Encyclopedia of Products & Industries—Manufacturing

Encyclopedia of Products & Industries—Manufacturing

VOLUME I
A–L

Patricia J. Bungert
Arsen J. Darnay

EDITORS

GALE
CENGAGE Learning™

Detroit • New York • San Francisco • New Haven, Conn • Waterville, Maine • London

Encyclopedia of Products & Industries—Manufacturing

Patricia J. Bungert and Arsen J. Darnay, Editors

Project Editor
Julie Gough

Editorial
Virgil L. Burton III, Miranda Ferrara, Louise Gagne, Peggy Geeseman, Lauren Haslett, Holly Selden, Justine Ventimiglia

Manuscript Editors
Alice Bernard, Claudia Bickel, John Coxeter, Pavlina S. Goodman, Mike Gruener, Andrew Grzeskowiak, Kevin Hile, Kristin Kahrs, Robert S. Lazich, John Magee, Monique D. Magee, Margaret Moser, Sheila Nason, Joyce P. Simkin, Jennifer Srnka, Angela Woodward

Product Management
David Forman

Indexing Services
Linda Mamassian

Product Design
Jennifer Wahi

Manufacturing
Rita Wimberley

For product information and technology assistance, contact us at Gale Customer Support, 1-800-877-4253.

For permission to use material from this text or product, submit all requests online at www.cengage.com/permissions.

Further permissions questions can be emailed to permissionrequest@ cengage.com.

Since this page cannot legibly accomodate all copyright notices, the acknowledgements constitute an extension of the copyright notice.

While every effort has been made to ensure the reliability of the information presented in this publication, Gale does not guarantee the accuracy of the data contained herein. Gale accepts no payment for listing; and inclusion in the publication of any organization, agency, institution, publication, service, or individual does not imply endorsement of the editors or publisher. Errors brought to the attention of the publisher and verified to the satisfaction of the publisher will be corrected in future editions.

EDITORIAL DATA PRIVACY POLICY
Does this product contain information about you as an individual? If so, for more information about how to access or correct that information or about our data privacy policies please see our Privacy Statement at www.gale.cengage.com.

LIBRARY OF CONGRESS CATALOGING-IN-PUBLICATION DATA

Encyclopedia of products & industries-- manufacturing / Patricia J. Bungert, Arsen J. Darnay, editors. -- 1st ed.
    p. cm.
    Includes bibliographical references and index.
    ISBN 978-1-4144-2983-0 (set hardcover) -- ISBN 978-1-4144-2984-7 (vol. 1 hardcover) --
ISBN 978-1-4144-2985-4 (vol. 2 hardcover)
    1. Manufacturing industries--Encyclopedias. 2. Manufactures--Encyclopedias.
I. Bungert, Patricia J., 1959- II. Darnay, Arsen. III. Title: Encyclopedia of products and industries--manufacturing. IV. Title: Products and industries-- manufacturing. V. Title: Manufacturing.
    HD9720.5.E53 2008
    338.003--dc22

                                                    2007019031

ISBN-13:

978-1-4144-2983-0 (set)
978-1-4144-2984-7 (vol. 1)
978-1-4144-2985-4 (vol. 2)

ISBN-10:

1-4144-2983-5 (set)
1-4144-2984-3 (vol. 1)
1-4144-2985-1 (vol. 2)

This title is also available as an e-book
ISBN-13: 978-1-4144-2986-1, ISBN-10: 1-4144-2986-X
Contact your Gale Cengage Learning sales representative for ordering information.

Printed in the United States of America
10 9 8 7 6 5 4 3 2 1

Contents

List of Graphics

Introduction

Encyclopedia of Products & Industries—Manufacturing (*EPIM*) is a compilation of essays on 120 major product categories across the entire spectrum of U.S. manufacturing activity. The purpose of this publication is to provide product information in structured format to students and analysts. While the focus is on products, the industrial context in which they are made and distributed is also carefully detailed.

SCOPE

The products covered in this work are presented in alphabetical order in two volumes. Essays are comprised of several standardized components so that important aspects of each product may be viewed from the same angles throughout the two volumes. This approach allows both similarities and differences between products to emerge. The rubrics used to organize the content are detailed below. Many essays feature graphical presentations for clarity; sales histories, product types, and market shares are typically charted. A bibliography follows each essay, and references to other closely linked topics in the *Encyclopedia* are provided in a *See Also* reference that follows the bibliography, where appropriate.

Topic Selection. An analysis of the national accounts that culminate in the government's reporting of the Gross Domestic Product (GDP) shows that ordinary consumers buy 70 percent of all manufactured goods. Industries and institutions, public and private, buy the other 30 percent of industrial output. In line with this broadly based distribution *EPIM* principally features consumer goods. These in turn tend to be categorized as durable goods such as automobiles, refrigerators, and televisions, or non-durable goods such as food, pharmaceuticals, clothing, and cosmetics.

Under the North American Industry Classification System (NAICS), the U.S. Bureau of the Census divides manufacturing into twenty-one major groups. *EPIM* offers essays on products manufactured in eighteen of these groups; those omitted, however, are covered indirectly. Within the major industry groupings, coverage also reflects what might be called the product densities present in each grouping. Some groups have a very extensive product array, while others are dominated by just a few categories. The largest number of essays are related to Food Manufacturing (18) followed by Chemical Manufacturing (15), Computer & Electronic Product Manufacturing (14), Miscellaneous Manufacturing (13), Transportation Equipment (11), and Machinery (9). The remaining forty essays deal with products in other sectors. Miscellaneous Manufacturing deserves special comment as this is the industrial grouping used by the Census Bureau to cluster industries that do not handily fit into other major categories. Included here is an important grouping of instruments—laboratory, medical, optical, and musical—as well as writing instruments and toys.

To the maximum extent possible throughout *EPIM* we have attempted to avoid very broad product categories. A generic category (e.g., Furniture) is easier to present as an industry than as a product line. In trade, commercial, and government reporting, data tend to blur the identities of products and the discussion becomes too general. In choosing products to feature, we have selected consumer goods over industrial products unless we saw a way of covering both in sufficient detail under the same heading. To provide some balance to the presentation, however, thirteen essays are devoted to industrial goods. Some of these anchor major transportation industries (Airplanes, Jet Aircraft, and Ships); some underpin agriculture (Fertilizers) or building activities (Construction Equipment); and some represent important developing or expanding areas (Satellite Communications Equipment, Security Equipment, and Turbines: Wind and Hydropower). Machine Tools are included because they underlie virtually all manufacturing activity, Medical Instruments because they are transforming health care, Industrial Finishes & Coatings because they protect virtually all surfaces, and Office Furniture because many of us spend so much of our time behind a desk. We have included products specifically intended for use in military or defense activities only in cases where the product category also has extensive or dominant private-sector uses. Such is the case in essays on Boats, Helicopters, Jet Aircraft, Ships, Satellite Communications, and Security Equipment.

ORGANIZATION

The *Encyclopedia of Products & Industries—Manufacturing* is a two-volume collection of essays presented in alphabetical order by product. Users will also be able to find products by industry grouping in the NAICS or the SIC indexes. These indexes present the contents arranged by industry groupings and, within these, by NAICS or by SIC codes. The alphabetical General Index is another tool for locating topics that are covered throughout the *Encyclopedia*.

Each essay is divided into twelve sections referred to as rubrics. The rubrics and their contents are:

- *Industrial Codes.* Codes assigned by the U.S. Bureau of the Census to the product under consideration are provided in two categories. Listed first are the codes of the North American Industrial Classification System (NAICS) currently used in the United States, Canada, and Mexico to classify industries and their products. These are six-digit codes rendered in the *Encyclopedia* with a hyphen between the first two and the last four digits (e.g., 31–1811 for Retail Bakeries). The hyphen is intended to help the user manage the long number more easily in copying it into notes for later use in additional research. The earlier Standard Industrial Classification (SIC) coding is still widely used outside of government. SIC codes are shown separately; they are four-digit numbers (e.g., 2051 for Bread, Cake, and Related Products). Finally, more detailed NAICS product codes are listed, which may help the user look deeper into the subject using statistics from the Census Bureau.

- *Product Overview.* The text of each essay begins with a general introduction to the product under review. Some products require minimal descriptions, while others require more extensive and technical descriptions. Extensive technical treatment is sometimes appropriate to show the ways in which categories within different types of the same product differ from one another. Technical treatment is also required to deal with manufacturing, performance, and pricing issues. The Overview also typically presents a history of the product and the range of its end uses. A fair number of product categories are under regulatory control by the government. The Product Overview at minimum summarizes regulatory issues and their implications. The general intent, under this rubric, is to position the product so that most of the discussion that follows is properly grounded.

- *Market.* This section provides the statistical data necessary to put the product into a broad commercial perspective. Measurements of the market are presented in dollars and, if appropriate, in units. In each case an attempt has been made to provide as extensive a history as possible on sales or shipments in order to highlight trends. In the United States domestic consumption of a product is often significantly higher than domestic production. In these cases imports account for a large part of consumption. Where this is the case, import and export data are presented. Forces influencing the market are also highlighted.

- *Key Producers/Manufacturers.* Under this heading brief capsule descriptions of important companies participating in the market are provided. Industrial concentration varies by product categories. In some cases just a few companies dominate, while in others, participation by many hundreds of companies is common. In all cases, however, we have endeavored to identify leading producers, both domestic and foreign. In some industries, such as dairy products, associations and cooperatives play a role very similar to the role normally played by key producers. In such cases, these associations and cooperatives are also listed in the Key Producers/Manufacturers section.

- *Materials & Supply Chain Logistics.* In some industries unique materials requirements and/or geographic factors play an important role. Such factors influence the producing industries in important and different ways. Generally geographic factors are highlighted in this rubric, but from a supply perspective. In those industries where natural resources are involved, a look at the occurrence of the resources in the United States and globally may be highlighted.

- *Distribution Channel.* The most common form of product distribution involves: (1) a manufacturer selling to a wholesale distributor, (2) the distributor supplying a retailer, and (3) the retailer selling to the consumer. Variations on this pattern abound. Distributional arrangements also tend to change over time. The Internet introduced e-commerce, yet another layer of distribution, which is often a hybrid of wholesale and retail distribution with courier services providing the transportation. Under this rubric the dominant and secondary forms of distribution used to bring a product to the market are outlined.

- *Key Users.* Users fall into many diverse categories, which are delineated and discussed under this rubric. Broadly speaking users are either consumers or institutions. Other delineations result from age and income stratifications, male or female predominance in use, and geographic considerations.

- *Adjacent Markets.* This rubric highlights product categories that represent alternatives to the product under consideration or products that move up or down in tandem with the product category under consideration.

- *Research & Development.* All industries spend money on research and development (R&D) at least in the form of new product development, the most common kind of R&D practiced by industry. In highly technical and advanced fields, both basic and advanced research tends to be funded. Breakthroughs can entirely transform an industry. Therefore, R&D activities are important indicators of future trends. Low levels of R&D expenditure, or spending on product enhancements alone, tend to signal loss of momentum in a field. In industries where breakthroughs are close to commercialization, changes loom ahead. Such issues are highlighted under this rubric.

- *Current Trends.* In many essays trends identified under other rubrics are summarized here and put into an overall context. In other essays underlying forces impacting a product or its industry are presented for the first time. A discussion of trends is most fully developed for those industries faced with one of the following challenges: resource scarcity, a changing regulatory environment, and/or a disproportionate reliance on a particular demographic segment of the population.

- *Target Markets and Segmentation.* This section highlights the manner in which producers position their products in order to appeal to what they view as different market segments. In some product categories producers sell the same product to every market segment, but differentiate the product by using different marketing approaches or packaging, or by making minor changes to the product itself, such as using different colors. In other cases products are substantially different depending on the market targeted, and may have been developed specifically for a particular clientele. The Key Users rubric and this one have commonalities. In some essays we enlarge upon user group findings by emphasizing complementary marketing approaches. In others we have used this section to discuss product differentiations, but with a marketing emphasis.

- *Related Associations and Organizations.* This rubric generally presents a listing of trade, professional, and user organizations that directly deal with the product or represent the industry that manufactures the products under consideration. Web addresses of the organizations listed are also presented. Such organizations are often important sources of additional insight into the topic and deserve close scrutiny by users wishing to go beyond the presentation in these essays.

PURPOSE, CONTENTS, AND TRENDS

Many encyclopedias function as repositories of information on universal and largely static facts and/or on past events. Manufactured products and their industries, by contrast, represent a dynamic flow of activity. In commerce the past is always prologue, and the present powerfully points to the future. The purpose of *EPIM* is thus two-fold. As an encyclopedic enterprise, it aims to present a thorough and comprehensive picture of what has been. As an accurate report on its topics, however, the *Encyclopedia* also attempts to identify the dynamics of the environment in which products exist. In every essay we have endeavored to describe trends that illuminate future directions. Some trends are quite apparent, while others vaguely point at something just emerging. The purpose of these volumes, therefore, is to deliver to students and analysts accurate pictures of a product and its industry from a point in time: the latter half of the first decade of the twenty-first century. The trends identified point at the second decade of the twenty-first century and, in some cases, beyond. The essays in these volumes should give the user a solid grounding for understanding the current status of a product and provide resources for additional investigation.

To achieve these ends the design of the *Encyclopedia* errs in the direction of comprehensiveness over such things as product detail and elaborate characterization of rapidly shifting technological, regulatory, distribution, and marketing arrangements. In many essays discussions of Key Producers/Manufacturers show the turbulence that has affected producing companies—turbulence that, viewed over even a brief period of a few years, has caused companies to merge, to spin off, to go private, to reemerge as new public entities, or to leave the market altogether. Such changes have been duly noted, occasionally emphasized. The contribution of *EPIM*, however, is to show the process itself and where it tends to point. The choice of rubrics used to cover each topic has been motivated by the goal of showing as many different aspects of the industry as possible, any one of which, if neglected, would result in a partial and possibly a misleading view. Within the rubrics themselves, the analysts writing these essays have, guided by the literature in the field, emphasized technology, regulatory activity, demographics, foreign competition, competing sectors, and resource issues. In virtually every product category, one or two trends stand out. Covering these extensively, while remaining aware of the total environment, has been an important objective of *EPIM*.

Although excessive generalization is never useful, especially not in dealing with products, seven broad trends are worth noting because they surfaced during the compilation of many essays, usually in groups. A brief summary follows with a general caution added: trends observed at any one time tend to be temporary and may rapidly reverse. The seven trends in brief are:

1. **U.S. Manufacturing Decline.** Since the mid-1990s the manufacturing sector in the United States has been declining, with more and more production taking place beyond U.S. borders. While this trend is common knowledge, finding it confirmed in detail and with the documentation of statistics, and occasionally with the documentation of multiple domestic producer bankruptcies, makes a more serious impression on the industrial analyst. The impact of this trend on the corporate community is often muted because the offshoring activity of well-known domestic leaders is not one of their widely touted advertising messages. To discover that many companies have gradually transformed themselves into domestic headquarters for foreign production is at minimum bad news for young people seeking careers in manufacturing.

2. **Brand and Producer Split.** Due in large part to offshoring activities, the direct relationship between a brand and its producer has been severed in many cases. It is no longer safe to assume that two similar items carrying the same brand name were made by the same company, in the same country, or that they are even being marketed by the same outfit. The sale or licensing of brand names has developed into a unique business and the buying public continues to associate genuine value with a brand because, in the past, quality associated with the brand gave it its power in the market.

3. **The Retailer as Manufacturer.** Many large retailers have discovered that their vendors buy their products from overseas manufacturers whom the vendors neither own nor control, except peripherally. Some retailers have decided to deal directly with third-party manufacturers in Asia and Eastern Europe, in many cases the same subcontractors with whom their previous manufacturing vendors also deal. In doing so the retailers are bypassing their traditional manufacturing vendors altogether in order to offer store brands at lower prices. Such developments have been particularly notable in the apparel industry but have emerged elsewhere as well.

4. **Emergence of China.** By the middle of the first decade of the twenty-first century China had come to replace Japan as the country most frequently mentioned as the chief competitor of U.S. manufacturers, as the source of goods sold at low cost by major merchandising firms. Somewhat less generally remarked, but certainly noted in many essays in the *Encyclopedia*, is the emergence of China as a market for goods as it gradually establishes a modern infrastructure. China is also emergent as a very substantial competitive buyer of natural resources, especially petroleum. Rarely noted in the general concern over China as a competitor is its own political history. China's history has been punctuated by major periodic upheavals of which the communist revolution dating to Mao Tse-Tung's emergence in the Long March (1934) and the Cultural Revolution (which began in 1969) are examples. Another period of uncertainty may change the face of China again and influence trends now seen as inevitable.

5. **Rising Commodity Prices.** A repeating theme in many essays is the rising price of commodities. In this context petroleum is the leading product. Precisely because crude oil prices influence both fuels and plastics, and because all industry is so highly mechanized and therefore energy-intensive, rising oil prices have been producing complementary increases in the costs of agricultural products, minerals, metals, and synthetic and natural fibers. Extended logistical systems, drawing raw materials and products from distant regions, have also added costs. Rising material costs have indirectly intensified efforts to curb expenditures on labor, accomplished in part by technological solutions (rising productivity) and by outsourcing labor-intensive operations to low-wage markets, thus diminishing domestic purchasing power.

6. **Aging Population.** A major demographic trend not subject to easy intervention, except perhaps by drastically modified policies of immigration, is the increase in the

older segment of the population and the resultant relative decline of children and young people as a percent of all people in the United States. Since the end of World War II, the Baby Boom has been the most important demographic fact in the United States. The boomers are now moving on, retiring, or reaching retirement age. As they do, their numbers are a transformative factor. Some markets are shrinking, while others are growing as the Baby Boom generation ages.

7. **Adverse Health Trends.** The last notable trend we wish to highlight, its impacts visible in a multiple cluster of products, is the growing incidence of obesity and the increasing number of overweight individuals. This trend is markedly present in children as well as adults and is associated with a rising incidence of Type II diabetes. The steady increase in these conditions, apparently highly resistant to efforts at public education and in part due to changes in lifestyle, has produced both real and cosmetic responses by participants in multiple branches of the food industry, radiating backward into agricultural commodities.

SOURCES OF INFORMATION

Each essay in *EPIM* was compiled from publicly available information sources. Major categories include: (1) periodical literature, especially trade journals, (2) industry and trade association Web sites supplemented by direct contact and correspondence, (3) professional association Web sites to obtain technical information, (4) Federal Government sources for statistical and regulatory information, (5) company Web sites to gain information on products, pricing, and distribution arrangements, (6) corporate annual reports, including Form 10Ks filed with the U.S. Securities and Exchange Commission, (7) academic sources and think tanks that offer studies, compendia, and handbooks, (8) books and reference materials accessed at libraries, and, to a limited extent, (9) direct field investigation in some cases to confirm information on pricing, displays, and to look at actual products in stores and showrooms.

Statistical Data. Where possible, federal data were used to determine the size of the total market for the products covered and for the industries that produced them. The two principal sources for this purpose are the *Economic Census* and the *Annual Survey of Manufactures* (*ASM*), both conducted by the U.S. Bureau of the Census. The *Economic Census* provides data for years ending in two and seven, collected by the Census Bureau after those years end and published with approximately a two-year lag. *Economic Census* data are the most accurate because they are based on a 100 percent survey of companies. Data for all other years are collected using a partial sample from which total industry data are projected. The data from the *ASM* survey also typically appear at a two-year lag. Thus in 2007, *ASM* data for 2005 were available. Combining these two sources, time series for the years between 1997 and 2005 were usually available for the essays presented in *EPIM*. The industrial classification system used by the Census Bureau changed in 1997 from SIC to NAICS; consequently data for years prior to 1997 were not used. Data from the Census Bureau on exports and imports were also widely used to make sense of U.S. markets, which, over time, have come to be dominated by imports.

The major limitation of the two sources highlighted above, apart from their time lag, is that reporting in some industries is not detailed enough to show the performance of product categories at the desired level. Shipment data on broad product groupings are usually reported, but the groupings hide details required to see specific products with any kind of precision. Another but more limited data series, the *Current Industrial Reports* (*CIR*), also published by the Census Bureau, provides somewhat more recent data (2006 data in 2007, for instance) and also provides very good product detail. *CIR*, however, does not cover manufacturing uniformly. The series concentrates on activities of highest interest to the public only with significant foreign trade aspects. Where possible, *CIR* data were used in the essays because they were more current.

Statistical information relevant to the analysis of many products is also collected by the U.S. Department of Agriculture (agricultural products, foods, livestock, and fish as well as lumber products), the U.S. Department of the Interior (minerals and fuel stocks), and the U.S. Department of Energy (fuels, electrical power, lighting, and related technologies). Valuable information on pharmaceuticals is available from the Food and Drug Administration. The Department of Housing and Urban Development publishes data on housing. In that it monitors many industrial processes for purposes of pollution control, the U.S. Environmental Protection Agency is a good source on process descriptions. The U.S. Department of Labor (DOL) reports on employment trends, which are useful for industry analysis alongside Census data. The DOL also collects information on prices and compiles the Consumer Price Index vital for tracking inflationary trends. The U.S. Department of Health and Human Services is a good source for information on health-related subjects and, occasionally, to provide perspective in certain product categories. The U.S. Department of Commerce, through its Census Bureau, is also the chief provider of demographic data widely used in the essays of this *Encyclopedia* to identify or to confirm trends in product sales.

In dealing with many products, federal sources of information do not provide sufficient resolution of product detail. In such cases we have made use of private sources of statistical data collected by market research firms. Most of these data are sold at high prices to those who subscribe to the reporting services. Summaries, however, are published and are made available to the general public. Similarly, some trade associations collect data from companies directly. In select essays, such data have been obtained and have been reproduced with permission of the associations.

ACKNOWLEDGEMENTS

A work of the scope and complexity of *Encyclopedia of Products & Industries—Manufacturing* relies on the inputs of many skilled and dedicated people. The editors are grateful to the contributors for their dedication and hard work. Contributing editors were Alice Bernard, Claudia Bickel, Patricia J. Bungert, John Coxeter, Arsen J. Darnay, Pavlina S. Goodman, Mike Gruener, Andrew Grzeskowiak, Kevin Hile, Kristin Kahrs, Robert S. Lazich, John Magee, Monique D. Magee, Margaret Moser, Sheila Nason, Joyce P. Simkin, Jennifer Srnka, and Angela Woodward.

The editors would like to recognize the contributions of Monique D. Magee, Vice President of Editorial Code and Data, Inc. Monique coordinated the complex effort, contributed essays, and produced all of the graphics. Helen S. Fisher bore a large share of the copyediting functions. Joyce P. Simkin contributed editorial assistance and invaluable technical and computer support to the entire enterprise. Last but not least, Julie Gough, as Project Editor, contributed overall guidance wisely and steadfastly to steer this project to a successful completion.

COMMENTS AND SUGGESTIONS

Comments on *EPIM* or suggestions for improvements of its usefulness, format, and coverage are always welcome. Although every effort is made to maintain accuracy, errors may occasionally occur. The *EPIM* editors will be most grateful if these are called to their attention. Please contact:

Project Editor
Encyclopedia of Products & Industries—Manufacturing
Gale Cengage Learning
27500 Drake Road
Farmington Hills, MI 48331–3535
Toll-Free: (800) 347-GALE
Fax: (248) 699-8069

Airplanes

INDUSTRIAL CODES

NAICS: 33–6411 Aircraft Manufacture

SIC: 3721 Aircraft Manufacturing

NAICS-Based Product Codes: 33–64111 through 33–6411100, and 33–64113 through 33–64113021

PRODUCT OVERVIEW

No one knows how long humans have dreamed of flying, but this concept informed some of Western culture's earliest stories. Around the time BC became AD, the Roman poet Ovid told the tale of Daedalus and his son Icarus. The first storytellers as well as their audiences could see snow persisting on the highest mountain peaks during warm weather; clearly, air did not grow warmer with altitude, and the sun would not have melted the wax holding Icarus's feathers to his wings.

A modern audience could note how the genius of Daedalus, using observation to mimic birds' wing shapes, was rendered irrelevant by his son's reckless behavior. Passion defeated by reality was a lesson that would-be aeronauts would discover repeatedly, frequently at a high price. This essay summarizes the history of heavier-than-air, fixed-wing powered aircraft.

Orville and Wilbur Wright had many self-taught engineer-inventor contemporaries who worked through repeated failures. The Wright brothers were partly inspired by the glider experiments of the German Otto Lilienthal and the Scottish engineer Percy Pilcher. Although these persons died while testing gliders of their own design,

both thought one secret of flight was the curved wing, and they published their findings. The Wrights based their design on that principle, created a steering mechanism, and added a 12-horsepower motor.

On December 17, 1903, their 750-pound plane launched from a railroad track at less than seven miles per hour, attained an altitude of perhaps ten feet, and landed after 120 feet. That beginning led to a new world in which gravity, while impossible to ignore, could be successfully challenged and even, less than seventy years later, overcome as humans traveled to land on the moon.

This flight was challenged for decades in the press and the courts, since wealth and fame would follow those who were actually first. Gabriel Voisin, who along with his brother, Charles, was the first to build aircraft in France on a commercial basis, scoffed at the influence of the Wright brothers and other aviation pioneers. Nonetheless, the early trickle of aviation pioneers soon grew to a swarm.

A company founded by Glenn Curtiss made the first commercial sale of an aircraft in 1909 to the Aeronautic Society of New York. In the same year, one of the Wright brothers succeeded in meeting U.S. Army specifications for an aircraft and sold it to the government for $30,000. The Wrights promptly sued Curtiss for patent infringement, virtually freezing the aircraft industry in the United States until World War I.

Despite several impressive demonstration flights the Wrights performed for the U.S. Army, its officials were unmoved. The Wrights took their show to Europe, where they flew for the German, French, and British armies. In the process, they prompted interest with such European aviation pioneers as Louis Blériot, Willy Messerschmidt, Anthony Fokker, and Marcel Dassault. Aviation made for good entertainment. Blériot and others such as Louis Paulhan built their own airplanes and began touring flying circuses in the first decade of the twentieth century.

With the outbreak of war in Europe in 1914, France and Germany were quick to apply aviation to the battlefield, producing the world's first aces, Roland Garros and Manfred von Richthoven. The United States Army embraced air power in 1914 by creating an aviation group within the Signal Corps. By 1918 the government showed its interest in aviation through expansion of an air squadron and active intervention in the industry. Having seen the effect of air power in Europe during World War I, the government resolved not to see American air power stunted by legal wrangling or patent hoarders. What emerged was a loosely policed competition for government contracts, primarily military and later airmail business. Hundreds of airplane builders emerged from garages and warehouses.

The aircraft industry received a tremendous boost in 1927 when Charles Lindbergh completed the first successful trans-Atlantic flight using a modified Ryan Aeronautics tri-motor. Lindbergh's daring and nearly fatal stunt so strongly revived interest in aviation that investors began pumping millions of dollars into aircraft companies.

In August of 1929, Allan Loughead and Fred Keeler sold the Lockheed Company to a group of automotive investors calling themselves the Detroit Aircraft Company. The company drew tremendous investor interest after aviatrix Amelia Earhart crossed the Atlantic with one of the company's Vegas aircraft. One month later, world financial markets were buffeted by the stock market crash that plunged the nation into the Great Depression. Aviation company stocks, valued at more than $1 billion on total earnings of more than $9 billion, were decimated.

The Depression would have destroyed the aircraft industry were it not for government support. It became official policy to award contracts to an increasingly privileged club of manufacturers, so that their expertise could be preserved and developed for military purposes. American aviation was controlled by three huge vertical monopolies, each maintaining huge airframe and engine manufacturing facilities and airline services.

While military preparations were stepped up in the 1940s, the Japanese attack on Pearl Harbor on December 7, 1941, sparked tremendous growth in the aircraft industry. Huge amounts of government money were poured into engineering and production facilities. President Roosevelt ordered 60,000 aircraft in 1942 and 125,000 the year after.

Emerging from the war with tremendous manufacturing capacity and engineering talent, the Douglas, Boeing, and Lockheed companies dominated the commercial aircraft industry. Competitors, including Curtiss, Martin, and Convair, were forced to exit the market in rapid succession, taking refuge in the more protected military businesses. Hughes Aircraft, famed for its massive Spruce Goose amphibian freighter, failed to break into the production market.

With fixed-wing designs having reached the peak of their development in the 1960s, we can glance back to the early Renaissance genius Leonardo da Vinci, who drew many flying machine designs. Whether any of these devices were constructed remains a matter of debate, but it is clear from his notebooks that human-powered flight fascinated da Vinci. However, the only materials he had access to were hardwoods and linen, and they were too dense to support this dream. Almost five centuries later, and bringing matters neatly full-circle, in 1977 the Gossamer Condor, built mostly of lightweight plastics, enabled the first human-powered flight.

MARKET

American aircraft companies build and sell airplanes for three distinct markets: the military, commercial aviation, and general aviation, which includes business aviation. From the end of World War II until the collapse of the Soviet Union in 1989, the American military services had an unending appetite for sophisticated aircraft, which American firms attempted to satisfy. The end of the Cold War, which reduced military spending around the world, provided the greatest challenge for American aircraft manufacturers, who had grown accustomed to lucrative contracts from the U.S. Department of Defense (DOD).

Developing commercial aircraft posed far greater risks than those of military aircraft. The development process for a passenger airliner capable of carrying several hundred people was lengthy and costly, requiring manufacturers to anticipate the needs of airlines far in advance and to gamble large amounts of money on the product's success. For this reason, Boeing canceled its development of a super jumbo aircraft. Manufacturers found a more stable market by designing new or modifying existing aircraft in response to the demands of carriers, who typically asked for improved fuel efficiency and more seating. By the early 1990s industry estimates showed that a new medium-sized airliner cost more than $2 billion to develop, with engines costing another $1.5 billion.

Due to the risks involved, commercial aircraft manufacturers tended to modify existing airframes rather than reinventing; most existing commercial airliners changed little in the last half of the twentieth century. However, some exciting new aircraft developments occurred in the areas of speed, range, capacity, and fuel efficiency. Many manufacturers by the early 2000s worked cooperatively, jointly developing a design and dividing work among partners if the design was successful. The merger of Boeing Company of Seattle, Washington, and the McDonnell Douglas Company of St. Louis, Missouri in 1997 resulted in economies of operation in many areas. Boeing concentrated on long-range, fuel-efficient planes with a slightly higher passenger capacity, justifying this with the ratio of development costs. Airbus continued to pursue the super jumbo concept, announcing its new model in 2007.

American manufacturers historically produce approximately 60 percent of the world's general aviation aircraft and 30 percent of the helicopters. The major U.S. manufacturers of general aviation aircraft are the Beech Aircraft Corp., Fairchild Aircraft Inc., the Cessna Aircraft Co., Gulfstream Aerospace, and Learjet Inc.

Based on data published in the *Statistical Abstract of the United States: 2007*, Figure 1 provides an overview of U.S. aircraft sales by military aircraft and civilian aircraft from 1990 through 2005. Annual fluctuations were greater for civilian aircraft sales than for military aircraft sales which saw slight declines in the early 1990s followed by a flattening out and slow increments thereafter.

Most aircraft manufacturers derive much of their profits from producing replacement and upgrade parts for their airplanes. Since large commercial jets represent such a large investment—a new twin-engine passenger jet may cost several hundred million dollars—airlines try to keep them in the air for many years. Moreover, the Federal Aviation Authority (FAA) sets stringent guidelines on repair and replacement procedures for passenger aircraft.

FIGURE 1

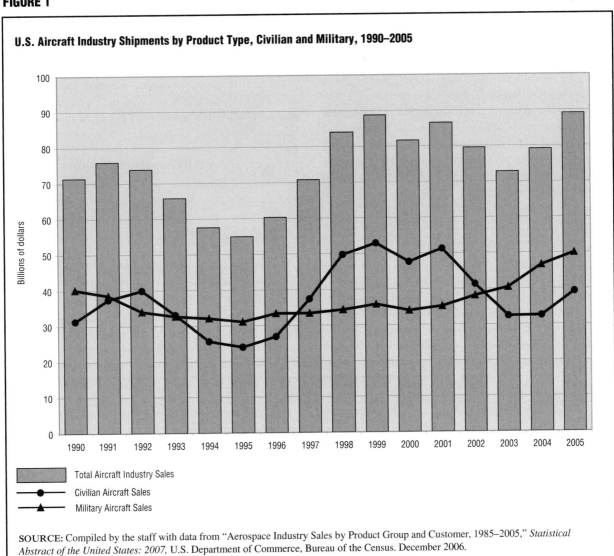

U.S. Aircraft Industry Shipments by Product Type, Civilian and Military, 1990–2005

SOURCE: Compiled by the staff with data from "Aerospace Industry Sales by Product Group and Customer, 1985–2005," *Statistical Abstract of the United States: 2007*, U.S. Department of Commerce, Bureau of the Census. December 2006.

By 2003 the aircraft industry was struggling in the wake of downturns in the air transportation market. The leading U.S. airlines lost more than $7 billion in 2001 and more than $3 billion through the first half of 2002. number of factors, including a slack economy, a decline in travel following the attacks of September 11, and heightened competition from discount airlines, contributed to the air transportation sector's woes. In December of 2002, United Airlines—which accounted for some 20 percent of U.S. flights—filed for bankruptcy after losing $4 billion over two years and laying off 20,000 employees.

U.S. manufacturers shipped about 4,088 units of complete civilian aircraft (fixed wing, powered craft; helicopters; and non-powered types of civil aircraft) in 2002, valued at approximately $34.7 billion. In terms of unit shipments, this figure represented a decrease from 2001, when the industry shipped 4,541 units valued at 41.8 billion, and from 2000 when shipments numbered 5,162 civil aircraft valued at $38.6 billion.

The Aerospace Industries Association forecasted that shipments of complete civil aircraft would total 2,751 in 2003, with an estimated value of about $25 billion. Some 275 airliners were expected to account for the majority of this total ($18 billion). According to *Standard & Poor's Industry Surveys*, Avitas, Inc. expected aircraft orders to fall from an estimated 816 in 2001 to 561 in 2002, after which levels would steadily improve, reaching a forecasted 973 by 2005. During the same timeframe, Avitas expected aircraft deliveries to fall from an estimated 1,148 in 2001 to 941 in 2002 and 707 in 2003. After 2003, deliveries were expected to improve slowly through 2005, when levels were forecast to reach 829.

KEY PRODUCERS/MANUFACTURERS

Boeing Company. In 1997 Boeing Company was the world's largest manufacturer of commercial jetliners and military aircraft, and was NASA's leading contractor. The company employed more than 200,000 people in 1999 in more than 60 countries worldwide. However, by 2002 it had reduced its workforce to 166,000. Company revenues were $54.1 billion in 2002, which represented a drop since 2001 of nearly 83 percent. During the late 1990s, production problems resulted in lost aircraft orders from companies such as British Airways, United Parcel Service, and Airbus Industries. These problems cost Boeing millions of dollars and threatened its standing as the top manufacturer.

Airbus S.A.S. Airbus surpassed Boeing's orders in 2001 and 2002. According to *Air Transport World*, Airbus reported net orders of 233 planes in 2002, compared to Boeing's 176. However, Boeing was still the market leader in deliveries, with 56 percent of the total market share.

Airbus began as a five-nation European consortium named Airbus Industrie. It was conceived as a European answer to America's domination of the large commercial transport market. By the early 2000s Airbus was restructured into a corporation named Airbus S.A.S. The company's majority shareholder (80 percent) is the European Aeronautic Defense & Space Co. BAE Systems of the United Kingdom holds the remaining 20 percent interest.

Northrop Grumman Corporation. Based in Los Angeles, California, Northrop Grumman Corporation this company employed 96,800 people in 2001. The company is responsible for the design, development, and manufacture of aircraft (including the less-than-perfectly-concealed Stealth Bomber), aircraft sub-assemblies, and electronic systems for the military. In addition to building ships, the company also designs, develops, operates, and supports computer systems. Sales totaled $17.2 billion in 2002. That year, Northrop Grumman saw its net income fall 85 percent, reaching $64 million.

Beech Aircraft Corporation. Best known for its line of Beechcraft propeller and jet airplanes, Beech Aircraft Corporation is one of several American manufacturers of small aircraft. Beech competes with Cessna, Piper, and Lear for shares of such markets as private pilots, small air taxi services, corporate customers, and military forces. Beech also manufactures a variety of aircraft parts and special systems for larger companies, principally McDonnell-Douglas.

In 1990 Beech recorded its best year, turning out 433 aircraft and collecting $1.1 billion in sales. Also in 1990, the new Starship model won certification. In 1992, Beech's 60th anniversary year, the company's 50,000th aircraft rolled out of the factory. That same year, however, a sales slump, attributed to a ten percent federal luxury tax, caused the company to cut back production and lay off 180 administrative staff.

Due to its 1900 and Jayhawk projects, Beech remains the largest of the small aircraft manufacturers, though Cessna builds more private aircraft. It offers a complete line of advanced aircraft, from the single-engine Bonanza, to the twin-engine Baron and Super King Air series, to the futuristic Starship. The bulk of Beech's more recent success, however, lies with its Beechjet and 1900 airliners. Barring any severe depression in small aircraft markets, Beech is likely to retain its leading position in this sector of the aviation industry.

Cessna Aircraft. Cessna Aircraft is one of the most famous names in small planes. A subsidiary of Textron, Cessna manufactures business jets, utility turboprops, and small single-engine planes. Best known for its small prop planes, Cessna is also a leading maker of business jets; it makes nine variations of its popular Citation jet. Its utility turboprop plane, the Caravan, has freight, bush, amphibious, and commercial (small connecting flights) applications. Cessna's single-engine planes are typically used for personal and small-business purposes. As it prepared to enter the 21st century, Cessna remained the largest private aircraft manufacturer in the United States. With its line of cargo craft and advanced private jets, including the new Citation X, Cessna still offered the broadest product range in the industry. With the company's relationship to owner, Textron, on solid ground, Cessna looked certain toward the end of the first decade of the 2000s to remain America's leading small aircraft manufacturer.

MATERIALS & SUPPLY CHAIN LOGISTICS

The number of units produced per year by the U.S. civilian aircraft industry is approximately one-thousandth that of the automotive industry. The U.S. aerospace industry shipped a total of 4,068 civil and 450 military aircraft in 2005 according to the U.S. Department of Commerce's International Trade Administration. On average, commercial transport aircraft cost $300 per pound. The use of relatively advanced materials combined with the production of intricate component forms without net-shape processes contribute to this higher cost per pound. The airframes of commercial aircraft are currently largely aluminum (70–80%) with smaller weight fractions of steel, titanium, and advanced composites. The gas turbine engines that power these aircraft use alloys of nickel (~40 percent), titanium (~30%), and steel (~20%), with the balance being advanced composites and aluminum.

A combination of composite materials and computer design has grown from the occasional application for a nonstructural part (such as baggage compartment doors) to the construction of complete airframes. For military applications, these materials have the advantage of being less detectable by radar.

Some aircraft of composite materials began to appear in the late 1930s and 1940s; these were usually plastic-impregnated wood materials. The largest and most famous example of this design is the Duramold construction of the eight-engine Hughes flying boat, popularly known as the Spruce Goose. A few production aircraft also used Duramold materials and methods.

Fiberglass, fabrics made up of glass fibers, were first used in aircraft in the 1940s and became common by the 1960s. *Composite* is the term used for different materials that provide strengths, light weight, or other benefits not possible when these materials are used separately. They usually consist of a fiber-reinforced resin matrix. The resin can be vinyl ester, epoxy, or polyester, while the reinforcement might be any one of a variety of fibers, ranging from glass through carbon, boron, and several other proprietary types.

To these basic elements, strength is often gained by the adding of a core material, essentially making a structural sandwich. Core materials such as plastic foams (polystyrene, polyurethane, or others), wood, honeycombs of paper, plastic, fabric or metal, and other materials, are surrounded by layers of other substances to create a structural sandwich. This method has been used to create, for example, Kevlar, used in aircraft panels, and Lucite, superior to glass for aircraft windows and canopies.

One advantage of composites is the ability to form them into a wide variety of shapes, accomplished by various methods. The simplest is laying fiberglass sheets inside a form, infusing the sheets with a resin, letting the resin cure, and polishing the result; this is how synthetic canoes are constructed. More sophisticated techniques involve fashioning the material into specific shapes with complex machinery. Some techniques use molds; others employ vacuum bags that allow atmospheric pressure to force parts into the desired shape.

Composite materials allow aircraft engineers to design lighter, stronger, and cheaper streamlined parts simply not possible when using metal or wood. Composites use has spread rapidly throughout the industry and will probably continue to be developed in the future.

DISTRIBUTION CHANNEL

Aircraft parts manufacturing can be seen as predating the invention of powered aircraft. The Wright Brothers' first airplane, little more than a propeller-driven kite, was equipped with cables, chains, and an engine built by others. In one sense, Orville and Wilbur Wright invented nothing; they merely designed and assembled their aircraft from existing parts. However, this view is too simplistic, since the Wrights put these parts together in a way no one had done before.

The American aircraft industry can be divided into four segments. In one segment, manufacturers such as Boeing and Lockheed Martin Corp. build the wings and fuselages that make up the airframe. Meanwhile, companies such as General Electric and Pratt & Whitney manufacture the engines that propel aircraft. The third segment covers flight instrumentation, an area where the most profound advances in aviation have taken place. But the fourth segment, broadly defined by the industrial classification *aircraft parts not otherwise classified,* includes manufacturers of surface control and cabin pressurization systems, landing gear, lighting, galley equipment, and general use products such as nuts and bolts.

Aircraft manufacturers rely on a broad base of suppliers to provide the thousands of subsystems and parts that make up their products. There are more than 4,000 suppliers contributing parts to the aerospace industry, including rubber companies, refrigerator makers, appliance manufacturers, and general electronics enterprises. This diversity is necessary because in most cases it is simply uneconomical for an aircraft manufacturer to establish, for example, its own landing light operation. The internal demand for such a specialized product is insufficient to justify the creation of an independent manufacturing division.

There is a second aspect to this distribution tier, since aircraft manufacturers have found it cheaper and more efficient to purchase secondary products from other manufacturers, who may sell similar products to other aircraft companies, as well as automotive manufacturers, railroad signal makers, locomotive and ship builders, and a variety of other customers. For example, an airplane builder such as Boeing, Grumman, or Beech might purchase landing lights from a light bulb maker such as General Electric. Such subcontractors supply a surprisingly large portion of the entire aircraft. On the typical commercial aircraft, a lead manufacturer such as McDonnell Douglas may actually manufacture less than half of the aircraft, though it is responsible for designing and assembling the final product.

When a major manufacturer discontinues an aircraft design, as Lockheed did with its L-1011 Tristar, a ripple effect is caused that affects every manufacturer that supplied parts for that aircraft. Therefore, parts suppliers that make up the third tier of distribution, strive to diversify their customer base to ensure the decline of one manufacturer will be tempered by continued sales to others. Given the unstable nature of the industry, parts manufacturers also attempt to find customers outside the aircraft business.

KEY USERS

The initial purposes of the Wright Brothers were identical to those of the thousands using balloons, gliders, and even kites in the centuries preceding them: to rise above the ground, stay aloft as long as possible, and to conclude with a landing from which one could walk away. Fascination led to experimentation, and the first widespread use of fixed-wing airplanes was entertainment, with exhibitions performed for a rapidly-growing audience. World War I shortly followed, and pilots moved quickly from enemy observation to techniques including dogfighting, naval attack bombing, and the perfection of strategic bombing tactics that would in World War II become carpet bombing.

Private concerns carried mail by aircraft before the war, and then the U.S. government established its own airmail service. However, The Kelly Airmail Act of 1925 returned airmail service to private bidders after a series

of crashes by the government's air service. Postmaster William Folger Brown encouraged the formation of large airline companies by carefully awarding profitable airmail contracts. Boeing acquired numerous private airmail companies and their lucrative contract rights; in 1928 they were banded together to form the National Air Transport Company. The following year, Boeing and Rentschler merged their airframe engine businesses to form the United Aircraft & Transportation Company. By the end of 1929, the company had taken over two propeller makers as well as Northrop's Avion company and laid out an air transportation network that later became United Air Lines.

Most people in the twenty-first century take air travel for business and especially for pleasure as a normal part of life, but the first air passenger services did not begin until 1937, when the emergence of the DC-3 and Electra enabled airlines to make money from passenger services alone and end their reliance on airmail. Many of the early passengers became airsick, so for decades, flight attendants could only be hired after completing nursing training.

Air travel for business and leisure continued to grow rapidly in the 1960s and the transition to jets was virtually complete by the middle of the decade. Yet the potential market remained largely untapped; as late as 1962, two-thirds of the American population had never flown. In 1961, Eastern inaugurated an hourly, unreserved shuttle service connecting New York, Washington, D.C., and Boston. There were major safety advances, although traffic growth strained the air traffic control system, and there was growing conflict between increasing volume and safety. Busy airports experienced conflicts between scheduled and business aircraft operations.

Despite financial pressures, air transportation was a critical part of the economy at the end of the twentieth century. Domestic traffic, 321 million passengers in 1984, rose to 561 million by 1998. United had a fleet of more than seven hundred airliners, and American and Delta had more than six hundred each by 2000. Perhaps the strongest indicator of the importance of the industry came after the hijacking of four airliners on September 11, 2001, in terrorist attacks on the United States. The federal government cancelled all air traffic for several days. Even after resumption, traffic dropped sharply, since much of the public was reluctant to fly again. Congress immediately appropriated approximately $15 billion in direct grants and loan guarantees to scheduled carriers, since many tottered on the brink of financial collapse. Manufacturers also suffered, because many airlines postponed or cancelled orders in response to the drop in passenger traffic and reduced schedules. Flight delays, overcrowding, overbooking, and cancellations are some of the incidents that still traumatize passengers. Nevertheless, long-term

FIGURE 2

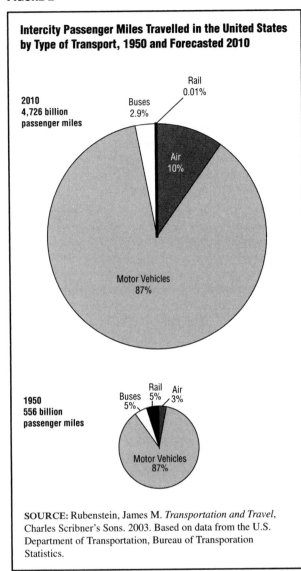

Intercity Passenger Miles Travelled in the United States by Type of Transport, 1950 and Forecasted 2010

2010
4,726 billion
passenger miles

Rail
0.01%

Buses
2.9%

Air
10%

Motor Vehicles
87%

1950
556 billion
passenger miles

Buses
5%

Rail
5%

Air
3%

Motor Vehicles
87%

SOURCE: Rubenstein, James M. *Transportation and Travel,* Charles Scribner's Sons. 2003. Based on data from the U.S. Department of Transportation, Bureau of Transporation Statistics.

growth in the early twenty-first century looks favorable and though various carriers continued to dodge in and out of bankruptcy, by 2007 air traffic neared pre-9/11 levels.

ADJACENT MARKETS

Even a single-engine, single-wing airplane contains thousands of parts whose peak functioning is essential to a safe landing. Indeed, a significant part of all aircraft control panels are instruments indicating whether the other instruments are working correctly; it's not as though, when something goes wrong, the pilot can pull over to the nearest cloud. The categories considered in this section include instrumentation systems, engine instruments, and fuel. Products produced by these industry sectors are necessary

to getting aircraft off the ground, keeping them in the air, and landing them at controllable speeds.

Guidance and Control Instrumentation. The products of this industry relevant to this essay include radar systems, navigation systems; flight and navigation sensors, transmitters, and displays; gyroscopes; airframe equipment instruments; and speed, pitch, and roll navigational instruments.

Main suppliers of search and navigation equipment are the same contractors who supply the larger U.S. aerospace and defense industry. Although not necessarily the most prolific producers of search and navigation instruments, many of the largest and most recognizable corporations in the United States are contributors, including AT&T, Boeing, General Electric, General Motors, and IBM.

A substantial majority of the industry's product types fall into the avionics (aviation electronics) classification, which includes aeronautic radar systems, air traffic control systems, and autopilots.

Historically, the primary customer for industry products has been the U.S. government—in particular the Department of Defense and the Federal Aviation Administration.

Search and detection systems and navigation and guidance systems and equipment ($29.1 billion worth of shipments in 2001) constitute 91 percent of the total search and navigation market and include the following product groups: light reconnaissance and surveillance systems; identification-friend-or-foe equipment; proximity fuses; radar systems and equipment; sonar search, detection, tracking, and communications equipment; specialized command and control data processing and display equipment; electronic warfare systems and equipment; and navigation systems and equipment, including navigational aids for aircraft,

During the 1970s the Global Positioning System satellite network first came under development. Inertial navigators using digital computers also became common on civil and military aircraft.

Industry shipment values for the above products totaled $31.9 billion in 2001, an increase over 2000 levels of $29.9 billion. In 2001, the industry's employment base of 153,710 workers was an increase from the previous year's count of 145,990 workers. Capital investment, which totaled approximately $1 billion in 2000, has remained relatively constant since 1997.

Aircraft Engine Instruments. The main customers of the aircraft engine instruments segment are General Electric, United Technologies, Rolls Royce, and other aircraft manufacturers. The sector shipped measuring devices for temperature, pressure, vacuum, fuel and oil flow-rate sensors, and other measuring instruments. Growth in this market is linked to aircraft production.

Through the first decade of the 2000s, the Measuring and Controlling Devices Industry was projected to grow at an annual rate of 3 percent. Aircraft engine instruments are predicted to be one of this industry's faster growing segments. Furthermore, the addition of software and services will contribute to overall industry growth, as will further expansion into overseas markets. The top five export markets in the late 1990s were Canada, Mexico, Japan, United Kingdom, and Germany; these five countries were also the top import countries. Looking into the 2000s, estimates indicate that 33 percent of measuring and controlling instruments product shipments will be exported, while 25 percent of U.S. production will be imported.

Aviation Fuel. The term gasoline is commonly used in North America where it is often shortened in to gas. This should be distinguished in usage from truly gaseous fuels used in internal combustion engines such as liquefied petroleum (which is stored pressurized as a liquid but is allowed to return naturally to a gaseous state before combustion). Mogas, short for motor gasoline, separates automobile fuel from aviation gasoline, or avgas. In the United States, avgas is known as 100LL (100 octane, low lead) and is dyed blue.

Gasoline fuels, called *white* products in the petroleum industry, comprise the lighter end of the refining process, usually about 20 percent of the total yield. Most automobile gasoline produced in the 1930s had an octane reading of 40, with aviation gasoline at 75–80. Aviation gasoline with such high octane numbers could only be refined through a process of distillation of high-grade petroleum. By 1945 American aviation fuel was commonly 130 to 150 octane, where it has largely remained since.

RESEARCH & DEVELOPMENT

In the wake of the September 11, 2001 terrorist attacks, the U.S. federal government ordered airlines to ensure that existing cockpit doors on commercial aircraft would be locked at all times and secured with extra bars and barriers. It also developed a standard redesigned, reinforced cockpit door that airlines were required to install on all aircraft by 2003.

With an annual research budget exceeding $1 billion for its aeronautical division, the U.S. National Aeronautics and Space Administration (NASA) contributes substantially to advances in aircraft technology. NASA has assisted the general aviation industry in the United States in such areas as developing new wing and blade designs—including the civil tilt-rotor project—and cockpit technology for business and commuter aircraft.

The Kyoto Protocol to the United Nations Framework Convention on Climate Change developed in 1997 was the first international treaty to set standards for greenhouse gas emissions—primarily carbon dioxide—by countries ratifying it. Although as of December 2006, the United States had yet to accept this treaty's limitations, NASA plans to develop aircraft that meet those environmental and safety standards.

Boeing is another leading aircraft technology researcher. Each year the company devotes between $1.5 billion and $1.8 billion for research and development. In the mid-1990s, the majority of the company's research funds went to developing its 777. In the first decade of the 2000s, with the delivery of its 777s, Boeing turned to refining its existing aircraft and designing new planes. In cooperation with NASA and several universities, Boeing began to develop a blended-wing-body (BWB) plane. The BWB's advantages include superior fuel economy, lower production costs, greater capacity, and greater range than the conventional aircraft of the 1990s. The BWB's capacity comes from the design of the wings, which hold seats for passengers. Researchers estimate that the plane could be ready by 2015. Meanwhile, Boeing plans to meet the fast-approaching requirements for environmentally friendly aircraft with its 717–200, which features reduced emissions and lower noise levels than its rivals. Test flights of the 717–200 began in early 1998.

In the mid-1990s, United States and Russian researchers jointly studied how to develop new supersonic civil aircraft. Although supersonic projects had largely ended in 1978, both countries renewed their interest. The U.S. component of the research team consisted of NASA, Boeing, Rockwell-Collins, Pratt & Whitney, and General Electric. The Russian component of the team included Tupolev, the developer of the Tu-144 supersonic jet. The collaborators went to work rebuilding the plane's engine to use the aircraft for studies of the ozone layer and sonic-boom problems.

In contrast to the huge government-sponsored research programs of the aerospace conglomerates, the research and development (R&D) efforts of the makers of ultralights and *kit planes*, designed to be assembled by the user, were lean but smart. The popular kit designs offered by Burt Rutan and others in the 1970s offered advanced materials such as exotic composites, plastic foams, and fiberglass and epoxy laminates. Also featured in these designs were canards, small wings placed at the nose of the aircraft, and winglets, fins at the end of the main wing, both of which increased efficiency and stability. Computer modeling enabled designers to incorporate advanced wing shapes into designs the ordinary enthusiast could build at home. At least one company has adopted these technologies to produce an inexpensive, six-passenger business turboprop (less than $1 million, compared to $3 million and up for competitors).

A significant experimental aircraft was the Gossamer Condor. In 1977, it enabled the first human-powered flight. Perhaps the greatest achievement in 1986 was Rutan's Voyager, the first aircraft to circle the world without refueling. By the early 2000s, Rutan's conceptions of lightweight craft with intercontinental range had found a military application in the U.S. armed forces—highly capable drones, used effectively during the hostilities in Afghanistan. High-altitude drones with extended range were also expected to acquire satellite-like global or regional communications roles in the new century. Rutan's designs and principles have found their civil application in the Beech Starship, a small business turboprop, and in a small jet fighter/trainer.

Instrumentation is another area of continuing research. A computerized display of flight information, the Electronic Flight Information System (EFIS), has promised to improve the decision-making abilities of pilots by providing an integrated, improved display of navigational, meteorological, and aircraft performance information in the cockpit. State-of-the-art airliners and business craft such as the Boeing 757 and 767, the Airbus A-310, and the Beech Starship are equipped with this system.

The Global Positioning Satellite (GPS) system, first developed for use by the U.S. military in the early 1970s, relies on groupings of satellites to provide extremely precise location information (including altitude) to receiving units within airplanes—some small enough to be handheld and inexpensive enough to be used by the general aviation market. However, the units' small size and accuracy have caused concern about their potential misuse in armaments.

CURRENT TRENDS

Civil aircraft production is controlled by the commercial market, supplying the jets and turboprops used by the world's passenger and cargo airlines. As of the middle of the first decade of the 2000s, just two manufacturers—Boeing in the United States and Airbus S.A.S. in France—have controlled nearly the entire market for commercial aircraft for more than a decade. This lead was secured by manufacturing medium and large jets for 100 or more passengers, the industry's most lucrative and capital-intensive segments. Aircraft manufacturers noted the increasing demand for large-capacity, wide-body planes and expected that the average number of seats per plane would increase to 240 by 2015. They expected Asian countries would help drive this trend with a 356-seat average capacity per plane by 2015. Airbus was set to deliver its 555 passenger A380 in 2008.

The huge costs and risks of aircraft manufacturing encouraged business consolidation and a proliferation of international joint ventures in what has been termed a borderless industry. Few countries could be considered self-sufficient in production, and even for those that could, most competitors in the industry pursue multiple cross-border ventures in order to keep costs down and draw on the special competencies and efficiencies of firms around the globe.

Globally, the industry experienced continued growth in the early 2000s. Boeing's World Air Cargo Forecast predicted an annual expansion rate of 6.2 percent through 2023, tripling the levels of overall air traffic. Strong growth was reported in international trade, with the most reported in the Asia-Pacific region. Traffic in North America and within Europe was expected to see below average increases. The U.S. firms Cessna Aircraft Co. and Raytheon led the continuing strong surge in sales in the general aviation segment. The U.S. industry reached $147 billion in 2003 sales. That year, Boeing, with about 280 units, and Airbus, with about 300 units, produced a combined $33 billion in aircraft. These had a per-unit value of $50 million or more, according to *Fortune*. For the first half of 2004, the companies delivered a combined 312 aircraft. According to researchers from the Teal Group, $421 billion in aircraft will be built by 2012.

Environmental groups in the United States, Europe, and Australia have focused on noise pollution. At the turn of the century, the U.S. Airport Noise and Capacity Act of 1990 required U.S. airlines to make their fleets meet quieter noise specifications. Smaller business jets were exempt from this rule. The International Civil Aviation Organization (ICAO) imposed similar standards.

Heavily congested airports have suggested the need for 600–800 seat, ultra-high-capacity aircraft (UHCA), or very large commercial transport (VLCT). Airbus began research on such a project, estimated to cost between $6 billion and $8 billion. Boeing also began research for its proposed UHCA, the 747-X. The potential market for these aircraft was projected at between 400 and 500 aircraft by 2010. In the early 2000s, Boeing studied development of smaller capacity, but higher speed, transports than the proposed UHCAs.

A concept for a 300-seat supersonic airliner, dubbed the Orient Express, has been the subject of a study group comprised of engineers and others from Boeing, Aerospatiale, British Aerospace, Japan Aircraft Development Corp., Tupolev, and Alenia. Traveling at Mach 3, or three times the speed of sound, the aircraft would cut travel time between Tokyo and Los Angeles to 4 hours, from the current 10. Fares were projected to fall eventually to a level just 20 percent higher than those for conventional flight.

Two types of vertical takeoff and landing (VTOL) aircraft also were being developed to serve inner-city airports. Ishida Corp. of Japan (in collaboration with U.K. and U.S. firms) is developing the 14-passenger TW-68. With wings that rotate 90 degrees, the craft would allow vertical takeoff and landing. Due to traffic congestion, Boeing projected a need for thousands of civil tilt-rotor aircraft (such as the Bell/Boeing V-22) in the first few decades of the new century.

TARGET MARKETS & SEGMENTATION

The opportunity to fly for business and the right to fly to increase the exotic possibilities of vacations have led the aircraft industry to a simple, if perhaps immodest goal: to get everyone in the air at some point in life, and the more often the better. Beginning in the late 1990s, online travel agencies such as Orbitz, Priceline.com, Travelocity, Expedia.com and others, used highly complex algorithms to scan millions of possibilities for discount hotel rooms, car rentals, and of course air flight times and days that would, they universally claimed, provide the consumer with the absolute cheapest available services, accommodations, and travel.

However, these algorithms have been analyzed, with the mildly disturbing conclusion that thorough examination of all possible variables would take longer than the probable age of the known universe. The agencies ignore this massive fact by simply choosing what look to be the top few possibilities and presenting them in a few seconds on the monitor screen, begging for the user to reach for a credit card. Nevertheless, airlines contribute to this process partly because competition constantly increases and because, thanks to various technology and new ways of ordering data, flights from almost anywhere to almost anywhere else are increasingly possible. Distance is no longer the barrier it was as little as fifty years ago.

Those parents of the baby boomers who flew in or before the mid-twentieth century recall a different travel experience than post-9/11 airports offer: friends and relatives could walk you to the gate and welcome you there when you returned; you could smoke on planes and even bring your own drinks; and you dressed up as if for a business meeting; meals were served and they often tasted good. Ironically, it was the success and safety of air travel, plus lowered costs and quicker journey times, that led to lowering the presence of gentility in the flight experience.

On the other hand, despite delays, congestion, and sometimes embarrassingly intimate searches for the sake of security, most people who fly usually get where they intended to go in pretty much the time promised. It's cheaper than a train, quicker and safer than driving. Airlines work ceaselessly to ensure that, sooner or later, everyone will fly.

RELATED ASSOCIATIONS & ORGANIZATIONS

Air Mail Pioneers, http://www.airmailpioneers.org

Aviation Safety Alliance, http://www.aviationsafetyalliance.com

Flying Apache Association, http://www.piperapacheclub.com/about_us.html

International Aviation Women's Association, http://www.iawa.org

National Gay Pilot's Association, http://www.iawa.org

Order of Daedalians, http://www.daedalians.org

Royal Air Force Historical Society, http://www.raf.mod.uk/history_old/rafhis.html

Swedish Aviation Historical Society, http://www.sff.n.se/engelska.htm

United Flying Octogenarians, http://www.unitedflyingoctogenarians.org

World War I Aeroplanes, Inc., http://www.aviation-history.com/ww1aero.htm

BIBLIOGRAPHY

"2001 Commercial Results Consolidate Airbus' Position as World's Leading Aircraft Manufacturer." Airbuss S.A.S. Available from <http://www.airbus.com>.

"Aerospace Industry Sales by Group and Customer, 1985–2005." *Statistical Abstract of the United States: 2007.* U.S. Department of Commerce, Bureau of the Census. December 2006, 646.

"Airbus Sees Strong Cargo Market." *The Journal of Commerce Online.* 30 March 2004.

"Boeing Commercial Airplanes." *Airfinance Journal.* April 2005.

"Bright Outlook in 2004." *Airline Business.* 1 February 2004.

"China to Buy 400 Aircraft in Five Years." *Alestron.* 3 July 2001.

"Component Tracking." *Flight International.* 13 April 2004.

Flores, Jackson. "Forecasts." *Flight International.* 27 April 2004.

Global Market Forecast 2004–2023. Airbus S.A.S. 2004–2005. Available from <http://www.airbus.com.>.

"India Will Buy 570 Aircraft—Forecast." *Airline Industry Information.* 22 June 2005.

Kingsley-Jones, Max. "Production." *Flight International.* 13 December 2004.

Lazich, Robert S. *Market Share Reporter.* Thomson Gale, 2004, Volume 1, 493–496.

Materials Research Agenda for the Automobile and Aircraft Industries. National Materials Advisory Board (NMAB). 1993.

Napier, David H. "2001 Year-End Review and 2002 Forecast: An Analysis." Aerospace Industries Association. 2002. Available from <http://www.aia-aerospace.org>.

Pattillo, Donald M. "Air Transportation and Travel." *Dictionary of American History,* 3rd ed. Charles Scribner's Sons. 2003, Volume 1, 82–86.

Taylor, Alex. "Lord of the Air." *Fortune.* 10 November 2003.

"Teal Group Analysts Predict that 6,743 Commercial Aircraft, Valued at $421 Billion, Are to Be Built Between 2003 and 2012." *Airfinance Journal.* September 2003.

"U.S. Firms Start Year Well." *Flight International.* 3 May 2005.

SEE ALSO *Helicopters*

Antiperspirants & Deodorants

PRODUCT OVERVIEW

Antiperspirants and deodorants are products used to prevent underarm odor. Their manufacture constitutes a major category of the larger cosmetics and toiletries industry, an industry that consists of other such personal care items as perfume, hair and shaving preparations, cosmetics, and creams and lotions, along with toothcare and bath and shower products. In 2005 the sale of antiperspirants and deodorants contributed $12 billion to a cosmetics and toiletries market valued at $253.3 billion worldwide. As Karl Laden notes in his 1999 book *Antiperspirants and Deodorants*, with over 90 percent of U.S. consumers using an antiperspirant or deodorant product regularly at the close of the twentieth century, these products were rivaled in the health and beauty sector only by toothpaste in frequency of daily use.

Though the natural odors of the human body came to be stigmatized in many parts of the world during the twentieth century, such an attitude has not always been the case. In Elizabethan England, lovers placed peeled apple in their armpits, later exchanging the sweat-soaked pieces of fruit as a way of sharing their affection. "Will be home in three weeks. Don't wash," the emperor Napoleon

famously wrote to his wife Josephine. Yet, in an admonition that could have been taken from an early American advertisement for soap, the first century poet Ovid, writing about the art of love, warned his readers of the difficulty of attracting members of the opposite sex when "harbouring a goat under the arms."

While the distinctive smell of underarms has been around for millennia, our understanding of what causes that smell had its genesis only as recently as 1833, the year the Czech scientist Johannes Evangelista Purkinje discovered sweat glands.

The human body has two types of sweat glands, the *eccrine* glands and the *apocrine* glands. Eccrine glands, the more numerous of the two, cover almost the entire body, whereas apocrine glands are located only in areas of concentrated hair growth such as the armpit, the genitals, and the scalp. Both glands produce a discrete form of sweat, neither of which is responsible for the smell associated with the odor emitted by a sweaty underarm. Instead, that smell occurs when bacteria naturally present on the skin react with proteins and lipids found in the secretion of the apocrine glands. The human underarm, both an area with high densities of bacteria on the skin as well as an area relatively free of air flow and therefore potentially moist and dark, provides an ideal environment for bacterial growth.

Eccrine glands are active in the human body from birth. They are linked with the autonomic nervous system, which means that they function involuntarily. They secrete a substance that consists of water, amino acids, electrolytes, and minerals, and they provide the body with a way to prevent reaching temperatures that might otherwise negatively affect well-being. Apocrine glands, on the other hand, develop during puberty and are largely activated in response to stress, emotions, or sexual arousal.

In the two centuries following Purkinje's discovery, scientists explored the chemistry of the underarm. They identified C19 androgen steroids and short chain fatty acids as sources of underarm odor and, in 1990, the discovery of 3-methyl-hexenoic acid as another source of odor was heralded in newspapers with such headlines as "Science sniffs out culprit in old mystery" and "Stink-tank scientist reports body odor breakthrough." While science made continual progress in understanding the chemical causes of what is commonly referred to as body odor, the history of human reaction to that odor and our efforts to control or mask it has been about as straightforward as our understanding of the very emotional states that lead to the production of sweat from the apocrine gland itself.

Social Context. There is no chronological advancement to be found from the past to the present with regard to personal cleanliness, only cycles of belief, social custom, and technological development. Attitudes toward personal cleanliness varied across cultures and throughout time according to practical factors such as the availability of water for bathing and the nature of living conditions as well as less tangible factors such as religious dictates and cultural norms. It appears, in addition, that smell is relative. As the American scientist W.T. Sedgwick conceded in a public health textbook at the turn of the nineteenth century, cleanliness is "doubtless an acquired taste."

There have been periods in history such as the medieval period in Europe and the early colonial era in North America, when bathing was rare. There was no running water, heating materials were expensive or required a great deal of time and energy to gather, sewage systems did not exist, and people often lived in close proximity with their animals. According to Charles Wysocki, an American physiobiologist who studies the chemistry of human olfaction, what is at first an unbearable odor can become commonplace after only a few days. This fact highlights the role that social norms play in dictating standards of personal hygiene and cleanliness.

For instance, while ancient societies such as those of the Greeks and Romans enjoyed highly developed water-delivery systems and elaborate bathing rituals, later societies such as those of early Christian Europe, not only lacked similar technological resources but even eschewed bathing and concerns with personal cleanliness as licentious and vain, activities to be avoided for moral as well as pious reasons. Meanwhile, in Islamic cultures during the same period, public steam baths called *hammams* were extolled as places of spiritual and physical purification.

The rise of concern with personal cleanliness in post-colonial America paralleled the rise of a public health movement and an increasing awareness of the importance hygiene plays in preventing disease. As in England, industrialization in the United States had been accompanied by an increasing number of people living in cities where a lack of plumbing and sewage systems exacerbated the incidence of illness. Mortality statistics from the 1800s show high rates of infant mortality due to diarrhea, the greatest cause of which was the communication of intestinal bacteria from adults who returned to children from the toilet without having washed their hands.

Early Market. A greater understanding among politicians and health reformers alike of the civic importance of plumbing and the significance of personal hygiene was accompanied by two factors closely linked with the formation of the antiperspirant and deodorant market: the growth of the soap industry and the beginnings of modern advertising.

Along with cereals, baking ingredients and foodstuffs, soap was one of the first products to be advertised nationally in the United States. Some of the earliest companies to make their start as soap producers and advertisers were Procter & Gamble, the Lever Brothers (now the Anglo-Dutch conglomerate Unilever) and Colgate-Palmolive; all three became global leaders in the cosmetics and toiletries industry.

The primary slant in early soap advertisements was the relatively utilitarian message of health reformers, thus emphasizing hygiene and the eradication of germs. By the early twentieth century, however, soap companies, with the help of advertising agencies such as the J. Walter Thompson Company, began to tap into a more psychological approach that linked cleanliness with social acceptance and self-esteem.

The term B.O. was first coined in 1926 print ads for Lever Brothers' Lifebuoy soap. The campaign ran for many years and depicted body odor as a social disgrace that could lead to ostracization, lost business ventures, and ruined romances. In one ad, a woman is pictured sitting on a man's lap. While he looks up at her imploringly, she looks away with a discouraged expression on her face. "Please tell me what is wrong dear!" reads a bubble caption that sits over the picture while below the picture in large bold print is written, "but she hadn't the courage to tell him he'd grown careless about B.O."

Other soap ads from this period, for products such as Palmolive and Procter & Gamble's Camay focused on soap's beautifying qualities and promoted smooth youthful complexions, in this way diverging from a message synonymous with the need for deodorants. However, while advertisements continued throughout the century to emphasize the ability of soap to leave bathers smelling good, the role of fighting the dreaded body odor fell increasingly to antiperspirant and deodorant products.

Product Development. In 1888 an unknown inventor from Philadelphia introduced the first ever trademarked deodorant. Called *Mum*, it was and marketed and distributed in the United States through his nurse. The primary active chemical in Mum was zinc oxide. The deodorant, which was sold in short jars, was a waxy cream that was messy to apply and potentially irritating to the skin. Though manufacturers continued to experiment with deodorants in the decades that followed, only after the first aluminum based compound was produced in 1942 did the antiperspirant and deodorant industry begin.

Aluminum compounds such as aluminum chloride, aluminum chlorohydrate, and aluminum zirconium glycine are key active ingredients in antiperspirants and deodorants. They are referred to as aluminum salts and are valued for their ability to temporarily plug sweat ducts, thus preventing the build-up of sweat in the underarm that leads to bacterial activity and odor.

In the late 1940s an employee at Mums came up with the idea for a new deodorant applicator. Inspired by the suggestion of a colleague, Helen Barnett Diserens developed a deodorant container based on the concept of a similarly new invention, the ball point pen. The new applicator was tested in 1952, and later marketed as Ban Roll-On.

By 1965 when the first antiperspirant aerosol deodorant was launched, the building blocks of an industry founded on technological innovation, scientific research, and savvy marketing had been set.

At the turn of the twenty-first century, when Americans were spending approximately $2 billion on deodorants and antiperspirants yearly, these products were produced in a wide variety of applications including pump sprays, squeeze sprays, aerosols, roll-ons, solid sticks, soft sticks, gels, and creams. While aluminum compounds form the primary active ingredient in most products, the development of compositions that would be clear upon application as well as the incorporation of hypoallergenic ingredients and milder, more natural fragrances were among the issues being explored in producing ever more effective and marketable products.

Industry Regulation. The terms antiperspirant and deodorant are often used interchangeably, but the two products are distinct. Deodorants reduce or prevent underarm odor by killing bacteria that cause odor and masking underarm smell through the use of fragrance. Antiperspirants inhibit the body's sweating process so that the underarm remains dry and bacteria have nothing to feed on in the first place. Despite this distinction, products marketed as deodorants most often also perform antiperspirant functions and antiperspirants, likewise, commonly contain fragrances meant to mask odor. In addition, the aluminum salts that make antiperspirants effective are known to have antibacterial qualities.

Prescription antiperspirants exist for the treatment of sweating disorders such as hyperhidrosis, excessive sweating. Dermatologists at the American Academy of Dermatology estimate the number of Americans suffering from excessive sweating to be approximately 8 million. Hyperhidrosis, when it occurs in the underarms, is most often treated with antiperspirants such as Drysol that contain high dosages of the active aluminum chloride compounds found in over-the-counter antiperspirants. In 2004 the drug Botox (botulinum toxin type A) administered as a shot to the underarm was approved for treatment in cases in which the condition could not be adequately managed by a topical agent.

In the United States, the Food and Drug Administration (FDA), following the dictates of the Federal Food, Drug, and Cosmetic Act, regulates products according to their intended uses. It considers any product that is designed to affect bodily structures or functions to be a drug. Since non-prescription antiperspirants are developed to inhibit the body's production of sweat, they are classified as over-the-counter drugs and are regulated by the FDA.

The FDA first began monitoring aluminum and aluminum zirconium containing products in 1977. Instead of reviewing each new antiperspirant that is produced, the agency publishes a monograph that lists the active antiperspirant ingredients that it has approved. As part of its process of approval, it is not only concerned with the safety of ingredients, but also with the veracity of any claims made about the product on its label. For instance, in a rule issued in 2003, the FDA allowed antiperspirant manufacturers to promote their products as being effective for up to 24 hours. In 2004 it reopened its ruling in order to test the efficacy of a new product claiming to provide sweat reduction for up to 48 hours.

If a product that is strictly a deodorant is found to cause harm once it is on the market, it is in the FDA's mandate to investigate that product's ingredients. Otherwise, because deodorants are classified as cosmetics by the FDA, they are not subject to the same kind of pre-market safety regulations as antiperspirants.

In 1976 the Cosmetic, Toiletry, and Fragrance Association, a trade association that represents the cosmetics and toiletries industry, established the Cosmetic Industry Review (CIR) with the support of the Food and Drug Administration and the Consumer Federation of America. The CIR gathers information about ingredient safety and provides that information to manufacturers. Improved consumer confidence in the cosmetic and toiletries industry stemming from the review of cosmetic ingredients and a resulting increase in manufacturer accountability was a notable factor in the industry's improved sales in the latter years of the first decade of the twenty-first century.

MARKET

According to data published by the U.S. Department of Commerce, Bureau of the Census, the U.S. Toilet Preparation Manufacturing industry posted a shipment total of $39.5 billion in 2005, up from $35.8 billion in 2004, $31.7 billion in 2003 and $33.2 billion in 2002. While the industry grew by an average of more than 5 percent between 1996 and 2000, its growth slackened in the early 2000s, slowing to a rate of only 2 percent between 2001 and 2002. By 2005 with a market value reaching $54 billion and an increase in sales of 3.3 percent, it had begun once again to pick up speed.

In 2002 the Census Bureau classified the deodorant and antiperspirant industry as a category of the larger toilet preparation manufacturing industry. It described toilet preparation manufacturers as those establishments that are engaged in the preparation, blending, compounding and packaging of personal care products such as perfumes, hair and shaving preparations, and creams and lotions. Statistics for the antiperspirant and deodorant product shipments within the toilet preparation manufacturing industry are classified as underarm deodorants, aerosol and spray, and underarm deodorants, roll-ons and solids.

Product shipment statistics published in the Census Bureau's *2002 Economic Census*, showed that shipments for aerosol and spray products were $1.4 million in 2002, down by approximately 30 percent from $2 million in 1997. Meanwhile, roll-ons and solids saw an increase in shipments of approximately 44 percent for the same period, rising from $1.2 billion in 1997 to $2.1 billion in 2002. The change reflected a trend that began in the 1990s with the introduction of new variations on the traditional stick antiperspirant/deodorant. The new products were considered invisible solids because they were smooth and dry upon application and did not leave a white residue. They were called, variously, clear gels, soft solids, sheer solids, and clear sticks, and developed in accordance with consumer preferences as revealed in market research conducted by antiperspirant and deodorant manufacturers.

The bar graph in Figure 3 presents the change in market share of antiperspirant and deodorant products by type in the United States from 1991 to 1998. It reflects the introduction of the new gel and soft solid products.

The U.S. cosmetics and toiletries market is one of the largest in the world. It is a also a mature market, which means that it has already reached a very wide base of

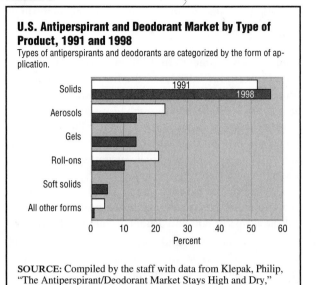

FIGURE 3

U.S. Antiperspirant and Deodorant Market by Type of Product, 1991 and 1998

Types of antiperspirants and deodorants are categorized by the form of application.

SOURCE: Compiled by the staff with data from Klepak, Philip, "The Antiperspirant/Deodorant Market Stays High and Dry," *Global Cosmetics Industry*. November 1999, page 24.

consumers and cannot count on tapping large sections of the population in order to achieve growth as an industry. Likewise, with most Americans already using an antiperspirant or deodorant daily, the manufacturers of these products must rely on technological innovation, product enhancement, and market research to find news ways of appealing to consumers.

Leading manufacturers of antiperspirants and deodorants in the United States such as Procter & Gamble, Mennen Co., and Helene Curtis are either multinational corporations themselves or are subsidiaries of multinational corporations such as Unilever, Colgate-Palmolive, and Henkel KGaA. Like the larger cosmetics and toiletries industry of which it is a part, the antiperspirant and deodorant industry constitutes a global market.

The North American and Western European markets were the two most highly developed markets worldwide in 2005, with North America accounting for 21 percent of global sales and Western Europe accounting for 30 percent. At the turn of the twenty-first century, however, the growth in the industry globally was increasingly being generated by activity in emerging markets such as those of Latin America, Eastern Europe and Africa, and the Middle East. With a market size of $28.8 billion in 2005, Latin America saw a jump of 11.3 percent in its share of the global market over 2004 figures. Equally impressive was Eastern Europe's increase of 9.6 percent over 2004, and an increase of 7.1 percent for Africa and the Middle East.

The Asia-Pacific region placed second globally in 2005 with a market valued at $63.1 billion, approximately $10 billion more than that of North America's for the same year. At an increase of 5 percent over 2004 earnings, the region's growth was largely due to the dynamism of the industry in China, where rising disposable incomes and growing product awareness contributed to the country's potential for market penetration.

One of the challenges for the makers of antiperspirants and deodorants seeking to expand into foreign markets is that consumer preferences vary from country to country. Since the early 1990s, for instance, Americans showed a consistent preference for antiperspirants over deodorants. French consumers, however, have been wary of the concept of stopping the body from sweating and showed a preference for deodorants over antiperspirants well into the early 2000s. Similarly, although aerosol products fell out of favor with U.S. consumers in the late 1970s, aerosol products were still a mainstay in the British market in the early 2000s.

Among emerging markets, Latin America posed a significant potential for continued growth in the first decade of the twenty-first century both as a global supplier of new cosmetic ingredients such as plant and fruit extracts and as an only partially tapped consumer base.

KEY PRODUCERS/MANUFACTURERS

At the close of 2005 Procter & Gamble was the world's leading manufacturer of cosmetics and toiletries. It had bumped major competitor and former leader L'Oréal Group from the top spot as the result of its acquisition of The Gillette Co. in October of that year. Euromonitor International's data for 2005 showed Procter & Gamble with a 12.8 percent share of the global market, followed by L'Oréal Group with 10.2 percent, Unilever Group with 7.5 percent, and Colgate-Palmolive with 4.1 percent. Each of these is profiled in order of market share.

According to *Market Share Reporter 2007*, Procter & Gamble also came out on top of the U.S. deodorant and antiperspirant market in 2005, taking a 29.2 percent share. The Gillette Co. was in second place with 18.9 percent, followed by Mennen Co. at 12.5 percent, Helene Curtis at 9.6 percent, and Church & Dwight at 6.3 percent.

Procter & Gamble. This market leader is a multinational consumer goods company. With operations in nearly 80 countries, Proctor & Gamble derives more than half its revenues from overseas and markets its nearly 300 brands in more than 110 countries. It divides its business into two main global units, health and beauty, and household care. For a period following its acquisition of The Gillette Co., it had a third category which it called Global Gillette. Its Gillette unit was incorporated into its two other units in July 2007. The company had sales of $68.2 billion for

the fiscal year ending June 2006 and employed 138,000 workers.

Formed in Cincinnati, Ohio, in 1837, Procter & Gamble began as a partnership between William Procter, a candle maker, and James Gamble, a soap maker. Both men were immigrants, Procter from England as an adult and Gamble from Ireland as a child. They had married sisters Olivia and Elizabeth Norris and their father-in-law urged them to form a business together. Though candle making and soap making might seem an odd coupling, both products at the time relied on the same key ingredient, lye. Lye was made from animal fat and wood ashes, and Cincinnati, home to a booming pork trade, provided ample resources for pork fat.

In 1875, the company hired its first full-time chemist, and in 1878 it introduced a new soap product which it called White Soap, later renamed Ivory. Affordable and yet of a quality equal to expensive Castile soaps, Ivory soap became the first of what would be many innovative products introduced by Procter & Gamble. Some of these products were Crisco vegetable shortening (1911), the first synthetic detergent (1933), Tide (1946), Crest toothpaste (1955), and Pampers disposable diapers (1961). Procter & Gamble introduced its Secret deodorant and antiperspirant line in 1960.

Procter & Gamble aggressively acquired companies in the mid-1950s, purchasing the Charmin Paper Company, the Clorox Chemical Company, and the Nebraska Consolidated Mills Company all by the end of that decade. In 1988 it took its first step into the cosmetics industry with its purchase of Noxell Corporation, producers of Cover Girl cosmetics and Noxema products. Other cosmetics and toiletries purchases included the Old Spice line of fragrances, skin care products, antiperspirants, and deodorants from the Shulton Company in 1990; the Max Factor and Betrix brands from Revlon Inc. in 1991; and the Clairol hair preparations business from Bristol-Myers Squibb in 2001.

The Gillette Company. In the late 1890s, King Gillette, already a successful salesman, inventor, and writer, set out to bring an inexpensive and effective disposable razor to market. In 1901 William Nickerson, a machinist who had been educated at the Massachusetts Institute of Technology, took an interest in developing the product with Gillette. That same year, Gillette formed the American Safety Razor Company to raise money for the development of the razor. He renamed the company the Gillette Safety Razor Co. in 1903 and in 1904 the company received a patent for its new razor.

The Gillette Company, based in Boston, Massachusetts, became a world leader in men's grooming products. Though its business remained grounded in the production of razors and blades, which it marketed to both men and women, it was equally successful as a manufacturer of toiletries. Three mainstays in the antiperspirant and deodorant market, Right Guard, Soft & Dri, and Dry Idea, were Gillette products.

In October 2005 Gillette became a subsidiary of Procter & Gamble, which purchased the company for approximately $57 billion in stock. The European Union and the United States Federal Trade Commission approved the merger with the condition that overlapping products be divested. Accordingly, in 2006 Gillette's Right Guard, Soft & Dri and Dry Idea product lines were sold for $420 million to The Dial Corporation, a subsidiary of Germany's Henkel KGaA.

Mennen Company. The Mennon Company, producers of the highly successful Mennen Speed Stick line of antiperspirants and deodorants for men, became a subsidiary of the Colgate-Palmolive Company in 1992. Mennen had been launched earlier in the century with a talcum based powder invented by its founder Gerhard Heinrich Mennen, a German immigrant. Originally based in New York, the company moved its headquarters to Morristown, New Jersey, in 1954. Among other products manufactured by Mennen were its Baby Care products, a line of toiletries for babies, the aftershave lotion Skin Bracer, and Lady Speed Stick, a line of antiperspirants and deodorants for women.

Colgate-Palmolive Company. The Colgate-Palmolive Company, based in New York, New York, is one of the most powerful consumer goods companies in the world. Its international presence includes operations in over 200 countries and it reported sales of over $12 billion for the fiscal year ending December 2006. Its brands include such mainstay products as Colgate toothpaste, Palmolive and Irish Spring soaps, Fab laundry detergent, and Ajax cleanser. It is also a major player in the manufacture of pet care products.

The company began as a manufacturer of soap, candles, and starch. Founded by William Colgate in 1806, it was originally incorporated as the Colgate Company. Upon the death of William Colgate in 1857, the founder's son changed the company's name to Colgate & Company. The company's name changed again in 1928, when it merged with soap manufacturers Palmolive-Peet to become Colgate-Palmolive-Peet. In 1953 the company dropped Peet from its title.

The Colgate-Palmolive Company was the first manufacturer to produce toothpaste in tubes. It was also a pioneer among U.S. companies in expanding operations abroad, creating a Canadian subsidiary in 1913

and a French one in 1920. Following throughout the 1920s were operations in Australia, the United Kingdom, Germany, Mexico, the Philippines, and Argentina, among others.

Major acquisitions in the personal care category for the company have included the purchase of cosmetics manufacturer Helena Rubenstein in 1973 and the purchase of Mennen Co. in 1992. In 2006, the company announced its intended acquisition of Tom's of Maine for $100 million.

Helene Curtis Industries Inc. This wholly owned subsidiary of the Anglo-Dutch consumer products giant Unilever is a personal care products company based in Chicago, Illinois. Helene Curtis' best-selling toiletries include shampoos, conditioners, antiperspirants and deodorants, and hand and body lotions.

The company was first incorporated as the National Mineral Company in 1927. Its founders, Gerald Gidwitz and Louis Stein, soon recognized that their one product, a facial mask made of clay mined in Arkansas, would not be enough to sustain them in an increasingly competitive market for women's toiletries. They turned to haircare products and in the 1930s introduced the first ever mass-produced hair-waving pads for the creation of permanents as well as one of the first detergent-based shampoos ever to be manufactured in the United States.

Innovation became a hallmark for Helene Curtis, which was renamed as such after Louis Stein's wife and son. In the 1940s and 1950s the company's Suave shampoo brand took the market by storm, followed by an aerosol deodorant, Stopette, that was a bestseller for several years. It was also during this period that Helene Curtis coined the term hairspray with the introduction of its aerosol hair product Spray Net.

Throughout the 1960s and 1970s, Helene Curtis expanded on the success of Suave shampoo, introducing creme rinses and wave sets under the Suave name. In the deodorant and antiperspirant category, it launched Secure, a powder deodorant, which was followed by a Suave brand roll-on. The 1980s saw the company begin producing skincare lotions and make two new highly successful haircare launches, Finesse conditioner and the Salon Selectives line of products.

The launch of the Degree brand of antiperspirants and deodorants in 1990, garnered the company a large share of the antiperspirant and deodorant market. However, the 1990s found Helene Curtis struggling to keep up with larger competitors such as Procter & Gamble and Unilever, and in February 1996 it announced that it would be sold to Unilever for approximately $770 million.

MATERIALS & SUPPLY CHAIN LOGISTICS

The primary active ingredients in deodorants and antiperspirants are antibacterial agents and aluminum compounds. Whereas such antimicrobials as ethanol, citricidal, triclosan, and zinc phenolsulphonate are effective in killing the bacteria associated with the formation of underarm odor, aluminum compounds such as aluminum chloride, aluminum chlorohydrate, aluminum sulfate, and aluminum zirconium glycine dissolve on the skin and hydrolyze in sweat ducts, thus preventing build-up of the sweat on which the bacteria feed.

One of the keys to a successful antiperspirant product is that its aluminum compound should be present at strong enough levels so as to prevent perspiration and yet not so strong that it irritates the skin. To achieve this balance, manufacturers calibrate the compound's acidity, which is the quality that makes it effective at blocking sweat ducts, with its base, the quality that neutralizes its potentially harmful effects. Early aluminum-chloride based antiperspirants suffered from just such an imbalance, not only causing skin rashes but also damaging clothing.

The non-active ingredients present in deodorants and antiperspirants depend largely on the form of application the product takes. In the early years of their development, antiperspirants and deodorants were available primarily as creams, powders, or roll-ons. In 1965 the Gillette Company became the first manufacturer to produce an antiperspirant in an aerosol dispenser when it introduced Right Guard. By the late 1960s aerosol antiperspirants and deodorants stormed the marketplace.

What gave the aerosol products their propulsion were fluorocarbons, which were favored for the soft dry spray they produced. However, in 1978 the U.S. government banned the use of fluorocarbons due to research that identified the gaseous compound as detrimental to the ozone layer of the earth. Manufacturers later replaced fluorocarbons with alternatives in aerosol products; however, negative publicity had a strong effect on the market. By 1982 U.S. sales of aerosol antiperspirants and deodorants dropped to a 32 percent share of the market from a high mark of 82 percent during the 1970s.

First formulated in 1934, solid stick products quickly filled the gap left by aerosols. They consisted of an active antiperspirant/deodorant ingredient suspended in a waxy base. Common base materials were stearyl alcohol, cetyl alcohol, hydrogenated castor oil, and glyceryl stearate. Base materials were blended with lubricating oils and emollients such as cyclomethicone, a volatile silicone compound.

The last two decades of the twentieth century saw new developments in the effectiveness of standard aluminum chlorohydrate and aluminum zirconium actives.

Though chemically indistinguishable from their standard forms, the new actives, referred to as enhanced efficacy aluminum compounds, were processed so that they contained lower weight molecular polymers, a property that enabled them to provide sweat prevention for longer periods of time.

One of the key suppliers of materials to the antiperspirant and deodorant industry is the specialty chemical manufacturer Reheis Inc., a subsidiary of GenTek. Based in Berkeley Heights, New Jersey, Reheis produces a range of 45 to 50 aluminum compounds including aluminum chlorohydrates and aluminum zirconium chlorohydrates. The company played a major role in the research and development of aluminum compounds. Some of its products are Rezal 36 GP, Rezal 36 GP SUF, Reach 301, and Reach-AZP 908.

Other inactive ingredients in antiperspirants and deodorants are fragrances, colorants, and skin care agents. Powders such as talc or starches enhance products esthetically, as do alkanolamides and propoxylated alcohols, which provide emolliency to sticks, and silicones, which act as both effective lubricants and anti-whitening agents. Glycerol esters keep metal agents such as aluminum zirconium stable in stick formats.

Some of the companies supplying inactive materials to antiperspirant and deodorant manufacturers are Dow Corning, ICI Specialty Chemicals, and Witco.

DISTRIBUTION CHANNEL

Traditionally, cosmetics and toiletries were classified as prestige products and mass market products. Prestige products were sold at department stores and select boutiques while mass market products could be found at supermarkets and drugstores. Unlike such cosmetic and toiletry categories as perfume, cosmetics, and creams and lotions, antiperspirants and deodorants never held the glamour of a prestige brand. They were more of a workaday product, a toiletry that is considered a necessity more than a prestige indulgence.

In the United States during the 1990s, the presence of large discount stores such as Wal-Mart began to reshape the retail landscape for cosmetics and toiletries. In 2001 figures from ACNielson showed that discount stores expanded their share of antiperspirant and deodorant retail sales from 35 percent of the market in 1995 to 42 percent in 2001. As of 2001 discount stores accounted for a majority of the volume in sales of antiperspirants and deodorants in the United States. Conversely, supermarkets sales of antiperspirants and deodorants over the same period decreased from 41 percent to 36 percent, and drugstores sales declined from 24 percent to 22 percent.

In the early 2000s consumers were willing to pay more for cosmetics and toiletries if these products provided good quality. This phenomenon led to the coining of the term masstige. Masstige brands referred to products that might have previously been considered prestige brands and therefore sold at a relatively exclusive location such as the cosmetics counter at a department store, but that were now being sold as mass market products in larger outlets such as discount stores and supermarkets.

Consumer interest in quality products and increased awareness of potentially harmful cosmetic chemicals led the manufacturers of antiperspirants and toiletries to develop products that contained many of the qualities associated with a prestige brand, such as ingredients that are good for the skin and possess a natural, non-synthetic smell. Such products were increasingly being sold at health food stores and by environmentally conscious retailers.

KEY USERS

Key users of deodorants and non-prescription antiperspirants are those individuals who regard underarm smell as something to be avoided. The desire to avoid emitting a body odor is often a subjective one. Physiological variations may lead some individuals to generate more underarm odor than others. Social and cultural norms also play a role in the decision to use an underarm product. Therefore, key users are most likely consumers who live in countries and societies that equate body odor with bad grooming, and bad grooming with low social as well as economic status, low self-esteem, and a general lack of respectability.

ADJACENT MARKETS

Antiperspirants and deodorants are closely adjacent to other scented toiletry items like perfumes and colognes, hair preparations, and creams and lotions. While these products contain fragrances that enhance their appeal to consumers, they are not developed to provide the preventative action of underarm antiperspirants and deodorants, nor do they claim to.

More directly adjacent to the function of underarm deodorants at the turn of the twenty-first century were body sprays and deodorizing soaps. While deodorizing soap brands such as Lifebuoy and Dial had been in the market since the early days of the toiletries industry, body sprays were introduced in the 1980s, taking off in Europe before making a successful entrance into the U.S. market in the first decade of the twenty-first century.

Body sprays are products that function between a perfume or cologne and a deodorant. As with a cologne, fragrance is a key selling factor for body sprays and brands are marketed in a variety of scents. Like a deodorant, these products claim to keep those who use them free of body odor and smelling fresh all day. Unlike deodorants, body sprays are meant to be applied all over the body, not just to the armpits.

Unilever kicked off the market for body sprays in the United States when it introduced its Axe brand in 2002. Axe, which is called Lynx in the United Kingdom and Australia, successfully targeted male consumers between the ages of 12 and 24. Following Unilever were Gillette with Right Guard Xtreme Sport body spray and Procter & Gamble with Old Spice High Endurance body spray. By the latter part of the first decade of the twenty-first century, body sprays were produced and marketed for women as well as men.

The information industry is adjacent to the antiperspirant and deodorant market. Information about product ingredients and industry ethics is accessible to consumers over the Internet. In the first decade of the twenty-first century consumers held companies accountable for such issues as the testing of ingredients on animals and the use of toxic substances in their products. An organization called the Campaign for Safe Cosmetics is an example of the adjacent information industry.

RESEARCH & DEVELOPMENT

Manufacturers of antiperspirants and deodorants continuously look for ways to make products more effective at preventing body odor. The main discovery for the antiperspirant and deodorant market was in 1942 when aluminum cholorhydrate was first produced in the United States. Since then, further research has led to the development of enhanced active aluminum compounds that provide longer lasting protection.

In the 1990s the development of products such as clear gels, soft solids, and invisible solids, satisfied the demand for antiperspirants and deodorants that apply clear and do not leave white marks on skin or clothing. Numerous patents during this and the following decade also addressed the need for improved deodorant actives. Some of the methods explored included the use of a chemical referred to as DTPA (diethylenetriaminepenta-acetic acid) that was found to bind to the iron contained in sweat, thus disabling bacteria that are known to feed on iron; a method that resulted in the inhibition of androgen receptor expression, which is known to carry human odor; and the use of a cellulose fiber product that was found to be an excellent deodorizer.

In the first decade of the twenty-first century, fragrance and the use of non-toxic substances were increasingly important to consumers of antiperspirants and deodorants. Research and development focused on the development of new, more natural and less synthetic smelling products as well as the inclusion of skin-friendly ingredients.

CURRENT TRENDS

At the turn of the the twenty-first century, cosmetics and toiletry manufacturers capitalized on well-known brand names to create lines of products referred to as mega brands. For the antiperspirant and deodorant industry, tried and true names such as Secret, Old Spice, Degree, and Soft and Dri were produced in different application forms from soft solids and clear gels to body mists and more. Within each form category, each product was manufactured in approximately four to six different scents. Marketing encouraged consumers to mix and match form types and fragrances to come up with a deodorant that spoke to their individual tastes and needs.

Another trend in underarm products was skin-friendly marketing. Products were introduced that contained such well-known skincare ingredients as aloe vera and vitamin E. They also allied themselves with brand names that were mainstays in the skincare industry. For instance, one of the products under the Secret brand was Secret Platinum and Olay, which made use of the famous skincare brand Oil of Olay. The Dove brand, originally a soap known for its moisturizing and gentle qualities, branched out into antiperspirants and deodorants in 2001, and marketing of these products highlighted their moisturizing and gentle effects on underarm skin as well as their strength at preventing odor.

Paralleling the interest in organic foods and health products in the two decades surrounding the turn of the twenty-first century, was the desire for cosmetics and toiletries that beautify and nurture with natural, non-toxic ingredients. Companies such as Tom's of Maine, Jason Natural Products, and Burt's Bees produced deodorants that were aluminum-free and made use of botanical ingredients such as hops, lemongrass, coriander, and lavender. L'Oréal acquired The Body Shop and Colgate-Palmolive acquired Tom's of Maine—evidence that traditional players in the industry saw value in the growing base of consumers interested in products containing natural ingredients.

TARGET MARKETS & SEGMENTATION

In the first decade of the twenty-first century the antiperspirant and deodorant industry targeted the male consumer. Penetration into the male market had traditionally been lower than that of the female market. A rise of interest among men in cosmetics and toiletries in general led to their increased use of deodorants and antiperspirants. Sales to young men between the ages of 12 and 34 drove growth in the industry as whole.

Scent proved to be a leading factor in a man's choice of product, and body sprays such as Axe Body Spray Deodorant, Old Spice High Endurance, and Right Guard Sport, though not antiperspirants, appealed to young men with fragrance names such as Aqua Reef, Arctic Force, Voodoo, Vitality, and Tsunami. Marketing campaigns centered around sports-related promotions as well as the selling of image—men who use the product being appealing to women, men who use the product being risk-takers and adventurers. Within the male market, Mennen Speed Stick and Degree continued to be first choice among men between the ages of 35 to 44.

In continuing to target the female market, manufacturers promoted the skincare benefits of their products. Campaigns focused on the esthetics of products, such as their smooth, dry feel upon application, the freedom from worry provided by their long-lasting protection, and their conditioning qualities for delicate underarm skin. Within the women's category, age was also a factor. Dove Pro-Age, a line of toiletries, included antiperspirants and deodorants targeted to mature women. The Secret Sparkle collection targeted teenage girls.

RELATED ASSOCIATIONS & ORGANIZATIONS

The Campaign for Safe Cosmetics, http://www.safecosmetics.org

Cosmetic, Toiletry and Fragrance Association, http://www.ctfa.org

The European Cosmetic Toiletry and Perfumery Association, http://www.colipa.com

Synthetic Organic Chemical Manufacturers Association, http://www.socma.com

Cosmetic Ingredient Review, http://www.cir-safety.org

BIBLIOGRAPHY

Abrutyn, Eric. "Global Underarm Technical Review 2002: A Market Looking Toward Differentiation." *Soap & Cosmetics.* November 2002, 25.

Alexander, Philip. "Solid Answers to Sticky Problems." *Manufacturing Chemist.* August 1992, 32.

Davies, Briony. "State of the Industry." *Global Cosmetic Industry.* June 2006, 28.

———. "State of the Industry." *Global Cosmetic Industry.* June 2007, 38.

Klepak, Philip. "The Antiperspirant/Deodorant Market Stays High and Dry." *Global Cosmetics Industry.* November 1999, 24.

Klingler, Laura. "Deodorants." *MMR.* 3 September 2001, 46.

Laden, Karl, ed. *Antiperspirants and Deodorants,* 2nd ed. CRC Press, 1999.

Lazich, Robert S. *Market Share Reporter 2007.* Thomson Gale, 2007. Volume 1, 298–303.

Parekh, Joe. "Enhancing Efficacy." *Soap Perfumery & Cosmetics.* July 2001, 22.

Rados, Carol. "Antiperspirant Awareness: It's Mostly No Sweat." *FDA Consumer.* July-August 2005.

Sivulka, Juliann. *Stronger Than Dirt: A Cultural History of Advertising Personal Hygiene in America, 1875 to 1940.* Humanity Books, 2001.

Short, John. "You Don't Have to Smell Like a Goat: A Review of Deodorant Research." *Soap Perfumery & Cosmetics.* July 1996, 36.

"Statistics for Industry Groups and Industries: 2005." *Annual Survey of Manufactures: 2005.* U.S. Department of Commerce, Bureau of the Census. Novermber 2006.

Stuller, Jay. "Cleanliness Has Only Recently Become a Virtue." *Smithsonian.* February 1991, 126.

"Toilet Preparations Manufacturing: 2002." *2002 Economic Census.* U.S. Department of Commerce, Bureau of the Census. December 2004.

SEE ALSO *Perfumes, Soaps & Detergents*

Athletic Shoes

—————◆—————

INDUSTRIAL CODES

NAICS: 31–6219 Other Footwear Manufacturing, 31–6211 Rubber and Plastics Footwear Manufacturing

SIC: 3149 Footwear (except Rubber), not elsewhere classified, 3021 Rubber and Plastics Footwear

NAICS-Based Product Codes: 31–621903, 31–621904, 31–62190Y, 31–62111, 31–62114, and 31–6211W

PRODUCT OVERVIEW

For almost a century, many Americans have worn canvas shoes with rubber soles in their youth. They were commonly called tennis shoes. Sometimes the same type of shoe was made of leather and rubber. Tennis shoes were referred to interchangeably as tennis shoes or sneakers. The English call sneakers *plimsolls*, a generic term for lightweight canvas shoes with rubber soles first produced in England. In Germany, the type of shoes casually referred to in the United States as sneakers are called trainers. These were casual shoes for playing outside, worn mostly by children.

A variation on the canvas and rubber shoe is the high top tennis shoe. Converse Rubber Shoe Company, headquartered in Malden, Massachusetts, introduced canvas high top shoes in 1917. Foreshadowing the later close connection between sports heroes and athletic shoe marketing, Converse classic high tops were made famous by Chuck Taylor. Taylor was a former basketball player for the Akron Firestones who joined Converse as a player

endorser in 1921. For thirty-five years he led basketball clinics across the country wearing Converse high top basketball shoes. In 1931 Converse started stitching the Chuck Taylor name on its canvas high top basketball shoe and renamed them the All Star. At this time, the celebrity athletic shoe era was in its infancy, where it remained for nearly fifty years.

During World War II, Converse manufactured shoes and apparel for the troops. Soldiers were outfitted in Converse Chuck Taylor All Star shoes for basic training. After the war, Converse All Stars remained the standard government issue athletic shoes for military and physical training. In the 1950s Hollywood icon James Dean wore "Chucks," as these shoes came to be called.

Along with James Dean, Converse high tops became a symbol of counterculture. A white T-shirt, blue jeans, black leather jacket, and high tops became the uniform of youth rebellion. In the 1970s punk rock became the symbol of counterculture. Early punk bands, such as the Ramones, wore Chucks at their performances in New York City and in the United Kingdom at Roundhouse in London. Converse continued to produce the shoes and athletes on the U.S. male basketball team won gold at the 1984 Olympics in Los Angeles wearing Converse basketball shoes.

Athletic shoes are built for high performance and are considered part of an athlete's gear along with clothing and equipment. Athletic shoes provide sports-specific levels of cushioning, flexibility, stability, traction, and durability.

- Cushioning minimizes the force of impact. Cushioning systems are designed to protect the knee and ankle from impact and provide rebound, spring, or lift to the foot and leg muscles.

- Flexibility is the ability to yield and bend. It is maximized or minimized in a shoe, depending on the sport. For example, runners need flexible outsoles and walkers require stiff, inflexible outsoles.

- Stability is the capacity to resist forces which cause a rapid change of motion and possible injury. High top shoes provide extra ankle support to prevent sprains.

- Traction is friction between the sole and the surface that helps the shoe to grip. Traction needs vary. Court shoes should not grab the surface and stop so suddenly that ankles are sprained.

- Durability is important. Athletic shoes should endure and continue to perform over time while maintaining cushioning, flexibility, stability, and traction.

An athletic shoe has four basic components. The layers are, from top to bottom, called the upper, insert, midsole, and outsole. Each component has a purpose. Sports-specific shoes vary the components for different functions.

The upper layer is the material—generally canvas or leather—that covers the top of the foot. Some contain nylon or mesh inserts for ventilation. The upper encloses the foot to snug the soles to the bottom of the feet. The shoe insert supports the bottom of the foot. The insert positions the arch support, which is ideally suited to the wearer's arch type.

The midsole contains the cushioning system that not only distributes impact force to prevent the full transfer of that force to the ankle and knee but also provides a rebound or lift to foot and leg muscles. Known as the heart of the shoe because it performs so many functions, the midsole also contains stabilizing features like a stiff rear heel cup for added support. The outsole provides traction and reduces wear on the midsole to increase durability. Typically made of carbon rubber or blown rubber, outsoles have a grade or slope from heel to toe.

Manufacturers invest significantly in the technology of athletic shoes. In order to help the athlete run faster, jump higher, and endure longer, designs for athletic shoes consider specific movements involved with different sports. Some of these movements are sprinting, pivoting, jumping, rapid starts and stops, and side-to-side shuffling. Athletic shoes are designed as sporting equipment that helps improve performance.

Categories of Athletic Shoes. The May 2007 issue of *Clinical Reference Systems* defined five major athletic shoe categories.

Running Shoes. Running shoes are lightweight and flexible. They are designed for anterior (forward) and vertical (up and down) motion. Running shoes have cushioning to absorb the impact of each stride, plus extra shock absorption in the heel.

Walking Shoes. Walking shoes are rigid and are designed to roll on a smooth tread from heel to toe. Walking shoes have extra cushioning at the ball of the foot. The rigid outsole rocks to encourage the natural roll of the foot during the walking motion.

Basketball Shoes. Basketball shoes have a thick stiff sole, and are designed to support the ankle for side-to-side shuffling and to provide for sudden starts and stops. Basketball shoes are subject to heavy abuse.

Tennis and other Racquet Sport Shoes. Court shoes have heavy traction to assist in rapid starts and stops. Posterior (backward) motion is more common in court sports.

Field Sport Shoes. These usually have cleats, spikes, or studs and include shoes for track and field, golf, bicycling, soccer, and baseball. While spike and stud formations vary from sport to sport, they are generally replaceable or detachable and affixed into nylon soles.

High performance sports-specific athletic shoes have become high fashion. Nike, Inc. is generally credited with blurring the line between performance and fashion. The classic example is the Air Force 1. A limited edition shoe rolled out by Nike in 1982 on the National Basketball Association courts that became so popular that Nike reissued the $89.95 white-on-white original and began to introduce colorful limited-edition versions. Enthusiasm for Air Force 1s spread.

Nike does not deserve sole credit. The Zeitgeist of the 1980s facilitated the popularity of high-fashion, high-priced athletic shoes. The 1980s were the era of hip hop. Early rappers, Run-D.M.C., had the first rap single played on MTV, and the first rap album to go gold. Run-D.M.C. established urban fashion trends by performing in leather pants, leather jackets, adidas sneakers—always with the shoelaces removed—and fedora hats. They performed at Live Aid in 1985. They were the first rappers on the cover of *Rolling Stone*. They wore athletic shoes designed by adidas and even had a hit song called "My adidas."

Nike and adidas were not operating in a vacuum, however. The National Aeronautics and Space Administration (NASA) takes credit for the development of high-tech athletic shoes. After NASA developed a process called blow rubber molding to produce helmets, a former NASA engineer pitched Nike the idea to use the new molding process to create hollow athletic shoe soles designed to be filled with cushioning materials. Nike developed cushioning consisting of interconnected air cells. In 1987 Nike Air shoes were released. Athletes and non-athletes bought them for the benefit of high-tech cushioning.

In the shoe industry trends constantly change and new models are rolled out up to four times each year. Yet some styles, such as the Air Force 1, remain popular. The Air Force 1 Web site chronologically showcases all 929 models including the still popular black-on-black model rolled out in 1993. Air Force 1s for women are often a fashion statement. For men, Air Force 1s often function as dress shoes that complete the outfit.

Athletic shoes are no longer worn only for fitness and sports; they are also a fashion statement. This expands the pool of potential customers eager to pay for the look even if they never go near a basketball court or grassy soccer pitch. The design of high-tech, high-performance athletic shoes has permeated the artistic community as well. In 2000 the San Francisco Museum of Modern Art presented "Design Afoot: Athletic Shoes 1995–2000." The exhibit featured more than 150 shoes from companies such as adidas, Converse, and Nike. The exhibit highlighted the evolution of athletic footwear design and demonstrated the blurred line that exists between fashion footwear and high performance athletic shoes.

Athletic shoes are considered nondurable consumer goods, which are goods that are purchased for immediate or almost immediate consumption and have a life span ranging from minutes to three years. For active people, athletic shoes last barely a year. For instance, running and walking shoes last for 300 to 500 miles. Basketball shoes take a lot of abuse; it is estimated that they last six months if the wearer plays twice a week for one hour. Because nondurable goods are destroyed by their use, consumers need to replenish their supply repeatedly. Generally, this equates to a large variety of style and price choices in the marketplace.

MARKET

Athletic shoes are built for high performance and are considered part of an athlete's gear. They make up just one part of the sporting goods market typically thought to include footwear, equipment, and clothing. The total retail sporting goods market was valued at approximately $50 billion in 2005 according to estimates made by the National Sporting Goods Association. Footwear made up one-third of this total.

The National Sporting Goods Association has tracked consumer spending for athletic shoes since at least 1988. Its twenty-six categories of athletic shoes include aerobic, baseball/softball, basketball, boat/deck, bowling,

FIGURE 4

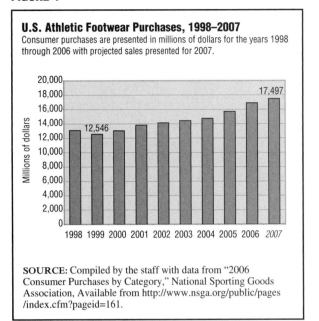

U.S. Athletic Footwear Purchases, 1998–2007

Consumer purchases are presented in millions of dollars for the years 1998 through 2006 with projected sales presented for 2007.

SOURCE: Compiled by the staff with data from "2006 Consumer Purchases by Category," National Sporting Goods Association. Available from http://www.nsga.org/public/pages/index.cfm?pageid=161.

cheerleading, cross training, cycling, fashion sneakers, fitness, football, golf, gym shoes/sneakers, hiking, hunting, jogging/running, skateboarding, soccer, sport sandals, tennis, track, trail running, volleyball, walking, water sports, and wrestling. To gather its data, the National Sporting Goods Association surveys 20,000 households twice a year for a total of 40,000 households.

Figure 4 depicts a ten-year sequence of consumer purchases of athletic shoes from 1998 to 2007. Consumer spending on athletic shoes steadily increased from a low of $12 billion in 1999 to a high of a projected $17.5 billion in 2007. Consumer spending increased during the period by almost 25 percent. The market grew in part because Americans buy more shoes. In 1980 each American purchased 5.7 pairs. By 2005 per capita purchasing was 7.71 pairs per year, a 40 percent increase.

While consumer spending on athletic shoes steadily increased, U.S. production steadily decreased. According to U.S. Census Bureau reports on footwear from their *Current Industrial Reports* series, in 1992, U.S. shoe manufacturers shipped products valued at $4.8 billion and within a decade shipments fell more than 50 percent, down to $2.2 billion in 2002.

The American Apparel and Footwear Association has tracked U.S. production and U.S. imports of shoes since at least 1990. Rather than focusing on consumer spending as does the National Sporting Goods Association, the American Apparel and Footwear Association counts numbers of shoes in pairs. For athletic shoes, 100 percent are imported as of 2005, the year for which most recent data are available. Of imports, 78 percent of athletic shoes come from China.

The National Sporting Goods Association conducts twice yearly surveys of footwear retail sales. Retail sales of athletic footwear rose 1.0 percent during the 6-month survey period of October 2006 through March 2007. Half of the sixteen types of athletic shoes covered in the National Sporting Goods Association 2006–2007 survey showed sales declines. Among the most popular athletic shoe categories—those with projected sales of five million units or more—only gym shoes/sneakers, fashion athletic shoes, and walking shoes increased in dollar sales. Fashion sneakers led the dollar increase, up 10 percent. Gym shoes/sneakers were up 2 percent. Walking shoe sales rose 6 percent.

Basketball shoe sales had the largest decrease among major athletic shoe categories, with a decrease of 4 percent. This decrease in sales of basketball shoes—a high-tech, high-priced segment of the shoe market—does not bode well for key producers of athletic shoes.

KEY PRODUCERS/MANUFACTURERS

In the U.S. marketplace adidas and Nike are essentially a duopoly. Since 100 percent of athletic shoes are imported, all key U.S. players have manufacturing operations abroad. For example, in 2004 Nike had 900-plus supplier factories—none of them owned by Nike—in fifty countries. In August 2007 the National Sporting Goods Association estimated that Nike controls almost half of the market for athletic shoes, jerseys and clothing.

Nike, Inc. Based in Beaverton, Oregon, Nike designs, markets, and distributes athletic shoes, clothing, and equipment. Nike reported $16.3 billion in worldwide revenue for the year ending May 31, 2007. Bill Bowerman (1911–1999), a University of Oregon coach, and Phil Knight, a University of Oregon accounting student and a middle-distance runner under Bowerman, each chipped in $500 in 1964 to start importing low-priced, high-tech athletic shoes from Japan.

From the beginning their goal was, in part, to dislodge the German domination of adidas in the U.S. athletic shoe market. By 1971 Bowerman invented a new shoe based on treads inspired by his wife's waffle iron. The waffle trainer became a bestseller and in 1971 the two founders took the name Nike, the goddess of victory, and adopted the swoosh logo. The Nike 1980 initial public offering involved 2.4 million shares at $11 each. After several splits, the stock traded at approximately $55 in August 2007.

Nike's 1982 success with Air Force 1s precedes its 1985 launch of the Air Jordan, endorsed by celebrity athlete Michael Jordan of the Chicago Bulls basketball team. Jordan signed a five-year contract in 1984 for $2.5 million

to represent the first marquee basketball shoe. Close to twenty-five years later Air Jordan's remain hip; forty-four Air Jordan models are available at prices of $175 and less.

In 1994 Nike entered the soccer market, an area historically dominated by adidas. Ten years later its share of the European soccer shoe market was 35 percent, exceeding adidas' 31 percent for the first time. To gain market share in Europe, Nike built on its celebrity athletic shoe marketing savvy. It paid the prestigious Manchester United soccer club in the United Kingdom $450 million over 14 years to outfit its players and to run its merchandising operation.

Nike began a series of acquisitions in 1988 by buying Cole Haan, the manufacturer of upscale shoes for $880 million. At the height of the inline skating fad in 1995, Nike paid $409 million for Bauer, the maker of ice and in-line skates. Some say Nike overpaid for Bauer since the in-line fad quickly faded. Seven years later it bought Hurley International, a surfing and skateboarding clothing company, for $95 million. It acquired Exeter Brands Group LLC in 2004 for $43 million, which makes value-priced footwear and apparel.

Nike's most successful acquisition was Converse in 2003 for $305 million. Converse has renovated its 60,000 square foot corporate headquarters in North Andover, Massachusetts, where it designs authentic presentations of the classic Chuck Taylor All Star. In 2006 Converse issued hundreds of variations on its moderately priced Chucks.

Nike founder, Knight, is chairman of the board, the largest shareholder, and chief executive officer. He has a net worth of $7.3 billion and is the world's seventieth-richest person according to *Forbes*. Knight gave the University of Oregon $100 million in August 2007 to create the Oregon Athletics Legacy Fund. Knight sold nearly $50 million of his stake in Nike since August 2007 but remains the majority shareholder, owning 93 percent of outstanding Class A shares as of May 31, 2007. *Fortune* recognized Nike in 2006 and 2007 for employee benefits like paid sabbaticals, on-site fitness centers and child-care facilities, and its 50 percent discounts on company products.

Adidas North America, Inc. Until 2003 adidas was the top European athletic shoe maker. It sold the top-ranked shoe for the top sport: soccer—known outside the United States as football. Following its $3.7 billion acquisition of Reebok in 2006, adidas had a 28 percent share of the world sporting goods market, not far behind Nike's 31 percent. Adidas and Reebok together make a worldwide sporting goods industry powerhouse. Reebok headquarters are in Canton, Massachusetts, near Boston. Adidas AG, based in Herzogenaurach, Germany, has its North American headquarters in Portland, Oregon.

Adidas made its name in soccer shoes. In the 1920s Adolph Dassler, a shoemaker in Herzogenaurach, Germany, decided to concentrate on athletic shoes and founded a business with his brother Rudolph. In 1936 Jesse Owens wore adidas when he won four Olympic gold medals in track and field in Berlin, Germany. The brothers later formed separate companies. Adolph formed adidas named after himself—Adi from Adolf and Das from Dassler—and Rudolph formed Puma in 1956, also headquartered in Herzogenaurach. Adidas made its name in soccer, with its famous three stripes logo developing from three support leather bands used to bolster the sides of soccer shoes. Adidas remains since 1954 the German soccer federation's official supplier beyond 2010. In 1974 the adidas-wearing German team won the World Cup.

After Adi Dassler died in the 1980s his wife and kids took over the company. By the 1990s new management started to move from a manufacturing to a marketing focus. The company went public in 1995. By 1996 adidas—with its famous three stripes logo—equipped 6,000 Olympic athletes from thirty-three countries.

Adidas acquired Salomon Group with brands Salomon, TaylorMade, Mavic, and Bonfire in 1997. Salomon in-line skates helps to explain the acquisition, but the fad had already begun to fade. The new company went by the name adidas-Salomon AG but the deal lasted only 6 years. By 2005 adidas sold Salomon Group, retaining only TaylorMade Golf Company Inc., to return to its core athletic shoes and apparel market. In 2006 it reverted to the legal name adidas AG.

The company unveiled the adidas 1, a $250 running shoe, in 2005. The shoe has a built-in 20MHz microprocessor computer chip that automatically adjusts the fit as the wearer runs. When the adidas 1 heel strikes the ground, a magnetic sensor measures the amount of compression in its midsole and the microprocessor adjusts firmness during the seconds the shoe is airborne. Adidas spent an estimated $20 million on the rollout. Film director Spike Jonze created several cinematic, big-budget TV spots with the theme "Impossible is Nothing."

Since Nike moved into the number one spot in European soccer shoe sales, adidas successfully attacked the U.S. basketball market, where Nike controls 60 percent of sales. Adidas signed three NBA all-stars: Tracy McGrady, Tim Duncan, and Kevin Garnett. Each will have his own sneaker. Adidas signed an 11-year deal with the National Basketball Association (NBA) in 2006 that makes it the official uniform provider for the league. The deal includes providing uniforms and other products for the Women's National Basketball Association and the NBA Development League, starting with the 2006–2007 season. Reebok, meanwhile, is the global marketing partner of the NBA with the power to create NBA branded

footwear. In 2007 adidas signed college contract after college contract, including luring Michigan and Texas A&M away from rival Nike.

Others. Sporting goods maker Puma AG was acquired for $7.3 billion in July 2007 by the French company that owns Gucci Group, Balenciaga, and Yves Saint Laurent, among other luxury brands. Puma is owned by Sapardis S.A., a nearly 100 percent subsidiary of PPR S.A. Puma is Europe's second largest sporting goods company and quadrupled sales since 2000.

In 1969 quarterback Joe Namath led the New York Jets football team to Super Bowl III wearing Puma shoes. In 2000 supermodel Christy Turlington and Puma jointly launched the Nuala yoga collection. Nuala was an instant, high-end success. In 2006 Puma was, with twelve teams, the dominant kit supplier at the World Cup in Germany. Puma is a premium brand.

Puma has a design center in Boston, Massachusetts that employs upward of 115 people. In August 2007 Puma North America paid $11.5 million for a 105,000 square foot office building in Westford, Massachusetts for its North American headquarters where it employs more than 250 people.

New Balance researched how running impacts the foot during the 1960s and developed an orthopedic running shoe with a rippled sole and wedge heel to cushion shock. Running grew in popularity and New Balance innovated with lightweight nylon to replace the heavy canvas and leather materials previously used in running shoes. New Balance has sales estimated at $1.5 billion per year, making it one-tenth the size of Nike.

MATERIALS & SUPPLY CHAIN LOGISTICS

For the U.S. footwear industry, the largest expense in production is for design, which is done in the United States, while the production of the shoes is done abroad. The materials needed to design shoes are different than materials needed to make shoes. The U.S. athletic shoe industry uses a variety of high-tech methods to facilitate design: specialized laboratory machines, objective athletic tests, and high-speed photography, film and video. Scientific techniques designers employ include motion analysis (kinematics), ground reaction forces and loading rates (kinetics), foot-pressure measurement (in-shoe and external), ankle range of motion (ROM), foot morphology, and electromyography. Designers need computer-aided design and drafting software and the state-of-the-art computer equipment to operate it.

Designers need materials and equipment to fabricate prototype athletic shoes that build on advances in three primary areas: biomechanics, the study of human movement and related forces; physiology, the study of the integration of the body's energy systems and responses to environmental stresses; and sensory/perception, the subjective evaluation of product attributes such as cushioning, flexibility, stability, traction, and durability.

Design prototypes are fabricated in-house. Prototypes are laboratory tested for cushioning, flexibility, stability, traction, and durability. Athletic shoe companies buy machines to test seam strain, adhesion, heat absorption, and water permeability. Durability tests are important as they provide a benchmark, a method to determine under identical conditions how one shoe compares to another. Durability tests are repeatable, and data acquisition systems can accumulate performance information. Athletic shoe companies buy machines to simulate specific conditions such as toe drag on a tennis shoe. Gait equipment is used to study the overall performance of the athlete in the shoe and the forces the lower extremities encounter. In addition to laboratory tests, prototypes are tested by actual athletes before shoes go into production overseas.

Nike has a nearly 13,000 square feet design department referred to informally as its Innovation Kitchen on its 175-acre headquarters campus. One wall displays every Air Jordan model ever made. Nike design materials include high-speed video cameras that capture soccer kick data at 1,000 frames per second and a scanner that produces a perfect 3D digital image of any human foot in just seconds. Nike testing surfaces include a huge section of regulation maple basketball flooring, artificial soccer turf, and a 70-meter running track. In 2001 Nike used the equipment to design and introduce Nike Shox, a system of columns of engineered foam that provide superior cushioning.

Other materials needed by the U.S. athletic shoe industry are computers to track the complicated worldwide supply chain. Since 100 percent of athletic shoes are imported, primarily from China, all key U.S. players have manufacturing operations abroad. For example, in 2004 Nike had 900-plus supplier factories in fifty countries. All players need materials to build top-flight information systems to handle logistics and to manage the supply chain.

Nike, for instance, designs and launches 120,000 products in four cycles per year. Nike spent $500 million in 2004 to modernize its technology system to track supply chain operations as it moves goods from its 900-plus factories to retailers. The $500 million helped Nike get products to customers faster and cheaper. Lead time for getting new sneaker styles to market was cut from 6 to 9 months.

Because 100 percent of athletic shoes are manufactured abroad, and because Nike controls an estimated 50 percent of the U.S. sporting goods market, in the late 1990s Nike was criticized about sweatshop labor condi-

FIGURE 5

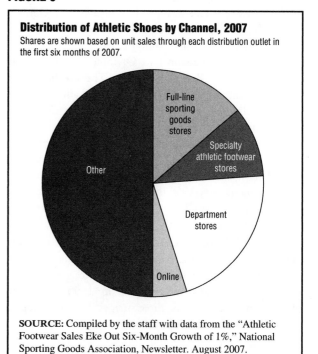

Distribution of Athletic Shoes by Channel, 2007

Shares are shown based on unit sales through each distribution outlet in the first six months of 2007.

SOURCE: Compiled by the staff with data from the "Athletic Footwear Sales Eke Out Six-Month Growth of 1%," National Sporting Goods Association, Newsletter. August 2007.

tions in China and Vietnam, the low-wage countries that produce 65 percent of its footwear. *Business Week* reported that by 2004 Nike had a staff of close to 100 inspectors to visit factories, grade them on labor standards, and work with managers to improve problems. Nike's staff performed 600 factory audits between 2002, when it built up its in-house monitoring staff, and 2004, including repeat visits to those with the most problems.

DISTRIBUTION CHANNEL

The athletic shoe retail distribution channel primarily involves full-line sporting goods stores, specialty athletic footwear stores, department stores, and online/internet stores. The distribution channel was described and quantified in unit sales by the National Sporting Goods Association in its August 2007 newsletter. Figure 5 depicts the percentage of unit sales controlled by the type of store in the distribution channel for the six-month survey period in 2006–2007.

Full-line sporting goods stores accounted for 14 percent of unit sales in athletic footwear, an increase from 12 percent over 2005. Specialty athletic footwear lost market share, claiming 10 percent of units in the athletic footwear market versus 12 percent the previous year. Department stores were even at 21 percent. Online/Internet sales were 5 percent of the athletic footwear market in units, up slightly from 4.6 percent the previous year.

Foot Locker, headquartered in New York, is the leading specialty athletic footwear store in the United States, operating approximately 1,300 primarily mall-based stores. It owns Lady Foot Locker, Kids Foot Locker, Footaction, Footquarters, and Champs Sports. Its direct-to-customer channels include Footlocker.com and Eastbay, a catalog retailer. Nike has also established retail space within some Foot Locker locations, often known as a store within a store. Nike represents an estimated 50 percent of Foot Lockers total sales, making Foot Locker a de facto retail division for Nike. This explains why Nike announced it will open jointly with Foot Locker as many as fifty House of Hoops stores in the United States by 2010. The stores are designed to boost sales of high-tech, high-priced, marquee basketball shoes. The first House of Hoops store will open in Harlem in New York, followed by stores in Los Angeles, Chicago, Houston, and Las Vegas.

Specialty athletic footwear stores lost market share. Perhaps as a result, Foot Locker announced in May 2007 that it reduced previously planned store openings from 170 to 125, including thirty-five Footquarters postponed store openings. In July 2007 Foot Locker announced it will close up to 250 stores in 2007, approximately twice the planned closures. Foot Locker also retained Lehman Brothers as an advisor to evaluate strategic alternatives, including inquiries received from private equity firms.

Major players Nike, adidas, and Puma own stores in prime locations. Nike announced in February 2007 that it will open an additional 100 of its own stores by 2010. The first Nike Town opened in 1990 in Portland, Oregon. The second was opened in 1992 in Chicago, Illinois, on Michigan Avenue. In 2007 there were 14 Nike Town stores. Nike Town New York on 57th and 5th was a travel destination recommended by Frommers. Nike.com lists 12 Nike Women stores, plus eighty-six Factory Stores.

Adidas operates 258 adidas Originals stores worldwide. Recent openings include Amsterdam, Los Angeles, London, Milan, New York, Berlin, Toyko, and Osaka. Its 1,750 square meter store in Paris on the Champs Elysées features Stella McCartney-designed skiwear and faux fur vests from rapper Missy Elliott. Adidas plans more stores in major cities worldwide, including one in China. Adidas operates 20 adidas Originals stores in the United States, mostly on the coasts; there are no adidas Originals stores in the Midwest.

Adidas and its new partner, the NBA, plan to increase the number of products at the NBA Store in New York. The league will create a store within a store adidas concept shop to serve as a prototype for future stores worldwide. The company will sell NBA products at its own shops, including its planned store in China, home of Houston Rockets player Yao Ming, who wears shoes designed by Reebok, which adidas acquired in July 2007.

Puma operates 40 Puma Stores in the United States. It opened its third store in New York on Union Square in Manhattan in October 2006.

KEY USERS

Athletic manufacturers divide their market into segments in two ways, by demographic characteristics and by athletic activity pursued while wearing the product. In terms of demographics, young people dominate this market and in particular, young men.

The key user of athletic shoes is male. Users of the marquee high-tech, high-price basketball shoe are young urban males. In January 2007 *PR Week* estimated that $2.9 billion worth of basketball shoes are sold per year, with males 12–20 years of age being the biggest consumer demographic. Manufacturers of athletic shoes market heavily to this demographic segment.

Women, too, wear athletic shoes but a far larger percent of female athletic shoe users wear those shoes primarily when they are participating in an athletic activity. Men are more likely to wear athletic shoes while doing their day-to-day activities, as a general-purpose shoe.

ADJACENT MARKETS

The primary adjacent market for athletic shoes is sporting goods. Since athletic shoes are built for high performance and are considered part of an athlete's gear, they make up just one part of the sporting goods market, typically thought to include footwear, equipment, and clothing. The total retail sporting goods market was valued at approximately $50 billion in 2005 according to estimates by the National Sporting Goods Association. Footwear accounted for 30 percent of this total. Clothing and equipment markets were the other two closely adjacent markets that together made up the rest of the $50 billion U.S. retail trade in sporting goods.

The U.S. athletic shoe producers also have clothing and equipment lines. Nike, adidas, and Puma depend heavily on brand recognition and brand loyalty. For example, those with a twenty-five year love of Air Force 1s might buy Nike clothing and sports equipment, or Nike sunglasses—which retail for $79 to $169. Those who identify with hip urban trends may buy adidas apparel and gear. When a brand is successful, its success spreads across adjacencies to increase sales.

RESEARCH & DEVELOPMENT

Current research and development related to athletic shoes is focused on redefining interactive marketing. R&D efforts involve how to best take advantage of the Internet as a medium for commerce, brand extension, and relationship building. Athletic shoe companies want to maintain a consistent brand image across retail and direct distribution channels, and across their adjacent clothing and equipment markets. Nike uses the interactive marketing of the World Wide Web to extend its brand in a sports-specific fashion. In addition to Nike.com, it operates Nikebasketball.com, Nikewomen.com, Nikerunning.com, Nikesb.com (skateboarders), Nikefutbol.com (Spanish), and countless others. Nikeid.com allows visitors to customize shoes. Its newest site is Nikeplus.com.

Nikeplus.com won a 2007 Industrial Design Society of America award in the consumer products category. The Industrial Design Society of America explained:

> Thanks to a unique partnership between Nike and Apple, your iPod nano can do more than play music. It can now be your coach, your personal trainer and your favorite workout companion. A small sensor inside the Nike+ shoe communicates with your iPod nano, and your iPod communicates with you. Real-time data, such as running pace, time and distance elapsed, and calories burned, are spoken over customized music tracks. Users can follow and manage workout data through a dedicated web site that offers goal-driven training programs and graphic results. The web site also allows you to create run-specific playlists, such as mellow songs for your warm-up and cool-down, and more intense tunes for the run in between.

Nike also won a 2005 Industrial Design Society of America award in the digital media category for its Nikeid.com Web site. The site transforms the visitor into a designer, allowing them to apply their own choice of colors and materials to shoes and gear. The design interface includes visual call-outs and audio cues to guide users through the design-build process.

As part of its success with Nikeid.com and Nikeplus.com, Nike increased its Internet spending more than tenfold since 2002, according to TNS Media Intelligence. Nikeid.com and Nikeplus.com both created experiences for consumers. During an era when most companies use the Internet primarily to tell their story and create their mythology, merely hoping it will become as legendary as the mythology surrounding Nike's founding, Nike transformed its Web presence from a mere storytelling platform to a digital world of customer connections. The iPod partnership demonstrated how a brand can market itself by offering something useful to a community rather than just communicating its assets. As a result, Nike shook up the entire advertising industry.

In 2007 Nike put its lead agency of the past twenty-five years, Wieden+Kennedy, into review for a longstanding assignment. Even though Wieden+Kennedy helped Nike create the iconic "Just do it" TV and print campaigns that catapulted the company into a commanding position, Nike is tellingly looking elsewhere. Wieden+Kennedy did not provide design services for the successful Nikeid.com (created by R/GA) and Nikeplus.com sites.

When Nike starts researching innovative ad agencies with interactive, digital and community-building capabilities and developing relationships with them, it has the power to shake down the entire advertising industry. Nike recently moved the advertising account for its high-profile running market segment to Crispin Porter+Bogusky. Wieden, whose relationship with Nike goes back to its founding, remains its primary creative ad agency.

Where Nike goes, others soon follow. In August 2007 New Balance placed its estimated $15–$20 million ad account in review. It had long used independent Boathouse Group Inc. in Waltham, Massachusetts, as its lead shop. In November 2006 the company awarded its interactive account and other nontraditional marketing chores to independent Boston, Massachusetts, shop Almighty.

CURRENT TRENDS

The newest trend in athletic shoes is the affordable basketball shoe. Many suggest that the day of the high price, high tech marquee basketball shoe is over. Sales of basketball shoes dropped 16 percent in 2006 according to *Business Week*. Furthermore, these sales dropped 4 percent over a single 6-month period in 2006–2007 according to the National Sporting Goods Association.

A McGill University, Montreal, Canada, study found that high price athletic shoes were overrated. The report, based on a 1987 Swiss study of 5,000 runners, showed a 123 percent higher incidence of injury in people wearing shoes priced at more than $95 than those wearing shoes costing $40 or less. McGill researchers tested fifteen men to find out if more expensive shoes changed athletic performance. They concluded that people tended to step harder on shoes that were priced higher, possibly causing more injuries. The conclusion, published in 1997 in the *British Journal of Sports Medicine*, was that no evidence available proved that less expensive shoes were worse than expensive shoes for sporting activities.

A high-profile example of a low-priced athletic shoe is the Starbury One; it sells for $14.98 at discount clothing chain Steve & Barry's. The Starbury One is endorsed by New York Knicks guard Stephon Marbury. Changing the celebrity athletic shoe style of marketing, Marbury's up-front endorsement fee was $0. Marbury makes money only from sneaker sales. As part of the introductory promotional tour in August 2006, Marbury traveled to forty cities in seventeen days selling an estimated three million pairs. The Starbury II came out on April 1, 2007, and Marbury visited thirty-eight cities in seventeen days. The

Starbury II also sells for $14.98. The expanded Starbury line supports twelve shoe styles, each available in a range of colors, along with basketball and lifestyle apparel. Marbury grew up poor with seven siblings and said $200 shoes were not an option for him.

Basketball shoes like the Starbury and other low-price athletic shoes have the potential to grow according to researcher NPD Group Inc. The low-cost shoe market—defined as sneakers under $50—grew nearly 9 percent over the past two years. Low-cost athletic shoes in the under $50 segment represent an estimated 50 percent of the $17.5 billion U.S. athletic shoe market.

As part of the low-cost trend, Payless introduced Dunkman athletic shoes in 2004. The shoes were endorsed by professional basketball superstar Shaquille O'Neal and sold for under $40. Dunkman shoes are available in several low top and high top basketball styles for men, women, youth, and toddlers. The line features full-grain leather, professional basketball team colors, and a Dunkman logo depicting Shaq doing his signature slam

dunk. The high-tech shoes perform similarly to shoes that cost five times the price. Shaquille O'Neal explained "These shoes are unbeatable—amazing styles, great quality and performance technology for under $40. They are professional basketball worthy."

TARGET MARKETS & SEGMENTATION

The major U.S. athletic shoe players segment the market into sports-specific categories. Categories at Nike are called action sports, baseball/softball, cardio, fitness dance, golf, lacrosse, running, sports culture, track and field, walking, all conditions, basketball, cycling, football, air Jordan, soccer, tennis, training, and yoga (yoga being a very small category of flexible shoes). Reflecting this sports-specific segmented approach, Nike has separate Web sites for various segments.

The National Sporting Goods Association segments the market in a similar manner. It tracks consumer spending in twenty-six categories of athletic shoes. Figure 6

FIGURE 6

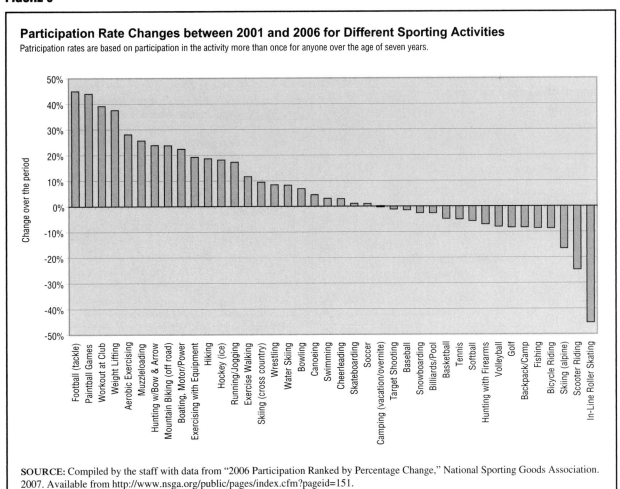

Participation Rate Changes between 2001 and 2006 for Different Sporting Activities

Patricipation rates are based on participation in the activity more than once for anyone over the age of seven years.

SOURCE: Compiled by the staff with data from "2006 Participation Ranked by Percentage Change," National Sporting Goods Association. 2007. Available from http://www.nsga.org/public/pages/index.cfm?pageid=151.

shows changing participation rates for sports-specific categories between 2001 and 2006.

Recognizing the importance of targeting segments, Nike reorganized internally in 2006 to focus on six major areas. The six main divisions, expected to generate 75 percent of future growth, are: running, basketball, soccer, women's fitness, men's training, and sports culture. Nike previously focused on the three traditional segments of footwear, clothing, and equipment.

Adidas is segmented toward distinct groups of sports-oriented consumers and the affiliated lifestyle. Men's athletic shoes at adidas are categorized by baseball, basketball, football, golf, outdoor, running, soccer, tennis, training, and wrestling. Reflecting their adjacencies in clothing and gear, adidas also has collections called Dunk the Impossible, Run the Impossible, Mexico Federation Collection, +F Series Collection, Award Winning Shoes, Men's Test Run 28 Shoes, Thrill & Destroy Collection, Beckham Collection, and the L.A. Galaxy Collection named after the soccer team where Beckham plays. Shopadidias.com has personal stories on video of athletes including soccer star David Beckham.

Adidas is ahead of Nike in targeting women. Adidas by Stella McCartney is a full line with fifty items that includes eight models of sleek high-fashion athletic shoes.

Since its 2006 acquisition by adidas, Reebok has targeted women. Its Run Easy campaign extols the virtues of running while chatting with a friend. Actress Scarlett Johansson developed a line of nine shoe and apparel items introduced in 2007.

RELATED ASSOCIATIONS & ORGANIZATIONS

American Apparel and Footwear Association, http://www.apparelandfootwear.org

Footwear Distributors and Retailers of America, http://www.fdra.org

National Shoe Retailers Association, http://www.nsra.org

The National Sporting Goods Association, http://www.nsga.org

BIBLIOGRAPHY

About Nike. Nike, Inc. Available from <http://www.nike.com/nikebiz/nikebiz.jhtml?page=5>.

adidas History. adidas North America, Inc. Available from <http://www.adidas-group.com/en/overview/history/default.asp>.

"Adidas to Supply German Soccer Post 2010." *CNN Money.* 22 August 2007.

"Apollo's Small Steps are Giant Leap for Technology." *NASA Science and Technology.* 21 June 2004. Available from <http://www.nasa.gov/missions/science/f_apollo_11_spinoff.html>.

"Athletic Footwear Sales Eke Out Six-Month Growth of 1%." The National Sporting Goods Association *Newsletter.* 13 August 2007.

"Athletic Shoes: Sports Medicine Advisor 2007." *Clinical Reference Systems.* 31 May 2007.

Bernstein, Aaron. "Nike's New Game Plan for Sweatshops." *Business Week.* 20 September 2004. Available from <http://www.businessweek.com/magazine/content/04_38/b3900011_mz001.htm>.

Cassidy, Hilary. "Nike Rules, and Everyone Gets Cool." *Brandweek.* 18 June 2007, S47.

"Fact File: Starbury Leads the Way as Low-cost Sneakers Make Big Noise." *PR Week.* 29 January 2007, 11.

"Footwear Production: 2003." *Current Industrial Reports.* U.S. Department of Commerce. Bureau of the Census. October 2004.

Gianatasio, David. "New Balance Goes Into Review." *ADWEEK.* New England Edition. 7 August 2007.

"Gucci Owner gets Reins of Puma with $7.3b Bid." *The Boston Globe.* 18 July 2007.

Holmes, Stanley. "Changing the Game on Nike; How Budget Sneakers Are Tripping Up its Basketball Business." *Business Week.* 22 January 2007, 80.

Holmes, Stanley and Aaron Bernstein. "The New Nike: No Longer the Brat of Sports Marketing, It Has a Higher Level of Discipline and Performance." *Business Week.* 20 September 2004. Available from <http://www.businessweek.com/magazine/content/04_38/b3900001_mz001.htm>.

Kamenev, Marina. "Adidas High Tech Footwear." *Business Week.* 3 November 2006.

King, Danny. "Shares of Nike Jump on U.S. Clothing Sales; 4th-Quarter Profit Up 32%." *The Seattle Times.* 27 June 2007, C2.

Moore, Matt. "Adidas Signs 11-Year Deal With the NBA." *USA Today.* 11 April 2006. Available from <http://www.usatoday.com/money/advertising/2006-04-11-adidas-nba_x.htm>.

Nagel, Kyle. "Nike Still Winning the Shoe WarsBut adidas is Courting College Athletics to Make up Some Ground." *Dayton Daily News.* 4 August 2007, B3.

"Nike Founder Gives $100M to U of Oregon." *Portland Business Journal.* 20 August 2007. Available from <http://www.bizjournals.com/pacific/stories/2007/08/20/daily9.html>.

O'Leary, Noreen. "When Great Ads Just Aren't Enough." *ADWEEK.* 19 March 2007.

O'Malley, Gavin. "Who's Leading the Way in Web Marketing? Nike, Of Course; Marketer Breaks Down Walls, and In the Process Sees Its Share Increase 2%." *Advertising Age.* 16 October 2006, D-3.

Sammon, Lindsay E. "Foot Locker Faces Tough Q1, But Serra Promises Comeback." *Footwear News.* 28 May 2007, 14.

"ShoeStats 2006." American Apparel and Footwear Association. July 2006.

Thomaselli, Rich and Alice Z. Cuneo. "Can Any Shop Run with Digital-obsessed Nike? Beaverton Giant Looks for Agencies to Help it Improve in Area it Already Dominates: Interactive." *Advertising Age.* 19 March 2007, 3.

"Trends: An Annual Compilation of Statistical Information on the U.S. Apparel and Footwear Industries." American Apparel and Footwear Association. 2005.

SEE ALSO *Shoes, Non-Athletic*

Auto Parts

———◆———

NAICS: 33–6310 Motor Vehicle Gasoline Engine and Engine Parts Manufacturing, 33–6311 Carburetor, Piston, Piston Ring, and Valve Manufacturing, 33–6312 Gasoline Engine and Engine Parts Manufacturing, 33–6320 Motor Vehicle Electrical and Electronic Equipment Manufacturing, 33–6321 Vehicular Lighting Equipment Manufacturing, 33–6322 Other Motor Vehicle Electrical and Electronic Equipment Manufacturing, 33–6330 Motor Vehicle Steering and Suspension Components (except Springs) Manufacturing, 33–6340 Motor Vehicle Brake System Manufacturing, 33–6350 Motor Vehicle Transmission and Power Train Parts Manufacturing, 33–6360 Motor Vehicle Seating and Interior Trim Manufacturing, 33–6370 Motor Vehicle Metal Stamping, 33–6390 Other Motor Vehicle Parts Manufacturing, 33–6391 Motor Vehicle Air-Conditioning Manufacturing, and 33–6399 All Other Motor Vehicle Parts Manufacturing

SIC: 3714 Motor Vehicle Parts Manufacturing

NAICS-Based Product Codes: 33–63111 through 33–63997554

PRODUCT OVERVIEW

A Global Enterprise. The auto parts industry has evolved over the last century from corner hardware stores supplying nuts and bolts for inventors such as Karl Benz, Armand Peugeot, and Henry Ford, to a global industry that supplies everything from screws, springs, and brake pads to total vehicle systems and in some cases entire automobiles.

In the early twenty-first century this industry produced parts, components, and systems for the world's car and truck producers projected to reach, according to the U.S. Department of Commerce, $1.1 trillion in goods by 2010. Like vehicle manufacturing, the auto parts industry is truly global. Suppliers operate on every continent except Antarctica.

Globalization has radically reshaped the industry especially in nations where domestic manufacturers have been under intense competitive pressure from offshore producers. As the Original Equipment Suppliers Association has put it: "More than ever, automakers are drawing on suppliers around the globe, shuttling parts across borders in search of lower prices and higher quality."

In earlier times independent parts suppliers were physically closer to their customers, rarely more than a day's drive away. In the global economy of the twenty-first century distance is not a barrier if your product is low-cost, meets industry quality standards, and can be delivered at the agreed-upon time. In fact, more than 20 percent of the auto parts produced in the world are exported from their country of origin to customers in other markets around the globe, primarily the United States, Western Europe, and Japan.

During the period from 2001 to 2005, auto parts exports across the world grew at an average annual rate of 12.7 percent reaching $220 billion by 2005 and more than 20 percent of global auto parts production. Emerging economies—Mexico, Brazil, Romania, Slovakia, Morocco, Saudi Arabia, Tunisia, India, and Taiwan—accounted for 29 percent of 2005 exports, their sales growing at a much faster pace (20.1%) than exports from established industrial nations.

This trend has had a pronounced impact on the domestic parts industry in the United States. The *Detroit Free Press* noted in a front page report (May 7, 2006): "Federal data found that vehicles built by Detroit automakers (Chrysler, Ford, and General Motors, the domestic 'Big Three') have steadily increased their proportion of parts from outside the United Sates and Canada. By the same measure, vehicles built in North America by Japan's largest automakers increasingly use U.S. and Canadian Parts."

Based on figures assembled by the International Trade Centre (ITC), foreign competition has had particularly negative impacts on U.S. auto parts producers. Until 2003 the United States was the world's leading exporter of auto parts. By 2004 it was second to Germany with Japan close behind and France, Canada, Italy, and Spain coming on strong. ITC is a joint technical cooperation agency of the United Nations Conference on Trade and Development (UNCTAD) and the World Trade Organization (WTO).

At the same time, U.S. auto companies dramatically increased their parts imports. Japan is the leading vehicle and parts producer; but unlike Germany and the United States, which are leading parts exporters as well as importers, Japan is only fourteenth on the list of importers. Japan relies more heavily on its domestic parts industry largely a result of its *keiretsu* structure under which manufacturers maintain exclusive relationships with their independent suppliers. According to the more comprehensive import/export figures of the U.S. Office of Aerospace and Automotive Industries (OAAI), U.S. imports of automotive parts were $95.2 billion in 2006. Exports totaled $58.9 billion—producing a trade deficit of $36.3 billion. The 2006 deficit was lower than the year before ($37.1 billion) but still triple the $11.7 billion deficit reported in 1999.

This reflects the continuing difficulties of the domestic auto parts industry, as outlined in the March 2007 U.S. Automotive Parts Annual Assessment of the OAAI, as their major customers continue to lose market share; costs of raw materials keep rising; the domestic Big Three (Ford, Chrysler, and General Motors) demand price and cost cuts; and foreign competition grows. "However," observed the report, "as transplant automakers (U.S. operations of foreign manufacturers) increase their presence in the United States, foreign-affiliated suppliers also increase their presence to supply the automakers, creating equipment and jobs in the U.S. economy."

Shrinking Domestic Sector. Employment in the U.S. auto parts industry has been eroding. Parts producers employed 920,000 in 2000 and 721,900 in 2006 according to data provided by the Bureau of Labor Statistics (BLS), a part of the U.S. Department of Labor. The number of participating companies has also been declining. In fact, as OAAI reports, "industry analysts predict that, of nearly 800 major suppliers in 2000, fewer than 100 will be left by 2010 as a result of bankruptcies, mergers and acquisitions, and migration to other industries." In 2005, for example, there were thirty-two mergers and acquisitions, up from twenty-six in 2004. In 2006 another eight major suppliers filed for bankruptcy. The employment figures are especially troubling in view of the fact that "Automotive suppliers are directly and indirectly reported to account for more jobs and provide more economic well-being to more Americans than any other manufacturing sector," according to the OAAI.

In some respects, the auto parts supplier industry is repeating the history of the industry it serves, but in a different form. A report in the May 1996 issue of *Ward's Auto World* presaged this trend in recounting historical highlights of the auto parts industry: "In the beginning, suppliers such as Henry M. Timken, Arthur Oliver Smith, Albert C. Champion, and the Dodge and Fisher brothers sold parts to the early automakers that they designed and manufactured themselves. Later, the automakers bought out some of these suppliers so they could control the parts that went on their vehicles. Meanwhile, other suppliers joined forces to create larger and more capable companies.

> In the 1990s, automakers are returning design and engineering responsibilities to suppliers for the components and systems they provide. Will automakers eventually return to vertical integration? That's not likely, say industry watchers, but the trend by larger suppliers to acquire smaller companies to give them systems capability and global presence closely resembles (automotive) industry history.

The report goes on to quote David E. Cole, director of the University of Michigan's Office for the Study of Automotive Transportation, to predict that "there won't be a wholesale return to vertical integration," although consolidation among Tier 1 suppliers takes the place of vertical integration from a historical perspective, which is essentially what has been transpiring. This has been a major factor in the decline of the number of major U.S. suppliers.

Industry executives and investors participating at Reuters Autos Summit in Detroit in September 2006, added their own take in a conference report: "Consolidation is inevitable among U.S. auto parts suppliers after two years of bankruptcies and declining Big Three vehicle production, but mega-mergers may not be in the

FIGURE 7

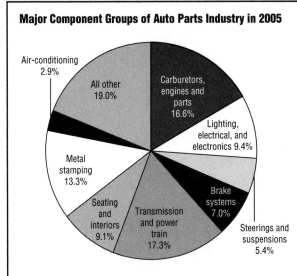

Major Component Groups of Auto Parts Industry in 2005

Air-conditioning 2.9%

All other 19.0%

Carburetors, engines and parts 16.6%

Lighting, electrical, and electronics 9.4%

Metal stamping 13.3%

Brake systems 7.0%

Seating and interiors 9.1%

Transmission and power train 17.3%

Steerings and suspensions 5.4%

SOURCE: Compiled by the staff with data from "Value of Product Shipments: 2001," and "Value of Product Shipments: 2005," *Annual Survey of Manufactures,* U.S. Department of Commerce, Bureau of the Census. January 2003 and November 2006.

cards." The report went on to observe that "consolidation in North America does not necessarily mean building a few very large suppliers, which have not fared as well as focused medium or small suppliers (companies focused on specialty technologies)," that have been much stronger than larger suppliers.

Major Product Categories. The auto parts industry produces a wide range of products, in effect all components of an automobile except its body and its tires. The sector is so diverse, in fact, that in reporting on it the U.S. Bureau of the Census breaks it apart into eleven separate industries which, in this presentation, we treat as nine major product groupings. The product array is presented in Figure 7, displaying category shares as percent of the sector's shipments in dollars.

Ignoring the All Other category, which includes a great multiplicity of parts, the largest category in 2005 was transmissions and power trains, the smallest automotive air-conditioning systems. In the combined carburetors, engines, and parts industry, carburetors represent 7 percent, engines and parts 93 percent of the total. Within the lighting, electrical, and electronics category, automotive lighting is 14 percent of the industry and all other electrical and electronic components 86 percent.

Shares of the components have remained roughly the same over time with small changes between 1997 and 2005. In 1997, for example, transmissions and power

trains were 17.2 percent, slightly smaller than in 2005. Similarly the carburetor/engine category was 15.9 percent in 1997 and 16.6 percent in 2005.

The All Other category includes filters, exhaust systems, wheels, bumper assemblies, automotive frames, fuel tanks, radiators, doors, sunroofs, air bag assemblies, and many other componentry that do not fit readily into other major categories.

MARKET

The North American auto parts market is not alone in experiencing difficult times in the early years of the twenty-first century. "Fully one-third of auto suppliers, globally, are in financial distress, with 41 percent in the Americas, 24 percent in Europe, and 32 percent in Asia," reported Neil DeKoker, president of the Original Equipment Supplier Association (OESA), in a presentation made in August of 2006 at a conference on "Rationalizing the Automotive Supplier Industry: Carving Out Profit from M&A [monitoring and evaluation] Activity." In its 2006/2007 Industry Review, the OESA reported a 7.3 percent decline in the total world original equipment parts market in 2005 ($781.7 billion) compared to 2004 ($843 billion) after several years of growth, including a 10 percent increase in 2003. Over the longer term, the world market is expected to experience growth, and is expected to exceed $1 trillion annually by 2010.

Domestic shipments of automotive and truck parts were reported by the Census Bureau as valued at $200.3 billion dollars, up from $174.6 billion in 1997. The growth rate in this period was 1.73 percent annually, but as shown in Figure 8, sales are cyclical. In the 1997 to 2005 period, the highest shipments were realized in 2000. Shipments dropped sharply in 2001 as a brief recessionary period set in. By 2005 the industry had again almost reached its 2000 peak in this period.

The U.S. new car and truck market (as contrasted to parts) was one of only two major markets in the world to lose ground, with sales fading 2.5 percent from 16.95 million vehicles in 2005 to 16.52 million vehicles in 2006. Japan was the other declining market. Japanese demand fell 2.5 percent from 5.73 million vehicles in 2005 to 5.59 million in 2006. Western Europe, the largest automotive market in the world, managed to grow slightly during 2006—0.8 percent from 16.52 million units in 2005 to 16.65 million in 2006. The emerging markets of the world, on the other hand, experienced more robust growth—other European countries were up a combined 8.1 percent; Brazil and Argentina, 13.3 percent, and the other markets of the world, 14.6 percent.

These anemic sales figures for new vehicles plus increasing pressure from auto parts suppliers in emerging economies were key contributors to the slow growth

FIGURE 8

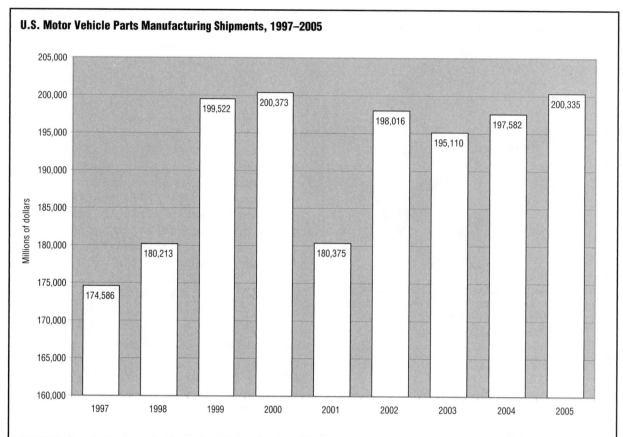

U.S. Motor Vehicle Parts Manufacturing Shipments, 1997–2005

SOURCE: Compiled by the staff with data from "Value of Product Shipments: 2001," and "Value of Product Shipments: 2005," *Annual Survey of Manufactures,* U.S. Department of Commerce, Bureau of the Census. January 2003 and November 2006.

shown by domestic auto parts suppliers. Exports from emerging economies countries have been growing at an annual rate of 20.1 percent, well above the export growth rate from more industrialized countries (12.7%). Automotive parts trade with China is an example of the impact imports from emerging economies can have. This trade has grown significantly. The United States imported $1.6 billion worth of auto parts from China in the year 2000. Seven years later, in 2006 the United States imported $6.9 billion worth of auto parts from China, an increase of 313 percent or nearly a 50 percent increase annually during the first years of the twenty-first century.

KEY PRODUCERS/MANUFACTURERS

There are thousands, if not tens of thousands of independent companies supplying the global automotive industry with everything from nuts and bolts to entire systems. More than 100 of these firms have sales of $1 billion or more based on global sales of original equipment. In 2006 the top 100 suppliers were dominated by companies in Germany, Japan, and the United States; both Japan and the United States had twenty-seven top 100 companies

and Germany twenty-three. The largest auto parts builder in the world in 2006 was Germany's Robert Bosch, GmbH, noted primarily for fuel injection systems, chassis systems, energy and body systems, and automotive multi-media and electronics.

Bosch. Bosch was established in 1886 when Robert Bosch opened the Workshop for Precision Mechanics and Electrical Engineering in Stuttgart, Germany. While its first projects were electrical equipment, including telephone systems, the business took off when Bosch produced his first magneto and adapted it to ignite an internal combustion engine, "solving one of the greatest technical problems faced by the automotive industry in its formative years," according to "Bosch History At A Glance" on the company's Web site.

The company has since evolved into its present preeminent position in the automotive world, producing not only an array of auto parts but providing diagnostic and other car service systems and advanced automotive technology developments as well. Corporate-wide, it operates in more than fifty countries with some 260,000 employees. Bosch has been number one since 2004 when it supplanted the Delphi Corporation, the U.S.-based auto parts giant that was spun off from General Motors in 1999, for the lead.

Delphi Corporation. Delphi, broadly based as a supplier of steering, chassis, energy, engine management, thermal management, interior, electronic systems and in-vehicle entertainment systems, had dropped to fourth in global sales by 2005 but regained second place in 2006 as it struggled to recover from Chapter 11 bankruptcy, for which it filed in 2005. The corporation has approximately 170,000 employees, including 17,000 engineers, and operates thirty-three technical centers and 162 wholly owned manufacturing sites in thirty-four countries. Since losing the top spot Delphi has been in an ongoing seesaw scramble with Denso Corporation of Japan and Magna International of Canada for the second, third, and fourth positions in the auto parts manufacturing industry.

Visteon Corporation. This company, which was the second largest auto parts producer in the world when it was spun off from Ford in 2000 and still ranked fourth globally as recently as 2004, had dropped to fourteenth by 2006. Visteon, like Delphi, is a broadly based supplier with an extensive array of products, notably chassis systems, climate controls, powertrain controls, electronics, lighting, engine management and fuel systems, exterior/interior systems, and cockpits. Visteon had sales revenue of $11.4 billion in 2006 and employed 43,000 people.

Denso Corporation. Japan's largest auto parts producer and another supplier with an extensive product portfolio, Denso has shown steady growth over the years, challenging Delphi as the number two global supplier and eventually Robert Bosch for number one. Should growth trends established in the first five years of the twenty-first century continue, Denso will be well-placed to continue growing.

Magna International, Inc. This Canadian auto parts leader is noted for interiors, exteriors, body and chassis systems, seats, mirrors, closures, electronics, engines, transmissions and drivetrains. Magna was also one of the first major suppliers to manufacture entire vehicles for customers. In 2007, for example, Magna assembled the BMW X3, and Chrysler 300, 300C, and Jeep Commander.

MATERIALS & SUPPLY CHAIN LOGISTICS

"The auto industry is one of the key oligopolistic, dynamically complex and networked global industries. To produce a car, which consists of some 20,000 parts and components, a producer needs to orchestrate the logistics and assembly of various kinds of input factors such as steel, glass, rubber and plastic, semi-assembled components though many a manufacturing technologies that are spatially distributed and located internationally."

A Look at the Spatial Distribution of the Automotive Industry reflects the realities facing the global auto parts industry in the twenty-first century as individual companies strive to survive in an extremely competitive environment.

"More than ever, automakers are drawing on suppliers around the globe, shuttling parts across borders in search of lower prices and higher quality," according to the Original Equipment Suppliers Association.

A May 7, 2006, Detroit Free Press report noted that "federal data found that vehicles built by Detroit automakers have steadily increased their proportion of parts from outside the United States and Canada. By the same measure, vehicles built in North America by Japan's largest automakers increasingly use U.S. and Canadian parts."

The report found that the value of parts coming from within the United States or Canada declined for all of the Big Three from 1995 to 2005. General Motors U.S./Canada-sourced parts in 1995 accounted for 91 percent of those used in vehicles sold that year and only 81 percent in 2005. Ford Motor Company U.S./Canada-sourced parts accounted for 86 percent in 1995, down to 82 percent in 2005, and Chrysler U.S./Canada-sourced parts accounted for 89 percent in 1995, down to 76 percent in 2005.

For that same period, the top three Japanese automakers increased their share of parts produced in the United States and Canada: Toyota with 49 percent in 1995 and 75 percent in 2005; Honda with 47 percent in 1995 and 68 percent in 2005; and Nissan with 42 percent in 1995 and 57 percent in 2005. However, the report also pointed out that "when considering all vehicles sold in the United States (in 2005), including those made in Japan, the share of parts value from U.S. or Canadian sources fell dramatically: Toyota, 49 percent; Honda, 58 percent; and Nissan, 48 percent."

The United States imported auto parts valued at $95.2 billion in 2005, an increase of more than 50 percent since 1999 when auto parts imports totaled $61.6 billion. During that same period, exports of auto parts from the United States grew only 18 percent, from $49.9 billion in 1999 to $58.9 billion in 2006, tripling the U.S. trade deficit in auto parts.

Mexico is the largest exporter of car parts to the United States with $26.4 billion in parts in 2006; Canada is second with parts exports valued at $20.4 billion; Japan is third largest with $15.4 billion in exports to the United States. Together, these three countries accounted for approximately two-thirds of all car parts imported into the

United States ($62.2 billion). Mexico and Canada, partnered with the United States in the North American Free Trade Agreement (NAFTA), exported auto parts valued at $46.8 billion to the United States in 2006. They also were the destination for 76 percent of all U.S. auto parts exported that year ($44.7 billion).

Most parts for vehicles built in the United States and Europe are supplied by independent producers since two of the largest supplier companies were spun off from their parent companies. Delphi Automotive Systems, a former subsidiary of General Motors, became independent in 1999 and Visteon Corporation, formerly part of Ford, became independent in 2000. They were among the 1,500 to 2,000 major Tier 1 automotive suppliers (companies selling components and systems directly to the vehicle builders) estimated to be serving the global automotive industry in 2004. This number is expected to shrink to between 500 and 700 major suppliers, and only about fifty of these survivors will be system-integrators dealing directly with the vehicle manufacturers, according to the Boston Consulting Group. The trend is for the major suppliers to provide larger and more complete vehicle systems or segments rather than simply supplying hundreds of parts and components, with some—like Magna International—building complete vehicles..

Japan, one of the leading global producers of automobiles, has traditionally operated under the *keiretsu* system in which an association of companies agrees to work together, often at the exclusion of firms outside the group. Automotive keiretsus sometimes have several hundred companies associated with a particular vehicle manufacturing company, usually in a somewhat exclusive arrangement. These agreements in essence bar non-members from obtaining supplier arrangements with the vehicle manufacturer. However, these keiretsu relationships are tending to dissolve as cross-sourcing of components among the suppliers increases with industry globalization.

The global automotive industry is a major consumer of a host of commodities worldwide. With vehicle production at more than 64 million units each year, it is a major consumer of various grades of steel, for example, which accounts for 55 percent of an average vehicle's content by weight, over 1,700 pounds in a mid-size sedan. Other materials consumed include cast iron, averaging more than 400 pounds, primarily for engine and suspension components; plastics and composites, nearly 250 pounds; aluminum, about 190 pounds; plus rubber, glass, copper, zinc, other metals, fabrics, and various fluids and lubricants. These materials are sourced from all over the world, the location depending on price, availability and accessibility on the part of the provider of parts and/or components supplying the vehicle manufacturer.

DISTRIBUTION CHANNEL

Auto parts are distributed through several different networks, according to their destination: as original equipment (OE) for assembly on vehicles in production; as original equipment service (OES) parts destined for the vehicle manufacturers' service networks; or as aftermarket parts and accessories, sold through an extensive network of independent jobbers, wholesalers, and retail outlets. The aftermarket parts are built or remanufactured to replace damaged or worn OE parts, while the accessories are parts sold after the original sale of a vehicle intended to add to the comfort, convenience, performance, safety or customization of that vehicle.

OE parts, which account for an estimated two-thirds to three-fourths of total automotive parts production, may move via several routes on their way to the vehicle assembly plant, where they may be delivered in bulk or timed to arrive shortly prior to installation on the vehicle—known generally as Just-In-Time (JIT) delivery. These parts, components and systems destined for final assembly have three basic points of origin: Tier 1 suppliers that sell finished components and systems directly to the vehicle manufacturers; Tier 2 suppliers that sell parts and materials for the finished components to the Tier 1 suppliers; and Tier 3 suppliers which supply raw materials. The Tier 3 companies may sell directly to the vehicle manufacturer, as well as to the Tier 1 and Tier 2 suppliers.

Auto parts are shipped using both truck transports and railroads to move vehicles to dealerships, although air freight may be used if critical parts are needed as soon as possible in order to keep an assembly line moving. Offshore exports and imports are handled by ship. In addition to OE and OES parts, shipments also may include auto parts destined for the various national and regional parts distribution networks serving the aftermarket segments of the world.

OES parts, like automobile sales, have traditionally been handled through the franchised independent dealership networks in the major markets of the world. In the United States, the National Automobile Dealers Association (NADA) represents some 20,000 franchised new car and truck dealers holding nearly 43,000 separate franchises, both domestic and international (many dealerships represent multiple nameplates, including those from more than one manufacturer). In addition the American International Automobile Dealer Association (AIADA) represents 11,000 international nameplate automobile franchises.

The independent aftermarket distribution networks are much more complex. There are approximately 45,000 companies engaged in the wholesale and retail auto parts industry with combined revenues of $135 billion according to an industry overview by Hoover's, a business information resource firm. The aftermarket has traditionally consisted of wholesalers who purchase parts and components from the manufacturers; jobbers, intermediaries between wholesale distributors and retail operation; and the retail outlets.

A wholesaler typically buys from 200 to 300 vendors, including manufacturers who may also be producers of OE and OES parts, and from many of the thousands of smaller companies that make parts specifically for the aftermarket, according to Hoover's. A wholesalers' distribution center may carry 300,000 parts for virtually every conceivable vehicle on the road.

Retailers range from small, single-location operations to network operators such as Genuine Parts, AutoZone, Advance Auto Parts, CSK Auto, and Pep Boys. Hoover's reports that AutoZone has 3,200 stores and Genuine Parts fifty-eight distribution centers and 900 retail outlets. Some retailers sell to consumers, repair shops, and commercial installers—gas stations, fleet operators and car dealer service departments; some also operate their own repair departments.

KEY USERS
The biggest consumers of the auto parts industry are the vehicle manufacturers, accounting for an estimated two-thirds to three-fourths of total automotive production. The balance is sold to the service, replacement parts, and accessories markets. A small user group is the do-it-yourself consumer, who repairs his own vehicles.

ADJACENT MARKETS
The market for auto parts covers cars, trucks and related four-wheeled motor vehicles, but there is a wide variety of other transportation vehicles served by their own parts-producing and distributing networks, including golf carts and motorcycles, construction equipment and commercial vehicles, trains, ships, aircraft and mass transit systems.

In the United States, according to statistics compiled by the U.S. Census Bureau in its *Annual Survey of Manufactures: 2005* and earlier editions, transportation equipment manufacturing, which included complete vehicles, aircraft, ships and railroad rolling stock, delivered a grand total of goods valued at $687.3 billion in 2005 and employed 1.56 million workers paid nearly $84 billion. These employees included 1.1 million production workers who earned more than $53.3 billion.

Within this Transportation Equipment Manufacturing Group, Motor Vehicle Parts Manufacturing, not including motorcycles, accounted for goods valued at more than $206.3 billion, employing 613,218 men and women, including 478,673 production workers. This represented 30 percent of the total value of goods shipped by transportation equipment manufacturers throughout the nation and 30 percent of that industry's total work force.

Aerospace product and parts manufacturing, which includes guided missiles, produced goods valued at $137.1 billion in 2005, of which about $48.5 was for aircraft engines, engine parts and other aircraft parts and auxiliary equipment. Overall, that industry employed 382,660 workers, including 157,055 in engine, engine parts and other aircraft parts manufacturing. Railroad rolling stock manufacturing produced goods valued at $9.2 billion while shipments of ship and boat building manufacturers were valued at $25.5 billion. Those industries employed 25,853 and 140,627 workers, respectively. All other transportation equipment produced in the United States in 2005 had a total value of $16.1 billion and those transportation equipment manufacturers employed 41,025 workers.

RESEARCH & DEVELOPMENT
The R&D efforts of the automotive industry are primarily focused on technologies that will make the automobile as environmentally compatible, as economical, and as safe as possible—goals which are the focus of both the vehicle manufactures and many of their leading suppliers. In fact many innovations introduced in cars and trucks have come from supplier companies, including electronic fuel injection, anti-lock braking systems (ABS), auto-dimming mirrors, electronic stability control, and a host of others. One of the keys to survival for any medium- to large-size auto parts supplier is innovation, either an improvement or significant refinement to an existing system or introduction of a totally new concept or technology.

A major emphasis in the early years of the twenty-first century was to make vehicles more environmentally compatible by reducing exhaust emissions and improving fuel mileage. This has included not only refining the venerable internal combustion engine but seeking alternative fuels and powertrains as well, including hybrid, all-electric, and fuel cell systems. Advanced safety systems are also a major focus, including everything from occupant protection systems ranging form "smart" air bags and seat belt systems to accident avoidance systems.

While the vehicle manufacturers take the lead in much development work, they rely heavily on their parts suppliers to develop and refine many of the technologies and devices required to make the new systems possible. These efforts include advanced batteries capable of long-range touring without frequent recharging; bio-fuels for both diesel and gasoline engines; inexpensive catalysts for fuel cells; and gasoline and diesel engine refinements to make these powertrains even quieter, more fuel efficient, and capable of higher power and torque outputs.

Electric motor advances are being pursued to make these powerplants practical as either the primary motive force or as part of a hybrid system mated with an internal combustion engine. In hybrid versions, the internal combustion engine may be used in any of three ways: as part of a generating system to charge the batteries; as an auxiliary to the electric motor providing extra power when needed as well as recharging the batteries; or as the main motive force while the batteries are used to power all other systems in the vehicle, reducing demands on the internal combustion engine.

A great deal of R&D effort by both the vehicle manufactures and engine suppliers is being devoted to refining the diesel engine. Diesel may wind up being the best solution to an efficient, economical and environmentally friendly powertrain. The diesel engine suffers from a negative image. It is thought to be noisy, dirty, hard to start, and expensive. Research, as reported in a May 2004 report in *Motor* magazine, has largely addressed these issues but ongoing R&D among diesel engine builders and fuel system suppliers is focused on improved engine management systems, common rail fuel systems, direct injection, high pressure injectors, multiple spray patterns, turbocharging, particulate filters, and new biomass fuels.

In the longer term, fuel cells are considered to be the next revolution in automotive powertrains. Fuel cells convert hydrogen and oxygen by using electrochemical devices. Such systems are inherently clean in that water is their by-product. Research into the development of viable fuel cell vehicles (FCVs) is directed at reducing the cost of fuel cells, improving their performance, and developing effective and efficient ways to produce and store hydrogen and other fuels. The effort is spearheaded by Freedom-CAR, a cooperate venture between the U.S. Department of Energy (DOE) and the U.S. Council for Automotive Research (USCAR), and CaFCP, the California Fuel Cell Partnership formed to encourage private companies and government agencies to work together to move FCVs toward commercialization.

CURRENT TRENDS

Since auto parts suppliers respond to the needs and demands of the automotive industry as dictated by global markets, supplier trends are a direct reflection of the factors determining the direction of the industry overall. These trends can be summarized as:

- The blurring of the line between cars and trucks as truck-like vehicles become increasingly popular.

- The drive to make vehicles more environmentally friendly by cleaning their emissions, making them more recyclable and lighter, and eliminating their consumption of non-renewable resources.

- The effort to make vehicles as safe as possible by improving their crash avoidance capabilities, crash worthiness, and occupant protection.

Within each of these trends, the driving force is the continuing application of advanced technologies, notably the continuing increase of mobile and in-vehicle electronics, expected to rise to $9.6 billion in 2007, an increase of more than 11 percent; automotive-grade semiconductors, experiencing year-on-year growth of 10 percent, reaching $18 billion in 2007; and automotive telematics and navigation, also expected to have strong growth in several world regions, generating about $38.3 billion in revenues by 2011. Most of these products and/or their components will come from auto parts suppliers.

The blurring of the line between cars and trucks is the result of the growing popularity of light trucks as multi-purpose passenger vehicles, resulting in a demand for vehicles that incorporate the comfort and performance of the automobile with the utilitarian benefits of a truck. The result is a new breed of vehicles that is a cross between a car and truck, requiring different steering, suspension, chassis, seating and other systems, most being sourced from the auto parts suppliers.

The same is true for the new breed of powertrains on the horizon in the first decade of the twenty-first century, notably direct-injection clean diesel engines, advanced gasoline engines with electronic valve timing, electrical and hybrid powertrains, and fuel cell-powered vehicles. Advanced safety systems, from smart air bags and seat belt systems that sense the size and positioning of passengers to more sophisticated electronic stability control systems and all the associated componentry will largely originate from innovative suppliers.

The auto parts industry is under intense pressure, both from increasing global competition and from many of its customers that demand lower costs and higher quality in the face of rising material, labor, and transportation costs.

The OAAI identified five issues facing the U.S. Automotive Parts Industry in its 2007 Annual Industry Assessment, which summarize the course the industry will likely be taking through much of the first decade of the twenty-first century:

1. Higher energy, plastic and steel costs, heavy debt, cash flow problems and overcapacity created by

production cuts at Ford, GM and Chrysler.

2. High legacy costs, employee wages and benefits with touch negotiations between suppliers, automakers and labor unions.

3. Continued depletion of industry ranks as a result of bankruptcies, mergers and acquisitions, plus migration to other industry—industry analysts predict that over nearly 800 major suppliers in 2000, fewer than 100 will be left by 2010.

4. Mergers and acquisitions of supplier companies. In 2005 there were thirty-two mergers, up from twenty-six in 2004, with private equity firms responsible for 53 percent of the deals.

5. Relations between the domestic Big Three and their suppliers, which are improving slightly, but remain poor compared with those of Japanese competitors.

TARGET MARKETS & SEGMENTATION

The target markets for auto parts are clearly defined, the primary target being the vehicle manufacturers of the world who account for anywhere from two-thirds to three-quarters of all auto parts purchases. The balance of auto parts production enters the service/replacement parts distribution network with most (approximately 70%) destined for commercial installers—auto repair shops, gas stations, fleet operators, and car dealer service departments. The remaining 30 percent goes to the retail market serving do-it-yourselfers.

The retail market can be divided into four basic segments: hard parts, such as brakes, mufflers, batteries, starters, alternators and pumps; maintenance items, such as oil, oil filters, lubricants, additives, spark plugs, fuel injectors, lights wipers, paints waxes and hoses; tools, including diagnostic equipment, wrenches and screwdrivers; and accessories, such as trim, hub caps, audio systems, and navigation systems.

RELATED ASSOCIATIONS & ORGANIZATIONS

ABS Education Alliance, http://www.abs-education.org

Alliance of Automobile Manufacturers, http://www.autoalliance.org

American International Automobile Dealers Association, http://www.aiada.org

Association of International Automobile Manufacturers, http://www.aiam.org

Automatic Transmission Rebuilders Association, http://www.atra.com

Automotive Aftermarket Industry Association (AAIA), http://www.aftermarket.org

Automotive Communications Council, http://www.acc-online.org

Automotive Engine Rebuilders Association, http://www.aera.org

Automotive Industry Action Group (AIAG), http://www.aiag.org

Automotive Maintenance and Repair Association, http://www.4amra.com

Automotive Parts Rebuilders Association, http://www.apra.org

Automotive Service Association, http://www.asashop.org

Automotive Trade Policy Council, http://www.autotradecouncil.org

Automotive Warehouse Distributors Association, http://www.awda.org

Brake Manufacturers Council, http://www.brakecouncil.org

Car Care Council, http://www.carcarecounciol.org

Council of Fleet Specialists, http://www.cfshq.com

Equipment and Tool Institute, http://www.etools.org

Filter Manufacturers Council, http://www.filtercouncil.org

Fire Apparatus Manufacturers Association, http://www.fama.org

Heavy Duty Distribution Association, http://www.hdda.org

Motor and Equipment Manufacturers Association (MEMA), http://www.mema.org

National Association of Trailer Manufacturers, http://www.natm.com

National Auto Auction Association, http://www.naaa.com

National Automobile Dealers Association, http://www.nada.org

National Automotive Technicians Education Foundation, http://www.natef.org

National Truck Equipment Association (NTEA), http://www.ntea.com

Original Equipment Suppliers Association, http://www.oesa.org

Recreational Vehicle Industry Association, http://www.rvia.org

Specialty Equipment Market Association (SEMA), http://www.sema.org

Tune-Up Manufacturers Council, http://www.tune-up.org

United States Council for Automotive Research, http://www.uscar.org

BIBLIOGRAPHY

"Automotive Parts: Trends in Global Production and Trade," U.S. Department of Labor. Available from <http://www.bls.gov>.

Bailey, David. "U.S. Parts Supplier Consolidation Inevitable." *The Boston Globe.* 14 September 2007.

"Career Guide to Industries—Automobile Dealers." U.S. Department of Labor. Available from <http://www.bls.gov>.

"December Global Light Vehicle Sales." *Global Monthly Sales Report.* JD Power and Associates. Available from <http://www.jdpowerforecasting.com>.

"Global Vehicle Production and Sales by Manufacturer," Global Market Data Book. *Automotive News Europe.* 26 June 2006.

Hill, Kim, Debbie Menk, and Steven Szakaly. "Contribution of the Motor Vehicle Supplier Sector to the Economies of the United States and Its 50 States." Center for Automotive Research, Economics and Business Group. January 2007.

Hyde, Justin and Michael Ellis. "Foreign? American? Auto Parts Go Global." *Detroit Free Press.* 7 May 2006, 1.

"Industry Overview: Automobile Parts Wholesale-Retail." Hoovers, a D&B Company. Available from <http://www.hoovers.com/automobile-parts-wholesale-retail>.

"International Trade Statistics by Country and Product." International Trade Centre, a Joint Technical Cooperation Agency of the United Nations Conference on Trade and Development (UNCTAD) and the World Trade Organization (WTO). 2007.

Keenan, Tim. "Supplier Sequel? Some See History Repeating Itself—Relations Between Parts Manufacturers and Automotive Producers—U.S. Automotive Centennial." *Ward's Auto World.* May 1996.

Klier, Thomas H., and James M. Rubenstein. "The Supplier Industry in Transition—The New Geography of Auto Production," Federal Reserve Bank of Chicago. August 2006, Number 229b.

"Rationalizing the Automotive Supplier Industry; Carving Out Profit from M&A Activity" Original Equipment Suppliers Association (OESA) and PricewaterhouseCoopers. 23 August 2006. Available from <http://www.oesa.org>.

"Statistics for Industry Groups and Industries: 2005." *Annual Survey of Manufactures.* U.S. Department of Commerce, Bureau of the Census. November 2006.

"Top 100 Global Suppliers." *Automotive News.* 25 June 2007.

Tsuge, Hiroto and Frank Bartels. "FDI Promotion Strategies in Developing Countries: A Look at the Spatial Distribution of the Automotive Industry." SAIS Working Paper Series. Johns Hopkins University, Paul H. Nitze School of Advanced International Studies. October 2003.

"U.S. Automotive Parts Exports, 1999–2006," U.S. Department of Commerce, International Trade Administration, Office of Aerospace and Automotive Industries' Automotive Team. Office of Aerospace and Automotive Industries' Automotive Team. 2007. Available from <http://www.ita.doc.gov>.

"U.S. Automotive Parts Imports, 1999–2006." U.S. Department of Commerce, International Trade Administration, Office of Aerospace and Automotive Industries' Automotive Team. Office of Aerospace and Automotive Industries' Automotive Team. 2007. Available from <http://www.ita.doc.gov>.

"U.S. Automotive Parts Industry Annual Assessment." U.S. Department of Commerce, International Trade Administration, Office of Aerospace and Automotive Industries. March 2007.

Verespei, Mike. "Detroit Needs A Different Driver." *Purchasing Magazine Online.* 7 April 2005. Available from <http://www.purchasing.com>.

Watkins, Thayer. "The Keiretsu of Japan." San Jose State University. Available from <http://www.sjsu.edu/faculty/watkins>.

SEE ALSO *Automobiles, Construction Machinery, Trucks*

Automobiles

———————————■———————————

INDUSTRIAL CODES

NAICS: 33–6111 Automobile Manufacturing, 33–6112 Light Truck and Utility Vehicle Manufacturing

SIC: 3711 Motor Vehicles and Passenger Cars Bodies

NAICS-Based Product Codes: 33–6111 through 33–61110100 and 33–6112 through 33–61120100

PRODUCT OVERVIEW

The automobile, which first appeared commercially in the nineteenth century and went on to revolutionize personal transportation during the twentieth century, is evolving through a series of fundamental changes early in the twenty-first century in order to achieve two somewhat incompatible goals. On the one hand, there is a growing mandate among many governments to minimize the automobile's impact on the environment and non-renewable natural resources. On the other hand, it must maintain its utility and affordability if it is to remain attainable by the general public.

Its evolution includes structural changes in two critical areas: (1) the continued blurring of the line between cars and other light-duty vehicles such as pick-up trucks, vans, and sport utility vehicles, and (2) a serious effort to eventually replace the internal combustion engine with an alternative powertrain and energy source, possibly the space age version of the fuel cell, in order to minimize or eliminate undesirable exhaust emissions. There has also been an ongoing emphasis on increasing the safety of all

light vehicles, triggered in 1965 by Ralph Nader's book, *Unsafe At Any Speed.*

The blurring of the roles of cars and light trucks is a reminder that the basic definition of an automobile needs to be either broadened or discarded. Typical modes of personal transportation are no longer limited to a "four-wheeled automotive vehicle designed for passenger transportation and commonly propelled by an internal-combustion engine, using a volatile fuel" per *Webster's New Collegiate Dictionary.*

In today's world, personal transportation includes pickup trucks, some capable of carrying up to five passengers, as well as sport utility vehicles (SUVs) and vans, both mini and full sized. Pickups and SUVs also come in compact and full-sized models. Since the story of both cars and light trucks is so interwoven and complex, for the purposes of this essay we will explore the automobile in the broader context of being any light vehicle used for passenger transportation.

As the name implies, a pickup is a small truck with an open cargo-carrying bed. Traditional SUVs function like station wagons with passenger and cargo-carrying capabilities but are built on pickup truck platforms. A new breed of downsized compact SUVs appeared early in the twenty-first century based on car rather truck platforms. These more car-like SUVs are an example of the blurring of lines between cars and trucks and will be covered later in this essay. Full-sized vans include commercial-style passenger vans and van conversions that provide living room comfort for its passengers, while minivans are primarily factory-built smaller versions of van conversion designed for active families with accommodations for both passengers and/or cargo.

Forerunners of the modern automobile actually first appeared in the eighteenth century—notably a steam-powered three-wheeled vehicle invented in France in 1769 by Nicholas Joseph Cugnot—but self-propelled vehicles didn't become commercially viable until the introduction of the internal combustion engine in the nineteenth century. Etienne Lenoir, a Belgian inventor, developed the first internal combustion engine, which he demonstrated in Paris in 1862. Then in 1878, Nicholas Otto, a German inventor, developed a four-stroke coal-gas engine that was quieter and smoother running. In 1885 Germans Karl Benz and Gottlieb Daimler built the first gasoline-powered vehicles, and in 1889, Frenchman Armand Peugeot built the first automobile for commercial sale. American brothers Charles and Frank Duryea began production in 1896 of the first commercially available gasoline-powered car in the United States.

Prior to the introduction of the internal combustion engine, steam engines and electric motors were used to power the so-called horseless carriage. In fact, by the close of the nineteenth century, some 30 manufacturers in the United States were offering an array of vehicles powered by gasoline, steam or electricity with electric vehicles outselling all other types of cars. Steam engines, however, proved too heavy to be practical for road vehicles, plus they had long start-up times and their need for plenty of water limited their range, so they soon faded from the scene. Electric vehicles of that era also had limited range and were very expensive and slow, which ultimately led to the declining popularity of electrically powered cars.

Early gasoline-powered vehicles also had their drawbacks, including noise, the smell of the fuel, complicated shifting requirements and difficult hand cranking needed to start the engine. Several developments in the early years of the twentieth century propelled the gasoline internal combustion engine to prominence and made possible the personal transportation revolution, most notably the invention of the electric starter by Charles Kettering in 1912, which eliminated the hand crank and initiated the mass production of the gasoline engine by Henry Ford.

It was the introduction in 1908 of the very affordable Model T by Henry Ford with its $950 price tag that really marked the beginning of the personal transportation revolution. That price dropped as low as $280 when Ford revolutionized automotive production in 1913 with the constantly moving assembly line and $5 per day wages in 1914. These developments made cars affordable for the average person, gave factory workers increased buying power, and triggered the explosive growth of the global automotive industry, which by 2006 was selling nearly 64 million vechicles annually.

The history of pickup trucks generally follows a timeline similar to cars. Gottlieb Daimler, the German automobile pioneer, built the first pickup truck in 1896, a four-horsepower belt-driven vehicle with somewhat limited capability. The first truck company to go into business in the United States, the Rapid Motor Vehicle Company, was opened in Detroit in 1902 by two brothers, Max and Morris Grabowski. The previous year they had designed and built a single-cylinder chain driven dray machine that was basically a motorized version of a horse-drawn wagon. Rapid and the Reliance Motor Company, also of Detroit, which also began building trucks in 1902, were purchased in 1908 and 1909, respectively, by William C. Durant, founder of General Motors Corporation (GMC).

Both Rapid and Reliant trucks were big gas-powered machines designed to replace horse-drawn wagons, so in the early years GMC used electric vehicles for light-duty delivery. However, in 1916 GMC converted everything to gasoline engines. Light-duty Chevrolet trucks appeared on the scene in 1918 and GMC-badged trucks were introduced in 1927, a product that originated with Pontiac but was badged a GMC to avoid the prospect of three General Motors-branded trucks. The other top-selling American pickup truck pioneers, Dodge and Ford, introduced their pickup truck models in 1918 and 1925, respectively.

Compact pickup trucks, which today are among the best-selling vehicles in the world, first appeared in the United States in 1959 when the Nissan Motor Company, a small Japanese manufacturer, exported the Datsun 1000, which had a load capacity of only a quarter-ton and a 1000 cc, 37-horsepower engine. While initial models only sold a few hundred vehicles per year, sales jumped to more than 15,000 in 1965 with the importation of the Datsun 520 pickup. Nissan eventually phased out the Datsun model name and badged all subsequent models as *Nissan*. Toyota jumped into the compact pickup fray in the U.S. market in 1964 by exporting its Stout, followed in 1969 by the Hi-Lux.

The popularity of the compact pickups caught the attention of the domestic Big Three (General Motors, Ford, and Chrysler) in the United States and they countered in the early years with imports of their own: Chevrolet with the LUV from Isuzu Motors Ltd. in 1972; Ford with the Courier from Mazda at about the same time; and Dodge from Mitsubishi in 1979. These models evolved into the domestically built Chevrolet S-10, Ford Ranger, and the Dodge Dakota, introduced in the mid-1980s as the first mid-size pickup trucks.

Both SUVs and station wagons trace their heritage to the 1920s when cars called depot hacks or suburbans were used to carry passengers and their luggage from train stations. Chevrolet and GMC applied the name to a utility vehicle introduced in 1936 as a passenger-carrying vehicle based on a commercial panel truck. Another early SUV forerunner was the Willy's Jeep Wagon introduced in 1940 as a utility vehicle for the family and eventually redesigned as the Jeep Wagoneer in 1963. The British Land Rover, another icon of the SUV industry, was inspired by the World War II Willy's Jeep and introduced by Rover Company Ltd. at the 1948 Amsterdam Motor Show.

While SUVs evolved as truck-based vehicles, the classic station wagon is a rear-wheel-drive car with a stretched wheelbase to accommodate a cargo-carrying compartment accessible via a rear tailgate. The first production station wagon was the 1923 wood-bodied Star built by the Star Motor Company, which had been purchased by the Durant Motor Company. Ford made the station wagon affordable to the general public with the introduction of a mass-produced Modal A version in 1929.

Passenger vans have been part of the light vehicle automotive scene since the 1920s but minivans did not emerge until they were introduced in 1983 by Chrysler Corporation, the market leader in full-size vans. The company recognized the desirability of a downsized van that fits in a typical garage and offers all the comforts and handling ease of a station wagon with a lot more room for passengers and cargo. Ford and General Motors followed suit in 1985 with the Aerostar and Astro/Safari, respectively.

MARKET

The global automotive market has been experiencing weak demand growth in the early years of the twenty-first century, but growth nonetheless. The worldwide market for cars and light trucks increased 3.4 percent from 61.69 million vehicles in 2005 to 63.81 million in 2006, according to J.D. Power Automotive Forecasting.

The U.S. market, which has been under intense competitive pressure from imports, was one of only two major markets to lose ground with sales fading 2.5 percent from 16.95 million in 2005 to 16.52 million in 2006. Japan was the other market experiencing declining demand, falling 2.5 percent from 5.73 million in 2005 to 5.59 million in 2006. The Toyota-Daihatsu manufacturing group, however, is projected to experience the strongest growth for the near future, increasing 3.8 percent from 2004 to 2009 with increased penetration in the North American and European markets, according to Dave Liggett in the just-auto.com Management Briefing global market review of car sales, March, 2005. No other major manufacturer is expected to experience any significant growth, according to the report. In fact, most are predicted to lose market share.

FIGURE 9

Production of the Top 20 Global Vehicle Manufacturers in 2004 and 2005

Company	2004	2005
General Motors	7,959,838	8,338,073
Toyota Motor Corporation	7,548,600	8,232,100
Fort Motor Company	6,636,329	6,631,718
Volkswagen AG	5,093,181	5,219,478
DaimlerChrysler AG	4,617,700	4,810,000
Hyundai-Kia Automotive	3,181,394	3,693,277
Nissan Motor Company	3,207,217	3,508,005
Honda Motor Company	3,181,624	3,409,991
PSA/Peugeot-Citoen SA	3,405,100	3,375,500
Renault SA	2,471,676	2,515,728
Suzuki Motor Corporation	1,986,749	2,124,584
Fiat S.p.A.	2,099,780	2,056,600
Mitsubishi Motor Corporation	1,413,403	1,362,673
BMW Group	1,250,345	1,323,119
Mazda Motor Corporation	1,134,421	1,146,145
AutoVaz	717,985	721,492
Isuzu Motor Ltd.	566,238	626,305
Fuji Heavy Industries Ltd.	592,676	588,331
China FAW Group Corporation	519,515	464,953
Chongquing Changan Automobile Co.	421,438	460,074

SOURCE: Compiled by the staff with data from the *Automotive News Europe*. 2006

Western Europe, the largest automotive market in the world, managed to grow slightly during 2006: 0.8 percent from 16.52 million units in 2005 to 16.65 million in 2006. The emerging markets of the world, on the other hand, experienced more robust growth: other European countries were up a combined 8.1 percent; Brazil and Argentina, 13.3 percent, and the other markets of the world, 14.6 percent.

In terms of production, the biggest producers are vehicle manufactures in the Asia-Pacific region, building a total of 24.83 million vehicles in 2005. Japan is the leading producer there, building 9.02 million cars and 1.78 million trucks that year. European manufacturers are the second largest vehicle producers with Germany the leader, producing 5.35 million cars and 407,523 trucks. In North America, the United States is the dominant producer, building more than 12 million cars and trucks in 2005 with General Motors the leader, building 1.15 million cars and 2.15 million trucks.

Worldwide, General Motors and Toyota are the biggest players, each building over 8 million vehicles annually with Toyota vying with GM for the number one spot. Ford is number three worldwide but is losing ground midway through the first decade of the twenty-first century. World rankings, based on global production in 2004 and 2005, are presented in Figure 9. Through the rest of the decade and as the century unfolds further, the rankings of all the manufacturers are very apt to change significantly

FIGURE 10

Global Production and Sales of Automobiles and Light Trucks by Region, 2005

Global Production

Region	Cars	Trucks	Total
Africa	413,015	237,874	650,889
Asia-Pacific	18,075,585	6,752,812	24,828,397
Central/South America	2,276,149	776,364	3,052,513
Europe	17,984,183	2,827,263	20,118,446
Middle East	1,441,313	564,663	2,005,976
North America	6,785,540	9,589,130	16,374,670
Total	46,975,785	20,748,106	67,723,891

Global Sales

Region	Cars	Trucks	Total
Africa	781,202	354,582	1,135,784
Asia-Pacific	12,105,398	5,818,859	17,924,257
Central/South America	2,335,205	776,158	3,111,363
Europe	17,376,760	3,019,281	20,396,041
Middle East	2,172,494	710,472	2,882,966
North America	9,541,213	11,155,510	20,696,723
Total	44,312,272	21,834,862	66,147,134

Truck production and sales figures reflect light, medium, and heavy commercial vehicles and buses.

SOURCE: Compiled by the staff with data from Automotive News Data Center and R.L. Polk Marketing Systems GmbH. 2006

as emerging players—notably vehicle builders in China and Korea—challenge the traditional leaders.

A comparison of global automotive sales with production in each of the world markets illustrates one of the major problems confronting domestic automakers in Europe and North America, who are being challenged by imports from Asia-Pacific producers, most notably in Japan and Korea. Nearly seven million vehicles built in 2005 in Asia, including more than 933,000 trucks, are exported, primarily to the United States, although 182,000 more trucks were sold in Europe than were built there that year. Figure 10 presents production and sales data for 2005 by region. Note the disparity between domestic production and sales in North America: 2.76 million more cars and 1.57 million more trucks were sold than were built there. Car sales also included exports from Europe, which produced 607,423 more cars than were sold in those countries.

The two largest, most developed automotive markets of the world—the United States and Western Europe—are both extremely competitive markets. In the United States, for example, the traditional Big Three are under unrelenting pressure from a host of new entrants from Asia and Europe, many of whom have established production facilities in North America. General Motors and Ford, number one and two in light vehicle sales, respectively, lost market share consistently through the first five years of the twenty-first century while Toyota,

FIGURE 11

U.S. Light Vehicle Sales History by Manufacturer, 2001–2005

Company	2001	2002	2003	2004	2005
General Motors	4,862,661	4,820,017	4,714,782	4,655,459	4,454,385
Ford Motor Company	3,962,659	3,623,221	3,477,444	3,319,767	3,153,875
DaimlerChrysler	2,479,846	2,418,671	2,346,168	2,427,634	2,529,254
Toyota Motor Company	1,741,254	1,756,127	1,866,313	2,060,049	2,260,296
American Honda Motor Company	1,207,639	1,247,834	1,349,847	1,394,398	1,462,472
Nissan Motor Company	703,308	739,517	794,481	985,988	1,076,669
Hyundai Group	569,962	612,464	637,692	688,670	730,863
Volkswagen of America	439,683	424,397	389,544	336,422	310,915
BMW Group	213,127	256,622	277,035	296,531	307,402

General Motors includes Buick, Cadillac, Chevrolet, GMC, Hummer, Oldsmobile, Pontiac, Saturn, and Saab.
Ford Motor Company includes Aston Martin, Ford Division, Lincoln, Mercury, Jaguar, Volvo, Land Rover, and the data for 2002 and 2002 includes Think.
DaimlerChrysler includes Chrysler Group, Mercedes Benz, and the data for 2003–2005 includes Maybach
Toyota Motor Company includes Lexus, Toyota Division, and in 2003–2005, Scion.
American Honda Motor Company includes Acura and Honda Division.
Nissan Motor Company includes Infiniti and Honda Division.
Hyundai Group includes Hyundai Division and Kia.
Volkswagen of America includes Volkswagen Division and Audi, in 2001 and 2002 Rolls Royce/Bentley, in 2003–2005 Bentley.
BMW Group includes BMW Division, in 2002–2005 Mini, and in 2003–2005 Rolls Royce.

SOURCE: Compiled by the staff with data from the Automotive News Data Center, available from http://www.autonews.com. 2007.

American Honda, Nissan, and Hyundai Group consistently increased market share. Figure 11 presents five years worth of U.S. light vehicle sales figures broken down by leading automobile manufacturers.

KEY PRODUCERS/MANUFACTURERS

There are six automotive production regions in the world: Africa, Asia-Pacific, Central/South America, Europe, Middle East, and North America. The 15 major producers—those building more than one million vehicles annually—are concentrated in three regions: Asia-Pacific, Europe, and North America. Since the automotive industry is a truly global enterprise with some 50 vehicle builders scattered throughout the regions of the world, many of the major manufacturers have production facilities in more than one region, often many more, as is the case with General Motors, Toyota, and Ford.

General Motors, for example, has manufacturing operations in 33 countries located in all six regions of the world. It employs about 324,000 people globally, about half in the United States, and had revenues of $192.6 billion in 2005. Toyota has manufacturing operations in some 26 countries in North and South America, Europe, Africa, and Asia-Pacific, has some 289,980 employees worldwide, including 38,340 in North America, and revenues of $174.6 billion in 2005. Ford, the third largest automotive producer, operates plants in 23 countries, also located in North and South America, Europe, Africa, and Asia-Pacific. Worldwide employment is about 300,000 and 2005 revenues were $176.9 billion.

The top 15 global manufacturers, including their affiliates and subsidiaries, account for 85 percent of the world vehicle production. These companies and their global affiliate and subsidiary operations are:

- General Motors, which includes Daewoo (Korea) and Holden (Australia)

- Toyota Motor Corp., which includes Daihatsu (Japan) and Hino (Japan)

- Ford Motor Company, which includes Aston Martin (UK), Jaguar (UK), Land Rover (UK) and Volvo Car Corp. (Sweden)

- Volkswagen AG, which includes Audi (Germany), Bentley (UK), Bugatti (Italy), Lamborghini (Italy), Skoda (Czech Republic) and Seat (Spain)

- DaimlerChrysler AG, which includes Chrysler Division, Dodge, Jeep, Mercedes-Benz (Germany), Smart (Germany), and Commercial Vehicle Division, EvoBus GmbH (Germany), Freightliner (USA), and Mitsubishi Fuso Truck and Bus Corp. (Japan)

- Hyundai-Kia Automotive, which includes Hyundai Motor (Korea) and Kia Motors (Korea)

- Nissan Motor Co. (Japan)

- Honda Motor Co. (Japan)

- PSA/Peugeot-Citroen SA (France)

- Renault SA (France), which includes Dacia (Romania) and Renault-Samsung Motors (Korea)

- Suzuki Motor Corp (Japan), which includes Maruti Udyog Ltd. (India)

- Fiat S.p.A. (Italy), which includes Fiat Auto (Italy), Ferrari (Italy), Maserati (Italy), and Iveco (Italy)

- Mitsubishi Motor Corp. (Japan)

- BMW Group (Germany) which includes Rolls-Royce (UK)

- Mazda Motor Corp. (Japan)

In North America, General Motors is the leading vehicle builder, producing Buick, Cadillac, Chevrolet, GMC, Hummer, Isuzu, Oldsmobile, Pontiac, Saab and Saturn vehicles. The corporation operates facilities in 28 states. Ford plants produce Ford, Lincoln and Mercury vehicles. DaimlerChrysler builds Chrysler, Dodge, Jeep and Mercedes-Benz vehicles.

DaimlerChrysler, the number five producer worldwide and number three in the United States, operates 33 plants throughout North America, including 24 in 7 of the U.S. states, 5 in Mexico, and 4 in Canada. Globally, the corporation has 382,724 employees in 17 countries in Europe; North and South America; Asia-Pacific, including Australia; Africa; and the Middle East. Revenues in 2005 were $177.36 billion.

The other major vehicle producers in the U.S. include Toyota Motor, which builds Toyota, Lexus and Scion vehicles at its plants in Cambridge, Ontario, Canada; Georgetown, Kentucky and Princeton, Indiana; and Tijuana, Mexico. Honda produces Acura and Honda vehicles at plants in Alliston, Ontario, Canada; East Liberty and Marysville, Ohio; Lincoln, Alabama; and El Salto Jalisco, Mexico. Nissan operates plants in Canton, Mississippi and Smyrna, Tennessee, in the U.S., and in Aquascalientes, and Cuernavaca, Mexico, producing Nissan-branded vehicles plus providing North American production for Renault in its Mexico plants.

Other companies building vehicles in the United States include: Volkswagen which builds the Beetle, Jetta, and VW medium trucks; BMW which builds the BMW Z4 roadster and X5, Subaru which produces both the Subaru Baja and Legacy, as well as the Isuzu Axiom, Rodeo and B9; Hyundai which makes the Hyundai Sonata; Misubishi which builds the Eclipse, Galant and Endeavor as well as the Chrysler Sebring and Dodge Stratus for the

Chrysler group of DaimlerChrysler; and AM General which builds the Hummer H1 and Hummer H2 for General Motors.

NUMMI, a joint venture between GM and Toyota in Fremont, California, was established in 1984 to build vehicles for the two automakers. In 2005 this plant produced the Pontiac Vibe, Toyota Corolla and Tacoma. The AutoAlliance is a joint venture between Ford and Mazda in Flat Rock, Michigan, building the Ford Mustang and Mazda 6 sedan, hatchback and wagon. CAMI, a joint venture between GM and Suzuki Motor Corporation in Ingersoll, Ontario, Canada, produces the Chevrolet Equinox, Pontiac Torrent and Suzuki Vitara.

MATERIALS & SUPPLY CHAIN LOGISTICS

The automobile industry is not only global in the extent to which finished vehicles are distributed for sale worldwide from their point of origin, but many of the materials, parts, assemblies, and systems for the majority of vehicles sold commercially are sourced from throughout the world as well. "More than ever, automakers are drawing on suppliers around the globe, shuttling parts across borders in search of lower prices and higher quality," states the Original Equipment Suppliers Association in a recent report.

For example, the report cites an analysis by the *Detroit Free Press*, noting that "federal data found that vehicles built by Detroit automakers have steadily increased their proportion of parts form outside the United States and Canada. By the same measure, vehicles built in North America by Japan's largest automakers increasingly use U.S. and Canadian parts."

Mexico, which builds more than one million vehicles annually for U.S., European, and Japanese car companies, is also a major component producer and the largest exporter of car parts to the United States—$22 billion in 2004. Canada accounted for $19 billion worth of parts exported to the United States that year and Japanese suppliers accounted for $14 billion, together these two countries accounted for 72 percent of all parts imported into the United States for use in auto production. While Mexico and Japan accounted for 55 percent of U.S. auto parts imports, over three-quarters of the $44 billion in U.S. auto parts exports were headed for auto plants in those two countries in 2004.

Overall, the United States has a trade imbalance in auto parts. In 2005 the imbalance reached $37 billion with exports of $55 billion and imports of $92 billion, including $46.9 billion from North American Free Trade Association (NAFTA) partners Mexico and Canada. Those two countries were also the destination for $42.6 billion in U.S. auto parts exports.

FIGURE 12

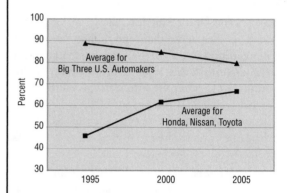

Percent of Auto Parts Used in U.S. Product Sourced in the United States or Canada by Automaker, 1995–2005

The percentages present represent the percent of automotive parts used in the production of vehicles in the United States that were themselves produced in the United States or Canada. If the percentage is 85, the remaining 15 percent of auto parts were sourced from outside the United States or Canada.

Percent by Company

Automobile Manufacturers	1995	2000	2005
DaimlerChrysler	89	80	76
Ford	86	87	82
General Motor	91	87	81
Honda	47	70	68
Nissan	42	58	57
Toyota	49	57	75

SOURCE: Compiled by the staff with data from Hyde, Justin and Michael Ellis, "Foreign? American? Auto Parts Go Global," *Detroit Free Press.* May 7, 2006, 1A.

Independent suppliers account for much of the content in vehicles built in the United States and Europe since the two largest suppliers in the world—Delphi Automotive Systems and Visteon Corporation—were each spun off from their parent companies, General Motors and Ford, respectively, and became independent in 1999 and 2000. They were among 1,500 and 2,000 tier one major system suppliers (suppliers selling components and systems directly to the vehicle builders) serving the industry in 2004, a number expected to eventually shrink to between 500 and 700, of which only about 50 will be system integrators dealing directly with the vehicle manufacturers, according to the Boston Consulting Group. This reflects changes in the role of the major suppliers that are being asked to provide larger and more complete vehicle systems or segments rather than simply supplying hundreds of parts and components.

The leading global suppliers include Delphi, Visteon, Lear Corporation, Johnson Controls, Inc., TRW and Dana Corporation all with world headquarters in the United States. Magna International, Inc. has its headquarters in Canada; Robert Bosch GmbH is based in

Germany; Valeo SA in France; and Denso Corporation, Aisin Seiki, Co. Ltd. and Yazaki Corporation, have headquarters in Japan.

In the United States, automotive suppliers employ more workers than any other manufacturing group, directly employing more than three-quarter million and contributing to 4.5 million jobs nation wide, including nearly 2 million of those indirect employees in support industries such as steel, plastic, and technical services. Suppliers' total annual payroll, including benefits, is $253 billion, or an average of $45,790 per worker annually, according to a 2007 report from the Center for Automotive Research.

Japan has traditionally operated under the keiretsu system in which an association of companies is organized that cooperates with and works with each other. In the automotive industries, supplier keiretsu sometimes have several hundred companies associated with a particular vehicle manufacturing company. Usually, these agreements are somewhat exclusive, obliging the two parties, the keiretsu companies and the auto manufacturer, to deal with each other and not others. This, in essence, bars non-members from obtaining supplier arrangements with the vehicle manufacturer. These keiretsu relationships are tending to dissolve as cross sourcing of components among the suppliers increases with industry globalization.

With the global automotive industry producing more than 64 million vehicles each year, it is a major consumer of various grades of steel, which accounts for some 55 percent of an average vehicle's content by weight—over 1,700 lbs. in a mid-size sedan. Other materials consumed include cast iron, averaging more than 400 lbs., primarily for engine and suspension components; plastics and composites, nearly 250 lbs.; aluminum, about 190 lbs.; plus rubber, glass, copper, zinc, other metals, fabrics and various fluids and lubricants. These materials are sourced from all over the world, the location depending on price, availability and accessibility on the part of the provider of parts and/or components supplying the vehicle manufacturer.

DISTRIBUTION CHANNEL

Automobile sales traditionally have been handled through franchised independent dealerships in the major markets of the world. These independent dealerships act as the retailing bridge between the vehicle manufacturers and the consumer. In the United States, the National Automobile Dealers Association (NADA) represents some 20,000 franchised new car and truck dealers holding nearly 43,000 separate franchises, both domestic and international (many dealerships represent multiple nameplates, including those from more than one manufacturer). In addition the American International Automobile Dealer Association (AIADA) represents 11,000 international nameplate automobile franchises.

Multiple nameplate representation often covers both domestic and foreign brands. For example, General Motor's automotive brands are Buick, Cadillac, Chevrolet, GMC, Holden (Australia), HUMMER, Opel (Germany), Pontiac, Saab (Sweden), Saturn, and Vauxhall (UK). In some countries, the GM Group distribution network also markets vehicles manufactured by GM Daewoo (Korea) and Isuzu, Fuji (Subaru), and Suzuki (Japan).

Since dealerships accept trade-ins as part of most new car sales, the dealerships also offer used cars, although new car sales typically account for more than half of all sales revenue for a dealership. The balance comes from used car sales, repair and service, and aftermarket parts and component sales. Vehicle leasing is also offered by dealerships as a financing option for consumers. In addition to traditional marketing techniques to attract buyers to their showrooms, automobile dealers are increasingly using the Internet to market new and used cars.

New car distribution systems use both truck transports and railroads to transport vehicles to dealerships; offshore exports are handled by ship. In the case of railroads, vehicles are shipped to distribution centers, then delivered to dealerships via truck. Ford, for example, employs an innovative vehicle distribution system with the Norfolk and Western Railroad in which vehicles sourced from Ford's 20 North American assembly plants are routed through a mixing center network and from there to Ford's dealership network. The mixing center network is a hub and spoke distribution system in which vehicles are shipped to one of four strategically located centers serving the U.S. and Canada. At each mixing center, the vehicles are unloaded, sorted and reloaded for delivery to a common destination, reducing time spent waiting at a plant until a railroad car is loaded with that plant's vehicles headed for a common destination.

In addition to new car dealerships, used cars are also sold by independent dealers, ranging from small, one-location stores or lots to large nationwide superstores.

KEY USERS

Next to a new home purchase or paying the rent, the personal car or light truck is the most essential expense for most members of the public throughout much of the world, accounting for a large majority of new vehicle sales. In the United States, for example, 25 to 30 percent of all new car sales are sold directly to the end consumers. Fleet sales—government agencies, businesses, and police departments—account for the balance.

ADJACENT MARKETS

In addition to automobiles and light trucks, transportation equipment includes a wide variety of powered vehicles, everything from golf carts, motor scooters, and motor cycles to commercial vans, medium and heavy trucks, recreational vehicles, other commercial vehicles such as buses, and heavy-duty construction equipment.

For the majority of the American public—with the exception of those adventurous enough routinely to hop on a bicycle or motorcycle to commute to work every day—markets adjacent to new automobiles are represented by used, or what are often referred to as previously owned, cars and mass transit. Viewed from the retail level, as shown by *Manufacturing & Distribution USA*, the used car market in 2007 represented dealer sales of $60.7 billion against new car dealer sales of $775.6 billion. Previously owned cars were thus 7.3 percent of all auto purchases in that year, not counting exchanges, which occurred between individuals. Historical patterns suggest that used cars are more and more of a choice (and perhaps a necessary choice) for a segment of the public. Used cars in 1987 were 3.7 percent of retail auto sales, in 1992 4.6 percent, in 2002 6.9 percent—suggesting that economic pressures may be inducing a growing number of people to select one of the adjacent markets to get their transportation.

Based on data collected by the Bureau of Transportation Statistics, an element of the U.S. Department of Transportation (DOT), just over 4 percent of all workers make use of mass transit of one kind or another (bus, streetcar, subway, or elevated train) in their daily commute. In 1989 4.9 million workers (4.6% of all commuters) reached the job by mass transit every day. In 2001, 5.6 million (4.7% of workers) were using transit. By 2005 numbers had declined to 5.4 million workers (4.4 % of all those commuting). People using bicycles and motorcycles, incidentally, represented 0.7 percent of commuters in 1989 and 2001 and 0.6 percent in 2005.

In quite a real sense, the growing trend of working from one's personal residence, made possible by the computer and the Internet, represents an interesting *adjacent* way of *avoiding* transportation—at least to the place of work. DOT's surveys also capture this phenomenon and show that it is growing. Some 2.7 million people worked at home in 1989 (2.6% of all workers). The numbers have steadily increased since, to 3.4 million in 2001 (2.8%) and 4.1 million in 2005 (3.4%).

Adjacent markets for long distance travel are trips by air, by bus, by train, or by ship. DOT surveys indicate that the automobile (no distinctions made between new and used cars) is the dominant form of transportation for taking trips. In 2001, 89 percent of all trips were made by using personal vehicles; all other forms came in a distant second: 7.4 percent of trips were made by air; 2.1 percent

by bus; 0.8 percent by train; 0.1 percent by ship, boat, or ferry; and the remainder were not classified by DOT. These results were somewhat less skewed in favor of autos/pickups if, instead of counting trips, we count *person-miles traveled,* to use DOT's terminology. Using that category, travel by mode in 2001 shows a different pattern: personal vehicles, 55.9 percent of miles; air travel, 41 percent; bus travel, 2 percent; train travel, 0.8 percent; and water-borne travel, 0.3 percent.

RESEARCH & DEVELOPMENT

The Research and Development (R&D) efforts of the automotive industry are primarily focused on technologies that will make the automobile as environmentally compatible, as economical to operate, and as safe as possible, goals which in many ways are mutually exclusive. Safety and emission reductions all add cost, while efforts to improve fuel economy by reducing weight can make the vehicle less safe and crash-worthy.

To make vehicles more environmentally compatible, the major emphasis has been on emission reduction, although there is increasing interest in making the vehicle as recyclable as possible. The latter is being improved by using recyclable materials as much as possible and making it easier to disassemble and extract recyclable materials at the end of its useful life.

The primary efforts, however, have been on powertrain technology, with the emphasis on cleaning up emissions and reducing fuel costs and consumption. This has included constantly improving the venerable gasoline internal combustion engine so it runs as cleanly as possible without an undue sacrifice of power and performance. But by the end of the twentieth century, it had become apparent that further improvements were providing only marginal emission improvement. Alternative power and fuel sources have been the primary goal ever since.

Ironically, one of those alternative sources is a reintroduction of the electric motor, either as the primary motive force or as part of a hybrid system mated with an internal combustion engine. In hybrid versions, the internal combustion engine may be used in any of three ways: strictly as part of a generating system to charge the batteries; as an auxiliary to the electric motor providing extra power when needed, as well as for recharging the batteries; or as the main motive force while the batteries are used to power all other systems in the vehicle, reducing demands on the internal combustion engine. Hybrids have been a major research effort by virtually all of the major manufacturers for use in both cars and trucks.

Electric vehicles reappeared in the 1990s with conversion vehicles from Ford, General Motors, Daimler-Chrysler and Toyota, plus two cars built from the ground up as electric cars, General Motors' EV1 and Honda's EV

Plus. Cost and short range imposed by the limitations of battery technology has kept pure electric vehicles out of the mainstream, although promising developments with advanced lithium ion battery packs and General Motors' development of the Chevrolet Volt, an electric car concept vehicle introduced at the North American International Auto Show in 2007, could bring electric vehicles back into a more prominent position. The Volt uses General Motors' patented E-Flex Propulsion System that consists of an electric drive system, lithium ion battery and an onboard generator powered by a one-liter turbocharged internal combustion engine to keep the batteries charged.

Yet another powertrain technology that dates to the early days of the automobile and has been around ever since may in the end be the technology to trump all the others. That is the lowly diesel, an internal combustion engine used extensively in Europe and Asia in small cars, yet largely looked down upon in the United States because of the perception that it is inherently noisy, dirty, hard to start and expensive to buy, although more fuel efficient and longer-lasting than gasoline engines.

A further plus of the diesel is its ability to operate on non-petroleum based fuels. Diesel owners have the option of using biodiesel, a domestically produced renewable fuel that reduces U.S. oil dependence and contributes to the national economy. The Diesel Technology Forum observes that "American consumers are turning to diesel-powered vehicles to help them save money on fuel costs without having to sacrifice the power and performance drivers have come to expect."

A great deal of R&D effort among diesel engine builders has been focused on addressing the negatives associated with classic diesel engines, including improved engine management systems, common rail fuel systems, direct injection, high pressure injectors, multiple spray patterns, turbocharging, particulate filters and new biomass fuels, according to a May 2004 report in *Motor* magazine.

Long-term, fuel cells are considered to be the next revolution in automotive powertrains. Fuel Cells 2000, an online fuel cell information resource, defines a fuel cell as "an electrochemical device that combines hydrogen and oxygen to produce electricity, with water and heat as its by-product…the process is clean, quiet and highly efficient—two to three times more efficient than fuel burning." Fuel cell vehicles (FCVs) are propelled by electric motors, but unlike battery-electric vehicles that use stored electric energy, FVCs create their own. They can be fueled by pure hydrogen gas stored onboard in high-pressure tanks or by hydrogen-rich fuels such as methanol, natural gas or gasoline which are converted by an onboard device called a reformer, according to the U.S. Department of Energy (DOE).

Research into the development of viable fuel cell vehicles (FCVs) is being spearheaded in the United States by FreedomCAR, a cooperate effort between the DOE and the U.S. Council for Automotive Research (USCAR, a consortium of Ford, General Motors, and DaimlerChrysler). FreedomCAR was formed to promote research into advanced automotive technologies with the potential of dramatically reducing oil consumption and environmental impacts.

Another organization involved in FCV development is the California Fuel Cell Partnership (CaFCP) a collaboration of auto companies, fuel cell providers, fuel cell technology companies and government agencies demonstrating fuel cell electric vehicles in California under day-to-day driving conditions. Its goals are to test and demonstrate the viability of FCVs and related technology under real-world conditions.

FCV R&D efforts are directed at reducing the cost of fuel cells, improving their performance, and developing effective and efficient ways to produce and store hydrogen and other fuels, according to the DOE. FreedomCAR and CaFCP were formed to encourage private companies and government agencies to work together to move FCVs toward commercialization.

Fuel cell technology is also being applied to hybrids. Ford has developed a HySeries drive system that is a battery-powered plug-in electric hybrid that also uses an onboard fuel cell to recharge the batteries once the batteries have discharged about 60 percent of their energy, extending the range of the vehicle.

CURRENT TRENDS

The automotive markets of the world have undergone a number of significant changes throughout the long history of the automobile in styling, technology and consumer tastes. Three dominant trends that are determining its growth and direction in the twenty-first century are:

- The blurring of the line between cars and trucks as truck-like vehicles become increasingly popular.

- The drive to make vehicles more environmentally friendly by cleaning their emissions, making them more recyclable and lighter, and eliminating their consumption of non-renewable resources.

- The desire to make vehicles as safe as possible by improving their crash avoidance capabilities, crash worthiness, and occupant protection.

Within each of these trends, the driving force is the continuing application of advanced technologies, notably the continuing increase of mobile and in-vehicle electronics, expected to rise to $9.6 billion in 2007, an increase of more than 11 percent; automotive-grade semiconductors, experiencing year-on-year growth of 10 percent, reaching $18 billion in 2007; and automotive telematics and navigation, also expected to have strong growth in several world regions, generating about $38.3 billion in revenues by 2011.

As noted, truck production in the United States significantly exceeds car production, the only market in the world where this situation exists. In every other market, truck production is only a fraction of the number of cars built. In Europe, for example, more than six times as many cars as trucks are built. Car and truck sales reflect this same preference. The sale of light trucks to the motoring public has always been strong in North America but is getting ever stronger as more than half the vehicles sold in this market have been light trucks since early in the twenty-first century. This clearly illustrates Americans preference for pickup trucks, SUVs, minivans and van conversions, a trend that is showing signs of taking hold in other markets of the world.

This demand for vehicles that incorporate the comfort and performance of the automobile with the utilitarian benefits of a truck has created a new type of vehicle built from the ground up as a cross between car and truck with unibody construction, relatively high seating positions, two-, four- or all-wheel drive, and capability of carrying up to eight passengers plus reasonable cargo space—the only missing ingredient is an ability to go off-road. This new breed of personal transportation vehicles are known generically as crossovers—built on car chassis with car-like suspension systems and powertrains, yet with the ruggedness, storage and utilitarian features of SUVs, vans and pickup trucks. The new breed of downsized SUVs mentioned earlier is an example of this trend.

The suburban vehicles of the 1920s and 1930s were the forerunners of this movement, especially as embodied in the early Chevrolet and GMC Suburban models. The modern crossover emerged late in the twentieth century, experiencing quick acceptance among the motoring public in the United States, growing 62 percent from 1999 to 2003 and predicted to continue at a 10 percent rate into the future, according to some forecasters. Among the early competitors in this rapidly growing market were the Acura MC, BMW X5X, Buick Rendezvous, Cadillac SRX, Chrysler Pacifica and PT Cruiser, Chevrolet HHR, Ford Freestyle, GMC Acadia, Honda Pilot and CRV, Infinity FX35, Lexus RX330, Mitsubishi Endeavor, Nissan Murano, Toyota Highlander, VW Touareg, and Volvo XC90.

Although the development of the internal combustion engine was a critical factor in the evolution of the automobile into a commercially viable product, it also sowed the seeds of the major problems associated with the automobile in the late twentieth and early twenty-first

centuries: pollution, consumption of a finite resource and, perhaps even more significantly, political issues associated with dependence on imported oil, much of which had to be purchased from unfriendly or unstable foreign governments. These political concerns were especially prevalent in Europe, North America, and Japan, the largest automotive markets in the world.

The diesel engine was invented in the late 1880s by Rudolph Diesel, a German looking for an alternative to steam power. He developed an internal combustion engine based on compression ignition principles capable of running on biomass fuel—peanut oil in initial demonstrations. Early diesels were too large and heavy for use in vehicles and it was not until the 1920s when smaller, lighter versions were introduced for lorries in Europe. Mercedes began using diesels in cars in 1936 and such use in cars has grown ever since. By the early twenty-first century, for example, Europeans were buying diesel-powered cars 35 percent of the time—45 percent if you include light trucks.

The only time drivers in the United States purchased diesel powered cars in any significant number was during the OPEC oil embargo during the nineteen seventies and then only in limited numbers. But advances in diesel engine technology have largely corrected the problems that kept American motorists away. According to the Diesel Technology Forum in a 2005 report, "advanced clean diesel technology offers American consumers a fuel-sipping alternative that does not sacrifice power or performance." Annual registrations of diesel-powered passenger cars in the United States increased 80 percent from 2000 to 2005, growing from 301,000 to nearly 550,000 vehicles, a trend that is expected to continue with diesel sales approximately tripling in the next 10 years, accounting for more than 10 percent of U.S. vehicle sales by 2015.

Vehicles with hybrid powertrains using some combination of an electric motor with a gasoline or diesel engine have been gaining popularity since 1997 when they became commercially available with the introduction in the Japanese market of the Toyota Prius and when Audi began volume production of the A4 Avant-based Duo in Europe (The Duo, which mated a gasoline engine with an electric motor, was not commercially successful so European automakers focused their efforts on advanced diesels).

The first hybrid car to be sold on the mass market in the United States was the two-door Honda Insight, introduced in 1999, followed in 2000 with the importation of the Toyota Prius, the first hybrid four-door sedan to enter the U.S. market. Subsequently, Honda introduced a hybrid version of the Civic in 2002 and Toyota released the Prius II in 2004, the same year Ford introduced the Escape Hybrid, the first American-built commercially available hybrid and the first SUV hybrid, according to the "History of Hybrid Vehicles" from HybridCars.com.

By 2005 hybrid auto sales had reached approximately 212,000 vehicles, 1.3 percent of all light vehicle sales, according to ConsumerAffairs.com, By 2012 hybrid sales were forecast to reach 780,000 vehicles, or 4.2 percent market share, according to that report.

The long-term trend toward more environmentally friendly powertrains is expected to lead to fuel cells. General Motors developed the first operational fuel cell-powered vehicle in 1968 but it wasn't until the 1990s that growing environmental and energy use concerns prompted increased industry and government investment in fuel cell research. By the end of that decade, DaimlerChrysler introduced NECAR IV, the first hydrogen fuel cell-powered commercial automobile, according to the World Fuel Cell Council. General Motors also introduced a drivable fuel cell concept, an Opel Zafira minivan, at the Paris Motor Show.

The twenty-first century could see the fuel cell emerging as a major player in automotive powertrains. "Light-duty automotive applications are by far the largest market opportunity available to fuel cell technology," reports the World Fuel Cell Council, "and have been the focus of intense development effort. All major automakers now have fuel cell vehicle programs. Most have either launched prototype cars or announced their intention to do so." Ford, for example, in early 2007 showcased a fuel cell-powered version of its Edge crossover (an SUV built on an automobile platform).

Weight reduction in order to improve fuel economy is the other environmental improvement target. Steel and cast iron have traditionally represented about two-thirds of the total weight of a typical car, weight reduction research has concentrated on lightweight steels, other lightweight metals such as aluminum and magnesium, plastics, carbon fiber, ceramics and other exotic materials.

Vehicle safety, which became a paramount issue following the 1965 publication of Ralph Nader's *Unsafe At Any Speed*, has focused on five areas: crash avoidance, pre-crash preparation, occupant protection, post crash measures, and security. Crash avoidance improvements are technical advances in systems that enhance a driver's ability to maintain control of the vehicle despite adverse driving conditions. These have included blind-spot detection systems, tire-pressure monitors, and enhanced vehicle suspension systems, including electronic stability control systems that use sensors to anticipate impending loss of control by the driver and electronically control brakes, throttle, and steering to help the driver guide the vehicle out of danger.

TARGET MARKETS & SEGMENTATION

The auto market can be broken down into a wide variety of segments based on a number of different criteria. Manufacturers analyze these segments very closely when they are designing new cars or planning new promotional campaigns.

A standard breakdown of the market is by the primary motivator involved with a new buyer's decision on which automobile or light truck/SUV to purchase. For many people this decision is made first based upon price range. Consequently, most automakers offer a range of vehicles within broad price ranges so as to offer some variety. There are, of course, automakers who specialize in only the high-end market, like Aston Martin and Rolls-Royce, but these are the exception.

New car buyers are also making the decision about what vehicle to purchase based on the functionality of the vehicle and consumers needs and desires. Attempts are made to create a profile of the features desired by the average member of a particular type or group of customers. Are, for example, most minivan buyers residents of the suburbs with school age children? Do these suburbanites wish that they could control the heating and air conditioning by zone so that while traveling in the vehicle alone a driver is able to focus these functions into the front of the vehicle? Knowing the answers to questions like these allows an automaker to both design for the likely buyer and also market their new features to the most receptive audience.

The market segment for which price is of little concern is often a segment for which style and prestige play an important role in their buying decision. Designing for and selling to this segment of the market requires a different approach. It is a smaller segment of the market than the other two discussed but it is a potentially lucrative one as the high-end vehicles provide automakers with a much higher profit margin. They are more expensive vehicles to make and the money involved in building up the brand recognition that must go along with the high-end vehicle is also a costly undertaking.

As the auto market has matured and the penetration rate of automobiles has climbed, so too have the ways in which the market is sliced and diced by analysts to try and find ways to appeal to a particular segment within it.

RELATED ASSOCIATIONS & ORGANIZATIONS

Alliance of Automobile Manufacturers (AAM), http://www.autoalliance.org

American International Automobile Dealers Association (AIADA), http://www.aiada.org

Automotive Aftermarket Suppliers Association (AASA),
http://www.aftermarketsuppliers.org

Heavy Duty Manufacturers Association (HDMA), http:
//www.hdma.org

Motor and Equipment Manufacturers Association
(MEMA), http://www.mema.org

National Automobile Dealers Association (NADA), http:
//www.nada.org

Original Equipment Suppliers Association (OESA),
http://www.oesa.org

Overseas Automotive Council (OAC), http:
//www.oac-intl.org

BIBLIOGRAPHY

"2006 Worldwide Fuel Cell Industry Survey." PriceWaterhouse Coopers LLC. November 2006.

Adler, Barry, Kathleen Lancaster, and Sharon Slodki. "Beyond Cost Reduction." The Boston Consulting Group. Executive Summary. March 2004.

"Auto Dealer Glut Hurts U.S. Makes." *Detroit News.* 3 February 2007.

Bairley, Susan. "USCAR and U.S. DOE to Invest Up to $195 Million in Lightweight Materials and Batteries Research." *USCAR.* 14 July 2005.

———. "U.S. Automakers Work to Maximize Vehicle Recycling Through USCAR and CRADA." *USCAR.* 8 January 2006.

Bunn, Don and Paul McLaughlin. "The Pickup Truck Chronicles—A History of the Pickup Truck in America." PickupTruck.com. Available from <http://www.pickuptruck.com>.

"Career Guide to Industries—Automobile Dealers." U.S. Department of Labor. Available from <http://www.bls.gov>.

"Cars, Trucks & SUVs" Diesel Technology Forum. Available from <http://www.dieselforum.org>.

"December Global Light Vehicle Sales." *Global Monthly Sales Report.* JD Power and Associates. Available from <http://www.jdpowerforecasting.com>.

"Dismantling the Keiretsu: Nissan Leads Shakeup of Japanese Supplier Network." *Ward's Auto World.* 1 May 2001.

"Europe Dealerships." American International Automobile Dealers Association. Available from <http://www.aiada.org>.

"Fuel Cell Vehicles." U.S. Department of Energy and U.S. Environmental Protection Agency. Available from <http://www.fueleconomy.gov>.

"Global Vehicle Production and Sales by Manufacturer," Global Market Data Book. *Automotive News Europe.* 26 June 2006.

Hill, Kim, Debbie Menk, and Steven Szakaly. "Contribution of the Motor Vehicle Supplier Sector to the Economies of the United States and Its 50 States." Center for Automotive Research, Economics and Business Group. January 2007.

"A History of the Diesel Engine." Yokayo Biofuels. Available from <http://www.ybiofuels.org>.

"History of Hybrid Vehicles." HybridCars.com. Available from <http://www.hybridcars.com>.

"Hybrid Sales Expected to Grow 268 Percent by 2012." ConsumerAffairs.com. 6 December 2006. Available from <http://www.consumeraffairs.com/news04/2006/01/hybrid_sales.html>.

"Industry Series: Historical Statistics for the Industry—2002 and Earlier Years." U.S. Department of Commerce, Bureau of the Census. July 2006.

Ingstad, David. *Crossovers, News.* NADAguides, 2006. Available from <http://www.NADAguides.com>.

Klier, Thomas H., and James M. Rubenstein. "Competition and Trade in the U.S. Auto Parts Sector" The Federal Reserve Bank of Chicago. January 2006.

Lazich, Robert S. *Market Share Reporter 2007.* Thomson Gale, 2007, Volume 2, 473–483.

Leggett, Dave. "The Manufacturers. (Global Market Review of Car Sales—Forecast to 2009)." Just-Auto.com. March 2005.

Lewin, Tony. "Nanjing Asks MG Rover Dealers to Stay Loyal; Chinese Carmaker Pleads for Patience." *Automotive News Europe.* 12 December 2005, 26.

"Nature of the Industry" National Automobile Dealers Association. Available from <http://www.nada.org>.

"North America Car and Truck Production History and Forecast." *Automotive News 2006 Market Data Book.* 22 May 2006, 7.

"North America Light-Vehicle Sales History and Forecast." *Automotive News 2006 Market Data Book.* 22 May 2006, 23.

"Rationalizing the Automotive Supplier Industry; Carving Out Profit From M&A Activity." Original Equipment Suppliers Association (OESA) and PricewaterhouseCoopers. 23 August 2006. Available from <http://www.oesa.org>.

"Safety Mandate/Government Set to Require Stability Control Technology On Autos." *Cincinnati Post.* 14 September 2006.

"Statistics for Industry Groups and Industries: 2005." *Annual Survey of Manufactures.* U.S. Department of Commerce, Bureau of the Census. November 2006, 33–34.

Stein, Jason. "Saab EU Dealer Expansion to Continue; GM Division Adds 40 Dealers in Europe; More to Come." *Automotive News Europe.* 23 January 2006, 8.

Stodolsky, F., A Vyas, R Cuenca, and L. Gains. "Life-Cycle Energy Savings Potential from Aluminum-Intensive Vehicles." Argonne National Laboratory, Transportation Technology R&D Center. Conference Paper: 1995 Total Life Cycle Conference and Exposition. October 1995.

"U.S. Automotive Parts Exports, 1999–2006." U.S. Department of Commerce, International Trade Administration, Office of Aerospace and Automotive Industries' Automotive Team. Available from <http://www.ita.doc.gov/td/rbdc/MFG_index.html>.

"U.S. Automotive Parts Imports, 1999–2006." U.S. Department of Commerce, International Trade Administration, Office of Aerospace and Automotive Industries' Automotive Team. Available from <http://www.ita.doc.gov/td/rbdc/MFG_index.html>.

Walczak, Jim. "The History of the Sport Utility Vehicle." Your Guide to 4-Wheel Drive/Offroading. Available from <http://www.4wheeldrive.about.com>.

Walsh, Brian, and Peter Moores. "Auto Companies On Fuel Cells." Breakthrough Technologies Institute. Available from <http://www.fuelcells.org>.

Washington, Frank S. "What the Hell Is a Crossover Vehicle?" *Inside Line.* 2 February 2006. Available from <http://www.edmunds.com>.

"Where Will Fuel Cells be Used?" World Fuel Cell Council. Available from <www.fuelcellworld.org>.

Wortham, April. "Study Shows Suppliers Manufacturer Most Jobs." *Automotive News.* 22 January 2007, 16.

SEE ALSO *Auto Parts, Trucks*

Bakery Products

———■———

INDUSTRIAL CODES

NAICS: 31–1811 Retail Bakeries, 31–1812 Commercial Bakeries, 31–1813 Frozen Cakes, Pies, and Other Pastries

SIC: 2051 Bread, Cake, and Related Products, 2053 Frozen Bakery Products, Except Bread

NAICS-Based Product Codes: 31–18110, 31–18121, 31–18122, 31–18123, 31–18124, 31–18125, 31–18127, 31–1812A, and 31–1812D

PRODUCT OVERVIEW

In modern industrial practice, bakeries are establishments producing bread, rolls, cakes, pies, and other sweet goods; however, those producing cookies and crackers are not. Restaurants and eateries that bake their own goods for consumption on the premises are also excluded from the bakery category.

The U.S. Census Bureau groups the following industries into the broader Bread and Bakery Product Manufacturing industry: Retail Bakeries, Commercial Bakeries, and producers of Frozen Cakes, Pies, and Other Pastries. A single company can and sometimes does participate in all three of these segments simultaneously. Retail bakeries, the original of all bakeries, were making a major comeback in the United States in the early twenty-first century. In these establishments bread and rolls are baked on the premises from flour (not from purchased dough) and are sold in the shop up front. If such an organization packages some of its product and delivers it to grocery stores for sale there, it must report that portion of its business as commercial baking. If some of the company's sweet products are frozen for later sale, the company is also participating in the frozen industry.

The U.S. commercial bakeries category is made up of approximately 2,400 companies operating nationally or regionally, producing packaged, branded products and selling them through grocery stores and supermarkets. These companies also make most of the frozen cakes, pies, and pastries. By contrast, the U.S. retail bakery category comprises nearly 7,000 companies, which share a much smaller market. They sell locally or, at best, serve a major metropolitan area. They specialize in more expensive products and serve a predominantly affluent market, although that pattern was changing in the early 2000s. Large grocery chains may also participate in retail baking when they bake bread, rolls, and cakes on the premises. A hybrid form, partial baking or par-baking, exists as well, where, for example, a distant bakery specializing in making French baguettes, will produce the dough, partially bake the bread, and ship this product to distant retailers in frozen form. The par-bread is put in the oven on the grocery store's premises and baking is finished on site for sale as freshly-baked bread to the customer.

Beginning in the 1990s and continuing in the first decade of the twenty-first century, the bakery products sector was experiencing shocks and undergoing transformations owing to demographic, health-related, and life-style changes in the United States. For this reason it is illuminating to look at bakery products in the context of other food industries that also convert basic grains into food. Eight major industries are involved, with bakery products representing three of the eight. In terms of shipment dollars in 2005 bakery products represented more than half (52%) of the eight-industry group. The others include breakfast cereals (17.6%), cookies and crackers (17.4%), flour mixes and dough manufacturing (6.6%), dry pasta manufacturing (3.2%), and tortillas (3.2%). To provide perspective the performance of bakery products will be compared to the products of these other industries.

Depending on how it is manufactured and packaged, pizza is sometimes included under Flour Mixes and Dough Manufacturing, sometimes under Frozen Foods Manufacturing (which excludes frozen sweet bakery products), and other times under Food Manufacturing, Not Elsewhere Classified. Doughnuts are included under commercial bakeries under a category known as Other Sweet Goods. In the adoption of the North American Industry Classification System (NAICS) in 1997, the United States introduced a brand new category, Snack Food Manufacturing. After that date hard pretzels were classified as snacks but soft pretzels continued to be part of bakery products produced either by commercial or retail bakeries.

A Look at Grains. Grains are the chief source of energy in food. As our cars burn hydrocarbons to move us around, our bodies burn carbohydrates. Both of these fuels are rich in carbon and hydrogen, but carbohydrates also incorporate oxygen. Ideally 55 to 75 percent of food energy should reach us in the form of carbohydrates, and grains are rich in this form of energy.

Wheat is the principal grain in the U.S. diet; however, since the middle part of the twentieth century the American diet has undergone a change. In 1967 wheat represented 80 percent of all grain consumed in the United States, corn 9 percent, and rice 6 percent. Contrast these numbers with 2005 and wheat still tops the list at 70 percent, though with a 10 percent decline, while corn consumption rose to 16 percent, rice rose to 11 percent, oats, barley, and rye round out the remaining 3 percent.

Grains are complex packages devised by nature in which the species that produce them carry their unique genetic codes. The three parts of all grains are: (1) the outer hull or bran, (2) the inner structure that holds the DNA, the germ, and (3) the intermediate mass of the

FIGURE 13

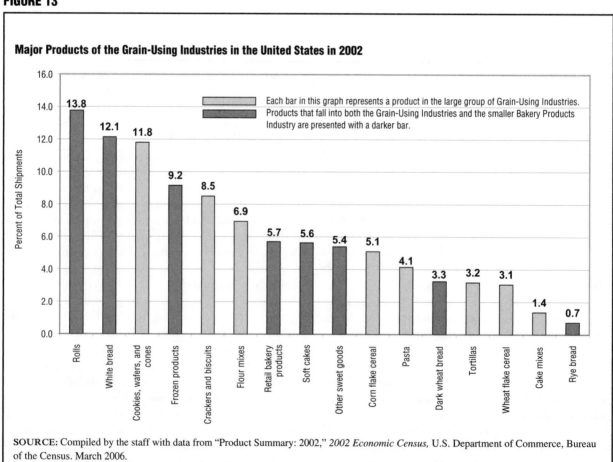

Major Products of the Grain-Using Industries in the United States in 2002

Each bar in this graph represents a product in the large group of Grain-Using Industries.
Products that fall into both the Grain-Using Industries and the smaller Bakery Products Industry are presented with a darker bar.

SOURCE: Compiled by the staff with data from "Product Summary: 2002," *2002 Economic Census*, U.S. Department of Commerce, Bureau of the Census. March 2006.

grain, called the endosperm or kernel. The bran is rich in minerals, fiber, and three vitamins: thiamin, riboflavin, and niacin (B1, B2, and B3). The germ carries the same vitamins plus vitamin E, proteins, and fat. The kernel holds most of the grain's mass in the form of carbohydrates and proteins. In obtaining refined flours producers remove the bran and the germ.

White flour comes from refined grains. The word enriched is sometimes used because vitamins removed in processing are added back in. Bakeries refined flour in efforts to improve the taste, texture, and appearance of bread and other baked products. Whole grain flours tend to be darker, less sweet, and slightly bitter. The nutritional content of refined flour, and consequently the products baked from it, is lower than that found in whole grains and also higher in carbohydrate content. Nutritional and medical experts promote use of whole grains in an effort to increase nutritional content, including fiber, which promotes good digestion. Grain refining increases the relative proportion of carbohydrates in the flour and reduces the total protein content.

Major Products. Considering the grain-using industries as a group, sixteen major product categories represented nearly 81 percent of total shipment dollars in 2002. Of these sixteen, eight were produced by the bakery industries, including white bread, dark wheat bread, rye bread, rolls, cakes and pies, other sweet goods, and the retail baking category as a whole. Bakery products accounted for 50 percent of group volume, indicating the dominant position of the baking sector in the distribution of grains.

The largest category, rolls, represents 13.8 percent of the group's shipments. Nearly half of the products in this category are hamburger and hot dog buns, which because of their role in the fast-food industry, explains their prominence. Bagels and English muffins are the largest product line, making up the other half of the rolls category.

The second largest category is white bread (12.1%). Cookies, wafers, and cones (not included in the bakery products industry) follow closely behind (11.8%). Dark breads (3.3%), including the nutritionally richer whole-wheat breads, rank twelfth of sixteen and rye bread (0.7%) rank last.

All of the products in the Retail Bakeries industry are 5.7 percent of the industry group's shipments. Within that industry the three largest product groupings are breads and rolls, cakes, other sweet goods including sweet rolls, coffee cakes, doughnuts, and cookies.

Nutritional and Consumption Issues. Data available from the USDA for 1970 and 2004 show per capita consumption of carbohydrates and proteins by major food groups.

FIGURE 14

Percent of Carbohydrates and Proteins Obtained Per Capita from Food Groups, 1970 and 2004

	Carbohydrates			Proteins		
	1970	2004	Change	1970	2004	Change
Meat, poultry, fish	0.1	0.1	0.0	39.7	40.3	0.6
Dairy products	6.9	4.5	-2.4	22.3	19.0	-3.3
Eggs	0.1	0.1	0.0	5.5	4.0	-1.5
Fats and oils	0.0	0.0	0.0	0.2	0.1	-0.1
Fruits	6.0	6.1	0.1	1.2	1.2	0.0
Legumes, nuts, soy	2.1	1.9	-0.2	5.3	6.1	0.8
Vegetables	10.5	8.7	-1.8	6.0	5.5	-0.5
Grain Products	**33.9**	**39.8**	**5.9**	**18.2**	**21.8**	**3.6**
Sugars and sweeteners	39.1	37.3	-1.8	0.0	0.0	0.0
Miscellaneous	1.3	1.5	0.2	1.6	2.0	0.4
Total consumed in grams per capita per day	394	481	87	98	113	15

SOURCE: Compiled by the staff with data from the USDA/ Center for Nutrition Policy and Promotion, February 15, 2007.

These data are displayed in Figure 14 and indirectly point to interesting dynamics in food consumption over a nearly 40-year period.

In 1970 the leading source of food energy, carbohydrates, was the sugars and sweeteners category (39.1% of total carbohydrates consumed) followed closely by grain products (33.9%) and then, at a distance, by vegetables (10.5%), dairy products (6.9%), and fruits (6%). By 2004, grain products had become the largest source of carbohydrates (39.8%), representing the largest gain in share, 5.9 percent, of any food group. People changed what they ate in response to very complex communications from public and private authorities concerning what is healthy and what is not. In the same period, however, people increased how much they ate, bumping its daily per capita consumption of carbohydrates from 394 to 481 grams, a 22 percent increase. Grain processors, including the bakery products industries, benefited from both the shift to grains and from increased consumption.

In 1970 the three largest sources of protein were: meat, poultry, and fish (39.7% of total protein consumed); dairy products (22.3%); and grain products (18.2%). By 2004, meats had fractionally increased their share (to 40.3%), but grains had become the second-highest source of protein (21.8%) up nearly 4 percent, while the dairy category (19%) slipped more than 3 percent. Because of the shifts between food groups and increased consumption, people began to eat more protein—up from 98 grams in 1970 to 113 grams in 2004, an increase of 15 percent. Grain processors benefited from this shift as well.

MARKET

The eight industries that form the grain-processing group had total shipments in 2005 of $65.1 billion. Of this total the three bakery products industries were 52 percent—retail bakeries with $3.0 billion, commercial bakeries with $25.5 billion, and frozen cakes, pies, and pastries industry with $5.4 billion for a total of $33.9 billion in shipments.

Industry Growth. From 1995 to 2005 the grain-processing group as a whole experienced compounded growth in shipments at the rate of 2.7 percent per year. This gain was the result of some of the component industries growing energetically, some showing only moderate advances, and one industry losing market share.

The largest industry, commercial bakeries, grew at an annual rate of 2.5 percent, thus slightly under-performing the industry group as a whole. Retail bakeries did much better, growing at 4.3 percent per year. These two industries achieved growth of 2.7 percent annually matching that of the total group. Taking into consideration frozen products, the composite of the three bakery products industries was 3.4 percent per year. Frozen cakes, pies, and pastries had very strong growth, 7.8 percent yearly, thus lifting bakery products above the eight-industry average.

Tortilla manufacturing saw the most rapid growth among the eight industries, with 8.2 percent per year. This was still a relatively small industry in the first decade of the twenty-first century ($2.1 billion in 2005), but it is a growing industry. Frozen products ($5.5 billion) came second in overall growth, at 7.8 percent. Third in annual growth rate was the breakfast cereal industry ($11.5 billion), which grew at a rate of 2.9 percent per year. In total size the cereals industry, was second only to commercial bakeries. Cookie and cracker manufacturing ($11.4 billion) grew 2.7 percent annually, matching the performance of the two bakery categories combined. Dry pasta manufacturing ($2.1 billion) had growth of 1.8 percent per year. The flour and flour mixes industry ($5.4 billion) lost shipments volume at the rate of 3.2 percent annually. The decline in the flour and flour mixes industry reflects broad trends in food toward more purchasing of finished products and away from purchases that demand labor in preparation. This contributed to the strong advance of frozen bakery products.

Looking at eight-year growth trends across industries as a whole, bakery products had moderate growth—below the 5 percent most industries hope to achieve. Retail bakeries outperformed commercial bakeries because companies that comprise the retail industry are smaller, are vertically integrated into direct sales, and have more rapid feedback, so are better able to discern consumer trends rapidly and to respond accordingly.

Product Growth. Of the sixteen large product groups, frozen sweet goods had the most rapid growth in the 1997–2002 period (8.2% per year), tortillas came next (7.1%), and retail-bakery products came third (4.7%). White bread, while the largest of the product segments, experienced relatively small growth for its size (1%). The second-largest product segment—cookies, wafer, and cones—fared slightly better (1.5%). Rye bread and crackers and biscuits turned in the lowest levels of positive growth (0.2% each). At the bottom of the listing were flour mixes and cake mixes, which lost market share (-6.3% and -6.5% per year respectively).

Bread sold by commercial bakeries grew slowly from 1997 to 2005. Mass produced bread is a mature industry receiving pressure from competing products including rolls, the products of retail bakers, cookies, and frozen cakes and pies. Growth in the bakery products industry comes from these products, not from bread. Tortillas, though a relatively small industry also represent a growing competitor to bread.

The two largest product lines of the breakfast cereal industry, corn flakes and wheat flakes, were also growing slowly in this period at 1.2 percent and 0.7 percent per year respectively; however the cereal industry was growing more rapidly in other sectors of their industry, including breakfast bars and other non-traditional breakfast products that have not yet forged a distinct category of their own.

KEY PRODUCERS/MANUFACTURERS

Throughout the latter half of the twentieth century, the leading bakery products manufacturer in the United States was Interstate Brands Corporation (IBC), producer of Wonder Bread (the top-ranked brand), Home Pride (second), Merita (eighth), and nine other bread brands, four major cake brands, and many other baked products. IBC began in the 1930s and is headquartered in Kansas City, Missouri. In 2004 IBC went into voluntary bankruptcy and by mid-2007 had not yet emerged. IBC filed for Chapter 11 citing declining sales, a high fixed-cost structure, excess industry capacity, rising health and pension costs, and increasing cost of ingredients. Sales of IBC stock on the New York Stock Exchange had been suspended. In 2006 the company reported sales of $3.06 billion. Despite IBC woes, *Milling & Baking News*, citing statistics from Information Resources, reported that IBC continued as the nation's leading vendor of bread.

Until 1985 Sara Lee Corporation was known as Consolidated Foods. It had acquired Kitchens of Sara Lee in 1956 and later took Sara Lee as its name. The company was ranked second in bread production in 2006. A $15.9 billion company, its bakery operations accounted for $1.87 billion of its total sales. Sara Lee is a diversified food company with major meat, coffee, and sweet product brands.

A close third in total bread sales in the United States in 2006 was George Weston Bakeries, Inc., a part of George Weston Ltd., which is a major Canadian bakery, founded in 1882, with total sales of Can$32 billion. George Weston sold its bakery operations in the western part of the United States to Bimbo USA in 2002, but continues to sell products in the rest of the country. Weston's top brand, Arnold, is the fourth ranked bread brand in the United States. It also sells Brownberry, Dutch Country, Freihofer's, and Stroehmann brands of bread, Entenmann's pastries, and the Thomas brand English muffin line.

The fourth-ranked bread producer in the United States in 2006 was Flowers Foods Bakeries Group, founded in 1919 in Thomasville, Georgia. The company had sales of $1.89 billion and owned the third- and sixth-ranked bread brands, Nature's Own and Sunbeam. Early to recognize opportunities presented by a growing health consciousness in the United States, Flowers was first to introduce an entirely sugar-free version of Nature's

Own and, in 2003, also introduced a low-carbohydrate, high-fiber version of the brand. The company also makes and sells Mi Casa tortillas, Cobblestone Mill rolls, and a line of frozen specialties.

Bimbo USA is a major regional baked products company serving twenty-two western states from 14 bakeries. Bimbo USA is owned by Mexico-based Grupo Bimbo, which is the third-largest bakery giant in the world and serves Latin America, North America, and Europe. Bimbo USA's 2005 sales were $1.27 billion. In addition to brands acquired from George Weston (Oroweat, Mrs. Bairds, Entenmann's, and Thomas), Bimbo USA's major brands include Bimbo, Francisco, Tia Rosa, and Marinela.

In 1937 Margaret Rudkin founded Pepperidge Farm, Inc. She started the company because she was trying to create bread for one of her sons who had allergies, preventing him from eating breads with preservatives. Rudkin's labors produced a natural bread of such high quality that her friends and neighbors urged her to try to make and sell it. Pepperidge Farm, named after pepperidge (sour gum) trees growing on the farm where the Rudkins had settled, was acquired by Campbell Soup Company in 1960.

FIGURE 15

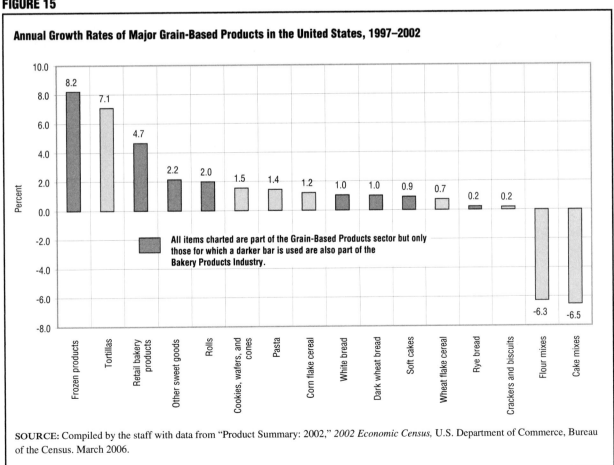

Annual Growth Rates of Major Grain-Based Products in the United States, 1997–2002

SOURCE: Compiled by the staff with data from "Product Summary: 2002," *2002 Economic Census,* U.S. Department of Commerce, Bureau of the Census. March 2006.

Pepperidge reported sales of approximately $1 billion in 2001. In addition to lines of bread, the company also sells cookies, crackers, and frozen meals and baked products.

Engelbert and Joseph Franz are credited with inventing the modern form of the hamburger bun early in the twentieth century. The company they founded, United States Bakery, sells under the brand names of Franz, Williams', and Snyder's. The company's 2006 sales were $275 million. United States Bakery is also a producer of pastries and cookies. Other brand names by which the bakery's products are known are Aunt Hattie's, Bay City, Seattle International, Svenhard's, and New York Bagel Boys.

Los Angeles-based La Brea Bakery, owned by the IAWS Group, PLC, an Irish corporation, is the leading supplier of partially baked goods that it prepares and ships to restaurants and retail bakeries. La Brea perfected par-baking when it discovered that partially baked bread, cooled and rapidly frozen, would turn out perfectly when baked the rest of they way. Par-baked bread is 80 percent baked initially. La Brea calls itself an artisan bakery because it specializes in baguettes, batards, rolls, sourdough breads, and extra-large loaves that look and taste as if just produced in a European bakery. La Brea makes and ships virtually all fresh-baked products supermarkets and other baker-sellers offer to their customers.

MATERIALS & SUPPLY CHAIN LOGISTICS

Bakery products are typically produced in close proximity to the markets in which they are bought and consumed. Baked goods were produced at 10,360 locations in 2002, for example, and of these 1,769 employed twenty or more people. In 2005 large metro areas, those with population of 250,000 or more, numbered 171 in the United States, suggesting that there are more than enough large bakeries to supply every town of some size and several small bakeries to supplement the output of the large ones, with some exceptions. According to Census Bureau data, in 2002 Idaho, Montana, Wyoming, North and South Dakota, Kansas, Mississippi, and Maine did not have any bakeries; their bakery products reached them by truck. In 2002 more than 440 flour milling operations supplied the nation's bakeries.

DISTRIBUTION CHANNEL

Unless the product is sold directly to the consumer by a retail bakery operating its own store, most products of this industry reach the customer by means of routes—organized truck deliveries from the bakery to retailers. Bakeries may own routes—Interstate Brands Corp. for example, owns 52 bakeries supported by 9,000 routes, while Bimbo USA has 14 bakeries and operates 3,000 routes. Direct Store Delivery (DSD) entrepreneurs may also in-

dependently own routes and may combine a number of products on a single route or serve one bakery exclusively. Frozen bakery products require refrigerated route deliveries. Par-baked bread moves through the frozen bakery products distribution system.

Bakery products also have a specialized retail segment, the baked goods stores. These operations, in distinction to retail bakeries, do not make the products they sell but specialize in selling breads, rolls, cakes, pies, and pastries. In 2002 nearly 4,500 such stores operated across the country. These retailers, like grocery stores and supermarkets, received their product from routes operated by bakeries. Special distribution arrangements are typical in the case of fast food chains whether the product is a bun (hamburger, hot dog, sub-sandwich) or doughnuts. Dunkin Donuts, for instance, operates its own wholly-owned distribution centers to supply its retail outlets.

KEY USERS

Nearly everyone eats bread. Consumption issues turn around how much and what kind of bread to eat. The difficulties experienced by parts of the industry, the relatively strong performance of retail bakeries, the introduction and promotion of whole wheat, sugar-free, and low-carb brands indicates that people are favoring newer brands, more traditional types of bread (even at higher prices), and are at least weakening the sales of white bread made with bleached flour.

ADJACENT MARKETS

Adjacent to bakery products are other markets that use grains, including the two largest, pasta and tortillas. Tortillas are rapidly growing as a category whereas pasta is largely flat. Both industries, however, are under the same pressures as bakery products and changing in similar ways, most prominently by replacing refined flour with whole wheat or other whole grains in making their products. The cereal industry, while not growing in its traditional segment, is growing in the new-product segment of breakfast bars and breakfast pastries. Cookie and cracker manufacturing is also considered an adjacent market to the bakery products industry.

RESEARCH & DEVELOPMENT

The central focus of research and development in the bakery products sector is on product transformation—changing popular products to make them more nutritious without losing customers by altering the product's taste and texture. Interstate Brands introduced whole-wheat versions of its flagship brand, Wonder Bread. The company's aim was to preserve, as much as possible, the taste and soft texture of Wonder Bread while making the bread more nutritious by using whole wheat. George Weston created X-Treme Wheat bread hoping this high-nutrition brand would please children who ate it while assuring parents who bought it that it was healthy.

R&D efforts in formulating new types of high-nutrition breads by a combination of selecting ingredients, modifying production methods, and intensely testing prototype products in so-called sensory panels are, in effect, the continuation of R&D efforts along lines that have characterized the commercial bakery industry since its emergence in the 1920s and 1930s. These efforts have resulted in highly engineered food products. The new and modified brands emerging from the lab are similarly formulated.

La Brea Bakery's innovations in par-baking represent another approach—making use of modern technologies and transportations systems to deliver traditional bakery products more efficiently. The combination of baking and freezing supports another trend in the industry—toward satisfying a public demand for freshness. People in supermarkets holding a French baguette still warm from the oven little suspect that the product was partially baked days earlier a thousand miles away. Additional R&D is aimed at developing products that can be microwaved to produce fresh taste. In these cases technological innovations to enhance microwave ovens themselves to deliver crispness in baked goods is part of the R&D thrust.

Much of the advanced R&D effort in this industry is aimed at ingredient formulations to produce additives that act as fat mimetics, which produce the taste of fats without actually being fats, such as sweeteners, or flavor enhancers. Routine R&D is devoted to nutritional analysis, shelf-life studies, process improvements.

CURRENT TRENDS

Significant trends for the bakery products sectors (and equally influential across the entire spectrum of the food industries) include the growing incidence of obese and overweight individuals in the population; demographic changes producing an age structure heavily weighted toward the elderly; and pressures on people's time due to increasing hours of work, particularly in families where husband and wife both work. These trends, as they are evolving, sometimes produce paradoxical results: time pressures on dual-earner couples disrupt meal planning, foster eating out, and lead to snacking. Eating out and snacking may contribute to weight gain and obesity. Gaining weight impels many people to diet, which leads to shifts between food groups, and the food industry must react to satisfy changing demands. Dieting individuals want to lose weight but are reluctant to give up pleasurable taste sensations. Food producers remove sugar but

substitute fats, or remove fats but substitute sugars to hold their customers.

Since 1963 the Centers for Disease Control and Prevention (CDC) has tracked obesity and excess weight information via the National Health and Nutrition Examination Survey (NHANES). In the survey (NHANES II) conducted between 1976 and 1980, 47.1 percent of the adult population was overweight or obese, with those termed obese representing 15 percent. In the 2003–2004 Survey, 66.2 percent of adults were overweight, with those deemed obese representing 32.9 percent of the population. In these same two periods, children with excessive weight increased as well. According to the period from 2003 to 2004, 13.9 percent of children aged 2 to 5, 18.8 percent aged 6 to 11, and 17.4 percent aged 12 to 19 were overweight, showing 8.9, 12.3, and 12.4 point jumps, respectively, over the 1976 to 1980 data.

Each survey has shown a growing trend in obesity. Excessive consumption of carbohydrates, sugars, and fats, and neglect of vigorous physical activity, are the principal causes of this trend. CDC data on physical activity indicate that only 48 percent of people engage in sports or other activities at recommended levels. While feedback from the healthcare bureaucracies does not seem to curb excessive eating or encourage people to exercise more, obesity findings sometimes cause people to shift consumption between food groups. From time to time dieting trends seem to threaten specific food clusters. The popular low-carbohydrate diets, which promote limiting the amount of carbohydrates people eat in the form of bread, pasta, and pastries, have caused many in the baked goods industry to find ways to counter the adverse effects these diets have had.

The American population is getting older. Census Bureau projections to the next population census year, 2010, show that four of six age groups under age 45 will decline as a percent of total population by then. The exceptions are the group aged 14 to 17, which will remain the same as in 2000, and the group aged 18 to 24 which will increase by a slender 0.5 percent. All of the older age groups, from 45 on up, show projected increases, with the largest increase occurring in those aged 45 to 64, which is projected to grow by 26.5 percent by 2010. The 65 and older group is expected to increase by 13.2 percent.

The growth in the group age 45 to 64 has particularly influenced the rapid growth of retail bakeries catering to the most value-conscious but also the most affluent age group in the population. The elderly, those aged 65 and older, may be a factor in the growth of the frozen segments of bread and pastry products in that this age group seeks convenience and smaller portions. Health-awareness increases with age and may become intense as people begin to have less energy or begin experiencing mobility challenges. The elderly generally consume less food, favor lighter foods, and are much more discriminating students of food labels.

In the early part of the twenty-first century it was commonplace to say that life is more hectic. The Families and Work Institute (FWI) provided some data to support this perception. In a 2002 study on changes in the work-force, the Institute reported that the work hours of couples where both people work increased from 81 to 91 between 1977 and 2002. In that same period, households with children headed by a single parent increased from 9 million to 17.6 million. All households increased 47 percent in that period, while single-family households increased by 100 percent. Because people are working more and are pressed for time, they spend ever more of their food budgets on meals eaten in fast food outlets and in restaurants. According to data assembled by the USDA's Economic Research Service, food expenditures away from home have been climbing. In 1965, 30 percent of a family's food budget was spent on eating out and steadily increased to 36 percent in 1975, 41 percent in 1985, 46 percent in 1995, and 48 percent in 2005.

Growing fast food consumption is clearly implicated in growing sales of rolls and tortillas. Pressures on people's time, similarly, favor convenience foods with increasing demand for frozen goods. At the same time they inhibit the urge to purchase food categories that require labor in the kitchen; hence a decline in the sale of flour and flour mixes.

TARGET MARKETS & SEGMENTATION

Bakery products appear, even in industrial classifications, as well-defined and separate markets—rolls, bread, cakes, pies, and pastries. Rolls taken as a whole divide into buns, bagels, breakfast, and dinner rolls all with their characteristic eating situations. In the last decades of the twentieth and in the first decade of the twenty-first century bread has divided into very clear market segments: traditional breads, mass-produced packaged breads, and, within that segment, the refined and whole grain categories. Traditional breads began to be aimed at the affluent segment of the market but are beginning to be widely purchased by everyone. Refined bread categories were in sharp decline at the beginning of the twenty-first century—to the point of driving their largest producer into bankruptcy. Whole grain breads appear to be the new mass segment aimed at routine, day-to-day consumption in sandwiches. The

frozen segment, which includes bread, rolls, and sweet products, is growing energetically in response to lifestyle changes that put pressure on people's time.

RELATED ASSOCIATIONS & ORGANIZATIONS

American Bakers Association, http://www.americanbakers.org

American Institute of Baking, https://www.aibonline.org

Independent Bakers Association, http://www.mindspring.com/~independentbaker

International Dairy-Deli-Bakery Association, http://www.iddba.org/default.htm

Retail Bakers of America, http://www.rbanet.com

Tortilla Industry Association, http://www.tortilla-info.com

BIBLIOGRAPHY

"Carbohydrate Counter." Carbohydrate-counter.org. Available from <http://www.carbohydrate-counter.org/cereal/search.php?cat=Wheat&fg=2000>.

Darnay, Arsen J. and Joyce P. Simkin. *Manufacturing & Distribution USA,* 4th ed. Thomson Gale, 2006, Volume 1, 115–122.

"Food CPI, Prices and Expenditures." U.S. Department of Agriculture, Economic Research Services. Available from <http://www.ers.usda.gov/Briefing/CPIFoodAndExpenditures/Data/table1.htm>.

"General Summary, 1997 Economic Census." U.S. Department of Commerce, Bureau of the Census. June 2001.

Lazich, Robert S. *Market Share Reporter 2007.* Thomson Gale, 2007, Volume 1, 121–126.

"National and State Population Estimates." U.S. Department of Commerce, Bureau of the Census. 22 December 2006. Available from <http://www.census.gov/popest/states/NST-ann-est.html>.

The National Study of the Changing Workforce. Families and Work Institute. 2002.

"Prevalence of Overweight and Obesity Among Adults: United States 2003–2004." U.S. Department of Health and Human Services, National Center for Health Statistics. Available from <http://www.cdc.gov/nchs/products/pubs/pubd/hestats/overweight/overwght_adult_03.htm#Table%201>.

"Product Summary: 2002." U.S. Department of Commerce, Bureau of the Census. March 2006.

"Whole Grains: High in Nutrition and Fiber yet Low in Fat." MayoClininc.com. Available from <http://www.mayoclinic.com/health/whole-grains/NU00204>.

Batteries

—■—

INDUSTRIAL CODES

NAICS: 33–5911 Storage Battery Manufacturing, 33–5912 Primary Battery Manufacturing

SIC: 3691 Storage Batteries, 3692 Primary Batteries, Dry and Wet

NAICS-Based Product Codes: 33–59111, 33–59117, 33–59112, 33–5911Y, 33–5911W, 33–59121, 33–59122, 33–59123, and 33–5912Y

PRODUCT OVERVIEW

A battery is one or a series of electrical cells or containers in which controlled chemical reactions take place. The energy generated by these reactions is captured as electromagnetic power. The atomic composition of the battery itself remains unchanged throughout its life—no elements are added or taken away—but atoms in it are rearranged. The battery has two terminals surrounded by an electrolyte. Electrolytes are, by definition, substances in which dissolved chemicals disassociate into charged ions; electrolytes are solutions of salts, acids, or bases. One terminal is made of substances that, in ionizing, build up a surplus of electrons; the other is made of a material that develops a deficiency. When a wire connects the two terminals, one yields and the other one gets electrons. Both loss and gain cause chemical reactions that kick in as soon as a circuit is established (the on-switch is pushed) but also stop when the circuit is interrupted. The electrons moving through the wire carry the energy of the chemical reactions. A device wired into the circuit (a bulb, say) captures the energy to do its job (say shine a beam).

Surpluses and deficits of electrons are possible only because the battery's internal environment is ionizing. Charged particles easily form inside it. Ions come in two varieties. Anions have more electrons than protons, cations more protons than electrons. This comes about because some part of the terminal material dissolves in the electrolyte and swims away, but by doing so it leaves part of its electrons behind and thus becomes a cation. The electrons left behind cause other atoms in the terminal to have too many; having too many, they become anions. Materials of different chemical composition have different reactivities, also called electrode potentials. Some are more and some are less easily ionized. By analogy, when a crisis erupts in social life, some people become quite hyper, others remain more steady.

In effect, for a battery to work, its terminals must have different electrical potentials. The difference is due to their different chemistries and rates of ionization. The means of naturally balancing this difference, however, must be inhibited. If not, the battery will not last long. When two terminals are suspended in an electrolyte, the electrolyte itself acts as the insulation. This inhibition is never total, one reason why batteries left lying for too long eventually lose their charge. Separators between terminals are also used to inhibit the flow of electrons. But in contrast with the electrolyte itself, a metal wire is an ideal highway for electron traffic. Atoms of copper in a wire, for example, are very loosely associated with their atomic cores. The cores are stationary but bathed in an agitated sea of electrons. In non-metals electrons are stickier; they cling to their atomic cores more firmly. For this reason a strand of silk will not conduct electricity at all but a copper wire will. When the two battery terminals are connected, the terminal with the high potential to give electrons sends them running to the terminal with a need to receive them. A current is established. For this reason batteries are sometimes called electron pumps.

The source of electrons in a battery is called the *anode*. It is associated with anions. Anodes are usually metals because metals easily give up their electrons, but any element similarly structured will do as well. One type of battery, for example, has a nickel-hydrogen chemistry, with hydrogen acting as the anode. Hydrogen is difficult to supply to a battery. The gas has to be kept under pressure, limiting its use to expensive stationary applications. A very popular nickel-hydrogen system, therefore, is the nickel-metal-hydride battery. It overcomes this difficulty and results in batteries one can hold in the hand. The hydrogen in such batteries is held as part of a metallic compound. Hydrogen can act as an anode because, under ionizing conditions, an atom of hydrogen is ready and willing to give up its sole electron.

The other terminal, called the *cathode*, is the receiver of electrons. It has a predominance of cations. Again, the cathode is usually a metal, but it must have a deficiency of electrons. Other substances meet this requirement as well if properly put together. Cathodes may be carbon. Indeed, they might be air. Air contains oxygen which, ionized, can and does act as a cathode. Some batteries use air as their cathode or, more precisely, the oxygen in the air. What matters chemically is that anodes must have too many and cathodes too few electrons.

In the standard lead acid battery familiar to us in cars, the anode is pure lead (Pb) and the cathode is lead dioxide (PbO_2). The electrolyte is a dilute solution of sulfuric acid ($2H_2SO_2$). As soon as a circuit is established between these two terminals, one atom of the anode, lead, transforms into lead sulfate ($PbSO_4$) and, in this process, gives up two electrons. At the same time one molecule of the cathode, lead dioxide, reacts with sulfuric acid to produce lead sulfate too— plus water. It gains two electrons in this reaction. Thus the electrons given up by the anode are absorbed by the cathode. The energy produced in the reaction is used by the car for starting the engine. Notice that at both terminals, one part of each has been chemically transformed into lead sulfate. The terminals have changed, if only a little. When both terminals have become lead sulfate, the energy potential in the battery has been exhausted. The electrolyte will have lost all of its sulfur and become pure water. The same atoms will still be present, but the battery will be dead.

Primary Batteries. Batteries are made of one or more electric cells. They come in two varieties. Primary cells are those that cannot be recharged. Storage cells, also called secondary cells, are those that can be charged many times. The first battery ever made was a primary cell fashioned by the Italian scientist, Alessandro Volta. Volta stacked sheets of zinc and copper plates separated by blotting paper soaked in salt-saturated water, arranging the plates so that the top and the bottom were different metals. Touching a wire to the top and the bottom of the voltaic pile produced a current in the wire, felt as heat. Two wires, one attached to the top, one to the bottom, touched to one another, produced a spark. Volta later discovered that a combination of zinc and silver plates yielded the best results. This invention, made in 1800, predated the invention of the first practical generator by about 70 years. Until the 1870s, the voltaic pile was the principal means of generating electricity for early experimentation.

All primary batteries are variations on this theme. They include dry cell batteries using zinc and carbon as the terminals and a mixture of ammonium chloride and zinc chloride as the electrolyte. The carbon terminal is surrounded by manganese dioxide. This battery dates back to the 1860s and was invented by George Leclanché. Another widely used type is the alkaline battery It deploys zinc and manganese-oxide as the terminals; the electrolyte, as the battery's name implies, is an alkaline paste (potassium hydroxide). The mercury battery has zinc and mercury oxide electrodes also encased in a potassium hydroxide. It can be fashioned into tiny disks and is thus suitable for powering hearing aids, watches, and other small devices. There are yet other combinations, including expensive but light silver-zinc devices used in aeronautics and space applications and very tiny zinc air batteries also used in hearing aids. This last category uses oxygen in air as its cathode. When the hearing aid is turned on, the switch actually uncovers a tiny opening to let the air in.

The major limitation of primary batteries is that, with use, chemical reactions change the composition of the two terminals and the power extracted to do work cannot be restored without destroying the structure and reprocessing the metals and the spent electrolyte. Indeed, primary batteries are usually discarded after use.

Storage Batteries. The internal arrangement of secondary or storage batteries is such that simply reversing the flow of current in them can charge them back up again. Instead of drawing current from the negative pole, current is applied to that pole. In a lead acid battery the nearly exhausted anode will gradually turn from lead sulfate back into pure lead; the exhausted cathode, also lead sulfate, will by repeated chemical reactions, triggered by the charging current, become lead dioxide again. The water surrounding the terminals will gradually turn into an electrolyte again as sulfur is reintroduced into it and chemically forms sulfuric acid dissolved in water. The energy of the incoming current will become chemical energy ready for extraction. The batteries are usually recharged before they are completely dead.

According to a listing published by PowerStream, a subsidiary of Lund Instrument Engineering, Inc., some thirteen different types of rechargeable batteries are on or nearly on the market. They are based on different metallurgies, including (in alphabetical order) iron-air, iron-nickel, iron-silver, lead, lithium-ion, manganese-titanium, nickel-cadmium, nickel-metal hydride, nickel-sodium, nickel-zinc, sodium-sulfur, zinc-manganese, and electrodes formed by metallic liquids. Other sources list more varieties or sub-varieties of those already named. The variety of combinations illustrates the great diversity of products the word *battery* constellates. Each type provides special advantages in energy density, useful life, cost, weight, safety, and other aspects crucial in different uses. Applications range from critical applications as in heart defibrillators and pacemakers, to emergency lighting and power, to power sources for vehicles and spacecraft, with less critical applications like portable telephones, cameras, laptops, tools, and many other devices occupying a middle ground.

Lead Acid. The largest category is the lead-acid battery, the mainstay of automotive starting, favored by low cost and relatively long life if the battery is treated right and not permitted to discharge entirely. In automotive applications the batteries are referred to as SLI batteries, the acronym standing for *starting, lighting, and ignition.* In industrial applications lead batteries are referred to as SLAs (for *sealed lead acid*) and as deep-cycle batteries. They have thicker lead plates and can thus be discharged to a deeper level whereas SLIs, when discharged below about 50 percent capacity, tend to get damaged. SLAs are popular as backup systems for computers and networks in case of power outages. Three other major categories of rechargeables have substantial sales volumes.

Nickel/cadmium (NiCad). These batteries are widely used in calculators, cameras, defibrillators, electrical vehicles, and in space applications. They are light, perform well in cold temperatures, and are excellent sources of continuous electrical power. They cost more than lead acid batteries for the same output. Their chief negative is the memory effect, to be discussed more fully under the subheading Problems and Issues below.

Nickel/metal hydride (NiMH). This battery—used in telephones, power tools, laptops, and electric vehicles among other applications—is a relatively recent type of battery developed to overcome the limitations of otherwise excellent nickel-hydrogen cells, the hydrogen acting as an anode, thus providing electrons. Nickel-hydrogen cells had to be kept under pressure. In the NiMH battery,

hydrogen is provided by various kinds of metal alloys able to hold hydrogen. The battery has more capacity and higher power densities than its competitor, the NiCad. It was also thought to be resistant to the memory effect—of which more later.

Lithium-Ion (Li-I). Perhaps the most important type, in light of future uses in electric vehicles, is the lithium ion battery. The category is widely used in laptops, cell phones, and electric vehicles, but is famous for powering the Mars Rover Spirit. Li-ion batteries have a cathode made of a graphite (carbon) matrix and an anode made of lithium oxide. The battery is light, has high energy density, and a long life. A tabulation comparing the four major categories of storage batteries is presented in Figure 16, drawn from a paper by Vulkson and Kelley with the Air Force Laboratory. These data indicate the more significant differences between the different technologies.

Li-ion batteries have superior energy density by weight and volume, total voltage delivered, and a reasonably high life, permitting 2,000 charges. Cycle life is usually calculated as charges applied after 80 percent of the battery charge has been exhausted. Most batteries must be recharged before they are fully drawn down. Battery costs for these types of batteries increase as one goes down in the table. Weight per unit of power decreases. Lead batteries are the oldest and lithium-based products the most recent technologies. In the industry lithium is viewed as the most important candidate to power a post-petroleum era. The technology, however, despite widespread use in electronic devices, is still relatively new and has technical and safety issues to be touched on later.

Other Names and Categorizations. In the North American Industry Classification System (NAICS), batteries are divided into two industries as above, but in the reverse order, thus storage batteries come first, primary batteries

FIGURE 16

Characteristics of Different Storage Battery Types

Type of Storage Battery	Watt-hours per kg	Watt-hours per liter	Voltage per cell	Cycle Life
Lead Acid	35	80	2.0	400
Nickel Cadmium	35	80	1.2	>1,000
Micket Hydrogen	55	60	1.2	>10,000
Lithium Ion	150	300	>3.6	>2,000

SOURCE: Compiled by the staff with data from Vulkson, Stephen P. and Michael Kelley, "High-Energy-Density Rechargeable Lithium-Ion Battery," U.S. Air Force Research Laboratory. February 2004.

second. Classification systems, of course, indirectly reflect the importance of the categories. Storage batteries are the larger industry. Within that industry, the Census Bureau divides the world between lead acid batteries and all others. Census divides the All Others category further into nickel cadmium batteries and all the rest in combination. Within the primary battery industry, the Census recognizes round and prismatic batteries as the larger and button and coin batteries as the smaller part. The prismatic type refers to rectangular 9 volt batteries with both of their terminals sticking out on top in the form of tiny crowns; this type of connection is known as the PP3 (patch panel); prismatics first appeared to power transistor radios. Round batteries have the positive pole extending above the surface of the battery on top; their bottoms are the negative pole. Within the round/prismatic product subdivision, the Census Bureau distinguishes between alkaline, zinc carbon, mercuric oxide, and lithium batteries, hiding the rest under all other. Button and coin batteries describe the smallest of these products. Both are round in shape but coins are much thinner. Here the Census recognizes by name silver oxide, alkaline manganese, zinc air, and lithium batteries. Another all other hides other button-coin metal combinations from detailed statistical view. As we march through this hierarchy of batteries, production numbers get ever smaller—at least in the context of U.S. manufacturing.

Battery Prices and Cost of Electricity. The energy capacity of batteries is stated in milliamps per hour, abbreviated as mAh. A milliamp is one-thousandth of an ampere, and an ampere is a measure of the flow of current. Batteries are also rated by voltage. A volt is the measure of the electromagnetic force between a battery's terminals, thus the battery's ability to pump electrons. This can be thought of as pressure. The milliamp then measures the amount of current that such pumping generates in a unit of time. The mAh and voltage, both as reported, can be used to calculate the electrical power delivered by a battery in a unit of measure most people understand—because they get electric bills in which usage of electricity in kilowatt hours is visible on the bill. A watt is an amp multiplied by volt, thus a composite measure of flow and force combined.

In 2003 electrical costs from the utility grid averaged from a low of 5 cents to a high of around 17 cents per KWh. The average residence consumed 29 KWh a day. To determine how many kilowatt hours a battery actually holds, one divides the milliamp capacity by 1,000 to get amperes, multiplies by voltage to get watt hours, and then divides that result by 1,000 to get kilowatt hours. Just to have some sense of this unit of measure, consider that 1 KWh is equivalent in energy to 860 calories, about 43 percent of a healthy adult's diet per day. To get the energy

of a gallon of gasoline, we would have to use around 33 kilowatt hours.

A triple-A battery (AAA) with an mAh rating of 1,100 and 1.25 volts can thus be shown to provide 0.0014 KWh. This is a tiny amount, of course. The battery will cost around $1.60. Dividing that price by the fractional kilowatt hour provided by the battery will produce a price per KWh of $1,164, or nearly 7,000 times the highest cost electrical power coming from the grid. The triple-A battery and the utility company live in different worlds. In order to detach from the utility's cable and still be able to use electricity to listen to that radio on a stroll in the countryside, we pay a high price but, using so little electricity, we don't really notice it.

Using such calculations we can determine, roughly, how primary batteries stack up. Tiny zinc air batteries used in hearing aids and coin batteries used in watches do best. They cost around 90 cents each and deliver electricity at the around $1/KWh, the lowest cost of any battery on the market; their uses, of course, are limited. Other major types are AAA batteries ($1,160/KWh), 9 Volt prismatics ($760/KWh), AA batteries ($520/KWh), C batteries ($200), and D batteries ($100). Prices for these five types of batteries are approximately $1.60 (AAA), $3.20 (9 Volt), $1.60 (AA) and $1.80 (Cs and Ds) per unit. They are, of course, usually sold in multiples.

According to *Market Share Reporter*, which in turn cites Information Resources Inc., the largest category of household batteries by size is AA (41% of batteries). Others in rank order are AAA (16%), 9 Volt (11%), C (9.1%), D (8.1%). All other kinds represent 15.3 percent of the market.

Storage batteries, of course, are rechargeable. For this reason their output in kilowatt hours is measured differently. Their lifetime output in KWh is calculated by multiplying the KWh of a single charge by the number of charges the battery will sustain before failing. The cost of the unit is divided by lifetime KWh to calculate cost per kilowatt hour. In this process analysts typically ignore the cost of electricity required to charge the battery. KWh costs, therefore, are somewhat understated.

The most cost-efficient battery is the ordinary lead-acid SLI (starting, lighting, and ignition) used in automobiles. SLI batteries are, by design, intended for brief draw-downs of electricity. They cost around $160 per unit and deliver electric power for around $2.22/KWh. Other types have higher costs. Lithium-ion batteries cost around $100 per unit and produce power at a cost of $23/KWh. Other examples are SLAs (sealed lead acids). They cost around $50 and deliver electricity at $8.30/KWh. NiCad rechargeables also cost around $50 and produce electricity at $7.40/KWh. Their chief competitor, the NiMH costs $20 a unit and produces electricity for $4.40/KWh.

These examples are, of course, intended merely to give some perspective. They are a snapshot in time (2007). The estimates are rough because each type of battery is available in multiple implementations and designs, prices vary widely; many batteries are sold at deep discounts. Technology in the field is also rapidly changing.

Problems and Issues. One of two performance problems associated with batteries is the so-called memory effect, experienced with NiCad batteries. The other is the overheating, leading to explosions and fires, associated with lithium batteries. Disposal of batteries holding toxic metals is a potential environmental problem. Other issues fall into categories one might call annoyance or unpleasant surprise.

The memory effect is best described as a temporary loss of battery capacity. For example, a battery recharged when it is still 30 percent charged will, after that recharge, only deliver 70 percent of its capacity, not the 100 percent for which it is rated. This problem was first encountered with NiCad batteries in satellite applications in which batteries were routinely charged at the same time. The loss of capacity can be restored by letting the battery discharge several times fully before being recharged, but battery life is shortened. Since the problem first appeared, changes to the product have largely corrected this difficulty, but the memory effect has, so to say, stuck in the popular memory. NiMH batteries are promoted as being free of the effect, but can also suffer from it under rare circumstances. In general technical solutions have essentially dealt with the problem, but the issue is still frequently mentioned. What some perceive as the memory effect is sometimes due to faulty charging equipment or charging regimes.

Lithium-ion batteries have problems when overheated, be that from external sources or a combination of charging, heavy use, and outside temperatures. Overcharging the battery or manufacturing defects in the insulation separating anode and cathode can cause the battery to explode. Microscopic metal residues inside them can also cause heating up and fires. In 2006, for instance, Sony recalled 10 million batteries used by Apple, Dell, Fujitsu, Hitachi, Lenovo/IBM, Panasonic, Sharp, Sony, and Toshiba laptops due to metallic contamination of the battery. The problems are serious, if rare, and signaled that the relatively new Li-ion technology was still under development at the end of the first decade of the 2000s. Efforts to forestall costly and dangerous product defects were, of course, actively pursued by changing the batteries' internal components and operational chemistries, but such changes also pointed in the direction of reduced electrical capacity in the safer devices under development.

Nickel, cadmium, mercury, and lead acid batteries have significant amounts of hazardous materials by con-

tent. Most jurisdictions require their collection separately from ordinary household waste. To make things simpler, all batteries are treated the same way for disposal purposes. The degree to which the public actually participates in separating spent batteries from other wastes is not well documented.

The lifespan of lithium-ion batteries begins immediately after they are manufactured, not when they are placed into use, and this lifespan is between two and three years. This limitation of the product is not exactly featured (or even mentioned) by producers. Lifespan has no relation to number of charges; it simply means how long a time the battery will remain operable. The individual who has bought a device powered by a Li-ion battery may thus have an unpleasant surprise when the device loses life so rapidly and requires an expenditure of $100 or more to keep on functioning. Rechargeable devices, like cell phones, are designed to announce when their batteries need charging, but the user often fails to get the message—a source of annoyance.

MARKET

Information about the U.S. market for batteries at the retail level is only spottily available, but data on industrial shipments of products made in the United States provide a picture of trends. Such data are available from the U.S. Census Bureau, with some detail for 1997 and 2002, Economic Census years. Only total industry shipments, in dollars, are available for other years and only up to 2005. These figures come from the *Annual Survey of Manufactures*. We will look first at U.S. production and then enlarge the view to the battery market the world over. Two contrary impressions will emerge. One shows domestic production in decline; the other reveals energetic expansion of domestic demand. Imports explain the difference.

In 2005 U.S. domestic battery shipments of $6.66 billion were recorded by the Census Bureau. These divided into primary batteries with 43 percent of the dollars and storage batteries claiming 57 percent. The year was not a peak, however. Eight years earlier, in 1997, the industry had shipped goods worth $6.9 billion. The decline came because storage batteries declined from a level of $4.3 to $3.8 billion between the two years. Primary battery shipments increased, from $2.6 to $2.9 billion, but not enough to compensate for the loss experienced by storage batteries. Total industry shipments thus declined at the rate of 0.2 percent per year, storage batteries declined at an annual rate of 1.5 percent, and primary battery shipments grew at an anemic rate of 1.7 percent per year, just a shade over the 1.3 percent growth of the population.

In contrast to these rather lackluster trends in domestic production, the world market grew at a much healthier 4.3 percent per year in the 1997–2005 period. World storage batteries grew at 2.5 percent and primary batteries at 7.3 percent per year. The total market was $49.6 billion in 1997, an estimated $68.0 billion in 2005, and $77.5 billion in 2007. Note, however, that these values are not directly comparable to U.S. production shipments. World data are retail sales. Estimates of U.S. retail sales are available for two years only. In 2007 the U.S. share of the world retail market was $15.5 billion, thus around 20 percent. U.S. retail sales, however, were also growing, up from $11.7 billion in 2002, a 5.8 percent increase in that period. That rate of growth was lower than world growth in the same period (7.6% annually) and in sharp contrast to U.S. domestic shipments, as noted above. The bottom line is that all the growth in the U.S. retail sector came from imports. Data on world markets and U.S. retail sales just given are based on information published by the Freedonia Group, a market research firm, as quoted in "Factors Affecting U.S. Production Decisions: Why Are There No Volume Lithium-Ion Battery Manufacturers in the United States," a 2006 report issued by the U.S. Department of Commerce (DOC).

To summarize the market picture as it appeared in the 1997–2007 period, both the total world and U.S. retail markets were growing at a fairly high rate, particularly for what is a rather mature industry. The most rapid growth has been in the primary battery category. The U.S. retail market advanced but its domestic segment was, at best, flat.

Factors Influencing the Market. The domestic industry's performance in the early 2000s was such that it caused the DOC to commission a special study to look into it, the study just cited above. It was carried out under the auspices of DOC's Advanced Technology Program. The study focused on Li-ion batteries because these represent the most advanced products powering electronic devices like cell phones, laptops, and music players. The study's conclusions highlight important new factors influencing the evolving battery market.

DOC found that domestic manufacturers opted out of volume manufacturing of Li-ion batteries for three principal reasons. They saw low return on investment compared to their existing operations, long lead times and high costs before commercial products could be produced, and because this product category required establishing technical sales organizations to reach Japanese producers who, in turn, would be the dominant original equipment manufacturers (OEMs) buying the product. Enlarging on this general conclusion, one might say that the shift in manufacturing of electronic devices of all kinds to Japan, Korea, and Southeast Asia has made it necessary for a U.S. manufacturer to take part intensively in that industry overseas in order to obtain the technical information

FIGURE 17

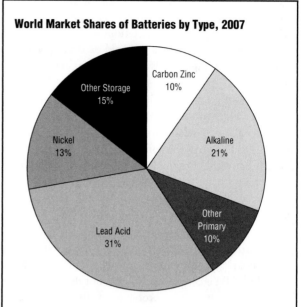

World Market Shares of Batteries by Type, 2007

Other Storage 15%
Carbon Zinc 10%
Nickel 13%
Alkaline 21%
Lead Acid 31%
Other Primary 10%

SOURCE: Compiled by the staff with data from Brodd, Ralph J. "Factors Affecting U.S. Production Decisions: Why Are There No Volume Lithium-Ion Battery Manufacturers in the United States," U.S. Department of Commerce, National Institute of Technology, Advanced Technology Program. December 2006.

and to keep up with development effectively enough to tailor-make products powering devices largely made overseas. In effect, U.S. manufacturers have chosen to concentrate on the mature segments of the battery market leaving the emerging and developmental segments to others.

World market shares of batteries by type, as estimated by Freedonia Group, are shown in Figure 17. The pie chart, based on total dollar sales, is divided into segments representing storage batteries, marked by dots, and primary batteries, marked by lines. Storage batteries represent 59 percent, and primary batteries 41 percent of the total market. Lead acid batteries, overwhelmingly automotive SLIs, are the largest segment followed by alkaline batteries, the second largest and the dominant segment in the primary battery category. Together with nickel batteries (which include both NiCad and NiMH unit sales) and carbon zinc, these segments together represent the old industry. New battery technologies are hidden under the other label in Freedonia's estimates. The fact that they represent major departures from the traditional segments becomes clear when we look at growth patterns from 1997 to 2007.

Figure 18 displays annual growth rates of each segment and of the Primary and Storage group for the ten year period. It is worth noting that in both categories, the highest rates of growth have been achieved by the *other* categories, that cluster growing at 12.3 percent per year in the primary and 15.7 percent in the storage battery grouping. Behind these very dramatic rates of growth is the world-wide expansion in the use of cells phones, laptops, and all manner of other portable electronic devices. The chief beneficiaries of this growth appear to be lithium-ion battery producers. The DOC study reports that Li-ion cell production increased from a level of 32 million cells in 1995 to 770 million in 2002, a growth of 57 percent per year. Lithium batteries have come to dominate a number of applications, including personal digital assistants (PDAs), digital cameras, movie cameras, cell phones, and laptops and other portable computers. Nickel-based batteries have been dominant in portable audio-visual equipment, cordless telephones, and power tools. In these latter segments NiCad batteries have been dominant. NiMH batteries, however, have a minority share of these applications and also participate in digital cameras, cell phones, and PCs where NiCad is not used. Beckoning from the more distant future is the prospect of using advanced storage batteries in transportation applications.

KEY PRODUCERS/MANUFACTURERS

Johnson Controls, Inc. (JCI). This company, based in Milwaukee, Wisconsin, is thought to be the largest manufacturer of lead acid batteries in the world, selling around 110 million units per year. The company entered this market in 1978 when it acquired Globe Union Inc. Since then JCI has expanded by acquisition. In 2002 it acquired the German Hoppecke Automotive Gmbh, in 2003 Varta Automotive Gmbh and 80 percent of Autobatterie Gmbh, and in 2005 the company purchased Delphi Corporation's worldwide battery business. Varta Automotive also owned dry cell battery lines, including button and coin batteries. That portion of the business was sold to Rayovac. JCI sells batteries to auto companies (OEM sales), replacement batteries, marine and industrial

FIGURE 18

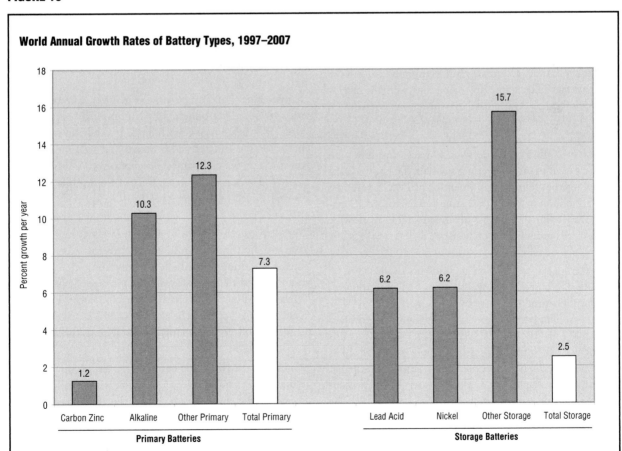

World Annual Growth Rates of Battery Types, 1997–2007

SOURCE: Compiled by the staff with data from Brodd, Ralph J. "Factors Affecting U.S. Production Decisions: Why Are There No Volume Lithium-Ion Battery Manufacturers in the United States," U.S. Department of Commerce, National Institute of Technology, Advanced Technology Program. December 2006.

batteries, and batteries used in telecommunications. Johnson Controls has 110 operations in twenty-eight countries in Asia, Europe, North America, and in South America. In 2006 the company had sales of $32.2 billion of which 11 percent ($3.5 billion) was in batteries.

Exide Technologies. Another world leader in the lead acid segment is Exide Technologies, based in Alpharetta, Georgia. The company, whose chief business is battery manufacturing, had sales in 2006 of $2.8 billion. Its products are destined for automotive, marine, motive power, and telecommunications applications. Motive units are used in forklifts, for instance. Telecommunications products include stand-by power products for computer networks, among others. The company began in 1888 with the commercialization by W.W. Gibbs of what was first known as the *chlorine accumulator*. The company was originally called Electric Storage Battery Company. Its name was originally a brand—the shortening of the phrase *excellent oxide*.

Duracell. Two companies dominate the alkaline battery market, Duracell and Energizer. Duracell had its origins in the 1920s when Samuel Ruben, an inventor, joined the P.R. Mallory Company. The company was later known as Mallory Battery Company and eventually as Duracell International. The company name was initially one of Mallory's brands, transferred to the entire corporation in 1964. Duracell merged with Gillette in 1996 and then became part of Procter & Gamble when P&G acquired Gillette in 2005. P&G reports on Duracell together with Braun, a razor manufacturer it also owns. These two companies together represented 4 percent of P&G's revenues of $68.2 billion in 2006, thus around $2.7 billion. Duracell is the leading producer of alkaline batteries worldwide but also sells other types of batteries.

Energizer. Energizer Holdings, Inc. began in 1896 as National Carbon Company (NCC) when it introduced the first mass-produced battery to the market, called The Columbia. The battery was six inches high and powered telephones. In 1914 NCC acquired a company called American Ever Ready. This organization, which began as American Electric Novelty and Manufacturing Company, created the flashlight. NCC thus became the first company to sell a device and the battery that operated it. The Energizer name arose as the rebranding of the Eveready battery. The Eveready name still continues in use in Asia. In 1917 NCC became part of Union Carbide. Ralston Purina acquired Energizer in 1986 but spun it off as an independent, publicly traded company in 2000. Energizer's sales in 2006 were $959.2 million, of which $723.7 million were batteries; and of that total 64 percent were alka-

line cells. Whenever batteries are mentioned in the United States, thoughts of the Energizer Bunny tend to arise, the Bunny having been one of the most successful advertising icons of the twentieth century. Worth noting is that the Energizer Bunny has been going since 1989—and is still going!

Spectrum Brands, Inc. Rayovac Corporation changed its name to Spectrum Brands, Inc. in 2005. The company is the leader in rechargeable NiMH batteries used in many portable devices in competition with lithium batteries. In 2006 Spectrum had sales of $2.55 billion; of this total batteries represented 34 percent or approximately $870 million. Spectrum's two major brands are Rayovac and Varta, the Varta brand being the primary battery lines of the German Varta Automotive mentioned above, the lead acid lines of which were acquired by Johnson Controls.

Asian companies dominate the lithium-ion battery market, with Japanese companies having an 80 percent share of total market. Among them the top three are Sanyo Electric Co., Ltd. with 27 percent, Sony Corporation with 21 percent, and Matsushita Battery Industrial Co., Ltd. (MBI) with 10 percent of world market share. These names, of course, have become very familiar to U.S. consumers in the context of electronic products generally.

Sanyo. This company's total sales in 2006, converted from yens, were $18.9 billion. Of that total, its battery sales were 18.7 percent or $3.5 billion. In addition to lithium batteries, the company also makes and sells nickel-based products. The company's orientation is industrial but Sanyo plans on introducing a line of consumer batteries as well.

Sony. In total size Sony is a much larger company, with $63.9 billion in sales in 2006. Of this total 64.3 percent was in electronics, in which category Sony includes components, but the company does not categorize batteries as such. Based on Sony's market share, however, its participation would appear to be at the level of $2.7 billion. Sony is best known for supplying battery systems for laptops.

Matsushita, MBI. This company is the producer of the Panasonic brand of batteries, available in alkaline, manganese, oxide, and lithium chemistries. The company also makes button and coin primary alkaline batteries. MBI's rechargeable batteries cover the range. They include lead acid, NiCad, NiMH, and lithium batteries, including a line of automotive batteries intended as primary propulsions for electric vehicles (EVs) and hybrids. MBI is part of the much larger Matsushita Electric Industrial Co., Ltd., a company with $76 billion in sales in 2006.

MATERIALS & SUPPLY CHAIN LOGISTICS

Major metals—and their future availability—play the most important role in the expansion of battery-based power. Zinc, nickel, and cadmium are central in the production of primary batteries. Storage battery technology is based principally on lead and nickel. Advanced storage batteries make use of lithium. Using summaries provided by the U.S. Department of the Interior in its Mineral Industry Surveys (MIS), the following snapshot emerges for the year 2006:

Zinc. The metal is available in large deposits all over the world and is also heavily recycled. Most zinc is used in the steel industry as an alloying component. World production in 2006 was 10 million metric tons with reserves at 220 million and reserve base at 460 million metric tons. The reserve base, as defined by the U.S. Geological Survey (USGS), includes reserves readily and economically extractable as well as other resources that require more development. One might think of it as deep reserves. When zinc is mined, the three largest co-products of its extraction are lead, sulfur, and cadmium.

Nickel. The metal was in high demand in 2006 with world consumption running at the rate of 1.55 million metric tons, two-thirds of which was consumed in making stainless steel. Growing demand came from economic development in China and India. Prices had more than tripled since 2002 running at $10.83 per pound in 2006, up from $3.07 in 2002. The largest reserves in the world were located in Australia (24 million metric tons) and in Russia (6.6 million metric tons). Well proved reserves suggested 41 years supply; deep reserves were sufficient to last 90 years, when more difficult underwater sources will have to be tapped.

Cadmium. Most cadmium mined (82%) was used in NiCad batteries. According to the USGS cadmium consumption declined 14 percent between 2002 and 2006 as substitutes were replacing it, presumably the NiMH technology. Very large reserves of sphalerite ores, the principal sources of cadmium, exist. Sphalerite is zinc and sulfur and mined for its zinc content; cadium is a by-product.

Lead. In the United States most of the lead consumed in 2006 was satisfied from recycled lead batteries. Consumption was 1.55 million metric tons and scrap collections 1.13 million metric tons (73% of consumption). Most of the lead consumed, 88 percent, was used for SLI-type auto batteries, 3 percent for ballast and counterweights, and 9 percent for ammunition, pipe, and industrial products. Mine production worldwide ran at the rate of 3.4 mil-lion metric tons in 2006. Reserves were 67 million metric tons, sufficient for a 20-year supply, and deep reserves amounted to 140 million metric tons.

Lithium. Lithium reserves across the world in 2006 were 4.1 million metric tons with mining drawing down 21,100 metric tons, suggesting supplies, at current rates of use, of nearly 200 years. Deep reserves were 11 million metric tons. Chile with 3 million metric tons and China with 540,000 metric tons had most of the reserves. U.S. reserves were around 38,000 metric tons. Lithium use in batteries amounted to 21 percent of consumption. Recycling of lithium was beginning to be practiced using spent batteries as the source of the metal in a development analogous to lead recycling.

Based on this snapshot in time, the metal under most pressure at the end of the first decade of the 2000s was nickel, not only because it is heavily used in making stainless steel but also because it was the basic raw material for NiMH rechargeable batteries in electric vehicles (EVs), the most common type used in experimental and developmental EVs for their inherent safety and long life. The deployment of lithium batteries in EVs is eventually likely, but at the end of the first decade of the twenty-first century lithium batteries were not ready for depolyment.

DISTRIBUTION CHANNEL

Alkaline batteries used in ordinary household applications have a three-tier distribution system in which producers sell to retailers through wholesalers, the retailer selling the product to the ultimate consumer in hardware, drug, grocery, convenience, and mass merchandising outlets. Rechargeable batteries typically follow a double path. They reach the consumer as part of a product and are therefore sold by the producer to original equipment manufacturers. The consumer, however, will also buy replacement batteries at retail. Where such purchases take place depends on the type of battery being replaced. Automotive products are sold by garages and auto supply houses. Batteries for electronic equipment are sold by stores that specialize in retailing computers, telephones, cameras, and audio-visual equipment. In that batteries suitable for a device are readily discernible by examining the equipment powered and/or its accompanying literature, many battery sales are taking place over the Internet, ordered online and delivered by mail or other carrier.

KEY USERS

The use of batteries is universal. In combination with satellite systems circling the globe, battery powered radios and telephones are providing information and communications even in areas of the globe once considered to be remote wilderness. Batteries are crucial to those who use them for immediate, personal, and physical support. Thus the smallest of batteries are the most important to selected individuals—those requiring heart pacers or using hearing aids. People required to use powered equipment far from the electric grid rely heavily on battery powered equipment in the field, not least of which are scientific workers, foresters, surveyors, rescue teams, fire fighters, and military and police forces. Batteries play a key role in health and security situations, thus in providing critical alternative power in medical institutions for equipment and lighting. Battery systems are installed in tandem with emergency generating equipment. Battery-based backup systems are routinely deployed to prevent the loss of information on single computers and on network servers. Battery based power, with the recharging role fulfilled by solar panels, is also the principal source of energy in space applications and other remotely-sited monitoring installations.

ADJACENT MARKETS

Batteries are by their very nature component parts of other things—indeed the crucial component parts. A large flashlight without batteries might serve, perhaps, as a handy weapon to club an intruder, but it has few other uses. The batteries without the flashlight might serve, perhaps, as paper weights. For this reason the concept of adjacency ill-fits this product category.

One way to look at batteries, in this context, is by functionality. In the past lighting was produced by candles and small motions, thus those inside of clocks, for instance, by wound springs or weights combined with gearing. In appliances that require very little power, such as calculators, solar power is used to run the device. Battery-driven power tools can be replaced by others employing small internal combustion engines. In larger devices used for lighting, for instance, adjacent products are those by-passed by time, such as kerosene lamps. In large applications the combustion engine reappears again powering an emergency generator.

In effect batteries represent the extension of a modern technology, electricity, into areas where the wire, which delivers electricity, cannot reach. To replace this new convenience, we have to dust off older solutions that predate Edison or reach ahead in time to imagine the perfection of arts still not quite measuring up, such as effective, energy-dense, portable, and miniaturized solar energy devices.

RESEARCH & DEVELOPMENT

A look at battery technology, as at least lightly touched upon in all that has been said so far, reveals that it is rooted in very sophisticated chemical engineering approaches to holding and releasing electromagnetic force by combining elements and compounds. The chief issues in improving

this technology and extending it further into even more applications include:

- Increasing energy density while decreasing weight
- Extending the inherent life of the device
- Removing danger and risk
- Reducing cost

We are living in an era where the ultimate exhaustion of hydrocarbon fuels for transportation is becoming a matter of decades rather than centuries. At the same time, coal resources and electrical energy from nuclear, hydro, and geothermal sources have much longer time horizons. For this reason a great deal of R&D in this field is directed toward development of effective batteries for electric vehicles. The best candidate for EV batteries as of the end of the first decade of the 2000s is lithium technology because it has the highest densities and the lowest weight. Lithium batteries, however, have a very short life and are extremely sensitive to tampering, contamination, and external temperatures. Battery failure can be very costly, indeed dangerous. Lithium batteries also cost a great deal. Research is being devoted especially to advanced, indeed the most esoteric, materials combinations aimed at powering electronic devices and automotive propulsion (rather than starting and ignition) batteries.

CURRENT TRENDS

Two major trends act like weather systems that create the climate for the battery industry. One of these is the expansion of computer intelligence into every conceivable device to provide convenience, unattended operation, transportability, and off-wire communications powers. This technological development leads to miniaturization and concomitant requirements for miniaturized sources of power, ideally supplied by batteries. Tiny, hand-held personal digital assistants with computing power matching those of once massive computers are an example but, if trends continue, only the tip of a future iceberg.

Looming future energy shortages, particularly in the transportation sector, already sketched briefly in the section on R&D, represent the second trend. In the long term liquid fuels from crude petroleum will disappear but electric power will continue to be available. The best means of transition between the two is represented by battery technology.

TARGET MARKETS & SEGMENTATION

This very mature industry—yet with a very new and hi-tech leading edge—has well established target markets based on end use. These have been largely outlined above. In summary, lead acid technology is aimed principally at automotive starting applications using thin-plate devices. SLAs, with thicker plates, capable of deeper discharge, are used in industrial and commercial backup systems in stationary applications. Alkaline batteries are aimed at consumer uses in portable devices. Miniaturized devices, disposable and rechargeable, are used in watches and in medical devices. Lithium technology dominates the rapidly growing demand produced by new electronic devices, competing with NiMH batteries. These same technologies compete for the other still gestating future deployment of batteries to move cars and trucks down the highways of the world.

RELATED ASSOCIATIONS & ORGANIZATIONS

Battery Council International, http://www.batterycouncil.org

National Electrical Manufacturers Association (NEMA), http://www.nema.org

Portable Rechargeable Battery Association, http://www.prba.org

BIBLIOGRAPHY

Beaty, William J. "How Transistors Work? No, How Do They *Really* Work." Amasci.com. Available from <http://amasci.com/amateur/transis.html>.

Brodd, Ralph J. "Factors Affecting U.S. Production Decisions: Why Are There No Volume Lithium-Ion Battery Manufacturers in the United States." U.S. Department of Commerce, National Institute of Technology, Advanced Technology Program. December 2006.

Cheng, Jacqui. "Sony Issues Global Li-ion Battery Recall." Ars Technica, LLC. 28 September 2006. Available from <http://arstechnica.com/news.ars/post/20060928-7858.html>.

Cringley, Robert X. "Safety Last." *The New York Times.* 1 September 2006.

"Household Batteries and the Environment." National Electrical Manufacturers Association (NEMA). June 2002. Available from <www.nema.org/gov/ehs/committees/drybat/upload/NEMABatteryBrochure2.pdf>.

Lazich, Robert S. *Market Share Reporter 2007.* Thomson Gale, 2007, Volume 1, 468.

Mineral Commodity Summaries. U.S. Department of the Interior, U.S. Geological Survey. January 2007. Available from <http://minerals.usgs.gov/minerals/pubs/commodity/>.

"Power Stream Battery Chemistry FAQ." Lund Instrument Engineering, Inc. Available from <http://www.powerstream.com/BatteryFAQ.html>.

"Rayovac Announces 1-Hour NiMH Charger." *Digital Photography Review.* 14 February 2001.

Vulkson, Stephen P. and Michael Kelley. "High-Energy-Density Rechargeable Lithium-Ion Battery." U.S. Air Force Research Laboratory. Available from <http://www.afrlhorizons.com/Briefs/Feb04/PR0306.html>.

Beer

——————•——————

INDUSTRIAL CODES

NAICS: 31–2120 Breweries

SIC: 2082 Malt Beverages Manufacturing

NAICS-Based Product Codes: 31–21201 through 31–21201231, 31–21204 through 31–21204291, 31–21207 through 31–21207121, and 31–2120A through 31–2120A151

PRODUCT OVERVIEW

> Without question, the greatest invention in the history of mankind is beer. Oh, I grant you that the wheel was also a fine invention, but the wheel does not go nearly as well with pizza.— Dave Barry

Beer is believed to be the world's first alcoholic beverage. Its origins date far back into human prehistory. All beers are fermented undistilled beverages derived from grains. They are usually carbonated and have been hopped for flavor. Beer comes in a truly staggering variety of costs and styles. Within the beer section of an average market the consumer can now find an ever-increasing variety of brews ranging from inexpensive mass-produced brews to high-end craft brews, from light lagers to dark ales, and from light beers to very high-alcohol content malt liquors. The growing variety found among commercially distributed beers is the most important trend in the modern brewing industry.

The Brewing Process. Beer is created when yeast ferments a grain mash. For most beers, which are barley based, the process usually begins when raw barley is malted. In the malting process the starches in the barley are converted to sugars when the grain germinates after being soaked in water. The malted barley is then roasted in an oven, a process that carmelizes the sugars and adds flavor to the barley malt.

After roasting the malted barley is boiled in water to create a sugar-laden mash known as a wort. Hops, a flowering herb, are also added to the boiling wort. The hops impart additional flavor and aroma, and also act as a preservative in the finished beer. Once the wort has cooled, yeast is added. Yeast converts sugar into alcohol and carbon dioxide. The wort is then left to sit while the yeast converts the sugars into alcohol that remains in the beer and carbon dioxide that is vented from the wort container. Once all of the sugar in the wort has been converted to alcohol by the yeast, the wort is then decanted, filtered, and bottled or kegged.

Home brewers and some traditional craft brewers add carbonation to the final beer via a secondary fermentation stage in which a bit more sugar is added to the beer at the time of bottling. The remaining live yeast then creates a small additional amount of alcohol and carbon dioxide, which is now contained and allowed to build pressure within the bottle. However, the vast majority of modern commercially brewed beers add carbonation to the final beer by simply adding carbon dioxide gas at the time of bottling.

Beer is a fresh consumable. Once it has been bottled or kegged, the quality of a beer begins to deteriorate. This deterioration is speeded by exposure to light, so most brewers minimize the exposure of bottled beers to light by using green or brown glass instead of clear glass. High temperatures or sudden changes in temperature can also diminish beer quality, so much of the manufacture and distribution chain uses refrigerated trucks and storage.

History. Though nobody knows exactly when mankind began brewing beer, the history of brewing dates back more than ten thousand years. The long history and importance of beer is reflected in the English word beer itself, which derives from the Latin word meaning to drink, *bibere*, the same root that led to the French word biere and the German word bier. The Spanish word for beer, cerveza, not to be outdone for historic origins, traces its roots to the Roman goddess of agriculture Ceres.

The first beers were almost certainly accidental brews, produced when stored cereals spontaneously fermented by naturally occurring yeasts when the stored grain was accidentally moistened. The prehistoric origins of brewing coincide with the prehistoric origins of agriculture. While some believe that the cultivation of grains necessarily preceded the origins of brewing, other experts believe that mankind pursued farming to provide a steady source of grain for brewing. The written histories of Ancient Egypt and Mesopotamia record brewing and the distribution of beer, and the *Epic of Gilgamesh* describes beer's role in the transformation of primitive man to civilized man.

Beer was an important part of ancient societies, just as it is in the twenty-first century. Five-thousand years ago the Mesopotamians and Egyptians brewed at least nineteen types of beer that are still known about today. Beer drinkers used straws to consume these cloudy and unfiltered brews, so as to avoid consuming the bitter yeast residues that settled in the bottom of those early beers. Compensation for the laborers that built the ancient pyramids in Egypt and ziggurats in Mesopotamia often included beer. At times high officials received a beer ration as great as five liters per day, while laborers received a lesser quantity of perhaps two liters. The Greek philosopher Plato is believed to have said, "He was a wise man who invented beer."

Brewing continued through the Middle Ages and the Renaissance in northern and eastern Europe and all social classes drank beer. Wine consumption outpaced beer consumption in southern Europe due to the ready availability of grapes, but beer dominated the northern and eastern parts of Europe. People often preferred alcoholic beverages because the alcohol served as a disinfectant and made beer safer than the oft impure sources of drinking water available during the Middle Ages.

Brewing was primarily a household activity during the Middle Ages and it was not until the fourteenth and fifteenth centuries that artisans began brewing beer for a larger consuming body. Monasteries played an important role in the development of beer production on a larger scale. To this day beer varieties such as Trappist are still brewed by monasteries.

The production and consumption of beer was such an important issue during this period that in 1516 William IV of Bavaria adopted the Reinheitsgbot, a law to safeguard the purity and consistency of beer by regulating its production. German brewing in the twenty-first century is still regulated by this German Beer Law of 1516. The Reinheitsgbot restricts beer to four ingredients: barley, hops, yeast, and water. Any beer produced in or imported into Germany must meet the Reinheitsgbot. Some craft breweries and micro breweries in the United States use the German Beer Law and adherence to it as a marketing tool, a way to differentiate their own beers from their competitors'. Many large breweries today produce beers that can not be exported to Germany because they do not meet the standards set by the Reinheitsgbot.

Prohibition. During the early twentieth century the prohibition movement swept across the globe and took deepest root in North America and northern Europe. The movement stemmed from nineteenth-century temperance movements and the Protestant wariness of alcohol and its ill affects. The duration of these periods of prohibition varied; in the United States prohibition lasted from 1920 to 1933 while in Canada prohibition varied by province but lasted from 1900 to 1948 on Prince Edwards Island.

The United States passed the Eighteenth Amendment on January 16, 1920. This constitutional amendment prohibited the sale, manufacture, and transport of alcohol, but did not prohibit the possession and consumption of alcoholic beverages. In 1933 the Twenty-first Amendment to the U.S. Constitution repealed the nationwide prohibition. The amendment gave individual states the right to restrict or ban the sale of alcohol, leading to a patchwork of state and local laws designed to regulate the sale and consumption of alcohol. Many states continued to enforce prohibition after the Twenty-first Amendment was passed. In 1966 Mississippi became the last state to repeal prohibition

Post Prohibition Rebuilding. A growing number of problems created by prohibition itself drove the repeal of prohibition at the national level: a black market for high potency distilled alcohol, the high cost of enforcing the law, and a sharp increase in crime rates. The desire to create jobs during the Great Depression of the 1930s also encouraged the repeal of prohibition. Many people continued to brew and distribute beer and other alcoholic beverages during prohibition, but all this economic activity happened on a black market basis. The government wished to legitimize the production of alcoholic beverages so that it could be regulated and taxed.

The passing of the amendment repealing prohibition went a long way toward solving the crime problem. The economic losses produced during prohibition took much longer to fix. Half of the breweries that operated in the United States prior to prohibition (1,345) went out of business and did not re-open after the repeal. By 1939 only 672 breweries operated in the United States according to the *United States Brewers Almanac*.

During the post-prohibition era great consolidation took place within the U.S. brewing business. The largest brewers had maintained operations during prohibition by producing "near beer," root beer, ginger ale, and other malt beverages with under one percent alcohol contents. They grew rapidly because they were more easily able to pick up production than those brewers who had discontinued production during prohibition. They also lobbied heavily after repeal for legislation that favored the large brewers by making production of beer legal only in large quantities for commercial distribution. In the decades following repeal the largest breweries grew and consolidated their market position while the variety of beer they provided to consumers decreased in variety. Before Prohibition a rich variety of local styles with roots in the many origins of American people themselves characterized American brewing. After the repeal of Prohibition light pilsener-style lagers with mass-market appeal dominated the American market.

In 1978 President Jimmy Carter signed legislation that legalized small-scale home brewing. This change led to a rebirth in the U.S. brewing industry. Home brewing expanded rapidly in the late 1970s and 1980s, and many of these small home breweries grew beyond the home base into small commercial breweries, microbreweries and pub breweries. Microbreweries grew between 5 and 7 percent of the total U.S. beer market by the turn of the century. These microbreweries have had a disproportionate impact on the U.S. industry despite their small size by bringing diversity back to American beers in the form of specialty and craft beers that featured new flavors and tastes. The big-three brewers, Anheuser Busch, Miller, and Coors all began introducing boutique brands and specialty beers of their own in the late 1990s and 2000s to compete with domestic microbrews and the growing numbers of imported beers.

Globalization. Beer markets had always tended to be local or regional until the end of the twentieth century. Although beer had always been imported and exported in the United States it was during the 1990s that foreign producers began to make deeper inroads into this vast market. By the early 2000s imports accounted for close to 12 percent of the U.S. market, and the United States had become the largest beer importing country in the world. In return U.S. producers established exporting networks through licensing agreements abroad and had made great inroads of their own by the early 2000s.

In 2002 the large African beer maker South African Breweries (SAB) bought Miller Brewing Company, the second largest U.S. beer maker and a long-time industry leader. The early 2000s marked another change in the global beer landscape as China surpassed the United States as the country with the highest total beer consumption. Neither country ranks in the top ten on a per capita basis, but the United States had long been the leader in total consumption.

Beer Types. Beers are generally categorized as ales or lagers by the behavior of the yeast used in the fermentation process used to brew the beer. Ales use top-fermenting yeasts that ferment the sugars of the wort more quickly and at a higher temperature than lagers. Ales are often brewed at a temperature above 55 degrees Fahrenheit. This warmer temperature tends to create a faster brewing process that brings more of the secondary flavors out of the barley malt, creating beers with fuller, more complex flavors. The faster brewing times and more complex flavors make ales a favorite of brewpubs and microbreweries.

Lagers use bottom-fermenting yeasts that ferment the sugars of the wort at a lower temperature than lagers. Ales are often brewed at a temperature well below 55 degrees, sometimes as low as 40 degrees. Lager yeasts take longer to consume the sugars in the wort at these lower temperature, and this slower brewing process creates a crisper, cleaner tasting beer. The crisp and light flavor of lagers make them a favorite of mass-market breweries.

Spontaneous fermentation with wild yeasts produces lambic beers. In tightly controlled brewing, wild yeasts can create a variety of *off* flavors, but those flavors are valued in the unique taste of lambic beers, which often have a sour flavor. The wild nature of these yeasts lead to very brewery-specific tastes for lambic brews.

In addition to the differences in taste imparted by different yeasts and brewing processes, brewers create a wide spectrum flavors for specific beers through their choice of the roast of the malted barley, the use of adjunct grains, the sugar content of the wort, and the variety and quantity of hops. Darker beers such as porters and stout use a very deeply roasted barley malt that has been roasted to the point of charring and imparts a dark color and a burnt flavor similar to coffee, cocoa, or caramel. Some brewers use caramel to darken and flavor their dark beers. Lighter roasts create more lightly flavored beers such as pilsner lagers and summer ales.

Brewers also use adjunct grains such as wheat, corn, and rice to create a lighter flavor in a beer, or simply to reduce the cost of the final brew by increasing the quantity of sugar in the wort without using more expensive malted barley. The mix of malt and water determines the sugar content of the wort, which then determines the alcohol content of the beer. Contrary to popular belief, darker beers do not have an inherently higher alcohol content. The renownedly dark Guinness Stout Draught clocks in with a relatively light 4.0 percent alcohol content while the much lighter-tasting Miller High Life has an alcohol content nearly a third greater at 5.5 percent.

The type and quantity of the hops impart a great deal of the final flavor of a beer. Hops act as a preservative in beer, but the oils and flavors from the hops also give a beer much of its odor and its distinctive bitter zing. Most domestic lagers are lightly hopped for a smoother taste, while many ales and darker lagers tend to have a stronger hop taste. India Pale Ale (IPA) is an especially hoppy variety of beer. Its basic recipe was created when British brewers added extra hops to kegs of ale being shipped to India to ensure that they were well-preserved. Some modern microbrewers take hopping to the extreme with brews such as the Hopslam from Bell's Brewing, a double-hopped IPA that touts is extreme hoppiness in its name.

The basic recipe used by most mass market brewers is a pilsner lager based on the style of lager brewed in Pilsen, Bohemia. The Pilsner Urquell lager from Pilsen that is often cited as the first true pilsner lager was first brewed by brewer Josef Groll in 1842. This imported full-barley beer can now be found in stores across the United States. The modern American pilsner shares its light golden color and high carbonation, though most mass-market brewers use adjunct grains such as wheat, corn, and rice in addition to barley to brew their version. Domestic pilsners such as Budweiser, Coors, and Miller Genuine Draft usually average around 5 percent alcohol content.

MARKET

The world beer market in 2003 measured 1.484 million hectoliters, equal to 39.2 million gallons. This represented an increase of 2.3 percent over the 2002 world sales figure. The dollar value of beer sales worldwide in 2003 rose at a greater rate than did volume sales, increasing by 10.2 percent over value shipments in 2002. Beer got a bit more expensive.

Beer shipments by U.S. brewers grew steadily in the early 2000s, increasing from $16.5 billion in 2000 to $20.8 billion in 2005 according to data published by the U.S. Census Bureau in its *Annual Survey of Manufactures.* These shipments were for sales by breweries to wholesalers and distributors and did not include imports.

The popularity of imported beers also grew in the early 2000s as did the value of imports to the United States. In 2000 the U.S. imported $9.4 billion worth of beer, equal to 11.8 percent of industry shipments that year. In 2005 the U.S. imported $10.5 billion worth of beer, 13.1 percent of industry shipments. Per capita consumption rates of beer, however, have decline slightly during the early 2000s.

In the late 1960s per capita consumption of beer stood at 16.5 gallons annually. Consumption rose through the 1970s and peaked at 24.6 gallons in 1981. The 1980s and early 1990s saw per capita consumption rates of beer in the range of 22 to 24 gallons. In the mid-1990s beer consumption began to fall slightly and in 2005 stood at 21.3 gallons per person per year. Figure 19 presents these data graphically.

Figure 19 also shows the median age of the U.S. population which rose steadily over the period shown, 1966 through 2005. This period coincides with the coming of age of the Baby Boom generation, those born after World War II through the early 1960s. Since consumption of beer is highest for those in their 20s the peak of per capita beer consumption logically occurred when the large Baby Boom generation was reaching and pass through its 20s and early 30s. Starting in the early 1980s, per capita beer consumption rates began to fall, if slowly.

At the same time the number of microbreweries and specialty breweries in the United States began to rise. According to data complied and published by the Beer Institute there were a total of eight specialty breweries operat-

FIGURE 19

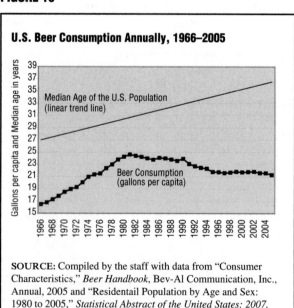

SOURCE: Compiled by the staff with data from "Consumer Characteristics," *Beer Handbook*, Bev-Al Communication, Inc., Annual, 2005 and "Residentail Population by Age and Sex: 1980 to 2005," *Statistical Abstract of the United States: 2007*, U.S. Department of Commerce, Bureau of the Census.

ing in 1980. By the turn of the century just shy of 1,500 such breweries were registered in the United States. As the per capita consumption of beer fell slowly, the types of beer being consumed began to shift towards a more varied beers and more expensive beers.

A total of 1,367 brewers operated in the United States in 2005. The Beer Institute characterized twenty-one of these brewers as traditional brewers and the remaining 1,346 as specialty brewers. The traditional brewers are the large, industry leaders by volume, while the specialty brewers include microbreweries and brewpubs, most of which produce specialty beers and craft beers. This large number of specialty brewers highlights the dramatic changes that have taken place in the U.S. beer brewing market during the last half of the twentieth century. In 1960 there were only 175 brewers in the United States.

KEY PRODUCERS/MANUFACTURERS

Three companies—Anheuser-Busch, Coors Brewing, and Miller Brewing—dominate the beer industry in the United States. Together they accounted for nearly 80 percent of the market in 2005. Of the many small breweries in the marketplace, a small number stand out as a second tier of key players. Microbreweries are by definition small and thus no single one of them appears as a particularly large player in the market, although their influence on the industry as a group is significant. Craft beers have been traditionally thought of as beers produced by microbreweries. As many of the microbrewers grew beyond the bounds of this category they did not change their beers and thus craft beers are now produced by beer makers of every size.

Anheuser-Busch. Started in 1852 as a Bavarian Brewery in St. Louis, Missouri, this company had a U.S. market share in 2006 of just under 50 percent, making it the largest U.S. beer maker, a position it has held since 1957. Eberhart Anheuser acquired the Bavarian Brewery in 1860 and took his son-in-law, Adolphus Busch, as a partner later that decade. In 1879 the name of the company was changed to Anheuser-Busch Brewing Association and by 1901 the company reached the threshold production figure of one million barrels of beer in a single year.

Today, Anheuser-Busch operates 12 breweries in the United States and has operations around the world. It produces more than 60 beers and flavored drinks. Among these are leading brands Budweiser, Bud Light, Budweiser Select, Busch, Michelob, Natural Light, and Ultra Amber to name but a few. Its net income in 2006 was $1.97 billion on gross sales of $17.96 billion. It is worth noting, however, that these financial results include all divisions of Anheuser-Busch, including those involved in the transportation and entertainment sectors.

Coors Brewing Company. This company dates back to the early nineteenth century and is a leading beer maker that employed over 11,000 people worldwide in 2006. It operated 10 breweries in three countries. Its operations are divided into three divisions which correspond to the countries in which it operates; Canada, the United Kingdom, and the United States. It marketed more than 40 brands of beer in 2006, including Carling, Coors, Coors Light, Blue Moon, Keystone Light, Killians Irish Red, and Molson Canadian. Coors Brewing Company had net income in 2006 of $361 million on net sales of $5.85 billion.

Miller Brewing Company, a division of SABMiller. The Miller Brewing Company was the United States' second largest brewing company in 2002. The company was formed by Fredrick John Miller in 1855 in Milwaukee, Wisconsin. It grew and expanded, managing to stay in business through prohibition and expanding again after the repeal of prohibition.

Miller Brewing Company was purchased in 1969 by the Philip Morris Corporation but continued to operate under its well-known name. It launched the first nationally available brand of low calorie beer, Miller Lite, in the mid 1970s. Philip Morris Corporation, renamed Altria in the 1990s, sold the Miller Brewing Company in 2002 to the large, international beer maker South African Breweries (SAB). The resulting company was named SABMiller plc.

The Miller Brewing division of SABMiller employed 6,300 people in 2006. Its activity accounted for 24 percent ($3.67 billion) of the SABMiller group's total gross revenues of $15.3 billion. Miller brands include Miller Lite, Miller Genuine Draft, Ice House, Miller High Life, Milwaukee's Best, Peroni Nastro Azzurro, Pilsner Urquell, Leinenkugel's, Henry Weinhard's, Olde English 800, and Mickey's.

Second Tier Market Leaders. Most of the brewers that make up the second tier of the industry started off as microbreweries and grew to be brewers too large to be categorized as micro. Some of the prominent players in this second tier of the industry are names familiar to beer drinkers by their company names and others are known only by their brand names. The Boston Beer Company is an example of the latter. It makes the Sam Adams line of beers. Sierra Nevada Brewing Company is an example of the former category, marketing a line of beers whose brand name includes the Sierra Nevada prefix. New Belgium Brewing Company, headquartered in Colorado is a second tier brewer that markets such craft beers as Fat Tire, Sunshine, and Blue Paddle. The Texas based Spoetzl Brewing Company produces a line of beers called Shiner:

Shiner Light, Shiner Bock, Shiner Blonde, and Shiner Kolsch. The F.X. Matt Brewing Company is another second tier brewer whose company name may be unfamiliar but whose products are well known in the northeast. It produces a line of beers which share the Saranac name: Saranac Lager, Saranac Black Forest, and Saranac Pale Ale to name a few.

The second tier brewers tend to focus on regional markets, although a few of these brewers have achieved national distribution during the later years of the twentieth century. Of those listed above, only Boston Brewing Company and Sierra Nevada Brewing Company have national distribution. The others market regionally, as does the Dixie Brewing Company out of New Orleans, Louisiana, maker of such provocatively named craft beers as Dixie Jazz and Blackened Voodoo. To rank this diverse group of brewers is a task the undertaking of which, although enjoyable, would take years.

Leading World Producers. In 2006 the three largest brewers in the world were InBev of Leuven, Belgium, SABMiller headquartered in London, England, and the U.S. market leader Anheuser-Busch which makes its home in St. Louis, Missouri. InBev global brands include Stella Artois, Beck's, Leffe, and Brahma. Other large brewers on the world stage are Heineken and Carlsberg, both of Germany, Grupo Modelo of Mexico, Tsingtao of China, Guinness of Ireland, and Kirin of Japan.

This list is incomplete and includes primarily brewers whose beers are exported beyond their national boundaries. Within most countries there are brewers of all sizes whose products are consumed in large quantity but are not seen on the global landscape because of their focus on domestic markets. Globalization is influencing the beer industry as it is most others but local, regional, and national beers are still important players within this age-old trade.

MATERIALS & SUPPLY CHAIN LOGISTICS

The basic ingredients of beer—and in the minds of some, the only legitimate ingredients—are water, barley, hops, and yeast. Flavorings and preservatives are sometimes added to beer as well but the four basic ingredients make up the bulk of the materials consumed in making beer.

Water. The quality of the water used to brew beer is perhaps the most important of these ingredients. By volume in the end product it is certainly dominant, but its quality too, is an essential part of the brewing process. The mineral components in water, which vary from one source to another, have an effect on the taste and appearance of beers made with that water. The water available in an area

dictates the type of beer that can be produced locally and has contributed to the evolution of different beer styles. Stouts are famously produced in Ireland, specifically in and around Dublin. The reason for this has to do with the hard water supplying the Dublin area. By contract, the pilsner beers made in Pilsen, Czech Republic are renowned for their X flavor which is a direct result of the soft water available in the City of Pilsen. Water from either city could not be used to produce the other city's beer.

Barley. While many beers are true to the Reinheitsgbot (the German Beer Law) requirement that beer contain only the four basic ingredients, many beers are brewed with what are referred to as fillers. These are used in order to replace the relatively more costly ingredient of barley with other starches that will ferment and produce alcoholic. These alternate ingredients—corn, rice, and other grains—also produce different flavors. Sugar is often added in other forms as well in order to speed the production process.

Hops. The use of hops in brewing contributes primarily bitterness, aroma, and flavor to the beer. However, hops also provide other attributes to the resulting beer such as better head retention, fewer unwanted proteins, and some anti-bacterial properties. By removing unwanted proteins, the use of hops can contribute greatly to achieving a clearer beer, favored by many Americans. Their potential to give beer a bitter taste makes the use of hops a matter of getting the right balance in order to gain the benefits without tipping the scale on bitter taste.

Yeast. Yeast is the catalyst in the beer brewing process as it is yeast that is responsible for converting fermentable sugars into alcohol, carbon dioxide, and other byproducts. There are hundreds of varieties and strains of yeast; the ones most commonly used in the brewing of beer are those in the Saccharomyces cerevisiae class of yeasts. Yeast often known as ale yeast is a top-fermenting type, while lager yeast is a bottom-fermenting type of yeast. Top-fermenting yeasts are typically used to brew ales, porters, stouts, and wheat beers. Bottom-fermenting yeasts are used to brew Pilsners, Dortmunders, Märzen, Bocks, and American malt liquors.

Packaging Materials. Other than the cost to compensate employees and workers, the largest single category of material inputs to the brewing business is the cost of packaging materials used to prepare the beer for distribution. Cans are by far the dominant method for packaging and distributing beer in the United States where 90 percent of beer is sold in cans. Consequently, the cost of metal cans, boxes, and other containers (not including glass), represented 9.8 percent of all inputs into the U.S. beer brewing industry at the turn of the century according to the Census Bureau.

DISTRIBUTION CHANNEL

The distribution of beer in the United States is a three-tier system. The system was established through legislation after the repeal of Prohibition in 1933. U.S. regulations require that brewers sell to wholesalers or distributors who control the distribution of beer to restaurants, bars, and stores. Thus, the three tiers of the system are producers, distributors or wholesalers, and retailers. Some beer wholesalers have locations throughout the United States but most are regional. Wholesalers are either brand exclusive or independent multi-brand wholesalers. The brand exclusive wholesalers are often partially owned by the breweries whose beers they distribute. In 2002 there were 2,561 beer and ale merchant wholesalers operating in the United States who together generated sales of $44.35 billion according to the Census Bureau's *2002 Economic Census.* Firms were active in every state of the Union and employed slightly more than 100,000 people.

The Beer Institute provides data on the number of beer distributors that are operating in the United States and their figures differ slightly from those of the Census Bureau, due to differences in how each defines a distributor. Nonetheless, the trends toward consolidation in the ranks of beer distributors is noted by both data series. According to the Beer Institute's *Brewery Almanac 2007,* there were just over 5,000 beer distributors operating in the United States in 1980. By 1990 that number had fallen to just shy of 3,000, and in 2004 the Beer Institute reported a total number of 1,907 beer distributors. Their numbers are shrinking but the volumes of product moving through these distributors has been growing, as has been seen in per capita consumption rates. The complexity of federal, state and local level regulation of the alcoholic beverage market is part of what has encouraged distributors to consolidate.

Individual states within the United States have their own regulations related to the consumption, distribution, and sale of alcoholic beverages within their borders. States control the licensing of establishments that sell alcoholic beverages, establish the hours in which it is permissible to sell these beverages, and regulate the distribution of these beverages as well. Some states even exert control over the sale price of alcoholic beverages and others operate state-owned distributorships. Distributors must manage this web of different regulations.

Distributors play an important role in the channel. They have the ability to monitor the sale of beer from the time the beer leaves the brewery until it arrives at a licensed retail outlet. This makes them an ideal point in the distribution system at which to collect state taxes and many states find it easier to collect taxes from a limited number of federally licensed beer distributors than from the far more numerous retail establishments that sell to the public. Other states collect taxes from the final retail outlet.

Managing the distribution channel in order to complement marketing efforts and assure that a steady supply of product is always available to the public is a challenge, as in the case of any perishable product.

Beer distributors sell their beer to a wide variety of final outlets, including bars, restaurants, stadiums, mass marketers, supermarkets, liquor and package stores, drug stores, convenience stores, and other retailers. The ability of an individual retail channel to sell beer and other alcoholic beverages is controlled by state law. Beer is generally more widely available than wine or liquor in many states, but the availability of beer, wine, and liquor in retail markets varies by state.

KEY USERS

Beer consumers tend to be young and male. According to the 2005 *Beer Handbook* from Bev-AL Communications, beer consumers are 61.7 percent male and 38.3 percent female. Men tend to prefer heavier, imported, and craft beers. The percentage of male consumption rises to 69.2 percent for regular beers, 68.0 percent for high-alcohol ice beers, 65.3 percent for imported beers, and 69.5 percent for craft beers. Women tend to prefer lighter, low-alcohol beers. The percentage of women consumers rises to 40.3 percent for light beers and 41.5 percent for non-alcohol beers. Surprisingly, since women otherwise prefer lower-alcohol beers, women also account for more than their usual share of beer consumption in the malt liquor segment. Women consumed 41.1 percent of malt liquors, which tend to be higher in alcohol content than the average beer. This may be an artifact of advertising and marketing that presents these beers as an alternative to beer, which is itself an historic artifact of some state labeling laws that banned presenting these high-alcohol brews as beer.

While women trail behind men in beer consumption there are several closely related markets in which they close the gap and even surpass men. Women consume 45.9 percent of hard ciders and lead men 60.5 percent to 39.5 percent in the consumption of malted beverages and prepared cocktails. The gap widens to 69.1 percent to 30.1 percent for wine coolers. These beverages have often been designed and marketed specifically for female consumers as an alternative to beer. Figure 20 presents selected rates of consumption by gender from Bev-AL's *Beer Handbook.*

Young consumers lead all beer consumers. Those in the 21 to 24, 25 to 34, and 35 to 44 age groups all account for significantly above-average beer consumption. Consumers aged 45 to 54 tend to consume beer in proportion to their overall population numbers, while beer consumption drops for those over the age of 55.

Beer consumers are also distinguished by race and many beers are marketed to target ethnic groups. According to 2003 statistics from the Simmons Market Research Bureau, caucasians purchase more light beer and micro/specialty beers than do consumers in general. African Americans purchase three times as much malt liquor and nearly twice as much ice beer as do consumers in general, while Asians and Hispanics purchase more imported beers than the general consumer. Light beers find less favor among African Americans and Asians, who purchase 1/3 to 1/2 less light beer than the overall market. African Americans also purchase just one-third as much micro/specialty beer as consumers in general. These trends are reflected in marketing campaigns. Malt liquor and ice beer advertising often aims at an urban target, while Asian beers and Mexican beers are aggressively marketed at those ethnic communities.

Retail beer purchases follow the age and gender consumption trends. According to statistics gathered by the Convenience Store News Shopper Panel in 2003, the average beer buyer between the ages of 21 and 29 accounted for 54 percent more retail beer purchases than the average adult, buyers aged 55 and above accounted for 41 percent fewer retail beer purchases than the average adult, and men accounted for approximately 10 percent more retail beer purchases than women. Beer is often a convenience purchase both for these young male consumers and beer consumers as a whole. The same study showed that nearly 2/3 (66%) of beer purchases were made through convenience stores, liquor/package stores, or drug stores, while just slightly more than 1/3 (34%) of beer purchases took place through destination shopping locales such as supermarkets or mass merchandisers.

Brewers target this young, male demographic and design their advertising and marketing campaigns accordingly, which is why most beer advertisements feature sports, attractive women, and humor. However, the rapid proliferation of beer varieties has also been accompanied by a proliferation of target marketing, with brands and marketing campaigns designed for all sorts of beer drinkers. All major brewers also offer beverages such as ciders, malt beverages, prepared cocktails, and wine coolers to be sure to appeal to all segments of the alcohol-consuming public.

FIGURE 20

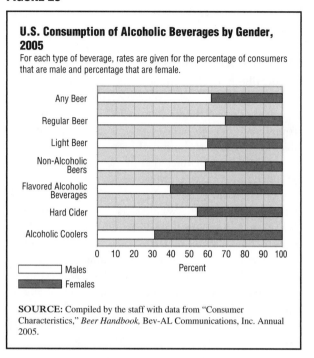

U.S. Consumption of Alcoholic Beverages by Gender, 2005

For each type of beverage, rates are given for the percentage of consumers that are male and percentage that are female.

SOURCE: Compiled by the staff with data from "Consumer Characteristics," *Beer Handbook,* Bev-AL Communications, Inc. Annual 2005.

ADJACENT MARKETS

The most obvious markets adjacent to the beer market are those for other alcoholic beverages. In the United States beer has always held a strong market position relative to other alcoholic beverages based on per capita consumption. Beer has consistently accounted more than 80 percent of all alcoholic beverage consumption in the United States based on per capita consumption figures. For the period 1966 through 2005 it averaged 86.5 percent according to the U.S. Department of Agriculture's Economic Research Service. However, over this period, wine gained market share, increasing its per capita consumption from 1 gallon in 1966 to 2.4 gallons in 2005.

Wine is an alcoholic beverage produced from the fermentation of unmodified grape juice. Grapes have a natural chemical balance which allows them to ferment without the addition of sugars, enzymes, or other nutrients. The top wine producing countries include, in order of 2005 production, Italy, France, Spain, Argentina, and the United States. Wine sales in the United States surged beginning in 1970 and were then flat or falling during the 1980s. The declines were attributed to the nationwide increase of the drinking age to 21, health and fitness concerns that discouraged alcohol consumption, label warnings for sulfites and for pregnancy, and lowered state blood-alcohol levels in driving-under-the-influence laws.

Beginning in 1994, sales began to climb steadily again and were continuing to grow in early 2007, when the report was written. The council attributed the renewed growth to a number of factors, including new reports that moderate wine drinking is good for health, increased and improved marketing campaigns, and the fact that the large millennial generation (those aged 21-30 in 2006), children of the Baby Boomer generation, was adopting wine drinking at a younger age than earlier generations.

Wine is still a distant second by volume within the three-product alcoholic beverage category. By value of shipments, however, it is much closer to beer, generating nearly half the domestic, wholesale shipment value that was generated by beer shipments in the United States in 2005. The value of domestic wine shipments in 2005 was $10.5 billion, according to the Census Bureau, while beer shipments were $20.8 billion, exclusive of imports for both products. Brewers are watching the growth in popularity of wine closely.

Distilled Spirits. The third product category within the larger alcoholic beverage sector has been declining. The consumption of distilled spirits or hard liquor in the United States has been on the decline for several decades. In 1966 the per capita rate of consumption of hard liquor was 1.6 gallons, exceeding wine's one gallon per person annual consumption rate in the United States that year. Hard liquor per capita consumption rose during the 1970s to a high of 2 gallons, and then began to decline. As of 2005 the annual consumption of distilled spirits in the United States stood at 1.4 gallons per person, well below the 2.4 gallon level of wine.

New flavors are being used to try and attract a new and younger consumer base for hard liquor. A new and younger consumer is seen as a way to counter act the decline in consumption seen in recent decades and often attributed to the aging of the population and concerns about health generally. Home distilling is also a trend that promises to increase demand for hard liquors as more people experiment with making craft liquors and acquire a taste for the drink.

Containers. The glass bottle and aluminum can markets are also important adjacent markets to the beer market. For producers of both of these types of containers, brewers continue to be important buyers. In fact, glass beer bottles have so far been a notable holdout in the transition of beverage bottles from plastic to glass.

Other Adjacent Markets. Beer advertisements are an important revenue stream for broadcasters, print, and Internet media organizations that in turn support a wide variety of entertainment activities, most notably professional sports. Anheuser-Busch alone spent $466 million on television advertisements in 2006, according to ACNielsen Monitor-Plus.

RESEARCH & DEVELOPMENT

Though it may seem there is little room for innovation with a traditional product such as beer, both manufacturing pressures and competitive markets have led to continuous innovation by brewers. As a manufacturing industry, large-scale brewers have taken advantage of many of the same sorts of technological innovations that have benefited other manufacturers. Where once a brewmaster relied on just his senses and experience to brew a batch of beer, now increasingly sophisticated control systems monitor everything from the germination rate of malting barley to the specific gravity of worts, allowing the brewmaster to precisely adjust the brew throughout the brewing process to ensure a more consistent product. Meanwhile, craft breweries and microbreweries consistently show an unending desire to try new and different beer recipes, experimentation which constantly introduces new tastes into the market.

The St. James Gate Brewery, maker of Guinness Stout, is a good example of the combination of tradition and innovation to be found among large-scale brewers. The St. James Gate brewery was founded in 1759 and in 1901 they established a research laboratory of their own to improve the manufacture and delivery of its beers, and the success of that laboratory has created a global market for Guinness Stout. Whereas once a *true pour* of their stout could only be found in Ireland, improvements in the taps used to pour Guinness Stout mean that the modern consumer can now find a good pint nearly anywhere in the world. A similar smooth and creamy pint of stout can also now come from an individual can or bottle, thanks to a small nitrogen pressurization widget that Guinness introduced into cans in 1988 and later to bottles in 1999. Other manufacturers such as S&N, Bass, Courage and Carlsberg-Tetley soon followed the 1988 introduction of Guinness's nitrogen widget with widgets of their own, and the technology can now be found in a wide variety of stouts, porters, bitters, and other draught-style beers.

As with widgets, much of the innovation in the beer industry that is most apparent to the consumer comes in the form of retail packaging designed to improve the delivery of an individual beer to a consumer. In 2005 Heineken introduced a portable draught keg that contains 4.75 liters of beer under pressure from carbon dioxide. Once opened, the keg stays pressurized for up to 21 days, which means that individual glasses of beer can be poured at a consumer's leisure. Many brewers have also introduced dates to their labeling. Because the quality of beer deteriorates over time, labeling that indicates the age of an individual bottle can be of great help to the consumer seeking the freshest possible beer.

Glass-lined cans have been another important development for the beer industry. Designed by the Ball Corporation, glass lined cans produce a similar taste to bottled beers, thus unaffected by its aluminum container. The glass-lined cans allow producers to package their beer in a package like aluminum to allow for a lower breakage rate than bottles, while protecting its taste. Consumers are also able to bring beers camping, hiking, and boating that they would not normally be able to due to the packaging. Oskar Blues Cajun Grill and Brewery in Lyons, Colorado, became the first brewpub to brew and can its own beer in 2002. The brewpub utilizes glass-lined cans produced by the Ball Corporation in Golden, Colorado.

CURRENT TRENDS

The U.S. craft brewing renaissance has had a significant affect on the market. Craft brewing is the fastest growing segment of the alcoholic beverage industry. While 2004 proved to be a difficult year for the industry as a whole, the craft beer segment grew at a rate of 7 percent. Moreover, craft brewers have been able to grow at the expense of importers during 2000–2004.

In 2006 microbreweries made up 10 percent of the U.S. beer market, with over 1,300 microbreweries. Sales increased by 16 percent from 2005 to 2006. Brewpubs, restaurants and pubs with an on premises brewery that serve the majority of their beer to customers on site, are gaining popularity and growing along with the beer industry in the twenty-first century renaissance. The craft beer industry, such as microbreweries and brewpubs, grew by almost 12 percent in 2006. In recognition of this growing segment, beginning in 2000, Sam Adams brewery started sponsoring a contest for the best home brewed beer. Each year since then there have been three winners whose beer has been put into production as a special brew and distributed by Sam Adams.

Large-scale brewers have countered an ever-growing number of small competitors by introducing new and different beverages as well as by taking popular brands and extending the brand across multiple products. Where Budweiser once stood alone as "The King of Beers," the king now has an increasingly crowded court that includes Bud Light, Bud Ice, Bud Dry, and Budweiser Select. In this increasingly crowded market, brewers must also fight for market share amid a declining overall share of the alcoholic beverage market. Beer sales fell from 56 percent of the alcoholic beverage market in 1999 to just 53.2 percent of the market in 2004, according to the Distilled Spirits Council of the United States. Total sales of beers and flavored malt beverages increased by just 1 percent from 2003 to 2004 according to the National Beer Wholesalers Association, and beer sales by themselves were actually down in that year.

For the foreseeable future brewers will continue to extend brands and introduce new beverages to compete in this increasingly diverse marketplace.

TARGET MARKETS & SEGMENTATION

As with many products and services, the next big, national market on the horizon for the beer industry is China. In fact for the first time ever, China consumed more beer than any other country in the world starting in 2003. Its enormous population accounts for the top ranking, but the rise of beer consumption in China also speaks to a developing taste for the drink. In 2003 China consumed 25 million kiloliters compared with 23.8 million kiloliters consumed in the United States. China's consumption rose to 28 million kiloliters in 2004 while the U.S. consumption figure remained reasonably steady at 24 million kiloliters.

On a per capita basis, neither the United States nor China even rank in the top ten by country. The Czech Republic, Ireland, and Germany ranked first, second and third in terms of per capita beer consumption in 2004. Their respective rates of per capita consumption were 157 liters, 131 liters, and 116 liters. The United States had a per capita consumption of 81.6 liters in 2004 and China's rate was 22.1 liters.

RELATED ASSOCIATIONS & ORGANIZATIONS

The Beer Institute, http://www.beerinstitute.org

Brewers Association, http:// www.beertown.org

British Beer and Pub Association, http://www.beerandpub.com

EBCU—European Beer Consumer's Union, http://www.ebcu.org

Mothers Against Drunk Driving, http://www.madd.com

National Association of Beverage Importers, http://www.nabi-inc.org

National Beer Wholesalers Association, http:// www.nbwa.com

BIBLIOGRAPHY

"Average Number of Monthly Beer Purchases at Different Retail Channels." *Beverage Aisle.* 15 May 2004, 1.

"Beer Consumption by Country." Kirin Company. Available from <http://www.kirin.co.jp/english/ir/news_release051215_2.html>.

Beirne, Mike. "Special Report: Beer." *ADWEEK Online.* 30 April 2007.

"Breweries: 2002." *2002 Economic Census.* U.S. Department of Commerce, Bureau of the Census. January 2005.

"Brewers to Get New Technology to Rate Malting Barley." *The Kiplinger Agriculture Letter.* 24 June 2005.

"Consumer Characteristics." *Beer Handbook.* Bev-AL Communications, Inc. 2005, 9.

"The Demographics of Beer." *Beverage Dynamics.* July-August 2005, 2.

Dennis, Mike. "What's Brewing in Widgets?" *Super Marketing.* 30 June 1995, 30.

"France: Heineken Launches Portable Draught Keg." just-drinks.com. 26 April 2005.

Gander, Paul. "Chapter 3 Glass Packaging." just-food.com. April 2003, 6.

"A History of Beer." Birmingham Beverage Company. Available from <http://www.alabev.com/history.htm>.

"Residential Population by Age and Sex: 1980 to 2005." *Statistical Abstract of the United States: 2007.* U.S. Department of Commerce, Bureau of the Census. December 2006, 12.

SEE ALSO *Distilled Spirits, Wine*

Bicycles

——————◼——————

INDUSTRIAL CODES

NAICS: 33–6991 Motorcycles, Bicycles and Parts Manufacturing

SIC: 3751 Motorcycles, Bicycles and Parts

NAICS-Based Product Codes: 33–69911 through 33–69911122

PRODUCT OVERVIEW

A bicycle is a vehicle with two wheels. The bicycle rider provides the energy to propel the bicycle. The use of foot pedals attached directly to one of the wheels, or indirectly through a chain and geared mechanism drives the bicycle forward. All bicycles—also called bikes for short—have handlebars to aid in steering, brakes of some sort, and most provide a seat for the comfort of the rider.

The history of the bicycle begins with Baron Karl Drais von Sauerbronn. The German inventor is credited with a number of inventions, including the meat grinder, the stenotype, and the 25 key typewriter. In 1817 he invented a two-wheeled device known as a Laufmaschine, *running machine.* The contraption was made of wood, had an upholstered seat, and weighed roughly 50 pounds. It lacked pedals, however; a rider had to push his or her feet along the ground in order to move forward. Drais received a patent for the device a year later.

The Laufmaschine device was only briefly popular, but it inspired inventors across Europe. It became known as a *Draisine* in Germany and a *Draisienne* in France. In England the device was known as a *hobby horse,* after the child's toy, or the *dandy horse,* after the foppish, affluent men who treasured novelty and were the primary owners. Despite its curiosity this primitive bicycle caused problems. Riders found their boots wore out prematurely from the constant pushing. Streets were in poor condition in the nineteenth century, which made for an unpleasant ride and the seating could not be adjusted and caused a number of accidents between riders and pedestrians. The devices were also hard to steer. By 1819 the device's popularity had faded in London.

Inventors continued to experiment with designs. Scottish blacksmith Kirkpatrick Macmillan is credited with inventing the bike pedal in 1840. Several historians suspect the pedal was attached to three- and four- wheeled vehicles, however; it remains unclear if it was attached to a two-wheeled vehicle. The velocipede (*fast feet* in French) was introduced at the start of the Victorian Age. Clearly inspired by the carriage, the device typically had two, three, or four wheels and levers to work the wheels and brakes. Its frame was made of iron and its wooden wheels varied in size. These vehicles also gained some popularity. But there were still speed and suspension issues. The two-wheeled version was known as the Boneshaker in England, a nickname that offers some insight into the ride enjoyed by Victorian cyclists.

Pierre Lallemont, who worked as a baby carriage and wheelchair maker in France, is thought to have made designs for a two-wheeled bicycle in 1863 or 1864. He sold the designs to the Olivier Brothers, who in turn engaged Ernest and Pierre Michaux to begin production. The Michauxs, it should be noted, are believed to have played some role in the new design; indeed, a few early sources actually credit them with inventing the modern bicycle.

They are believed to have drawn upon their expertise in wagon and carriage building and improved the vehicle's suspension, for example. Lallemont left for the United States, where he was given the first U.S. patent for a bicycle in 1866. Lallemont returned to France two years later to find Europe in the midst of a bike craze.

The Modern Bike. More design innovations followed. Frenchman Eugene Meyer's model, known as the first high wheel bike, was produced in 1869. This was a modification of the Boneshaker design, with a large front wheel (limited only by the length of the rider's leg), a small rear wheel, and a lighter metal frame. Such bikes were known as ordinary or penny farthing bikes in England. The nickname came from coins used at the time. The penny was a large coin and the farthing significantly smaller one; held next to each other, they resembled a side view of Meyer's bike. The bike was first produced in the United States in Boston in 1878. While this new model enjoyed popularity, it also had problems: it was difficult to mount, dismount, and ride, and because of its large front wheel a rider could easily lose his balance and fall.

James Starley was a prominent British bicycle maker. He was in the sewing machine business, but was also an inventor, and is credited with the invention of the tricycle. In 1868 his company Turner and Starley started making penny farthing bikes. Starley's nephew John Kemp Starley went to work for him. In 1885 the younger Starley built his Rover bike. Highly influential, Starley's Rover is seen as the first modern bicycle. It had two 26 inch wheels (still a standard bike wheel size), a diamond shaped tubular frame, tangential spokes, ball bearings in the wheels and cranks, and pneumatic tires. It also held another crucial element: the chain drive. Chain drives consist of a chain moving over a sprocket gear; the device allows for power to be evenly distributed on the wheels of bicycles, motorcycles, and other vehicles. The principle was not new, but Starley was the first to use it on the bicycle.

The bicycle started to be mass produced in the United States about this time. According to government and trade association estimates, bike production jumped from 800,000 in 1895 to two million in 1897; 1.1 million bikes were produced annually from 1899 to 1903. The growth of the bike market sparked more innovations. Dr. John Boyd Dunlop began to explore commercial applications for the pneumatic tire, which had been developed in 1845. The first aluminum bicycle, far lighter than steel framed bikes, was made in France in 1890. William Reiley invented the two-speed mechanism in 1896; his work would be used to create the three-speed, which followed in 1902. New types of pedals and toeclips were patented. Peugeot produced the first recumbent bicycle in 1914. A recumbent bicycle is one that is designed to provide the rider with a more upright position from which to sit and peddle the vehicle.

The bicycle market faced major challenges from the automobile and motorcycle as America's preferred mode of transportation. There were 300 firms involved in bicycle production in the late 1890s; by 1905 there were only 12 manufacturers left. Over the coming decades, children would become the primary users of bikes. During this period bicycles were mass produced and quite similar in design. The Schwinn Bicycle Company set out to make a new kind of bike in the 1930s, emphasizing style and quality. Their new bikes were very popular and helped stimulate the market. The industry would see another surge in interest in the 1960s and 1970s when people became more interested in recreation, fitness, and environmental causes. Bike production climbed from 6.8 million in 1970 to 15.2 million in 1973. The interest in fitness continued into the 1980s. Mountain biking and cycling became popular sports activities; spandex bike apparel even became part of mainstream fashion by the end of the decade. The bike industry became much more specialized in the 1980s. Some bikes were designed for road use, others for racing, and others were meant for rough mountain terrain. Various advances in plastics, composite metals, and materials processing were used to achieve these new lines of bikes. Companies such as Specialized, Cannondale, and Trek all got their start in niche markets during this period.

MARKET

Early Production History. Colonel Albert A. Pope of Boston was a great admirer of the high-wheeled bike when he first saw it in the late 1870s. The businessman and philanthropist worked with Englishman John Harrington to import bikes into the United States. Pope soon became interested in building his own bikes, however. Pierre Lallemont of France registered a patent for a two-wheel bicycle in the United States in 1866. He formed a partnership with James Carroll but was unable to make his venture successful. Carroll's patent was sold several times and was eventually split between carriage maker Richardson and McKee and the Montepelier Manufacturing Company. Pope acquired the patent rights from these companies. For the actual production of his bicycle, however, he turned to the Weed Sewing Machine Company of Connecticut. The company and indeed this region of Connecticut had earned a reputation as specialists in metallurgy and the mass production of metal parts.

Pope's company produced 50 bikes in 1878 under the brand name Columbia. The Rover, or safety bicycle, was produced in England in the late 1880s (design improvements made this bicycle safer than its predecessors). The various design improvements to the Rover helped boost the bicycle market in England and France. Pope produced his first safety bike in 1888; his factory was now producing far more bicycles than sewing machines.

The number of bike makers in the United States climbed from 27 in the early 1890s to 300 in 1896; a third of these were located in the Chicago area. Pope had also been an early advocate of better roads so that bicycle riding could be more enjoyable. In the 1890s new bike owners, setting out on the bumpy, unpaved roads for the first time, joined his cause. Women became bike riders, granting them greater freedom from the home. A bike was a significant purchase in the late nineteenth century. Pope's Columbia was priced at $95 in 1888; a racing bike from Schwinn cost $150 in 1902 (the equivalent to more than $27,000 in current dollars). But Adolph Schoeninger of the Chicago Western Wheel Works used sheet metal works and other innovative production methods to produce a bike that would be much more affordable for the average family.

The emergence of the automobile in the 1920s was another challenge for bicycle makers. In fact, many bicycle designers and engineers moved to the automobile industry as it developed. According to government and trade association figures, annual bike production fell from one million units in the 1890s to a low of 250,168 in 1927. In the 1930s Schwinn created a department to focus solely on improving bike and motorcycle design and overall quality. Schwinn is credited with reinvigorating the industry, and would become a market leader.

There were improvements to the bicycle in the 1940s and 1950s. The three speed gearing system, hand brakes, headlamps, kickstands, and reflectors all became more common bike features. Many bikes were produced for children, a growing segment of the population. According to the Census, there were 47.3 million children 0–17 years of age in 1950; the figure climbed steadily to 64.5 million in 1960. With the post World War II move to the suburbs, bicycles got another boost as they offered children greater mobility and were thus quite popular.

During this period, President Eisenhower relaxed tariffs on specific products. Bicycle makers started hearings with the U.S. Tariff Commission in 1954 to ask for higher tariffs and quotas on foreign bikes. Their alarm was understandable. Bicycle imports climbed from 67,789 in 1950 to 1.2 million in 1955, according to government and trade industry sources. Imports represented approximately a third of the market during this period. This figure dropped during the bike boom of the 1960s to just under a quarter of the market. As bike makers began to move their operations outside the United States in the 1990s, the U.S. market has become overwhelmingly composed of imports.

The Bike Boom. Many historians describe the 1960s and 1970s as a bike boom. Sales did indeed increase. There were 3.7 million bikes sold in 1960. In 1973, bike sales climbed to 15.2 million. There are a number of reasons for this boom in sales. There were a growing number of young people in the country. There was also an increased interest in physical fitness. The President's Council on Youth Fitness was formed in 1956. President Kennedy changed the organization's name to the President's Council on Physical Fitness in 1963 to emphasize the need for all Americans, not just children, to be physically fit. Schwinn released the Sting-Ray brand in 1963 and sold approximately 40,000 in the first year. The 10 speed already existed in Europe but became increasingly popular in the United States. Americans were also becoming more conscious of their environment. In 1970 the first Earth Day was held to call attention to various environmental concerns; bike riding was seen as one way to be environmentally friendly. For some Americans, bike riding became a necessity. The Oil Embargo of 1973 brought on gas shortages and higher prices and forced many Americans to find alternate transportation methods.

Certain sectors of the market received greater attention in the 1980s. Amateur cycling became more popular. In 1984 women's cycling was added to the Olympics for the first time. That same year the first La Grande Boucle Feminine Internationale was held in France; it is sometimes called the Tour de France for Women. The origin of BMX (bicycle motocross) racing actually begins in 1971. In that year the motocross documentary *On Any Sunday* was released. In its opening scenes boys are shown on their Sting-Ray bicycles emulating the stunts performed by motocross riders. BMX racing is thought to have initially existed in California. The industry appears to have developed quickly; the first professional race was held in 1975. In 1982, the young heroes of *ET: The Extra Terrestrial* were seen riding BMX bikes and are seen riding in silhouette against the moon in that film's highly iconic shot. This film is given some credit for moving the BMX bike into the mainstream. Mountain biking became popular in the 1980s. Some people had used bikes on mountain trails as early as the 1930s. But Joe Breeze is credited with really moving the mountain bike industry forward in 1977. His Breezer bikes had rugged frames and fat, durable tires that were well suited to difficult terrain. The industry developed into the 1980s and 1990s along with many extreme sports.

The Bike Market in 2006. The U.S. bicycle industry was a $5.8 billion industry in 2006, including the retail value of bicycles, related parts, and accessories through all channels of distribution, according to research funded by the National Sporting Goods Association. Bikes represent $2.6 billion of this total. Sales in the industry have been fairly consistent with previous years. The industry saw estimated sales of $5.3 billion in 2002, $5.4 billion in 2003, $5.8 billion in 2004, and $6.1 billion in 2005 (an all-time high). In 2006, 18.2 million bikes were sold in the United States, a number that has seen little change since 1991.

Approximately 2,000 companies were involved in the manufacturing of bikes and related accessories in 2006. These companies market approximately 200 brands across the industry's specialized sectors. Sources estimate that nearly 99 percent of the U.S. market consists of imports. Of the 18.1 million bikes sold during the year, only 195,000 were made domestically. Approximately 60 percent of these bikes were made in China and 27 percent were manufactured in Taiwan. Some firms do maintain small domestic manufacturing operations. Also, there is a small but growing base for the manufacturing of custom bike frames.

The Gluskin Townley Group and the *Cycling Consumer of the New Millennium Report* survey the state of the bicycle industry each year. Their reports published in 2006, quoted in *Bicycle Retailer and Industry News*, note that Trek was the largest producer and importer of bicy-

FIGURE 21

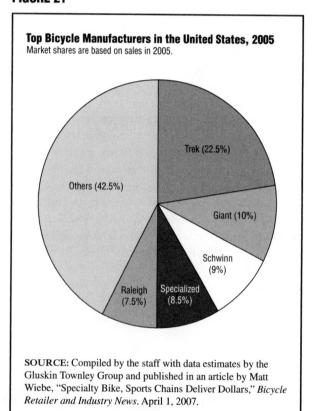

Top Bicycle Manufacturers in the United States, 2005
Market shares are based on sales in 2005.

SOURCE: Compiled by the staff with data estimates by the Gluskin Townley Group and published in an article by Matt Wiebe, "Specialty Bike, Sports Chains Deliver Dollars," *Bicycle Retailer and Industry News*. April 1, 2007.

FIGURE 22

Specialty Bicycle Sales in the United States by Type, 2004–2006
Market shares are shown as percentages based on unit sales.

Type of Bike	2004	2005	2006
Mountian Bike	32.9	28.8	28.5
Comfort Bike	15.5	14.2	14.0
Hybrid/Cross Bike	12.5	12.5	15.0
Cruiser	3.9	5.5	6.0
Road Bike/700C	10.8	16.4	17.0
Youth Bike	24.0	18.8	16.5
Other	2.0	2.4	2.0

SOURCE: Compiled by the staff with data coming from the U.S. Department of Commerce and estimates prepared by the Gluskin Townley Group and published in the National Bicycle Dealers Association's *Industry Outlook 2006,* available from http://nbda.com/page.cfm?PageID=34.

cles in the United States in 2005. It sold 575,165 vehicles and had a 22.5 percent market share. Much of Trek's operations are still in the United States. The country is its major market; the company has less than 4 percent of the global market.

Giant was the next largest company. It sold 255,629 bikes and had a 10 percent market share. Schwinn sold 230,066 bikes, which gave it a 9 percent share. Specialized sold 217,284 bikes and had an 8.5 percent market share. Raleigh rounded out the top four with sales of 191,722 bikes and a 7.5 percent market share.

Bikes in the United States are purchased primarily for recreational purposes; according to trade industry figures, only 5.5 percent of bikes were purchased for uses other than recreation, such as racing and transportation. The categories shown in the table require some explanation. A road bike is a bike intended for use on a paved road. Types of road bikes include racing, utility, and touring. A hybrid bicycle is a cross between a mountain and road bike. A comfort bike is seen as a subset of this category. Both hybrid and comfort bikes might have hub rather than derailleur gears, front suspension forks, and seat post suspension. A hybrid bike typically has larger wheels than a comfort bike. These are relatively new, ill defined categories, and there is overlap between the categories. Mountain bikes are the most popular type of bike in the United States. They have lost some market share. Part of the reason may be the loosely defined categories—some hybrid bikes are also mountain bikes, for example. But foreign competition is a major problem. Mountain bikes from China and Vietnam touched off price wars in some areas.

KEY PRODUCERS/MANUFACTURERS

Pacific Cycle. Chris Hornung founded Pacific Cycle Inc. in 1977. The company began importing bicycles made in China and other Asian countries into the United States. Pacific developed a reputation as a distributor of high quality low-cost bikes. The company broadened its product lines and in the 1990s began distributing its bicycles and related merchandise at Wal-Mart, Target, Toys R Us, and other mass merchandisers. In 2000 the company acquired the Roadmaster and Mongoose brands. Mongoose was a popular brand with young BMX bike riders, and Pacific's distribution relationships moved this brand into the mass market category.

In 2001 Pacific acquired the Schwinn and GT brands. Schwinn is perhaps the most recognizable bike brand on the market. The brand had been a market leader for some time, but had failed to capitalize on the BMX bike movement. As well, the brand had suffered by being available only at independent bike dealers. Pacific introduced Schwinn to the mass market and GT to the sporting goods channels for the first time in 2002. In 2004 Pacific was acquired by Dorel Industries Inc., a global supplier of juvenile and home furnishings. Brands owned by Pacific include Schwinn, Mongoose, GT, Schwinn Motor Scooters, Kustom Kruiser, Roadmaster, Pacific, Dyno, Powerlite, InSTEP, and Pacific Outdoors. Pacific Cycle is located in Madison, Wisconsin. Dorel is headquartered in Montreal and employs approximately 4,700 people in 15 countries. The company reports annual sales of $1.8 billion.

Trek Bicycle Corporation. Richard Burke and Bevel Hogg established the Trek Bicycle Corporation in 1976. The company began producing stylish framesets that incorporated European and American production methods. Their first framesets sold for $275 and were only available in independent bike stores. This helped the Trek name develop a sense of exclusivity, and the company received good word of mouth among customers. Trek's focus had been on the road bike market. In 1983 the company took some preliminary steps into producing mountain bikes. It developed an aluminum bonded bike that was very popular in the 1980s. At the same company leaders put a greater emphasis on the company vision and added dealers and strengthened their customer service policies. By 1996 the company was the leading bike maker in the specialty channel. Mountain bikes represented 80 percent of its product line. It was servicing 1,500 dealers. Trek became a sponsor for bike racing and in 1997 it signed Lance Armstrong. Armstrong became the first American to win the Tour de France in 1999. The model he was riding, the Trek's OCLV Carbon 5200, saw a jump in sales

after Armstrong's win. The company is based in Waterloo, Wisconsin.

Giant Manufacturing Company Ltd. Giant Manufacturing was established in Taiwan in 1972. The company began manufacturing bikes to be sold under other company brand names. It became the leading bike maker in Taiwan in 1980. In 1986 it began making bikes under its own brand name. It manufactures bikes for competition, exercise, recreation, and transportation. It also makes biking apparel and accessories. The company's products are sold in bike shops in more than 60 countries and it operates 20 of its own stores. Giant Manufacturing also makes bikes for other popular brands through its four manufacturing facilities in Taiwan, China, and Europe.

Schwinn. Ignaz Schwinn started the Arnold, Schwinn Bicycle Company in 1895. After the end of the bike boom at the turn of the century, Schwinn bought up the operations of some failed bike makers. The company initially made some motorcycles. But Frank W. Schwinn, Ignaz's son, decided to start making bikes that distinguished themselves from the competition. Schwinn became an industry leader through its innovative designs. Schwinn introduced the bike balloon tire in 1933, which quickly became the industry standard. In 1946 it introduced built in kickstands. In 1963 the company introduced its popular Sting-Ray. Roughly 40,000 Sting-Rays were sold in the first year and by 1968 and other bike makers copied the bike's high rise handlebars, banana seat, and stickshift. Schwinn ventured into other recreational equipment markets; in 1965 it introduced the first in-home workout machines.

Schwinn continued to make heavy steel bikes; the growing market, however, was in the lightweight Asian bikes being imported into the country. The company lost market share. In 1951 Schwinn had a 22.5 percent share based on dollar sales. By 1961 its share had dropped to 12.8 percent. Then, in the 1960s Schwinn made some controversial management decisions. In 1967 the U.S. Supreme Court ruled that Schwinn had engaged in price fixing and other illegal dealings. Schwinn set up its own warehouses and distribution process after the ruling in the hopes of better controlling the way its products were sold and marketed. Many analysts feel this move alienated them from market trends. The company failed to embrace the mountain and BMX market of the 1980s, for example. Schwinn filed for bankruptcy in 1993. In the mind of a young bike shopper a Schwinn bike is libel to be considered the bike his or her parents rode, and the company is working to come up with more engaging designs to attract the attention of this important customer—a young biker.

MATERIALS & SUPPLY CHAIN LOGISTICS

The standard bicycle frame, often referred to as a diamond shaped frame, can be broken into two major sections. The front of the frame holds the top, seat, down, and heat tubes. The rear holds the chainstays, seatstays, and rear wheel dropouts. The first bike frames were made from wood. Manufacturers later made the frame from steel. This increased the bike's durability, but the heaviness of this metal added considerably to the bike's weight. It was not until the 1970s that manufacturers were able to produce steel alloys that were both light and durable. The brakes, chains, and wheels are generally made from stainless steel. They are purchased separately by the bike maker and then attached to the frame.

How a Bike is Made. The production of a bike begins with the manufacturing of flat steel. The flat steel is then shaped around a tube and welded together. The thickness of the tube typically varies in places. It is thicker at joints and stress points to provide greater durability. The thickness may be reduced in places; this reduction helps reduce the bike's weight. The tubes then go through a number of processes to make sure the tubes are of the appropriate strength and weight. Joints are then sealed, and the frame goes through a process of alignment before being painted. The various components such as the seat, handlebars, stems, brakes, and headsets are attached to the frame. Their placement varies depending on the type of bike. The gear shift, for example, may be attached to the down tube, the handlebars, or the stem. These parts are typically made of an aluminum or steel alloy. The wheels are attached last.

DISTRIBUTION CHANNEL

Bicycles are sold primarily through three main distribution outlets: mass merchandisers, specialty bike chains, and chain sporting goods stores. Companies market specific brands to each of these channels and there is very little crossover between them; certain specialty bikes, for example, are unlikely to be found at a mass merchandiser. Trek's bikes are found only in specialty stores and are not available for sale online.

Mass Merchandisers. Large, mass merchandisers such as Kmart, Target, Toys-R-Us, and Wal-Mart sold 13.9 million bikes, or three quarters of all bicycles sold in the United States in 2006. These retailers controlled such a large market share in part because their bikes were inexpensive. The average price of a bicycle at a mass merchandiser was $72 in 2006, far below the average $450 price at a specialty bike retailer or the $225 price at a sporting goods store. Approximately 80 percent of children's bikes are sold through the mass merchandising channel.

Specialty Bicycle Stores. There were 4,500 independent bike dealers (the local bike shop) in 2006. This figure includes performance retail stores. These stores sold 2.5 million bikes in 2006, representing 14 percent of industry unit sales. The National Bike Dealers Association (NBDA) estimated there were approximately 8,000 independent bike dealers in the early 1980s. The bicycle retail industry loses as many as 1,000 retail shops each year, but the loss is tempered by a significant number of start-up operations that enter the industry annually. The NBDA estimates that the average bike dealer needs a 36 percent profit margin to cover his or her costs and break even. The average profit margin falls far below 36 percent, standing at something just shy of 5 percent in the middle of the first decade of the 2000s according to the NBDA.

According to an NBDA report titled *2004 Cost of Doing Business*, the average specialty bicycle retailer had gross sales of $550,000 per year. Roughly 91 percent of them had one location. The average store size was 4,822 square feet. Revenues for most bicycle dealers were generated on the sale of bicycles (47%) although bike parts and accessories represented 35 percent of their revenues. The remaining 18 percent of revenues were generated by bicycle repair services (11%), bicycle rentals (2%), and by the sale of other product holdings such as fitness equipment (5%). The average store sells approximately 650 bicycles

FIGURE 23

Number of U.S. Specialty Bicycle Shops, 2000–2007

Legend:
- Number of specialty bicycle retailers (left axis)
- Total number of bicycles of any size sold (right axis)

SOURCE: Compiled by the staff with data coming from the U.S. Department of Commerce and estimates prepared by the Gluskin Townley Group and published in the National Bicycle Dealers Association's *Industry Outlook 2006*, available from http://nbda.com/page.cfm?PageID=34.

per year, carries five bicycle brands, and numerous accessories. Approximately 80 percent of adult bikes are sold through this channel.

Sporting Goods Chain Stores. The large sporting goods chains such as Dick's Sporting Goods and the Sports Authority sold 1.1 million bikes in 2006. Their bike sales were worth $246 million. While these figures are substantial, sporting goods chains represented the smallest portion of the overall bicycle sales in the United States, just 6 percent of unit sales and 9.3 percent of dollar sales. Sporting goods chains typically stock a wide range of bikes and related equipment; their selection is generally smaller, however, than the specialty retailer.

Approximately 550,000 bikes were sold through all other channels in 2006, including the Internet, mail order, hardware stores, and local/regional retailers. These locations represented just 3 percent of unit sales. Bike sales in these outlets totaled $178 million, or 6.7 percent of retail sales.

KEY USERS

According to the National Sporting Goods Association (NSGA), bike riding was the eighth most popular sports activity in 2006. A total of 35.6 million Americans age 7 or older reported riding a bike at least once during the year (walking and swimming were the most popular sports). Women represented 45 percent of this total. More people reportedly rode a bike than played basketball (26.7 million), played golf (24.4 million), or went jogging (28.8 million). When broken down by age one sees that 16 million cyclists were younger than 18 in 2004. Approximately 18 million were 18 to 34 years of age, 6 million were 35 to 44 years of age and 7.8 million over 45 years old. Figure 24 presents bicycle ridership figures in the United States by age group.

Lance Armstrong brought visibility to the sport in much the same way as Tiger Woods did for golf and Michael Jordan did for basketball. After Armstrong's Tour de France win in 1999 more people briefly took up the sport. However, bike riding is on the decline. According to NSGA figures, the number of people bike riding has fallen from 53.3 million in 1996, despite the fact that a good number of Americans developed a real love of the activity. The number of licensed racers reportedly doubled to 5,400 between 2002 and 2006. As of early 2007 there were 1,112 cycling clubs in the United States. Bike tours were a popular part of many vacationers' travel plans.

Interest in cycling grew in specialized categories such as mountain biking or BMX riding. According to the NSGA, 7.3 million Americans reported participating in mountain biking at least once in 1996; that figure climbed to 8.5 million in 2006. Such extreme sports are often seen

FIGURE 24

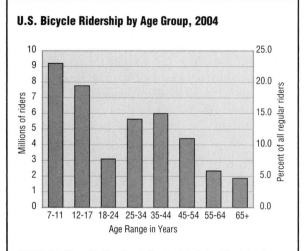

U.S. Bicycle Ridership by Age Group, 2004

SOURCE: Compiled by the staff with data from "Participation in Selected Sports Activities: 2004," *Statistical Abstract of the United States: 2007*, U.S. Department of Commerce, Bureau of the Census. December 2006.

as the domain of children or teenagers. This is not so. The average age of a frequent mountain biker (25 or more outings per year) is 30 years. Nearly 40 percent of regular mountain bikers reported having graduated from college. The average BMX cyclist was 24.7 years of age.

ADJACENT MARKETS

The bike industry is closely aligned with several other industries. Nearly half of the overall bike market is related to sales of bike accessories such as tires, helmets, apparel, pedals, and footwear. The National Sporting Goods Association considers bicycles a subset of the recreational transport industry, which had an estimated $36.7 billion in sales in 2005. This industry includes pleasure boats, snowmobiles, and recreational vehicles. It has more than doubled since 1990, when it registered sales of $14.5 billion. The recreational transport sector, in turn, is part of the sporting goods market, which includes gear and footwear, and had estimated sales of $86.7 billion in 2005.

Bicycles offer wide opportunities for large urban areas. Many cities, for example, are sending police officers on bike patrol. Having law enforcement officers on bicycles is a way to increase their visibility and help them to access pedestrian areas more easily while encouraging a gentler, kinder image of the police in general.

Bicycle delivery services are another business that builds on bicycles by using them to maneuver the busy roads of a populous urban center. There were roughly 300 bicycle delivery firms working in New York City in the early 2000s. At the industry's height in the 1980s there were believed to be as many as 4,000 commercial cyclists in the city. The fax machine, email, and firms like Federal Express have replaced much (but clearly not all) of the business; the threat of bike accidents with pedestrians or vehicles also helped chip away at the business. According to Chicago's *Bike 2015 Plan*, that city's bicycle messenger companies employed more than 300 bicyclists who made an estimated 1.1 million deliveries annually. City officials planned to boost the use of bike messengers by 25 percent by 2015.

RESEARCH & DEVELOPMENT

Manufacturers have been adding more rider friendly features to bikes in the hope of attracting new riders. Automatic shifting has been around since the early 1990s. The technology had become more sophisticated in 2006 and 2007. In automatic shifting, a computer chip is built into the bike near the pedal. The chip measures wheel speed and the pedaling of the rider; it then automatically shifts into a higher or lower gear. Some have the ability to adapt to the rider's skills and shift gears during difficult turns.

In early 2007 bike makers released models offering clever storage features, improved brakes, coatings that can be peeled off to change the color of the bike, and provocative styling. Beginning in the 1990s businesses became more conscious of the need to offer products that were more comfortable for human use—a design concept known as *ergonomics*—and bike makers have adopted this design principle as well. In 2006 and 2007 bike makers released models with wider, more comfortable seats and adjustable handlebars with better designed grips. Companies returned to some of their old styles in the hopes of persuading Baby Boomers to return to the bikes of their youth.

CURRENT TRENDS

Global Trends. World bicycle production increased by nearly 9 percent to some 101 million units in 2003. China is a major market for bikes and indeed is a world producer, making 58 percent of all bikes sold worldwide. However, the country has begun to ban the bikes from their roads. With the country's booming economy, many Chinese purchased their first car in the early 2000s. Bike lanes were being cleared to make way for these and other vehicles. Shanghai, China, for example, had an urban population of 20 million in 2003; there were thought to be 9 million bikes in the city. The number of private vehicles in Shanghai nearly doubled to 142,801 by the end of 2003, according to China's National Bureau of Statistics. Bike usage dropped from 33 percent of trips in 1995 to 27 percent in 2000. According to the Cycling Association of China, each household owned, on average 1.43 bicycles in 2002, compared with 1.82 in 1998.

Certain populations have embraced bicycle riding wholeheartedly. The Germans owned 66 million of the 150 million bikes estimated to exist in Western Europe in 2006. Riders used the bikes for recreation as well as transportation. Many cities in Western Europe struggle to relieve traffic congestion and some governments have pushed cycling as one possible remedy. The Dutch, for example, were some of the most active cyclists in the region. The average citizen there traveled 2.4 miles a day on his or her way to and from work. In a working population of 6 million, 1.2 million people commuted to work by bicycle in 2003. Interestingly, a bike from Dahon, a U.S. company, was named Dutch bike of the year in 2006.

U.S. Trends. The bicycle is used primarily for recreation rather than transportation in the United States. According to the Census, 0.4 percent of workers used a bicycle to commute to work in 1990, 86.5 percent of workers took a car or carpooled, while the rest walked or took public transportation. In 2004 87.8 percent of workers commuted by car or carpooled. Only 1.4 percent reported using "some other means" to get to work other than public transportation. Few Americans have incorporated biking into their daily commute. This is largely due to the fact that the average U.S. worker lives far from his or her place of work. According to the Census Bureau the average U.S. commute lasted 25 minutes in 2004, suggesting a distance long enough to make commuting by bicycle an unrealistic goal for most people.

With the rise of gas prices that began in the early 2000s, and increased attention brought to environmental issues, bike use may increase. Organizations such as the League of American Bicyclists have lobbied state, local, and federal governments to enact policies that are more bike friendly. Their efforts appear to be succeeding. On August 10, 2005 President George W. Bush signed the Safe, Accountable, Flexible, Efficient Transportation Equity Act: A Legacy for Users (SAFETEA-LU). It authorizes as much as $4 billion in new federal spending on bike paths, trails, and related programs through September 2009.

TARGET MARKETS & SEGMENTATION

In the last century the bike market experienced several periods of strong interest: immediately before the turn of the century and in the 1960s and 1970s. The industry saw a smaller boom in sales in the 1980s when the market became more highly specialized, leading to the development of the BMX, racing, and mountain biking sectors. During its history manufacturers have made the bike lighter, more durable, and more attractive. There is little bike production in the United States. In 2006 nearly all bikes were imported. The decline in the market also seems to be reflected in bike riding in the United States, which has declined almost steadily from 1996 to 2006. However, there is some hope on the horizon for the bike. The aging population appears to see bike riding as a good source of exercise. The bike may also be further incorporated into plans to curb traffic problems and promote sound environmental policies.

RELATED ASSOCIATIONS & ORGANIZATIONS

Bicycle Product Suppliers Association, http://www.bpsa.org

National Bicycle Dealers Association, http://www.ndba.com

National Sporting Goods Association, http://www.nsga.org

BIBLIOGRAPHY

"Bike Ban for Shanghai." *CNN Online.* 9 December 2003. Available from <http://www.cnn.com/2003/TRAVEL/12/09/china.bike.ban.ap>.

"Bike Cult Book Global Bike Counts." *Bike Cult.* 22 February 2003. Available from <http://bikecult.com>.

"Economic Decision Making—Colonel Albert Pope." *New England Economic Adventure.* 7 July 2007. Available from <http://www.economicadventure.org/decision/pope.pdf>.

Hook, Walter. "China Rocks Global Bike Industry." *Sustainable Transport.* January 2003.

Hudson, William. "Myths and Milestones in Bicycle Evolution." Jim Langley—Bicycle Aficionado. Available from <http://www.jimlangley.net>.

"Industry Overview 2006." National Bicycle Dealers Association. Available from <http://nbda.com/page.cfm?PageID=34>.

Maus, Jonathan. "Portland Bicycle Industry Worth $63 Million." Bike Portland Organization. Available from <http://www.bikeportland.org>.

Norman, Jason. "End of Road Surge Contributes to Specialty Retail Sales Decline." *Bicycle Retailer and Industry News.* 1 April 2007.

Ritfeld, Jennifer. "The Netherlands: Bicycle Industry." U.S. Department of Commerce, International Trade Administration. February 2007. Available from <http://www.export.gov>.

"Table 1079: Commuting to Work by State: 2004." *Statistical Abstract of the United States: 2007,* 126th ed. U.S. Department of Commerce, Bureau of the Census. December 2006, 689.

"Table 1233: Participation in Selected Sports Activities: 2004." *Statistical Abstract of the United States: 2007,* 126th ed. U.S. Department of Commerce, Economics and Statistics Administration, Bureau of the Census. December 2006.

"Table 1235: Sporting Good Sales by Product Category: 1990 and 2005." *Statistical Abstract of the United States: 2007,* 126th ed. U.S. Department of Commerce, Economics and Statistics Administration, Bureau of the Census. December 2006.

"Ten Year History of Selected Sports Participation." National Sporting Goods Association. Available from <http://www.nsga.org>.

Van Der Meer, Ben. "Bicycle Retailers Adapt to Meet Needs of Aging Boomers." *Reading Eagle.* 30 June 2007.

Wiebe, Matt. "Specialty Bike, Sports Chains Deliver Dollars." *Bicycle Retailer and Industry News.* 1 April 2007.

SEE ALSO *Fitness Equipment, Motorcycles*

Blinds & Shades

————————◆————————

INDUSTRIAL CODES

NAICS: 33–7920 Blind and Shade Manufacturing

SIC: 2591 Blinds and Window Coverings

NAICS-Based Product Codes: 33–79202, 33–79204 and 33–79208

PRODUCT OVERVIEW

Blinds and shades are used to cover windows. They are part of the window coverings industry, which is part of the broader home textiles and home furnishings industry. The trade journal *Home Textiles Today*, a leading print and Web publication, discusses window coverings in terms of whether they are soft or hard. Soft coverings encompass drapes and curtains. Traditional drapes are used in living rooms and bedrooms, designed to match the furniture or bedding. They are usually floor-length, extending from near the ceiling to the floor and are made of heavy, lush fabric that was pleated, tuck-pointed, gathered, and/or layered. Curtains are used in less formal rooms as well. In kitchens they are often short, lace or net panels. Industry observers reported that soft window coverings are losing market share to hard coverings for windows in the middle of the first decade of the twenty-first century.

Blinds and shades are considered hard window coverings. Even if they feature soft, billowing Roman shades or soft, honeycombed slats, they are referred to as *hard* because they are operational. Blinds are multi-part, slatted panels, typically horizontal, operated by a string and pulley system contained in a headrail that lifts slats up

from the bottom of the window to reveal the view. Some modern shades are designed to affix to the bottom of the window frame and roll up instead of the more traditional top-to-bottom format. Vertical, slatted blinds operate from side to side. Shades are distinguished from blinds in that they are simpler—generally one panel that may be billowy or flat and is pulled up or rolls up from the top. Blinds and shades are functional products designed to filter light and provide privacy. While privacy is valued, many people also desire a view. Blinds and shades allow both, since they are easy to open fully, partially, or close completely.

The desire to cover a window or enhance a view is nothing new. Thousands of years ago Egyptians living along the Nile River wove wetland reeds together for use as both floor and window coverings. The Chinese used bamboo for similar purposes. In Renaissance Europe, wooden shutters covered the outside of windows. Later, in Colonial America, wooden shutters were installed inside. Everything old is new again and natural blinds from wood, grass, and bamboo are growing market share. Desire for natural materials emerged in direct relation to the decades-long reign of aluminum mini-blinds.

Aluminum, slatted mini-blinds gained market share slowly but steadily after they were invented in the late 1940s by Hunter Douglas, a Dutch group with roots in the United States. Hunter Douglas developed the technology and equipment to cast and fabricate aluminum and it first used the new material in blinds. Aluminum is a lightweight metallic element that is silvery white when uncoated. Aluminum is mined from the bauxite ore that makes up approximately 7 percent of the Earth's crust, and is most commonly used for soft drink cans. The new lightweight metal product became universally known as the mini-blind since its slats were narrower than one inch. Prior to the introduction of mini-blinds by Hunter Douglas, blinds were commonly known as Venetian, a name derived from European-style exterior slatted wooden shutters that were adjustable.

Mini-blinds grew in popularity until they represented approximately 70 to 80 percent of the U.S. hard window covering market in the early 1980s. Many manufacturers entered the market and made mini-blinds available in a vast array of colors because aluminum holds color coatings well. Mini-blinds were popular due in large part to the inherent qualities of aluminum. It is strong yet lightweight, flexible yet resistant to warping, durable yet low maintenance. Special anti-static finishes helped keep the narrow slats reasonably dust-free. Even though they were a relatively recent product introduction, mini-blinds soon took on an aura of timelessness. Since they were not constructed of fabric, fabric patterns and choices no longer dated a home interior.

As the market grew, manufacturers made blinds and shades ever more multifunctional. The 1970s and 1980s were a period of increased concern over energy efficiency. Besides the obvious dual function of providing privacy while enhancing views, blinds and shades were designed to insulate against hot and cold and thus contribute to a home's overall energy efficiency. In 1985 Hunter Douglas developed its Duette Honeycomb shade. Instead of flat slats, it featured three dimensional honeycomb slats. The six-sided fabric slats had advanced insulating properties due to pockets of air trapped within the honeycomb structure. The Hunter Douglas honeycomb design was copied by other makers. For instance, Levolor markets a similar product as a cellular shade. Energy efficiency is just one of the many functions of blinds and shades.

Another function built into blinds and shades involves highly specified light control. Fabrics developed to meet this need range from room-darkening to opaque to sheer. Highly specialized solar fabrics were developed that block the ultraviolet rays that damage home textiles and home furnishings. Cellular blinds and solar shades were not improvements over wood, which offers the best sun protection because it acts as a total sunblock. Such ever expanding functionality caught the attention of The International Designers Society of America. It honored shade makers seven times since 1991 with its industrial design excellence award. The 1991 award honored Levolor for yet another function: sound control. Levolor St. Tropez Blinds were recognized for a design that included rubber endcaps and a rubber bottom rail bumper to stop the irritating sound of a mini-blind clacking in the breeze.

While within the industry, window coverings are often separated into hard and soft segments, in practice

consumers combine both components. Many blinds and shades are accessorized with a fabric top treatment to soften the hard edges of a window frame. In practice, a consumer can choose from many types of blinds. According to The British Blind & Shutter Association, there are eight major types of blinds. Typically blinds are assumed to be horizontal slats unless specifically stipulated to consist of vertical slats.

American manufacturers tend to categorize blinds and shades as mini-blinds, wood blinds, faux wood blinds, roller shades, solar or sheer shades, Roman or pleated shades, woven natural product shades, cellular shades, and vertical blinds. Award winning designs blurred the line between categories of blinds and shades as they combined aesthetics and function.

The International Designers Society of America awards demonstrate this blurring, at which Hunter Douglas especially excels. Hunter Douglas dominated The International Designers Society of America awards, winning four of only seven awarded to blind and shade makers since 1991. Headquartered in Dulles, Virginia, The International Designers Society of America represents the 1965 merger of several organizations with roots in the late 1930s such as American Designers Institute, Industrial Designers Education Association, and Society of Industrial Designers.

After Hunter Douglas earned its initial recognition for design excellence in 1993 for its Silhouette window shades, it again won for its Luminette Privacy Sheers in 1998. Luminette has 3.5 inch vertical rigid flat fabric slats that rotate for varying degrees of light control. When closed the rigid fabric slats block 99 percent of ultraviolet rays, yet when open they stack discretely. Luminette combines the advantages of vertical blinds with elements of drapes, proving vertical blinds are not just for patio doors anymore. The fabric is 100 percent polyester for durability and easy maintenance (i.e., vacuuming).

The Hunter Douglas Duette UltraGlide honeycomb shade was honored for its slick operating system in 2001; no matter how high it is raised, the cord remains at the same safe distance from the floor, keeping it out of the reach of children and pets and maintaining a clean, uncluttered look. The Hunter Douglas Trio Convertible Shade was honored in 2004 for its energy-efficient, highly insulating cellular construction that offers clear views, complete privacy, and a high level of UV protection. Trio Convertible Shades use three-dimensional 1.14 inch hexagonally-shaped, fabric slats that compress and expand to let in more or less light. Each slat can open and close individually, filling up with air or flattening out to let light in. Fully flattened fabric vanes allow a perfect view without raising the blind. Fully expanded fabric vanes completely block the view while letting in light.

Blinds and shades perform many functions yet must also be aesthetically pleasing. Innovations in the market effectively trained customers to expect—even demand—more functionality in blinds and shades. The effect is a steadily growing market.

MARKET

The U.S. blind and shade industry is a mature market. Figure 25 shows nine years worth of industry shipments ranging from a low of $2.4 billion in 1997 to a high of $3.0 billion in 2005, for an overall increase over the period of 23 percent. The industry shipment data for both 1997 and 2002 were derived from the U.S. Bureau of the Census *Economic Census.* For all the other years, non-census years, data were gathered from the Census Bureau's *2005 Annual Survey of Manufactures.*

The production of blinds and shades experienced uneven growth over the period, slowing during periods of general economic slowdown and increasing again as the economy rebounded. The market for blinds and shades tends to follow, with a slight delay, the market for new construction—both residential and commercial—and this market was strong through the period presented in Figure 25. Declines occurred in the building sector during late 2005, 2006, and 2007 which will likely be reflected in the shipment data for blinds and shades in these years once those figures are available.

The U.S. Consumer Product Safety Commission and the Window Covering Safety Council jointly estimate that roughly 85 million window blinds are sold each year. The U.S. Census Bureau reports in "Blind and Shade Manufacturing: 2002," that American factories shipped $2.8 billion worth of blinds, shades, and parts in 2002, up 17 percent from $2.4 billion in 1997. The main Census Bureau product categories are: (1) window shades plus accessories and rollers, (2) Venetian blinds including components and parts, (3) blinds including natural bamboo, rattan, reed, and wood, plus curtain and drapery hardware, and (4) other.

Shipments of products in the first category (window shades plus accessories and rollers) were up 36 percent between 1997 and 2002, growing from $370 to $513 million. Venetian blinds including components and parts were down 13 percent from $1.1 billion to $982 million. Blinds including natural bamboo, rattan, reed, and wood, plus curtain and drapery hardware were up 56 percent from $494 million to $772. The other catch-all category also grew 20 percent.

Of the three categories in the blinds and shades sector, only one experienced declining manufacturers' shipments during the 1997 to 2002 period—Venetian blinds. The fastest growing category of blinds and shades was

FIGURE 25

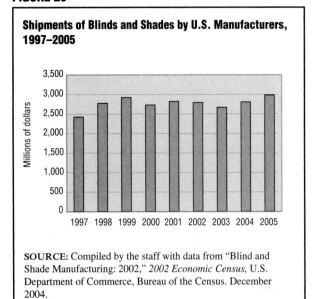

Shipments of Blinds and Shades by U.S. Manufacturers, 1997–2005

SOURCE: Compiled by the staff with data from "Blind and Shade Manufacturing: 2002," *2002 Economic Census,* U.S. Department of Commerce, Bureau of the Census. December 2004.

the one including products made with natural materials such as bamboo, rattan, reed, and wood as well as curtain and drapery hardware. Curtain and drapery hardware is sometimes referred to as the suspension sector. The simplest suspension system favors poles. For years, drapery hardware was designed to stay hidden. Hardware can be manufactured from lightweight aluminum, heavyweight cast iron, steel, bronze, brass, and wood, and is meant to be on display. Attention is drawn to the ends of the suspension poles with finials. Amber glass knobs and raffia interwoven with brass threads are the types of finials recommended by HGTV.

The International Trade Administration tracks U.S. exports of blinds and shades. It compiles tariff and trade data from the U.S. Department of Commerce and the U.S. International Trade Commission. In 2006 U.S. exports of blinds and shades totaled $46 million. Canada and Mexico were the main source of these exports. As Figure 26 shows, the United Kingdom, Hong Kong, Korea, and Australia were among the top recipients of U.S. exports of blinds and shades.

KEY PRODUCERS/MANUFACTURERS

According to *Market Share Reporter 2007,* leading U.S. manufacturers were Newell Rubbermaid with 21 percent of the market and Springs Window Fashions with 18 percent of the market. On the global scale, Hunter Douglas stands out as the leading maker of window coverings.

The Hunter Douglas Group. With its world headquarters in Rotterdam, the Netherlands, Hunter Douglas

FIGURE 26

Leading Importers of Blinds and Shades Exported from the United States in 2006

Countries are listed by the value of their imports of blinds and shades in thousands of dollars.

All others $14,064

Canada $13,414

Mexico $10,879

Australia $1,587

Hong Kong $2,067

South Korea $2,320

United Kingdom $2,338

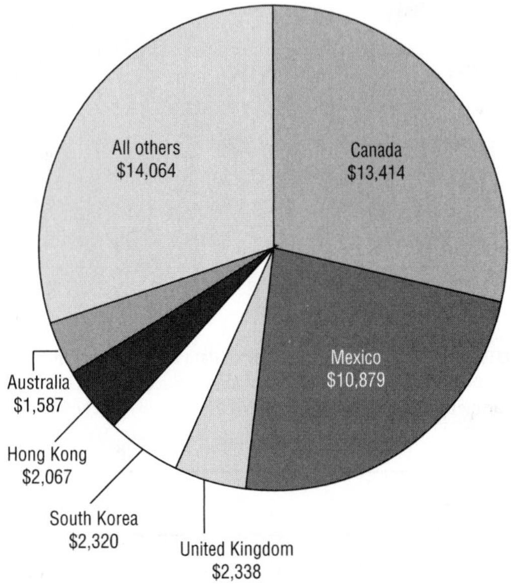

SOURCE: Compiled by the staff with data from "NAICS 337920 Blind and Shade Manufacturing: FAS Value of All Countries, U.S. Domestic Exports," U.S. Department of Commerce, International Trade Administration, available from http://www.ita.doc.gov/td/ocg/exptab.htm.

had worldwide sales of $2.6 billion and about 20,000 employees in 2006. It is the world's largest manufacturer of window coverings, which accounted for 87 percent of its 2001 revenue. In North America, Hunter Douglas had sales of $1.2 billion and 8,900 employees in 2006. Hunter Douglas donated custom window coverings for more than 5,000 new Habitat for Humanity homes built during 2006.

With roots going back to Düsseldorf, Germany in 1919, the forerunner to Hunter Douglas was established in the Netherlands in 1933 as a machine tool operation. In 1940 it moved to the United States as Douglas Machinery Co. In 1946 Douglas Machinery Co. established a joint venture with Joe Hunter, who developed technology and equipment for the continuous casting and fabrication of aluminum. This led to the production of lightweight aluminum slats for Venetian blinds. Hunter Douglas was formed. Hunter Douglas aluminum blinds led the American market. It also led the business model for distribution of blinds and shades. Hunter Douglas developed a network of more than 1,000 North American independent fabricators. In 1956 and 1976, respec-

tively, the U.S. business was sold and reacquired. In 2001 Hunter Douglas moved to its newly renovated 230,000 square foot manufacturing facility in Cumberland, Maryland, and took 500 plus employees with it after it outgrew its facility in Frostburg, Maryland.

All companies say they innovate. Hunter Douglas actually does. In 2006, for the 11th consecutive year, Hunter Douglas swept the annual Window Coverings Manufacturers Association competition. It won 20 out of 34 total awards. It introduced a record number of new products, including Duette Architella, yet another proprietary honeycomb shade design with even higher energy efficiency combined with superior aesthetics.

Newell Rubbermaid. Headquartered in Atlanta, Georgia, Newell Rubbermaid had 2003 global sales of $7 billion and 33,000 employees worldwide. Newell Rubbermaid products are organized into four segments: office products, tools & hardware, home & family, and cleaning, organization & decor. It owns Sharpe, Paper Mate, Parker, Waterman, Rubbermaid, Calphalon, Graco, and others. In 1993 Rubbermaid acquired Levolor Kirsch Window Fashions. Headquartered in High Point, North Carolina, Levolor was founded in 1914 and its name was for a while synonymous with one inch mini-blinds. Levolor makes blinds and shades in aluminum, wood, and woven wood products in styles it calls Roman, cellular, wood, natural, vertical, soft vertical, roller, panel track, and metal. Levolor also sells decorative drapery hardware. Its High Point headquarters include a 50,000-square-foot office and a 193,000-square-foot plant. Employing approximately 2,300 associates, Levolor Kirsch Window Fashions' other locations include: Athens, Georgia; Garden Grove and Westminster, California; Rockaway, New Jersey; Salt Lake City, Utah; South Holland, Illinois; and Mexico. Levolor closed its plant in Shamokin, Pennsylvania, in 2002, and in 2003 closed its factory in Ogden, Utah. Levolor moved its operations to Mexico resulting in a significant loss of jobs in the United States. Newell Rubbermaid, which owns Levolor and Kirsch, plans to close one-third of its 80 U.S. factories by 2008, resulting in a loss of approximately 5,000 U.S. jobs.

In June 2007 Levolor Window Fashions announced a new cordless Roman Shades collection. The exclusive patent pending cordless lift system can be operated either top down or bottom up. Cordless shades eliminate dangling lift cords, which are hazardous to children and pets, and provide a clean, uninterrupted window silhouette.

Springs Window Fashions. Headquartered in Middleton, Wisconsin, Springs Window Fashions has two customer service locations in Pennsylvania (Montgomery and Williamsport) and a centrally located distribution center in Groveport, Ohio. Four manufacturing plants are in Grayling, Michigan, Reno, Nevada, and Reynosa and Tijuana, Mexico. Springs Window Fashions manufactures blinds, shades, and drapery hardware under three separate brands acquired over a 15 year period: Bali (acquired in 1989), Graber (acquired in 1979 for $38.5 million), and Nanik (acquired in 1995), and for private label accounts. Each addresses a particular segment of the market.

Bali started in 1906 as Carey-McFall, producing blinds for automobiles and railroad cars. In the 1960s it introduced the Bali custom mini-blind. The custom—or more appropriately, the cut-to-fit—blind is its focus. Bali targets the do-it-yourself customer. It sells cellular shades; faux, wood blinds; vinyl and aluminum horizontal blinds; and natural shades from materials like bamboo, sisal, grass, jute, and even straw. It touts the classic roller shade as a retro window treatment; Roman shades with a soft appearance; and solar shades that filter light to reduce glare and damaging UV rays. Bali's Web site provides information so customers can select the right custom-made (cut-to-fit) window treatment.

Graber blinds are sold only by professional dealers and decorators. Graber began in 1939 with the invention of a special bracket to hold drapery panels over Venetian blinds. Initially focused on drapery hardware, by the late 1960s, Graber added pleated shades and verticals to its line. Graber developed patented designs to keep pleats uniformly spaced, with no unsightly sags. It makes energy efficient cellular shades; wood shutters from North American basswood with sleek cordless controls; and decorative hardware like valances and cornices. Graber's version of the Hunter Douglas Honeycomb shade is called CrystalPleat. CrystalPleat's multifunctions include insulation, light control, and sound control. Crystals are often six-sided structures, so it may be assumed that the Crystal-Pleat line is a direct nod to Hunter Douglas.

Nanik handcrafted, custom, wood blinds are a premium product. Nanik makes only wood blinds. Its distinction is that it forms and punches slats prior to finishing them with stain or paint for optimum coverage and protection. Its naturals line is made from North American basswood hand-selected for its mineral marks, knots, and cross-grain patterns. Basswood is used exclusively on 2 inch blinds, available in 13 standard colors. Nanik is a real, custom shop capable of making specialty shapes for unusual applications, such as arches, circles, trapezoids, and cutouts. The custom Nanik color palette consists of 50 colors ranging from paints to stains, with an option to custom color match at no extra charge.

Vertically-integrated window treatment company 3 Day Blinds announced on June 25, 2007 that it would manufacture and retail a new line of Laura Ashley Sheer Window Shadings in North America. The sheer shade is designed to disappear into the headrail when raised to maximize the view.

MATERIALS & SUPPLY CHAIN LOGISTICS

American factories shipped $2.4 billion worth of blinds, shades, and hardware in 2002, up 11 percent from $2.2 billion in 1997. To do so, American factories purchased $1.2 billion of total materials in 2002, up 8 percent from $1.1 billion in 1997, according to the U.S. Census Bureau report titled "Blind and Shade Manufacturing: 2002." That report discusses materials consumed by kind. The three kinds of materials needed to manufacture blinds and shades are plastics, metals, and all other materials, a category that includes polyester. Together, these three categories represent 60 percent of materials consumed in the U.S. production of blinds, shades, and window covering hardware.

Plastics consist primarily of plastic coated fabric and shade cloth along with plastic sheets, rods, tubes, film, and other shapes. Metals consist of four classes of materials: fabricated metals including forgings; steel shapes and forms (excluding fabricated metals and forgings); aluminum; and aluminum and aluminum-based alloy sheet, plate, foil, and welded tubing. The other category includes polyester.

Polyester reigns. It is the most-used material in the manufacture of blinds and shades. The synthetic material is popular with manufacturers because of its strength, versatility, and durability. For years, people looked down their noses at polyester because it was a synthetic. Kim Kiner, vice president of product design for Hunter Douglas, explained that polyester is a material that has transformed its reputation over the years and slowly grown into a dominant force. "This is a fabric that never had a great reputation in years past," Kiner said. However, polyester material fares better under ultraviolet rays, unlike natural materials whose color tends to fade and eventually break down. Polyester can be made to look like silk, suede, or any other material.

DISTRIBUTION CHANNEL

Ever since the introduction of aluminum mini-blinds in the 1940s, blinds and shades were distributed via two distinct channels. Channels were developed for standard, ready-make merchandise and for custom-made merchandise. When the internet became popular, blind manufacturers and retailers both were early adaptors of it as a sales distribution channel. Popular electronic retailers of blinds and shades include 3 Day Blinds and Blinds Galore, among others.

The mass merchandise channel sells value priced blinds and shades. These are a ready-made or cut-to-fit product, sometimes referred to as custom but a more apt description is customized. Common outlets in the value distribution channel include home improvement centers like Lowe's, The Home Depot, and Menards; department stores like JC Penney, Sears, and more recently Bed Bath and Beyond; and mass merchandisers such as Kmart, Target, and Wal-Mart. In June 2007 when Levolor announced its new Roman Shades collection, it described its distribution channel as including only EXPO Design Center, Home Depot, JCPenney, Lowe's, Menards, and Sherwin Williams. During the summer of 2007, Target Stores had blinds and shades for sale ranging in price from $25 to $200 including Woolrich, microsuede, cotton polyester, Roman shades; natural jute and bamboo shades; 100 percent cotton, Roman shades with acrylic foam thermal backing; and solar roller shades constructed from 71 percent polyvinyl chloride fabric and 29 percent polyester fabric.

The custom merchandise channel is characterized by high-end window treatments. For example, at the highest end, Nanik sells its exceptional custom-made wood blinds exclusively through independent design consultants and showrooms. Independent retail shops sell the majority of high-end blinds and shades. In order to stay solvent, many independent retail shops are affiliated with high-end manufacturers. Hunter Douglas is a good example. Its 2006 corporate report details how it expanded its network of affiliated independent retailers, which it refers to as a dealer alliance. In 2006, Hunter Douglas expanded its *Showcase* dealer alliance program from 500 to over 800 dealers in North America, and its *Gallery* dealer alliance program from 320 to 350. It also expanded to 300 the Centurion Club, a dealer alliance category designed to keep its product firmly in the high-end channel as distinct from the mass merchandise channel. In the Centurion Club, participating Gallery and Showcase alliance dealers commit 100 percent of their business to Hunter Douglas branded products.

A part of the distribution of window coverings involves the use of product sample books, which contain samples of both fabrics and finishes. Hunter Douglas sold more than 140,000 new product sample books to retail dealers in 2006. This was a record number and interesting in that in other industries, such as the paper industry, product sample books are free. Maintaining a smoothly operating distribution channel also includes training for dealers, retailers, affiliated stores, or members of an alliance program. Hunter Douglas trained more than 28,000 retail dealers in face-to-face professional customer education programs and an additional 8,000 through remote training sessions made available over the Internet in 2006.

Levolor uses product seminars conducted by its sales representatives at home improvement centers to educate retailers and customers about the proper use of its products. Levolor also has training videotapes at stores such as EXPO design Center, Home Depot, JCPenney, Lowe's, Menards, and Sherwin Williams.

KEY USERS

Users of blinds and shades are all those who occupy structures that include windows. In fact, blinds and shades are sometimes used within a structure to separate one space from another as a sort of temporary wall or divider so even those in structures without windows are at least potential users of blinds and shades.

If we were to segregate the users of blinds and shades into categories, they would fall into three: the residential user, the commercial user, and the institutional user. Each is defined more by the function carried out in the buildings in which the blinds and shades are used than by any difference in the way in which those blinds and shades are used. They do differ, however, in the ways in which they buy blinds and shades.

ADJACENT MARKETS

Blinds and shades are initially purchased and installed as new buildings are brought into use. Consequently, the building sector is an important adjacent market to the blinds and shades market. In fact, construction starts serve as an indicator of future business for manufacturers of blinds and shades. In 2005 the U.S. housing market, which had been strong through the first five years of the new century, began to slow and the slowdown continued into 2006 and 2007. Hunter Douglas reported that its 2006 sales and operating profits were negatively affected by this slowdown.

The purchase of replacement blinds and shades is also an important factor in the overall market for these products. Consequently, the business of home decorating is an important adjacent market to the market for window coverings, blinds and shades included. Trends in home decorating strongly influence the selection of window coverings by establishing fashions and thus the fashionable use of blinds, for example, over curtains, or roman shades over horizontal blinds. These style trends influence the selection of window coverings for new installations as well as replacement installations.

During large-scale redecorating projects consumers are often willing to make a substantial dollar investment in window coverings. Blinds and shades are considered a durable good expected to last seven years on average and this helps decorators promote the upgrading of blinds during a redecorating project. Consumers expect newly purchased blinds and shades to perform many functions yet also to be aesthetically pleasing and compliment the style of the room.

A 2003 book titled *The American Demand for Household Furniture and Trends* presents an analysis of the U.S. furniture market. Examining the period from 2001 to 2011, it reported that as a result of faster growth in the population among those over 40 years of age, the number of households is expected to grow by close to 11 percent, a pace slightly faster than the 8 percent growth expected for the total population. As the number of U.S. households grows, household furniture spending will grow by 23 percent or from $64 billion a year to $79 billion a year (in constant 2001 dollars), it says. Blinds and shades are part of the broader home textiles and home furnishings industry.

RESEARCH & DEVELOPMENT

Research and development efforts in this field tend to focus in two areas, the technical and functional or the fashion and trend oriented. Research and development in the fashion and trend sphere focuses primarily on color and pattern changes in the home textiles and home furnishings industry. In April 2007 Levolor predicted trends such as glam high-gloss finishes and dramatic color combinations; classic colors like plum, jade, gold, silver, and dark grey; casual handcrafted and hip; eco-chic natural fabrics; and global fusion wood and woven wood blinds and shades.

Highly technical research and development efforts have resulted in many advances over the years, including photosensative materials, automation of the opening and closing functions, and the integration of remote control devices with large-scale or high-end installations.

Blinds and shades can be operated by remote control, either by battery power or electricity. Graber motorized operational systems promise smooth operation with headrails known for being the best in the industry. Battery-operated are the least expensive remote control systems and can be installed without ripping up the walls. More technologically sophisticated electronic systems can be controlled from a touch panel that also controls, lights, thermostat, and the audio/visual system. Electronic systems can be programmed so shades rise and fall at precise times of day to protect furniture from ultraviolet rays.

Highly technical research and development resulted in sun control fabric. One fabric was developed by Freduenberg Nonwovens in Durham, North Carolina. Called the Pellon Wonder-Shade, it was developed for custom roller shade applications. Decorator fabric can be ironed on to the fabric's fusible backing, trimmed, and attached to a roller. The result is a room darkening shade with a smooth white backing. DuPont developed Sontara, a nonwoven fabric designed to create enhanced softness and drapability. The fabric is used by Graber in its CrystalPleat cellular shades.

CURRENT TRENDS

Increasing energy efficiency has been a trend in most fields during the first decade of the twenty-first century as the cost of energy has risen. Systems of all sorts have been retrofit in order to increase energy efficiency. Window treatments are no exception. Energy efficient window treatments are one way to increase energy conservation. By using the sunlight coming through a window to heat the internal space in the winter, the cost to heat a building is reduced. In the summer, effectively blocking the incoming sunlight can greatly reduce the heat buildup in a structure and thus reduce the energy needed to cool the building. Commercial buildings with large expanses of glass—common in both high-rise urban areas and suburban office parks—are prime candidates for the installation of energy efficient window blinds and shades. Window treatment manufacturers are paying attention to this trend and are designing new products that offer greater and greater protection from glare, heat gain, and UV rays.

Natural fibers have been another trend during the first decade of the twenty-first century. During the 1950s through the 1970s, woven shades made with natural materials were relegated to the porch. During the last decades of the twentieth century they migrated into the home with improved construction and enhanced operational systems. Natural materials are aesthetically pleasing and filter light naturally. These natural material-based window treatments were also a popular theme in home decorating fashions during the later half of the first decade of the twenty-first century. Bamboo blinds and sea grass shades help to create an environment that projects a nature friendly and green image.

TARGET MARKETS & SEGMENTATION

The do-it-yourself builder is an important target market for manufacturers of blinds and shades. This target market is accessible through home improvement stores and e-commerce Web sites. Bali, for examples, targets do-it-yourselfers through its Web site by offering a great deal of information to consumers about styles and color trends. Their tagline, "your home, your style," gives customers the feeling they are customizing their home and not following the broader, more beaten path.

Interior decorators are another target market for blind and shade manufacturers. This audience is accessible through professional trade shows as well as through advertising. Prime media for such advertisements are interior design magazines and any one of the numerous cable television networks dedicated to shows about home improvements, decorating, design, and gardening.

Builders and commercial property managers are another audience targeted by makers of blinds and shades. These customers are often able to make decisions that influence very large purchases, for such things as a commercial office building, a hospital, a hotel chain, or an apartment complex. Consequently, these buyers are an important market segment for makers of window treatments, although one that is generally able to command a very competitive price.

RELATED ASSOCIATIONS & ORGANIZATIONS

The British Blind & Shutter Association, http://www.bbsa.org.uk

The International Designers Society of America, http://www.idsa.org

US Consumer Product Safety Commission, http://www.cpsc.gov

Window Coverings Association of America, http://www.wcaa.org

Window Coverings Manufacturers Association, http://www.wcmanet.org

The Window Covering Safety Council, http://windowcoverings.org

BIBLIOGRAPHY

"Blind and Shade Manufacturing: 2002." *2002 Economic Census.* U.S. Department of Commerce, Bureau of the Census. Available from <http://www.census.gov/econ/census02>.

DeKorne, Clayton. "Upselling Window Treatments: Profitability in this Specialized Arena Requires Expertise." *Remodeling.* September 2005, 128.

"The Domestic Window Coverings Market UK Report 2006." AMA Research Ltd. Available from <http://www.amaresearch.co.uk/Windowcoverings06s.html>.

Girard, Lisa. "Natural Window Shades of the '70s are Back— With a Twist." *Home Channel News.* 15 April 2002, 14.

"Hard Window Sales Cover $3.9 Billion of Biz." *Home Textiles Today.* 18 March 2002, 15.

International Directory of Company Histories, Volume 19. St. James Press, 1998.

"Laura Ashley Window Due from 3 Day Blinds." *Home Textiles Today.* 25 June 2007, 14.

Mancini, Rosamaria. "Polyester Leads the Pack: Versatility and Durability Make the Material a Popular Choice in Window Coverings." *HFN The Weekly Newspaper for the Home Furnishing Network.* 1 August 2005, 12.

Meyer, Nancy. "Special Treatment: Retailers are Stressing Shades and Blinds." *HFN The Weekly Newspaper for the Home Furnishing Network.* 26 August 2002, 14.

"Pellon Wonder-Shade." *Nonwovens Industry.* Feb 2000, 38.

Rothstein, Shari Lynn. "Dupont's Sontara: Made in the Shade." *HFN The Weekly Newspaper for the Home Furnishing Network.* 9 August 1999, 20.

Trucco, Terry. "Technophobia Window Treatments: Motorized Shades and Blinds." *House Beautiful.* November 2006, 78.

SEE ALSO *Windows*

Board Games

———————■———————

INDUSTRIAL CODES

NAICS: 33–9932 Game, Toy, and Children's Vehicle Manufacturing

SIC: 3944 Games, Toys, and Children's Vehicle Manufacturing

NAICS-Based Product Codes: 33–99327 through 33–99327226

PRODUCT OVERVIEW

A board game contains several parts. One essential part is a flat surface on which the game's activities are played, the board. Games may involve the role of dice, the spin of a wheel, or the answering of questions. In these types of board games the players move pieces on the board as part of the competition. Other games require players to rid themselves of property or acquire the property of their opponents. Some board games require skill and knowledge on the part of the player while other games are driven primarily by luck.

Early Games. Early versions of chess, checkers, and backgammon were all played in ancient civilizations. These games were created in Egypt, Greece, India, or China and became popular with established rules. Games were introduced to cultures through trade; the game of Wei Qi, for example, appears to have made its way across the Asian continent along the early trade routes of the region. Board games were also introduced through exploration and military campaigns. The Vikings carried the game *Hnefatafl* on their travels to Greenland, Iceland, Ireland, and Great Britain. Translated as King's Table, this game involved moving pegs around a wooden board. Several sources credit Christopher Columbus with bringing chess and cards to the New World. Roman soldiers brought early versions of backgammon with them on their travels. As a game was introduced into a culture the players typically made it their own, changing the name of the game and perhaps slightly altering rules or board designs.

Backgammon. Senet is the oldest known board game. It has been found in ancient digs of burial grounds in Egypt and dates to at least 3500 BC. More than 40 versions of the game have been recovered. Players used sticks, bones, or other materials as player pieces, which they moved across the 30 square board. Romans played a similar game called Ludus Duodecim Scriptorum, the *game of 12 lines*. Romans introduced the game to Britain in its first century conquest. Turkey, Spain, Germany, and China were only a few of the countries that had similar versions of the game. They were known by other names, of course, and there were subtle differences. In the game of Nard, for example, which developed in Asia in 800 AD, two dice were used; the Arabic version of the game used three dice. The first use of the word *backgammon* was in 1645, with the name coming from the Saxon or Welsh languages. By the 1930s the rules to the game were more firmly established, and the game has remained unchanged since then. In the 1960s the first tournaments were held. The game saw a brief surge in popularity in the 1970s.

Chess and Checkers. The game of Chaturanga originated in India and is considered by many to be the earliest form of chess. The game dates to at least as early as the sixth century. The name literally means having four limbs or parts, and seems to reflect the four military divisions used in the game. Chaturanga is played on a square board of eight squares. The pieces were the Raja (the king) and his Mantri (the counselors), the gaja (elephants), the asvas (horses), the ratha (chariots) and the Pedati (infantry). There are clear similarities between this game and chess, including the arrangement of pieces on the board and the name of the pieces (the ratha is an early rook piece, the pedati an early pawn). The game of Shatranj is quite similar to Chaturanga. It is referred to in Persian writing in 600 AD. The game made its way to Greece and across Europe; it is believed to have been in France by 760 AD. By 1400 the game of chess was firmly established in Europe.

Ancient civilizations played a game called Alquerque, an early version of checkers. Alquerque boards and pieces have been found at Egyptian digs dating from 600 BC; Egyptian artwork suggests the game may have existed as early as 1400 BC. Both Homer and Plato mention the game. The Moors introduced a version of the game to Spain about the tenth century. While the name given the game varied from culture to culture (the Arabs called the game El-Quirkat) certain aspects of the game were retained. The game was played on a board with two players and twelve pieces. Each player starts play with either six white pieces or six black pieces. The game was enjoyed throughout Medieval Europe.

Historians believe the combining of Alquerque with the 8 x 8 chessboard took place in France in the middle of the twelfth century. This game was initially known as Fierges. The game became known as Checkers in the United States. It is known as International Draughts in other countries. Rules vary slightly, and the game is played on a 10 x 10 or 12 x 12 board depending on the country. English draughts is played on an 8 x 8 board and is the checker game's closest relation. Chinese Checkers, it should be noted, is played on a hexagon shaped board. It was actually invented in the United States and not China; the adding of the word Chinese was thought to give the game a more exotic feel.

One game the Chinese can take credit for is Wei Qi, meaning surrounding pieces. It is first mentioned in writing in 548 BC, although it existed for centuries before this. Wei Qi's origin is unclear. Several emperors have been credited with its creation, as have members from that nation's military. The game appears similar to chess or checkers at first glance. However, it is quite different. The game is played with black and white stones on a board; the object of the game is to move the stones strategically to take as much territory as possible. The game became known by many names as it moved along Asian trade routes (*Shogi* in Japan and later *Baduk* in Korea, *Go* in the West). A game that clearly emphasizes strategy, Wei Qi has a reputation for being very challenging because of the countless moves one can make on the board.

During the thirteenth century Alfonso X, king of Leon and Castile, commissioned the *Libro de los Juegos*, the Book of Games. The *Book of Games* is the first encyclopedia of games in Europe and was completed in 1281. It contained the rules for the early games of chess, backgammon, dice, and other table-based games. Table and board games were mainly popular with the aristocracy, although certainly not exclusively. Egyptian cave paintings suggest that early members of the lower class enjoyed chess and backgammon games as well. It was the upper class who had developed a particular love of their leisure time, however, and board games offered them a pleasant diversion. Their favorite games are thought to be those that emphasized chance rather than skill.

The 1800s Begin. This concept of leisure time would come much later for those in the lower economic brackets. The average American child in the 1700s and early 1800s had little time for leisure and game playing. Children were expected to work outside the home or contribute to the running of the daily household. Sensibilities changed during the Victorian period, when the typical household became much more child oriented. The middle class began to develop in the United States about this time as well. Some families were wealthy enough to bring nannies and other servants into the home. More time could be devoted to leisure activities.

With the increasing affluence, the increase in leisure time, and the development of a middle class, the board game industry began to flourish. Chess and checkers were still popular pursuits for families, but other games were developed that could entertain and provide moral instruction. In 1843 W. & S.B. Ives published and released the game Mansion of Happiness. The game is the first commercially produced board game in the United States. In the game, created by Anne W. Abbott, a player encounters virtues such as honesty and temperance and vices such as cruelty and ingratitude along the 64 spaces on the board. A virtue sent a player ahead spaces, while a vice sent him back several spaces. The winner was the first one to the Mansion of Happiness.

At the turn of the century wealthy men such as John D. Rockefeller and Andrew Carnegie were helping to shape the country. The economy was growing during the Industrial Revolution, and certain segments of the population began to notice growing gaps between the rich and poor. Henry George was a writer and economist of the late nineteenth century. He believed that land is a fixed asset that belongs to the people. As a society develops, rents increase and the landlords unfairly reap the benefits from the land they do not own. Elizabeth Magie was one of his most ardent disciples. In 1904 she invented a game called the Landlord's Game. The game has a striking resemblance to Monopoly. There are 40 spaces on a square board; they include railroads, utilities, a water company, and rental properties. Magie explains in the game's rules that the game is intended to entertain but also to show how "under the present or prevailing system of land tenure, the landlord has an advantage over other enterprises."

She took the game to Parker Brothers in 1924. They rejected the game. Copies of the game reportedly made their way to the economics departments of the University of Chicago and University of Philadelphia (Magie lived near each of them for periods of time). Reportedly, it developed a fan base there. The game's rules were tweaked and it developed the informal name of Monopoly. Knapp Electric issued a new version of the game called Finance. Heater salesman Charles Darrow began making homemade versions of the game to give to friends. His version of Finance strongly resembled the modern version of Monopoly. He took the game to Milton Bradley and Parker Brothers. Both rejected the game as too complicated; Parker Brothers famously told him the game had "52 fundamental playing errors." Parker Brothers later changed its mind when it learned of the game's strong sales in Philadelphia. Monopoly was sold in 80 countries and 26 different languages in the early twenty-first century and has sold more than 200 million copies.

The 1950s and 1960s. The mass production of board games truly began in the post war period in the United States. Plastics, metals, and other materials that had been needed for the war effort were now available. By the 1960s manufacturers would have greater understanding of plastics, molds, and precision cutting devices. As mass production developed it stimulated adjacent industries. The country's advertising industry became more robust. Manufacturers marketed board games in attractive ways and emphasized their entertainment value. Historians have noted certain trends that took place in the late 1940s. A soldier returned home and got married. These new couples started families and what followed was the baby boom. Many families then moved to the suburbs. The country enjoyed economic prosperity and leisure time increased.

Some very popular board games make their first appearance during the 1950s and 1960s. Alfred Butts first tried to market his word game Scrabble in the 1930s. In 1948 he met James Brunot. They tried to market the game together. In 1949 the Brunots made 2,400 sets and lost $450. The president of Macy's discovered the game in 1950 and started selling the game in his stores. Production soon could not meet demand. By 1953 a million copies of the game had been sold. Candyland, a popular game

with children, was published in 1949. Anthony Ernest Pratt invented the game of Clue in 1944. Parker Brothers bought the game in 1948 and started marketing it. Yahtzee, published in 1956, is regarded as the grandfather of the modern dice game. The Game of Life was published in 1960. Charles F. Foley and Neil W. Rabens patented Twister in 1969.

1970s and Beyond. Some board games in the 1970s became much more sophisticated. In 1974 Tactical Studies Rules published the first edition of Dungeons and Dragons. The game has some of the elements of a traditional table-based game, with players, play pieces, and dice. However, this fantasy game marks the beginning of the role playing game genre. In the game a Dungeon Master acts as the game's director leading the other players through the dungeon. These players assume a particular character. They roll multi-sided dice to determine that character's attributes—his strength or agility, for example. Early games featured manuals to provide rules for the game, and miniature figures (orcs, elves, etc.) to represent the characters as they moved through the dungeon. The supernatural aspects of the game were troubling to some parents, and subsequent editions of the game have been modified.

Scott Abbott and Chris Haney invented Trivial Pursuit in 1979. Approximately 20 million copies of the game were sold in North America during the peak of the game's popularity. Nearly 88 million games had been sold in 26 countries and 17 languages by the end of 2004. In 1980 the game Civilization was published. Each player starts with a single population token, representing 7,000 people, and grows and expands his empire over the course of turns. The goal of the game is to be first to advance to the final age on the Archaeological Succession Table. The game has a loyal following because of its intelligent handling of its subject matter. The board game ceased publication in the United States. Intriguingly, it moved into a PC game format. The video game and home computer markets were developing quickly in the late 1970s and early 1980s. Some board game makers thought them a fad while others viewed them as serious threats. Pictionary was the last popular game of the 1980s. Published in 1986, the game is played in teams. Players must draw certain objects to be identified by their teammates. Correct answers move them along a game board.

A few games were popular in the 1990s—Cranium, for example, published in 1998, went on to win several awards from trade associations. Marketers released board games based on favorite, classic characters or those that received brief cultural interest (The Smurfs, Captain Planet). Games in other formats became popular (Magic: The Gathering was released in 1993). Americans were quickly obtaining Internet access during this period and this new media would become a dominant focus of American's shrinking leisure time. A 2001 estimate suggested the average American had 26 hours of leisure time a week in 1973; by 1997, the figure had fallen to 17 hours.

MARKET

Board Game Basics. Most board games are one of two types: race games, like Candyland and Parcheesi, or strategy games in which a player must rid himself of tokens, gain territory from his opponent, or the like. The board game industry is a reliable performer. There are several reasons for this stable performance. Among these reasons are the relatively low cost of board games and their ease of use. An interactive board game (Trivial Pursuit, for example) is typically priced from $24.99 to $49.99. A game made of cardboard or plastic sells for $11.99 to $19.99. Children are still the primary consumers, but the industry has evolved to include more games for adults and families.

Board game sales tend to increase near the end of the year as the holiday season approaches. The industry may not be as high profile as video games or other toy categories but it is just as competitive. As many as half of all new board games fail in their first year; of those remaining, another half fail in their second year. Creating a successful board game is clearly challenging. One must create a game with a minimal number of pieces, and with rules that are easily understood and require minimal math. The game should be completed in a reasonable amount of time, be played with 2 to 8 players and be able to fit in a box. This needs to be produced at a reasonable cost by the manufacturer. It also, of course, needs to be fun. Many inventors work each year to come up with the next hot game. Some game makers accept submissions but others do not. A 2002 report noted industry leader Hasbro receives 1,600 submissions for new games each year.

The Modern Board Game Market. The decade of the 1980s marks the arrival of two powerful competitors into the entertainment market: home computers and video games. The Commodore 64 was a powerful home computer that reached the market in 1982. Offering superior speed and graphics at the time, the computer reportedly sold 17 million units from 1982 through 1994. The Nintendo and Playstation 2 were two of the more popular gaming consoles. Super Mario and Pac-Man became pop culture references as common as Mr. Moneybags from the Monopoly game.

During this time manufacturers and advertisers became more keenly aware of the power of licensing and the economic power of young people. Certain characters starred in their own board games. The Cabbage Patch

dolls, for example, enjoyed only a brief popularity with young girls in the 1980s; it was long enough, however, to get the dolls a board game of their own. Shipments of Cabbage Patch related materials (games, stickers, etc.) totaled $1 billion in 1984. Board games were also released based on *Captain Planet*, the *Goonies* movie, and other properties.

This form of marketing helped keep sales strong in the board game industry. In 1988 board game sales increased 14.5 percent from $810 million in 1987 to $950 million in 1988, according to the Toy Manufacturers Association. Unit sales increased 193 million to 205 million over the same period. The low cost of many games helped keep the market performing as well. The average price of a board game in 1988 was $9.67. Board games have always been popular with children, of course. In the 1980s analysts noticed that more families seemed to be playing them together. As well, a new market sector developed: board games for adults. The popular Trivial Pursuit game essentially created this category. Popular games from 1988 were Pictionary; Outburst; Jenga; Win, Lose or Draw; and Balderdash.

Perennial favorites like Monopoly drive the market as do games that become hot. In 1989 classic games such as Monopoly, Clue, and Twister were on the best seller list. That same year, Donald Trump, business executive and entrepreneur, launched the Trump game, in which players buy and sell real estate properties. Trump claimed the first edition of this game sold 1 million copies. Board games leveled off slightly in 1990 as the country entered a recession. Sales were strong enough however that the NPD Group, a leading industry tracking firm, decided to start tracking adult game sales separately for the first time.

In the mid-1990s the toy industry saw increased interest in educational toys. Some toys have always been marketed as such, of course, but one can not help but wonder if such games appeared in response to popular but mindless video games. The company Leapfrog was founded in 1995. The company quickly became a leader in the market for electronic toys that help children learn to read and write. The board game industry seems to have followed the trend. In 1994, We the People, a game set during the Revolutionary War, was published. It has been credited with starting the war game genre that would be adopted by video games. In 1995 Mayfair Games pushed its English language version of the Settlers of Catan (the game was first published in Germany). Players must build roads and erect cities in this game. In 2006, according to Mayfair Games, more than 11 million copies of the game have been sold since its debut. The game Cranium was published in 1998 and requires its players to engage in tasks that draw on their language, performance, and art skills.

Americans quickly embraced DVD technology when it first arrived in the late 1990s. Manufacturers began to adopt this technology into their games in an effort to stimulate the category. Hasbro released the Twister game with a DVD of dance moves. Screenlife LLC released its SceneIT? Movie Edition in 2002. The game incorporates the DVD technology with the board game technology; players must answers questions based on film clips to move around the board. The company has sold nearly 10 million copies and publishes 20 different titles. This new DVD game genre generated annual sales of $200 million from 2003 to 2006. Other titles exist based on American Idol, Harry Potter, and James Bond.

Incorporating DVDs into board games is seen as a major innovation to the board game market. This happened just as some other games were showing their age. By 2002 Trivial Pursuit sales had fallen to $15 million, a 98 percent drop from the game's 1984 sales peak. The only players who seemed to be left, one *Business Week* article noted, were those "who do crossword puzzles in ink." Hasbro chose to release pop culture versions of the game. The marketing plan succeeded; unit sales of the game jumped from 500,000 in 2001 to 2.4 million in 2004. This was good news for Hasbro, but it was seen as troubling by many who thought it a sign of a culture that too often appears to favor the less intellectual. While a player once had to know something about Julius Caesar or French history, contemporary players must now draw on knowledge of celebrities and pop music.

FIGURE 27

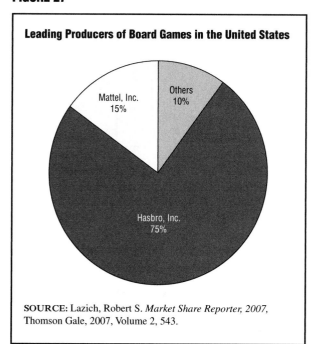

Leading Producers of Board Games in the United States

- Mattel, Inc. 15%
- Others 10%
- Hasbro, Inc. 75%

SOURCE: Lazich, Robert S. *Market Share Reporter, 2007,* Thomson Gale, 2007, Volume 2, 543.

Some analysts believe that more Americans were staying closer to home after the attacks of September 11, 2001. Board games saw a small increase in shipments. Manufacturer shipments of nonelectronic games and puzzles increased from $393.7 million in 2002 to $483.6 million in 2003, according to the U.S. Census Bureau. In 2003 retail sales of board games totaled $870 million; another estimate suggested the industry topped $1 billion. About 34 establishments made board games in 2002 as their primary product. Other toy companies made board games as side ventures; the Census reports that there were 742 establishments engaged in the overall toy, game, and children's vehicle market in this year. According to *Market Share Reporter 2007*, Hasbro's acquisitions of Milton Bradley and Parker Brothers gave it a 75 percent share of all board game manufacturing in the United States in 2005. Mattel was second in the market with a 15 percent market share. Board game sales totaled $870 million in 2006, up from $706.2 million in 2005, according to the NPD Group.

KEY PRODUCERS/MANUFACTURERS

Hasbro, Inc. Hasbro began in 1923, when two brothers founded a small store in Providence, Rhode Island. Henry and Helal Hassenfeld's store was known as Hassenfeld Brothers Inc., and they started off selling textile remnants. These cloth remnants were used to line hats and pencil boxes. They soon decided to start making the pencil boxes and school supplies themselves. From there, the eight company employees, all family members, ventured into toy making. Their first offerings included wax crayons, paint sets, and doctor kits.

The company was primarily a toy maker by the early 1940s. The company was on its way to producing lines of toys that have become classics. In 1952 Hasbro produced Mr. Potato Head, the first toy to be advertised on television. In 1954 it became a licensee for Disney characters. In 1964 the company introduced the G.I. Joe action figure. The company would cease production of G.I. Joe in 1975. There were several reasons the action figure was discontinued. The increased cost of plastics in the mid-1970s made producing the G.I. Joe unattractive. Concerns with violence were another. Consumers were increasingly concerned with violent, military-themed toys and arcade games. Disquiet with the Vietnam War was fresh in the minds of many parents as well. The G.I. Joe was reissued in 1982.

In 1968 the company changed its name to Hasbro Industries Inc.; it had actually produced toys under the trade name Hasbro for some time. In the 1970s the company expanded into some ill conceived ventures. In 1970 it expanded into the nursery school market through the Romper Room brand name. It also produced cookware associated with the Galloping Gourmet. The company deemed its Javelin Darts toy unsafe. Some of its toys simply performed poorly in the market.

The 1970s also mark the beginning of some key acquisitions for Hasbro. In 1977 it bought the rights to license the Peanuts characters. It purchased assets from Warner Communications in 1983, including the rights to Raggedy Ann and Andy. In 1983 it acquired GLENCO Infant Items, the world's largest maker of bibs. In 1984 it acquired Milton Bradley, which gave the company more of the board game industry's biggest titles: The Game of Life, Candy Land, Scrabble, and Chutes and Ladders. In 1989 it acquired Coleco Industries, maker of the Cabbage Patch dolls. In 1991 Hasbro acquired the Tonka Corporation, including its Kenner Products and Parker Brothers divisions. The acquisition brought Play-Doh, Easy Bake Oven, Nerf, Monopoly, and some Star Wars and Batman properties into Hasbro's holdings.

Mattel, Inc. In 1945 Ruth and Elliot Handler and Harold Matson created Mattel Creations, which made dollhouse furniture and picture frames. In 1959 the company introduced Barbie to the world. Reportedly Ruth created the doll when she noticed her daughter Barbara's preference for adult dolls over her baby dolls; the name for Ruth's new doll was also inspired by their nickname for their daughter. The doll was a hit and it remains the best-selling toy of all time as of 2006.

Mattel continued to introduce popular toys through the 1960s, such as the Chatty Cathy doll and the Hot Wheels cars. Like Hasbro, it also began rapidly acquiring companies, absorbing a dozen companies before 1970. The company made some missteps and neared bankruptcy in the 1980s. New management helped turn the company around, as did its 1993 merger with Fisher Price, the world leader in infant and preschool toys. In 1998 it reached an agreement with Walt Disney to produce toys for the company (the company had sponsored the *Mickey Mouse Show* in 1955). The company, headquartered in El Segundo, California, employed 32,000 people and had revenues of $5.6 billion for the year ended December 2006.

Fundex Games. Fundex was founded in 1986 and is based in Indianapolis, Indiana. Fundex is the producer of a number of games and has received many awards from trade groups and parents organizations for their products. Its bestsellers include What's in Ned's Head, Booby Trap, and Gnip Gnop. It has an exclusive relationship with the Professional Domino Association to market its products. From 1996 to 2005 the company grew from eight to 50 full time employees and spends $200,000 annually

marketing its products. The company had an estimated 3 percent market share in 2005.

Cranium Inc. Richard Tait and Whit Alexander invented the game of Cranium in 1998. The game was launched through Starbucks, Barnes & Noble, and Amazon.com. These were unusual distribution points—Starbucks and Amazon.com had just arrived on the American retail scene, but the game found a following among the customers of these companies. Cranium Inc. has developed games for fifteen international markets and has sold some 6 million units in the United States alone. Cranium won the Game of the Year awards at the American International Toy Fair, the industry's annual trade show, in 2001 and 2002. The company has made a version for children. Its other games include Hoopla and Ziggity, released in 2003 and 2004 respectively. The company is based in Seattle, Washington.

MATERIALS & SUPPLY CHAIN LOGISTICS

The materials used in a board game vary depending on the game, of course. Most game pieces are made from plastic using an injection molding process. Various polymers, iron oxides, and dyes are used in the coloring process. Dice are also made from plastic. Their manufacturing process is similar to game pieces. The plastic used must be easily colored, heat stable, and durable. The gameboard is typically made of chipboard with a laminated cover featuring some sort of graphic (the spaces a player must move to, for example). The laminated coating also increases the gameboard's durability. The packaging is an important part of the production process. The box features color photos on its lid. Often this photo is of a group of children or adults enjoying themselves as they play the game. This artwork must be designed, photographed with models, and printed.

DISTRIBUTION CHANNEL

The toy industry is adapting to a changing market in which children are putting down traditional toys at earlier ages for more high-tech gadgets. This means board game makers must fight for shelf space at Wal-Mart and other mass merchandisers. Traditional toy stores also must compete for shoppers, who often turn to discount stores and online stores to make their purchases. U.S. toy sales moved up to $22.3 billion in 2006.

According to *Market Share Reporter 2007*, 55 percent of all toys and games were sold at mass merchandisers and discount stores in 2006. Toy stores represented 18 percent of sales. Another 6 percent of toys were sold online. Other markets such as food stores, drug stores, and department stores represented the other 21 percent of the market.

FIGURE 28

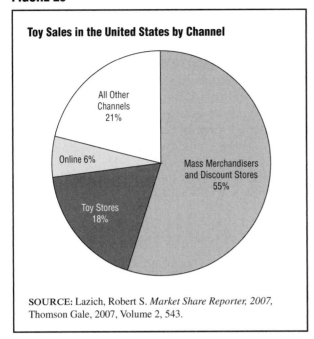

Toy Sales in the United States by Channel

All Other Channels 21%

Online 6%

Toy Stores 18%

Mass Merchandisers and Discount Stores 55%

SOURCE: Lazich, Robert S. *Market Share Reporter, 2007,* Thomson Gale, 2007, Volume 2, 543.

KEY USERS

Both children and adults play board games. Families often have game nights, which provides them with entertainment at a far cheaper cost than a trip to the movies or a sporting event in a big city. Psychologists have long pointed to the positive features of board games. Such activities promote critical thinking and good sportsmanship. Games once were used to provide moral instruction and still are. A market for religious games first blossomed in the 1980s. There are Christian versions of Trivial Pursuit and Monopoly; in the latter, churches rather than hotels are built. The National Scrabble Association sanctions over 150 Scrabble tournaments and more than 200 clubs in the United States and Canada every year. Monopoly, reportedly played by more than 750 million people since its creation in 1935, has its own tournaments each year attended by thousands.

ADJACENT MARKETS

Just as Hasbro and Mattel lead in the board game market, they are also leaders in the toy market. As stated, the two firms produce or license some of the most well known brands in the toy market: Hot Wheels, Sesame Street, Fisher Price, Star Wars, Batman, and Barbie. Both are publicly traded with Mattel reporting sales of $5.1 billion and Hasbro $3.1 billion in their most recent fiscal year.

The toy industry is dominated by four major product categories: electronics, vehicles, arts & crafts, and infant/preschool products. These four categories represented $9 billion in retail sales in 2005. Even as this industry enjoys

strong sales, its manufacturing base has moved overseas. This trend in production began in the 1990s. In 1993, according to government statistics, there were 42,300 workers in the industry. By 2005 that figure had dropped to 17,400 workers.

RESEARCH & DEVELOPMENT

Board game makers continually bring updates and innovations to their products. Hasbro introduced a new version of Monopoly in 2006 that dispensed with Boardwalk and other game-board spaces named after sites in Atlantic City, New Jersey, and landmarks from U.S. cities. DVDs are being incorporated into games to make the experience more interactive and exciting. But others are looking at board games in more serious ways. Psychologists have long seen such games as revealing of human behavior, from how humans handle competition to our likelihood to cheat. Scientists have only just begun to see how the mental activity involved in game playing might delay the onset of Alzheimer's and other diseases. Computer programmers use games as useful ways to test the limits of a computer system. In 1997 a computer famously beat chess champion Gary Kasparov. In 2005 it was announced that a computer had solved the notoriously challenging game Go for a 5x5 playing board.

CURRENT TRENDS

There has been a growing chorus of voices noting how Americans are overworked, stressed out, and short of time. Some people like the idea of board games, but dislike the time necessary to sit down and actually play them to completion. Recognizing this, Hasbro has released streamlined versions of some of its games. The Game of Life that uses a Visa card rather than cash and electronically keeps track of points—both keep the game moving. In 2007 *Express* versions of some classic games were released: Monopoly Express, Scrabble Express, and Sorry Express.

TARGET MARKETS & SEGMENTATION

Considering the popularity of DVDs, video game consoles, and online games, one may be surprised that the board game industry continues to see strong sales. When such new media appeared, more than a few parents and toy store owners feared the demise of classic games that had been popular during their early years.

Will Manley, a columnist in a 1995 *Booklist* article predicted that board games would be obsolete in 20 years. He laments that children don't have the patience for board games because they are too hooked on the violence and speed of video games. He speaks for many when he extols the pleasures of board games: "There's a certain leisurely comfort involved with popping a big bowl of popcorn, pouring everyone a steamy cup of hot chocolate, and spreading out an oversize playing board on the kitchen table during a frosty February afternoon. Likewise, nothing beats heading to the beach on a sunny day in July with a transistor radio, a cooler full of Squirt, and a game of Scrabble. The best feature of board games, of course, is the fact that they are slow. They take the hurry out of life. There are no unnerving beeps to nudge you along and no blinking screens to raise your blood pressure. If you want to conquer the world or even just Atlantic City, you can take your own sweet time doing it." Setting aside Manley's reference to a transistor radio, his description feels just as timely and relevant as readers must have found it then. In 2015, the year Manley predicted board games would go the way of the dodo bird, the Census Bureau predicts there will be about 49.5 million children under 12 years of age. One hopes they too will discover the simple pleasures of which Manley speaks.

RELATED ASSOCIATIONS & ORGANIZATIONS

American Specialty Toy Association,
   http://www.astratoy.org
Game Manufacturers Association, http://www.gama.org
Hasbro, http://www.hasbro.com
National Scrabble Association,
   http://www.scrabble-assoc.com

Toy Industries Association, http://www.tia.org

BIBLIOGRAPHY

Bandahl, Brian. "Dice Comes Up Good for Miami Board Game Maker." *South Florida Business Journal.* 8 June 2007.

"Busy Lives, Short Attention Spans Prompt Speedier Board Game for Americans." *International Herald Tribune.* 29 March 2007.

DeMott, John S. "Booming Sales in Toyland." *Time.* 10 December 1984.

"Hasbro Has Kids Hopping." *Business Week.* 15 August 2005.

Jensen, Jennifer. "Teaching Success Through Play: American Board Games and Table Games 1840–1920." *Magazine Antiques.* December 2001.

Keller, Lauren. "Games: A Family Affair." *Playthings.* July 1997.

Lazich, Robert S. *Market Share Reporter 2007.* Thomson Gale, 2007, Volume 2, 543.

Manley, Will. "Games." *Booklist.* 1 March 1995.

"SceneIt Soars to No. 1 Again!" *CNW Group.* 15 February 2006.

Schulman, Milt. "Steady Sales Keep Board Games on the Map." *Playthings.* March 1993.

"Table 1: Value of Shipments for Product Classes: 2005 and Earlier Years." *Annual Survey of Manufactures.* U.S. Department of Commerce, Bureau of the Census. November 2006.

"Table 5: Industry Statistics by Primary Product Class Specialization: 2002." *Game, Toy, and Children's Vehicle Manufacturing: 2002.* U.S. Department of Commerce, Bureau of the Census, Economics and Statistics Administration. January 2005.

Boats

—■—

INDUSTRIAL CODES

NAICS: 33–6612 Boat Building

SIC: 3732 Boat Building and Repairing

NAICS-Based Product Codes: 33–66121, 33–66123, 33–66125, 33–66128, and 33–6612W

PRODUCT OVERVIEW

People think of ships and boats as water-borne craft and use size to distinguish between the two. When Jane Doe talks about a sailing ship she means an old-fashioned clipper cruising the oceans with multiple masts carrying clouds of sail. But when Jane talks about a sailboat she means something smaller and intended for recreation. She would think of yachts as boats but feel that they were also something in between, especially if the yacht routinely sailed the oceans.

The U.S. Census Bureau distinguishs between products manufactured by the shipbuilding industry (NAICS 33–6612) and the boat building industry (33–6611). The Census Bureau does not use size alone to make its divisions. It follows a functional definition in which end use plays a role. Thus the Bureau explicitly defines barges, crew boats, ferryboats, fireboats, patrol boats, and tugboats as objects made by the shipbuilding industry. But the Bureau assigns lifeboats to boat building, divides hydrofoils between the two industries by size, and it explicitly excludes from both ships and boats inflatable craft made of plastics or rubber; but these last categories are classified as products of the plastics and rubber products industries.

Types of Boats. The vast majority of boats are used for recreation, but a small proportion of the boat industry's output is sold to commercial, institutional, and military users. Some of these boats are especially made for such applications; most are ordinary pleasure craft outfitted with other equipment after purchase for uses in patrolling rivers or for utility purposes.

Boats are classified, first of all, by means of propulsion. They are motorized, move under sail, or they are rowed. Motorboats are subdivided into four major classes. Outboard motorboats, the largest category, carry their propulsion system affixed to the back of the boat. Both the engine, its gearing, and the propeller are outside the boat. The boats are designed to be moved by an engine and may be sold as a unit, with a motor, or without one; in the latter case the owner can buy his or her preferred brand of outboard motor or use one he or she already owns.

Inboard motorboats come with an engine built into the vessel itself and thus taking up some of the vessel's space. When the power plant is located toward the front or center of the boat, a shaft runs from it back to the rear of the craft; it is tipped by the driving propeller just beyond the stern of the boat. This type of arrangement is known as a direct-drive inboard arrangement—a single shaft going straight back. An alternative arrangement, called the V-drive, is used to create more cabin space in the front of the boat. The letter V is used as a designation because this type of inboard engine uses two shafts, with gearing at the bottom of the V. The engine is placed in the back of the boat, thus out of the way, and oriented so that the first shaft points toward the front of the boat—thus in the wrong direction. The gearing transfers power from the engine to the second shaft, which points at the back or stern of the boat. The rear-pointing shaft carries the propeller.

Sterndrive motorboats, also called inboard/outdrive boats, present an alternative to inboard arrangements for all boats of some size. Sterndrive engines are located under the deck of the boat at the very back of the vessel. A short shaft carries power to gearing and propeller located outside the boat and under water, the shaft passing through a sealed opening in the transom (the back end) of the boat. Sterndrive boats are growing in popularity. They are powerful and provide maximum room inside the boat for cabin space and at the stern of the boat for fishing without the interference of an outboard motor.

Jet boats are inboard motor boats propelled forward by a jet of water. The water is drawn in from beneath the boat by an inlet, its flow accelerated by an engine moving pumps or by a turbine-style engine. The water is shot out of the boat at its stern but just above the water line. Jet boats originated in New Zealand in the 1950s and were originally designed for use in shallow waters in which rocks tended to damage propellers. Jet boats, however, also have their limitations in that mud or debris can block the flow of water and stop the engine. This system of propulsion gave rise to the most popular category of boat, or boat-like recreational vessel, the personal watercraft or PWC. While people ride inside a jet boat, which looks like any other boat, people straddle PWCs like a horse or motorcycle.

Sailboats are classified as sailboats not because their sole method of propulsion is by wind but because their principal function is to move under sail. Larger sailboats typically come equipped with inboard or sterndrive motors. Midsized sailboats all use outboard motors to come in if becalmed or to maneuver easily near the docks on exit or reentry. The smallest category of sailing vessel is a surfboard equipped with a mast and a sail. The largest is a yacht quite capable of ocean crossings with a crew and passengers. Catamarans have double hulls separated by a platform that supports the mast, the steering, the passengers, and all their gear.

Two-thirds of all boats in 2005 were motorized. The remaining third were made up of rowboats, canoes, and kayaks. The Census Bureau does not include inflatable boats under NAICS 33–6612 because their methods of manufacture do not resemble construction of rigid boats. At the retail level, however, boats are boats and the industry considers inflatables as part of its normal product spectrum. Inflatables may be large enough to require motors for their propulsion, may be rowed, and some are also equipped for sailing. Many people convert boats designed for rowing into motorboats by attaching an outboard motor to them. Canoes and kayaks are narrow vessels; canoes carry two or three people, kayaks one or two.

Materials of Construction. Boats are also classified by their material of construction. People speak of fiberglass and of aluminum boats. The largest materials category, based on boats in use, is fiberglass, although the fiberglass itself is actually only a component of so-called fiberglass boats. These are really plastic boats. The formal designation used is fiberglass-reinforced plastics (FRP). The mass of fiberglass boat hulls is made with styrene polymers into which fiberglass matting is embedded to give the composite material greater strength.

In 2005, 58 percent of motor and 45 percent of non-powered boats were made of fiberglass. Metal, predominantly aluminum, formed the hulls of 37 percent of motor and 33 percent of other boats. Humanity's oldest boat-building material, wood, was the hull material of 1.4 percent of all motor boats and of 4.1 percent of non-powered boats. Hulls may also be formed from composites of metals, wood, canvas, rubber, plastics, and sealants. Indeed trends in the direction of composites may develop in countering the environmental problems of forming FRP hulls. Motor boats with composite or treated canvas hulls were 2.5 percent of such boats and 16.4 percent of non-powered vessels in 2005. Inflatables, made of plastics and rubber, were 0.9 percent of mechanically propelled and 1.9 percent of non-powered boats. The source of these data is the National Marine Manufacturers Association (NMMA) which publishes an annual statistical abstract of boating.

Size Classes. NMMA also provides insight into the sizes of mechanically propelled boats in use. The majority of such boats (51.8%) fell into the 16 to less than 26 foot category followed in second place by boats under 16 feet (43.5%). Large boats ranging from 26 to under 40 feet represented 4 percent of boats. Large yachts, 40 feet long and larger had the same proportion to all boats as the very rich have to the population as a whole, 0.6 percent.

Average Unit Costs. According to NMMA data, in 2005 those wishing to participate in recreational boating at the lowest possible cost could buy themselves a kayak retailing at an average price of $478. In ascending order were canoes ($627), inflatables ($1,912), personal watercraft ($9,495), outboard motor boats ($15,006—but $25,118 with the outboard motor added), jet boats ($25,108), sterndrive boats ($35,592), inboard ski boats ($40,297), sailboats ($44,926), inboard cruisers ($399,815), and houseboats ($720,210). People with smaller boats normally stored at home in or just outside the garage were likely to have to add another $1,846 on average for a trailer. On the whole, however, it is evident from these data that the boating industry offers products for all income classes, from the most modest on up to the most lavish and demanding.

MARKET

In 2005 boat building was an $11.4 billion industry, having more than doubled in the eight years from 1997, a year in which it was $5.2 billion in size. Boat building enjoyed a 10.5 percent annual growth rate between 1997 and 2002, the two full economic survey years for which data were available at the time of writing, and again between 2002 and 2005. But this strong and sustained growth was temporarily reversed in 2001 as the economy turned down: shipments fell from $8.3 to $7.2 billion between 2000 and 2001 but then grew again to $8.5 billion by 2002, absorbing both the shock of an economic turndown and the mood change introduced by the 9/11 terrorist attack rather quickly.

As the Census Bureau sees this market, measuring its size in dollars rather than in units, outboard motorboats were the largest segment in 2002 (34% of shipments, valued at $2.6 billion). Next in order were inboard motorboat (30%, $2.4 billion), sterndrive boats (20%, $1.6 billion), and every other kind, including sailboats (16%, $1.3 billion). The Census Bureau reports on various sizes of sailboats but its legally imposed obligations to protect the identity of producers, which can be inferred from data for certain sizes of boats, causes it to suppress enough data so that detailed analysis of this "all other" category is not possible. Slight changes in the share of these categories took place between 1997 and 2002. Outboard motorboats increased (from 32% in 1997) as did inboard motorboats (from 25% in 1997). Sterndrive boats lost share of shipments (down from 28% in 1997) and the all other category gained a point (up from 15% in 1997).

Product category data are not available from the Census Bureau for 2005, but changes in boat registration data indicate patterns of change, in the 2002–2005 period. Personal watercraft enjoyed the sharpest gain in registration, growing at 4.7 percent per year, followed by non-powered boats, growing at a 1.8 percent rate. Next came sterndrive boats, growing at 1.7 percent annually. Inboard boats had a 1.0 percent and outboard motorboats a 0.7 percent annual growth in registrations. Sailboats lost registrations in this period at the rate of 0.8 percent per year. Registration data, of course, are not market data. They represent net additions to a list of boats in active use calculated by reducing registrations by annual retirements and adding new registrations. What these registration data do show, however, is a broad trend: the market is growing most rapidly at the bottom (relatively inexpensive boats and toys) and at the top (expensive sterndrive and inboard vessels).

Based on registration data, approximately 1 percent of all boats are in the commercial category. Registrations, to be sure, capture only units, not their values. The commercial boat sector represents a much higher percentage of

dollars for the simple reason that such boats are of the largest sizes and are also the most sophisticated in construction and internal furnishings. They are passenger vessels, commercial fishing boats, and are used in various research and industrial deployments.

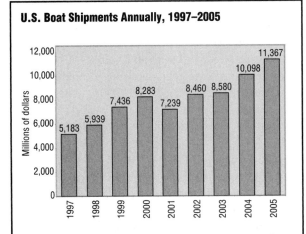

FIGURE 29

U.S. Boat Shipments Annually, 1997–2005

SOURCE: Compiled by the staff with data from "Boat Building 2002," *2002 Economic Census*, and "Statistical for Industry Groups and Industries: 2005," *Annual Survey of Manufactures*. January 2005 and November 2006.

FIGURE 30

U.S. Boat Production by Type in 2002

Percentages are based on a total value of boat shipments equalling $7.85 billion.

- Boat building not specified by kind 10%
- Sailboats and all other 6%
- Sternboats 20%
- Outboard Motorboats 34%
- Inboard Motorboats 30%

SOURCE: Compiled by the staff with data from "Boat Building 2002," *2002 Economic Census*. January 2005.

In 2005 boating was supported by an estimated 71.3 million participants in boating activities. This population waxes and wanes in relation to income and employment trends. In the period 1989–1997, the population of participants was steadily growing to a peak of 78.4 million people reflecting a vigorously growing economy. Participants declined in the period 1998–2001, went up again in 2002, declined to an all time low in this 1989–2005 period, to 68.7 million in 2003, and then showed growth again to 2005. In 2005, 17.95 million boats were in use, suggesting roughly four participants per boat in the water. This total tally of boats is made up of 12.98 million registered and 4.97 million non-registered boats.

Trends in household income are perhaps the best indicator of what has driven growth in the boating industry. Household income in constant dollars grew 9.5 percent between 1986 and 2005. But the distribution of this income by fifths (quintiles) of households is the more telling indicator. In the 1986–2005 period, the two lowest quintiles of households saw income growth of 8.6 and 8.4 percent respectively. The middle quintile's income grew at 9.6 percent. The two highest fifths of households, however, enjoyed growth of 14.5 and 32.5 percent respectively. The rich were getting richer faster. Such a pattern of skewed income growth creates markets for recreational goods of the more expensive kinds—personal watercraft, speedboats, and motor cruisers.

KEY PRODUCERS/MANUFACTURERS

Some 1,060 manufacturers share the $11.4 billion (2005) pleasure boat market distributing its products through approximately 5,000 dealers. Entry into the boat-building business has always been and continues to be relatively easy, accounting for the large number of producers and the fact that a major component of boats, the outboard motor, is produced by just a handful of major companies. A very strong consolidation-trend has been underway in boating, suggesting that in due time just a handful of companies will emerge into dominant positions, but this is not yet the case. The two largest companies control 35 percent of unit sales. No other company has even a 5 percent share of units. And the leaders display a rather bewildering array of brand names, suggesting that buyers of boats continue to support a great diversity of offerings.

The largest boat company is Brunswick Boat Group, a company with $2.8 billion in sales in 2005 and an 18 percent share of units. To illustrate brand diversity, Brunswick's brands are Albemarle, Baja, Bayliner, Boston Whaler, Cabo Yachts, Cresliner, Harris Kayot, Hatteras, Laguna, Lowe, Lund, Maxum, Meridian, Palmetto, Princecraft, Sea Boss, Sea Pro, Sea Ray, Sealine, Triton,

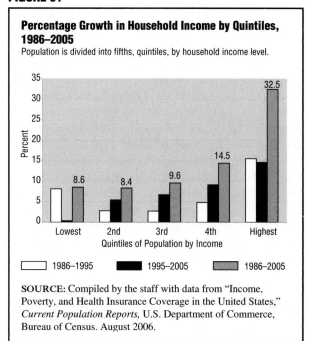

FIGURE 31

Percentage Growth in Household Income by Quintiles, 1986–2005

Population is divided into fifths, quintiles, by household income level.

SOURCE: Compiled by the staff with data from "Income, Poverty, and Health Insurance Coverage in the United States," *Current Population Reports,* U.S. Department of Commerce, Bureau of Census. August 2006.

and Trophy. Brunswick acquired four of these brands (Ablbemarl, Sea Boss, Sea Pro, and Triton) in 2005 alone. In dollar volume, Brunswick is estimated to hold 30 percent of the market.

Second-ranking Genmar Holdings, Inc., a privately held company, estimates its sales at $1 billion. Like Brunswick, Genmar has a diversity of products and many brands: Carver Yacht, Champion, Four Winns, Glastron, Hydra-Sports, Larson, Marquis, Ranger, Seaswirl, Stratos, and Wellcraft. The company's unit share of the market is 17 percent.

Third-ranked, with 4 percent of units sold, is Tracker Marine Group. The company specializes in fishing boats. Its brands are Kenner, Mako, MYacht, Nitro, Seacraft, and Sun Tracker.

Backing up these and all the other companies are outboard motor companies. Some seventeen producers offer motors. Of the motor producers six are Chinese, four are Japanese, two Italian, one Korean, and four are North American. Bombardier, a Canadian company, owns the Evinrude and Johnson outboard motor brands. The ownership is Canadian, but the boats are manufactured in the United States. Mercury Marine, also manufactured in the United States, belongs among the top three U.S. brands alongside Evinrude and Johnson. In the United States, Briggs & Stratton also makes small gasoline and electric outboard motors. Yamaha, the leading Japanese motor producer, also sells a line of boats; the boats represent an estimated 3 percent share of total units.

MATERIALS & SUPPLY CHAIN LOGISTICS

Production facilities of large pleasure craft—those difficult to transport by road—tend to be sited at or very near bodies of water where they will be used. Smaller boats, however, including quite sizeable vessels readily transported by truck, are produced all over the country. Thus, for instance, Brunswick's large offshore sports fishing boat, the Albemarle, is produced in Edmonton, North Carolina, on the Albemarle Sound, but its Lund brand of aluminum boats are made in New York Mills, Minnesota, not far from Fargo, North Dakota. Tracker had its origins in Lebanon, Missouri, not far from the Lake of the Ozarks but situated along Interstate 44 with no water in sight. And Genmar began in Minnesota. The company continues to draw some of its product, the Larson line of boats, for instance, from that state. Larson, to be sure, is produced in Little Falls on the banks of the Mississippi.

The principal manufacturing materials for metal boats are aluminum and aluminum-base alloys, delivered in sheets. Fiberglass boats depend on plastic resins delivered in granules or pellets and on glass fiber arriving as mats. These two categories of materials amount to around 10 percent of total materials consumed by the boating industry measured in dollars. The major category of purchased equipment takes the form of diesel and gasoline engines, accounting for 20 percent of inputs. All of the rest is accounted for by steel or plastic forgings, castings, shapes, and extrusions, hardware of a great diversity of types, lumber, plywood, and purchased furnishings, pumps, canvas products, paints and coatings, piping and fittings, and nautical instruments and communications equipment.

DISTRIBUTION CHANNEL

Most boat producers utilize a two-tier distribution system in which the producer sells to a dealer and the dealer sells directly to the ultimate consumer. In this business—unlike others that involve motorized recreational products (snowmobiles, motorcycles)—the wholesaler/distributor is absent. With some 1,000 producers and 5,000 dealers characterizing the U.S. distribution system, on average every producer delivers goods through five dealers. Averages, of course, do not apply, but the numbers indicate that many producers deliver their products through a handful of local dealers only, whereas big brands command large dealer chains.

The large boat dealers, however, operate very sizeable business. This is evident from data published by *Boating Industry* magazine as part of its Top 100 program. Data for 2004, averaging results from 314 major dealers, indicated that such dealers had, on average, sales of $17 million, $6.9 million per store, and employed 18 people per store, 43 per dealer.

A major institutional aspect of boat distribution in the United States and elsewhere in the world is the annual boat show. These are typically held in large conference centers in the cold season but may also be held on lakes or ocean shores in the open during the warm months. Local and regional dealers come to display their wares for the inspection of people pondering the purchase of a boat. More than a hundred such shows are held in the United States every year. Depending on location, the first are staged each year in February. The majority of boat shows take place in March and April when Spring is near but the weather is still inclement or blustery and potential customers' thoughts turn to the warm season ahead. Most major cities have a boat show. Shows continue to be held through August.

KEY USERS

With some 70 million participants in boating, virtually the entire population participates in recreational boating at one point or another, at one time or another, either through ownership, rental, or enjoying a little water cruising with good friends. Boat registration and expenditure data indicate that key users are people who live close to large bodies of water. Thus the top five states in total boat registration are Florida, Michigan, California, Minnesota, and Texas. Florida is surrounded by ocean, Michigan by the Great Lakes. Most of California's population is on its Pacific coast. Minnesota touches Lake Superior and proclaims itself, on its license plate, the Land of 10,000 Lakes. Texas looks at the Gulf of Mexico and is drained by several great rivers. Not surprisingly, the next five states in rank are Wisconsin, New York, and Ohio, adjoining the Great Lakes, South Carolina on the Atlantic, and Illinois on the Great Lakes. Surprisingly, perhaps, Hawaii is next to the last in boat registrations although it is an island state; perhaps Hawaii's people are glad to be on solid ground. Last is the District of Columbia.

These raw ranks of states by boat registration do not account for the significantly different population sizes of these states. If the rankings are restated as number of boats per 1,000 of population, the leading state is Minnesota, with 171 boats per 1,000, followed by Wisconsin (117), South Carolina (103), Michigan (97), and Florida (60).

The top ten states in expenditures on boating are shown next in rank order, but to indicate the ability of the states' populations to afford boats we also show each state's rank in per capita income in the nation in parentheses. For instance, Florida spends the most money on boating despite the fact that it is 23rd in per capita income. Location and climate are obviously important aspects in boating. The top ten: Florida (23rd) California (12th), Texas (32nd), New York (5th), Washington (11th),

North Carolina (37th), Michigan (22nd), Minnesota (8th), Wisconsin (21st), and Virginia (10th).

ADJACENT MARKETS

From a functional point of view, less so from a manufacturing perspective, the market most adjacent to boat building is the shipbuilding industry. In 2005 shipbuilding was a $14.2 billion industry in the United States, thus somewhat but not significantly larger than boat building. The industry produces floating platforms like oil drilling rigs, oceangoing commercial and naval vessels, as well as ships and barges used in lake and river traffic. The shipbuilding industry's output is two-thirds military and one-third commercial vessels (or their refurbishment). Shipbuilding and boat building are quite different activities. Very differently organized and structured companies participate in the business—but both produce craft that float on the water.

In recreational boating, and viewed from the perspective of the participant, other motorized toys represent adjacent markets, including four-wheeled all-terrain vehicles, motorcycles, and snowmobiles. Viewed more broadly, all recreational markets compete with one another. The couple heading out on a tour of famous golf courses could, instead, invest in a splendid yacht. And that long-desired second home built out of redwood logs, up on the hill, overlooking the lake, costs a pretty penny too. But down on the shimmering water, by that little dock, there may very well be a boat for the evening excursion to see the sunset.

In recreational boating, the boat itself may be the central focus but the enjoyment of the sport requires a substantial infrastructure and other equipment as well. Trailers to transport boats to water are an important market. The NMMA's estimate of the retail market for trailers for 2005 was $247 million. In the services sector, marinas and boatyards are important by providing slips, buoy moorings, fuelling, and security services. Although many marinas and boatyards are publicly owned and operated or operated by private clubs for their own members, *Boating Industry*, in a 2006 survey, reported that more than 70 percent of marinas are commercial, for-profit operations.

Very expensive houseboats, found in use along major rivers and in coastal regions, represent an alternative to real estate. Such boats typically spend their lives moored along docks. For some they are the primary residence, for others they are second homes. Houseboats see the wider reaches of water only on occasional cruises on special occasions.

Boat-making is also practiced as a hobby by many individuals and this hobby represents a unique adjacent market in materials, tools, and accessories. Canoes are the most commonly made vessels of this type, but hobbyists continue to make every other kind of boat as well, including very fast speedboats with unique engine designs and such improbable-sounding but very functional objects as sailing yachts, the hulls of which are made of reinforced concrete.

RESEARCH & DEVELOPMENT

A significant problem in the building of fiberglass-reinforced plastics (FRP) pleasure boats is emission of styrene to the atmosphere. Styrene emissions are controlled by National Emission Standards for Hazardous Air Pollutants. Emissions appear in open-mold FRP boat construction when the resin is heated in contact with air. Standards set limits on emissions during fixed periods of time; these can only be met by limiting production. Significant effort in R&D is devoted to this issue in the industry to bring emissions under control short of introducing closed-mold manufacturing technologies with capital requirements significant enough so that the technique is beyond the means of most smaller boat builders.

Alongside and as a consequence of emission problems in FRP boat construction, the industry is engaged in research aimed at developing composite hull designs using modern materials—and sometimes in combining traditional materials like wood and modern sealants. Such research is very costly because many years of product testing are required to detect negative consequences of what appear to be good designs. Engine R&D is actively pursued by the much larger outboard motor companies. The U.S. Coast Guard maintains a continuing R&D effort aimed principally at boating safety, but Coast Guard findings in turn result in design changes by boat manufacturers.

CURRENT TRENDS

Boating for pleasure—not merely for transportation, commerce, or military purposes—predates modern times and, no doubt, will flourish even if, as a consequence of political or climate changes, a very different future emerges. But in the current context, the most important driving force behind demand for boating is disposable income. It has been responsible for a general upgrading of purchases, exemplified by the rapid growth in personal watercraft, a relatively new line of products, and innovation in engine designs. Growth has been in powered craft at the expense of sailboats. More expensive boats have gained share.

Trends in disposable income indicate much more rapid growth in the income of the top two-fifths of the population and, for the immediate future (the first and second decades of the twenty-first century) this trend may continue to hold. Mitigating it is the predicted increase in fuel cost which is making the use of water craft more and more expensive. Gasoline or diesel purchased at marinas typically costs more than at filling stations, but buying fuel and transporting it to the boat is often impractical. Prices of fossil fuels are likely to trend up. In time this factor may have an impact on the kind and sizes of boats purchased.

Another trend, already noted, will be continuing consolidation in boat manufacturing, in part driven by environmental concerns, as already discussed above under research and development trends. As yet consolidation has not resulted in the reduction of actual manufacturing establishments. Establishments, in fact, have grown in number. But modern techniques in producing fiberglass boats may lead in that direction with possible reductions in the number of different kinds of craft likely to be available in the future.

TARGET MARKETS & SEGMENTATION

People purchase boats with a significant primary experience in mind and then consider, as well, secondary experiences that the boat may also provide for them. These experiences include the exercise of physical skills—and the opportunity to teach them to the young—and the desire to have these experiences may be the principal decision behind the purchase of sailing craft. People are drawn to sailing precisely because it puts them in a position to move across water using only nature's power and their own ingenuity. At the most basic level, such individuals can enjoy the experience and practice their skills by participating in wind surfing all by themselves. Those who buy sailing yachts can reproduce the oldest human exploratory experience, crossing the great water in search of adventure. The sailing yacht, of course, also provides a social experience. Sailboat racing is a sport in its own right, and boats specifically designed for racing are a distinct segment in this category. In marketing their sailboats, producers aim to stir up the desire for such experiences.

The experience of speed creates another special market in boating. It accounts for several distinct segments, including speed boats, boats designed for pulling water skiers, and personal water craft that, by exposing people to the water at fairly high speed, provide the exhilaration of speed boating but at relatively low cost.

Kayaks and canoes are targeted to that segment of the population that sees boats as a means to exercise in the watery environment at very close quarters. The boat is a means of testing one's skills while, at the same time, being in contact with the natural environment and, in the more extreme uses of such boats, experiencing the thrill of danger.

A third experience that draws industry's specific targeting is fishing—so much so that bass boats are a distinct category of boats. Boating accessories, such as virtually silent trolling motors, are supported by the fisherman's needs. Ocean-going fishing vessels represent the top end

of this market segment. A very large proportion of outboard motorboats are sold to men and women who enjoy many varieties of fishing.

Cruising at slow or modest speed—simply the enjoyment of the water itself and of the shoreline passing by—accounts for the remainder of all boating uses excluding commercial deployments. Rowboats can provide this experience at very modest cost—and afford some exercise as well. Pontoon boats and small cruisers are ideal for small lakes. Large cruisers are best suited for coastal settings and large bodies of water. Cruising boats of one sort or another offer sightseeing for the eyes, the wind for the skin, and the sound of water for the ears, and the companionship of others for entertainment, and the whole experience as a relaxing diversion from the stresses of the rat race.

RELATED ASSOCIATIONS & ORGANIZATIONS

American Boating Association, http://www.americanboating.org

American Power Boat Association, http://www.apba-racing.com

Boat Owners Association of the United States, http://www.boatus.com

National Marine Manufacturers Association, http://www.nmma.org

US Sailing, http://www.ussailing.org

BIBLIOGRAPHY

2005 U.S. Recreational Boat Registration Statistics. National Marine Manufacturers Association. 2006.

"Boat Building 2002." *2002 Economic Census.* U.S. Department of Commerce, Bureau of the Census. January 2005.

Boating Statistics–2005. U.S. Coast Guard. 31 August 2006.

DeNavas-Walt, Carmen, Bernadette D. Proctor, and Cherryl Hill Lee. "Income, Poverty, and Health Insurance Coverage in the United States: 2005." *Current Population Reports.* August 2006.

Kong, Emery J, Mark A. Bahner, and Sonj L. Turner. "Assessment of Styrene Emission Controls for FRP/C Boat Building Industries." U.S. Environmental Protection Agency. October 1996.

"More Consolidation Predicted for U.S. Recreational Boat Segment." Udenrigsministeriet Danmark Eksportrad, Trade Commission of Denmark. 19 January 2006. Available from <http://www.dtcatlanta.org>.

Pascoe, David. *High Tech Materials in Boat Building.* Marine Survey Online. Available from <http://marinesurvey.com/yacht/material.htm>.

"Share of Aggregate Income Received by Each Fifth and Top 5 Percent of Households (all Races): 1967 to 2000." *Current Population Survey.* U.S. Department of Commerce, Bureau of the Census. 21 March 2000.

"Top 100." *Boating Industry.* 21 June 2005.

SEE ALSO *Ships*

Bottled Water

————◆————

INDUSTRIAL CODES

NAICS: 31–2112 Bottled Water Manufacturing

SIC: 2086 Bottled and Canned Soft Drinks and Carbonated Waters

NAICS-Based Product Codes: 31–211201

PRODUCT OVERVIEW

The sale and consumption of water considered to be special in some way appears to be as old as civilization. In Europe under the Roman empire communities grew up around hot springs and mineral springs. These waters were considered to have healing powers and these sites developed into resorts frequented by the wealthy who bathed in the waters and also ingested mineral waters as drink. Centers of this nature have had continuous existence ever since they were settled and have carried names like *water, bath,* or *spring*—used in the singular or plural.

The oldest brands of bottled water illustrate the industry's origins. Probably the very oldest brand of bottled water, Evian, is produced on the southern shore of Lake Geneva, in France, near a place called Évian-les-Bains, literally Evian-the-Baths. Evian began to be sold in 1830 in earthenware jugs. Vittel Grand Source, which appeared in 1855, translates as Vittel Great Spring. Perrier is the name of a doctor who bought a mineral spring in France, but the name of the location is telling: it was then called Les Bouillens, literally meaning The Bubbling Forth, thus a spring. Perrier dates back to 1863, but its promoters don't fail to tells us that Hannibal, invading Europe from

Carthage, stopped at Les Bouillens to enjoy its waters. Whether or not the elephants partook of what later came to be called Perrier is merely speculation. Contrexéville, the home of the Contrex brand, is a thermal spring and health resort the French began to visit in the 1760s to cure kidney stones. The top Italian brand, San Pellegrino, is bottled at the location of the same name, a famous bath and health resort. The oldest brand of American water, Poland Spring, began to be bottled in Poland Spring, Maine in 1845. The inn built near the spring had by then become a well-known health spa. By 1904 the water had won international prizes at the World's Columbian Expositions and the World's Fair. All brands of spring, mineral, and sparkling waters that predate the growth of bottled water as a major product category, pre-1980s, had very similar origins.

The initial motivation for paying money for water was people's perception of a unique health benefit available from certain waters occurring in special places. The taste of the water came to play a greater role later. The purity of specially processed waters became important in regions of hard and/or brackish waters, after the introduction of water chlorination, and eventually as a replacement for soft drinks filled with potentially fattening sugar.

Categories. The U.S. Environmental Protection Agency (EPA) regulates tap or municipal water under the National Primary Drinking Water Regulations (NPDWR). Under its authority to oversee food along with drugs, the U.S. Food and Drug Administration (FDA) regulates bottled water under Title 21 of the *Code of Federal Regulations.* FDA classifies bottled waters into five categories of which two have to do with the source and the other three with content or processing.

The two sources are *Artesian* and *Spring* water. Artesian water is obtained from deep-lying aquifers. The water is held in rock formations and moves very slowly—at fractions of an inch per year. It is reached by drilling deep wells into the bedrock. Spring water bubbles to the surface naturally. The FDA also implicitly recognizes that the source of most bottled water sold in the twenty-first century comes from ordinary tap water, but the agency handles such water under one of its processing definitions.

The other categories are *Mineral*, *Sparkling*, and *Purified* water. Mineral water, which must be from natural sources, must have a minerals contents of at least 250 parts per million in dissolved form; the processor may not add any minerals to the water. Sparkling water, again from artesian or spring sources, must naturally contain carbon dioxide, but if some of this natural carbonation is lost in extraction or processing, it may be replaced. The FDA's purified water category implicitly recognizes tap water as a source. Such water is defined as having undergone distillation, deionization, or reverse osmosis. Distilled water is boiled and the vapors are then liquefied by cooling; mineral contents are left behind. Deionization removes magnesium, calcium, and salt in a process similar to but more intense than water softening. Reverse osmosis is a water cleaning process using engineered semi-permeable membranes; water is forced through the membranes at very high pressure; only pure water passes the membranes leaving a mineral brine on the far side.

Using FDA's definitions, bottled water must either be purified in the strict sense or must be natural without additions of any sort. Only antimicrobial agents (e.g., chlorine) and fluoride may be present. If the processor adds anything at all, the bottled water or mineral water designation must be augmented by the processor who must identify the additive and provide its magnitude on the ingredients list required to be part of the label. FDA's regulations thus assume that EPA regulations on drinking water are already met. EPA's approach is to set maximal content in the water of a wide spectrum of microbes, radionuclides, inorganics, inorganic synthetics, volatile organics, and disinfectants.

Bottled Water In Perspective. The soft drink industry is fond of repeating that people drink in response to three motivational need states: simple refreshment, rehydration, and nutrition. The terminology is marketing deep think but provides a lens for looking at trends in beverages—and how these have changed over time. Changes between 1980 and 2005, as reported by the U.S. Department of Agriculture's (USDA's) Economic Research Service serve as the underpinning.

In 1980 the categories were soft drinks (25.1% of total beverages), milk (20.6%), coffee (20%), beer (18.1%), fruit and vegetable juices (5.7%), tea (5.5%), bottled water (2%), wine (1.6%), and spirits (1.5%). That year per capita availability of purchased liquids was 134 gallons, translating into 33.6 gallons per person of soft drinks and 27.6 gallons of milk. Milk is refreshing but usually classified as a food product. At the opposite extreme distilled liquor falls into the "refreshment" category. The other products exist somewhere between these polarities. Bottled water is the obvious leading candidate in the rehydration category.

In 2005 the same products, again shown in rank order, were soft drinks (27.7% up from 1980), fruit and vegetable juices (16.6% up), bottled water (13.7% up), coffee (13% down from 1980), beer (11.5% down), milk (11.3% down), tea (4.3% down), wine (1.3% down), and spirits (0.7% down). This reordering in shares signals a massive transformation of consumer behavior. Fifth-ranking fruit and vegetable juices moved up to second rank; seventh-ranking bottled water moved up to third place. Only three categories increased their share. Of these bottled water had the greatest gains over 1980, increasing its share by 11.7 points, fruit and vegetable juices by 10.9 points, soft drinks by 2.6 points. Leading the losers were milk, coffee, and beer, declining by 9.3, 7, and 6.7 points respectively.

By 2005 consumption had increased to 186 gallons per person. Tea and wine, although losing share, showed increases in absolute consumption. Fruit juices and vegetables scored the greatest gains in volume. They advanced from 7.6 to 30.8 gallons between 1980 and 2005, a net increase of 23.2 gallons per capita. Bottled water came in second. It increased from 2.7 to 25.4 gallons per person, up 22.8 gallons. Soft drinks were next, increasing by 17.9 gallons. Milk exhibited the largest decline, moving from 27.6 to 21 gallons per person, a net loss of 6.6 gallons. Advances in fruit and vegetables and bottled water together account for nearly 90 percent of the increase in total beverage consumption per capita in this 25-year period.

These data suggest that the need state definitions are malleable rather than laws of taste. The public appears to be redefining nutrition to exclude fats (fruits and vegetables thus win out over milk), is in the process of replacing sugar-sweetened beverages with water, and is consuming much more purchased water than in the past. Other factors are also at work. The population aged 15 and under is decreasing as a percent of all people; the elderly population is increasing; more people are obese; lifestyles are more hectic. These trends have had their most profound impact on the soft drink industry which, as will be dis-

FIGURE 32

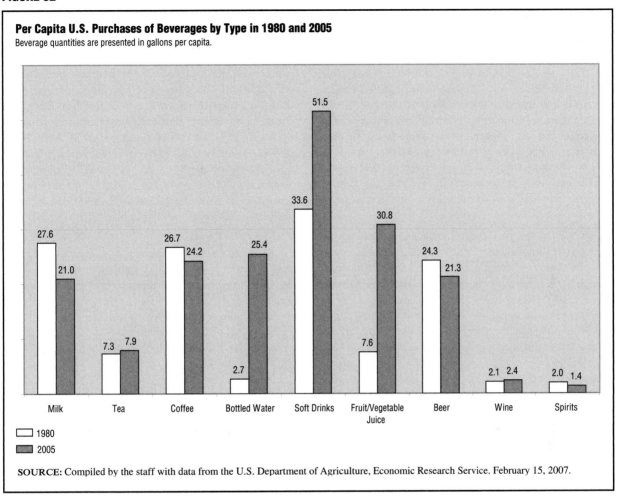

Per Capita U.S. Purchases of Beverages by Type in 1980 and 2005

Beverage quantities are presented in gallons per capita.

SOURCE: Compiled by the staff with data from the U.S. Department of Agriculture, Economic Research Service. February 15, 2007.

cussed below, is seeing its traditional product threatened by the rising sales of bottled waters and, to counter the consequences, has entered the bottled water market to forestall its own decline.

Energy and Sports Drinks. A discussion of bottled water would remain incomplete without special recognition of another growing category of beverages which occupy a place between carbonated soft drinks and bottled water. These are energy and sports drinks, sometimes also referred to as isotonics. This category had its start at the University of Florida in 1965 where Head Coach Ray Graves, working with professors at the university, stimulated the development of Gatorade to fight athletes' dehydration. Gatorade was initially produced for public distribution by Stokely Van-Camp, bought by Quaker Oats in 1987. PepsiCo acquired Quaker Oats in 2001 and has been selling Gatorade since. Coca-Cola's PowerAde is Gatorade's chief competitor. Other brands exist as well.

Sports drinks are beverages intended to replace salt lost in sweating. Salt (sodium chloride) contains positive and negative ions. Sodium and chloride, along with potassium, calcium, magnesium, bicarbonate, phosphate, and sulfate are important electrolytes in the body's chemistry. An electrolyte can be either a positively charged ion (a cation) or a negatively charged one (an anion). The body's cells use ions to generate electrical voltages to preserve their shapes and to signal required activities, as in sending nerve pulses. The loss of salt in muscle tissue leads to contractions and thus cramps for lack of electrolytes. Energy drinks are also called isotonics. The word means *same tone* or same tension. When the salt level outside the cells, thus in the blood or in the liquids between cells (the interstitial fluid) and in the cell are the same, a happy isotonic condition prevails—conditions are the same all over. But imbalances cause cell deformations. Drinking isotonics restores the balance. Sports drinks have a lot of salt. The trick in formulating them is to make them taste right despite this disadvantage. The drinks also carry proteins and sugars.

Bottled Water

Energy drinks appeared in the United States in the late 1990s and are based principally on caffeine as a stimulant, sugar as the energizer, and popular hype as the mystique. The best-known brands are Red Bull (Red Bull Inc.), Monster Energy (Hansen Natural Inc.), Rockstar (Rockstar International), Full Throttle (Coca Cola Co.), and SoBe No Fear (PepsiCo Inc.). Red Bull's sugar content is less than that of Coca-Cola Classic and its caffeine content, at 80 mg per 250 ml, is actually less than that contained in ordinary liquid coffee. Based on data from the Coffee Science Information Center (CoSIC), the equivalent quantity of instant coffee has 100 mg of caffeine and roasted coffee has 142 mg. The CoSIC is an element of the British Institute for Scientific Information on Coffee (ISIC).

As viewed by the Census Bureau energy and sports drinks are reported as parts of the Soft Drink Manufacturing industry under the Noncarbonated Soft Drinks category. Bottled water has its own industrial code. Census data on neither category is very useful. The Bureau largely suppresses details on energy and sports drinks in order to avoid disclosing the revenues of specific companies. In the bottled water category, product subdivisions by type are simply not provided. The analyst must consult private research services.

MARKET

Commercial developments in the beverage industry are well tracked by a number of organizations specializing in retail trade research generally and in trends affecting the beverage industries in particular. For purposes of this essay, the shape of the bottled water market has been developed using data cited in the trade literature which, in turn, uses the work of the Beverage Marketing Corporation, a major research firm serving all aspects of beverage manufacturing and distribution. Data on the soft drink industry specifically are quoted from *Beverage Digest's* annual press releases on the industry's condition.

Based on these sources, the U.S. bottled water segment in 2005 had estimated retail sales of $10 billion. The industry grew at a heady rate of 14 percent a year from 1997 (sales of $3.5 billion in that year) whereas the traditional soft drink industry, dominated by carbonated beverages, advanced in the same period at a rate of 2.8 percent a year (from $54.7 to $68.1 billion).

FIGURE 33

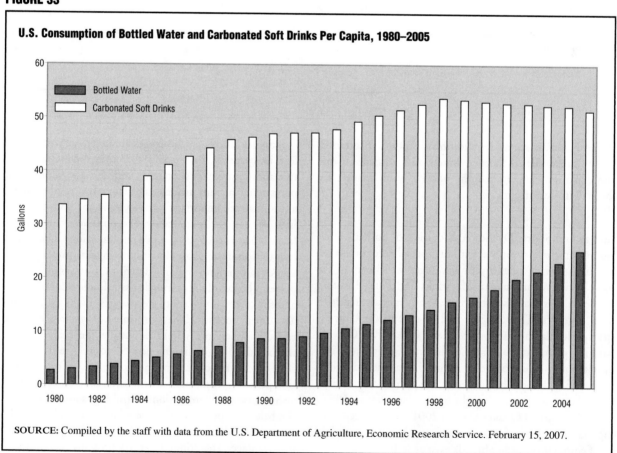

U.S. Consumption of Bottled Water and Carbonated Soft Drinks Per Capita, 1980–2005

SOURCE: Compiled by the staff with data from the U.S. Department of Agriculture, Economic Research Service. February 15, 2007.

124   ENCYCLOPEDIA OF PRODUCTS & INDUSTRIES—MANUFACTURING, VOLUME 1

These developments signal the ongoing transformation of the non-alcoholic beverage industries in the first decade of the new century. Carbonated soft drinks (CSDs) and bottled water combined were a $58.2 billion industry in 1997 with CSDs representing 94 percent of sales. Eight years later the combined industry was $78.1 billion but CSDs had slipped to a still dominant but contracted share of 87 percent of that. In the soft drink industry, which includes both CSDs and bottled water, the growth of the latter has been largely noted with gloom because water cuts into the growth of sweet sodas. Water's growth and the equally energetic growth of energy and sports drinks indicate the maturing of the traditional soda industry. It has slowing growth but retains a huge volume.

Consumption data over a 25-year period graphically illustrate the concerns of traditional carbonated beverage producers. Data show a flattening of per capita consumption of soft drinks beginning in the 1989–1993 period. A brief period of growth follows—then another flattening in the 1998–2005 period. Meanwhile bottled water consumption had an unwavering period of advance.

Beginning in 2005 and again in 2006, the weakness was not in absolute consumption of soft drinks but certainly in its growth translated into actual *decline* in the volume of shipments of the two leaders in the soft drink category. In 2005 Coca Cola's CSD volume dipped 0.6 percent, PepsiCo's dropped 1.2 percent. In 2006 Coke had a downturn in carbonated soft drink volume of 1.2 percent and Pepsi, 1.3 percent. Both companies, however, are major participants in the bottled water and in the sports drinks categories so that losses in one category were compensated by gains in the other.

Comparing consumption and sales data for the 1997–2005 period, it is evident that bottled water is improving its profitability. In 1997, 3.6 billion gallons produced $3.5 billion in sales (96 cents per gallon on average). In 2005, 7.5 billion gallons produced $10 billion in sales ($1.33 per gallon on average). Some of the high-priced brands of bottled water cost more than gasoline per gallon; and tap water reaches the consumer at less than a penny a gallon.

Product Categories. Beverage Marketing Corporation divides bottled water into Nonsparkling, Sparkling, and Imported categories. In 2005 these represented 95.1, 2.5, and 2.4 percent of total consumption in gallons. Nonsparkling water was also dominant in 1997 but gained share slightly over the other two categories. This major category grew at the rate of 10 percent a year, 1997–2005, sparkling water at a rate of 2.3 and imports at a rate of 2.6 percent a year. More precise product differentiations into the categories FDA uses for regulating bottled water are not available.

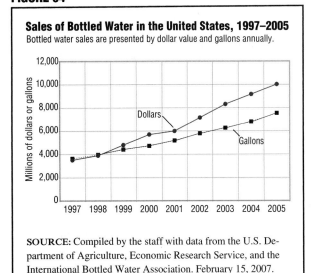

FIGURE 34

Sales of Bottled Water in the United States, 1997–2005

Bottled water sales are presented by dollar value and gallons annually.

SOURCE: Compiled by the staff with data from the U.S. Department of Agriculture, Economic Research Service, and the International Bottled Water Association. February 15, 2007.

A dollar-denominated look at these major categories gives some indication of price levels for each major category. Imports, the smallest volume category, represented 7.24 percent of dollar sales in 2005. Second-ranked sparkling water was 4.5 percent of the dollar market. The largest category, nonsparkling water, with 95 percent of physical volume, represented a smaller 88.2 percent of sales in dollars, indicating that this mass market for water was also lowest in price. Translating these numbers into per gallon indicators, Nonsparkling averaged $1.23, Sparkling $2.45, and Imported $3.97 per gallon. By way of comparison, carbonated soft drinks averaged $4.46 and beer $13.80 per gallon in 2005. Sports drinks cost about the same as CSDs, energy drinks cost at least a dollar more. Observers of the bottled water industry, particularly those critical of it for cultural reasons, tend to compare the costs of bottled water with that of tap water. But looking in the other direction, at bottled water's main competitors, water has a distinct price advantage.

Factors of Growth. Most industry observers explain rising consumption of bottled water by pointing to increasing interest in health issues by the public. Growing obesity and its accompanying increase in the less severe form of Type II diabetes are associated with fats and sugars. Milk and soft drinks are associated with one or the other. Behind this increasing concern with health, however, are broader demographic changes that, usually, underlie major shifts in public behavior. The demographic phenomenon of our times has been the very large population cohort called the baby boom (those born 1946 through 1964). In 2000 the youngest of this group was 36, the oldest 54. A second major development has been increased lifespan.

Changes in population structure between the last two census years, 1990 and 2000, are revealing. If we look at population in twenty age groups of five years, the greatest increases were in those aged 50–54 (up 54.9%), 45–49 (up 44.8%), and 90–94 (up 44.6%). Among all twenty age groups, nine of the older all had higher growth rates than the four groups 19 and younger. Those in the prime of life, 20–34, current and future parents, had actually declined in number. These numbers suggest that the population as a whole was getting older. Trends since 2000 have merely intensified this tendency. Consumption of soft drinks is associated with youth; in the first decade of the twenty-first century the young are a *decreasing* percentage of the population. An aging population, alerted to the potential dangers of sugar consumption, have shifted to diet drinks—of which the most obviously healthy variety is plain water.

KEY PRODUCERS/MANUFACTURERS

Nestlé Waters. The leading supplier of bottled water the world over is Nestlé Waters, an element of Nestlé S.A., the global food company. Nestlé is also the top-ranked producer of bottled water in the United States with a market share of 38.5 percent in 2006. The company has 72 brands worldwide and sales in the category of 9.6 billion CHF (Swiss Franks), approximately $8.5 billion (2006); Nestlé S.A. had total revenues of around $70 billion. Nestlé owns four of the five most famous brands around the globe—Aqua Panna, Contrex, Perrier, San Pellegrino, and Vittel. The company owns Poland Spring, the oldest U.S. brand and Nestlé's most popular consumer brand; other important brands in the United States are Ice Mountain, Deer Park, Ozarka, Arrowhead, Calistoga, and Zephyrhills. Nestlé Waters is also a dominant supplier of the bulk drinking water market in the United States with a 25.8 percent share. Office workers know this product as five-gallon jugs waiting, upside down, in coolers that dispense iced water into cups. This category is known in the business as "Home & Office Delivery" or HOD. Nestlé's HOD water brands are Ice Mountain, Zephyrhills, and Poland Spring. Nestlé has been in the water business since 1969 when it initially acquired a 30 percent share of Société Générale des Eaux Minérales de Vittel.

Groupe Danone. The French food company, Groupe Danone (best known in the United States for its yogurt brand, Dannon) owns the world's best-selling brand of bottled water, Evian. Danone's own origins and its acquisition of Evian coincided when two companies, BSN-Evian and Danone-Gervais merged in 1973 to create the company. Evian is sold in 125 countries around the world. Groupe Danone had total worldwide sales for all products of €13.7 billion in 2004 (around $18 billion). Evian

distribution in the United States has been controlled by Coca-Cola Company since 2005 when Coke bought out Danone's interest in a joint venture.

PepsiCo Inc. PepsiCo Inc, a leader in the soft drinks market, became involved in the bottled water market with its introduction of Aquafina in 1994. Since that time the brand has achieved top rank in the United States among bottled water brands although Pepsi remains in second position overall, behind Nestlé. Pepsi's other brands are Propel and Flavorsplash. PepsiCo had total revenues of $32.6 billion in 2006 of which 28 percent were in beverages, most of that volume in CSDs. PepsiCo is also the leading producer of sports drinks with its flagship brand, Gatorade, acquired in 2001.

Coca-Cola Company. Coca-Cola introduced the Dasani brand of bottled water in 1999. By 2006 Dasani was the second-ranking brand in the category and Coke the third-ranking seller of bottled water. As already noted, the company also sells Evian in North America. Earlier, in 1992, Coke began to bottle the sports drink, Powerade, a product that it had launched in 1990 as a fountain beverage. Coca-Cola had revenues of $24 billion in 2006.

Other Top Companies. Three other companies ranked among the top six producers in the United States by *Beverage World* magazine are DS Waters, Crystal Geyser, and Glacier Water Services Inc. DS Waters of America, Inc. began as a joint venture formed by Groupe Danone and Suntory International in 2003. The company assembled a number of regional bottled water companies with histories going much further back to form the company and also participates in bulk water sales. DS Waters brands, with their years of origin noted in parentheses, are Alhambra (1902), Belmont Springs (1900s), Crystal Springs (1920s), Hinckley Springs (1888), Kentwood Springs (1963), Sierra Springs (1950), and Sparkletts (1925). Danone eventually bought out Suntory's interest in the venture and sold the DS Waters to Kelso & Co., an investment firm.

Crystal Geyser Water Co. is a privately owned firm in business since 1997. The company specializes in spring waters acquired on Mount Shasta and Olancha Peak, both in California; in Benton, Tennessee; in Salem, South Carolina; and in Moultonborough, New Hampshire (in the Ossipee Mountain region). Crystal Geyser, with headquarters in San Francisco, also bottles juices and sells ready-to-drink tea.

Glacier Water Services Inc., an $87 million company in 2006, has a unique approach to the drinking water market: it sells pure water in 15,000 vending machines in

39 states. The company was founded in 1983 with headquarters in Vista, California.

While the discussion above captures leading producers and the better-known brands, more than one hundred companies participate in bottled water manufacturing in the United States. They operate 228 establishments, 78 of which had twenty or more employees. To the products of this industry must be added important brands much valued by their customers but sold in limited quantities. High-priced brands beyond Evian, Perrier, and Vittel, according to Sam Gugino, *Wine Spectator*'s *Taste* columnist, are Chateldon (France), Solé and San Faustino (Italy), Alpenrose (Switzerland), Tynant (Wales), and Highland Spring (Scotland)—all sometimes offered in upscale restaurants. At the other end of the scale are private label bottled waters like Kroger, America's Choice, which is A&P's brand, and Kirkland, Costco's brand.

MATERIALS & SUPPLY CHAIN LOGISTICS

Depending on the kind of water in the bottle, the cost of transportation or cost of processing may be a higher percentage of its cost. Evian, coming from the French/Swiss alpine region, crosses the Atlantic and then begins its trip home to a location in the United States. The well known Fiji brand of bottled water comes from the island cluster of that name next door to Australia. Products labeled spring or mountain water have their origins in specific geographies and have to reach their markets by means of paid travel too. The two leading brands, Aquafina and Dasani, are labeled purified. Their source is the municipal water at the places where they are bottled. The principal purification process used by both brands is reverse osmosis combined with other methods. Other brands carry the distilled water label showing that their process of purification is by distillation. All types of water have the same basic packaging costs.

The introduction of polyethylene terephthalate (PET) in beverage packaging in the 1980s created the most popular packaging solution in bottled water. This clear, strong, and flexible plastic is readily recycled back into PET bottles and many other products. When incinerated it emits no chlorine by-products like the earlier polyvinyl chloride. PET is widely available from some sixteen different producers (DuPont, 3M, Honeywell, Eastman Chemical, and others) at twenty-nine production sites across the nation.

DISTRIBUTION CHANNEL

Bottled water reaches consumers by three distinctly different channels—home and office delivery in bulk containers (usually 5 gallon glass or plastic bottles), vending machines (a very small percent of total volume), and through retail channels in single bottles or in multi-bottle packages. The retail channel itself is divided by the industry into grocery, mass merchandising, drug store, and convenience and gas outlets. Overall, including all food and beverages sold, not just water, grocery stores represent about 55 percent; mass merchandisers, including warehouse clubs, about 18 percent; and convenience stores around 16 percent of volume. Bottled water distribution most likely follows this pattern as well.

Two- and three-tier distribution systems coexist. In the two-tier system major producers (like the soft drink giants) sell directly to large retail chains and these chains sell directly to the customer. Bottled water wholesalers operate to serve as a third link in three-tier systems between the producer and the retail outlet itself. Large chains prominently feature—and dedicate substantial shelf-space—to their own private label brands. The chains deal directly with the bottlers to obtain this product, typically a price-leader in each store, often sold in bulk (1 gallon) containers as well as smaller bottles or packs.

KEY USERS

Catherine Ferrier, in "Bottled Water: Understanding a Social Phenomenon," assembles research on motivations for buying bottled water in the United States and elsewhere. Consumers in the United States in 1997 appeared to be motivated principally by concerns over the safety of tap water and by a desire to displace other beverages with something better. French respondents, surveyed in 2000, were motivated principally by the superior taste of water and water hardness in regions where such conditions prevail. Bottled water consumption has significantly increased everywhere, in the United States as well as Europe, in the time since the studies cited by Ferrier took place suggesting that other rationales have played a role as well. The substitution motivation in the United States is promoted by concerns with growing obesity, highly publicized in the late 1990s, and word-of-mouth promotion. Other surveys indicate that adults, and principally women, are primarily responsible for the growth in the use of bottled water. Men have been behind the equally explosive growth of sports drinks and, most recently, youths for the dramatic increase in the consumption of energy drinks.

ADJACENT MARKETS

A well-established alternative to buying bottled water is the production of such water at home using tap water as the raw material and purchased filtering equipment as the purification route. This market, manifesting as purchases of filtering devices, is very small in comparison with bottled water sales themselves but has shown substantial growth in the same period. The leading producer of such equipment is Brita GMBH, a German company, founded in the 1960s and now a global supplier of such equipment. Brita is sold in the United States by Clorox Company (the $4.6 billion diversified concern known best for its bleach product). Clorox bought Brita's U.S. operation in 2003. A competitor to Brita is Procter & Gamble's PUR, a functionally very similar purification and dispensing system. Procter & Gamble is a diversified, $68.2 billion home products producer. Small water purifiers from these companies are available for under $50. They utilize replaceable carbon-filtration cartridges. In many homes across the country householders fill PET bottles originally obtained holding one of the brands of bottled water but now serving to hold filtered water brewed at home.

In the beverage industry itself sports drinks, energy drinks, fruit "ades," and vegetable juice drinks are viewed as adjacent markets. In the dairy industry leading producers are formulating and promoting milk-based beverages aimed at the same market. Non-alcoholic beer products are yet another manifestation of reaction by a traditional industry to the pressure applied by a major, new, and ultimately threatening, product category.

RESEARCH & DEVELOPMENT

Research and development efforts in the industry touch every aspect of the business, from water treatment at one end of the spectrum to packaging at the other. Development of more effective treatment systems is process development. This work is typically focused on requirements unique to specific sources of water. Part of such R&D is of an environmental nature and involves study of water tables and regional hydrology in response to environmental opposition (further delineated below). Water purity is a major interest in the industry, underlined by Coca-Cola's recall, in 2004, of 500,000 Dasani bottles in the United Kingdom after bromate levels were found to exceed legally permissible levels. Bromate occurs when water is disinfected using ozone and reacts with naturally occurring bromide in water. Bromate is suspected of being a carcinogen.

A trend in the industry is the introduction of flavored waters with subtle tastes and a faint touch of sweetness—achieved using artificial sweeteners. Considerable research effort is expended on finding so-called water white flavors, meaning flavor-providing substances that do not cause the water to cloud up. Oxygenation of the water is yet another enhancement sought by some bottlers on the theory that oxygen in the water will oxygenate the blood (accomplished naturally by breathing). It is difficult to force much oxygen into water and the health effects are not well-established, but work continues along these lines and super-oxygenated waters are already on the market. Providing health benefits to the elderly (bone health, joint protection) by dosing water with minerals is under

investigation to develop new bottled water features. Finding the right acid-sweetener ratios is an aim of product development.

With rapidly growing volume, parts of the bottled water industry are coming to resemble the soft drinks industry. The latter is famously marketing-driven and the packaging is a very important aspect of the hype. An annual awards program offered by bottledwaterworld (Zenith International Publishing, of Bath, England) highlights packaging, labeling, and marketing concepts. Substantial materials research and development effort lies behind winning concepts in packaging. An example from the 2006 competition was Flavor Top, a screw cap closure that both stores and dispenses ingredients as the container is opened. Winners in the category of Best Cap were Innovation Fund and Nottingham Spirk. Innovation Fund is a technology development and licensing firm. Nottingham Spirk is a development company. The 2006 "Best Packaging Innovation" award was won by Nestlé Waters for a 3-liter PET bottle with interlocking features so that bottles can be joined into packs. Bottledwaterworld's 2007 competition is based on environmental themes—possibly in response to the environmental movement's less than warm embrace of growing bottled water sales.

CURRENT TRENDS

The dominant trend in the bottled water industry is strong growth now and more to come. The response to this welcome prospect takes somewhat divergent forms. Continued consolidation in the industry is likely as large companies continue to grow by acquiring small producers of spring and mountain waters. Not yet visible on supermarket shelves, crowded as they are with unflavored purified water brands and featuring two or three upscale brands (Evian, Perrier, Vittel, Fiji), is the growing category of so-called enhanced waters. These are flavored or nutritionally enriched products attempting on the one hand to resemble soft drinks and capture a market based on unique tastes and, on the other, attempting to capture shares of the energy and sports drinks markets. Product proliferation is likely—and likely also to be resisted by the retailers. Retailers prefer large volume achieved by a few popular brands to a forest of many competing flavors all vying for limited shelf space, already a feature of CSD and beer distribution. Trends in soft drinks and beer—will these product categories continue to decline?—will influence the shape of the market.

How to handle energy and sports drinks in this mix is also adding to the merchandising problems of the retail channel. As the first decade of the twenty-first century was drawing to a close, the two major soft drink giants were positioning their water products as healthy and pure beverages but aiming the product (at least as indicated by the imagery used) at the youth market. Bottled water, in actuality, is principally consumed by adults. Product development efforts by these companies were directed at enhanced waters thus enlarging the number of brands or choices within one brand. Others, Nestlé being the leading example, continued to concentrate on the high-priced end of the market with conspicuous corporate efforts made to communicate with environmental interests.

This last point is relevant because the rapid growth of bottled water as a category, and not least the growth of purified water—ultimately derived from municipal sources—has also generated a cultural reaction of a negative sort. This position is taken by environmental groups. These groups view drinking water regulations as not sufficiently rigorous and have problems with the health claims made for purified waters. They also worry about aquifers and future water supply. They anticipate problems if commercialization of water as a consumer commodity leads to lowering of water tables in specific areas.

TARGET MARKETS & SEGMENTATION

Bottled water has a prestige segment represented by leading imports and their American counterparts in traditional spring and mountain waters. The latter are sold both in the United States and overseas. These products are aimed at the discriminating palette. The other extreme is represented by purified waters sold in bulk under private label or in the Home & Office category. These products are aimed at the healthy-minded. The middle ground is occupied by regional spring and mountain brands, with loyal followings, and mass-market brands produced and promoted by the major soft drink companies as health-providing or recreational drinks.

RELATED ASSOCIATIONS & ORGANIZATIONS

American Beverage Association, http://www.ameribev.org/index.aspx

The Association of State Drinking Water Administrators, http://www.asdwa.org

Canadian Bottled Water Association, http://www.cbwa-bottledwater.org

International Bottled Water Association, http://www.bottledwater.org/public/statistics_main.htm

Water Quality Association, http://www.wqa.org

BIBLIOGRAPHY

Age: 2000. Census 2000 Brief. U.S. Department of Commerce, Bureau of the Census. October 2001.

Bellas, Michael C. "Motivational Need States and How We Need to Think." *Beverage World.* June 2006.

"Beverage Digest/Maxwell Ranks U.S. Soft Drink Industry for 2004." *Beverage Digest.* 4 March 2005.

"Bottled Water Proves It's a Big Fish." *Beverage Industry.* July 1999.

"Caffeine." The Coffee Science Information Center (CoSIC). Available from <http://www.cosic.org/background-on-caffeine>.

Ferrier, Catherine. "Bottled Water: Understanding a Social Phenomenon." The World Wide Fund for Nature. April 2001. Available from <http://assets.panda.org/downloads/bottled_water.pdf>.

Gugino, Sam. "Water, Water Everywhere." *Wine Spectator.* 30 April 2003.

"The History of Bottled Water." Nestlé. Available from <http://www.nestle.com/Our_Responsibility/Water/Case_Studies/The+history+of+bottled+water.htm>.

Landi, Heather. "Bottled Water Report: Back to double-digit volume growth, the bottled water market is flowing swift and steady." *Beverage World.* 15 April 2005.

Mastrelli, Tara. "Cold Beverages." *ID: The Information Source for Managers and DSRs.* June 2002.

Posnick, Lauren M. and Henry Kim. "Bottled Water Regulation and the FDA." *Food Safety Magazine.* Reprinted by U.S. Food and Drug Administration, Center for Food Safety and Applied Nutrition. August/September 2002. Available from <http://www.cfsan.fda.gov/~dms/botwatr.html#authors>.

Roberts, Michael Bliss Vaughan. *Biology: A Functional Approach.* Thomas Nelson & Sons Ltd., 230.

"Special Issue: All-Channel Carbonated Soft Drink Performance in 2005." *Beverage Digest.* 8 March 2006.

"Special Issue: Top-10 CSD Results for 2006." *Beverage Digest.* 8 March 2007.

Stevens, James M. "A Look into the Crystal Ball: From a commodity to a branded beverage, bottled water's best days are yet to come." *Beverage Industry.* August 1996.

"The US Bottled Water Market." *Beverage World.* 15 August 2004.

Boxes: Packaging & Packing

INDUSTRIAL CODES

NAICS: 32–2211 Corrugated and Solid Fiber Box Manufacturing, 32–2212 Folding Paperboard Box Manufacturing, 32–2213 Setup Paperboard Box Manufacturing, 32–2214 Fiber Can, Tube, Drum, and Similar Products Manufacturing, 32–2215 Nonfolding Sanitary Food Container Manufacturing

SIC: 2652 Setup Paperboard Boxes, 2653 Corrugated and Solid Fiber Boxes, 2655 Fiber Cans, Drums & Similar Products, 2656 Sanitary Food Containers, 2657 Folding Paperboard Boxes

NAICS-Based Product Codes: 32–221101, 32–221102, 32–221103, 32–221104, 32–221105, 32–221106, 32–22110Y, 32–221201, 32–221202, 32–221203, 32–221204, 32–221205, 32–221206, 32–22120Y, 32–221301, 32–22130Y, 33–22141, 32–22143, 32–2214W, 32–22131, 32–22133, 32–22135, and 32–2213W

PRODUCT OVERVIEW

Most products, excepting only large equipment and durable goods of substantial size, reach customers packaged in boxes. Boxes also hold goods temporarily when people or institutions move. Boxes are so overwhelmingly present in every area of life—the home, office, warehouse, retail store, hospital, school, and factory—that we tend no longer to see them. In a manner of speaking we look through them and see the products they hold. We acknowledge their utility by saving them in attics and basements for shipping a package or storing goods we do not immediately need. Boxes are made of natural and predominantly softwood fibers. They enjoy high rates of recycling; some categories of boxes are made predominantly of fibers recovered from old newspapers obtained by municipalities in separate waste collections.

As outer containers used in transporting goods in bulk, boxes function as packing. When boxes hold a single object or quantity of product, they function as packaging. We know packing containers as those brown boxes known as corrugated or kraft containers. Packaging boxes come in three categories of which a cereal box, a milk carton, and a small jewelry box holding a valuable ring are representative. The first is called a *folding box*. The cereal box itself is made of so-called food-grade paper, but its brethren used for textiles, tissue, shoes, toys, and a myriad of others are in the same category and use coarser grades of board. The milk carton belongs to the *sanitary food container* category under tighter regulatory control. The jewelry case is part of a category called the *setup box*—a type of stiff, highly decorated, and often expensive container. In addition to these objects, all of which have the requisite rectangular shape to merit the name box, we must add a category called *fiber cans, tubes, and drums*. Although cylindrical in form, they function the same way as their right-angled cousins.

Packaging, in a way, is a game of boxes within boxes—and the final item may be another kind of container yet. A plastic or glass bottle of aspirin may come in a folding box. Fifty or more of them will be packed in a corrugated box. The corrugated box itself will be mounted on a pallet with a number of others all sitting on a corrugated tray. The pallet itself, furthermore, may be turned into a single unit using metal or plastic strapping. Outer containers have a single function—to hold product and to carry minimal identification. Packages typically have a much more elaborate display and communication function. Some are also intended to store the product. Milk cartons are a good example as are software packages. Both sell themselves on store shelves. Milk containers store the product in the refrigerator. The software package ends up on a shelf in the office to hold disks and manuals after the software has been installed and is running on the user's personal computer.

Wood Fiber Basics. Between 40 and 50 percent of wood consists of cellulosic fibers, strong crystalline sugar polymers with no free water content and therefore not very reactive chemically. The rest of wood is hemicellulose and lignin. Both are polymers but have different structure. Chemists refer to them as diverse and amorphous polymers. They mean that these substances, unlike cellulose, cannot be collectively defined by predictable and uniform shape. When the wood comes from conifers, or evergreens, it will also have a high resin content. The paper maker must remove all three of these substances to get at the useful fiber. Cellulose is desirable because it is strong and chemically stable. Freed of binders, the fibers will align to form sheets that resist tearing or bursting, and will have fold-endurance and uniform optical properties. Fold-endurance means that a sheet, once folded, will not tend to spring back. Good optical properties means uniform coloration and ready acceptance of inks.

To make paperboard, the producer first grinds pulpwood into fine particles and then uses either mechanical, chemical, or a hybrid process to get the fiber out. All of the processes are wet, thus the fibers are turned into a liquid slurry. Once only fibers remain, they are formed into sheets in which the cellulose fibers are uniformly oriented, which helps with folding. Bleaching may be applied to turn the board white. Drying and compression under drums produces a board or a sheet of uniform color, fiber-orientation, and thickness. The simplest process is groundwood pulping used predominantly to make newsprint. Wood is ground, moistened, and agitated until fibers separate from binders. The process results in an inferior, coarse product with low strength and inferior optical characteristics. The three types of chemical pulping are sulfite, sulfate (or kraft), and semi-chemical pulping. Sulfites are ions made of a single sulfur and three oxygen atoms; sulfates have four oxygen atoms adhering to one sulfur atom. One or the other is used because incoming wood has different chemical characteristics. Semi-chemical pulping combines mild chemical treatment with mechanical agitation to get a reasonably clean fiber.

Sulfite Pulping. This process uses coniferous trees cooked in a solution of sodium sulfite. The pulp is next bleached. The process works well on woods low in resin like spruce, fir, or hemlock. The sulfite method makes good printing and tissue papers. Sulfite paper is also used to make inner and outer liner sheets for boxes.

Sulfate (Kraft) Pulping. The sulfate process is effective on trees with high resin content. Southern pines, the most widely used raw material in large containers, have such content. The resulting product is very strong, hence the name of the process, kraft, the German word for strength, introduced as a name for the method by the German chemist, C.F. Dahl. Dahl developed sulfate pulping in 1879. The kraft process is used for packing board and other grades of coarse paper. Bleached kraft is one type of board used in both containers and in some types of higher-strength folding boxes. The process is also employed to make printing grades of paper.

Semi-chemical Pulping. In this process wood fibers are pre-treated in a chemical bath rich in sodium sulfite to soften them. Producers then employ mechanical means to separate as much fiber as possible. The process has higher yields than kraft pulping (more of input comes out as board or sheet by weight), but the method produces a weaker fiber. It leaves more binder in the product, which accounts for binders weakening the structure but increasing the yield. Semi-chemical pulps are made from hardwoods, such as birch and waste paper. The corrugated sheet between two rigid outer sheets used in brown boxes is typically made by this process. The corrugated sheet is also called corrugated medium or simply medium.

Types of Boxes. Insider talk in the boxboard industry is sometimes confusing, so some definitions may be useful. *Containerboard* refers both to solid fiber materials and what might be called sandwiches of solid fiber and corrugated medium. In either case such board is intended as outer packing. The boxes may be brown or white. If they are white they are still kraft but have been bleached. Confusion arises because the industry also uses the term *liner-board* to mean the same thing, but when people talk about *liner* they mean thin sheets of white or colored paper used to laminate the outside or the inside of boxes—so that the cereal box may be decorated with the most recent cartoon character or baseball slugger.

Packing Boxes. These come in four categories. Solid fiber containers have a single stiff board. Corrugated boxes (sandwiches) come in: (1) single-wall, (2) double face, and (3) double-wall varieties. Single wall consists of a solid fiber laminated to an internal corrugated sheet. That sheet, in turn, is protected inside by a thin liner. Double face is a true sandwich, with two stiff boards separated by a corrugating medium. Double-wall boxes, the strongest, are formed by joining one or more double-face boards into a single structure. While the corrugated medium is used as a filler and is not as strong as the kraft board, its shape gives the box the ability to compress under pressure—the corrugations act as buffers.

In contrast to containerboard, which is invariably kraft paper, the material used in folding and setup boxes is called *boxboard*. This is confusing to industry outsiders because boxboard may be kraft. Boxboard, however, is never containerboard. Finally, bleached virgin fiber board used in liquid food containment is referred to as *sanitary board*.

Setup Boxes. These packages are delivered ready-made and ready-to-fill. They are stiff boxes usually in two pieces, the box and its lid, and are decorated by the maker to the user's specifications. The lid may or may not be affixed to the box. Setup boxes are made of joined pieces of thick boards, usually of coarse boxboard typically made of groundwood pulp containing waste paper. The box is lined inside and out and may feature internal padding or structure to hold product—jewelry or candy, for instance. In its report on the industry the U.S. Bureau of the Census always uses the word rigid in parentheses when referring to setup boxes. The industry is small. Boxes of this type always serve a purpose beyond containment. They present the product. The stiff box, attractively decorated, sends a message of quality. The largest users of such boxes are producers of high-end apparel and textiles, soap and cosmetics, and confectionery. Department stores and other retailers also use such boxes to package on site upscale goods that they have sold. Setup boxes should not be confused with light-gauge boxes used to hold textiles (like a blouse or sweater) easily folded from flat pieces into a boxy shape by retail clerks. Boxes of that type are folding boxes.

Folding Boxes. These boxes arrive at the buyer's facility as flat, pre-creased, shaped board ready to be folded into boxes on the user's machinery or, in retailing, by the retailer's employees. Measured in dollars, folding boxes are the largest packaging category whereas corrugated is the largest box category. Almost all folding boxes are laminated on the outside to sheets of printing grade paper. The boxboard may be made of virgin fibers or recycled paper depending on the end use. Food containers are invariably made of virgin fibers and conform to Food and Drug Administration (FDA) standards for food-grade paper; the fibers may or may not be bleached. Converters may print box blanks, die-cut them, and insert translucent windows into some to display the merchandise—common in the toy industry. The packager folds the flat blanks into the final shape and produces a closure

by folding precut tabs or applying an adhesive. The top five consumers of folding boxes are: packagers of dry foods (e.g., cereals), canned sodas and beer (the six-pack container); frozen foods; cosmetics and medicines; paper products, including book mailers; and packagers of bakery products, butter, and ice cream. Virtually every production sector, however, makes use of folding boxes.

Nonfolding Sanitary Boxes. These products are a relatively small portion of a larger industry called Nonfolding Sanitary Food Containers. Milk cartons and identical containers holding other liquid dairy products such as juices, juice drinks, and tea, are in the box category. Other industry products—to think outside the box for just a moment—are paper cups, cylindrical food containers, paper plates, trays, and miscellaneous other paper products that come in contact with food, such as soda straws and table cloths made of paper. Shipping such containers ready-made is not economical. For this reason paper companies ship sanitary board in large rolls. The actual milk or juice container is made at or near the filling operation on packaging lines. Liquid containers are coated with hot melts, a specially formulated coating of plastic and wax. Manufacturers use virgin fibers and adhere to FDA regulations. FDA rules set the maximum levels of bacterial contamination permitted, levels usually defined in number of different colonies of bacteria.

Round Boxes. Fiber cans, tubes, and drums are functionally identical to boxes used for bulk or final packaging. For this reason they are generally referred to as paperboard boxes. The cylindrical containers are typically composites. Their sides are made of kraft board or pressed fiberboard. The ends are made of wood, metal, or plastic. In effect, except for shape and their composite character, the materials used in cylindrical products match those of equivalent boxes. Mailing tubes are made of coarse groundwood fibers and waste paper; food-grade containers are made of virgin fiber. The largest markets for cans are frozen fruit juices and refrigerated dough products in the food category and motor oil in the non-food category. Fiber drums are widely used for bulk shipments, the three largest markets being dry chemicals, plastic compounds and resins, and dry food products shipped in bulk.

MARKET

The total market for fiber-based boxes, including cylindrical containers, was $44 billion in 2005 at the production level. These were shipments by the industry of finished products to the next level down, the final buyer or the wholesaler. Shipments in 1997 were $37.9 billion. In this eight-year period, sales increased 16 percent overall and advanced at the rate of 1.9 percent per year. The overall growth rate of the industry came in just over that of the population as a whole, which increased at the rate of 1.4

FIGURE 35

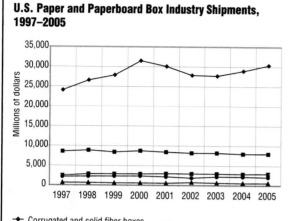

U.S. Paper and Paperboard Box Industry Shipments, 1997–2005

◆ Corrugated and solid fiber boxes
▲ Setup paperboard boxes
■ Nonfolding sanitary food containers
● Folding paperboard boxes
◆ Fiber cans, tubes, drums, and similar products

SOURCE: Compiled by the staff with data from "Value of Product Shipments: 2001," and "Value of Product Shipments: 2005," *Annual Survey of Manufactures,* U.S. Department of Commerce, Bureau of the Census. January 2003 and November 2006.

percent during this period. A visual presentation of sales performance is provided in Figure 35 by component industries.

Figure 35 shows that kraft container packaging is the dominant form that boxes take as measured in dollars and in tonnage. In 2005 corrugated shipments were $30.3 billion, folding boxes $8 billion, sanitary board $3 billion, fiber drums and related cylindricals $2 billion, and setup boxes $600 million. Of the sanitary board category, only 25 percent, approximately $750 million, were cartons for liquids; the rest of the shipments went into cups, plates, and trays. Figure 36 shows industry data in dollars for 1997 and 2005, together with annual rates of growth or decline.

The tabulation shows that only two of the industries participating in this sector had growth—corrugated and sanitary boxes. The others all experienced modest declines. Within the sanitary category, food cartons grew at a lower rate than the total industry of which they were a part.

Competition between materials on the one hand and shifts in international trade on the other explain the patterns we see displayed in the table. Corrugated boxes have only marginal competition from plastics. Plastic containers capable of delivering the same strength and durability cost much more than kraft-based products. Certain types of products can be effectively packed by putting them on cardboard trays and then shrink-wrapping the whole.

Examples are canned foods. Such techniques do not work well with other more fragile products. The growth of corrugated boxes, however, has been held down by growing imports. Imported products reach our shores in corrugated boxes too, but those are made outside of the United States.

Folding boxes are most vulnerable to competition from plastics. For instance, Plastic strapping hold six-packs of canned beer or soft drinks together. Shrink-wrapping provides advantages in showing the product but it eliminates most of the board except for one sheet which is used to hold the plastic tight. A spot-check of the histories of participants in set-up box manufacturing shows that most companies began producing their boxes from paperboard, but along the way, in moves toward diversification, they have also introduced plastic boxes to sell alongside fiber boxes. In the fiber can, tube, and drum business, the market decline has been least noticeable. A market decline of 0.4 percent per year, especially in a period marked by a recession, amounts to saying that sales have been flat. Sanitary liquid cartons have held their own against plastics because they stack more tightly and take up less space overall in shipping.

KEY PRODUCERS/MANUFACTURERS

Given the dominant role corrugated boxes play in this sector, it is not surprising that leaders in the industry are principally engaged in kraft pulping. The industry is forward-integrated. Those who produce pulp and board typically also produce boxes in wholly-owned conversion plants. The top five companies have nearly two-thirds of the market. There is, however, no single dominant firm several horse-lengths ahead of the field. The leaders are shown here not in rank order of total sales—which often involve a wide array of activities—but in order of their

FIGURE 36

Shipments and Growth Rate of Fiber Containers, 1997 and 2005

Product	Shipments		% Change Annually
	1997	2005	
Corrugated/solid fiber boxes	24,075	30,363	2.9
Folding boxes	8,564	7,998	-0.9
Nonfolding sanitary board	2,483	2,971	2.3
Food cartons	672	747	1.3
Fiber cans, tubes, drums	2,145	2,070	-0.4
Setup boxes	635	606	-0.6

SOURCE: Compiled by the staff with data from "Value of Product Shipments: 2001," and "Value of Product Shipments: 2005," *Annual Survey of Manufactures,* U.S. Department of Commerce, Bureau of the Census. January 2003 and November 2006.

shipments of corrugated containers in 2006. The leaders all participate in some aspect of folding box and sanitary board production.

Smurfit-Stone Container Corporation. This Chicago-based company was the result of a merger between Jefferson Smurfit Corporation and Stone Container Corporation in 1998. The company has grown by acquisition since that time—buying St. Laurent Paperboard Inc. and one of MeadWestvaco's container board plants—thus making Smurfit-Stone the leading producer in this industry with $7.2 billion in sales revenues in 2006. Like its competitors, the company operates a few mills to feed a large number of conversion plants. Smurfit-Stone reported operating four board mills and owning 39 consumer packaging establishments located near population centers.

International Paper Company. This company had sales in 2006 of $21.99 billion from producing printing papers (31.5% of total), industrial packaging (22.4%), consumer packaging (11.2%), distribution services (30.8%), forest products (3.5%), and specialty products (4.3%). Packaging was around one-third of International Paper's activity. The company operated 18 pulp mills and 94 conversion plants. Of the latter, 65 produced kraft containers. The company's combined sales in the industrial and consumer packaging categories were $7.3 billion. International Paper was founded in 1898 and is based in Memphis, Tennessee.

Weyerhaeuser Company. Founded in 1900 and based in the state of Washington, this company is also diversified with $21.9 billion in sales in 2006. Weyerhaeuser is engaged in paper production, timberlands management, wood products, and real estate. Twenty-two percent of Weyerhaeuser's revenues in 2006, $4.9 billion, came from paperboard production and box manufacturing. The company operated 13 container board plants and 100 box conversion facilities. Weyerhaeuser was also a leading supplier of liquid packaging board that it shipped to the food industry.

Temple-Inland Inc. This is a relatively new company, but one that is built of components with deep roots. Temple Industries, founded as Southern Pine Lumber Company in Texas in 1893, and Inland Container Corporation, founded in 1925, were merged in 1983 to create Temple-Inland. The company also has a real estate and a financial component. In 2006 Temple-Inland had sales of $5.56 billion, $2.98 billion of which it earned in its corrugated operations. These consisted of five containerboard plants, a corrugated medium mill, and 67 converting plants making and selling boxes.

Georgia-Pacific Corporation. This company began in 1927 as a hardwood lumber wholesaler in Atlanta, Georgia. Georgia-Pacific became a public company in 1949 and remained public until 2005 when it was acquired, and taken private, by Koch Industries, Inc. For 2004, its last full year as a publicly-traded company, Georgia-Pacific reported sales revenues of $5.6 billion. The company operates four corrugated board mills and converts between 70 and 75 percent of the output of these mills into boxes in 25 conversion plants, selling the rest of its output to independents. The company is also a major producer of bleached boxboard for the frozen food market. Its sales in corrugated boxes in 2004 were just under $3.0 billion, and its sales in bleached board $2.2 billion.

The companies listed account for the bulk of sales in corrugated containers and portions of folding and sanitary board as well. Packaging Corporation of America (PCA), with 2006 sales of $2.2 billion, is an important factor in folding boxes, operating one semi-chemical pulp mill, two linerboard mills, and 68 conversion operations making boxes. PCA also has around 5 percent of the corrugated market. Norampac Inc., a Canadian company with a substantial U.S. production presence, is an important participant with Can$1.3 billion revenues in 2006. The company has eight board mills and 26 conversion plants. Green Bay Packaging Inc., a privately held packaging company, describes itself as "Large enough to serve you, small enough to know you." Small enough, one assumes, refers to the company's estimated sales of around $790 million, based on data reported in *Manufacturing and Distribution USA*. Large enough perhaps refers to Green Bay's operations in 15 states through 30 divisions. The company is a leader in folding box production and also has a 2 percent share of corrugated boxes.

Setup box manufacturing is very much the domain of small, privately held operations. ThomasNet, the online presence of Thomas Register, the directory publisher, listed 120 manufacturers participating in the business in 2007. The first four companies in ThomasNet's listing provide a feel for the size and character of leading participants. Companies listed first by ThomasNet are advertisers and may be viewed as leaders, although no rankings are available. Brimar Packaging Inc., in Avon, Ohio, fell into the $5 to $9.9 million category and employed between 10 and 49 people. Boxit Corp. of Cleveland, Ohio, had 50 to 100 employees and had been operating since 1932. Capitol Box Corporation of North Bergen, New Jersey, had 10 to 49 employees and had been producing setup boxes since 1936. Jarrett Industries, Inc., a woman-owned corporation in Owings Mills, Maryland, had sales in the

$10 to $24.9 million range, employed between 10 and 49 people, and exported products to Western Europe, Africa, the Middle East, Asia, Australia, Europe, Canada, and Mexico. These producers, and others like them, typically offer boxes made of paper and plastic; some offer metal and wood boxes as well and provide special box interiors of fabric and foam familiar to all of us as the insides of jewelry boxes.

According to the Census Bureau statistics, the total U.S. fiber-box industry had 1,585 participating companies in 2002. They operated 2,664 establishments, of which 1,841 employed twenty or more people.

MATERIALS & SUPPLY CHAIN LOGISTICS

In 2006 the U.S. forestry industry harvested 15 billion cubic feet of roundwood for all purposes. Roundwood is forestry's jargon for what is generally referred to as trees. Of that total 4.4 billion cubic feet went into pulping (29.3%), the pulpwood itself divided into softwood species (60% of total) and hardwood species (40%). Virtually all containerboard and boxboard mills in the United States rely on softwood. Nearly 85 percent of softwoods seek the sun and are harvested in the South Atlantic and the South Central states, thus the region bounded to the west by Texas, to the east by Florida, Georgia, the Carolinas, and Virginia, and to the north by Oklahoma, Arkansas, Kentucky, and West Virginia. To achieve the best economies, pulp mills must be where the pulpwood is. The industry is therefore concentrated in the southern states, although some plants are present in the Northeast and in the Pacific Northwest. The most common softwood species used in box making are loblolly, shortleaf, longleaf, and slash pine. Western operations predominantly use Douglas fir and ponderosa pine.

Conversion operations, by contrast, are located in or near major population centers. As already noted, converters are very often wholly-owned operations of the paper company, but independent converters also participate in the business and large packagers also own their own conversion facilities, purchasing board directly from mills.

Recycling plays an important role in this industry. Between one-quarter and one-third of kraft mill raw materials arrive as bales of spent corrugated boxes. Temple-Inland, for example, reported that 36 percent of its inputs in 2006 came from waste products. Georgia-Pacific reported recycling rates between 25 and 30 percent. Other producers are also dependent on waste products and discuss the availability and price trends of paperstock (the industry term for paper scrap) in their annual reports. High rates of recycling are facilitated by the fact that most waste cardboard occurs in centralized locations—at retail stores and in warehouses. Recycled corrugated cardboard

has always commanded the highest prices in the wastepaper trade; flattened boxes are routinely bundled in large balers owned by the distribution sector, with converters or recyclers collecting bales and eventually returning them to the pulp mills.

The folding box industry is also a high-level user of waste paper, but the fiber it uses comes from old newspapers typically assembled by municipal waste collection systems. Shredded newspapers undergo de-inking and are then converted into the grey-colored board which forms the basic material of folding boxes used in non-food applications.

DISTRIBUTION CHANNEL

The general structure of distribution in this industry may have seven or more levels. The pulp mill sells to a converter directly or to a wholesaler/trading company then sells board-stock to a converter. Here we have two or three levels. The converter sells to a packager (level 4). The packager distributes goods to a wholesaler (level 5) who sells to a retailer (level 6). The final recipient, the retail consumer, is level 7. Packing containers are also used to deliver parts and components from one manufacturer to the other, introducing additional levels.

In practice only, a relatively small percentage of board production reaches independent converters through a wholesaler/trader; wholesalers serve smaller operations. In sanitary food packaging, the food producer is usually also the converter and buys directly from the pulp mill. In such cases the number of levels involved is smaller. Relations between converters and packagers are often very close. The converter functionally serves as an extension of the packager's manufacturing activity. The converter will, for instance, print the customers' images, messages, and codes on packages and packing materials as part of die-cutting and scoring the board into box blanks. Printing operations are a large component of conversion. Large packaging operations requiring a continuous flow of product will work under long-term contracts with the converter. The converter often maintains one or more dedicated lines to serve a client.

In the distribution of boxes, the packager is the ultimate buyer. Customers downstream from the level where the product is made do not buy boxes; they buy products. The final customer rarely sees the packing (as distinguished from the package) in which goods are delivered to the retail sector unless the goods arrive by mail or by a private carrier directly to the home. Sometimes people delivering appliances in large cardboard boxes leave them at the home; more frequently they unpack the product and leave the carton in the truck.

This industry also features a reverse-distribution system in which retailers, wholesalers, municipalities, and paperstock dealers participate to return corrugated board or to acquire used newsprint for repulping. About one-third of all kraft paper has a sort of homing instinct and returns to the mill that made it.

KEY USERS

The key user of boxes is the packager. It is the packager who needs the box to protect his or her goods in transportation or to sell the product on a shelf. Another important user category is represented by moving companies. They purchase corrugated boxes for packing households and offices. The U.S. Post Office buys boxes for resale to the public as shipping containers. Businesses also purchase storage boxes of every kind—corrugated, folding, and setup—used in archival applications. A small quantity of packaging board reaches the consumer directly for packaging gifts. Gift boxes are typically purchased during the holiday season.

ADJACENT MARKETS

Generally speaking, all other forms of packing and packaging are adjacent markets to paperboard boxes. Of these, pallets are most intimately associated with boxes as the platform on which the latter are assembled and strapped for easy handling by forklift trucks in transport and warehouse operations. Most pallets are made of rough wood planks, although more expensive composite or heavy plastic pallets are also found on the market.

Glass containers compete with sanitary board. Plastics do so as well but are principally in competition with folding and setup boxes. Woven baskets compete with setup boxes as attractive presentational packaging for goods. Textiles are used in sacking to carry produce. Metal and plastic strapping is an important adjacent market in that it is used to secure pallets and also to hold tight bales of recycled board. Gift markets—especially jewelry and fancy candy—are adjacent markets to setup boxes; the boxes are heavily used in the packaging of such goods and therefore perturbations of these industries affect setup boxes as well.

RESEARCH & DEVELOPMENT

Paperboard packaging and packing containers, boxes and cylindrical containers both, represent a mature business in which competitive forces have long ago produced an equilibrium so that competition between materials is minimal. R&D activity in the industry is focused on externalities to the product, thus on pollution control and protection of the silviculture that produces the raw material. Product research tends to focus on product improvements such as coatings and barriers to protect the product and its contents, on stiffening agents that promise stronger products

at reduced fiber consumption, and on decoration of the product, thus on printing inks. Package producers, in attempts to improve sales, work on innovative features, such as easy closures, to differentiate themselves. Improving efficiencies in conversion are also a focus of research.

Pulping wood and turning it into fiber has always been and still remains a dirty industry, threatening air and water resources. In that both sulfite and sulfate mills use sulfur as a major process ingredient, these industries produce bad odors. Discharged sulfur contributes to sulfur oxides in the air and to acid rain formation. Bleaching of fibers using chlorine produces dioxin in effluents, a poisonous chemical that cumulates in fish. For best economic results, waste combustion for process heat is common in the industry and, if uncontrolled, produces nitrogen oxides implicated in smog and sulfur emissions. Pressures to control pollution emanate from the U.S. Environmental Protection Agency (EPA) which conducts considerable R&D for monitoring purposes. The high costs of pollution control continue to stimulate industry expenditures on process improvements and on devising improved and more efficient control devices.

The southern pine forests, the principal source of fiber used by this industry, are subject to periodic attacks by the southern pine beetle, often rendered as SPB, *Dendroctonus frontalis Zimmerman*. Annual destruction can remove 0.2 percent of softwoods—not a very high number, to be sure, but major periodic outbreaks can be 30 to 40 times more destructive. Considerable research centered on SPB is taking place under U.S. Forest Service and private auspices aimed at improved early detection of outbreaks and effective management of forests to minimize beetle incursions.

CURRENT TRENDS

In this mature industry it is more apt to speak of *concerns* rather than of *trends*. Concerns—as noted in issues highlighted by industry associations and voiced by producers in their annual reports—include worries about the shrinking share of paper and paperboard in packaging; the cyclicality of the industry; rising costs for energy, raw materials, and chemicals; industry consolidation; disruptive events; and government regulation.

Shrinking market share is cited by many. Such contraction is substantiated indirectly by the fact that growth rates are lower in paper and board than in economic activity taken as a whole. To be sure, the erosion is taking place slowly. Furthermore growth may resume if plastics prices, which have been increasing sharply, continue on that trajectory because petroleum prices continue to spiral upward. The packaging industry has always been cyclical in that reduced consumption directly translates into reduced packing and package acquisitions. Cyclicality is visible in Figure 35, showing the significant downturn in corrugated shipments and smaller erosions in sales in all but the sanitary category from 2000 to 2001. The industry is highly dependent on fossil fuels for its pulping operations. Energy costs also effect the cost of its principal processing chemicals. Industry consolidation worries producers but, in comparison with other sectors, very little consolidation has taken place. Disruptive events, on the scale of the Katrina hurricane of 2005, worry the industry as do power failures, forest fires possibly due to climate change, and SPB infestations. Government regulations represent a perennial pressure on costs and therefore on profitability in a market participants view as very competitive; they see intermaterials competition driving down paperboard prices.

TARGET MARKETS & SEGMENTATION

Each major type of container is targeted at distinct segments, with packing boxes having the most universal use for containerization of goods for bulk shipment, folding boxes serving principally consumer goods in the food and textiles sectors, fiber drums serving bulk markets, and setup boxes claiming the luxury and gift market as principal outlets. With the exception of setup boxes, the other categories represent commodity markets where product quality must be—and essentially is—uniform. Converters, therefore, attempt to compete on services and minor innovations—better closures, more attractive printing—to differentiate themselves.

RELATED ASSOCIATIONS & ORGANIZATIONS

American Forest & Paper Association, http://www.afandpa.org

Association of Independent Corrugated Converters, http://www.aiccbox.org

Fibre Box Association, http://www.fibrebox.org

North American Packaging Association, http://www.paperbox.org

Paper Industry Association Council, http://www.paperrecycles.org/about_us/index.html

Paperboard Packaging Council, http://www.ppcnet.org/index.aspx

Reusable Industrial Packaging Association, http://www.reusablepackaging.org

Technical Association of the Pulp and Paper Industry (TAPPI), http://www.tappi.org/s_tappi/index.asp?pid&equals

BIBLIOGRAPHY

Darnay, Arsen J. and Joyce P. Simkin. *Manufacturing & Distribution USA,* 4th ed. Thomson Gale, 2006.

Darnay, Arsen J. and William E. Franklin. "Salvage Markets." U.S. Environmental Protection Agency. 1972.

"EPA Cracks Down on Polluting Pulp Mills." U.S. Environmental Protection Agency. Press Release. 20 April 1999.

Lazich, Robert S. *Market Share Reporter 2007.* Thomson Gale. 2007.

"Product Summary: 2002." *2002 Economic Census.* U.S. Department of Commerce, Bureau of the Census. March 2006.

"Setup Boxes." ThomasNet. Available from: <http://www.thomasnet.com/nsearch.html>.

Thatcher, Robert C. and Patrick J. Barry. "Southern Pine Beetle." *Forest Insect & Disease Leaflet 49.* U.S. Department of Agriculture, Forest Service. Available from <http://www.na.fs.fed.us/spfo/pubs/fidls/so_pine_beetle/so_pine.htm>.

"Two E.P.A. Studies Confirm Threat to Fish of Dioxin from Paper Plants." *The New York Times.* 27 August 2007.

"Value of Product Shipments: 2005." *Annual Survey of Manufactures.* U.S. Department of Commerce, Bureau of the Census. November 2006.

SEE ALSO *Cans*

Breakfast Cereals

—■—

INDUSTRIAL CODES

NAICS: 31–1230 Breakfast Cereal Manufacturing

SIC: 2043 Cereal Breakfast Foods

NAICS-Based Product Codes: 31–123011, 31–123012, 31–123013, 31–123014, 31–123015, 31–12304, and 31–1230W

PRODUCT OVERVIEW

Breakfast cereals are corn, wheat, oats, rice, and other grains processed, cooked, and shaped by flaking, puffing, or shredding. Flaked products are usually also toasted; shredded products are baked into biscuits. The typical single unit in a cereal bowl, the atom of the industry, you might say—the flake, kernel, or one shred of a shredded biscuit—is actually a single grain of corn, wheat, oat, or kernel of rice. In some varieties of cereals, however, the individual item may be produced as dough. The dough is then extruded into shape by pressure through a die. Additional toasting or puffing may follow. These ready-to-eat food products are the dominant segment of the breakfast foods industry. The more traditional and less processed cereals require cooking. Steamed rolled oats can be eaten raw with milk. Most traditional breakfast cereals are based on oats while corn holds the largest share of the ready-to-eat market.

Production Process. An outer membrane known as the *bran* protects the genetic code inside each grain. The bran itself is protected by the hull, a relatively hard and fibrous structure. The inner mass of each grain is mostly carbohydrate, a type of sugar (chemically carbon, hydrogen, and oxygen), and a smaller amount of protein. This part is the kernel, also called the *endosperm*. The endosperm surrounds the *germ*. The germ holds the DNA of the species. DNA is protein. The germ also holds vitamins and a small amount of carbohydrates.

In the process of making cereals, producers first employ cleaning and sorting screens to remove dust and debris. They also screen out defective grains. Doubled grains in which two grains have fused together are undesired; they have too much hull and not enough endosperm. Very small grains are also screened out for the same reason. Grains are de-hulled by propelling them at high velocity against hard surfaces. The impact fractures the hull. The lighter hull is separated from the heavier remaining mass of the grain by gravity and air suction. At this phase the industry refers to the grain as the *groat*. Grains that resist de-hulling are screened out and recycled for another round.

In processing grains of corn, the bran and germ are separated from the groat by a process called degerming. It leaves behind one-half to three-quarters of the original grain—most of the endosperm and little else. Wheat is processed further to prepare it for easy flaking. Producers use rollers that lightly crush the grains in a process known as bumping. Polished rice used in cereals does not need any additional processing. Grains reaching this stage are once again renamed, referred to as *grits* in the trade. Grits are next placed into a flavoring solution in which sugar, malt, and salt have been dissolved. Most grains have very little taste and need this enhancement to taste good. Then the grits are cooked. Cooking times vary with the grain; corn grits require two hours, rice around one hour, and wheat around one-half hour. The grits clump together during cooking into sometimes quite large (soccer-ball-sized) formations and undergo a delumping process next until the grits are once more separated and uniformly distributed on a moving belt. The belts next pass through drying chambers in which heat and humidity are carefully controlled. The last step before the serious business of making cereal begins is tempering. The dried grits are held for several hours until the moisture levels between individual grits and also within the grits reach equilibrium.

The serious business is the creation of flakes or shreds of cereal; puffing is done while the grits are still moist. Producers flake grains by using cooled rollers that simply flatten out the grits. Knives mounted over the surface of the rollers peel off the flattened items. Corn grits are moistened with a flow of steam to cause them to stick to the rollers first. In modern practice flakes are then toasted while suspended in very hot air rather than being brought into contact with hot surfaces. Packaging follows.

Shredded wheat involves compressing individual grits between a smooth drum and another which has grooves. In the shredding process the grits are pressed by the first drum into the grooves of the second and the grits are elongated in the process. Sequences of drums, arranged at right angles to each other, produce layer after layer of shreds which overlay each other like a network. Biscuit-forming machinery is fed these layers or mats of shreds. The machines employ relatively dull knives to separate out biscuits by pressure (which seals their edges) and cutting (which separates one biscuit from another). All grains can be shredded, but shredded wheat is the dominant form of this cereal.

In making puffed cereals, puffing immediately follows cooking, the grits still moist and hot. The process inserts them into vessels; the vessels are sealed and heated from the outside to increase the temperature and pressure in the chamber. Then, suddenly, the vessels are popped open. The rapid change in pressure causes the moisture inside the grits to expand, puffing them up. If the process is controlled precisely, most of the grits will puff and will not burst. The resulting objects are dried in fluid beds of hot air so that the weight of the grains will not cause individual puffed particles to collapse again. Producers then use sorting beds to remove un-puffed grains and the detritus left behind by those that burst.

Producers use very similar methods to flake or puff individual items initially made from dough that has been forced through dies. By using dough products, manufacturers can combine different grains and achieve unusual forms—tiny doughnuts, letters of the alphabet, shaped flakes, and so on. Coating the products with sugar and flavoring them with spices is optional and often used.

All of these processes are characterized by high precision and careful control of temperature, moisture, and dwell times. High engineering skill is required to make machines able to deal effectively with very small and delicate particles. Breakfast cereals, thus, are high-tech products made by advanced technological means—a fact easily missed or forgotten. The public experiences cereals primarily as a bewildering jungle of competing products or as advertising messages aimed at childish instincts or memories of pleasant tastes.

An Interwoven History. It is ironic, in a way, that a food product today appealing to our more hedonistic impulses originated as a health food embraced by religious pioneers. The inventor of the category was James Caleb Jackson, head of a vegetarian sanatorium in Dansville, New York. He served his patrons bran nuggets for breakfast. These hard grains had to be soaked overnight to soften them enough to eat. He called the product Granula. The year was 1863. Elen G. White was one of the patrons of the Dansville Sanatorium. She became the founder of the Seventh Day Adventist denomination. A member of the church she founded was John Harvey Kellogg, himself the head of another sanatorium in Battle Creek, Michigan.

John Kellogg formulated a health biscuit for the clients of his sanatorium. He also called his biscuit Granula, but, when sued by Jackson for using the name, Kellogg renamed his biscuit Granola. Further experiments conducted by John with his brother, Will Keith Kellogg, eventually, and by accident, produced the industry. One of their products was a cooked wheat meal they rolled into sheets to make a cracker-like food. A pot of the cooked wheat, left unprocessed overnight, was the accident in question. The following morning they rolled out the product, as usually. This time it did not turn into a sheet. Instead, every grain in the mixture separated as a single flake. These flakes, roasted and served to the sanatorium's customers, were an immediate hit. In 1906 W.K. Kellogg bought his brother's portion of patents. They had obtained these patents jointly. W.K. formed a company, and began to sell cereal on the open market. Kellogg Company has been in continuous existence ever since and is the dominant producer of breakfast cereals still.

Charles William Post was a customer of the very same Battle Creek Sanatorium where the Kelloggs' held sway. He came to the place in 1893 in order to recover from a nervous breakdown. In 1894 he formed his own sanatorium. There Post began to experiment with hickory coffee and his own cereals. The coffee, Postum, became a success. Post's attempts to make an easier-to-chew version of Jackson's Granula became Grape Nuts (1897). His own corn flakes, initially called Elijah's Manna, took off when he renamed the product Post Toasties (1908). The C.W.

Post Company eventually diversified and renamed itself General Foods in 1922; it became part of Philip Morris (which itself renamed itself Altria by way of gaining distance from tobacco). Philip Morris acquired Kraft in 1988 and merged the two acquisition into Kraft Foods. Kraft became independent in 2007 and, as an inheritor of Post, ranks third in the breakfast cereals business.

The year before Post introduced Elijah's Manna, thus in 1921, the breakfast of champions, Wheaties, was born in Minnesota. A health clinician accidentally spilled some bran gruel on a hot stove and produced some very crisp and nice-tasting flakes. By that time, of course, thanks to Kellogg's success in the market, the environment was more than ready for such accidents. The clinician took the idea to a milling company in the area called Washburn Crosby Company. Washburn Crosby was operating mills in what later became Minneapolis. It used the falling waters at St. Anthony Falls on the Mississippi for power. In 1928 Washburn Crosby became General Mills through a merger of several area millers. Wheaties, in the long run, turned General Mills into the second-ranking cereals company, a position it still held early in the twenty-first century.

Categories, Brands, and Trends. Americans consume more wheat than any other grain, but in the breakfast cereals category the leading grain is corn. It accounts for approximately 28 percent of all cereals as measured in dollar sales. Corn is followed by oats, the grain used in 17 percent of products. Wheat, with 16.5 percent of share, is a close third; rice, with 8.4 percent, is a more distant fourth. These data come from the *2002 Economic Census* which reports on major categories by grain. The Census Bureau classifies the remaining 30 percent of shipments without showing the grain used in the product. Thus the Census Bureau reports an additional 18 percent of shipments as breakfast preparations made of other or mixed grains, 8 percent as other breakfast foods, including infant and instant formulations and cereals intended to be cooked before consumption (the traditional categories); and the remainder were labeled as not specified by kind (nsk) because those reporting the shipments failed to provide a description. Virtually all serious growth in the breakfast cereals category between 1997 and 2002 was in this somewhat mysterious nsk category, a subject we shall look at more closely under the Market heading below.

Organizations such as Information Resources Inc. (IRI) and ACNielsen also produce reports on the cereals industry. These market research firms focus on brands of cereals (and many other products) collected as a service to manufacturers and retailers. Their data occasionally appear in trade and general circulation magazines. Based on IRI data cited in *Market Share Reporter*, the top brands in 2004 in rank order were Cheerios (General Mills, made of oats), Frosted Flakes (Kellogg, made of corn), Honey Nut Cheerios (GM, oats), Honey Bunches of Oats (Post, owned by Kraft), Cinnamon Toast Crunch (GM, wheat and rice), Frosted Mini Wheats (Kellogg), Froot Loops (Kellogg, made of corn, oats, and wheat), Lucky Charms (GM, oats), Kellogg Corn Flakes, and Special K (Kellogg, rice).

The leading brands accounted for less than 40 percent of all cereals sold in 2004, the rest of sales were shared by some 250 other varieties, at least as estimated by *Progressive Grocer* magazine. The impression of an overwhelming number of brands is, of course, confirmed by any visit to a decently sized supermarket. Cereal.com, an Internet site specializing in cereals, provides nutritional details on 135 brands. A look at the Kellogg's Web site showed 28 brands, and perusal of General Mills' site revealed 65 brands for these two leading producers in early 2007. All told, in 2006, the industry was using 2,235 unique Universal Product Codes (UPCs) according to ACNielsen—UPCs are read by the scanning machines built into grocery store checkout counters. The large number of UPCs is explained by the fact that each size category has its own code and any change in product merits requesting and using new codes.

Data from ACNielsen for 2006, cited by *Progressive Grocer*, indicated decline in the sales of branded and boxed breakfast cereals through most of the first decade of the twenty-first century. The survey was based on the sales of food, drug, and mass merchandising stores—but excluding Wal-Mart. This category (branded and boxed) accounted for 87 percent of all such sales. Bagged granola and granola-based and other natural breakfast bars, by contrast, showed energetic growth. Increasing awareness of health issues, combined with changes in life styles on the part of at least a portion of the U.S. population, appear to be the factors behind a slow but ever more visible transformation of this industry.

MARKET

Issues of Measurement. Measuring the breakfast cereals market is complicated by the fact that federal industry data are sometimes significantly different from federal product data. The U.S. Census Bureau conducts an Economic Census every five years (in years ending in 2 or 7) principally to support national measurement of Gross Domestic Product (GDP). Thus it measures all transactions taking place within an industry, not just final output. The GDP is a record of all economic exchanges. Duplication in the data is sometimes a consequence, the degree of it influenced by industry structure. Thus the Bureau also records transactions between manufacturers before products reach finished form. When Producer A sells Producer

B a semi-finished product classified as a breakfast cereal (e.g., corn grits or corn meal), that transaction is added to total industry shipments. When Producer B sells the same product in finished form to a wholesaler, that transaction is also added to the industry total. In the NAICS industry 31–1230, duplicate shipment counts have been high, about $1.8 billion in 1997 and $1.1 billion in 2002. Product shipments were lower than industry shipments by those two values in those two years, with the paradoxical consequence that industry shipments declined but product shipments increased.

In building a multi-year pattern, difficulties also arise because the Economic Census is a 100 percent survey of all producers, but only every five years. Producers are mandated under federal statute to provide information. In all other years, however, the Census Bureau conducts the Annual Survey of Manufactures (ASM). ASM data are obtained from a sampling of participants only and pro-

jected to the industry as a whole. ASM results are thus influenced by the sample—and more so in an industry like breakfast cereals where fewer than 50 companies make up the total. Furthermore, no product level data are acquired in the course of an annual survey.

Private sector measurements in this industry are usually built from the bottom up, thus from check-out counter scanning data using UPC definitions. These are not directly comparable to the government's NAICS product codes. Data from private sources have tended to be lower—and sometimes significantly lower—than Census Bureau estimates. These issues are always present in market size measurements, but more visibly in this industry.

Market Size. The casual reader on this subject will encounter published values ranging from $6 on up to $11-plus billion per year for the breakfast cereals industry in the early 2000s. The lower value reflects private surveys

FIGURE 37

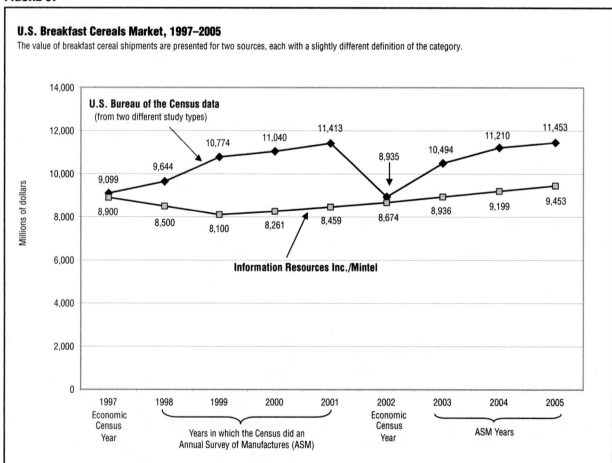

SOURCE: Compiled by the staff with data from "Historical Statistics for the Industry: 2002 and Earlier Years," *2002 Economic Census,* American Fact Finder, and "Statistics for Industry Groups and Industries: 2005," *Annual Survey of Manufactures.* U.S. Department of Commerce, Bureau of the Census, November 2006. *Prepared Foods,* November 2003 and August 2004.

that typically report on sales taking place in a subset of all merchandising channels, thus in grocery stores, supermarkets, and convenience stores, for instance. The highest values are based on the ASM. Figure 37 provides an illustration.

The data show Census Bureau estimates based on two Economic Census years, 1997 and 2002, and ASM data in the intervening years, 1998–2001 and 2003–2005. The two lowest points in this sequence were Economic Census years, all the higher estimates took place in ASM years. The industry appears to have declined from 1997 ($9.1 billion) to 2002 ($8.7 billion). The very sharp decline from 2001 ($11.4 billion) to 2002, a 21.7 percent drop, was not noted anywhere in the trade literature in 2002 or the year after, although weakness in the market has been noted by many observers from 2000 forward, albeit for the well-established branded-and-boxed ready-to-eat cereals sector only.

Shown on the same chart is a time-series developed by Mintel International Group, a market research firm. Mintel used data from Information Resources Inc. and its own estimates. These figures were reported by *Prepared Foods* magazine in two separate articles. Data for 2004 and 2005 represent Mintel's projections. Worth noting is the fact that federal and private figures for 2002 (an Economic Census year) correspond quite closely; the federal estimate is just $261 million above the Mintel estimate. If Mintel's data for 1999 and 1998 are projected backwards to 1997, the resulting value for 1997 would be within about $200 million of the Census value as well. ASM data for 2005, however, place the private estimate nearly $2 billion under the result reported by the Census Bureau possibly reflecting duplications in the ASM data.

These figures suggest that the size of the breakfast cereal market at the product level and at retail pricing was at least $9.5 billion in 2005, possibly higher. Data from the 2007 census, which will be collected in 2008 but will not likely be available until 2010, will reveal if growth is sharply up, as the federal data suggest (growing at 8.6% per year since 2002) or just moving upward modestly (at an annual rate of 2.9%) as suggested by Mintel's estimate singled out here for purposes of illustration.

1997–2002 Growth. If we take the Economic Census years as providing the most accurate data on the details of this industry in the United States, a look at changes in components will highlight trends in the industry as a whole. As noted, product shipments increased between 1997 and 2002 as measured by the Census Bureau although industry sales declined. In the 1997–2002 period total product shipments grew at a rate of 1.1 percent per year. Looking more closely at broad categories, we get the results shown in Figure 38.

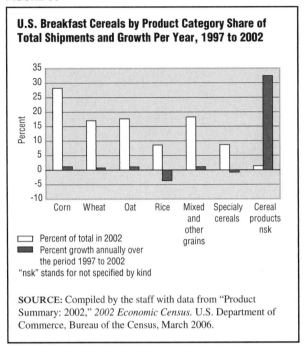

FIGURE 38

U.S. Breakfast Cereals by Product Category Share of Total Shipments and Growth Per Year, 1997 to 2002

Legend:
□ Percent of total in 2002
■ Percent growth annually over the period 1997 to 2002
"nsk" stands for not specified by kind

SOURCE: Compiled by the staff with data from "Product Summary: 2002," *2002 Economic Census.* U.S. Department of Commerce, Bureau of the Census, March 2006.

In the largest product categories the annual growth rate never exceeded 1.2 percent, achieved by corn-based products, and was actually negative in some categories (-3.7% for rice and -0.8% for specialties such as instant and infant cereals). At the same time, the industry's smallest category, comprising a mere 1.4 percent of total shipments in 2002, the growth was a rather spectacular 32.4 percent per year between 1997 and 2002. This is the category the Census Bureau labels, "Breakfast cereals and related products, not specified by kind." The latter part of this title—not specified by kind—calls attention to the evident difficulties that cereal producers have in classifying some of their products. It is under this category that they place all new products they have difficulty fitting into the NAICS classification scheme.

In this same period (1997–2002), the U.S. population increased at the rate of 1.3 percent annually. The obvious conclusion is that the industry as a whole was growing at a rate below that of the population, with only the most popular product category, the corn-based products, almost but not quite keeping up with population increase. The decline, however, has been slow. Most of the population still consumed a large amount of breakfast cereals in the middle of the first decade of the twenty-first century, but has changed what it buys. The growth within the industry itself appears to be in new product categories. A significant portion of that market consists of items nicely matched to growing segments of the food industry as a whole (e.g., snack foods, natural products, dieting, and health foods). Demographics, particularly fewer children as a percent of total people, play a role as well.

KEY PRODUCERS/MANUFACTURERS

In 1997 the industry had 48 companies operating 71 establishments; of these 47 had twenty employees or more. Five years later two companies had disappeared in a process of industry consolidation. Establishment counts had declined to 66 all told. In 2002 the large establishments (20 or more employees) still numbered 47. Employment in the industry had decreased from 14,700 to 13,000 people. Six companies dominated the market in the middle of the first decade of the twenty-first century.

Kellogg Company. The top producer of breakfast cereals, with a 34 percent share of the total market and sales of $10.9 billion (2006) was Kellogg Company, the original inventor of the category, still headquartered in Battle Creek, Michigan. The company operates in seventeen countries around the world and sells its products in 180 countries. Five of its brands in the United States rank in the top ten, its leading brand being Kellogg's Frosted Flakes (ranked second overall).

General Mills. A close second in the industry is General Mills, a diversified food company with roots in flour milling going back to the 1860s. The company had sales in 2006 of $11.6 billion, $8 billion in retail products of which breakfast cereals represented approximately one-quarter. The company has a 29 percent share of the market. General Mills' Cheerios is the top brand; three other brands are in the top ten as well. The company has a joint venture with the global food giant, Nestlé S.A., under which Nestlé sells GM products in 80 countries excluding North America.

Kraft Foods. The third ranked producer of breakfast cereal, Kraft Foods, is best known as a producer of cheese but ranks among the leading diversified food companies in the United States. Its two major brands of cereals are the Post family of brands and Nabisco's Shredded Wheat. Post Honey Bunches of Oats is the fourth-ranked cereal brand in the country. Kraft dates back to 1916 when its founder, J.L. Kraft invented the process for making processed cheese. As already mentioned earlier, Kraft was owned by Philip Morris for a period of years but has become independent (in 2007). While still under Philip Morris, Kraft acquired RJR Nabisco (in 1993).

Quaker Oats. Begun in 1901 when several leading oat millers formed the company, Quaker Oats developed into the fourth-largest breakfast cereal producer. The company holds the leading share in the traditional oat cereal market.

It operated as an independent producer for one hundred years before merging with PepsiCo, Inc. in 2001. At the time of the merger the company's sales were $5 billion. Quaker Oats offers twenty brands of cereals of which the best known is Quaker Oatmeal. The company holds approximately 6.5 percent of the total cereals market.

Malt-O-Meal. Malt-O-Meal is a privately held producer, based in Minneapolis, Minnesota. The company, originally called Campbell Cereal Company, began in 1919 as a producer of wheat-based cooked cereals, operated as a private label producer, and then entered the cold cereals market, renaming itself Malt-O-Meal in 1953. The company makes three hot and eleven cold cereals brands and has a market share of approximately 3.5 percent.

Weetabix, Ltd. An important European producer of breakfast cereals is Weetabix of the United Kingdom, operating in the United States as Weetabix USA in Clinton, Massachusetts. The company's best-known brands are Weetabix and Alpen.

Ralston Foods. The most important producer of private label breakfast cereals in the United States is Ralston Foods, one of the operating elements of Ralcorp Holdings Inc., a $1.85 billion private label producer. The organization is one part of the former Ralston Purina Company which had two components, one produced pet food and animal feeds and the other food products for humans. Ralcorp was the food element spun off in 1994. Nestlé acquired the pet food business in 2001.

In a historical footnote it may be useful to note that Ralston Purina once more illustrates the linkage between breakfast cereals, health, and new religious or reformist movements. The name Ralston in Ralston Purina was originally an acronym built of the words Regime, Activity, Light, Strength, Temperation, Oxygen, and Nature. These were the seven principles of Ralstonism, a nineteenth century utopian movement founded by Albert Webster Edgerly. This controversial and charismatic figure appeared to like *noms de plume*; he was known among his followers as Dr. Everett Ralston and wrote many books under the pseudonym, Edmund Shaftesbury. He was also a popular promoter of diet foods. According to Ralston Purina's company history, William Danforth, founder of the Purina Wholefood Company, invited Edgerly to endorse a cracked wheat product that Danforth had introduced in 1898. Edgerly agreed but required that the food be named Ralston Wheat Cereal. So popular was the product that Purina renamed itself Ralston Purina in 1902.

MATERIALS & SUPPLY CHAIN LOGISTICS

Breakfast cereals are made all across the United States but with concentrations in the Upper Midwest, the Midwest, California, and the North East. These clusterings are largely due to historical patterns in that this industry is still largely dominated by companies that invented the category. The leading companies began operations in Michigan and Minnesota; oatmeal producers began in Iowa and Ohio. The industry established additional operations in major centers of population.

Collectively, and measured in dollars of costs, whole grains (wheat, oats, corn, barley, and rice), partially processed grains (corn grits, flakes, and meal), and flour are the dominant input category, but most production plants do not use all of the grains that are turned into cereals. Although not intuitively obvious, the second major cost category is paperboard containers and corrugated paperboard, the paperboard typically originating in northern and corrugated in southern paper and board mills. Sugar is the largest single category of input (in contrast to any one of multiple types of grain) and is used by virtually all production plants. Fruits and nuts used as additives are the last major grouping. The bulk of these—raisins—originate in California.

The value added in manufacturing breakfast cereals is high. Value added is the difference between incoming cost and outgoing shipments. Value added was nearly 75 percent of total shipments for breakfast foods in 2002, compared to 48 percent for all manufacturing activities in that year—suggesting that this product category can readily sustain high transportation costs if necessary. Most of the product is destined for major population centers.

DISTRIBUTION CHANNEL

Cereals are distributed through independent grocery stores, grocery chains, and mass merchandising outlets. Data on distribution by channel, unfortunately, exclude data from Wal-Mart because this giant does not contribute information to the common pool of scanned data, as other retailers do, that market research firms mine to provide broad reports on retail product sales. For all practical purposes, however, Wal-Mart, in relation to this industry (indeed to all grocery products) can simply be seen as the nation's largest grocery store, representing approximately 17 percent of all grocery sales. Some of the product also moves through convenience and drug stores, but for purposes of looking at the distribution channel, these are also supplied using the same methods.

Grocery stores face the major problem of maintaining adequate stocks of many hundreds of items at all times, to avoid out-of-stock situations, and yet to minimize inventories, especially of perishable items. For this reason the industry relies on two forms of distribution of which the most important is the DC, an acronym for distribution center. DCs supply multiple stores usually within a radius of approximately 150 miles maximally. In modern practice computerized systems track levels of product in the stores a DC serves and of inventory within the DC itself. Orders flow out to suppliers when triggering levels are reached. Suppliers then replenish the DC. The second method used is the DSD; the letters stand for direct store delivery, namely routes maintained by suppliers or wholesalers. DSD routes are used for products that are replenished routinely and often daily, including fruits, vegetables, dairy products, meat, and bread. DCs and DSDs may specialize in refrigerated products. Large chains typically maintain their wholly-owned DCs and also use independent wholesalers. Wholesalers supply most independent grocers. Two examples of such wholesalers are Supervalu Inc. and McLane Company. Supervalu is itself a grocery chain but also operates 35 DCs from which it supplies 5,000 independent grocery stores. McLane operates 10 centers for groceries and supplies 19,000 convenience and grocery stores.

Given the very large number of brands maintained by cereals producers, these companies operate distribution centers of their own from which they supply other DCs or deliver directly using routes.

KEY USERS

Virtually everyone consumes breakfast cereals—or has done so in the past. The product is associated with childhood eating habits, and the industry's advertising and promotional effort aims to influence children and their parents.

A significant demographic trend in the twenty-first century, already visible in 1990, is the aging of the U.S. population, meaning that children and youngsters (19 and younger) will represent a smaller proportion of the public in 2010 than they did in 2000 or in 1990. The population aged 19-and-under was 28.6 percent of the population in 2000 and is projected to be 26.9 percent of the population in 2010. The total population increase between those two years is projected to be 9.5 percent while the increase for those 19 years old or younger is projected to grow only 3.3 percent. In part for this reason, the industry is engaged in product development and advertising to reposition breakfast cereals as a convenience and a snack food.

ADJACENT MARKETS

The chief alternatives to breakfast cereals are other foods typically consumed for breakfast, the traditional bacon-and-eggs with toast, bagels and cream cheese, or pastry products. Cereal breakfast bars, as opposed to cereal in bowls mixed with milk, are increasingly consumed but cannot be considered adjacent because their products are classified with cereals; they are made of the same components that go into the bowl. Also emerging are energy drinks specifically formulated for consumption in lieu of breakfast. Such flavored beverages contain cereal products but in liquid forms rather than as flakes or shapes. Also different from standard breakfast cereals are waffles and toasted products of which two examples come from Kellogg—Eggo waffles and Pop Tarts. These products, in effect, represent second-order processed foods thus incorporating greater convenience and more complex combinations of different components than the traditional cereal.

RESEARCH & DEVELOPMENT

The chief thrust of all R&D activity in this industry may be characterized as new product development. Considerable work is also underway on changing existing products to make them more appealing to a more demanding market. The leading companies all maintain substantial R&D facilities in which they develop products in test kitchens and prototype them in experimentally-sized miniature production facilities.

The chief thrusts aimed at traditional products include creating healthier cereals based on whole grains and their mixtures. Cereal breakfast foods contain considerable amounts of salt and sugar to give them flavor. Replacing all or some of these ingredients with newly formulated flavor enhancers, replacing carbohydrates with fiber, and adding additional vitamins to the product are strategies pursued by all the companies.

The industry is experiencing sluggish growth in its standard product despite what seem to be strenuous attempts to introduce modified products by dozens each year. Real growth, however, is associated with new products that may be classified as breakfast or as snacking foods but are well off the traditional reservation, the breakfast cereal. It is in these areas where most of the emphasis is being placed, at least as indicated by the companies' communications directed at their stockholders.

CURRENT TRENDS

Among the many trends influencing the breakfast cereals product category the most important is probably a growing public awareness that eating habits in the United States need correction. Data began to emerge as far back as 1963 with the first National Health and Nutrition Examination Survey (NHANES), conducted by the Centers for Disease Control and Prevention (CDC), showing that Americans are overweight, and high percentages overweight enough to be called obese. Since that time other NHANES have not only confirmed the early findings but have shown deteriorating trends. Excess weight is closely

correlated with the milder Type II diabetes. Obesity in children is increasing as well. This phenomenon has put a sharp flood-light on the food industry, not least breakfast cereals. These grain products have large amounts of carbohydrate (nature's natural sugar) as well as added cane and beet sugar and corn syrup. A segment of the population has responded to the warnings offered with more and more emphasis by the nation's health authorities. The consequences have been declining or flattening sales patterns in traditional processed foods.

Demographic shifts, already noted above, have produced lower growth in precisely those segments of the population that consume the largest amounts of packaged cereals, the young. Pressures on time have mounted as dual-earner couples have increased and, even more sharply, the number of single-parent households with children. This phenomenon has put pressure on available time and has caused the fraying of old habits. Fewer and fewer families have sit-down breakfasts. Mornings in families, with both parents and children all heading out, are most affected by vanishing time.

Another trend is cost pressures that began to plague the industry in the middle and end of the first decade of the twenty-first century. Hostilities in Afghanistan and Iraq and tensions mounting with Iran have put the issue of oil supplies in the future on the front page again. These events, remote from breakfast, in a way, have affected it by changing the view of corn. As a source of ethanol it loomed so large in the latter half of the first decade of the 2000s that virtually all corn-processing associations made ethanol the center-piece of their speculations about the future.

The industry's response was still in process of unfolding as the first decade of the new century was nearing its end. Participants in the industry have been transforming their products to make them more healthy in response to the public's heath concerns. Reacting to demographic changes, the industry has presented traditional cereals as snack foods to be eaten between meals and has introduced a substantial number of new products that can serve both the purposes of breakfast and the midday or afternoon snack. In response to rising costs, the industry has raised prices, albeit carefully. It has reduced package sizes. General Mills, which initially resisted industry-wide price increases raised them in 2007 in the wake of seeing its sales decline 1.2 percent in 2006. Up-trending prices had in the past resulted in shifts to private label brands and may further slow sales of traditional products. The new product categories, however, many of which have less or no corn content, may not be affected as much.

TARGET MARKETS & SEGMENTATION

At the dawn of the industry breakfast cereals were targeted at healthy-minded adults and vegetarians, but the industry rapidly evolved a strategy of aiming its marketing at children by appealing to them directly and persuading their parents to buy the product. These two modes of targeting the product coexist into the twenty-first century, the health-appeal returning again. Most products are aimed at children and are high in sugar content to reinforce that appeal. A growing subset of products continues to be marketed to the health-conscious. The most recently emergent trend has been, as already noted, to make all cereals more healthy by increasing fiber, reducing carbohydrates, lowering sugar levels, and using whole grains. The provision of this product in the form of ready-to-eat breakfast bars is aimed at a life-style segment of adults and youngsters on the move, in a hurry, with no time but to munch on the go.

RELATED ASSOCIATIONS & ORGANIZATIONS

American Corn Growers Association, http://www.acga.org

American Institute of Baking, http://www.aibonline.org

Grains for Life, http://www.grainpower.org/index.asp

National Corn Growers Association, http://www.ncga.com

North American Millers' Association, http://www.namamillers.org

Wheat Foods Council, http://www.wheatfoods.org

BIBLIOGRAPHY

Darnay, Arsen J. and Joyce P. Simkin. *Manufacturing & Distribution USA,* 4th ed. Thomson Gale, 2006, 35–38.

"The Early Days of Breakfast Cereal." MrBreakfast.com. Available from <http://www.mrbreakfast.com/article.asp?articleid =13>.

Fishman, Charles. "Corporate Secrecy is Bad for Business, Bad for Democracy." *Philadelphia Inquirer.* 9 June 2006.

"History/Timeline of Ralston Purina Company." Nestlé S.A. Available from <http://www.purina.com/company/History.aspx>.

Kiple, Kenneth F. and K.C. Ornelas. *The Cambridge History of Food.* Cambridge University Press, 2000.

Lazich, Robert S. *Market Share Reporter 2007.* Thomson Gale, 2006, Volume 1, 138.

Lempert, Phil. "Super Bowl: The New 'Snap, Crackle, and Pop' of Product Diversity in the Cereal Aisle Seem to be Working." *Progressive Grocer.* 1 March 2007.

Phillips, Bob. "Breaking with Tradition: Single-Serve Cereals and Breakfast Bars Provide Morning Fuel for On-the-go Consumers." *Convenience Store News.* 8 May 2006.

Roberts, William A., Jr. "Breakfast Cereals: A Vulnerable Position." *Prepared Foods.* 20 November 2003.

———. "A Soggy Cereal Market." *Prepared Foods.* August 2004.

Six, Janet. "Hidden History of Ralston Heights." *Archaeology.* May/June 2004.

Thompson, Stephanie. "Big G Yields on Cereal-price Cuts: Joins Rivals in Downsizing Boxes to Effectively Lower Consumer's Checkout Bill." *Advertising Age.* 19 March 2007.

SEE ALSO *Snack Foods*

Cameras

―――――――◆―――――――

INDUSTRIAL CODES

NAICS: 33–3315 Photographic and Photocopying Equipment Manufacturing

SIC: 3861 Photographic Equipment and Supplies

NAICS-Based Product Codes: 33–331511, 33–331512, 33–331513, 33–331514, 33–331515, and 33–3315Y

PRODUCT OVERVIEW

If we view photography as the recording of images, the camera is the device that collects and focuses light from a target and film is the medium that does the recording. Such, at any rate, was the natural division of functions until about the mid-1990s when digital photography spread from industrial and space applications to the public in general. After that time, two technologies for capturing light have been in competition: film and magnetic media. We get printed photographs from either source. Terminology also changed in the 1990s. The traditional camera loaded with film has come to be known as the analog and its new competitor as the digital camera. Analog cameras have a venerable history going back to the nineteenth century. They existed long before that time as well, thus before the development of image-storage chemistry, as curiosities and as devices used by artists. Analogs exhibit levels of sophistication not yet matched by digital devices, but the general trend in the industry suggests that digital machines will ultimately replace the analog camera except in special uses.

The basic elements of the camera are lenses that collect light through an aperture, a viewing mechanism that shows the user what the picture will look like, storage space for film, a mechanism for placing frames of film behind the lens one by one, and a mechanism to expose the film to light for a brief moment. Modern cameras do much more. They have motors to advance and to rewind the film, flash lights, and automatic focus and light control using optical sensors, computer chips, and mechanics. Cameras count the pictures taken and display the tally. They may be equipped so lenses can be exchanged; alternatively they will have a zooming lens. Finally, they have battery power to drive the mechanisms. Digital cameras use electronic means to translate light into digital signals; the digitized data are stored in built-in computer memory. Once memory is full, cameras have disks to store pictures. Disks may be offloaded to other computers. Digital devices do not need film or means of moving it through the camera.

There are four major types of camera named for the method by which the photographer previews the picture to be taken. Viewfinder cameras provide a special window arranged so that it more or less matches what the lens sees. The match is never perfect because the angle of view is different. This difference is the parallax error; the picture in the viewer and the print from the lens will be slightly different. Most amateur photographers are little bothered by the error, but it matters to professionals. Single Lens Reflex (SLR) cameras are the most popular. They show exactly what the lens sees, the image transferred from lens to user by a series of mirrors. The first mirror, placed at an angle in front of the film, is snapped aside when the camera is clicked. Twin Lens Reflex (TLR) cameras have two lenses; one takes the picture; another, above it, displays the same image to the photographer. The view is larger and sharper, but with two lenses used, a parallax error is still present. View cameras let the user see exactly what the lens is looking at, but no mirrors are involved. The view screen is placed directly behind the lens where the film is normally waiting. In view cameras the film is held elsewhere. When the picture is snapped, the film is first moved into place in front of the view screen for the exposure.

The image reaching the film or the electronic sensor from the lens is upside down and inverted from side to side (left is right and right is left). The last two types of cameras show the photographer this upside down view. In the SLR mirrors are so arranged that they put the picture the right way around again. In viewfinder cameras the view port is just an aperture; the user's eye is just looking at the scene.

Camera lenses come in standard, wide angle, and telephoto varieties. Focal length measured in millimeters is used to define these categories. Focal length is the distance from the center of the lens to the surface of the film or chip that will record the image. Standard lenses most closely match what human eyes see and are around 50 mm wide, ideal for 35 mm film. Wide angle lenses are less than 50 mm and are best for landscape work or wide pictures of a large family gathered for its picture in the backyard at reunion time. Telephoto lenses come in medium and long versions, the medium being 85 to 135 mm, very good for portrait photography. Long telephoto lenses are greater than 135 mm and are used in sports photography and taking pictures on safari to capture elusive or dangerous animals. These types of lenses are known as fixed focal view or as prime lenses. Zoom lenses are designed so that the focal length can be changed by the user without changing lenses. In handheld cameras the focal length of zooms will range from 35 to 100 mm. Zoom cameras are a compromise. A good intermediate solution is to use an SLR on which different lenses can be mounted as the situation requires it.

Digital cameras are defined by resolution, thus the number of dots or pixels the receiving sensor can support. A minimal resolution is 320 x 400 (128,00 pixels). Photo quality begins at 800 x 600 (480,000 pixels). A high-end resolution is 4064 x 2704 (11 million pixels). The more pixels the sharper the image—but the greater the memory capacity needed to store an album.

Historical Snapshot. The Chinese philosopher Mo-Ti, living in the fifth century B.C., first recorded a curious phenomenon. A tiny hole in the wall of a darkened space will produce an image of the outside on a wall—but upside down. Why? Light beams travel in a straight line. Assume that the tiny hole is in the door of an empty but

darkened chamber. Light from the highest point outside, the sky, for instance, will enter the hole at a downward angle, reaching the bottom of the opposing wall. Light from the lowest areas outside will angle up and reach the wall inside near the ceiling. Light from the right will end up on the left, light from the left on the right. Today we know that precisely the same thing happens when light enters ours eyes. What we see is upside down and the sides are reversed, but the brain conveniently switches things back the way they ought to be. People observed this phenomenon throughout time here and there. The Muslim sage Abu Ali Al-Haytham, working in the eleventh century in Egypt, laid the foundations of the science of optics. His extensive experiments involved, among others, working in a dark room with light. His major work, *Book of Optics* reached an awakened European public in Latin translation at the time of the Renaissance. The phrase "dark room" translated to Latin is *camera obscura*. Of that phrase today one word remains in common use, the camera. The idea of these projected images through tiny holes, be the chambers that hold them room-sized or just boxes, immediately fascinated people, and the fascination has not died. Leonardo da Vinci included a design for a camera obscura in his notes. Such objects were made for public entertainment and used by artists to project images that they traced on paper. All manner of small, portable cameras of this type were produced. Photography emerged when chemical means were found to capture the images.

Chemical pioneers were Joseph Nicéphore Niépce and Louis Daguerre in France. Niépce produced the first image on a pewter plate in 1826 using bitumen of Judea as his emulsion, an asphaltic substance that hardens under light. Daguerre made the first daguerreotype in 1839, a photograph on copper that had been treated with silver, which darkens with light. Chemistry based on the characteristics of silver came into general use thereafter. Negatives were made on coated glass, first by a very laborious process known as wet plate photography requiring the photographer to prepare the plate, expose it, develop it, and print it immediately on paper that, in turn, he first had to coat with an emulsion. George Eastman in upstate New York founded what later became known as Eastman Kodak by simplifying the wet process. He developed coated glass plates that could be stored before use and after exposure to be developed at the photographer's convenience. Dry plate came on the market in 1879. Eastman is also credited with the development of roll film when he substituted a cellulose substrate for glass in 1889. Equally momentous was Eastman's introduction of the low-cost Kodak camera for the masses. He named it that because he liked the sound of the letter K—and wanted a word that started with and ended in the same letter.

The basic chemistry underlying photography exploits a characteristic of silver-halides, a group of compounds of silver with bromine, iodine, and chloride. Silver-halides are distributed evenly on film. Light causes chemical bonds to break so that metallic silver remains; it darkens the negative. Silver-halides on photographic paper turn the paper black when touched by light. Developing either film or prints requires removing silver-halides that did not react with light. This description neglects the chemical complexities involved in film and paper preparation, developing, and fixing. Complexities are even greater where color is involved. Color requires layered emulsions for three primary colors and more complex chemistry yet. Early pioneers abandoned solid surfaces like Daguerre's copper in favor of translucent plates like glass because they wished to print the actual picture, not its mirror image. Thus negatives came into use as an intermediate transfer mechanisms. The developed negative is turned around so that its bottom becomes its top. Then it is turned front to back so that the side-to-side inversion is corrected. Everything restored to normal, it is then time to print on paper.

The evolution of the camera itself, the hardware as opposed to the wetware of chemistry, involved development of lenses, making the camera itself as compact as possible and eventually portable, adding view ports on the outside and mechanisms for film advance on the inside, adding artificial lighting by means of flash guns and then flash bulbs, and, finally, automating the functions of focusing by range finders and lighting by aperture and shutter control. Development of the modern camera took place predominantly in Europe, the United States, and in Japan. France had pioneered the category; Germany became dominant in camera development by, for instance, introducing the first SLR (the Ihagee Exacta), the first 35mm film and camera (Leitz's famous Leica) and the first TLR (the Rolleiflex by Rollei Gmbh). Eastman Kodak was a major factor across the board in introducing and developing mass-produced cameras and successive innovations in film, including color film. General Electric introduced the first flashbulb. Fuji and Nikon (Nippon Kogaku) were early participants in the field, Fuji predominantly in film, Nikon in cameras. Nikon, which originated as an optical firm, supplied the lenses for the first Canon cameras.

Digital Cameras. Digital cameras, a revolutionary departure from film-based devices, originated in 1969 with the development of a new integrated circuit technology at Bell Labs called CCD, abbreviating charge coupled device. CCDs can be built up into sensor arrays in which a single CCD can record the charge produced by light hitting one spot or pixel of a surface. The intensity of the charge will correspond to a color, a value that can be recorded in computer memory as a number. The tiny size of CCDs means that very high resolution sensors are possible. Earliest uses of CCDs were in video recording and

in digitizing photographic images by NASA for transmittal from the moon and from satellites to earth in digital formats. Texas Instruments developed a digital camera but did not commercialize it. The first effective digital camera aimed at the professional market was introduced in the United States by Kodak in 1991, the DCS-100. The first consumer cameras appeared in the mid-1990s introduced by Kodak, Casio, Ricoh, Sony, and others. Another sensor technology is CMOS, the acronym for "complementary metal oxide semiconductor." The advantages of CMOS include lower power consumption hence longer battery life. Sensors are cheaper to manufacture and thus bring costs down. Their pixel density is lower, however, and they pick up less light. Both technologies are evolving rapidly, both have supporters. More brands use CCD than CMOS technology as we come to the close of the first decade of the twenty-first century.

MARKET

Three approaches to sizing a market are to look at domestic shipments as reported by the U.S. Bureau of the Census, to calculate apparent consumption, and to obtain market research data at the retail level. The first approach is meaningful if the United States is a major producer. Examination of shipments then will show the product's contribution to the economy and job creation. The second approach is useful when substantial portions of the product made here are exported or, similarly, significant percentages of demand are satisfied by imports. The look at retail sales usually demands reviewing data collected by private research organizations. The Census Bureau's retail reporting is never detailed enough to capture the sales of a narrow product category like cameras. In this look at the market we will briefly examine results from all three methods by way of tracking the major transformation in the industry that has taken place in the first decade of the twenty-first century.

Domestic Production. From the viewpoint of national accounting, cameras are part of Photographic and Photocopying Equipment (NAICS 33-3315). Within that industry they are grouped as one category under Still Picture Photographic Equipment. Figure 39 provides the big picture (the entire industry as defined by the NAICS code 33-3315) for the period 1997 to 2005, the most recent available at time of writing. It shows an industry with shipments that were $8.4 billion in 1997, peaked at $8.8 billion in 1998, dropped precipitously from that peak to a low of $1.96 billion in 2002, and have been growing very slowly since that time to reach $2.3 billion in 2005, having shrunk to just over a quarter of the industry's size in 1997.

FIGURE 39

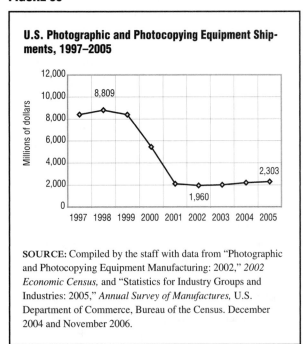

U.S. Photographic and Photocopying Equipment Shipments, 1997–2005

SOURCE: Compiled by the staff with data from "Photographic and Photocopying Equipment Manufacturing: 2002," *2002 Economic Census,* and "Statistics for Industry Groups and Industries: 2005," *Annual Survey of Manufactures,* U.S. Department of Commerce, Bureau of the Census. December 2004 and November 2006.

FIGURE 40

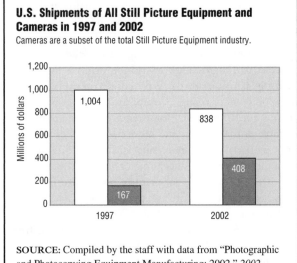

U.S. Shipments of All Still Picture Equipment and Cameras in 1997 and 2002

Cameras are a subset of the total Still Picture Equipment industry.

SOURCE: Compiled by the staff with data from "Photographic and Photocopying Equipment Manufacturing: 2002," *2002 Economic Census,* U.S. Department of Commerce, Bureau of the Census. December 2004.

Details beneath this major category are available, but only for two years: 1997 and 2002. The still picture equipment segment represented 11.9 percent of the industry in 1997 and 34.2 percent in 2002, meaning that other categories bore the brunt of the decline. Between those years the total industry shrank at an annual rate of 25.3 percent but the still picture category at the much more modest rate of 3.6 percent. These years saw the meltdown of the photocopying, microfilming, and motion picture equipment segments. Photocopiers and motion picture equipment went digital, domestic production was replaced by Asian manufacturing. Microfilming suffered from a similar transit from film to digital archiving. Still photography was the hold-out. Still cameras, a subsegment of the still picture equipment category, did even better, as shown by Figure 39.

Camera shipments actually advanced at an annual rate of 19.6 percent from $167 million in 1997 to $408 million in 2002. This healthy growth in the category was directly attributable to domestic production of digital cameras. The total market represented by this volume of shipments, however, is quite small when compared to other sources of data used to size the U.S. market as a whole.

Apparent Consumption. Somewhat marginal data are available on imports and exports from the International Trade Center (ITC), an organization jointly supported by the UN Conference on Trade & Development (UNCTAD) and the World Trade Organization (WTO). The data are marginal because ITC only reports on photo-graphic equipment, thus at a level roughly corresponding to NAICS 33-3315. Using such data, it would appear that apparent consumption in 2002 was made up of domestic production of $1.96 billion less exports of $1.2 billion plus imports of $3.3 billion, netting out to apparent consumption at the production/wholesale level of $4.1 billion. Imports, thus, accounted for 81 percent of total demand given the high levels of exports ITC reports. Corresponding values for 2005 suggest apparent consumption of $3.7 billion, a decline from 2002. Imports as a percent of demand also declined to 69 percent. Given that this was a period of economic growth, if not exactly at rousing levels, the data are puzzling until we recognize the fact that the costs of digital equipment have been dropping, certainly in the digicam category but in other branches of the industry as well. As we shall see, unit sales have been growing very rapidly. Foreign trade statistics have very low resolution, reminding one of faces fuzzed out on the screen to protect the innocent. They serve merely to indicate that exports represent high proportions of domestic demand.

Retail Estimates. Data of this type are collected by the Photomarketing Association International (PMA) and The NDP Group, to name two prominent market research firms. PMA reported unit shipments of digital cameras of 4.5 million in 2000, 7.0 million in 2001, 9.4 million in 2002, 13.0 million in 2003, 18.2 million in 2004, and 20.5 million in 2005. These data were published in *Market Share Reporter 2007.* NDP's estimates of unit shipments for 2005 were higher, 27 million, but the estimates made later. That company also put 2006 shipments at 29.5 million units. The PMA series indicates a growth rate from 2000 to 2005 of 35 percent per year. Very dramatic rates of growth are reported by all observers—matched by falling-off-the-cliff style declines in traditional analog cameras and film.

Dollar estimates are also made but are typically tightly held and made available by their producers to subscribers only. Data occasionally surface to view. NDP published estimates for 2005 and projections for 2006 in March 2006. According to NDP the market for digital cameras in 2005 was around $6.2 billion at retail. Point-and-shoot cameras with an average sales price of $199 represented 95 percent of units and 74 percent of dollar sales; more expensive digital SLRs costing on average $1,352 were 5 percent of units and 26 percent of dollars. NDP projected units growing by nearly 18 percent to 2006, revenue dollars 8 percent. Prices of both the low and high end cameras were projected to decrease.

In a report on the first quarter of 2007, PMA data showed that most rapid growth in unit sales was taking place in the high segments. PMA rates its segments by picture density, thus megapixels (millions of pixels, mpx). Greatest change in unit sales was in the 7 and greater mpx cameras, 235 percent. Next were 6–6.9 mpx cameras, 162 percent, and third the low end of the resolution categories, below 3 mpx, growing at 20 percent in the first quarter of 2007. All other segments showed declines. These data, for unit sales, were paralleled by dollar sales with one exception: 7+ mpx machines showed growth of 101 percent in dollars, 6–6.9 mpx machines growth of 66 percent, but the below 3 mpx units showed a decline in dollar sales of 27 percent. Prices were dropping over year-before prices but dropping the most in the low-end category. These patterns, although a snapshot in time only, reflect a widespread conviction in the industry that buyers tend to cluster in two groups, those who just want to snap a picture at a decent cost and the knowledgeable aficionados who will spend time at the shop, know all about the products, and don't mind spending serious money.

Worth noting in this look at digital cameras are data also reported by PMA showing unit sale declines of 45 percent and dollar sale declines of 50 percent for the analog camera segment—and similar but somewhat lower drops of 21 percent (units) and 20 percent (dollars) for one-time-use cameras, thus the throwaways. Also worth noting was that one category within the analog cluster resisted the general erosion. This category, Advanced Photographic Systems (APS), declined in units sold but showed a 167 percent growth in dollar shipments. Photography was going digital as the first decade of the

2000s was closing but the highest end of analog photography looked like a survivor.

KEY PRODUCERS/MANUFACTURERS

The universe of cameras is dominated by Japanese and German companies with significant roles played by Eastman Kodak in the United States, Samsung in South Korea, and well-known brands coming from Sweden and Switzerland as well.

The list of Japanese participants reads like a recital of electronics giants. Participants in alphabetical order are Canon, Casio, Fuji, Kyocera, Minolta, Nikon, Olympus, Panasonic (which is Matsushita), Pentax, Ricoh, Sanyo, Sharp, Sigma, Sony, and Toshiba. Observers of the field pick either Canon or Sony as the top camera producer worldwide. Canon, which entered the Chinese market and gained penetration toward the end of the first decade of the twenty-first century, is probably the leader by a small margin. Based on estimates by International Data Corporation Canon had top ranking in the U.S. market with a 17.1 percent share in 2004 followed by Sony with 16.7 percent.

German companies are best known for high end products serving the professional photographer or expert amateur. Like their Japanese counterparts, they have also entered the digital camera markets. Producers are Leica, Linhof, Minox, Rollei, and Zeiss AG. Minox is famed for producing very small cameras; until the early 2000s it was a division of Leica but regained its independent status in 2001. Zeiss is a leading global producer of lenses with operations in Germany as well as in Japan. It has its own camera brand, the Zeiss Ikon, but its chief business is manufacturing lenses for others. Among Zeiss's customers are Sony, Nikon, and Nokia in Japan, the last predominantly active in making telephones and therefore not listed above; other Zeiss customers are Rollei in Germany, Sinar in Switzerland, and Hasselblad in Sweden.

Eastman Kodak is ranked third in the United States with a market share of 11.8 percent. The company had sales in 2006 of $13.3 billion. Of this total the company's Consumer Digital Imaging Group earned $2.9 billon. The segment includes cameras, equipment sold to retailers for photo print production, photo printers, and the sale of CMOS imaging sensors to others.

Two Swiss companies participate in the market with well-known brands, Alpa of Switzerland and Sinar AG. Hasselblad is a Swedish firm with world-wide operations. A very small part of Samsung, the global electronics giant based in South Korea, is a line of digital cameras.

MATERIALS & SUPPLY CHAIN LOGISTICS

Camera production is more closely associated with technological expertise than raw materials used in the camera's manufacturing. Although any stable, translucent substance can be fashioned into lenses, glass is used exclusively for at least the outermost lens of cameras because glass resists scratching. All lenses used in high-end cameras are glass; cheaper cameras may use plastic lenses inside. Sand suitable for making glass must be high in silica content and low in iron, chromium, and cobalt; these latter color the glass, but clear glass is desired. Such sands are not uniformly available but common enough all over the world. The high value of lenses versus the raw material from which they are made means that lens manufacturing need not be located where the sand occurs. Flat glass producers, however, dealing in mass products, tend to locate where the sand is. Film used in cameras is now predominantly polyester-based (polyethylene terephtalate), thus a plastic, and widely available. Camera housings are made of plastics and metals; mechanical components are precision-fabricated metal. With digital sensors replacing film, semiconductors are becoming an important component of cost. These products are also high in value and low in weight; they do not influence location.

Cameras have always been high-tech products—even before that phrase emerged with electronics. Camera production has therefore tended to be concentrated where expertise resided, thus in technologically advanced areas like Europe, Japan, and the United States. Most producers in the latter years of the first decade of the 2000s were Japanese because Japan had become the predominant electronics producer in the world.

DISTRIBUTION CHANNEL

Producers use wholesale merchants to supply retailers; retailers sell to the consumer. Based on data provided by *Manufacturing and Distribution USA*, the wholesale business, including cameras, auxiliary equipment, and supplies was projected to be a $20.3 billion market in the United States in 2007, down somewhat from a $21.0 billion level in 2002, the decline most likely due to eroding film sales. Retailers divide into two major categories. The smaller of these, representing roughly 30 percent of total volume, are specialized camera stores typically serving the high end of the market. Sales of this sector were projected to be $2.98 billion in 2007, down from $3.1 billion in 2002—the shift downward attributable more to declining camera prices than loss of film sales in that camera stores are not major factors in film sales; drug stores are. Most cameras reach the ultimate consumer through department stores, mass merchandisers, and drug stores (low end and one-time-use cameras). Online sales are also a growing factor. Web sites provide means of displaying hundreds of different models—and, indeed, there are hundreds—with full information about features, thus enabling the shopper to engage in extensive and painless comparison-shopping at leisure from the home computer.

Camera stores attract the serious amateur photographer and the professional. Such retailers typically concentrate on a selection of brands only, provide expert in-store staff, and carry substantial inventories of auxiliary products. They also offer repair services on site or act as intermediaries to get damaged cameras repaired by the producer.

KEY USERS

It is axiomatic in the industry to say that key users come in two varieties, the casual user interested only in point-and-click and the serious user. The second category is divided into amateurs and professionals. Sales patterns tend to bear this out. Thus market reports tend to segment the industry into low-end and high-end with the middle barely mentioned in summary reporting. Product sales also cluster in the same manner. Product offerings by producers, however, suggest that demand exists for cameras across a wide spectrum of quality/performance and that the low–high division is more of a convenience than a fact. Otherwise, to take digital cameras as an example, no cameras in the middle range—greater than 3 and less than 7 megapixels—would be on offer.

One-time-use (OTU) cameras are purchased for occasions, thus a single outing or a vacation. Most households own low-end cameras for occasional picture taking. This base of customers supports the industry by buying the most product. The amateur—perhaps best defined as the person who becomes interested in photography as such, not just the product of it—begins above this level and may be divided into at least two segments. Beginners, typically young, will buy the best camera they can afford, thus the upper range of the low end. As they grow older they upgrade the equipment and purchase products in the midrange of performance. A subset of these amateurs graduates into serious photography, typically because individual skills are recognized by family members and friends or personal fascination and discrimination grow. As such amateurs get the means, they become buyers of high-end equipment. If they continue to participate, they enter the upper ranks of photographers.

At the highest end specialization begins to create markets for different kinds of cameras, thus, for instance, those ideal for high speed action, portraiture, landscape, nature photography, and other contexts. The distinctions between amateurs and professionals begin to blur when their skills converge.

ADJACENT MARKETS

Camera sales support a range of adjacent markets. Most prominent of these, but declining, is the market for film, photographic paper, and photo-developing services. This market is largely being displaced by others, including photo printers based on inkjet or laser printer technologies, toner and ink cartridges, and special paper. The services sector for digital image printing remains. Photomarketing Association International, for instance, reported that in 2006, 49 percent of digital prints were produced at home (or in the office on borrowed equipment), 42 percent were made by retail outlets, and 9 percent were ordered using online services. In the latter case, disks are shipped to a processor, or digital files are sent electronically, and prints are returned by mail.

Serious photographers are buyers of extra lenses, light sensors, tripods, and other auxiliary products. Projectors and screens for showing 35mm slides were a large market. Projectors are still used to show digital photographs in business presentations using computers and PowerPoint software. In the home new television sets based on plasma and liquid crystal display (LCD) technology are replacing screens, with the TV set's hardware readily putting photos from a disk up on the screen. Camera cases, picture frames, and photo albums are yet other adjacent markets ultimately created by the presence of the camera.

RESEARCH & DEVELOPMENT

Research and development in this industry has come to be focused powerfully on the rapidly growing digital sensor technology. CCD and CMOS technologies would both benefit from further miniaturization in order to lift possible picture resolutions. Broad trends in the semiconductor field promise yet tinier devices.

Transistors (switches) are now 65 nanometers (nm, for billionth of a meter) in size for the highest end computers. By 2008, 45nm transistors were in the offing and, beyond that, 32nm and 22nm transistors were envisioned by leading companies, including Intel, Advanced Micro Devices (AMD), Toshiba, and NEC Corporation. Sensor pixel densities set ultimate limits to the sharpness of images available in digital cameras. CCD technology is limited by energy demand and thus battery life. The most effective batteries, lithium ion, are plagued by over-heating problems; they sometimes cause fires. CMOS technology crowds circuits onto the surface of the sensor itself. This circuitry, while cooler, interferes with the reception of light. Tinier circuits would improve the ability of CMOS sensors to receive more light and thus sharpen the images they produce.

In that both CCD and CMOS technologies are relatively new, very significant improvements, including new kinds of sensors, are likely to emerge in the 2010s and beyond. Worth noting is that improvements will follow R&D trends in semiconductors rather than leading them.

CURRENT TRENDS

The trend in the camera industry is the digital revolution. It has yet to run its course, but all indicators suggest that other than in professional applications, film cameras will disappear from the market. At the end of the first decade of the twenty-first century prices of digital cameras were still falling, but industry observers were projecting a leveling off by 2010. The digital camera has stimulated transformations in the processing of images to paper.

Film processors have lost roughly half their old market for photo development and, with the rapid growth in photo printing, may lose more of this business yet. Color printing generally had, as of the middle of the first decade of the 2000s, just begun to nibble at the monochrome printer market, but color machines were growing and black-and-white versions were declining in share.

TARGET MARKETS & SEGMENTATION

To some extent markets and segmentation have been outlined under the Key Users heading. Market segments in the industry are usually rendered as disposable cameras aimed at special occasions, point-and-shoot cameras targeted at the casual user emphasizing ease of use and modest cost, mid-level cameras with higher resolution aimed at knowledgeable amateurs with marketing messages based on features and costs, and high end products sold for performance and further subdivided by end-use categories. At the highest end lens quality and the reputation of the producer are the chief selling features.

RELATED ASSOCIATIONS & ORGANIZATIONS

Consumer Electronics Association, http://www.ce.org

Business Technology Association, http://www.bta.org/i4a/pages/Index.cfm?pageID=2031

Electronic Industries Alliance, http://www.eia.org

National Electrical Manufacturers Association, http://www.nema.org

Photomarketing Association International, http://www.pmai.org/index.cfm/ci_id/1198/la_id/1.htm

Professional Photographers of America, http://www.ppa.com/splash.cfm

BIBLIOGRAPHY

Bellis, Mary. "The History of the Digital Camera." About.com. Available from <http://inventors.about.com/library/inventors/bldigitalcamera.htm>.

Burns, Paul. "The History of the Discovery of Cinematography." October 1999. Available from <http://www.precinemahistory.net/index.html>.

Darnay, Arsen J. and Joyce P. Simkin. *Manufacturing & Distribution USA* 4th ed. Thomson Gale, 2006, Volume 2, 1088–1092.

"The First Photograph." Harry Ransom Center, The University of Texas at Austin. Available from <http://www.hrc.utexas.edu/exhibitions/permanent/wfp/>.

Lazich, Robert S. *Market Share Reporter 2007.* Thomson Gale, 2007, Volume 2, 532–537.

"The NDP Group Focuses on Digital Camera Forecast Through 2010." Press Release, The NDP Group, Inc. 6 March 2006. Available from <http://www.npd.com/press/releases/press_060306.htm>.

"PMA Monthly Printing and Camera Trends Report." Press Release, Photomarketing Association International. 7 June 2007. Available from <http://www.pmai.org>.

Spehr, Paul. "Why Nitrate?" Conservation OnLine. 7 August 2003. Available from <http://palimpsest.stanford.edu/byform/mailing-lists/amia-l/2001/01/msg00018.html>.

Steffens, Bradley. *Ibn al-Haytham: First Scientist.* Morgan Reynolds Publishing. 2006.

Wilgus, Beverly and Jack Wilgus. "The Magic Mirror of Life: An Appreciation of the Camera Obscura." Available from <http://brightbytes.com/cosite/what.html>.

Candy

———■———

INDUSTRIAL CODES

NAICS: 31–1340 Nonchocolate Confectionery Manufacturing

SIC: 2064 Candy and Other Confectionery Products (nonchocolate confectionery)

NAICS-Based Product Codes: 31–13401001, 31–13401004, 31–13401007, 31–13401015, 31–13401021, and 31–13401026

PRODUCT OVERVIEW

While the demand for chocolate continues to be strong, non-chocolate candy also finds frequent favor with consumers seeking to satisfy a sweet tooth. The human love of sweet treats can be traced all the way back to cavemen who extracted honey from beehives, according to a timeline prepared by the National Confectioners Association (NCA). In Europe's Middle Ages, sugar candy was a delicacy available only to the rich, but by the seventeenth century boiled sugar candies were popular in both England and the American colonies. By the mid-1800s almost 400 American factories were producing candy, which usually took the form of penny candy sold loose from glass cases. Americans also enjoyed homemade hard candies such as peppermints and lemon drops. By the late 1800s the discovery of sugar beet juice and the development of more efficient mechanical equipment led to the introduction of such products as candy corn in the 1880s and Tootsie Rolls in 1896.

Other items on the timeline include: peppermint sticks, 1901; NECCO Wafers, 1901; Conversation Hearts, 1902; Life Savers, 1912; the chewy candies that would eventually be called Gummi Bears, 1922; Tootsie Roll Pops, 1931; Marshmallow Peeps, 1954; Atomic Fireballs, 1950; Starburst Fruit Chews with Vitamin C, 1960; and Lemonheads, 1962.

According to the NCA, the stories behind some of the most popular candies are as follows:

Candy Canes. Candy canes can be traced, or so goes the legend, to 1670 when the choirmaster of the Cologne Germany Cathedral handed out sugar sticks to his singers. It wasn't until the beginning of the twentieth century, however, that red and white stripes and peppermint flavor became the accepted norm.

Candy Corn. Candy corn was invented in the 1880s by George Renninger, who worked for the Wunderlee Candy Company. The revolutionary three-color design was very popular, but the candy was produced only seasonally from March to November. The machinery was needed for other products during the remaining years. Candy corn has remained a Halloween favorite.

Conversation Hearts. Conversation hearts, usually sold for Valentine's Day, can perhaps trace their origin to homemade candies made by American colonists, who scratched love notes on the surface for Valentine's gifts.

NECCO Wafers. NECCO Wafers were first produced in 1912 by the New England Confectionery Company, with the name derived from the company's initials. The company was formed by the merger of three candy companies, including Chase and Company, which had been producing mints in Canada since 1847.

Lifesavers. Lifesavers were developed in 1912 by Cornelius Crane, a chocolate manufacturer, as a product to be sold in summer when chocolate sales were down. A malfunctioning machine created holes in the centers of the candies, and the company kept the design, naming the new confection for it.

Cotton Candy. Cotton candy can be attributed to several sources. John Wharton and William Morrison received a patent for a cotton candy machine in 1899, and they took their invention to the St. Louis World's Fair in 1904. Thomas Patton received a patent in 1900 for a machine that used a different process to spin the cotton candy threads. A more reliable machine was introduced in 1949 by Gold Medal Products, which made the confection widely available at circuses, amusement parks, and fairs.

Gummi Bears. Gummi bears and other Gummi candies were invented in Germany in 1922 by Hans Riegel, who formed the Haribo Company. In 1981 the Herman Goelitz Company, which later became the Jelly Belly Candy Company, began making Gummi Bears in America. Gummi worms were introduced in 1981, and a wide variety of shapes followed. In 1985 Walt Disney introduced the cartoon show, *The Adventures of the Gummi Bears.*

Jelly Beans. The jelly center of jelly beans is generally traced to Turkish Delight, a Middle Eastern treat since Biblical times. The shell coating became possible in the seventeenth century when the panning process was invented in France to make Jordan Almonds. Objects such as almonds or the jelly centers are coated by being rocked in large rotating pans of syrup. Jelly beans were popular penny candies in the United States. They have been associated with Easter since the 1930s because of their egg-like shape.

Licorice. Licorice, which comes from the glycyrrhiza plant, was enjoyed as long ago as the time of the pharaohs. Crusaders brought the candy back to England from the east, and an English monastery eventually began making it. The colonists brought a number of licorice recipes to America.

Lollipops. Lollipops, hard candies on a stick, may have made their first appearance during the U.S. Civil War, when pieces of candy were placed on the ends of pencils for children. George Smith invented one version of the modern lollipop in 1908, putting hard candies on a stick and naming them after his favorite racing horse, Lolly Pop. The Racine, Wisconsin, Confectioners Machinery Company created a machine that could make 40 lollipops per minute. In 1916, Samuel Born, a Californian, invented another version that he called the Born Sucker machine.

Marshmallows. Marshmallows trace their history to ancient Egyptians, who made a special treat from the mallow plant that grows in marshes. In France in the mid-1800s, candy stores began painstakingly whipping the mallow sap into fluffy texture to fill molds. In the late 1800s, the starch mogul system was invented in which the candies were formed by machines in molds of modified cornstarch, and candy makers replaced the mallow root with gelatin. The Girl Scout Handbook in 1927 printed the first known recipe for s'mores, which combined toasted marshmallows, chocolate, and graham crackers. In 1948 an extrusion process was invented by Alex Doumak in the United States, and the marshmallow became very popular as an ingredient in many recipes in the 1950s. Americans are now the main consumers of marshmallows.

Taffy. Taffy has been popular in American since the late 1800s, and salt water taffy first appeared in Atlantic City, New Jersey, at the end of the nineteenth century. One version of the story traces the treat to an 1883 storm that flooded the boardwalk and led storekeeper David Bradley to jokingly offer his soaked candy as salt water candy. Many manufacturers of salt-water taffy, popular in seaside resorts, still add a little salt to the taffy.

MARKET

The United States is the world's biggest candy market. The U.S. Census Bureau reports on candy makers in a report titled *Confectionery: 2006.* This publication reported that candy producers shipped more than 1 million tons of non-chocolate confectionery in 2006, merchandise with a wholesale value of $4.733 billion. The National Confectioners Association (NCA) estimated that non-chocolate retail sales in 2006 totaled $8.9 billion, compared to $15.6 billion in retail sales of chocolate candy and $2.7 billion in sales of gum.

The *Confectionery: 2006* report broke down the $4.733 billion in industry shipments of non-chocolate candy as follows: hard candy, $1.32 billion; chewy candy, including granola bars, $1.47 billion; soft candy, $887 million; iced/coated candy, $33 million; panned candy, $756 million; and licorice and licorice type, $257 million.

FIGURE 41

U.S. Confectionery Market, by Industry Shipments and Retail Sales in 2006

The source for Industry Shipment data is the U.S. Bureau of the Census and the source for Retail Sales data is the National Confectioners Association.

Billions of dollars

■ Industry Shipments
▨ Retail Sales

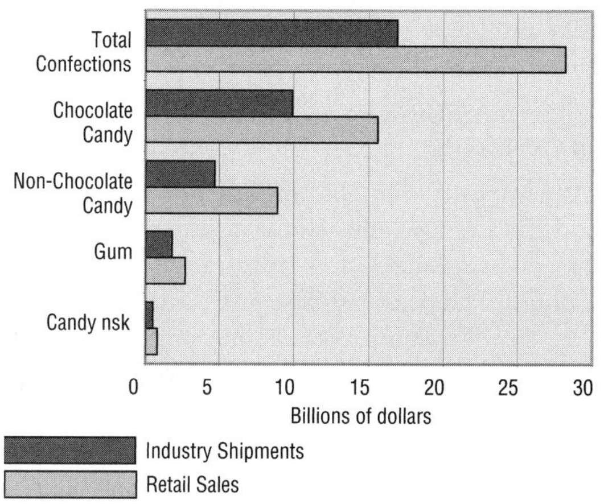

SOURCE: Compiled by the staff with data from *Confectionery Industry Reveiw 2006 Year End,* National Confectioners Association and "Confectionery: 2006," *Annual Survey of Manufactures,* U.S. Department of Commerce, Bureau of the Census. June 2007.

9 percent over 2005, in Eastern Europe 18 percent, and in Latin America 23 percent. Euromonitor figures placed 2006 global sales of chocolate candy at $74.1 billion and non-chocolate at $43.6 billion.

The candy category has a number of attributes that promote sales, *Confectioner* reported in 2006. One is impulse consumption, with a high percentage of consumers buying front-end display confectionery products. Another is the product's expandable nature, as opposed to items with finite usage such as toothpaste. Affordability is also important, with 79 percent of snacks that are sold through the food/drug/mass/convenience channels costing less than $1.

Andrew Lazar, a packaged foods analyst for Lehman Brothers, spelled out some of the challenges threatening the market in a keynote address at the 2006 NCA State of the Industry Conference in Orlando, Florida. He warned that costs of ingredients, energy, and packaging were going up, forcing price increases. He called consumers schizophrenic and demanding. "The consumer who will spend $4 on a latte is the same consumer who will stand in line at Costco to save on a pack of paper towels," he said.

Consumers were also variety-seeking, he said, making constant innovation necessary. He stressed that candy makers were not just competing with other candy companies. "Every one of you is competing with Procter & Gamble and Pepsico…because they all want that retail space, and they're good at getting it," he said. In another sense, candy companies are also competing against the whole snack market, with impulse buyers making such choices as candy versus chips for a quick break.

Another important factor in the U.S. candy market is seasonal sales. The NCA reported that non-chocolate candy sales increased 12.3 percent for Easter 2007 compared to 2006. Valentine's Day non-chocolate sales increased 11.9 percent for 2007. Halloween 2006 non-chocolate sales were 1.2 percent higher than 2005 sales. Candy sales at Christmas, however, were facing greater competition from gift cards and other purchases, the NCA said. While total candy sales were up 1 percent for Christmas 2006 over 2005, sales of non-chocolate candies were down 2.8 percent.

KEY PRODUCERS/MANUFACTURERS

The *2002 Economic Census* listed 475 companies in the United States that manufactured non-chocolate candy. These ranged from corporate giants to small owner-operated firms that reported a payroll for any part of the year. Figure 42 lists the top ten companies in the confectionery industry in the United States. A profile of the top four companies follows.

The U.S. International Trade Administration (ITA) calculates export value as the total transaction price at seaport, airport, or border ports of exportation, including transportation costs to the port, insurance, and other expenses. Using this method, the United States exported $356.5 million worth of non-chocolate candy in 2006, with $186.9 million of that going to Canada. Other export destinations included Mexico, $35.78 million; Australia, $14.65 million; Korea, $11.44 million; and the United Kingdom, $10.57 million.

The ITA reported $1.44 billion in customs value of imports of non-chocolate candy, including $553.48 million from Canada; $385.35 million from Mexico; $101.34 million from China; $43.78 million from Brazil; $40.15 million from Spain, and $31.96 million from Germany.

Candy Industry reported in June 2007 that while the United States remained the world's biggest spender on candy, other segments of the global market were increasing rapidly as consumers with rising incomes developed a taste for sweets. In response, manufacturers increasingly began supplying products catering to local tastes. The growth of a modern distribution and retail infrastructure in many countries also contributes to rising sales.

Using figures obtained from Euromonitor International, a London-based research firm, *Candy Industry* said that 2006 confectionery sales in Asia-Pacific rose

FIGURE 42

Top Non-Chocolate Candy Makers in the United States, 2006

Company	Sales in million dollars	Percent of market
Hershey Company	32.90	13.64
Wrigley Company	29.60	12.27
Masterfoods USA	23.70	9.82
Just Born, Inc.	17.60	7.29
Nestle USA, Inc.	16.10	6.67
Brach's Confections, Inc.	13.90	5.76
Tootsie Roll Industries, Inc.	6.50	2.69
Ferrero USA, Inc.	5.40	2.24
Jelly Belly Candy Company	5.09	2.11
Cadbury Adams USA, LLC	4.80	1.99

SOURCE: Lazich, Robert S. *Market Share Reporter 2007*, Thomson Gale, 2007, 582.

Hershey Company. Although The Hershey Company is generally thought of as a chocolate maker, Hershey was also the leading non-chocolate candy maker in terms of 2006 U.S. sales at supermarkets, drug stores, and mass merchandisers (excluding Wal-Mart). With sales of $32.9 million, Hershey accounted for 13.64 percent of the market.

On a dollar basis, Hershey's sales are about 75 percent chocolate and 25 percent non-chocolate products. In fact, the first product produced by Milton S. Hershey, the company founder, was caramel candy, not chocolate. Non-chocolate products now include breath mints Ice Breakers and Breathsavers, Twizzlers (which represents 87.8 percent of vending machine licorice sales), Good & Plenty, and Jolly Rancher. Located in Hershey, Pennsylvania, Hershey is the largest North American candy maker, and it employs more than 13,000 people worldwide.

Wrigley Company. Ranked second in the U.S. market is Wrigley with $29.6 million in non-chocolate sales and 12.27 percent of the market. Wrigley markets Life Savers, the top brand of non-chocolate candy and mints. Other products include Airwaves, Altoids mints, Crème Savers, Eclipse, and Orbit mints. Chicago-based Wrigley, which is the world's largest manufacturer of gum, has global sales of more than $4 billion.

Masterfoods USA. The third largest U.S. candy maker is Masterfoods with $23.7 million in U.S. candy sales and a 9.82 percent market share in 2006. Masterfoods is a division of leading candy maker Mars, which is privately held by the Mars family. The non-chocolate candies made by Masterfoods include Skittles and Starburst.

Just Born Inc. With 2006 U.S. candy sales of $17.60 million Just Born had a 7.29 percent market share. Just Born is a privately owned company founded in 1923 and located in Bethlehem, Pennsylvania. The company has more than 560 employees and markets its candy in more than 50 countries. Products include Marshmallow Peeps, its original product, as well as jelly beans and movie favorites Mike and Ike, Hot Tamales, and Zours.

MATERIALS & SUPPLY CHAIN LOGISTICS

The largest categories of ingredients used by the non-chocolate candy industry are sugar, corn sweeteners, dairy products, peanuts, and tree nuts such as almonds and walnuts. The delivered cost of the materials used in 2002 by the 475 companies in the United States that made non-chocolate candy was $1.9 billion, according to the *2002 Economic Census*. Figure 43 provides a detailed breakdown of the costs of materials used by U.S. candy makers in 2002.

FIGURE 43

Materials Consumed in the Production of Candy, 2002

Ingredients and Materials	Cost (thousands of dollars)
Ingredients	
Cane and beet sugar solids	194,000
Dextrose and conr surup	123,900
Essential oils and flavors	144,300
Sugar substitutes	94,600
Milk and milk products	31,700
Raw nutmeats	31,700
Fats and oils	27,800
Chocolate coatings	26,000
High fructose corn surup	21,300
Processed nutmeats	19,500
Fruits, fresh and dry	19,380
Cocoa	3,600
Unsweetened chocolate liquor	3,600
Dry fructose	918
Nuts in shell [1]	890
Packaging Supplies	
Paper and coated film	198,300
Paperboard containers	176,000
Aluminum foil	25,370
Plastic containers	11,000
Metal cans, tins, and lids	8,860
Glass containers	8,400
All other packaging materials	381,800

[1] Data are from the *1997 Economic Census*

SOURCE: Compiled by the staff with data from the *2002 Economic Census*, U.S. Department of Commerce, Bureau of the Census. December 2004.

As can be seen in Figure 43, sugar in various forms is a key ingredient for the candy industry, and the cost of sugar is a major problem. In 2005 Congress directed the Secretary of Commerce to report on whether jobs had been lost because of the differential between U.S. and world sugar prices. The resulting report stated that the U.S. used 17.8 billion pounds of refined sugar in 2003, down from 18.5 billion pounds in 1999, with about 85 percent of the sugar produced domestically and the rest imported. Employment in sugar-containing products decreased by more than 10,000 jobs between 1997 and 2002. The three main sugar-using industries were non-chocolate confectionery, chocolate and chocolate confectionery, and breakfast cereal.

In 2004 the price of U.S. refined sugar was 23.5 cents per pound compared to the world price of 10.9 cents, with the U.S. price kept high by price supports and quotas. As a result a number of U.S. manufacturers closed or relocated, with many going to Canada or Mexico. The report calculated that for every agricultural job saved by the price supports, three manufacturing jobs were lost. In addition, the loss of manufacturing was contributing to the trade imbalance, with imports of sugar-containing products growing from $6.7 billion in 1990, to $10.2 billion in 1997, and then to $18.7 billion in 2004.

The NCA has worked to bring this issue to the attention of legislators and has lobbied in support of trade agreements such as CAFTA (the U.S. Central American-Dominican Republic Free Trade Agreement) that allow U.S. manufacturers to purchase some sugar from Central American growers at world market prices.

Also of concern to the NCA are federal labeling regulations for food ingredients that are common allergens. The organization has initiated an education program for its industry members and works with the Food and Drug Administration (FDA) and its members to be sure that labeling is sufficient to ensure safety. About 5 to 8 percent of children have food allergies, according to the NCA. The ingredients that cause 90 percent of allergic reactions are milk, eggs, peanuts, tree nuts, soy, wheat, fish, and shellfish.

The growing interest in health issues has also affected the nature of materials purchased by the industry. Confectioners are using more fruits, nuts, and seeds, particularly ingredients such as blueberries and almonds, both known to carry high levels of antioxidants.

There also has been an interest in value-added ingredients that contain functional substances such as prebiotic and probiotics (substances that improve health through beneficial bacteria), calcium, antioxidants, and vitamins. These ingredients can create new manufacturing concerns such as avoidance of temperatures that break down important enzymes.

Candy Industry reported in 2006 that the increased demand for healthy and organic products had led to shortfalls and supply chain issues. The magazine quoted the Organic Trade Association's *2006 Manufacturer Survey*, which said that 52 percent of respondents had reported that a "lack of dependable supply of organic raw materials has restricted their company from generating more sales of organic products." The magazine predicted that the shortfalls would persist, with milk and almonds, organic ingredients of particular interest to confectioners, being in particularly short supply.

DISTRIBUTION CHANNEL

Candy is sold widely in venues ranging from super-size warehouse stores to concession stands at Little League baseball games and vending machines in hospital waiting rooms. Bed, Bath & Beyond sells candy that it markets for instant consumption, at home use, and gifts. Even Amazon.com now sells candy from suppliers around the world.

In its *Confectionery Industry Review 2006 Year End*, the NCA reported sales in the following channels for 2006, including the percentage increase or decrease over 2005: Supermarkets, $4.3 billion, +0.1 %; Wal-Mart, $3.2 billion, +6.7 %; Mass other than Wal-Mart, $1.3 billion, +1.5 %; Convenience Stores, $4.2 billion, +9.4 %; Drug no change; vending machines, $1.2 billion, +0.5 %; bulk sales, $1.5 billion, -0.4 %.

The wide range of sellers makes candy distribution networks particularly important. In 2005 an estimated 60.7 percent of candy was sold directly by manufacturers to retail stores, and 39.3 percent went through retail distributors known as brokers.

For the brokers, the first decade of the twenty-first century has been a time of economic turmoil as such major corporations as Hershey Foods, Amurol Confections Company (a subsidiary of Wrigley), and Willy Wonka Candy Factory took their sales in-house, removing business from brokers. In early 2004 *Professional Candy Buyer* quoted one estimate that nearly $1 billion had been sucked out of the candy brokerage business. Others noted, however that the business is cyclical. Masterfoods, for example, dropped its internal sales force in favor of brokers at the same time that Hershey took its sales force direct.

The companies that went to in-house sales said they were making the move to benefit retailers by cutting costs and offering a single point from which to order and distribute. Some brokers noted bitterly that they had built up small candy companies by distributing their products widely, only to see these companies go to direct sales. They also questioned whether in-house systems would overlook small retailers such as local ballparks and swimming pools. Brokers claim that the loss of sales through small retailers could have a long-term negative impact since sales through these local outlets are known to build children's brands.

Distributors, whether in-house employees or brokers, push for the maximum shelf space, the most effective displays, and valuable checkout-line space for their products. One issue of concern to these distributors is a growing tendency to install self-checkout lines in supermarkets. Customers busy scanning their groceries have little time to think about impulse buys. In addition, installation of self-checkout lines, which usually requires eliminating three or four traditional checkout lines, is generally done by the operations department with little consultation with the store's merchandising team.

Representatives of candy companies are working to convince stores that with a little creativity, it is possible to place effective display racks near self-checkout lines. They point out that candy, gum, and mints account for 32 percent of front-end sales and almost 34 percent of front-end profits.

One distribution issue that has been of concern to the industry is an October 2005 decision by the National Motor Freight Traffic Association to change the freight classification for candy. The price of sending less-than-truckload shipments is negotiated based on this classification. The Freight Association ruled that candy packages were less dense than indicated by their classification, which meant the product was taking up more space in trucks. The new rating resulted in higher shipping rates for the entire industry. The NCA appealed, and while it did not get candy back to its old classification, it obtained an agreement with the trucking industry on a sliding classification based on actual density. This meant that shippers of dense packages were charged less.

KEY USERS

Almost everyone eats candy. The NCA estimates that 99 percent of U.S. households purchase candy during the year. Children are an especially important market for non-chocolate candy makers, and many products are developed especially to appeal to younger customers. With a variety of sugar-free products, as well as organic candy and candy formulated to deliver nutrients such as calcium, even the health conscious, diabetics, and dieters can enjoy candy in moderation.

ADJACENT MARKETS

In addition to chocolate candy, non-chocolate candy competes with a large array of snack foods that are generally eaten between meals. In this sense, candy competes with ice cream bars, frozen fruit bars, fruit rolls, single-serving containers of yogurt, energy drinks, granola bars, cup

cakes, cookies, and even beef jerky. Many of these snack foods aim for the same checkout-line impulse buyers who drive up candy sales.

Although these products can be found under various headings in government reports, there is a NAICS (North American Industrial Classification System) code for snack food manufacturing that includes nuts, chips, pretzels, popped popcorn, and even pork rinds. In 2005 U.S. snack food producers shipped $20.8 billion in products, compared to $15.1 billion for all confectionery products except gum and $4.7 billion for non-chocolate candy.

The snack food industry, like the candy industry, has been affected by the growing interest of American consumers in health and diet, with many more products featuring whole grains and an expanding list of snacks available in 100-calorie individual serving packs.

RESEARCH & DEVELOPMENT

The NCA collects information and supports research of interest to both the chocolate and the non-chocolate candy industry. It also conducts education programs in areas of interest to its members. The NCA, for example, sponsored research presented at the 2006 All Candy Expo on techniques to grow confectionery sales. The organization sponsored a multi-year, in-store study by Dechert-Hampe, Inc., aimed at learning more about the category and consumers, establishing benchmarks on best practices, and identifying opportunities to increase sales in the overall category.

Much of the research undertaken by manufacturers is aimed at innovative new products and packaging. This is particularly important for companies targeting children, who as a group have substantial spending money to help them keep up with the latest innovations. Tung Toos, one such innovative product, creates temporary tattoos on the tongues of those eating the candy.

Batman Projector Pops actually project the Batman signal on a flat surface when the child pushes a button. The lollypop itself also lights up. Pops Rocks Candy Laboratory comes with a plastic test tube. The child puts the rocks in the tube and adds Secret Ingredient 1 to create a color change. Adding Secret Ingredient 2 sets off a foaming action. The child can then drink the citrus-flavored magic potion.

Manufacturers also seek new products to increase their share of the adult market. One report in 2003 listed more than 2,000 new candy, chocolate, and gum products that same year. Hershey reported in May 2006 that it had increased its share of the mints market category seven points to 35 percent over the past year by introducing new products. At the same time, Hershey acknowledged that too many new or special edition products can push a company past the point of diminishing returns. The flood of new products also creates problems for retailers who have only limited space.

Masterfoods tackled the problem of shelf space and display by sponsoring extensive research by Forbes Consulting on designing the most effective display for the 2006 Easter season. The research included consumer surveys, focus groups, and one-on-one interviews with approximately 1,000 people. The result was a visually appealing area, coated with Easter colors. The goal was to attract customers and keep seasonal purchases in the candy aisle.

The growing interest in organic and natural products has spurred research to bring such products to market. Research by color companies at the end of the twentieth century, for example, made the use of natural colors—colors created from fruits and vegetables—more practical and less expensive. These colors work particularly well in hard candies or gummy-type products.

CURRENT TRENDS

Major trends in the candy industry cater to the health-conscious and those on diets. The NCA reported on its Web site roundup of 2004/2005 trends that diet candy represented only about 3 percent of the overall candy market, but that the diet candy market segment was growing rapidly—90 percent for the twelve months that ended in April 2005.

Organic candies also cater to this health trend, including value-added organic alternatives among such favorites as lollipops, licorice, and gummies. *Candy Industry* reported in November 2006 that the organic market had grown 45 percent to $149 million for the year, compared to 1.8 percent for the conventional candy market. Manufacturers also continue to introduce bite-size and portion controlled products to appeal to those who want to limit their candy consumption.

Sales of licorice are growing, and manufacturers are also increasing the number of sour varieties they offer. SweeTart Gummy Bugs from Nestlé, for example, come in a variety of bug shapes and combine two sour flavors. In general, new candy flavors are stronger and more unique, according to the NCA, including such offerings as berry blasts, tropical twists, and sweet and sour. Jelly Belly's new lineup included Mint Trio and even Baked Beans. Super fruit flavors, derived with fruits believed to have antioxidants or other beneficial compounds, are gaining popularity. They include blueberry, pomegranate, black currants, guava, and lychee.

In another trend, candy makers are trying to take advantage of strong sales of candy for seasonal holidays by broadening the appeal of their products. In 2003 the company Just Born celebrated the 50th anniversary of its marshmallow Peeps with a marketing campaign featur-

ing a yellow Peeps chick dressed up for Valentine's Day, Halloween, and Christmas as well as for Easter. Jelly bean makers, too, are attempting to broaden the appeal of these traditional Easter candies for other holidays.

Meanwhile, private label non-chocolate products are expanding. This is in part because of the strength of Wal-Mart, which carries many of these private label products.

The results of many of these trends were seen in Professional Candy Buyer's 2005 "What's Hot" list, which included portable pocket-size packs, indulgent and natural, trans-fat-free, extreme cooling peppermint, bright packs in primary colors, and almonds. Its "What's Not" list contained bizarre flavor blends, unidentifiable ingredients, synthetically derived, spearmint and wintergreen, and walnuts.

TARGET MARKETS & SEGMENTATION

Although almost everyone eats candy, not all population groups are tempted by the same products. Candy makers aim their products at a number of target markets. The children's market is huge, with sales of non-chocolate kids' candy estimated at $500 million in 2006. With more than 40 million children between the ages of 6 and 14 in the United States, children's brands represented 18 percent of sales in 2005, according to ACNielsen data. Packaged Facts reported that the purchasing power of children tripled in the 1990s, with the direct buying power of children expected to exceed $51.8 billion in 2006. This group tends to favor stronger, more intense flavors than those chosen by adults. They also like sour flavors.

Candy aimed at children is marketed with advertisements in publications or television programs aimed at the young. Buzz marketing—word of mouth marketing—is often effective in groups vulnerable to peer pressure. Internet blogs, news groups, and chat rooms can also help spread the word about a new product.

Given the expense of licensed products (an average 10 percent of the wholesale price), candy makers try to choose a handful of the most popular evergreen and television characters, with perhaps an occasional family movie, to license.

The evergreens are particularly effective in appealing to mothers for products with a positive value—portion-control packaging or added nutritional and/or low calorie content, highlighted prominently on labels. These products must compete with organic low glycemic snacks and dried fruit wraps, also marketed to mothers.

Innovative interactive candy products are the fastest growing segment of the kids' candy market. Light pops, pops attached to tubular handles through which lights shoot when a button is pressed, are popular. Other innovations include a Star Wars liquid candy light saber, Pez dispensers that play music, lollipop lipstick, and a candy harmonica.

While candy makers are eager to market to children, they do not want to be blamed for the growing epidemic of childhood obesity. On its Web site, the NCA has created confectionery marketing guidelines for its members, which the organization says can bring balance to marketing materials and help educate children and adults about nutrition.

Other products are marketed to different target groups. The primary target age for breath freshener mints is between 18 and 34. One mint product aims to expand that to the teen and pre-teen—often called 'tween—audience by using packaging that can be attached to a key ring, backpack, or purse.

Meanwhile women aged 25 to 64, are the primary jelly bean consumers, but specialized colors and flavors are aimed at the kids and the teens markets. As has been seen, candy makers target the health conscious and the dieting with sugar-free and organic products and with candies said to contain healthful substances.

Not surprisingly candy designed for Hispanics is gaining market share, considering a 2000 Census that showed that the population of Hispanics grew from 22.4 million to 32.8 million from 1990 to 2000. Hispanics are expected to account for more than 15 percent of the population by 2010. The market is complicated, however, since Hispanics from different countries have different taste preferences.

Given the importance of the Hispanic market, companies such as Montes USA Inc., Bimbo Snacks USA, and Arcor—all with Hispanic roots—are finding retailers eager to handle their products. The packaging is generally bilingual, with Spanish used since the largest group of Hispanics in the United States comes from Mexico. Mainline companies such as Hershey are also aware of the trend, offering bilingual packaging for retailers in areas with large Hispanic populations.

RELATED ASSOCIATIONS & ORGANIZATIONS

American Association of Candy Technologists, http://www.aactcandy.org

National Confectioners Association, http://www.candyusa.org

Pennsylvania Manufacturing Confectioners Association, an International Association of Confectioners, http://www.pmca.com

Retail Confectioners International, http://www.retailconfectioners.org

BIBLIOGRAPHY

"Best Practices Boost Retail Sales." *Professional Candy Buyer.* July-August 2006, 36.

Brewster, Elizabeth. "Sweet Success for Sours: Non-chocolate Candies See Growth in Sours, Diet Products, All-Occasion Treats." *Confectioner.* May 2003, 62.

"Brokers: A Vanishing Breed? Millions of Dollars Are Disappearing from the Broker Community as Suppliers Go Direct." *Professional Candy Buyer.* January-February 2004, 50.

"Confectionery: 2006." *Annual Survey of Manufactures.* U.S. Department of Commerce, Bureau of the Census. Available from <http://www.census.gov/industry/1/ma311d06.pdf>.

"Confectionery Industry Review 2006 Year End." National Confectioners Association. Available from <http://www.ecandy.com/ecandyfiles/2006_Annual_Industry_Review_February_2007.ppt>.

"Confectionery Timeline." National Confectioners Association. Available from <http://www.candyusa.org/Classroom/timeline.asp>.

Covino, Renee M. "What's Up—and Down in the Candy Universe." *Confectioner.* April 2006, 42.

"Employment Changes in U.S. Food Manufacturing: The Impact of Sugar Prices." U.S. Department of Commerce, International Trade Administration. Available from <http://www.ita.doc.gov/td/ocg/sugar06.pdf>.

Fuhrman, Elizabeth. "Color My World: Candy Colors Send Consumers Signals Triggering Subconscious Responses." *Candy Industry.* May 2004, 56.

"In-aisle Strategy to Boost Seasonal Sales: Consumer Research Results in a Revolutionary Approach to Seasonal Merchandising by Masterfoods USA." *Professional Candy Buyer.* July-August 2005, 25.

"Kids' Candy Brands: Wooing Today's Youthful Confectionery Consumers Is More Challenging than Ever Before but – considering their significant spending power – Potentially More Rewarding As Well." *Confectioner.* January-February 2006, 32.

Kuhn, Mary Ellen. "Candy Plays Catch-Up in Self-Checkout Aisles: This Rapidly Growing Retailing Paradigm Requires Effective Front-End Merchandising to Avoid Big Losses in Confectionery Sales." *Confectioner.* March 2006, 52.

Lazar, Andrew. "Candy's Challenges Catalogued." *Confectioner.* April 2006, 10.

Lazich, Robert S. *Market Share Reporter: 2007.* Thomson Gale, 2007, 582.

"Licensed Candy: Hot Licenses Can Sell More Candy–but Retailers Have Become Much More Selective in Their 'Picks.'" *Confectioner.* January–February, 2006, 36.

"A Little Candy History!" National Confectioners Association. Available from <http://www.candyusa.org/Classroom/timeline.asp>.

"Nonchocolate Confectionery Manufacturing: 2002." *2002 Economic Census.* U.S. Department of Commerce, Bureau of the Census. Available from <http://www.census.gov/prod/ec02/ec0231i311340.pdf>.

"Processed Foods Index Page." U.S. Department of Commerce, International Trade Administration. Available from <http://www.ita.doc.gov/td/ocg/food.htm>.

Rehan, Kelly. "Bolstering Benefits: Functional Confectionery Continues to Thrive as Manufacturers Reach Out to the Growing Health-Conscious Consumer Base." *Candy Industry.* July 2007, 50.

———."Harvesting Health: Though Traditional Favorites Like Peanuts and Almonds Show No Signs of Going Out of Style, Many Manufacturers Have Taken Advantage of Consumers' New Willingness to Try More Unusual Offerings." *Candy Industry.* January 2007, 42.

Rogers, Paul. "One Sweet World: Demand in Developing Nations Drives Global Confectionery Revenues and Creates a Platform for Future Growth." *Candy Industry.* June 2007, 32.

Thompson, Stephanie. "Hershey Reaches Limits of Limited-Edition Candy; Innovation: CEO Lenny Slashes 25% of Portfolio, Pledges to Seek Platforms." *Advertising Age.* 2 May 2006, 12.

Vreeland, Curtis. "Special Report: Organic Confectionery Market." *Candy Industry.* November 2006, 34.

SEE ALSO *Chocolate, Snack Foods*

Canned Foods

——•——

INDUSTRIAL CODES

NAICS: 31–1111 Dog and Cat Food, 31–1225 Fats and Oils Refining and Blending, 31–1330 Chocolate and Confectionery, 31–142M Fruit and Vegetable Canning, Pickling, and Drying, 31–1514 Dry, Condensed, and Evaporated Dairy Products, 31–161N Animal (Except Poultry) Slaughtering and Processing, 31–171M Seafood Preparation and Packaging, 31–191M Snack Food Manufacturing, 31–1920 Coffee and Tea Manufacturing, 31–194M Seasoning and Dressing Manufacturing, 31–199M Food Manufacturing, not elsewhere classified, 31–211M Soft Drink and Ice Manufacturing, and 31–2120 Breweries

SIC: 2011 Meat Packing Plants, 2023 Dry, Condensed, Evaporated Products, 2032 Canned Specialties, 2033 Canned Fruits and Vegetables, 2047 Dog and Cat Food, 2066 Chocolate and Cocoa Products, 2068 Salted and Roasted Nuts and Seeds, 2076 Vegetable Oils, 2082 Malt Beverages, 2086 Bottled and Canned Soft Drinks, 2091 Canned and Cured Fish and Seafoods, 2095 Roasted Coffee, and 2099 Food Preparations, not elsewhere classified

NAICS-Based Product Codes: 31–111111, 31–111141, 31–111142, 31–13207, 31–14211, 31–14214, 31–14217, 31–1421A, 31–1421D, 31–1421G, 31–1421J, 31–1421W, 31–14221, 31–14224, 31–14227, 31–1422B, 31–1422W, 31–15145, 31–1611M, 31–191112, 31–19201, 31–194211, 31,194212, 31–19994, 31–1999P1, 31–21114, and 31–21201

PRODUCT OVERVIEW

The canning of food began early in the nineteenth century when Nicholas Appert, a French chef and candy maker, responded to a contest held by Napoleon Bonaparte. Bonaparte hoped to supply his troops with wholesome preserved foods when on campaign. Appert invented canning in glass containers in 1809. Britain, Napoleon's chief political adversary, rapidly responded. In 1810 the Englishman Peter Durand received a patent for food preservation. He used containers made of tin-coated sheets of iron hoping to produce a less breakable and more easily transported product. Durand was thus the inventor of the tin can, so called—although tin was only a coating intended to protect the underlying metal from the acids in foods. Canning expanded rapidly.

Can making commenced in the United States in 1812, brought from Europe by Thomas Kensett, an immigrant from England. The Civil War greatly accelerated use of canned foods. Just before the war began in 1861 production was around 5 million cans per year; after the war production had reached 30 million cans. In 1900, less than one hundred years after the basic technology had been invented, the so-called sanitary can made its debut. This was a can made with double-folded seams, the seams covering up the solder and thus preventing its migration into the food. By the early 1920s automated can making had been developed and cans were churned out at the rate of 250 cans per minute as compared to roughly 10 cans per day per laborer before. After that, technological improvements came in stages. Cans began to emerge in the beverage market in the 1950s and aluminum cans appeared in the 1960s. Lead solder gave way to other ways of seaming cans. Ever more sophisticated coatings were introduced to protect the can from the food and the food from the can. Dual-metal (steel-aluminum) closures were introduced for easy-to-open cans. Cans became lighter, their production more automated, and can making and canning operations more tightly integrated.

The purpose of canning is food preservation. From this perspective canning—whether the container is metal or glass—provides the same protection. Both types of containers are vacuum-sealed when holding perishable foods, accomplished by applying seals as the contents are submerged in boiling water. The phrase home canning refers to preserving fruits and vegetables in glass containers. Over time, however, canned food has come to mean food packaged in metal cans. The Census Bureau's reporting follows this convention. Until 1997 foods and beverages were a single industry category ("Food and Kindred Products") under the Standard Industrial Classification (SIC) system. After that date, with the introduction of the North American Industry Classification System (NAICS), Food Manufacturing was separated from Beverage & Tobacco Products. In this essay we follow the earlier convention and will discuss both canned food and beverage products under the general term Canned Foods.

Food preservation continues to be the principal reason for using metal cans as packaging. Canning, however, has expanded beyond its original purpose so that some foods are packaged in metal cans even when other packaging would do as well. An example is distribution of cookies, salted nuts, and roasted coffee in attractive metal cans. The great majority of foods and beverages delivered in cans, however, are in metal containers in order to protect them.

Although canned foods as a category is in common use both in popular and in commercial speech, no precise measurement of the total category is available from the Census Bureau. Put more precisely, the Bureau provides census data on a range of major industry subcategories in which the packaging mode is specified. Examples are some of the biggest categories like fruits and vegetables, soft drinks and beer, pet food, dairy products, and salted nuts. In other food categories, however, the mode of packaging is not provided and requires estimates based on such data as, for instance, the industry's consumption of different kinds of packaging materials. An alternative source of information is provided by the U.S. Department of Agriculture (USDA) which, in reporting on food consumption by the public, identifies quantities consumed in fresh or canned forms.

Data from the USDA dealing with food availability per person per year provide one view of canned food. Thus in 2005 just under 1,650 pounds of food were available for consumption per person, up from 1,634 pounds in 1995. Availability should not be confused with actual consumption because substantial quantities of food are lost in processing. These data also exclude beverages. Of this total, however, around 8.6 percent reached consumers in canned packaging in 2005, down slightly from the 1995 level of 8.7 percent. In that these data inevitably miss some of the food that ends up in cans—because no basis is provided for a good estimate—it may be reasonable to assume that approximately 10 to 12 percent of all food may come to us in cans of some sort. Major categories include:

- Fruits: 17% of all food, 6.1% canned
- Vegetables: 25% of all food, 25.3% canned
- Dairy: 17% of all food, 0.6% canned
- Meat and Poultry: 11% of all food, 2.5% canned
- Fish: 1% of all food, 15% canned
- Fats and Oils: 5% of all food, 12% canned
- Tree Nuts: 0.2% of all good, 15% canned
- Coffee: 0.6% of all food, 53% canned
- All Other: 23% of all food, the canned proportion unknown

The All Other category includes grains, dried legumes, sweeteners, eggs, and peanuts. It is possible to produce a rank order of canned foods from this listing by weighting the canned portion by the quantitative magnitude of the food category. Using that method, the top five categories of canned foods are vegetables, fruit, fats and oils (primarily vegetable shortenings), coffee, and fish (primarily seafood). Another way of looking at the category is from an industrial perspective. A look from that viewpoint will be carried out under the Market subheading below in which, alongside the look at food, beverage packaging in cans will be included as well.

The Canning Process. The principal object of all food preservation is to prevent bacterial activity in the food from causing it to spoil. The major approaches to this end have been drying, pickling, salting, smoking, and freezing. Bacterial activity requires water which, removed in drying, inhibits bacterial action. Pickling preserves food by creating an acidic environment intolerable to most microorganisms. Cheeses, for instance, are stabilized by the presence of lactic acids produced by certain bacteria feeding on milk-sugar, making it difficult for other pathogens to live. Salting causes cell dehydration and kills off bacteria; salting is a type of pickling. Smoking causes antioxidants to be present in food and these suffocate bacterial life. Freezing immobilizes the water that bacteria need to live.

The innovation introduced by canning depended on two factors. One was hermetically sealing of containers and the other was heat sterilization. Bacterial organisms are killed off by the application of high heat applied to a sealed container. The seal is tight enough that bacteria cannot reach the food again until the can is opened.

Using fruit and vegetable canning as examples, the process of canning itself may be envisioned as four major operations, the two central steps are unique to the canning process itself.

Food preparation and filling. Fruit canning begins with washing the fruit, a sorting and grading step, peeling and coring of products that require it, chopping and slicing if desired, and optionally cooking of the fruit before it is placed in a can. Vegetable processing is similar but with some of the steps arranged in a different order. Most vegetables are subjected to blanching after being washed, sorted, and cut. Blanching is intended to kill off bacteria by immersing the vegetable briefly in boiling water. Vegetables are peeled, if required, after blanching and/or cooking. Another wash cycle follows peeling.

Container sealing. Filled containers are vacuum sealed. This means that air is drawn out of the container immediately before it is sealed, accomplished either by sealing the cans while their contents are hot or sealing them in a vacuum chamber. The object is to create a pressure inside the can lower than atmospheric pressure to keep seals tight, to remove as much free oxygen from the can as possible, and to keep cans from bulging later if used at high altitudes. When hot water vapor condenses into steam in a closed chamber, a vacuum is created. If some of the air is also evacuated from the chamber, vapor condensation is even more effective. Steam tunnels or locally applied steam condensation is used in creating a vacuum during can sealing.

Heat sterilization. This crucial process in canning follows sealing. It is produced by heating the containers briefly to a boiling temperature. In many modern processes, sterilization of the product and of the container itself are separated. At the point of filling, a sterile product is placed in a pre-sterilized container.

Cooling, labeling, and storage. The hot product is next cooled and dried. After that labels are affixed, cans are placed in packing boxes, and the boxes aggregated onto pallets for transport to the warehouse.

Variations on this general scheme characterize all canning operations. Differences arise from the kind of product canned, its levels of natural acidity, and whether or not the process is continuous or handled in batches.

MARKET

The most recent year for which the total market for canned foods could be estimated at the time of writing was 2002. Data for that year were issued in the spring

of 2006 for the first time by the U.S. Census Bureau. Data for 2007 would be issuing no earlier than 2011. Thus details for 2002 are, in the slow-moving world of economic analysis, current data. These data, which follow, are based on dollar transactions rather than food weights, the basis of the earlier discussion above using USDA per capita data. In 2002 the U.S. market for canned food was $56.5 billion, down from $58.1 billion in 1997. By the very nature of the subject—in that canned foods are sold as part of a number of major food industries—these data were aggregated from different census reports and represent a composite picture. Thirteen major industries were included in the list. Shown in rank order based on food sold in cans, these industries are shown in Figure 44.

In the majority of these industries the Census Bureau's own reports identify those subcomponents of the total industry that come to the consumer packaged in cans. In a few cases no data are provided and the values shown in the table of Figure 44 were based on estimates. In the Coffee and Tea industry, it was assumed that all ground roasted coffee was delivered in cans but that no tea came in cans, probably overstating the canned market. In the Fats and Oils industry, it was assumed that 25 percent of vegetable shortenings were canned, based on the industry's consumption of packaging materials. In the Seasonings and Dressings industry, 15 percent of pepper and spices were assumed to reach the retail market in cans. Some industries were excluded for lack of data. Some chicken reaches the market in cans, but the amount is im-

FIGURE 44

U.S. Food Industry Shipment Totals and Percent Shipped in Cans in 2002

Industries are listed in order of their size within the canned food sector.

Industry	Total Industry Shipments ($Billion)	Shipments of Canned Product ($Billion)	Canned Shipments as Percent of Total (%)
Fruit and Vegetables	31.9	25.60	80.3
Breweries	17.6	9.10	51.8
Soft Drinks	34.1	8.70	25.5
Coffee and Tea	5.2	3.60	79.7
Pet Foods	10.7	3.30	29.3
Dried Condensed, and Evaporated Milk	9.5	2.10	15.0
Seafood Preparations	8.8	1.35	15.1
Red Meat	53.1	1.30	2.5
Fats and Oils	8.4	1.00	11.8
Snack Foods	16.8	1.00	4.8
Food Manufacturing nec	1.6	0.80	5.0
Seasonings and Dressings	11.1	0.20	1.9
Chocolate and Cacao	4.1	0.15	3.7

SOURCE: Compiled by the staff with data from "Product Shipment Details," *2002 Economic Census*, U.S. Department of Commerce, Bureau of the Census. March 2006.

FIGURE 45

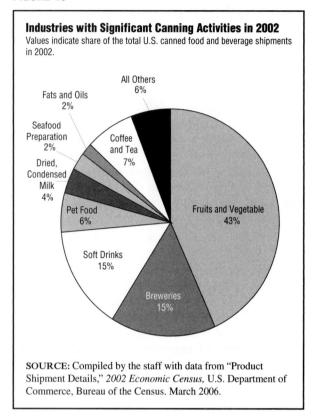

Industries with Significant Canning Activities in 2002

Values indicate share of the total U.S. canned food and beverage shipments in 2002.

- All Others 6%
- Fats and Oils 2%
- Seafood Preparation 2%
- Dried, Condensed Milk 4%
- Pet Food 6%
- Soft Drinks 15%
- Breweries 15%
- Coffee and Tea 7%
- Fruits and Vegetable 43%

SOURCE: Compiled by the staff with data from "Product Shipment Details," *2002 Economic Census,* U.S. Department of Commerce, Bureau of the Census. March 2006.

possible to discern from Census data. The same is true of baby food, cookies, flavorings and syrups, and other categories in which data on packaging materials consumption are either not reported or suppressed. The general picture presented, however, closely matches data reported by the Can Manufacturers Institute.

If all the canned food categories, thus excluding soft drinks and beer, are expressed as a percent of total food shipments in the United States, canned food represented 8.7 percent of dollar value in 1997 and 8.4 percent in 2002. These numbers are very much in the same magnitudes as those obtained from the USDA on quantities available in canned form to U.S. consumers per capita. If the food and beverage sectors are combined, those portions packaged in cans represented 12 percent of total in 1997 and 10.8 percent in 2002.

If we look at the eleven industries where most of the food canning takes place (excluding beverages), these industries in total, including all of the products that they ship, not merely canned goods, represented 44.5 percent of all food shipments in 1997, 38.3 percent in 2002, and 37.7 percent in 2005. Data for 2005 are aggregates with the canned portion not available. These industries were thus, as a group, losing share of total market. If we look at the two beverage industries, the pattern is somewhat different. These two industries taken as a whole—including

product bottled in glass and plastics as well as metal—represented 83.1 percent of the total beverage market in 1997, 78.5 percent in 2002, a drop in share, and then 80.4 percent in 2005, a gain over 2002. The result for these industries is mixed. The total beverage category also includes distillers and wine producers.

Examining growth trends between 1997 and 2002 for the thirteen industries and comparing their annual compounded growth rates, shows, as a whole, a negative growth rate of 0.9 percent. The canned portion also declined but at a slower rate, 0.5 percent per year. This negative growth rate was due principally to declines in the canned components of soft drinks and beer, declining at an annual rate of 3.6 percent. In the food sector, the canned components grew at a rate of 1 percent, thus below the rate of population increase. Among the industries taken individually, ten had growth in total shipments, three declined. In the canned portions, seven had growth in shipments and six declined. The details are shown in the graphic. The major issues underlying these trends are discussed under the heading of Current Trends below.

The general picture that market data present is that canned foods/beverage categories are components of thirteen very large and generally mature food and beverage industries in which they represent about a quarter of all shipments and about 11 percent in the total food and beverage industry. Canning growth in food is growing minimally and declining in beverages, more so in soft drinks than in beer.

KEY PRODUCERS/MANUFACTURERS

The twentieth century in the food industry was a period of corporate consolidation so that, in the twenty-first century, most of the largest food companies offer a very wide array of products and have been assembled by their managements by acquisition of leading brands, each one once offered by specialized companies. The key producers and manufacturers of canned food in the United States are consequently the leading food companies, and the key producers in beverages are the dominant companies in soft drinks and beer.

The leading producers in this major segment, arranged here in alphabetical order, are Anheuser-Busch, Campbell Soup Company, Coca-Cola Company, ConAgra Foods, Inc., Del Monte Foods Company, Dole Food Company, General Mills, Kraft Foods Inc., Nestlé S.A., PepsiCo, Inc., Procter & Gamble, and J.M. Smucker. Revenue data in the following company descriptions refer to 2006 or 2005 data unless otherwise noted.

Anheuser-Busch. With $18 billion in sales, Anheuser-Busch is the nation's top beer producer. The company is also a major producer of aluminum cans through a wholly-owned subsidiary (Metal Container Corporation), much of the production of which Busch itself buys for its own packaging.

Campbell Soup Co. With revenues of $7 billion, Campbell Soup is the dominant company in what the industry calls canned specialties. It is the leading maker of canned soups sold under its own and the Swanson brand. Campbell produces V8 canned juices and also sells SpaghettiOs, the second-ranking brand in canned pasta products.

Coca-Cola. This company, with sales of $24 billion, is the nation's and the world's top soft drink producer.

ConAgra. A diversified food manufacturer with approximately $12 billion in sales, this company is an important canner of tomato products (Hunt's), of canned beans (Van Camp's and Ro*Tel), owner of the leading canned pasta producer (Chef Boyardee), and is also the producer of a leading canned cooking spray brand (Pam).

Del Monte. This company, with sales of $3 billion, is the leading producer of fresh fruits and vegetables. These product lines represent Del Monte's traditional business. Like many other food companies, however, Del Monte has diversified and is a factor in canned seafood (StarKist), in soups, and in pet food (Gravy Train, 9 Lives).

Dole. This company had sales of $4.7 billion in 2003. Soon thereafter the company went private. Dole is also a leader in fruits and vegetables, but its canned products are principally oranges, pineapple, and tropical mixes.

FIGURE 46

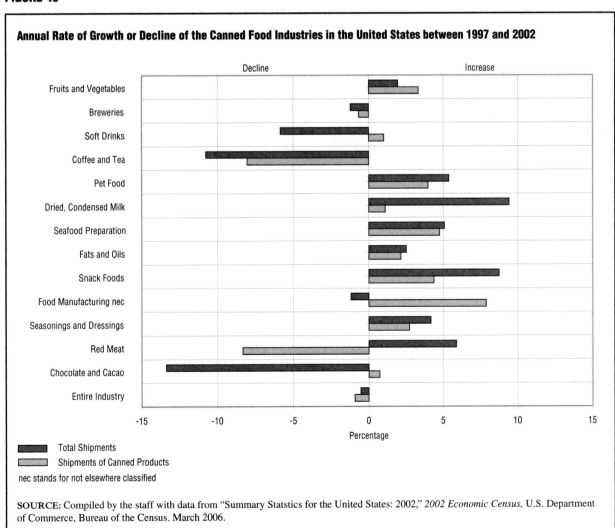

Annual Rate of Growth or Decline of the Canned Food Industries in the United States between 1997 and 2002

Total Shipments
Shipments of Canned Products
nec stands for not elsewhere classified

SOURCE: Compiled by the staff with data from "Summary Statstics for the United States: 2002," *2002 Economic Census*, U.S. Department of Commerce, Bureau of the Census. March 2006.

General Mills. This company which originated as a flour producer, is an $11.6 billion diversified food company participating in the canned food business principally through its Green Giant line of vegetables, Progresso soups, and Old El Paso canned bean specialties.

Kraft. This company, with sales of $34.4 billion, is known above all as a dominant producer of cheese products. The company plays a role in canned food by being an important coffee producer through its Maxwell House, Yuban, and General Foods coffee brands. Between 1988 and 2007 the company was owned by Philip Morris (later renamed Altria). Altria spun off Kraft as an independent corporation, with full independence reached in March 2007.

Nestlé S.A. The Swiss food giant, with revenues of approximately $75 billion worldwide, participates in canned foods in many areas but most notably in vegetables through the Libby brand, acquired in 1970, its own brands of chocolate syrups, the Carnation brand of condensed and evaporated milk, acquired in 1985, and in pet foods (Alpo, Mighty Dog, Purina).

PepsiCo. This company is the second-largest soft drink producer in the world with $32.6 in total sales of which 28 percent are related to beverages.

Procter & Gamble. This company's predominant activities are in the personal care, household cleaning, laundry detergents, prescription drugs, and disposable diapers markets. This $68.2 billion company, however, is a major factor in coffee production, selling the leading brand of coffee, Folgers. P&G also sells a line of pet food (Iams).

Smucker. This company, with sales of $2.15 billion, is best known for its fruit preserves and jams, sold in glass containers. Smucker, however, is the owner of the Crisco brand, the leading vegetable shortening product and of the Pet brand of condensed milk. Smucker also owns the Knudsen brand of fruit juices some of which reach the market in cans.

The full range of corporate participation in canned foods is partially indicated by Census data which show that, in 2002, 899 companies were engaged in the fruit and vegetable canning business alone. A substantial volume of canned food is produced by many participants in the food industry for private labels. Examples of such brands are America's Choice, the private label brand of A&P; the Kroger brand of fruits, vegetables, and canned specialties; the Kirkland brand owned by Costco; and Sam's Choice, Wal-Mart's brand used in foods. A leading producer of baked and canned beans is Bush Brothers

& Company, a privately held organization selling the Bush brand of canned products. The best-known canned meat product in the United States is Spam, produced by Hormel Foods, a $5.7 billion company. The second-ranking tuna brand is Chicken of the Sea, owned by Tri-Union Seafoods LLC, a $450 million company. The oldest brand of condensed milk is Borden. The product has been produced since 1997 by Eagle Family Foods, Inc.

MATERIALS & SUPPLY CHAIN LOGISTICS

With the largest category of canned foods being fruits and vegetables, products which easily spoil or lose their freshness soon after harvesting, the dominant logistical concern in canning is to carry it out as rapidly as possible after the raw product is obtained. This requirement, which also applies to such categories as fish, meat, and dairy, has concentrated canning activities in close proximity to the occurrence of the product to be canned, thus in agricultural and horticultural centers, near slaughtering operations, and close to seashores where fresh fish are delivered for processing. In continuous operations of any size, the manufacturing of the containers takes place very close to filling operations. The can factory is located adjacent to food preparation, the cans arriving ready-to-fill on conveyor belts from next door, as it were. The co-location of product processing and can-making is largely dictated by economics: it is more cost effective to deliver the raw materials for cans (coils of sheet steel, pre-fabricated ends and closures, and coatings in bulk) than empty cans.

DISTRIBUTION CHANNEL

Canned foods and beverages reach the ultimate consumer through grocery stores, convenience outlets, and grocery chains. A portion of soft drinks is distributed through vending machines. Typically distribution is two- or three-tier. Large retailers act as their own distributors; they buy directly from producers and sell to the consumer. Independent retailers usually rely on the services of food wholesalers; the latter buy from the producer, sell to the store, and the store to the customer in a three-tier distribution. The ultimate rationale for buying canned goods is product preservation on the shelf. For this reason institutional buyers like restaurants and hospitals also buy canned goods exactly like the ordinary shopper; they may buy the goods in larger cans, of course. Such buyers, however, buy canned goods in bulk quantities and, even in cases where the purchases are made from retailers rather than distributors, institutional buyers usual receive discounts.

KEY USERS

Virtually everyone buys canned goods and beverages. Certain product categories are available in no other form, an example being condensed or evaporated milk. In all other cases, the principal motivation is convenience and economy. Buying only fresh products imposes much more frequent shopping and the risk of higher spoilage rates. Buying only frozen products requires substantially higher refrigeration capacity. And in all cases more labor is required to reach the ultimate object—the steaming meal on the table.

ADJACENT MARKETS

Canned foods and beverages compete with products packaged in glass, plastics, and paper/plastic composites. This competition is visible across the board. All beverages and most foods are delivered in glass containers sealed with metal caps. Plastic bottles are a major competing product for aluminum cans in carbonated beverages and almost completely preempt metal cans even in oils; a very small fraction of edible oils still moves in cans, but this mode of packaging is disappearing in that category too.

Adjacent in another sense are markets offering food preserved using alternative methods. The most prominent method is freezing. Frozen foods offer many of the same advantages as canned foods if refrigeration space is available. Dried foods are another alternative but have the disadvantage of requiring reconstitution by adding water. The combination of freezing and drying has played an important role in coffee distribution, indirectly displacing canning of roasted coffee, which preserves the flavor and freshness of the product. Freeze-dried coffee, for example, was 11 percent of all coffee sold in 1997 and 13 percent of coffee sold in 2002.

A very important, difficult to measure, and still relatively new development in the food industry is the growth of prepared foods and ready-to-eat meals which represent an indirect competition to canning. This trend is aimed at transferring the labor of cooking from the kitchen to the factory—the food industry hoping that it can achieve growth in a mature industry by tempting the busy homemaker to pay a much higher margin for convenience. This development is indirectly aided by such improvements in kitchen tooling as advances in microwave ovens and innovations in food preservation by the exploitation of composite materials, including paper-plastics-metal composites. To the extent possible, canned food manufacturers are attempting to provide buyers with products that fit the ready-to-eat category and in this way they may be able to benefit from the trend.

RESEARCH & DEVELOPMENT

Research & Development in this broad category is principally centered on the packaging itself, thus the can, rather than on the broad range of products that are distributed in cans. Where the product is receiving a great deal of attention—an example being soft drinks where the industry appears to be engaged in what looks almost like a desperate search for gaining and holding the attention of customers with new flavors and formulations—packaging changes are driven by marketing motives. Here, too, an example comes from beverages where tapered cans, resembling bottles but cooling faster in the refrigerator, have been introduced to catch the eye.

Other areas of interest to researchers in the canned food and beverage industry are glass lined cans for use with beverages, in particular beer, and self-heating cans for use with food. Beer sold in cans far outsells beer sold in glass bottles, nonetheless, connoisseurs of beer have long complained that some beers take on a tinny flavor from the cans in which they are packaged. Can producers are working with microbreweries like KettleHouse Brewing Company in Missoula, Montana, to use cans lined with a glasslike, food-grade coating that helps to prevent any flavor deterioration.

Self-heating cans are another new type of can being tested for widespread use. The self-heating can looks like any other food can but it has a hidden interior chamber that contains lime and water. A mechanism is used to allow a user to cause these substances to mix when the user is ready to heat the contents of the can. So far, these cans are still being tested in trials while work is being done to improve their functionality.

An aging Baby Boom generation has inspired the introduction and improvement of easy-to-open closures so that people suffering from arthritis can avoid the effort of using can openers. Throughout the history of canning, coatings have improved. The better the coating the less likelihood exists that acids in the food will attack the can or that metallic tastes migrate into the food or beverage. Taking costs out of the package or the packaging process is receiving continuous attention. Not least, creation of new composite products is of great interest to the industry. One example is a coated, aluminum can with a non-metallic lid aimed at producing products that can be placed into the microwave directly. The canned food industry is interested in all product modifications that will allow it to compete head-on with the growing number of ready-to-eat meal products, many of which are designed for heating in a microwave oven.

CURRENT TRENDS

Canned Food. Looking first at trends in food packaging alone, leaving beverages to the side for a moment, statistical data for the 1997 to 2002 period indicate that the eleven major industries where canning takes place have declined in shipments, from $187.6 billion (1997) to $175.6 billion (2002), a decline in shipments of 1.3 percent per year. In this same period, the canned portion of these industries increased from $38.5 to $40.5 billion, a rate of increase of 1 percent per year. Thus canned goods increased in share of these industries from 20.5 to 23.1 percent in 1997 and 2002 respectively. Canning has thus held its place, indeed has increased its presence in these crucial industries, but the industries of which they are a part themselves have lost ground.

The largest absolute gains in canning were achieved in four traditional categories, fruits and vegetables, pet foods, and in condensed/evaporated milk. The largest percentage gains were seen in the last category, milk, and in snack foods, specifically in canned salted nuts. The largest absolute and percentage losses were in the coffee and tea category, an industry that, as a whole, shrank from $8 billion in 1997 to $5.2 billion in 2002. Based on later aggregate reporting on the coffee market, coffee as a category improved after 2002, but data on specific product lines were not available.

It is worth noting that in the 1997 to 2002 period shipments of steel cans themselves (steel is principally used in food), thus the packaging rather than the product, declined at the rate of 2.7 percent per year, indirectly suggesting that physical quantities of product sold declined as well, with the 1 percent annual growth in canned products, noted above, achieved by price increases and/or caused by shifts in the canned product mix from commodity-style goods to branded products, as the growth in snack foods in part illustrates. Snack foods as a whole, for example, grew at 4.4 percent per year whereas fruits and vegetables grew at a rate of 3.3 percent per year.

It is of particular pertinence to note that in the food group itself, the All Other food category (formally Food Manufacturing not elsewhere classified) showed the greatest rate of growth of all eleven industries that have some canned products. This industry grew at 7.9 percent per year. It has very little canned food content (just puddings and canned frostings). But it includes the novel new products the food industry envisions as the leaders into the future: prepared foods and ready-to-eat meals. The growth of this industry indirectly influences the slow growth in canned categories and hints at likely future developments provided that purchasing power in the nation continues to grow.

Canned Beverages. Trends in the beverage markets based on 1997 to 2002 data strongly hint at a decline in the canned segments of the two markets involved—soft drinks and beer. These two industries in combination had shipments of $50.6 billion in 1997 and $51.7 billion in 2002, advancing at a rate of 0.4 percent per year. In this same period, the canned portions of the industries declined from $21.5 to $17.8 billion or from 42.4 to 34.5 percent of these two beverage industries taken as a whole. The decline in the canned components was 1.2 percent per year in beer and 5.8 percent per annum in soft drinks. In the beverage categories competition from glass, and more intensely from plastics, is the major factor. Glass has always enjoyed superiority in preserving taste; glass is one of the most inert of packaging materials available—one reason why it is universally used in packaging wine. Plastics have been favored by superior economics. Trends in the beverage industry, powerfully influenced by marketing rationales, favor novel material combinations and shapes. The winner, if there ever will be one, is yet to emerge. Can producers hope it will be a bottle-shaped can, or a glass lined can.

TARGET MARKETS & SEGMENTATION

The major segments in this wide cluster of products are preservation and utility, exemplified by commodity-style food products like fruits and vegetables, beverage packaging in which strength (to hold in carbonated liquid pressures) and rapid cooling are advantages, and taste protection desired by coffee and nut canners. In the twenty-first century the can, as such, is no longer promoted as a segment-defining packaging feature except in those situations where the producer wishes to communicate prestige, exemplified by packaging cookies in beautifully decorated tins.

RELATED ASSOCIATIONS & ORGANIZATIONS

American Beverage Association, http://www.ameribev.org/index.aspx

The Beer Institute, http://www.beerinst.org

Can Manufacturers Institute, http://www.cancentral.com/index.cfm

Canned Food Alliance, http://www.mealtime.org

BIBLIOGRAPHY

"Chapter 9.8.1 Canned Fruits and Vegetables." *Compilation of Air Pollution Emission Factors.* U.S. Environmental Protection Agency. January 1995.

Christensen, Tyler. "KettleHouse to Introduce it's Double Haul IPA Beer in a Pop-Top Option." *Missoulian.* 31 May 2006.

"Complete History." Can Manufacturers Institute. Available from <http://www.cancentral.com/hist_overview.cfm>.

Darnay, Arsen J. and Joyce P. Simkin. *Manufacturing & Distribution USA,* 4th ed. Thomson Gale, 2006.

"Dawn of the Can." The Aluminum Association Inc. 24 October 2004. Available from <http://www.aluminum.org>.

"General Summary Tables." *1997 Economic Census,* and *2002 Economic Census.* U.S. Department of Commerce, Bureau of the Census. June 2001 and March 2006.

Lazich, Robert S. *Market Share Reporter 2007.* Thomson Gale, 2007.

"Long Live the Metal Can." *Food Logistics.* 15 October 2005.

Panjabi, Ghansham and Philip Ng. "Packaging Unplugged." Wachovia Capital Markets, LLC Publication. 22 December 2006.

"Product Summary: 2002." *2002 Economic Census.* U.S. Department of Commerce, Bureau of the Census. March 2006.

SEE ALSO *Cans, Frozen Foods*

Cans

———————■———————

NAICS: 33–2431 Metal Can Manufacturing

SIC: 3411 Metal Cans

NAICS-Based Product Codes: 33–24311 through 33–2431W

PRODUCT OVERVIEW

Metal cans are used in the packaging of foods, beverages, as well as for packaging household, institutional, and industrial cleansers, aerosol sprays, solvents, paints, lubricants, and related products. They are typically cylindrical containers but, in recent times, have also appeared in more complex shapes as promotional products, including in the form of tapered shapes resembling bottles.

Steel and aluminum, sometimes in combination, are the principal raw materials forming the can itself. Steel cans may carry a very thin layer of tin to protect the metal from corrosion and/or organically-based (plastic) coatings that serve the same purpose. Since 1995 soldered cans containing lead in the solder have been prohibited, but lead solder is still occasionally used in can-making overseas. Metal cans typically feature a printed paper wrapper adhering to the can by tiny dabs of glue. Aluminum cans, almost exclusively used in beverage packaging, typically carry a printed brand logo instead of being wrapped. The principal purpose of metal containers of either metal is preservation of food during distribution and storage over extended periods of time. In the industry itself closures are

a distinct product category that may be sold and even used separately, thus in cardboard containers with a steel lid.

Canning in metal appeared early in the nineteenth century—a product of war, as so many other innovations. Napoleon Bonaparte kicked off the process in 1795 by offering a prize of 12,000 francs to the inventor of a method for preserving food. Nicholas Appert, a chef and candy maker, won the prize in 1809. Appert invented canning, but his containers were sealed glass. He sealed his containers in boiling water, thus producing a vacuum. Britain at the time saw the Napoleanic expansion across Europe as a threat and responded by taking the next step toward canning. In 1810 Peter Durand in England received a patent for food preservation. Durand wished to make a less breakable container better suited for the rough-and-tumble of military life and used containers made of sheet iron coated with tin; the tin coating was intended to protect the iron from corrosion by acidic food. The tin can was born. For a century workmen made cans by hand, roughly ten per day per laborer. They cut the metal, bent it into shape, cut tops and bottoms, and soldered these on. The top lid, equipped with a small opening, received the food to be preserved. The cans were sealed by soldering the small hole shut.

Can manufacturing began in New York in 1812, introduced by an immigrant from England, Thomas Kensett. The industry developed rapidly so that by 1861, at the opening of the Civil War, 5 million cans were made annually—and after the Civil War 30 million. Innovations speeded up the process. Producers discovered that salting the water in which cans were cooked increased temperature and thus saved time. Producers introduced mechanized soldering of can seams and increased output from ten to sixty cans per day. The so-called sanitary can appeared in 1900 produced by creating double folds of metal at the seams. The folds were applied over the solder; this eliminated soot and minimized the migration of traces of lead into the food. By the early 1920s automated can body production appeared in the United States and production speeds of up 250 cans per minute became possible. Experiments to can beverages began in the 1940s but beverage cans, which required innovations to handle internal pressures generated by sodas and beer, were delayed by World War II and the Korean War that soon followed. Critical materials were scarce. This application emerged after 1953 when rationing of metals was suspended. Aluminum cans appeared in 1965. Technological developments, thereafter, centered on eliminating lead from solder, technological improvements in processing to speed production and to take out costs, dual metal containers (steel can, aluminum closure), and improved closures generally. Within the aluminum can system, cans were made with ever less aluminum.

The rise of environmental consciousness, symbolized by the first Earth Day celebration which took place in 1970, brought metal containers, particularly beverage containers, into the forefront of environmental concern. The industry became involved with recycling programs. Aluminum containers, which dominate the metal beverage industry, are an ideal product for recycling because aluminum is valuable as a metal and thus incentives for their return can be high; they are also light and easily compressed in recycling operations.

Cans made of either metal represent mature end markets. Alternatives to metals in liquid or moist food containments are limited but making headway through innovation; alternatives in beverage packaging are glass and plastics; both are competitors although aluminum dominates sodas and beer. Innovation in aluminum is centered on shapes and closures—thus on features that have a strong promotional and marketing aspect.

MARKET

Based on the Economic Census conducted by the Bureau of the Census, metal cans represented an $11 billion industry in 2002 and, by projection of past trends, was around $11.4 billion in the early 2000s. The concept of industry shipments underlies these numbers, thus the receipts of the manufacturers from the users of the can—companies that package the actual goods sold to the ultimate consumers. Still using 2002 census results, steel cans were $4.07 billion (37%) and aluminum cans $6.86 billion (62%) of the total, the remainder accounted for by miscellaneous parts and closures. In the five-year period before the census, metal can shipments declined at the rate of 5.7 percent annually, but trends in steel and aluminum

differed. Steel cans saw an annual decline of 2.7 percent between 1997 and 2002 whereas aluminum cans saw an annual growth of 0.9 percent.

The can manufacturing industry employed 38,600 people in 2002, down from 41,400 in 1997. In 2002 the industry had 51 participating companies, 36 engaged in steel and 15 in aluminum can-making. The total has declined from 57 companies in 1997, 41 in steel and 16 in aluminum. The decline in number of corporations is best explained by mergers and acquisitions in that consolidation has been widespread in this industry.

Cans serve three distinct markets which, in order of relative size, are beverage distribution, food distribution, and the distribution of selected chemical products. The beverage distribution market in turn divides into a larger and smaller market, the larger being cans for soft drinks. In the North American market, including Canada, just under 70 billion cans, virtually all made of aluminum, carried soft drinks to the market based on data from the Can Manufacturers Institute (CMI). Data from the same source but for North American beer sales indicate the use of around 32 billion aluminum cans.

The North American market for food cans accounted for just under 30 billion cans in 2005 as estimated by Wachovia Capital Markets, LLC based on company reports. The major segments of this market in order of size, based on CMI data for 2005, were vegetables (34%), pet foods (22%), soups and similar products (17%), dairy products (7%), fruits and meat and poultry (each 6%), seafood (5%), baby food (2%), and coffee (1%). The overwhelming majority of cans used in these applications are made of steel.

FIGURE 47

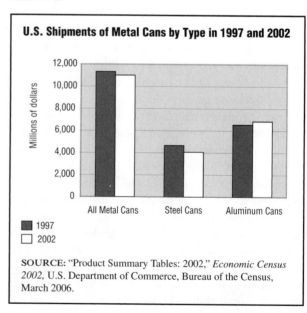

U.S. Shipments of Metal Cans by Type in 1997 and 2002

SOURCE: "Product Summary Tables: 2002," *Economic Census 2002*, U.S. Department of Commerce, Bureau of the Census, March 2006.

The remaining category of non-food cans, usually labeled general line cans, represent around 4.5 billion units of which aerosol cans account for 72 percent and all other kinds for the rest. These products typically carry chemicals, lubricants, solvents, and paints but may also be used for packaging novelty items. The category is heavily based on steel rather than aluminum.

As a consequence of supply logistics, cans made of either metal are manufactured very close to end users. For this reason the import and export of cans, if any, is not traceable in statistics. The raw materials that go into these products, of course, are traded on the international markets.

KEY PRODUCERS/MANUFACTURERS

The corporate structure of can manufacturing divides along lines indicated by the two basic component materials, steel and aluminum, both of which, represent major raw materials production activities in the background. Thus there is a distinct aluminum and a distinct steel can production industry, although leaders in one may be involved in the other as well. Behind these companies are major suppliers of steel and aluminum. Another distinct manner of participation in this industry—or the two metal-based portions of them—is by supplying specialized componentry or supplies, such as proprietary easy-open ends, valves (for aerosol cans), coatings, sealing systems, and inks.

It is also worth noting that in this industry, as in many others, physical as well as corporate consolidation has taken place. The transportation of empty cans is inefficient because a great deal of air has to be moved, translatable into cubage. Instead of moving empty cans producers make cans as close as possible to the point at which they are filled. Some end-product producers also own the packaging production function. This is the case with the beer company, Anheuser-Busch.

Aluminum. The three leading producers of aluminum cans are Ball Corporation, Metal Container Corporation, and Rexam Beverage Cans Americas. Of these three Ball makes both aluminum cans for beverages and steel cans for food packaging. The company's beverage business was by the acquisition of Jeffco Manufacturing Company (1969) and purchase of Reynolds Metal Company's domestic can making assets (1998). Reynolds was a major aluminum producer later acquired by Alcoa (2000). Ball entered the European beverage industry in 2002 by purchasing Schmalbach-Lubeca AG, the second largest such producers in Europe. Ball operates 20 beverage can production plants in North America (including Puerto Rico) and 11 in Europe.

FIGURE 48

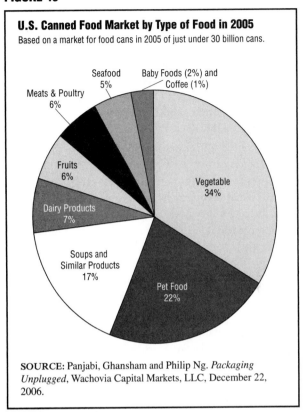

U.S. Canned Food Market by Type of Food in 2005

Based on a market for food cans in 2005 of just under 30 billion cans.

SOURCE: Panjabi, Ghansham and Philip Ng. *Packaging Unplugged*, Wachovia Capital Markets, LLC, December 22, 2006.

Metal Container Corporation (MCC) is a subsidiary of Anheuser-Busch Companies. The company makes aluminum containers and closures (tops), supplying roughly two-thirds of its parent's requirements of the former, three-quarters of the latter. MCC also produces cans and tops for other soft-drink beverage companies. The company was built by Anheuser-Busch from the ground up in 1973.

Rexam Beverage Cans Americas is the inheritor of a leading U.S. can producer, American National Can Company (a major supplier of Coca Cola Co., Coors, and Anheuser-Busch), which it acquired in 2000. Rexam began in 1923 as the Bowater's Paper Mills in England, specializing in newsprint. Later the Bowater Group renamed itself Rexam and transformed itself, principally by acquisitions, into a consumer packaging company, becoming the world's top beverage can producer in the process.

These companies, in their domestic and international operations, rely on the principal feedstock, aluminum sheeting, and on the major aluminum producers. These are Alcoa, Inc., ARCO Aluminum, Inc., Novelis Corporation, Wise Alloys LLC, and others.

Steel. The leading steel can producers include Ball, BWAY Corporation, Impress USA Inc., Silgan Containers Corporation, Sonoco Products Co., and Van Can

Company. Ball entered the food can business in the 1980s (after entering the aluminum can market for beverages) and expanded in the steel category as well by acquiring Heekin Can, Inc. in 1993. BWAY Corporation, earlier known as Brockway Standard, began in 1870 and grew by acquisition of several other companies and represents the leading producer of general line cans serving the paint and other related solvent producers, including Sherwin Williams, Glidden, DuPont, PPG, and Thompson MinWax.

Impress USA Inc. is the operating presence in the United States of Impress Group BV, a Netherlands-based packaging conglomerate originally formed by the merger of the French company Pechiney and the German firm Schmalbach-Lubeca (another part of which was sold to Ball Corporation). Impress is a leading can producer in Europe. Its U.S. division is a leader in seafood and petfood canning. Impress has proprietary easy-open closures and supplies these widely to others. Impress USA was formed when Impress acquired the canning assets of Starkist from H.J. Heintz Company Ltd. in 2000.

Silgan Containers Corporation is a U.S. company, a producer of food containers, its origins going back to the 1920. The company supplies approximately half of the entire food canning market and serves such major food companies as Campbell, Del Monte, General Mills, Hormel Foods, Nestle and other well-known brands. Enthusiasts for Spam (the edible kind) may note that Silgan has been packaging this Hormel product since 1927. The company also makes aluminum packaging and is a leader in easy-open ends, a feature of increasing importance in the metal packaging market.

Sonoco Products is a highly diversified packaging, packing, and product protection company with participation in many niches as a component supplier. The company's participation in the can business includes manufacturing of steel full panel pull out (FPPO) ends and several related products, including plastic end caps, all important features in the development of steel cans because they provide convenience for the consumer. The company began operations as the Southern Novelty Company in 1899 introducing the first ever paper yarn carrier, replacing more expensive wooden objects used before—a novelty for its time.

Van Can Company is a producer of steel cans for food distribution. The company emerged by degrees from the Ralston Purina Company. The latter sold its Van Camp Seafood property, Chicken of the Sea, along with the canning plant that served it, to an Indonesian investor. The investor first separated the seafood and the canning operations as separate entities and then sold Van Can Company to Komodo Enterprises Inc. Van Camp remains an important participant in seafood canning, pro-

ducing both cans and ends, and has diversified into other food areas as well.

These and other companies engaged in steel can production rely on steel sheet manufacturing by such producers (in alphabetical order) as Dofasco Inc., Mittal Steel USA, US Steel, and USS-POSCO Industries.

Specialists. In the twenty-first century can making has reached a very high level of technological sophistication reliant on a number of specialists in metal products, sealants, coating techniques, adhesives, inks, and more. Can makers look to such companies for specialized knowledge, products, and techniques which it would be inefficient for each of them to develop independently. This technological status of the industry has produced a number of suppliers who do not make cans but are integral to the industry.

Examples are DAREX Container Products, a specialist in sealants and coatings, and ICI Packaging Coatings, a leading coatings producer. Sealants guarantee product freshness. Coatings inhibit corrosion of the metal that they protect. INX International Ink Company combines coatings know-how with a specialization in suitable and compatible inks used to print can surfaces. PPG Industries (originally Pittsburgh Plate Glass) provides the same expertise and has a long history (back to 1883) as a specialist in sealants, coatings, paints, and ink. Other leaders are The Valspar Corporation and Watson Standard, both specialists in coatings.

MATERIALS & SUPPLY CHAIN LOGISTICS

Cans are typically manufactured in close proximity to the facilities in which they are later filled. The product is not completed until the closure is applied to the open can already holding the beverage, soup, vegetables, paint, or what have you. Thus the making of the can itself takes place in one operation but its final sealing in another. In most situations the two production operations are in close proximity. Open cans are conveyed on belts to the filling operations from within the same building or from an adjacent building. Relatively small canning operations—too small to justify the placement of a complex can-making facility next door—will receive open cans arranged on pallets from trucks. The general rule, however, is the co-location of canning and filling operations.

Canning plants typically receive their basic metal in huge coils—steel or aluminum. If simple closures are used—as in an ordinary soup can made of steel—the body of the can and its two ends are cut from the same sheet of metal. The bottom is joined to the body to form a cup; the top is applied after contents have been added. If the can uses an easy-open top, these specially engineered lids may be manufactured at a distant location and shipped to the can-making facility in bulk. In making two-piece beverage containers, the cup portion of the can is punched out of the incoming coil of aluminum sheet and the bottom, already part of the extrusion, is shaped in a mold. Virtually all beverage cans feature pull-tab openings which are frequently produced off-site by specialists. The general rule from a logistical point of view, is that the cup portion of the can is always made on site from coils of metal brought in by truck or rail but that the closures may come prefabricated from vendors unless the cans made are standard steel containers intended to be opened by a can-opener.

The other important components of cans, almost always acquired in bulk containers and applied in the manufacturing process are coatings, inks, varnishes applied over inked surfaces, and sealants. To achieve efficient operations in steel can production, coatings are applied to and decorative or informational messages are printed on the sheet itself before it is scored and cut. Varnishes are applied after the inks are force-dried in ovens. Internal coatings are applied to the cup itself after it is steam-cleaned before filling. Aluminum cans are printed as cups because the sheet itself is deformed in making the cup; all other operations, therefore, also take place with the cup already formed. Incoming closures, especially those of large surface area that can carry a message will arrive preprinted and varnished.

If paper labels are applied to cans these may arrive in pre-printed rolls ready to be installed into feeding machinery for cutting and application with dabs of adhesive. Similarly, if the cans feature a plastic cap—to be used by the consumer after opening the can—these caps are usually manufactured by a vendor and arrive in bulk.

From a logistical point of view, the most important aspect of can manufacturing is to locate the operation as close to the final user's plant as possible. Production of the cans is managed and controlled to feed the filler exactly what it requires as it requires it, eliminating inventories of cans waiting to be filled. Steel or aluminum, plus other supplies in bulk, are the can makers inventory.

DISTRIBUTION CHANNEL

Metal cans are part of the vast category of packaging and, as such, distribution has two distinct aspects. If viewed from the perspective of the seller of cans to a canner of goods, distribution is direct from the manufacturer to the buyer. The buyer, indeed, may be the manufacturer, as is the case for Anheuser-Busch and others. Two corporate entities may be involved, but in practice the can-making operation is usually established as part of the total canning enterprise as an on-site supplier. Long-term contracts are the rule. In an operational sense, the can maker coordinates its production schedules with those of the canner in the most detailed and intimate ways. The principal sales

event takes the form of an extended negotiation and planning cycle between a can maker and the canner. Thereafter price adjustments take place on the basis of terms fixed in the contract that, for example, may explicitly recognize factors over which the vendor, the can maker, has no control—such as the price of aluminum or steel.

Once cans are filled, sealed, and additionally packaged in cartons affixed to pallets, the can itself has, in effect, lost its identity as a product distinct from its contents. It will proceed thereafter in the distribution channel appropriate to the product it carries. Viewed from this perspective, the can itself invariably has minimally a three-tier distribution: from the seller to the canner, from the canner to the retailer, from retailer to the consumer. Some portion of virtually all products packaged in cans also pass through a wholesale level and cans thus often have a four-tier distribution.

KEY USERS

From the can producer's perspective, the key users of cans are producers who need a package capable of containing perishable liquid or semi-liquid products—products that need protection from organic processes of decay. Among key users are producers of pressurized liquids requiring the strength of metal to contain their goods as well as those selling products that, in the event of accidental spill, can cause harm directly to those who come in touch with the product or may result in fires. Cans provide long-term product protection and are thus attractive to producers of goods looking for long shelf-life. Metals, by reason of their durability, imply higher cost than flimsy paper or stretchable plastic and carry an aura of permanence and value. These features of canning attract users who wish to exploit their generic capacity to draw attention to their products or to give them value or visibility. A trivial example is the packaging of cookies in firm and beautifully decorated metal cans—despite well established alternatives that provide the same protection at significantly lower costs. But some buyers of novelty metal packaging, which falls under the Census Bureau's definition of a can, will package many products in metal even when these need no protection at all—jewelry, stationary, and even dolls.

Consumers are the ultimate users of cans. The overwhelming majority of canned goods are intended for the ordinary consumer. To be sure, industrial/commercial products are also distributed in cans, notably lubricants, industrial solvents, and commercial-grade paint products. Institutional markets received canned goods, for instance medical products intended for hospital use—and military supplies. But in terms of numbers of cans or billions of dollars, the largest single category is beverage distribution followed by food packaging.

ADJACENT MARKETS

As part of the packaging industry, the metal can stands on the grocer's or drug store's shelves side by side with glass containers, plastic bottles and jars, rigid and flexible paper containers, and composite packages made of different materials in laminated, coated, and nested compositions. As the history of canning indicates, the nearest adjacent market to the metal can is the glass container. When we speak of home canning the image of glass jars appears, the first type of container doing the job metal cans now do. Glass also forged the way in the distribution of beer and sodas in single-portion quantities. Long before the aluminum can appeared, the returnable glass bottle was king. Glass offers all of the core functionalities of the can. It adds the feature of making the contents visible to the users, and the inert nature of glass keeps it from even microscopic contamination of the material it holds. But glass has two major and one minor drawback. Glass is breakable. It is heavy and thus adds significant costs in distribution. The minor drawback of glass is that it takes longer to cool the contents. This aspect is important in the distribution of beverages, be they beer, soda, or milk: the product in a can cools much faster and thus provides a consumer benefit. Glass still remains the exclusive package for wine because glass least interferes with the taste of the product.

The history of packaging is the history of materials competing for dominance based on complex factors including inherent suitability, cost, weight, shelf-life, stackability, capacity to display decoration and messages, translucency, and other factors. Aluminum has achieved dominance in beer and soda packaging and will retain it unless, due to increasing aluminum prices, it comes under pressure. Such an eventuality is at least theoretically possible because the energy requirements for making aluminum from bauxite are very high. When and if that point arrives, aluminum's currently minor competitors, plastics and glass, will have difficulty competing as well. Plastics depend on petroleum, and an energy crisis driving aluminum prices very high will also make plastics expensive. Glass, similarly, is also an energy-intensive product in that it requires the melting of sand. In the very long run—unless humanity discovers some genuinely novel form of cheap energy—we may slide backward to the oldest package of all, the wooden keg.

One of the reasons that metal cans showed a history of flat growth or outright decline in the late twentieth and early twenty-first centuries is because its dominant segment is food packaging, an extremely diverse category in comparison with beverages. Alternatives abound. For example, soup in a can competes with soup packaged in sealed glass jars; it also competes with instant soup that may be packaged in sealed paper cups or laminated sacks combining paper, plastics, and a metallic outer layer. In addition, from time to time, canned soup may also compete with societal fashions, as for instance increased cooking at home from fresh ingredients. Such trends, however, are difficult to measure. Another example of indirect competition by product transformation is the sale of instant or freeze-dried coffee in composite paper-plastic containers in competition with ground coffee in cans.

Industry observers, such as Wachovia Capital Markets, LLC, discern as one of the chief factors causing sluggish growth in steel can production the rising competition from flexible packaging. Technological advances in materials emerged strongly in the sustained period of growth following World War II and have not abated. Flexible packaging is only likely to nibble at the edges of the steel can market. In many largely commodified applications such as soups, fruits, and vegetables, consumers are unlikely to abandon long-established habits, not least the ability to store and stack cans on shelves, in favor of products that will, at least initially, appear to be less obviously safe.

RESEARCH & DEVELOPMENT

Packaging in cans began as the consequence of a deliberate effort to stimulate R&D in France in support of military goals. The field has, since then, marched from innovation to innovation and continues to do so to this day. The major thrusts of development may be subdivided into basic materials research, production-oriented innovation, and product differentiation.

From the earliest times of canning in metal, a negative aspect of this form of packaging has been that some of the metal, if in microscopic quantities, has always migrated into the product itself. The human palate has extraordinary sensitivity, moreso in some than others. For some the metallic taste is still detectable despite the use of very effective coatings on the side of cans. This issue goes by the name of *manufacturing-induced taste factors* in the industry and is receiving continued and intensive attention; the work is focused on coatings, particularly the thoroughness with which suitable coatings cover the surface at the submicroscopic level. Taste panels are used by the industry to support this research, reminiscent of panels used in wine evaluation.

Research is also directed at the exterior of cans in the context of environmental factors that may affect the container in shipping and storage. This sort of R&D is focused on the performance of outer varnishes, deformations due to heat and cold and their effects on seals, and on the sealing technologies used in cans generally. Part of this work is active testing of products *to* failure and then the forensic tracing of the failure event and its consequences.

Research is also being applied at the manufacturing process itself intended to speed up can making, detecting difficult-to-see bottlenecks and events that cause them, and introducing automated controls and quality assurance measurements at more stages of the production process.

The can itself receives a great deal of R&D attention. Here the effort is driven by competitive forces and the work merges into new product and new application developments. A central focus remains on easy-to-open closures, a feature likely to give producers a competitive edge as the baby boom generation ages. This effort has been paralleled by a phenomenon associated with alternative markets, namely the development of ever better can openers. Within the aluminum sector, substantial work is being undertaken to achieve new shapes in aluminum, a leading shape being a bottle and, preferably, a reclosable bottle. Other shapes, providing consumers with an easier grip, are also emerging or still being perfected. The quest for new shapes, including multi-purpose cans, is also an important aspect of steel can R&D. Thus, for instance, a can that will double as container and soup dish or cup is a research goal. Steel itself prevents microwaves from penetrating into a can or can-like container, but with a composite can that features a plastic closure it appears likely that microwavable cans are in the offing as the twenty-first century advances. All of these efforts converge on providing the consumer new conveniences and to offer the canner innovative products that stand a chance to gain them market share.

CURRENT TRENDS

The major trends in this industry have to some extent been alluded to already. In the aluminum segment, the stronger of the two, innovations are trending toward new shapes. Whether or not bottle-shaped aluminum cans will be adopted widely in the market, and thus become commodified, widely accepted, and cheap remains to be seen. The bottle shape, while novel in aluminum, introduces inconveniences in storage while demanding a higher price. The major trend in steel has been absence of growth and losses of market share to other forms of packaging—stemmed or slowed by innovations in easy-to-use closures. Likely developments in the steel sector of canning will be the emergence of composite cans, with the lid or a portion of it being made of plastics, taking out weight and adding new features.

No discussion of trends in the can industry can be complete without noting the environmental issue. It is of long standing but continues to receive a great deal of attention inside the industry. The issue rose to prominence in the 1960s, already in a concern over beverage can litter and evolved into the more general concern with recycling. The industry, through its associations of producers, like the Can Manufacturing Institute, and buyers, like the American Beverage Association and The Beer Institute, are energetically backing recycling programs for beverage and other containers in efforts to avoid mandatory deposit legislation or outright bans on different types of containers. As a consequence of such outreach, promotion, and actual programs, substantial portions of aluminum are recycled. In that process, 95 percent of the energy consumed in making the original aluminum from bauxite is saved—achieving genuine environmental and economic savings.

TARGET MARKETS & SEGMENTATION

In the discussion of the market presented above, the major markets and segments have already been outlined adequately, this being an industry where the product itself is a container common to many specific markets. But aside from the broad structure of the industry, its division into two separate metals used for their unique characteristics, and then into the separate containment applications for which they are used, target markets and segments in another sense of the word are well worth noting. These really point to the leading edge of an industry that is, on the whole, the supplier of a well-established commodity.

In this special sense, producers and users of cans are clearly orienting their efforts of product differentiation at three different groups. One of these is the youth market, likely to respond to novelty and innovation. By accepting new products, these young users will then, the industry hopes, lead in the expansion of new products as their early acceptors grow and mature. Beverage can makers are aiming products at this market when they develop and introduce bottle-shaped aluminum cans. A second distinct segment receiving targeting by the producers are busy young adults looking for rapid and efficient accomplishment of routine tasks in the course of busy lives. Developments in the field aimed at producing reusable containers, a can that serves as a soup bowl, for example, and microwavable composite cans have young adults in mind. Specially shaped products that enable children to drink juices without spilling them belong in this category as well. The third segment is made up of the elderly challenged by arthritis or simple weakness in the hands. Easy-open cans have a special appeal to this segment and will cause its members to give preference to canned food products that are easy to open.

RELATED ASSOCIATIONS & ORGANIZATIONS

The Aluminum Association, Inc., http://www.aluminum.org

American Beverage Association, http://www.ameribev.org/index.aspx

American Iron and Steel Institute, http://www.steel.org/AM/Template.cfm?Section=Home

The Beer Institute, http://www.beerinst.org

Can Manufacturers Institute, http://www.cancentral.com/index.cfm

Canned Food Alliance, http://www.mealtime.org

Composite Can & Tube Institute, http://www.cctiwdc.org

National Paint & Coatings Association, http://www.paint.org

Packaging Machinery Manufacturing Institute, http://www.packexpo.com

Women in Packaging Inc., http://www.womeninpackaging.org

BIBLIOGRAPHY

"Complete History." Can Manufacturers Institute. Available from <http://www.cancentral.com/hist_overview.cfm>.

Darnay, Arsen J. and Joyce P. Simkin. *Manufacturing & Distribution USA,* 4th ed. Thomson Gale, 2006, Volume 1, 927–929.

"Firms Close to Creating Inkjet-Printed Cans." *Print Week.* 19 October 2006.

Fuhrman, Elizabeth. "Packaged to be Different: Package Material and Shape Matters For Every Beverage Category." *Beverage Industry.* May 2006.

Hartman, Lauren R. "Canning Sauces the Hirzel Way: Canning line at Hirzel Canning Co. & Farms in Northwood, Ohio." *Packaging Digest.* June 2005.

"Long Live the Metal Can." *Food Logistics.* 15 October 2005.

Panjabi, Ghansham and Philip Ng. "Packaging Unplugged." Wachovia Capital Markets, LLC Publication. December 2006.

"Product Summary: 2002." *2002 Economic Census.* U.S. Department of Commerce, Bureau of the Census. March 2006.

Spaulding, Mark. "Menu of Food Containers Shifts Toward Flexibles: Overall Demand Keeps Pace With Population Growth: Freedonia." *Converting.* November 2005.

Theodore, Sarah. "Container Demand Spans all Categories." *Beverage Industry.* April 2005.

Toto, DeAnne. "A Can-do Attitude: The Metal Can's Reputation For Recyclability and Energy Efficiency Keeps It a Popular Choice for Food and Beverage Packaging." *Recycling Today.* January 2006.

SEE ALSO *Canned Foods*

Carpets & Rugs

―――――■―――――

INDUSTRIAL CODES

NAICS: 31–4110 Carpet and Rug Mills

SIC: 2273 Carpets and Rugs Made of Textiles

NAICS-Based Product Codes: 31–41101 through 31–41105008

PRODUCT OVERVIEW

Carpets cover two thirds of floors in the United States and thus dominates the floor covering industry. Other floor coverings include ceramic tile, stone, laminate, vinyl, and wood. After carpet was declared an approved floor covering by the Federal Housing Administration in 1966 it rapidly took over the floor covering market. In 1973 shag carpeting was introduced and outsold other carpeting types for years thereafter. The popularity of carpet is due to its many benefits. It is soft, quiet, warm, and generally considered affordable. Special treatments can be applied to carpets and rugs so that they are stain, static, and soil resistant. New colors, patterns, and textures are introduced every year to keep up with fashion trends. Carpet is relatively easily replaced to update the interior design of residential and commercial properties.

The cushioning effect of carpet controls both airborne sound and impact noise. Airborne sound is noise like radios and voices that carries between rooms and levels. Impact noise is sounds like footfall and vibration from appliances that radiate through structural parts of buildings. Carpet on floors, and sometimes on walls, is an important acoustical component of public places like auditoriums, restaurants, and schools. How carpet performs depends on the type of material used to make it. The most common material for carpets is nylon. Polyester, polypropylene, and wool are also used. These materials are used to make long skeins of carpet yarn for the face fiber, or pile, in carpet and rugs.

Long skeins of carpet yarn are made in two ways. Carpet yarn is known as either spun staple fiber or bulked continuous filament fiber. Both create yarns that produce carpet and rugs with distinct characteristics.

Spun Staple Fiber. A series of short strands, typically 5 to 7 inches long, are spun together to form one continuous skein known as spun or staple fiber. Several skeins are twisted together to form a strand of yarn. When tufted into a carpet or rug, spun staple fibers bloom more, exhibiting a bigger hand finish.

Bulked Continuous Filament Fiber. A continuous strand manufactured as one long skein is twisted and heatset together to form yarn known as bulked continuous filament fiber. The continuous strands of synthetic fiber are texturized to increase bulk and strength. When tufted into a carpet or rug, continuous fibers allow a smooth fine finish.

The material used to make long skeins of carpet yarn determines the aesthetics and performance of the finished carpet or rug. Factors like available colors and textures, and resistance to stains, static, and soil are determined by the face fiber, or pile. The most common materials for carpet and rugs are nylon, polyester, polypropylene, and wool.

Nylon. Nylon was invented in 1939 by DuPont Company and first used for carpet in 1959. Nylon is the most common carpet material. It represent 65 percent of all U.S.-made carpet and rugs. Nylon is a petroleum derived product defined by the Federal Trade Commission as a manmade fiber produced from a long-chain synthetic polyamide. The polyamide is melted, spun, and drawn into strands after cooling to make both spun fiber and bulked continuous filament fiber. Nylon takes color well and can be dyed either during the melting and spinning phase or white skeins can be made to be dyed later.

Besides its excellent color affinity, the benefits of nylon include affordability, durability, and versatility. While higher priced than other materials, nylon pricing is favorable compared to wool, its closest competitor. It can be treated to be stain, static, and soil resistant. It is versatile enough for both residential and commercial uses. Nylon withstands rigorous use and can be readily cleaned.

Polyester. Sometimes referred to as PET, short for its chemical name polyethylene terephthalate, polyester is a petroleum derived product defined by the Federal Trade Commission as a manmade fiber produced from any long-chain synthetic polymer composed of at least 85 percent by weight of an ester of a substituted aromatic carboxylic acid. Polyester is made by combining ethylene glycol with terephthalic acid at high temperature and in a vacuum to achieve the high molecular weight needed to form useful fibers. PET is melt spun into primarily a spun staple fiber, although some bulk continuous filament fiber is produced.

Polyester fiber is noted for its luxuriously soft hand and bloom. Polyester has good color clarity and colorfastness. It resists water-soluble stains and retains its luster. Polyester is used for value-priced carpets and rugs. Its use as a face fiber material is growing, almost doubling between 2001 and 2006. In 2006 the three largest manufacturers introduced new carpets made from polyester because it is less sensitive to price increases than either nylon or polypropylene.

Polypropylene. Commonly referred to as olefin, polypropylene is a petroleum derived product defined by the Federal Trade Commission as a manmade fiber produced from any long-chain synthetic polymer composed of at least 85 percent by weight of ethylene, propylene, or other olefin units. Basically, olefin fibers are products of the polymerization of propylene and ethylene gases under controlled conditions. The fibers resist dyeing, so colored olefin fibers are produced by adding dye during melt spinning; this results in a somewhat limited color selection.

Olefin is generally used for bulked continuous filament fiber and is very lightweight. Olefin is preferred for commercial installations. It is favorably priced, colorfast, easily cleaned, and inherently resistant to chemicals, fading, mildew, moisture, stains, static electricity, and wear. Olefin is the fiber of choice for indoor-outdoor applications.

Wool. The most expensive carpet fiber, wool is noted for its luxurious hand and performance. It is soft, has high bulk, and takes dye more beautifully than any other fiber. It is a durable material with natural fibers that scatter light and reduce visible soil. Because it is a natural material, it is always spun, never bulked as a continuous spun filament. Wool is a style and trendsetter that other fibers are used to copy. At the Surfaces 2007 trade show in Las Vegas, the place of wool in the materials hierarchy was underscored. Traditionally nylon mills introduced carpets in wool, wool blends, or proprietary nylon technologies that create wool looks. Rising petroleum costs gave manufacturers the impetus to turn to wool. As the cost for petroleum derived synthetic fibers increases, the cost for all natural wool has leveled off, making wool relatively affordable.

No matter what material they are made of, carpets and rugs can be knitted, tufted, or woven. Knitted carpets are made using a method in which yarn is stitched to the backing and anchored by a plastics coating. Woven carpet involves a traditional manufacturing process where carpet is produced on a loom by which lengthwise and widthwise yarns are interlaced to form the fabric. The most common manufacturing method is tufted.

Close to 95 percent of U.S. carpet and rugs are tufted. The process involves several hundred needles (up to 1,200 across the typical 12 foot width) that thread carpet yarn through a lightweight backing, forming loops or tufts of the required height. An adhesive coating is applied to the back to anchor tufts in position then a second backing is applied for extra strength. The tufted manufacturing process is used to produce different carpet textures known as loop pile, cut pile, and cut loop pile.

Loop Pile. Loop pile is the basic look upon which cut pile and cut loop pile are built. Loop pile is formed by continuous rows of tightly spaced loops. The result can be textures from thick and nubby to smooth and plain. Either level loop pile or multi-level loop pile can be produced. Multi-level loop involves two or three varying pile heights to produce a high/low or sculptured effect. Level loop pile wears well because the tight pile bears the weight of the foot evenly. It also hides footprints and vacuum marks. Loop pile results in a carpet with a hand-crafted appearance that fits room styles from contemporary to country to cottage.

Cut Pile. After standard loop pile is constructed, all the tops of the loops are cut. The tufts of yarn stand up straight and form an even surface. Cut pile results in different looks depending on different twist levels of yarns. For example, cutting a yarn with low twist results in the smooth classic look of velvet or velour. Highly twisted yarn produces a hard twist pile affiliated with plush styles because a well defined twist gives the pile individual definition. Cut pile results in a thick, rich carpet suitable for bedrooms and living rooms.

Cut Loop Pile. After standard loop pile is constructed, a sculptured appearance is created by cutting some loops while leaving others uncut. Either level cut loop or multi-level cut loop can be produced. Cut loop combines the practicality of loop pile with the classic appearance of cut pile. The sculptural effects hide soil, stains, and vacuuming marks. Cut loop pile results in a carpet that can be casual or classic.

Whether the texture is loop pile, cut pile or cut loop pile, the word carpet generally denotes wall-to-wall floor covering while rug denotes area rugs, scatter rugs, and sometimes but not always bath mats. Rugs pose a special problem because they are a subcategory of other industries. Rugs are sometimes discussed as a category within the floor covering industry, the home furnishings industry, the textiles industry, and even the gift industry when it comes to bath mats or welcome mats. Commerce lumps rug statistics in with both carpet and textiles figures. This makes generalizing difficult within the rug subcategory of the carpet industry.

Premium high-priced hand tufted area rugs woven from wool are imported from places like Belgium, China, India, Italy, and Egypt. Rugs from manmade fabric are made in the United States. These can be categorized variously as area rugs, scatter rugs, bath rugs, door mats, and indoor-outdoor rugs. Common sizes are 3 by 5 feet, 5 by 8 feet, and 7 by 10 feet.

Indoor-outdoor rugs are becoming more common. This type of all-synthetic rug was typically referred to as utilitarian and inexpensive since "the words 'ugly' and 'cheap' are frowned upon in polite company," according to Lisa Wymann, editor of RugNewsBlog.com, writing in the December 25, 2006 issue of *Furniture-Today*. Retailing for under $100 in a 5 by 8 size, these rugs were sold out of cartons in discount stores and home centers. The newer breed of indoor-outdoor rugs is more decorative. They come in a broader range of styles and take advantage of olefin so that they are inherently resistant to chemicals, fading, mildew, moisture, stains, static electricity, and wear. This new breed of rug is used in outdoor kitchens, BBQ areas, decks, porches, boats, and even inside the home in kitchens, bathrooms, sunrooms, and garden rooms.

MARKET

Carpet and rug manufacturing in the United States grew during the early years of the twenty-first century. As part of its *Current Industrial Reports* series, the U.S. Census Bureau reported total product shipments from 1997 to 2006. Figure 49 depicts the growth of the carpet and rug industry during this 10 year period from a low of $10.3 billion in 1997 to a high of $14.3 billion in 2005 with a slight drop to $14.2 billion in 2006. Total cumulative growth between 1997 and 2006 was 28 percent. Figure 49 also provides estimates for 2007 through 2009 based on research conducted by New Horizons Marketing Inc. and reported in *HFN*.

In 2003 carpets and rugs controlled 68 percent of the U.S. flooring product market, down four points from 72 percent in 1997. Forecasts are for carpet to decline to 62 percent in 2009 due to the growth of other flooring choices like ceramic tile, laminate, vinyl, and wood. Wood flooring in particular doubled its flooring market share between 1997 and 2004, primarily at the expense of carpet. While continued erosion in carpet as a percentage of flooring by type was expected, researchers predict that carpet will fail to fall below 60 percent of the U.S. floor covering market.

Within the $14.2 billion carpet and rug industry, the main Census product categories in order of market size are tufted carpet and rugs, woven carpet and rugs, and other carpet and rugs. Tufted carpet and rugs dominates representing 95 percent. Tufted carpet and rugs were up

FIGURE 49

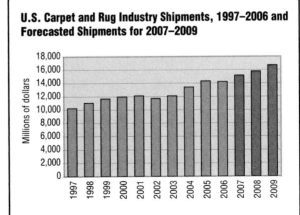

U.S. Carpet and Rug Industry Shipments, 1997–2006 and Forecasted Shipments for 2007–2009

SOURCE: Compiled by the staff with data from "Carpets & Rugs–2006 Annual Report," *Current Industrial Reports*, U.S. Department of Commerce, Bureau of Commerce, June 2007 and "Carpet One Predicts a 4.3 Percent Spurt by 2009," *HFN*, August 8, 2005.

25 percent between 1997 and 2006 from $9.969 billion to $13.445 billion. Of this $13.445 billion in 2006, Census tracked product shipments based on whether the yarn fiber was nylon, polyester, or polypropylene. Figure 50 is a pie chart that shows percentages of each. Nylon dominated with 69 percent. Polypropylene is 16 percent and polyester is 11 percent. All other is 3 percent.

The Census Bureau series of reports titled *Current Industrial Reports* highlight changes in major carpet product categories between 2001 and 2006. Tufted nylon product shipments were up 17 percent between 2001 and 2006, from $7.8 billion to $9.4 billion. Tufted polypropylene—common name olefin—product shipments were up 18 percent between 2001 and 2006, from $1.8 billion to $2.2 billion. Tufted polyester—common name PET—product shipments were up 42 percent between 2001 and 2006, from $876 million to $1.5 billion.

Rug sales are sometimes discussed as a subcategory within the floor covering industry, the home furnishings industry, the textiles industry, and even the gift industry when it comes to welcome mats. The U.S. area and bath rug business totaled $2.7 billion in 2004, a 9 percent increase over 2003 according to a *Floor Focus Magazine* 2005 survey. The survey also reported that rug imports were up nearly 12 percent in 2004 over 2003, with India up nearly 20 percent. China, Pakistan and Turkey were other countries whose rugs showed strong growth as imports in the United States during this period. Turkey was

up 41 percent. Based on research by HTF and cited in *Home Textiles Today,* the rug industry (defined as accent, scatter, and area rugs but not bath rugs) was worth $4.5 billion at retail in 2005.

KEY PRODUCERS/MANUFACTURERS

The top three U.S. manufacturers of carpeting are Shaw Industries, Inc., Mohawk Industries, Inc., and Beaulieu of America, in that order. Besides Shaw and Mohawk, who also make area rugs—which Beaulieu does not—Maples Industries is a top rug maker.

Shaw and Mohawk accounted for a hefty 71 percent of the carpet business. Shaw led with sales of $4.3 billion and a 38 percent market share. Mohawk was close behind with $3.8 billion in sales and a 33 percent market share. Beaulieu trailed at $1.1 billion and a 10 percent market share. Maple with its rug-only emphasis had $245 million in sales and a 10 percent market share. Sales estimates and market share figures are for 2004, the last year for which data on all four companies was available, based on material published in *Floor Focus Magazine* in May 2005. Each is profiled in alphabetical order.

Mohawk Industries, Inc. Headquartered in Calhoun, Georgia, Mohawk is a diversified floor covering manufacturer with 34,000 employees. It designs, manufactures, and markets woven and tufted carpet, area rugs, and hard surface flooring products. Included in Mohawk carpet and rug brands are: Aladdin, Alexander Smith, American Olean, American Rug Craftsmen, American Weavers, Bigelow, Galaxy, Harbinger, Helios, Horizon, Image, Karastan, Lees Carpet, World, WundaWeve, Custom Weave, Mohawk, and Mohawk Home.

Karastan is top of the line and Mohawk Home is a value priced brand. For instance, a rug such as Empress Kirman from the Original Karastan collection is woven from fully worsted New Zealand wool and retails for around $1,199 in a 5 foot 9 inch by 9 foot format. From the Mohawk Select Passport collection, rugs made from a nylon yarn system that blends shades of color into multiple tones with an ultra-soft pile retail for $299 in 5 foot 3 inch by 7 foot 10 inch format.

Mohawk Carpet Mills began in 1878 when four Shuttleworth brothers brought 14 second hand looms from England to Amsterdam, New York. After a 1920 merger with the nearby firm of McCleary, Wallin, and Crouse the name was changed to Mohawk Mills, Inc., derived from the Mohawk River Valley in eastern, upstate New York where Amsterdam is located.

In the 1950s Mohawk moved and constructed manufacturing facilities in Mississippi and South Carolina. In 1956 Mohawk merged with Alexander Smith, Inc., to

FIGURE 50

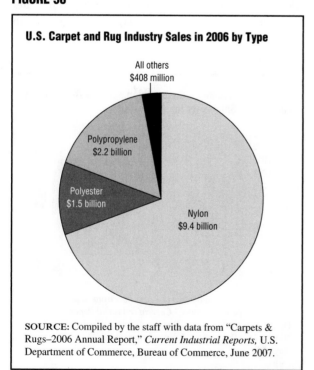

U.S. Carpet and Rug Industry Sales in 2006 by Type

All others
$408 million

Polypropylene
$2.2 billion

Polyester
$1.5 billion

Nylon
$9.4 billion

SOURCE: Compiled by the staff with data from "Carpets & Rugs–2006 Annual Report," *Current Industrial Reports,* U.S. Department of Commerce, Bureau of Commerce, June 2007.

form Mohasco Industries, the largest carpet manufacturer in the world. By 1989 Mohawk Carpet returned to its roots by purchasing the carpet manufacturing business from Mohasco. Mohawk grew to dominate the carpet and rug industry primarily through acquisitions: Horizon Industries in 1992; American Rug Craftsman and Karastan-Bigelow in 1993; Aladdin Mills in 1994; Galaxy Carpet Mills in 1995; certain assets from Diamond Carpet Mills in 1997; Newmark Rug Company, American Weavers & World Carpets/WundaWeve in 1998; Durkan Patterned Carpets and Image Industries in 1999; Alliance Pad in 2000; Dal-Tile and American Olean in 2001; and Lees Carpets in 2003.

Shaw Industries, Inc. Based in Dalton, Georgia, Shaw leads in both textile and carpets and has 30,000 employees. Shaw got its start in 1946 when Clarence Shaw, father of CEO Robert E. Shaw and J.C. Shaw, bought Star Dye Company, expanded dramatically, and started finishing carpet as Star Finishing Company. In the late 1960s, Shaw acquired Philadelphia Carpet Company and Star Finishing moved into carpet manufacturing.

Shaw Industries, Inc., went public in 1971 with $43 million in sales and 900 employees. In 1972 Shaw became vertically integrated with the acquisition of its first yarn plant, following up with the acquisition of its first continuous dye plant in 1973. It gained 100 percent control of its yarn supply when it acquired six yarn spinning mills in 1983.

By 1985 Shaw was listed on the Fortune 500 with more than $500 million in sales and close to 5,000 employees. Shaw became further vertically integrated when it acquired Amoco's polypropylene fiber production facilities in 1992. The acquisition also made Shaw the largest producer of polypropylene fiber in the world. The move helped Shaw capitalize on the tremendous growth of Berber carpet, the trend that displaced shag carpeting as the fashionable new look in the 1990s.

Like Mohawk, Shaw grew through acquisitions: WestPoint-Pepperell Carpet and Rug Division gave it the Cabin Crafts and Stratton brands and added 40 percent to its sales volume in 1987; the Armstrong World Industries, Inc. flagship Evans & Black brands added 30 percent to its sales volume in 1989; Salem Carpet Mills, Inc., added 25 percent to its sales volume with the Salem and Sutton brands in 1992. Because area rugs grew in popularity during this period, in 1993 Shaw formed its Shaw Rugs division. For instance, a Shaw Living area rug from the collection designed in collaboration with outdoors artist Phillip Crowe is machine woven of olefin and retails at $379 in the 5 foot 5 inches by 7 foot 11 inches size.

Shaw positions itself as a low-cost provider of carpets, a factor in its 1998 merger with Queen Carpets. Renowned financier Warren E. Buffett acquired a majority of Shaw through his Berkshire Hathaway, Inc., holding company in 2001 and by 2002 owned 100 percent, ending Shaw's tenure as a public company. In 2003 Shaw purchased the north Georgia operations of the Dixie Group, acquiring such brands as Carriage Carpets, Bretlin, and Globaltex.

Beaulieu of America. Headquartered in Dalton, Georgia, Beaulieu of America was established in 1978 out of Beaulieu Belgium by Mieke and Carl Bouckaert. It began as a producer of polypropylene area rugs. Beaulieu became vertically integrated into yarn extrusion in 1981 and diversified into tufting carpet in 1984, establishing a factory in Chatsworth, Georgia. In 1987 Beaulieu increased its ability to produce nylon polymers and nylon yarns with the addition of a facility in Bridgeport, Alabama.

In 1990 Beaulieu continued its vertical integration with the addition of another new facility in Bridgeport, Alabama, for the extrusion of polypropylene staple fiber. Like Mohawk and Shaw, Beaulieu grew through acquisitions of companies like Conquest Carpet Mills, Interloom, Coronet Industries, Grass More, D&W Carpets, Marglen Industries, Columbus Carpet Mills, Peerless Carpet Corporation, Princeton Rugs, and American Polycraft.

Its acquisition of Coronet and Peerless in 1980 helped establish Beaulieu of Canada. Beaulieu had 8,500 employees in the United States and Canada. Its acquisition of Sterling Carpet Mills helped establish Beaulieu of Australia in 1995 where, in addition to making rugs, it extrudes bulked continuous filament polypropylene and nylon fibers. In 2002 Beaulieu of America decided to concentrate strictly on carpet, and sold its rugs and hard surfaces divisions.

Maple Industries. Maple is a family affair. Founded in 1928 by John Maple, it remains privately held. Maple is a leading U.S. manufacturer of tufted accent and area rugs. In 1966 Maple opened a 30,000 square foot factory and then doubled its factory size in 1969. In the 1970s Maple innovated a three color cross-dyed process for its rugs. In the 1980s Maple introduced the first bulk continuous fiber nylon rugs for use in the bathroom, started using state-of-the-art computerized dyeing and computerized tufting, and pioneered the use of olefin accent rugs that could be machine washed. In the early 1990s Maple doubled the size of its New York City showroom and opened its own yarn mill.

Maple reported a compound annual growth rate of 17.5 percent between 1994 and 2004. Maple grew organically by investing at least 90 percent of its profits into capital improvements. Between 1997 and 2004 it invested over $35 million in its manufacturing facility in Scottsboro, Alabama, in both expansions and new equipment so that it led technological innovations in the industry. In 2000 Maple expanded its factory to a total of 500,000 square feet. Between 2000 and 2004 Maple grew enough to warrant the 2004 doubling of its factory space from the 500,000 square feet it built out in 2000 to over one million square feet. It spent $6 million on the 2004 expansion. Maple does not try to be all things to all segments in the floor covering industry. It concentrates on manufacturing tufted rugs.

Others. Nourison Rug Corp., a maker of carpet and rugs, opened a 305,000 square foot plant in Calhoun, Georgia, in June 2007. Initially it will make carpet and later rugs and bath products at the plant, which has the latest cutting and finishing equipment to produce roll runners and rugs, already averaging 30 to 50 cuts of roll runners per day. The company has the ability to expand the building with another 420,000 square feet when it is needed.

MATERIALS & SUPPLY CHAIN LOGISTICS

U.S. carpet and rug makers had shipments of $14.2 billion in 2002 and spent more than half—56 percent—that amount on the materials it consumes in the manufacturing process. In 2002 manufacturers spent $7.9 billion on materials, up 15 percent from $6.7 billion in 1997. Since most of the materials it consumes are petroleum derived, it is difficult for the carpet and rug industry to control costs.

Most of the $7.9 billion spent on materials was for nylon. More than half (61%) was spent purchasing nylon. In 2002 spending on nylon totaled $4.8 billion, up 20 percent from $3.8 billion in 1997.

Besides nylon, carpet and rug makers need materials to produce backing. In 2002 spending on backing was its next largest expenditure after nylon. Industry wide expenditures for backing materials, which are divided into primary and secondary backing materials, were $931 million. Primary backing materials are mostly polypropylene based. Secondary backing materials are mostly woven and nonwoven manmade materials, but also include jute, foam/rubber/latex, and plastics like vinyl and polyurethane.

Besides expenditures for highly valued nylon and all important backing materials, the next largest categories of materials expenditures are polyester and polypropylene. In 2002 total spending on polyester was $719 million and spending on polypropylene was $689 million.

Since such a high percentage of the raw materials used in the carpet and rug industry are derived from petroleum products—with the exception of comparatively small amounts of wool, silk, jute, and dyes—the top three car-

pet makers announced price increases effective June 2007. Shaw, for example, raised prices 5 to 6 percent because, as it told *Floor Covering Weekly*, "worldwide demand is driving costs on propylene, benzene, paraxylene and other raw materials impacting all carpet costs," and it received notice of increases from all its nylon, polyester, propylene, and latex suppliers. Mohawk raised prices 4 to 6 percent. Beaulieu increased prices 6 percent. The price of benzene, a key ingredient in the manufacture of nylon, increased 28 percent in the first five months of 2007 according to the May 14, 2007 issue of *Floor Covering Weekly*.

Since so many of the materials consumed in the carpet and rug industry are petroleum derived and cost control measures are difficult to implement, carpet and rug makers turned to recycling as a means of gaining control over costs related to petroleum derived raw materials. Mohawk, for example, owns a polyester recycling facility in Summerville, Georgia, known as Image. The facility recycles the long-chain synthetic polymer known as polyethylene terephthalate from carbonated soft drink and water bottles. Bottles are sorted, ground into fine chips, and then cleaned. Ground chips are melted, extruded into fiber, and spun into staple carpet yarn. Mohawk reports that plastic beverage bottles are made with top quality polyethylene terephthalate as required by the U.S. Food & Drug Administration, so its recycled spun polyester is superior to lower grades used by other makers.

Mohawk is one of the largest plastics recyclers in America. It has kept approximately 21 billion plastic bottles out of landfills since it acquired Image in 1999. Image recycles 3 billion bottles per year or 25 percent of all bottles collected in North America. The Image plant produces approximately 160 million pounds of recycled polyester spun fiber per year. Mohawk donated its EverStrand polyester carpet, made from fiber spun from recycled plastic bottles at Image, for the backstage celebrity greenroom at the 79th annual Academy Awards in 2007.

Shaw, too, is involved with recycling. It collects and recycles carpet nationwide and expects to collect 300 million pounds annually, the equivalent of a 12 foot roll of carpet more than 10,000 miles long. It has partnered with recycling companies to establish a collection network across the country. It uses the carpet at its Evergreen nylon recycling facility in Augusta, Georgia, to produce nylon 6, the raw material used to make nylon fiber.

DISTRIBUTION CHANNEL

The distribution channel for rugs and carpets is characterized by: (1) specialty/independent flooring stores, (2) home centers/building materials centers, and (3) building contractors. In April 2007, *Floor Focus Magazine* reported in detail on the distribution channel for over $54 billion of retail flooring—ceramic tile, carpet, laminate, vinyl, and wood—sold in the United States in 2006. Specialty/independent flooring stores led with 35 percent of retail sales. Home centers/building materials centers accounted for 24 percent. Reflecting the growing influence of the big box stores, Home Depot and Lowe's accounted for 16 percent of that 24 percent.

Floor Focus Magazine examined the total 2006 $54 billion retail flooring distribution channel in depth by breaking the industry into three distinct segments which it defined as the consumer residential replacement segment, the builder new residential segment, and the commercial mainstreet segment. A fourth segment called commercial-specified is a unique distribution channel handled by special distributors known as contract dealers.

In the consumer residential replacement segment, specialty/independent flooring stores led the distribution channel with 45 percent. Home centers/building materials centers accounted for 36 percent of retail sales. Home Depot and Lowe's accounted for 26 percent of that 36 percent.

In the builder new residential segment, specialty/independent flooring stores accounted for 38 percent of all flooring dollar sales, while building flooring contractors led the channel with 47 percent. Within the builder new residential segment, home centers/building materials centers accounted for 12 percent of retail sales. A shakeup in this segment occurred in July 2007 when Home Depot sold HD Supply, the largest builder flooring contractor in the country. In 2006 that firm alone accounted for 7 percent of all flooring sold to U.S. builders.

In the commercial main street segment, specialty/independent flooring stores once again led the channel with 38 percent of all flooring dollar sales, while home centers/building materials centers accounted for 20 percent.

The commercial-specified segment is somewhat different. Firms called contractors or contract dealers control 87 percent of flooring dollar sales in that sector. Other emerging nontraditional distribution channels are shop at home retailers.

While the Big Box stores have a growing influence on the market—Home Depot and Lowe's have a combined 3,325 U.S. stores—specialty/independent flooring stores led the distribution channel overall, controlling 35 percent of the total $54 billion retail flooring market. Specialty/independent flooring stores also led the distribution channel for residential replacement flooring with 45 percent of this important segment of the market. The position of the specialty/independent flooring stores within the distribution channel was strengthened in part by Abbey Carpet and Carpet One.

Abbey Carpet has 620 dealer franchises. In 2007 it changed its name to Abbey Carpet and Floor, reflecting the shift in the U.S. floor covering industry where the market share of carpet is diminishing due to the growth of other flooring choices like ceramic tile, laminate, and wood. Abbey also launched a new business called Abbey Floors at Home in direct response to the success of Empire Home Services, the shop at home retailer that grew from a $30 million local Chicago business in the 1990s to a $500 million national business by the middle of the first decade of the twenty-first century. Empire spends in the neighborhood of $10 million per year on its local television commercials that promise next day delivery on carpet and other flooring.

Carpet One started in 1985 as a 13 member co-operative in Atlanta, Georgia. By 2005 it had more than 1,000 stores that collectively sold $3.3 billion of flooring. The co-op was established to offer a helping hand to specialty/independent flooring stores tired of being dwarfed by the giants. The powerful co-op Carpet One is the single most important retail entity in carpet, giving it the muscle to negotiate prices and products with industry leaders Mohawk and Shaw.

Mohawk and Shaw dominate the flooring industry, together accounting for a sizeable 71 percent of the carpet business. Mohawk and Shaw each own distribution systems. Mohawk owns one of the largest distribution companies in America with a fleet of hundreds of trucks. Shaw created its own trucking subsidiary in 1982. By the turn of the current century it had close to 30 distribution centers.

A traditional distribution channel within the sizeable carpet and rug industry is the trade show. These include the International Rug Market in Atlanta, Georgia, the annual Surfaces show in Las Vegas, Nevada, and the Las Vegas Furniture Market show, also in Nevada. At the trade shows, the major players try to out-do one another with lavish parties and huge showrooms.

Rug sales are sometimes discussed as a category within the floor covering industry, the home furnishings industry, the textiles industry, and even the gift industry when it comes to welcome mats. For rugs, department stores are the largest distribution channel with a 25 percent share. Home centers/building materials centers were the second largest distribution channel with 16 percent.

KEY USERS

Key users of carpeting are those involved in either the process of updating the interior design of residential and commercial properties or those involved in building new such structures. The updated interior can be part of a remodeling or redecorating project. Other key users include large national or regional homebuilders such as D.R. Horton, the largest homebuilder in the United States, and Toll Brothers, the leading builder of luxury homes.

ADJACENT MARKETS

Remodeling, redecorating, and home improvement spending is an adjacent market to carpeting. The market for residential remodeling is driven by home values. As the value of a home increases, the owner's ability to obtain financing for home improvements generally increases as well. According to the U.S. Census Bureau, the U.S. median home value rose 32 percent from 2000 to 2005. This boom fueled consumer spending and much of that spending was for home improvements, including remodeling and redecorating to update the interior design. In 2006 and 2007 home prices dropped for the first time in years. Moody's Investors Service estimated that median existing home prices fell on average by 3.6 percent in the first half of 2007, and as much as 20 percent in some markets, notably the Detroit, Michigan, market.

New construction is another adjacent market to carpeting. Recognizing that new house construction drives carpet sales, in 1992 Shaw launched Home Foundations, a carpet program targeted at builders. The U.S. housing market was at record levels during the 2002 through 2005 period, buoyed by low interest rates and flexible lending practices. Housing starts—new residential construction—as well as sales of existing housing both set records during this period.

The market for wood flooring is also adjacent to carpet. The wood flooring market on a whole is dwarfed by the carpet and rug industry. U.S. wood flooring shipments represent 10 percent of the total value of U.S. shipments of rugs and carpets; in 2002 the Census Bureau reported wood flooring industry shipments of $1.56 billion compared to $11.75 billion for rugs and carpets. Forecasts were for carpet shipments to decline relative to other flooring choices like ceramic tile, laminate, vinyl, and wood. For instance, wood flooring doubled its market share at the expense of carpet between 1997 and 2004.

RESEARCH & DEVELOPMENT

Research and development on the part of the Carpet and Rug Institute resulted in the Carpet America Recovery Effort. The goal of Carpet America Recovery Effort is to divert 27 to 34 percent of used carpet from landfills by 2012. The Effort emerged in 2002 as a result of a Memorandum of Understanding for Carpet Stewardship, a national agreement signed by members of the carpet industry, representatives of state governments, the U.S. Environmental Protection Agency, and nongovernmental organizations.

To reach its goal of diverting up to 34 percent of used carpet from landfills by 2012, Carpet America Recovery Effort fosters market solutions and helps expand the used carpet collection infrastructure. It added 24 U.S. reclamation sites in 2006, building on its 16 preexisting collections site for a 2006 total of 30. Its June 2007 newsletter announced a total of 52 collection sites. Half of U.S. states have carpet reclamation facilities.

Carpet America Recovery Effort reported at its 2007 annual meeting that a total of 261 million pounds of used carpet was diverted from landfills in 2006. A decade earlier almost no carpet was diverted. The 2006 accomplishment represents a 16 percent increase in diversion over the previous year. In 2005, 225 million pounds were diverted, an increase of 108 percent over 2004. Of the 225 million pounds collected in 2005, 86 percent was recycled, while the rest was burned for energy.

It is anticipated that diversion for waste to energy will decrease as demand for recycled material increases. Used carpet can be reused in three ways. It can be used as carpet, generally as backing material or padding. It can be turned into something else such as composite lumber for use as decking, railroad ties, and marine timbers. Lastly it can be burned as fuel, with a Btu (British thermal unit) value by weight approximately the same as coal.

Recycled carpet is generally commercial and involves circumstances where the cost to haul used carpet to the reclamation facility is less than the landfill disposal fee. Often used carpet goes to a sorter where the backing is removed and fibers separated. The sorter sends it to a processor, where the material is cleaned and broken down so that it is pure and usable. Processors sell to manufacturers. Some businesses cover all three processes, such as LA Fibers, the largest carpet recycler in the world.

Interest in recycling carpet—a product consisting primarily of nylon—grew as oil prices increased. The year 2007 was pivotal. Reflecting the increased demand for the nylon in carpet, Shaw re-opened its Evergreen nylon recycling plant in Augusta, Georgia, which had been idled for almost five years due to lack of demand.

Evergreen was built in 1999 to reduce nylon to caprolactam, its raw material, and then make new nylon from the broken down elements with negligible loss. Shaw acquired 50 percent of Evergreen in its 2005 purchase of Honeywell's fiber business, then bought DSM Industries' 50 percent share of the plant in 2006. Shaw re-opened the plant in 2007 due to increased raw material costs and a marketplace with a better understanding of the value of recycled materials.

Since the plant operates as a sorter and a processor, Evergreen has to collect close to 300 million pounds of carpet to get 100 million pounds of nylon. Shaw gets double duty out of its trucking subsidiary that hauls new carpet out and reclaimed carpet it picks up from its dealers back to Georgia, the world capital of carpet. Its initial goal is to produce 30 million pounds of caprolactam annually, which will yield about the same volume of nylon fiber. Evergreen has a capacity of 100 million pounds. All of the nylon produced at Evergreen will be used by Shaw to make nylon carpet.

CURRENT TRENDS

The industry keeps up with fashion and interior design trends. It introduces new colors, patterns, and textures of carpets and rugs every year to keep up with trends. Wool is a style and trend setter that other fibers are used to copy. For that reason, the Wools of New Zealand annual carpet and rug color forecast is eagerly awaited. Its 2008 international carpet and rug color forecast is called *The Colour of Wine–Fine Living for a Fine Palette*. Colors for the 2008 palette for wool carpets were inspired by the art of viticulture. Colors were derived from vineyard landscapes, the various varieties of grapes, and the finest wines of New Zealand. Its collection of New Vintage Carpets for the Connoisseur includes new soft shades of white that were not cold and clinical, but rather both warm and welcoming, as well as funky and futuristic.

One industry trend is to offer coordinated products. Carpet makers are offering coordinated runners and area rugs. A consumer can purchase roll runners for stairs and area rugs that match newly installed carpeting. This allows consumers to coordinate carpet with other rugs in the house. Some manufacturers are taking this cross-category coordination one step further and include woven bedspreads, tapestries, pillows, throws, and window blinds.

Another industry trend is licensing agreements. Shaw Living led the licensing charge, with products for almost every type of consumer—from children to golf fans—through licensing agreements with Kathy Ireland, Mary-Kate and Ashley Olsen, Jack Nicklaus, Tommy Bahama, and many others. In October 2006 Shaw showed new rug designs for its licensed Tommy Bahama Collection called Squared Away.

TARGET MARKETS & SEGMENTATION

Mohawk and Shaw target young consumers with lower priced products with plans to capture loyalty and eventually move them into the higher end. Karastan, owned by Mohawk, introduced its Studio Collection of area rugs in synthetic materials—an area it normally eschewed—with the idea that early lower price point purchases will lead the consumer to aspire to eventually owning its high-end products. Shaw took a similar approach with its mary-kateandashley collection. The collection initially targeted

the juvenile market with bright and fun rugs. In 2005 it unveiled six new designs that target not only young adults but anyone interested in a fashionable new rug.

Manufacturers target commercial carpet segments like corporate, hospitality, education, and retail. The commercial segment is active because carpet has a well-recognized 5 to 7 year life span so it must be replaced relatively often.

Carpet companies target commercial carpet segments with modular carpet tiles. Modular carpet allows damaged or stained carpet tiles to be replaced quickly and cost-effectively, resulting in a better overall appearance and the ability to maintain an always-clean appearance. Because carpet tiles are not sold in rolls they are less flexible with a firmer backing material, making them better wearing. Shaw debuted its Create Your Own Designer Rug carpet tile line at the April 2007 textiles market in New York. The solid carpet tiles are available in 12 different colors.

Retail and grocery stores, lifestyle centers, enclosed shopping malls, and the education sector generally are all targeted as potential users of modular carpet. Teachers prefer carpet on classroom floors because it absorbs noise, making for quieter classrooms. Makers predict the education sector will be one of the fastest growing markets for modular carpet according to a 2005 survey by *Floor Focus* magazine.

RELATED ASSOCIATIONS & ORGANIZATIONS

American Fiber Manufacturers Association, http://www.fibersource.com

Carpet America Recovery Effort, http://www.carpetrecovery.org/index.php

The Carpet and Rug Institute, http://www.carpet-rug.org

Oriental Rug Importers Association, http://www.oria.org

World Floor Covering Association, http://www.wfca.org

BIBLIOGRAPHY

"1985–2005 Carpet One: Two Decades of Making a Difference in the Flooring Industry." *National Floor Trends.* July 2005, 17.

Bittner, Rick and Mehrdad Yazdani. "Carpet Elevates Image of 21st-Century Learning Environments." *School Planning and Management.* December 2005, 21.

"CARE Reclamation Effort Rocks." *Floor Covering Weekly.* 21 May 2007.

"Carpeting Success: Beaulieu Group has Become the Third-largest Carpet Manufacturer in the World." *US Business Review.* January 2006.

"Carpets and Rugs: 2002." *Current Industrial Report.* U.S. Department of Commerce, Bureau of the Census. December 2004.

"Carpets and Rugs: 2006." *Current Industrial Report.* U.S. Department of Commerce, Bureau of the Census. June 2007.

"Carpet Use Increasing: Retailers Find Shoppers Linger When they Have Happy Feet." *Chain Store Age.* July 2006, 96.

Corral, Cecile B. "Expectations Rise for Rugs." *Home Textiles Today.* 16 January 2006, 1.

———. "Expo Design Alters Rug Mix: Retailer Heeds Customers' Calls for Roll Runners and Larger Area Rugs." *Home Textiles Today.* 8 January 2007, 10.

"Focus 100 Manufacturers." *Floor Focus Magazine.* May 2005. Available from <http://www.floordaily.net/features/Feat100Man0505.htm>.

Frank O'Neill. "Distribution Evolution." *Floor Focus Magazine.* April 2007.

Gavin, Kimberly. "Nourison Opens Calhoun, Georgia-based Plant." *Floor Covering Weekly.* 4 June 2007.

Helm, Darius. *Focus 100 Retail.* November 2006. Available from <http://floordaily.net/focus_article.aspx?article=9985>.

"Homebuilder Sentiment Hammered." *Focus 100 Retail.* 17 July 2007. Available from <http://floordaily.net/newsarticle.aspx?article=10848>.

Mancini, Rosamaria. "A Tween and Hungry Look: Manufacturers are Tailoring their Products to Reach both Young and Old." *HFN The Weekly Newspaper for the Home Furnishing Network.* 8 August 2005, 30.

———. "Carpet Tiles are Again Flying Off the Shelves." *HFN The Weekly Newspaper for the Home Furnishing Network.* 6 June 2005, 16.

"Maples Adds to its Stainless Reputation in the Industry." *Home Textiles Today.* 4 October 2004, 10.

"Mohawk Closing Tufting Operations at Lavender Drive." *Floor Covering Weekly.* 4 June 2007.

"Mohawk to Acquire Wood Flooring Assets." *Home Textiles Today.* 9 July 2007, 13.

Quail, Jennifer. "The Polypropylene Paradox: Synthetic Rugs Helped Alter the Marketplace, but Economic Factors are Keeping Wool in the Loop." *HFN The Weekly Newspaper for the Home Furnishing Network.* 13 December 2004, 36.

"Report: Moderate Growth Ahead; Researcher for Carpet One Predicts a 4.3 Percent Spurt by 2009." *HFN The Weekly Newspaper for the Home Furnishing Network.* 8 August 2005, 32.

"Shaw Living Doubles Atlanta Space, Signs 9-year Lease." *Home Accents Today.* 1 August 2006, 14.

Torcivia, Santo. "The Home Center Challenge." *Floor Focus Magazine.* April 2007.

SEE ALSO *Resilient Floor Coverings, Vacuum Cleaners, Wood Flooring*

Caskets

———■———

INDUSTRIAL CODES

NAICS: 33–9995 Burial Casket Manufacturing

SIC: 3995 Burial Casket Manufacturing

NAICS-Based Product Codes: 33–99951, 33–999511, 33–99951101, 33–999512, 33–99951206, 33–99953, 33–999531, 33–99953101, 33–99953106, 33–99955, 33–999551 and 33–99955100

PRODUCT OVERVIEW

A casket, or a coffin, is a box that holds a deceased body. The term casket is thought to be a North American euphemism, for a casket properly defined is a box to hold jewelry. The word coffin has a similar meaning, taken from the Greek word for basket. The terms casket and coffin are often used interchangeably. However, they denote boxes of different design. Coffins typically have tapered edges and tend to be hexagonal or octagonal in shape. Caskets are more rectangular—a standard box. The lid is also split in such a way that allows the deceased to be viewed during a visitation, memorial service, or funeral if so desired by the family or in a will.

The casket industry has its roots in the woodworking traditions of the early 1800s. The funeral director was then known as the undertaker. The undertaker was also the local furniture maker. They had the skills and tools to manufacture caskets, and did so on an as-needed basis. By the late 1800s casket manufacturing developed as a

distinct business and the manufacturers devoted their efforts to the production and sale of coffins and caskets.

Caskets are typically made of wood, metal, or cloth. Cloth-covered caskets were the best-selling types of casket for decades. In 1950 more than half of all caskets sold were cloth-covered. Such caskets are made of particleboard or cardboard and are finished with a cloth covering. They are simple in construction and are shape and the least expensive form of casket available.

There are four major types of metal caskets: sealed steel, unsealed steel, copper/bronze, and stainless. Sealed steel (also known as *gasketed*) caskets are constructed from 16-gauge, 18-gauge, and 20-gauge metal (16-gauge is the heaviest, 20-gauge the lightest). A rubber gasket is used to seal the casket at the point of closure. Sealed caskets have continuous welds at seams and in the corners. Various glues and epoxies are used to guarantee a tight, durable seal. Thicker-gauge caskets may have more complicated shapes, such as rounded or urn-shaped corners. Lighter gauges are more likely to have traditional square edges. Sealed caskets represent approximately three-quarters of all steel unit sales.

Non-sealed steel gaskets are less expensive than their sealed counterparts. They are made of the thinnest-gauge steel. They do not have the continuous welds at seams and corners, as do the sealed gaskets. It is interesting to note that this is a product category in which the top two manufacturers do not compete head-to-head in all major product categories. York manufactures both sealed and unsealed gaskets. Batesville, the largest casket maker, produces only sealed models.

Stainless steel caskets are more expensive than regular steel caskets. They are almost always sealed. They may have rounded or square edges. Copper or bronze caskets are the most expensive types of caskets available, although some exotic woods are used to make caskets that are also very expensive. These caskets also allow for finishes with more luster than other metal versions.

Hardwood caskets include pine, cottonwood, maple, cherry, hickory, and mahogany. The wood has a glossy finish and may be hand polished. This, of course, adds to the cost. Wood caskets do not have a sealing mechanism and are known as nonprotective caskets.

MARKET

About 1.8 million caskets are sold in the United States each year. This represents approximately $1.4 billion in casket sales per year based on retail prices through funeral homes and casket stores.

The value of all shipments of caskets in the United States is an accounting of the wholesale value of caskets and does not include retail mark-ups or bundled funeral service fees which are included in the $1.4 billion annual sales estimate for all caskets. The value of manufacturer's shipments is provided by the U.S. Census Bureau in its *Annual Survey of Manufactures*. In 2005 shipments of burial caskets were valued at $860.2 million. The 2005 shipments were slightly higher than those recorded in 2004, when they were valued at $843.7 million. Total casket and coffin shipments have been declining slowly since 1997, when total shipments were valued at $1.17 billion.

Metal caskets and coffin shipments were valued at $490.5 million in 2005, or 57 percent of total shipments. In 2002 total shipments were $778.6 million and metal caskets were 66 percent of this total. Wood caskets and coffin shipments were valued at $184 million in 2005, or 21.3 percent of total shipments. Shipments of wood caskets and coffins were $273.2 million in 2002. Wood caskets and coffins represented 23 percent of total shipments in that year. Both of these categories refer to completely lined and trimmed caskets and coffins for adults. Other types of caskets and coffins shipments, valued at $125.8 million, include burial boxes and vaults (except concrete and stone), casket shells, casket shipping containers and cases, and caskets and coffins for children. This category has been increasing steadily since at least 1997, when shipments were $65.8 million.

Cloth-covered caskets were the most popular types of caskets for many years. In the late 1950s sheet metal became more readily available, thus production of steel caskets could be increased. In 1960, consumers began to favor steel caskets for the first time. According to the Casket and Funeral Supply Association, it was during

the 1960s that steel surpassed cloth-covered caskets, with 44 percent to 34 percent market shares, respectively. By 1980 the preference for metal caskets peaked. Steel took a 69 percent share of the market. Wood caskets seem to be gaining in popularity again. Hardwood was thought to have represented 17 percent of all caskets sold in 2000, up from 12 percent in 1980. Steel represented 60 percent of all sales in 2000, cloth covered 14 percent, copper/bronze 3 percent, and stainless steel 4 percent. Children's caskets accounted for 2 percent of sales since 1990.

Part of the reason for the rise in hardwoods is that consumers see them as more attractive looking. Such wood may also be perceived as more natural. Such perceptions are related to the rise in cremations. Noticing this trend, casket manufacturers have also marketed wood caskets specifically designed for cremations.

KEY PRODUCERS/MANUFACTURERS

The first casket producers were small companies serving local markets. It was not difficult for a company to manufacture the cloth-based caskets that were popular during this period. Such caskets could be manufactured with saws, drills and other general-purpose machinery. However, the metal caskets that became popular in the late 1950s were more difficult for companies to manufacture. Working with metal meant significant capital costs in metal stamping, bending, cutting, and grinding equipment. It also meant greater investments in painting and finishing systems.

These factors helped drive industry consolidation. Another factor influencing the industry was the fact that casket demand in North America has remained steady at approximately 1.8 million units in recent years. Some analysts even forecast a decline in sales as more people choose cremation. Without increased demand, the main opportunity for growth for a company was to take market share from other casket makers.

Data from the U.S. Census Bureau detail the change in the industry. In the early 1950s there were over 700 casket manufacturers with more than 20,000 employees in the industry. By 1967 the Census Bureau reported 523 establishments in the industry with 16,800 employees. In 1992 the establishment count was down to 211 and employed stood at 7,800 persons. Five years later there were 173 establishments employing 6,792 persons. Further declines were experienced thereafter and by 2002, the Census reported 164 establishments employing just 5,069 persons.

Batesville Casket Company. This company is the largest casket maker in the United States, representing almost half of all casket sales. It is headquartered in Batesville, Indiana. The company can trace its roots back to John

Hillenbrand who began making wood caskets in 1884. In 1904 Hillenbrand purchased the Batesville Coffin Company, and renamed it Batesville Casket Company. The company manufactures all types of cloth-covered, hardwood, metal, and copper/bronze sealed (gasketed) caskets. It also manufactures related products, such as urns, containers, and other memorialization products used in cremations. It is a division of Hillenbrand Industries, which reported having 9,800 employees and sales of $1,962.9 million for the fiscal year ending September 30, 2006.

The York Group. Second largest of the casket makers in the United States, The York Group began as The York Wagon Gear Company. It manufactured wooden carriages, but in 1892 switched to manufacturing wooden caskets. In the late 1990s the company acquired a number of casket companies as part of an aggressive expansion plan. They also acquired firms that manufactured goods related to the funeral industry, such as urns and memorials. The York Group's parent company, Matthew International, employs 4,000 people and reported sales of $715.9 million for the fiscal year ending September 2006.

Aurora Casket Company, Inc. This company is the third largest casket maker in the United States. It is also the largest privately owned casket company and is headquartered in Aurora, Indiana. The third largest casket maker employs 700 people, many of whom are descendents of previous workers. Colonel John J. Blackman founded the company in 1890. It manufactures approximately 150,000 caskets annually. It also makes urns and grave markers.

These three firms dominate the industry. Each has acquired competitors during the 1990s, a period of great consolidation within the industry. The York Group acquired Sacramento Casket Co. and Houston Casket Co. In 1998 the Aurora Group acquired Mountain States Casket Co. and J&B Casket Co. It acquired Clarksburg Casket Co. in 2000 and Hastings Casket Co. in 2003. In 1993 Batesville acquired casket makers in Canada and Mexico to lock in its control of the North American market. These companies also acquired urn, wood, and metal manufacturing companies.

The top three firms represent nearly 80 percent of casket sales based on units and dollars. Other manufacturers include Southern Heritage, Goliath Caskets, Astra Caskets, Freeman Metal, and Victoriaville.

MATERIALS & SUPPLY CHAIN LOGISTICS

According to the U.S. Census Bureau, casket makers consumed $360.4 million worth of materials in 2002, up from $325.6 million in 1997.

The most expensive materials consumed by casket manufacturers are the woods and metals needed for the production process. Bronze is the most expensive metal, and is preferred by the casket industry due to its strength and natural ability to resist rust. Copper is comparable to bronze, but is a less expensive material. Stainless steel has a higher tensile strength than either bronze or copper and is also a naturally rust resistant material. Metal is also necessary for handles and other hardware.

Rough and dressed lumber is needed for the construction of the casket or the outer shell. Wood is becoming a popular choice for caskets again. This will mean an increased need for hardwoods such as pine, mahogany, and oak. In the past, consumer selection of wooden caskets over metal caskets has been governed by regional preferences, with rural areas being more likely to purchase wooden caskets. Urban areas have traditionally had higher sales of metal caskets.

Manufacturers also use paints, stains, and lacquers on the outer shell of the casket to give it as attractive an appearance as possible. Some wood finishes are applied by hand, which adds to the cost. Casket hardware consists of cast and forged metals and formed plastics. Interior fabrics are usually made of materials such as cotton, satin, wool, or velvet.

DISTRIBUTION CHANNEL

For many years, funeral homes were the only source for caskets for the consumer. There were several important government regulations that opened the door for change.

In 1984 the Federal Trade Commission (FTC) passed a series of regulations known as the Funeral Rule. This set of funeral industry rules was established to protect the grieving families of the recently deceased from being easily taken advantage of by funeral homes. The rule required funeral homes to disclose their prices for goods and services with potential clients in a general price list. This allows consumers to compare prices for various services among funeral homes. Funeral homes must also disclose these prices over the phone. In 1994 the FTC also prohibited funeral homes from charging handling fees for caskets purchased from a source other than the funeral home. The new rules offered other protections for consumers, such as forbidding funeral directors from insisting that embalming is necessary for deaths (it is not) and from touting the preservative qualities of a casket (decomposition of the body will still take place).

The new legislation shed light on the fees charged by the funeral homes. In 2007 the average casket was $2,000, and represented half the cost of a modestly priced funeral. The funeral plot cost another several thousand dollars. The casket is often priced well above its wholesale cost to provide revenue for the funeral home. The industry drew criticism because many funeral homes charged high fees for handling a casket they had not provided.

With the new rules prohibiting such activities, the market was effectively opened to third-party casket retailers. There are an estimated 300 casket stores in the United States. These discount casket dealers may operate out of an actual storefront or an electronic storefront (a Web-based retail site). Casket retailers have names such as CasketXpress and Funeral Depot and sell caskets for a fraction of the funeral home cost. But even larger retailers are getting into the market. In December 2004 popular discount retailer Costco started selling caskets in its stores.

Many funeral homes have accepted the presence of these third-party retailers. They do not discourage consumers purchasing caskets from such outlets. Some funeral directors work with consumers, encouraging them to inspect the caskets they purchase through these outlets, for example. Often the funeral homes have prices that are competitive enough that the discount from the discount retailers is negligible. These discount retailers represent only a fraction of all caskets sold annually. In other words, most consumers still purchase caskets from funeral homes.

Because the casket industry is not seeing increasing demand, manufacturers have focused on customer service. The three major casket makers all have their own distribution networks. When a funeral director calls one of them with an order, the casket is typically delivered within 24 hours. The third-party retailers function in much the same way. These companies often partner within someone to aid in the transportation of the casket. A consumer browses the Web site of Funeral Depot, for example, and selects a casket. Funeral Depot calls one of its partners to verify the availability of the casket from the manufacturer. The manufacturer then drop-ships the casket to the partner firm. The partner firm delivers the casket to the funeral home for a fee or it holds the casket until Funeral Depot can arrange to pick it up.

Critics of the funeral industry have noted that many funeral homes participate in subtle but manipulative marketing techniques, such as presenting inexpensive caskets in unattractive colors in their showrooms. The more expensive caskets, in turn, are made to be more appealing with handsome finishes or are shown in better displays. The casket industry is, in short, highly competitive. In 2005 Hillenbrand Industries (owner of Batesville, the largest casket maker) was named as a co-conspirator in a price-fixing lawsuit. The lawsuit accused the top three funeral home operators (Alderwoods Group, Service Corp., and Stewart Enterprises) and Hillenbrand of conspiring to sell only Batesville Casket models. Customers were also forced to pay inflated prices. The lawsuit also alleged Hillenbrand was forced to ban casket sales to discount chains, independent stores and Internet retailers. The lawsuit, which had yet to make its way through the courts as of early 2007, was the first of its kind to pit funeral homes against discount casket dealers.

KEY USERS

Demand for caskets has been steady at aproximately 1.8 million units annually. Some analysts even estimate a shrinking demand.

There are several factors that help explain this. The number of deaths is increasing modestly, with annual deaths in the United States at 2.5 million. The population is aging, but is also tending to live longer. Health care has improved. Batesville noted in its most recent report that the most recent flu seasons have been mild and resulted in few deaths. Also total deaths each year include body donations and casketless cremations.

Demand is affected by an increasing preference for cremations over traditional burials. The cremation rate began to increase slightly in 1960. In that year there were 60,987 cremations in the United States, equating to 3.56 percent of all deaths, according to the Cremation Association of North America. In 1980 the rate had climbed steadily to 9.72 percent of all deaths. A year later it was over 11 percent of all deaths. By 1994 the rate had nearly doubled to 21 percent. In 2005 there were an estimated 778,025 cremations, or 32 percent of all deaths. The rate varies by state, of course. In Arizona, Hawaii, California, Montana, Oregon, Washington, and Colorado over half of all consumers chose cremation over traditional burial (Washington's rate was the highest at 64.01 percent in 2005). By 2010 the projected cremation rate in the United States will be 33.7 percent of all deaths.

For a direct cremation a cardboard box is normally used. Those who wish to have a funeral visitation (sometimes called a viewing) or traditional funeral service will use a coffin of some sort. But it is also possible to rent a regular casket for the duration of the service. These caskets have a removable bed and liner, which is replaced after each use. There is also a rental casket where there is an outer shell that looks like a traditional coffin. The deceased is placed in a cardboard box that fits inside the shell. At the end of the service the inner box is removed and the deceased is cremated inside this box.

ADJACENT MARKETS

There is, of course, a strong relationship between casket makers and the funeral industry. For a period of time some casket makers had their facilities close to cemeteries and funeral homes. But with demand for caskets basically static, the large casket makers have had to develop new relationships with the 22,000 funeral homes, 115,000 cemeteries, and 1,155 crematoriums in the United States.

Casket manufacturing companies have begun to manufacture other products related to the funeral industry. Many manufacture urns, memorials, display furniture, hardware and mortuary equipment. Matthew International, owner of the York Group, is the top cremation equipment maker in North America. It also builds mausoleums and is the country's top manufacturer of urns, bronze memorials, and commemorative plaques.

Funeral homes are the largest market for casket makers. Most funeral homes are family run operations and have been in operation an average for 47 years. Service Corp. International is the largest corporate provider of funerals in the United States. Alderwoods Group, Stewart Enterprises, and Carriage Services are the other leading funeral service providers. The top five chains represent approximately 20 percent of the market.

RESEARCH & DEVELOPMENT

The casket industry is very competitive, and new models and products appear regularly at annual trade shows. New models have increased in recent years, which is an interesting development, as models tend to stay in the market for 10 to 15 years.

Batesville recently advertised the merits of its new line of wood caskets. The veneer process offers improved structural integrity and eliminates the natural defects that are found in wood. The new veneer process also uses 95 percent of the wood, which cuts down on the waste associated with more traditional veneer processes.

The York Group recently got its new write-on casket model to market. The write-on model allows mourners to write final goodbyes to the deceased directly onto the casket. York's most significant development in recent years was its York Management System in 1997. Previously, consumers would select a finished casket from a funeral showroom. The new system featured partial sample caskets, displays of lining material, and a selection of handles and other hardware. Some of this was to ease what is a difficult task for consumers. But the prevailing theory was that when given a wide range of choices, consumers would gravitate toward the higher-priced goods.

This theory proved correct. One funeral director was quoted as saying it was the first new innovation in casket manufacturing in 75 years. Considering the sales figures, other funeral directors surely agreed with him. York's first 50 displays showed an average gain of $438 per sale under the new system. More than 500 funeral homes were using the York system at the end of the first decade of the twenty-first century.

CURRENT TRENDS

The most interesting development in the industry is the rise of green burials. The term generally refers to cremations or full-body burial with no embalming fluids and a biodegradable wooden box or shroud. Many proponents of green burials feel that the cost of a casket and traditional funeral are prohibitive. Some people choose to be buried in a meaningful spot. For example, some people have had their remains sunk into coral reefs. In another unusual example, James Doohan, "Scotty" of the television show *Star Trek*, elected to have his remains shipped into space.

A casket may be custom-designed. It might be airbrushed or have a specialty finish, for example. Some people are choosing caskets modeled into specific shapes, such as guitar cases. One company is experimenting with computer-generated murals placed on the casket. A few individuals have confessed to sleeping in coffins. Some of these people are being deliberately provocative, but there is among this group of people a small subset who genuinely believe themselves to be vampires.

In an increasing number of cases, specialized caskets are a necessity. As obesity rates increase in America, so have the demand for oversized caskets. For years, caskets were built with a standard inside shoulder width of 22 inches to 24 inches. Some companies are now producing caskets with widths from 28 inches to 44 inches. Batesville introduced a line of oversized caskets in October 2004 that they report have been very successful. Southern Heritage, which makes only steel caskets, reports oversized caskets now represent 20 percent of its sales. Goliath Caskets, who specialize in plus-size caskets measuring up to 52 inches in width, recently claimed to be "very, very busy." They manufacture 600–800 oversized caskets each year.

In short, more people are customizing their funerals. The funeral industry has been quick to respond to customers' desires for individualized memorial services. There are now funeral coordinators to organize and stage the service. Music that had meaning to the deceased is played during the service. One can arrange for photos and movies of a loved one to be put on a DVD and distributed to mourners. A funeral service might be broadcast live on the Internet for family and friends to watch if they can't be there in person. The Hollywood Forever Cemetery in Los Angeles, resting place of many Hollywood stars, is a high-profile example of changes in the industry. They host movie nights and walking tours of the grounds. They also offer *digital memorialization* packages—film, audio, and biographical information of the deceased—that range in price from $400 to over $4,000. As Brent Cassidy, one of the funeral directors notes, "Instead of just going to the cemetery and saying to your daughter, 'Well, here's your grandfather—I really wish you could have met him,' you can actually introduce her to him."

Even the image of funeral directors is changing. The television show *Six Feet Under* ran from 2001–2005 on HBO. It featured a family of funeral directors and received many Emmy awards. In old movies, the town undertaker was often seen as a sinister fellow dressed all in black. This has changed greatly. Approximately half of those graduating from funeral service programs were women in the first decade of the 2000s, and the average age of graduates was the late 20s. Two-thirds of graduates had no prior direct family relationship with the funeral service business. This new breed of funeral director even has sex appeal. The first *Men of Mortuaries Calendar*—featuring good-looking funeral home directors—was published at the beginning of 2007.

TARGET MARKETS & SEGMENTATION

Caskets still remain the preferred method of burial. This will be the case for some time to come. People still like the traditional funeral, with a coffin and flowers in the front of a church. Cremation is gaining in popularity, true. But some people do find it distasteful. It goes against the religious or cultural beliefs of others. The type of casket purchased by the consumer will vary. Casket makers may have to follow funeral directors and offer more personalized services: more high-end, expensive hardwood caskets, or caskets composed of more environmentally friendly materials in coming years.

RELATED ASSOCIATIONS & ORGANIZATIONS

Casket and Funeral Supply Association, http://www.csfa.org

Cremation Association of North America, http://www.cremationassociation.org International

Conference of Funeral Service Examining Boards, http://www.cfseb.org

National Casket Retailers Association, http://www.casketstores.com

National Funeral Directors Association, http://www.nfda.org

National Funeral Directors and Morticians Association, http://www.nfdma.

BIBLIOGRAPHY

"About the Casket Industry." Casket & Funeral Supply Association of America. Available from <http://www.cfsaa.org/about.php>.

Annual Survey of Manufactures 2005. U.S. Department of Commerce, U.S. Bureau of the Census. November 2006.

Barol, Bill. "Death & The Salesman Irreverence and Death Don't Mix." *CNN Money.* 1 March 2004.

"Burial Casket Manufacturing: 2002." *2002 Economic Census.* U.S. Department of Commerce, Bureau of the Census. December 2004.

"Caskets Growing as Waists Do Too." *QC Times.* 30 April 2005.

Chadderdon, Lisa. "The Customer is Always Dead." *Fast Company.* November 1999.

Donhardt, Tracy. "Casket Maker Named in Price Fixing Lawsuit." *Indianapolis Business Journal.* 9 May 2005.

Evanoff, Ted. "Hillenbrand Caught Up in Casket Suit." *Indianapolis Star.* 5 May 2005.

"Frequently Asked Questions." American Board of Funeral Service Education. Available from <http://www.abfse.org/faq.html>.

Graham, D. Douglas. "Overpaying is Not Dignified." *Multichannel Merchant.* 1 May 2003.

Lubove, Seth. "Six Feet Under." *Forbes Global.* 31 October 2005.

Rafater, David. "Big-Box Wholesaling." *Business Week.* 18 November 2004.

Saltzman, Steven and Joshua N. Rosen. "The Death Care Industry." *The Death Care Industry.* 9 January, 1998.

Valifria, Lori. "Green Burials Often Unique, Less Costly." *National Geographic News.* 9 September 2005.

Cellular Phones

———————■———————

INDUSTRIAL CODES

NAICS: 33–4210 Telephone Apparatus Manufacturing, 33–4220 Communication System and Broadcast and Studio Equipment Manufacturing

SIC: 3661 Telephone Apparatus Manufacturing

NAICS-Based Product Codes: 33–422012, 33–42201201, 33–42201204, 33–42201207, and 33–42201209

PRODUCT OVERVIEW

A cellular phone is a handheld, wireless receiver that is connected to a nearby radio transmitter and receiver station. This wireless connection allows the user to move about freely, and the user may place and receive telephone calls. More expensive versions of this device allow the user to check email, store photos, play music, and browse the Internet. Cellular phones (also called cell phones, mobile phones, wireless phones, and handsets) resemble the receiver on a standard, analog phone. There is a mouthpiece, ear piece, and keypad. Display screens are an important feature of the cellular phone used for inputting information to be recorded and for other data management purposes. Cellular phones are also much smaller than traditional telephone receivers.

How Cell Phones Work. Cellular phones operate with radio frequencies, a form of electromagnetic energy located on the electromagnetic spectrum between FM radio waves and the waves used in microwave ovens, radar, and satellite stations. The geographic region served by a cellular system is subdivided into areas called cells. Such cells are supposed to be uniform regions but in fact frequently overlap. Each cell has a central base station and two sets of assigned transmission frequencies; one set is used by the base station, and the other by mobile telephones. Each area or cell contains an antenna. When a person moves within the network area carrying a cell phone, the cell phone keeps contact with the local antenna.

Each cell has its own base station, which is identifiable by its transmitting and receiving antenna located on a tower at the top of a hill or building. When a call is placed from a cellular phone, a signal is sent from the cell phone antenna to that cell's base station antenna. A special frequency, the control channel, is used when cell phones and base stations communicate. The base station sends a System Identification Code (SID) to the cell phone over this control channel. This code, a unique five digit number assigned by the Federal Communications Commission (FCC), identifies the cell phone carrier in that cell. If the SID sent from the base station is the same as the SID programmed in that phone, then the phone is using its home carrier's cell. If the SIDs do not match, then the phone is roaming, using another carrier's cell.

In addition to the SID, the cell phone antenna transmits a registration request. This request is sent to the Mobile Telephone Switching Office (MTSO). The MTSO tracks the phone's location in a database. It then assigns a frequency pair in that cell with which the phone will communicate with another phone. When the phone moves close to the edge of a cell, the base station in the cell in which the phone is in identifies the signal as getting weaker and, through the MTSO, coordinates with the base station into which the phone is moving to signal to the phone to switch frequencies to those that will work in the new cell. This coordinated switching allows users of cell phones to communicate with each other while traveling from cell to cell and city to city.

History. The first cellular phone was available from Motorola in 1983. However, the first integration of radio and telephone communications took place shortly after the turn of the twentieth century. Bell Labs (which would one day be part of AT&T) created the first version of a mobile, two-way, voice based radiotelephone. The first ship-to-shore radiotelephones were used in 1919. A radiotelephone was installed in a Detroit Police Department police car in 1921; the technology would be introduced into other dispatch services in later years. In Europe, radio-telephone communications were also used on first-class passenger trains about 1926.

Bell Labs claims the first truly mobile telephone call was made in St Louis, Missouri, on June 17, 1946. The systems operated under the push to talk format. A user selected a channel, used the push button feature to speak, and had to be connected by an operator. By 1948, wireless telephone service was available in almost 100 cities and highway corridors. Customers included utilities, truck fleet operators, and reporters. There were only about 5,000 customers in the United States at the time making 30,000 weekly calls.

The wireless network could not handle large call volumes. A single transmitter on a central tower provided a small number of channels for an entire metropolitan area. A handful of receiver towers handled the call return signals. No more than three subscribers could make calls at one time in any city. The service cost $15 per month, plus 30 to 40 cents per local call, and the equipment weighed 80 pounds.

In December 1947 W. Rae Young and D. H. Ring, Bell Labs engineers, proposed a system to support a network of wireless phone systems throughout the country. They proposed regions be divided into a series of hexagonal cells; these regions would be comprised of multiple low-power transmitters spread throughout a city. There would be automatic call handoff from one hexagon to another and reuse of frequencies within a city. Phil Porter, also of Bell Labs, felt that the cell towers should be at the corners of the hexagons rather than the centers and have directional antennas that would transmit/receive in three directions. However, the technology at the time was unable to support such a move (that system, it should be noted, would eventually be adopted).

That same year, 1947, AT&T proposed that the FCC allocate a large number of radio-spectrum frequencies so that widespread mobile telephone service would become feasible. By doing so, the FCC would provide incentive for telecommunications companies (namely AT&T, historians have pointed out) to research the new technology. The FCC decided to limit the amount of frequencies available in 1947. Their limitations allowed only 23 phone conversations possible simultaneously in the same service area. These frequencies were used for emergency services such as police, fire, or medical emergency.

1950s through 1980s. By the 1950s, the electronics industry was starting to advance. Transistors and miniaturization had progressed to the point that a radio might be carried in the palm of one's hand, for example. In 1956 the first car phones were put into use in the United States (the system had been used in Sweden a few years earlier). The systems were bulky and an operator was needed to switch calls. A 40-pound transceiver was typically stored in the trunk. By 1964 customers could dial phone numbers directly from their cars, which removed the need for push-to-talk operators.

By 1967 mobile phone technology was available throughout the country; however, the user had to stay within one cell area. Base stations could not yet hand a phone call off from one cell tower to another. This problem was solved by Amos Edward Joel (again, an engineer from Bell Labs), who developed a method for a tower to hand off a phone call to another tower while keeping the user connected. Users could now move from cell to cell and not lose a phone call.

In 1968 the FCC initiated an inquiry on whether to allocate part of the TV UHF band (channels 70–83) to mobile services. AT&T submitted a proposal, detailing the hexagonal cell plan and describing how phone calls might be handed off from tower to tower as a caller moved about with his portable handset. The FCC proposal met with considerable resistance from television and private radio (police, ambulance) broadcasters, who were unwilling to lose part of their broadcasting spectrum. There were some lawmakers who were uneasy with AT&T's powerful influence in the market. The FCC would not resolve the matter for roughly another decade, when the FCC made cellular service possible to the public in March 1982. In that agreement 20 megahertz each were set aside for Wireline Common Carriers (WLCCs) and Radio Common Carriers (RCCs) (Baby Bells and smaller phone companies), 30 megahertz for private radio, and 45 megahertz were held in reserve for cellular service.

As the FCC sorted out the bandwidth issues, Bell Labs continued its research. In 1970 it filed a patent for a mobile communications system. In 1977 experimental systems were launched in Chicago, Illinois, and the Baltimore/Washington D.C. corridor. Meanwhile, the cellular industry moved forward in other countries. In

1979 the first commercial cellular system began operation in Tokyo, Japan. Cellular systems in the Nordic countries also entered service about the same time.

Dr. Martin Cooper of Motorola created the prototype of a cellular phone (Motorola and AT&T were frequent partners). On April 30, 1973 Cooper and John Mitchell, a fellow engineer, introduced the first cellular phone at the New York Hilton. The audience consisted of about 50 reporters. Cooper placed a call to Joel Engel, head of research at AT&T's Bell Labs, while walking the streets of New York City talking on the first Motorola DynaTAC prototype. According to Cooper, sophisticated New Yorkers gaped at the sight of someone moving around while making a phone call. The Motorola DynaTAC 8000x would not be available until 1983. It cost $3,995. The phone, dubbed the *brick*, had one hour of talk time and eight hours of standby.

Consumer demand was initially underestimated. The Motorola prototype had a waiting list of 25,000 people in 1980. Farley cites some of the industry forecasts. In 1980 consultants McKinsey & Company estimated that there would be 900,000 subscribers by the year 2000; the actual total was 109 million. Donaldson, Lufkin, Jenrette in 1990 guessed no more than 67 million by 2000. A 1994 estimate by Herschel Shostech Associates estimated 60–90 million by 2004; the actual total was 182 million. Nobody believed the average citizen would be interested in such a service or be able to afford it.

The FCC, frequently criticized for dragging its heels in the decision making process, was completely overwhelmed when it came to issuing licenses for the new cellular industry. In 1981 the FCC announced it would issue licenses to 306 Metropolitan Statistical Areas and 428 Rural Statistical Areas, a non-wireline company and a wireline company. It took the FCC nearly a year to process the 190 applications for the top 30 areas. It expected a similar period of time to process the 353 applications for markets 31 through 60. It estimated licenses for the next thirty markets not to be processed until the end of 1985. Another 567 applications existed for the next 30 markets. In 1986 the FCC switched to a lottery system to award the applications. It also allotted another 10 megahertz of spectrum. By 1987 the industry saw more than 2 million subscribers and more than $1 billion in revenue, according to the Cellular Telecommunications Industry Association.

1990s and Beyond. Technological developments with electronics in the last decade of the century were rapidly implemented by the mobile phone industry. Smaller, more powerful components were integrated into cell phones. The compression of digitized information also made it possible to make more efficient use of precious bandwidth. Calls could be multiplexed, which allowed

more than one call to be carried on the same line. Sound quality would also be improved. The handset could be smaller and even more portable. These technological advances helped usher in the second generation technology: GSM (Global System for Mobile Communications), TDMA (Time Division Multiple Access) and CDMA (Code Division Multiple Access).

In 1982 the Conference of European Posts and Telecommunications (CEPT) hoped to create a mobile standard in Europe. By 1987, the GSM standard was created based on a hybrid of analog and digital technologies. This allowed for fast bit rates and more natural-sounding voice-compression algorithms. GSM was launched in Finland in 1991. GSM is now the world's leading wireless phone format, with more than 2 billion users as of June 2006. It operates on the 900-megahertz and 1.8-gigahertz bands in Europe and on the 1.9-gigahertz band in the United States. GSM is the only service that permits users to place a call from either North America or Europe. The GSM standard was accepted in the United States in 1995.

The major difference between TDMA and CDMA is how they handle multiple users. TDMA chops the channel into sequential time slices. Each user of the channel takes turns transmitting and receiving signals. CDMA assigns a unique code to each conversation, allowing for some overlap along the spectrum. CDMA provides three to five times the calling capacity of GSM and TDMA and has become the standard in North America.

CDMA2000 is a slightly newer, second generation technology that provides voice and data capabilities between mobile phones and cell sites. It nearly doubles voice capacity over second-generation CDMA networks and supports high-speed data services. A few European carriers in the former Soviet bloc have rolled out CDMA2000, as have carriers in Japan and South Korea. Verizon and Sprint were the first to use this technology in the United States.

With increased processor power and lighter, better designed parts, manufacturers began creating phones with other functions integrated into them. Computer companies, it should be noted, about this time were creating the first personal digital assistants—handheld computers that could run a calculator, clock, alarm, and any number of personal organizing functions. The same could be done for phones. In the late 1990s, several efforts were made to integrate camera and cell phone functions. Philippe Kahn did so in 1999, as did Olympus with its Deltis VC-1100 model in 1994. In both cases, however, digital cameras and phones were linked by wires, which allowed for the uploading and transmission of digital photos. Kyocera and Sharp were the first companies to work on a handset with an integrated camera in 1999. Sharp released the J-SHO4 in November 2000. Camera phones started shipping to North America in 2002.

IBM and BellSouth introduced the first smartphone, the Simon Personal Communicator, in 1994. Smartphones are combination mobile phones and PDAs (Personal Digital Assistants). Such phones provide cellular service, email, text messaging, Internet access, multimedia services, and organizational capabilities. Most phones are so multifunctional that the distinction between smartphones and high-end handsets has blurred considerably. The devices are more popular in Europe than in the United States.

Third Generation Technology. The third generation (3G) technologies are those items that include handsets with increased capacity and high-speed data applications up to two megabits. Companies from South Korea introduced such phones to the United States in 2003. Third generation wireless employ wideband frequency carriers and a CDMA air interface. Networks must be able to transmit wireless data at 144 kilobits per second. One example of third generation technology is W-CDMA (Wideband Code Division Multiple Access), which uses 5-megahertz channels for voice and data with peak data speeds of 384 kilobits per second. Another is TD-CDMA (Time Division CDMA), which uses increments of 5 megahertz of spectrum, each slice divided into 10 millisecond frames containing fifteen time slots (1,500 per second).

In early 2007 the GSM Association launched a campaign called *3G for All*. The purpose of this program was to gain wider distribution of third generation phone technology. This would be attained through various major economies of scale in manufacturing, logistics, and marketing.

MARKET

Modern mobile technology has been introduced in various levels of sophistication known as generations. The exact years of these generations are open to debate in the industry. The first handsets shipped were analog phones in the 1980s. These phones are considered first generation. Digital phones, the second generation, were shipped in the 1990s. Third generation phones started to be shipped between 2001 and 2003. A more advanced level of phones, the 3.5 generation phones, started to ship in 2005. Analysts expect fourth generation phones to start shipping about 2009.

Market research firm Handelsbanken Capital Markets estimated that there were 415 million mobile phones shipped worldwide in 2000. Shipments fell slightly to 405 million in 2002. Industry shipments experienced double-digit growth through most of the next few years, climbing to 978 million in 2006. The company estimated

FIGURE 51

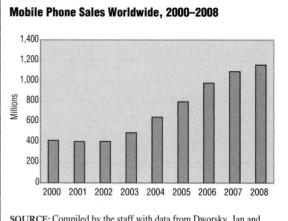

Mobile Phone Sales Worldwide, 2000–2008

SOURCE: Compiled by the staff with data from Dworsky, Jan and Karri Rinta, "Table 12: Volume Shipments by Vendor." *Telecom Equipment Handset Monitor, 4th Quarter, 2006,* Handelsbanken Capital Markets, February 8, 2007.

shipments of 1.1 billion phones in 2008 as can be seen in Figure 51. Other market research firms come up with slightly different figures when analyzing the same industry. For example, Informa Telecoms & Media estimated 942.7 million phones shipped in 2006 and anticipated shipments of 1.1 billion in 2008. Piper Jaffray estimated 982 million units shipped in 2006 and 1.2 billion in 2008.

According to Handelsbanken, Nokia took 30.9 percent of global shipments in 2000. It was forecast to take 36.8 percent of shipments in 2008. Motorola had an 18.1 percent market share in 2000; its share in 2008 was expected to be 22.9 percent. Samsung had a 6.2 percent share in 2000 and was anticipated to have a 12.6 percent share in 2008. Sony Ericcson had a 10.4 percent market share in 2000. It was expected to rank fourth in 2008 with a 9.3 percent share.

The handset industry is expected to see strong (but slowing) growth in the coming years. Wireless penetration is high in developed countries, and increasing in less developed ones. Figures vary, but over 70 percent of the U.S. and Japanese population use a cell phone, compared to 40 percent in Latin America and 90 percent in parts of Europe. Consequently, as penetration rates rise, an increasing percent of the sale of new phones will be generated by the need for replacement handsets rather than handsets for new subscribers. About 55 percent of handset sales were replacement models in 2002. By 2007 this figure was expected to increase to 57 percent.

GSM will be the dominant technology standard through the end of the first decade of the twenty-first century at least. In 2007 there were expected to be 1.02 billion phones shipped. About 808 million, or 78 percent, will be in the GSM format. Approximately 190 million phones were shipped using CDMA technology, which represents 18.5 percent of total shipments. Other technologies, including TDMA (Time Division Multiple Access), PDC (Personal Digital Cellular), PHS (Personal Handy-Phone Systems) and IDEN (Integrated Digital Enhanced Network) represented the remaining 3.5 percent. These technologies have a larger presence outside the United States.

Market Analysis. Piper Jaffray places total handset sales in North America at 170 million units in 2007. Motorola was the leader in the North American market that year. Gartner Dataquest and Piper Jaffray estimated that Motorola had 38 percent of the market in 2007, up from 23 percent in 2002. Nokia was the leader in 2002, with a 36 percent market share. The company lost share from 2002 to 2003 falling from 30 percent to 21 percent. In 2007 Nokia's share was forecast to be 15 percent. Samsung was third in the market; Samsung had a 10 percent market share in 2002 and was forecast to have 16 percent of the market in 2007. LG had an 8 percent share in 2002, which grew to 14 percent in 2007. Together, all other firms represented 23 percent of the market in 2002 and 16 percent in 2007. Figure 52 presents the Piper Jaffray market share estimates for 2007 graphically.

FIGURE 52

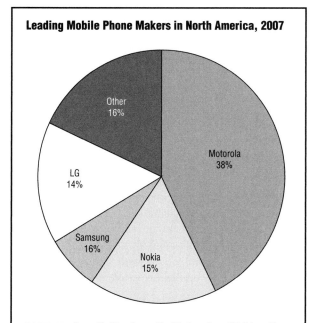

Leading Mobile Phone Makers in North America, 2007

SOURCE: Compiled by the staff with data from Walkley, T. Michael, Amir Kapur, Michael Sklansky, and Preetesh Munshi. "Exhibit 18: North American Market Share Trends," Global Wireless Handset Market, Piper Jaffray, February 2007.

Total handset sales in Western Europe were forecast to reach 185 million in 2007. Nokia had a 37 percent market share, followed by Motorola with 16 percent, and Samsung and Sony Ericsson with 15 percent each. Nokia's market share declined since 2002, when it led the market with a 48 percent share. According to Piper Jaffray, however, it still enjoys considerable brand loyalty. Sony Ericcson's line of third generation (3G) phones is expected to take market share as the 3G technology gains market penetration.

The Asia-Pacific region (excluding Japan) is expected to see double digit growth in 2007. Handset sales will exceed 350 million units. Much of this growth will be in China, the world's largest handset market. Handset unit sales grew 40 percent in 2006 from both new and replacement handset sales. In 2007 Nokia had an estimated 42 percent of the Chinese market, Motorola 19 percent, Samsung 11 percent, LG Electronics 5 percent and other firms 24 percent. Nokia brought its share up from 32 percent in 2002 to 42 percent in 2007 because of improved distribution channels. Motorola followed Nokia's lead; improvements to its supply and distribution structure brought its share up 8 percent from 11 percent in 2004.

Japan has a high handset penetration rate. Consequently, growth will come largely from replacement sales as wireless users switch to 3G technology. In 2007 Sharp had an estimated 19 percent of the Japanese market. Panasonic and NEC each had 12 percent.

Latin America saw strong growth in subscribers during 2004 and 2005. Growth was expected to continue in the double digits. Piper Jaffray noted that the Middle East had 20 percent of the world's population; the region is thus expected to represent a similar share out of the next billion subscribers worldwide. Handsets are important in Africa because of the unreliable telecommunications infrastructure on that continent. In all these regions, however, replacement sales will outpace new subscribers after 2008. Nokia was forecast to be the leading handset provider in Latin America in 2007. Its share was expected to take a third of the market. Motorola was expected to be close behind with a 31 percent share, Sony Ericsson a distant third with 9 percent. Nokia was anticipated to have nearly half of the market (48 percent) in Eastern Europe, Africa, and the Middle East. Samsung was forecasted to be second with 15 percent.

The Asia-Pacific region had 1.06 billion subscribers in 2007, or 44.5 percent of the 2.38 billion subscribers worldwide, according to Informa. Europe had 521.4 million subscribers, or 22 percent of the total. Africa/Middle East had 298.5 million subscribers and 12.5 percent of subscribers. Latin America had 259.3 million subscribers and 11 percent of the total. North America had 233 million subscribers and 10 percent of total subscribers.

FIGURE 53

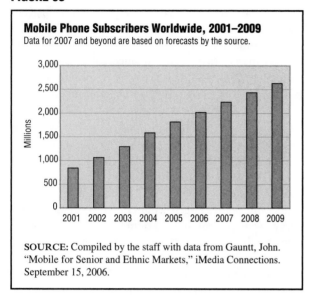

Mobile Phone Subscribers Worldwide, 2001–2009
Data for 2007 and beyond are based on forecasts by the source.

SOURCE: Compiled by the staff with data from Gauntt, John. "Mobile for Senior and Ethnic Markets," iMedia Connections. September 15, 2006.

The total number of mobile phone subscribers is expected to exceed 2.6 billion in 2009, according to PricewaterhouseCoopers. NTT DoCoMo, China Mobile, KDDI, Cingular Wireless, and Verizon Wireless were the leading carriers based on wireless revenues for the first half of 2006, according to Chetan Sharma Consulting. The United States added about 23 million new subscribers in 2006. The consultancy reports that Verizon and AT&T each had a 26 percent market share, Sprint had a 23 percent share, T-Mobile an 11 percent share and other firms 14 percent of the market.

KEY PRODUCERS/MANUFACTURERS

Nokia Corporation. This company operates in four segments: mobile phones, multimedia, enterprise solutions, and networks. The mobile phones segment offers mobile phones and devices based on GSM/EDGE, 3G/WCDMA, and CDMA cellular technologies. It has its operations in Europe, the Middle East, Africa, China, the Asia-Pacific, North America, and Latin America. The company was founded in 1865 and is based in Espoo, Finland. It employed 68,483 people and generated revenues of $55.2 billion for the fiscal year ended December 31, 2006.

Motorola, Inc. Providing wireless and broadband communication products worldwide, the company manufactures cellular infrastructure systems, associated software and services, application platforms, switching technologies, two-way radios, and a variety of voice and data systems. Motorola offers its products through direct sales, distributors, dealers, retailers, licensees, and agents. The company was founded in 1928 and is based in Schaum-

burg, Illinois. It employed 66,000 people and generated revenues of $48.2 billion for the fiscal year ended December 31, 2006.

Samsung. This company makes many kinds of consumer devices, including DVD players, big-screen televisions, digital cameras, computers, color monitors, LCD panels, printers, semiconductors, flash memory, and communications devices ranging from wireless phones to networking switches. Samsung generated revenues of $78.9 billion in 2005 and employs 128,000 people.

Sony Ericsson. This company was established to draw on the cellular technology of Ericsson, the world's leading maker of wireless infrastructure equipment, and Sony's expertise in developing popular consumer electronics. Sony and Ericsson each own half of the venture, which began operations in 2001. The company generated $8.6 billion in revenues in 2005 and employed 5,000 people.

MATERIALS & SUPPLY CHAIN LOGISTICS

Several market experts have noticed that while the industry was still performing well, growth in the number of subscribers will start to slow. This is to be expected as cellular penetration increases in developing markets and reaches near saturation levels in mature markets. iSuppli reports that global mobile phone subscriber growth increased an average of 25 percent in 2004, 2005, and 2006, but will only be at about 12.8 percent in 2007. Similarly, unit production grew an average of 19.3 percent from 2004–2006. In 2007 iSuppli forecast growth slowing to 9.1 percent, and 6.9 percent in 2008, 4.8 percent in 2009, and 3 percent in 2010. Industry growth will be sustained through replacement handset sales rather than through new subscribers. Consumers will replace their handsets as they fail or as they are encouraged to do so by attractive new features.

In short, the industry needs to make handsets that are attractive to the user. This has begun to change the structure of how handset makers design and build their products. Informa Telecom notes that a supplier previously shipped its goods to the handset maker. A chipset supplier put little emphasis on when or how the chips would be used. Some suppliers are now part of the design process, working more closely with handset makers on how to optimize handset construction.

A good example is in the construction of entry-level phones. One source of new subscriber growth will be in developing markets of Africa, Latin America, and the Asia-Pacific region. The populations in such regions typically have low incomes and so handsets shipped to these countries must be inexpensive, costing $30 or less. Suppliers and handset makers have worked to design such a phone. The handsets contain very basic processors. More circuitry is combined on one die, an efficient structure that dramatically cuts the bill of materials (manufacturing cost). Such low cost models may see sales of 28 million in 2007, according to iSuppli.

Higher-level phones present greater challenges to the handset manufacturer. Consumers are drawn to phones that are multifunctional. Such phones might have digital cameras, gaming capabilities, high-resolution screens, and GPS/Bluetooth functions. Such handsets also require powerful processors. These features drive up the cost of manufacturing a cellular phone and increase the complexity of both the hardware and software components.

The average bill of materials (BOM) costs were $61 for an entry level phone, $66.03 for low-end, $91.49 for a mid-range phone and $135.09 for a high end phone, according to iSuppli. Displays are the most expensive part of a phone, followed by digital logic chips and then memory. The percent of total cost varies depending on the type of phone, however. In a low-end phone, the baseband (digital signal carrier) represents 28 percent of the cost of materials, the bill of materials cost, followed by the RF/IF circuitry (integrates the radio signal with the processor) with 12 percent and displays 11 percent. In a high-end phone displays take 28 percent, followed by the baseband with 16 percent and the RF/IF circuitry with 4 percent. The battery made up 10 percent of the cost in the lower-end phone, but 3 percent in the high-end model. A camera made up 4 percent of the materials cost of the high-end phone. Generally, low-end phones do not include cameras. Other costs included memory, user interfaces, and power management applications. Various electrical components represented about 20 percent of the cost of phones on both ends of the cost spectrum.

Software is a vital part of high-end phones. Companies must pay royalties to software developers for operating systems, application platforms, browsers, and predictive text software. Such royalties are becoming an increasingly expensive part of handset manufacturing. Informa Telecoms & Media estimates that the percentage of intellectual property and software royalties will reach 20 percent of total cost for 2.5G (2.5 generation) phones and could be as high as 30 percent or more of total cost for 3G phones by the end of 2006.

Handset makers must distinguish themselves in a crowded market. There are numerous models on the market, many with similar features. Gartner reports that the number of new models in Japan increased 30 percent in 2005 as manufacturers rushed out models to appeal to every possible end user. Many experts are curious to see how Apple's iPhone, released in 2007, will ultimately transform the market. Apple has had great success with the iPod and many analysts expect similar success with the iPhone. The size of the demand for music-enabled phones is estimated to be large and manufacturers are attempting to fill this demand with hundreds of music-enabled cellular phone models. As many as 835 models are expected to be on the market in 2007. Roughly fourteen models have features that closely resemble the features on the iPhone.

DISTRIBUTION CHANNEL

With slowing subscriber growth has come declining revenues for mobile operators. The average voice per user revenue fell 7 percent in 2006 while the average data revenue per user increased 50 percent, notes Chetan Sharma Consulting.

According to iSuppli, revenues from mobile content (games, video, services, and music) will grow to $43 billion in 2010, up from $7.7 billion in 2005. Music is the most downloaded form of content, representing about 40 percent of sales. Music's market share may drop slightly by 2011 as mobile television and other video become more popular. Global text messaging service revenues are expected to increase from $60 billion in 2006 to $93 billion in 2011. Informa Telecom & Media anticipated that by 2011, just under half of all mobile subscribers worldwide will use mobile browsing.

The distribution of mobile content is available to the consumer from the manufacturer, service provider stores, or consumer electronics stores. New sites, such as mall kiosks, online vendors, and store-within-a-store players, have made the industry even more competitive. Mobile operators have controlled most of the distribution for mobile content, but there are signs that third party players are gaining ground in these mobile data sales. Handset makers are also making a stronger effort to deliver this content to consumers though direct marketing and Web sites.

According to Mintel, about 62 percent of phones and plans were sold through service provider stores in 2003. Cellular specialist stores selling more than one brand sold 10.2 percent and electronics stores had 7.3 percent, and Web sites had 3.1 percent. The balance was sold through department stores, drug stores, and other outlets. Among large retailers, Wal-Mart, Radio Shack and Best Buy were the most popular outlets from which to purchase a wireless phone.

KEY USERS

About 76 percent of U.S. households reported having a cell phone in the third quarter of 2006, according to research by ACNielsen. Recent research notes that cell phone owners are replacing their handsets sooner than in the past. Piper Jaffray reported that as of early 2007 the average cell phone user held onto his or her handset for 30 months.

Market tracking firm Mintel reports that among those 18–24 years of age, 54 percent had a cell phone in 2003. Among those 25–34 years of age, 60 percent had one. About three-quarters of those 35 to 44 years old did. As well, 62 percent in the 45 to 54 age group owned a cell phone compared to 52 percent in the 55 to 64 category. Forty-four percent of those 65 years of age or older reported having a cell phone.

The Mintel report did not address cell phone ownership among children and young teenagers. The company Firefly makes phones aimed at preteen users that have flashing lights and Mom and Dad on speed dial. Robin Abrams, the company's CEO, estimated that 10 percent of children between 8 and 12 years of age (about 23 million people in all) had a cell phone. Parents may find some merit in children having phones. Working parents often feel that such devices keep them better connected to their children, for example. Phones are also helpful during emergencies. But the phones can be a distraction for young people as well. School districts have been instituting cell phone limitations (or complete bans) in schools. Teachers cite the disruption phones bring to the classroom, and with text messaging functionality, the possibility of cheating.

ADJACENT MARKETS

A cellular phone without a service package—a phone number and connection to a service provider—is little more than a fancy looking toy for a child. Consequently, the most direct market adjacent to cell phones is the mobile phone service provider market. These two markets are, in fact, dependent on one another. Cell phones have, in fact, become a sort of loss leader for phone service providers. For a multi-year contract, a service provider in the latter years of the first decade of the twenty-first century will give the customer the actual phone itself.

The market for electronic components is another adjacent to cell phones. The transition to third generation technology will place demands on makers of semiconductors, transceivers, digital baseboards, power amplifiers, and memory makers. This demand helps stimulate various electronic component markets. Baseboards had sales of $13.4 billion in 2005, while amplifiers and transducers had sales of $6 billion, according to Gartner. However, these industries also face strong competition, the cost of research and development, and short life cycles of products. Such challenges do make it difficult to predict how well these markets might ultimately perform.

Again, mobile content providers will perform well. Sales of wireless applications processors increased 35 percent from 2004 to 2005, according to International

FIGURE 54

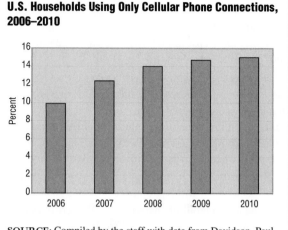

U.S. Households Using Only Cellular Phone Connections, 2006–2010

SOURCE: Compiled by the staff with data from Davidson, Paul. "Sprint, T-Mobile: Local Carriers Drag Feet," *USA Today*. January 15, 2007.

Data Corp. Texas Instruments, Intel, and Renesas are the leading makers of chips that help phones play games, multimedia applications, and music. According to the Semiconductor Industry Association cell phones are the second largest market for semiconductors behind computers. The increased need for memory to run large applications and store audio and video will benefit the $20 billion flash memory market. Piper Jaffray anticipates rising volume and density in the market with declining cost per bit.

Ringtones have proven a popular accessory for cell phones in the early 2000s, and, often an annoying addition to the public space. Ringtones do have a practical purpose; a personalized ring can help a user distinguish his or her phone in a roomful of ringing phones. Broadcast Media, Inc. estimated that the market was worth $68 million in 2003, $245 million in 2004, $500 million in 2005 and $600 million in calendar year 2006. The analysts at Broadcast Media, Inc. anticipated a drop in the market, with sales slipping to $550 million in 2007.

RESEARCH & DEVELOPMENT

Wireless service and device providers are teaming with the holders of music and video rights to offer downloads, and they will be building relationships with broadcasters to deliver mobile TV without burdening the mobile network. On the handset side, a rapidly growing number of collaborations will be formed between hardware and software companies to build more complete, more reliable, and more feature-rich cell phone platforms. Fewer companies will be able to do everything required to build a platform on their own. Cellular baseband experts, for example, will collaborate with providers of global positioning system solutions, for installation in these basebands, to provide value-added functionality. These collaborations will help accelerate time to market while minimizing research and development costs.

NTT DoCoMo has run some tests on fourth generation technology. Such devices would build on the 3.5 generation technology high speed downlink packet access (HSDPA), which is used primarily in Japan. A downlink is the transmission signal from a base station to a cell phone. HSDPA is a packet-based data service with data transmission ranging from 8 to 10 megabits per second over a 5 megahertz bandwidth in W-CDMA.

CURRENT TRENDS

Trends in the later half of the first decade of the twenty-first century show a leveling off in the number of new subscribers. One developing trend is the number of people who elect to dispose of their regular landline in favor of their wireless service. In 2003 the Federal Communications Commission (FCC) instituted stiff penalties against local phone companies that don't allow customers to keep their phone number when they switch to wireless service. However, in early 2007 wireless providers T-Mobile and Sprint Nextel complained to the FCC that many local phone companies are not allowing their customers to make this switch. Number transfers ideally should only take days. Some phone companies let the process drag out for months and some frustrated customers simply give up. About 10 percent of households had removed their landline in 2006, and Yankee Group expects this figure to climb to 15 percent by 2010. Just under half of those who are cell phone only are under 30, according to the Pew Research Center. They are also typically less affluent, young, and single.

TARGET MARKETS & SEGMENTATION

Vendors structure their marketing around various factors. They consider overall trends in cell phone usage, the frequency with which subscribers use phones, and the most utilized functions (checking e-mail or text messaging, for example). They consider basic demographics as well, such as gender, age, and geographical location.

Young people, for example, are typically drawn to technology and are willing to adopt new devices. They are often very brand conscious. They are typically drawn to a new phone that is multifunctional, allowing them to text message, play music, and browse the Internet. Manufacturers design cellular phone devices and services that cater to these features. Business users, on the other hand, typically have different needs. Those using cellular phones for business have needs that tend to involve handsets with a large screen, mobile e-mail functionality, and advanced calendar and synchronization capabilities. Manufacturers design devices accordingly for the business user and market them through different media outlets directed to a professional audience.

RELATED ASSOCIATIONS & ORGANIZATIONS

Federal Communications Commission, http://www.fcc.gov

Wireless Association, http://www.ctia.org

Wireless Communications Association International, http://www.wcai.com

Wireless Infrastructure Association, http://www.pcia.com

BIBLIOGRAPHY

"BMI Projects Downturn in 2007 Ringtone Sales." Broadcast Media, Inc. 12 March 2007. Available from <http://www.bmi.com>.

"Boomers Most, Teens Least Likely to Have Cell Phones." *Research Alert.* 19 September 2003.

Burns, Ed. "Cell-Only Population: Young and Tech Savvy." ClickZ News. 16 May 2006. Available from <http://www.clickz.com>.

"Coming Soon: A Cellular Radio System Near You." *Communications News.* February 1984.

Davidson, Paul. Sprint, "T-Mobile: Local Carriers Drag Feet." *USA Today.* 15 January 2007.

DeVoss, Mark. "Teardown Analysis: Nor Flash Retains Mobile Place." *EE Times.* 4 August 2005.

Dworsky, Jan and Karri Rinta. "Table 12: Volume Shipments by Vendor." *Telecom Equipment Handset Monitor Q4 2006.* Handelsbanken Capital Markets. 8 February 2007.

"Executive Summary." *Future Mobile Handsets.* October 2006.

Farley, Tom. "Cell-Phone Revolution." *American Heritage.* Winter 2007.

Ford, Dale. "The Three Ps of Mobile Handsets." Nokia Corporation. 30 November 2006. Available from <http://www.nokia.com>.

Gauntt, John. "Mobile for Senior and Ethnic Markets." iMedia Connections. 15 September 2006. Available from <http://www.imediaconnection.com>.

Roberts, Bill. "Cell Phone Market to Shift." *EDN.* 21 February 2007.

Terrell, Kenneth. "That Ringing Sound is a Cash Register." *U.S. News & World Report.* 12 December 2005.

Tyson, Jeff. "How Cell Phones Work." Verizon.com Learning Center. Available from: <http://www22.verizon.com/about/community/learningcenter/articles>.

"US Wireless Market Mid–Year Update." Chetan Sharma Technology & Strategy Consultants. 20 August 2006. Available from <http://www.chetansharma.com>.

Walkley, T. Michael, Amir Kapur, Michael Sklansky, and Preetesh Munshi. *Global Wireless Handset Market.* Piper Jaffray. February 2007.

Wolpin, Stewart. "Hold the Phone." *American Heritage.* Winter 2007.

Cheese

———◆———

INDUSTRIAL CODES

NAICS: 31–1513 Cheese Manufacturing

SIC: 2022 Natural, Processed, and Imitation Cheese

NAICS-Based Product Codes: 31–15131, 31–15134, 31–15137, and 31–15137

PRODUCT OVERVIEW

Cheese manufacturing is the second largest industry based on milk. The production of cottage cheese is reported by the Census Bureau as part of the fluid milk industry, the largest of the five. The other three are ice cream, butter, and dry, condensed, and evaporated products.

Most of cow's milk by weight is water (88%) followed by lactose (4.7%), fat (3.4%), protein (3.2%), and minerals (0.7%). The protein in milk is further subdivided into casein (about 80% of the protein) and the solids in whey (20%). The basic raw material of cheese is the casein in milk. The word is derived from the Latin for cheese (*caseus*). When young sucklings of any mammal species drink milk, enzymes in their stomachs cause casein to curdle into digestible solid clumps. Professor David Frankhauser of the University of Cincinnati, in a paper titled "Rennet for Making Cheese," provides a plausible scenario of how humanity must have discovered cheese making.

> Presumably [he writes] the first cheese was produced by accident when the ancients stored milk in a bag made from the stomach of a young goat, sheep or cow. They found that the day-old milk would curdle in the bag (stomach), yielding solid chunks (curds) and liquid (whey). Once they discovered that the curd-chunks could be separated out and dried, they had discovered a means by which milk, an extremely perishable food, could be preserved for later use. The addition of salt was found to preserve these dried curds for long periods of time.

The enzymes causing the curdling are called rennet, rennin, or chymosin. Initially rennet was obtained by slicing up stomach tissues and inserting them into containers of milk. In modern practice chymosin is produced by recombinant DNA techniques in cultured bacteria.

Natural Cheese Production. Natural cheese is produced by preparing the milk, coagulating it to produce curds, treating the curds, draining off the whey, knitting the curds, salting them, pressing them, and ripening the cheese mixture. Cheese may also be dried, grated, or shredded.

In the preparation stage raw milk is first adjusted to a predetermined fat content by skimming off cream or by using a centrifuging process to accomplish the same end. Once the desired fat content is reached, the milk is pasteurized, that is to say it is heated, but not boiled, and held hot long enough to kill off all pathogens harmful to humans. Most cheeses are coagulated using rennet, but certain kinds of cheeses (Ricotta and cottage cheese being examples) are coagulated using acid. With many types of cheeses a two-stage coagulation is common in which bacteria are used as starters and, after the curdling process begins, typically because lactose-consuming bacteria generate lactic acid, rennet is added for the full digestion of the casein. The process requires between 20 and 35 minutes depending on the cheese being made, longer if acid alone is used. Curd treatment consists of making the curds larger or smaller by cutting them apart or pressing them together. Cutting curds reduces the moisture content; creating larger formations causes these to retain their moisture. As part of treatment most cheeses undergo a heating process; the curds are scalded, cooked, or simply heated to influence their moisture content and texture.

In the draining process whey is removed from the cheese using a heat process, mechanical treatment like cutting, stirring, oscillating or pressing the mixture, or some combination of the two. Spores of penicillium molds are injected into interiorly mold-ripened cheeses like Roquefort, blue cheese, and the French Bleu d'Avergne during the draining process. There are numerous varieties of penicillium molds—not to be confused with penicillin, the drug, derived from species of mold other than those used in making cheese.

Throughout the cheese-making process producers aim at bringing about desirable chemical transformation known to take place if the environment is managed appropriately. After the whey has been drained, curd knitting takes place. Producers use mechanical devices to pull, knead, press, draw, and shape the cheese. Their motivation for going to these lengths is knowledge that lactic acids forming in the cheese must be evenly distributed in order to achieve desired cheese textures. The maker controls the curd's acidity in this process and, for certain cheeses, may also introduce additives. Salting of cheese is an important step. Producers apply salt to the outside of shaped blocks or dip blocks into salt-brine briefly. Salt draws residual whey components to the surface of the clump, slows down the chemical processes inside the cheese, acts as a preservative, and plays a role in crust formation. Finally, with the exception of cottage or cream cheese, most varieties of cheese undergo a period of ripening. Ripening may take a few weeks up to a year or longer. Cheese continues to form as it ripens, bacteria and enzymes continuing to work. They slowly transform the texture, taste, and chemistry of cheese. Cheese producers maintain temperatures and humidity ideal for each variety as it ripens.

Processed Cheese. The first step in making processed cheese is to make natural cheese in the laborious way described above. Processed cheese manufacturing is thus a form of secondary processing of natural cheese in an effort to transform it into a uniform, industrial product. The method originated in Switzerland but was adapted so widely in the United States that processed cheese is also called American cheese. American cheese is usually made of Cheddar or Colby. The producer begins with blocks of finished cheese. Rinds are cut off and the cheese is ground. Sampling determines its fat and moisture content. The ground cheese is melted until the fat and the protein portions (the serum) separate. Emulsifiers are mixed in, their quantities based on the earlier sampling. Emulsification causes uniform dispersion of fats throughout the mass and their suspension in the serum (the liquefied protein). This produces a high level of homogeneity absent from the original Cheddar or Colby. The molten cheese is rapidly cooled, shaped, and packaged in oxygen-free environments; in the absence of oxygen molds cannot survive. Processing takes place in a nitrogen gas atmosphere.

Cheese Classification. The most common classification of cheese is by its degree of hardness or consistency. Those inclined to make finer distinctions classify cheeses as soft, semi-soft, semi-hard, and hard. The semi-soft category is often reserved for processed cheeses. Others are content with three categories: soft, semi-hard, and hard. Using this last classification, examples are the following:

- Soft Cheese: Brie, Camembert, Feta, Gorgonzola, Roquefort, Mozzarella, Muenster.

- Semi-Hard Cheese: Brick, Gouda, Monterrey Jack, Provolone, Stilton.

- Hard Cheese: Asiago, Cheddar, Colby, Emmental, Gruyere, Jarlsberg, Parmesan, Romano.

There are hundreds of varieties of cheese, of course, and people have strong loyalties to the cheeses that they favor. With this in mind any listing of examples will invariable leave out someone's best-liked variety. To see a list of 103 varieties, the reader is referred to CheeseNet on the Internet (see references). Even that list, of course, will leave some seekers dissatisfied. The total count of varieties is said to exceed 2,000 the world over.

We can also classify cheese by the mammals from which the milk is obtained. Most common animals are cows, goats, ewes, camels, and buffalo. Cheeses are also classified as having smooth, holey, and veined surfaces. Cheese with holes (like Emmental and the American Swiss cheese) are made by introducing gas-producing bacteria in the cheese; veined cheeses are produced by introducing molds. Consumers also sometimes classify cheeses by their countries of origin. Wherever milk-producing mammals are raised, fine cheeses are also produced. Some of the best of these are eaten all around the world. In the United States cheese varieties from European countries are consumed in large quantities and are widely known. Brick cheese (originally from Wisconsin and so named because actual bricks were used to press the cheese), Colby, Monterey Jack, and Swiss—America's compliment to Emmental produced in Switzerland—are distinctly American cheese varieties as is, of course, processed cheese in a great variety of forms.

MARKET

Cheese-making in the United States accounts for much of the demand for cow's milk. Annual production consumes around 38 percent of all milk produced. In 2005 cheese shipments by U.S. producers were a $27.1 billion industry. Cheese production has been growing at a rate of 3 percent per year in the period from 1992 to 2005, but this period had one major hiccup. Between 1992 and 1998, the industry advanced at a rate of 3.5 percent a year but then declined sharply for two years between 1998 and 2000. Growth picked up again in 2001 and has been growing at 6.5 percent per year to 2005, the last year for which data were available at the time of writing.

Industry growth generally has been supported by growing consumption of cheese, but market fluctuations have been due to price issues in the dairy industry as a whole—most particularly wide swings in milk pricing beginning in the mid-1990s. Demand for cheese has been increasing steadily as shown by per capita consumption figures maintained by the U.S. Department of Agriculture (USDA). Cheese consumption per person stood at 25.9 pounds in 1992. By 2005 consumption had increased by nearly 6 pounds to 31.7 pounds per capita. Natural cheese consumption grew more energetically whereas processed cheese showed signs of weakness, particularly late in this period. Per capita consumption actually dipped between 2004 and 2005. In 1992 natural cheese accounted for 56 percent of consumption, in 2005 for 59 percent.

Average milk price in the period 1981 through 1995 was between $12 and $14 per hundredweight (cwt—equivalent to approximately 11.6 gallons of milk). In the 1996–2005 period, prices began to fluctuate more wildly, the average price falling into the $12 to $16 per hundredweight range. Average price changes year to year (and much more pronounced changes month-to-month) tell the story of increasing fluctuation: from 1992 to 1993 the change in average price was 29 cents per cwt, from 1993 to 1994 it was 17 cents, and from 1994 to

FIGURE 55

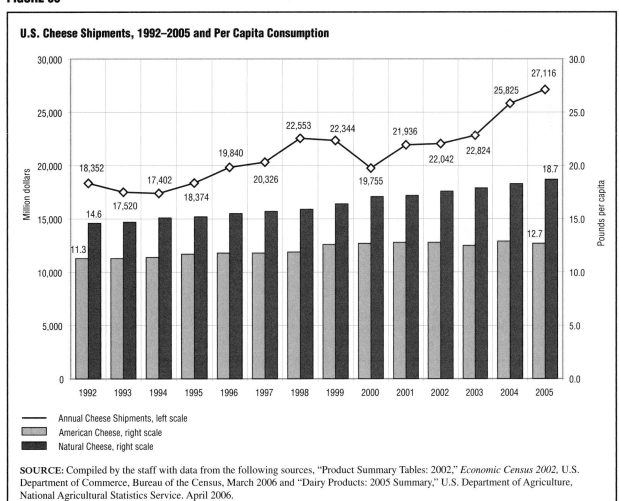

U.S. Cheese Shipments, 1992–2005 and Per Capita Consumption

Annual Cheese Shipments, left scale
American Cheese, right scale
Natural Cheese, right scale

SOURCE: Compiled by the staff with data from the following sources, "Product Summary Tables: 2002," *Economic Census 2002,* U.S. Department of Commerce, Bureau of the Census, March 2006 and "Dairy Products: 2005 Summary," U.S. Department of Agriculture, National Agricultural Statistics Service. April 2006.

1995 it was 23 cents per cwt. But the next four years set the pattern for the future: 1995 to 1996 showed a $2.14 swing in average price upward, 1996 to 1997 a $1.54 shift downward, 1997 to 1998 a $2.16 swing upward—and so on throughout the period to 2005. This was a period of decreasing dairy cow population and sharply increasing output of milk per cow. The price volatility, however, is more directly traceable to the 1996 Farm Act. It called for elimination of milk subsidies by 1999, later reversed; it also reorganized the subsidy schedules. The 2002 Farm Act ensured continued milk subsidies but subsidy schedules, simplified by the 1996 act, continued in place—as did swings in average milk pricing.

Production of cheese by variety is available from the National Agricultural Statistics Service (NASS), an ele-

FIGURE 56

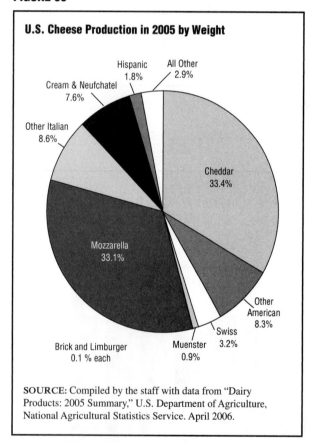

U.S. Cheese Production in 2005 by Weight

Hispanic 1.8%
All Other 2.9%
Cream & Neufchatel 7.6%
Other Italian 8.6%
Cheddar 33.4%
Mozzarella 33.1%
Other American 8.3%
Swiss 3.2%
Muenster 0.9%
Brick and Limburger 0.1% each

SOURCE: Compiled by the staff with data from "Dairy Products: 2005 Summary," U.S. Department of Agriculture, National Agricultural Statistics Service. April 2006.

ment of the USDA, based on weight of production. Total production of cheese in the United States in 2005 was 9,127 million pounds. NASS data indicate that Cheddar is the largest variety produced in the United States (33.4% of total) closely followed by Mozzarella (33.1%). Third place was held by other types of Italian cheeses (8.6%). Other American cheeses, including Colby and Monterey (8.3%) held the fourth place; cream cheese and Neufchatel

(7.6%) came in fifth place. Processed cheeses were made from Cheddar and the "other American" categories.

Varieties with the largest increase in production from 2004 were Hispanic cheeses (up 17.5% from 2004), Brick (up 9.8%), and Muenster (up 7%). Losing production tonnage were Limburger (down 10.1%), and cream cheese and Neufchatel (down 1%).

Connoisseurs of cheese may buy all kinds of imported cheeses, but imports represent a relatively small fraction of total U.S. cheese consumption—around 5 percent. In 2005 consumption was 9,409 million pounds and imports represented 461 million pounds of that total. Imports have been growing at an annual rate of 3.8 percent, but U.S. exports, while smaller in total tonnage than imports (128 million pounds in 2005) have been growing at a rate of 11 percent per year in the 1992 to 2005 period—suggesting that America makes high quality cheeses. Net imports, consequently, have been growing at a rate of 2.2 percent per year. Thus, in 1992, 4.3 percent of consumption was satisfied with imports contrasted with 5 percent thirteen years later.

Trends Supporting Cheese Consumption. People are cutting back on dairy foods if consumption is measured by weight of product on a per person basis. People are drinking less milk, eating less ice cream, consuming less cottage cheese, sour cream, buttermilk, yogurt, dried milk, evaporated milk, and also less whipping cream. Declining per capita consumption does not mean that the public is spending less. On the contrary. Expenditures are rising, but actual consumption of product is down. People are paying more for less quantity of dairy products. This trend had held for twenty years by the middle of the first decade of the twenty-first century. Only two categories defied the trend—butter and cheese. Why are these two products hold-outs? The USDA attempted an explanation in its 2001–2002 *Agricultural Fact Book*. After noting per capita consumption trends, the *Fact Book* authors come to the following conclusions regarding cheese:

> Lifestyles that emphasize convenience foods were probably major forces behind the higher consumption [of cheese]. In fact, more than half of our cheese now comes in commercially manufactured and prepared foods (including food service), such as pizza, tacos, nachos, salad bars, fast-food sandwiches, bagel spreads, sauces for baked potatoes and other vegetables, and packaged snack foods. Advertising and new products—such as reduced-fat cheeses and resealable bags of shredded cheeses, including cheese blends tailored for use in Italian and Mexican recipes—also boosted consumption.

Production figures for 1997 and 2002 underline the USDA's findings. These production figures distinguish between shipments purchased for consumption by consumers (packaged cheese shipments) and shipments purchased by the food industry for use in other food products (bulk cheese shipments). In both years, bulk shipments of natural cheese, the larger component of the market, accounted for 65 percent of shipments, the cheese thus reached industrial and institutional buyers. Nearly 40 percent of processed cheese also moved in bulk. Butter, similarly, is a product diet-conscious individuals avoid, but producers of ready-to-eat meals and snacks incorporate into the product because it delivers a taste very difficult to achieve using substitutes. More than half of all butter shipments were also bulk shipments.

KEY PRODUCERS/MANUFACTURERS

The leading producer of cheese in the United States is Kraft Foods. James L. Kraft began the company in 1903 as a cheese wholesaler in Chicago, Illinois. In 1914 the company began making its own cheese and supplied canned cheese to the military during World War I. Kraft pioneered processed cheese production in the United States and received many patents for its process, the first one in 1916. Kraft was also a pioneer in specialty cheese products, introducing Velveeta in 1928 and Kraft's Macaroni & Cheese dinners in 1937, a ground-breaking entry into the prepared food market. The company remained independent until 1988 when it was acquired by Philip Morris, the tobacco company. Philip Morris had already purchased General Foods (in 1981). In 1989 the company formed Kraft General Foods from the two acquisitions and, in 1995, renamed this element Kraft Foods. Philip Morris, renamed Altria, spun off Kraft Foods in early 2007. Kraft had revenues of $34.4 billion in 2006 of which cheese and dairy products represented $6.4 billion. The company's overall share of the cheese market in the late first decade of the twenty-first century stood at around 25 percent, lower in natural cheese and higher in the processed cheese categories.

The largest cheese market segment, representing nearly 40 percent of all cheese made, is the chunk or loaf market. Leaders in this segments, after Kraft Food, are Tillamook County Creamery, Lactalis USA, Land O'Lakes Inc., Cabot Creamery Inc., Cacique Creamery Inc., H.P. Hood LLC, and Saputo Cheese USA. Dairy cooperatives play an important role in the dairy industry as a whole—and cheese is no exception. Tillamook, Land O'Lakes, and Cabot are all cooperatives. Tillamook began in 1909 in the county of that name in Oregon and has been operating since that time as a farmers' co-op, currently representing 150 dairy families but distributing cheese products across the nation and other dairy products

regionally. Land O'Lakes is the nation's third-largest dairy cooperative and fourth largest producer of natural chunk cheese. The cooperative is based in St. Paul, Minnesota, and has 1,200 members. It was founded in 1921 and is the largest producer of butter in the United States. Cabot is the nation's fifth largest chunk cheese producer, a dairy cooperative based in Montpelier, Vermont, first organized in 1919. It is a producer and distributor for 1,500 farms, operating four plants.

The other leading participants in this segment of cheese manufacturing, in addition to Kraft Foods, are corporations. Lactalis USA is the representative in the United States of Groupe Lactalis, a very large multinational dairy company based in France. Groupe Lactalis is Europe's largest cheese producer with top market share rank in France as well as Italy. It holds third place in chunk natural cheese (fifth place in natural shredded cheese) in the United States. Lactalis built plants in Belmont, Wisconsin and Turlock, California and owns two other acquired operations. Cacique is a relatively recent corporate entry into cheese making in the United States. Established by a Mexican family in 1973 in California, Cacique began making just 100 pounds of cheese per day in a tiny facility, producing unique Hispanic cheese products. The company has achieved fifth-place ranking in chunk natural cheese and is the leader in the rapidly growing Hispanic cheese varieties. Cacique continues to be family-owned and operated. H.P. Hood, with sixth-place rank, participates in the category through Heluva Good Cheese Inc., a company it acquired in 2004 as part of its acquisition of Crown Foods. Heluva was established in 1925. It began and is still operating in Sodus, New York (near Rochester). Since then the company has expanded and operates facilities in the states of New York, Massachusetts, Pennsylvania, Virginia, and Tennessee. Saputo Cheese USA is the U.S. element of Saputo, Inc., a major Canadian multinational cheese company. Saputo had revenues of $4 billion (Canadian) in 2006. It has operations in Canada, the United States, Argentina, and Germany. Saputo USA has 13 cheese-making plants in the United States. In 2007 the company was negotiating with Land O'Lakes to acquire the cooperative's West Coast industrial cheese manufacturing business.

The leader in the second largest cheese segment, shredded natural cheese (about 35 percent of total production) is also Kraft Foods with an estimated share of 32 percent in 2004. The second largest producer in that year was Sargento Foods Inc. with a 12 percent share. Sargento was founded by the Gentine family in Plymouth, Wisconsin, a family that still owns the company. Leonard Gentine and Joseph Sartori founded the company by joining forces and their names, but the Gentines bought out the Sartori interest in 1965. Sargento was the first company ever to market shredded cheese and remains a strong presence in that market. The third ranked producer in the market is American Dairy Brands, an element of Dairy Farmers of America, the nation's largest dairy cooperative based in Kansas City, Missouri, but operating nation-wide. Lactalis, Saputo, and Tillamook come in fourth, fifth, and sixth place.

The seventh-place participant in shredded natural cheese, Schreiber Foods Inc., deserves special mention. The company is the second largest manufacturer of cheese in the United States overall but is rarely mentioned because it is predominantly a supplier of the fast food industry and the leading private label cheese producer. In virtually every market share compilation, the "private label" category has the largest share of sales; and the largest share of the private label market belongs to Schreiber. The company was founded in 1945 in Green Bay, Wisconsin. It is a privately held company with sales in excess of $3 billion. As a privately held company, it does not interest those in the media who track the performance of publicly traded entities. Schreiber claims to be the largest privately held cheese company in the world. Schreiber operates globally (in the United States, Europe, Latin America, the Middle East, and Asia). Schreiber also sells its own pro-prietary line of cheese processing equipment.

Although the eleven organizations mentioned above represent a substantial segment of total production (approximately 65%), cheese manufacturing was carried out by 366 companies in 2002, using Census Bureau data, suggesting that the market concentration is relatively low. The 366 companies operated 501 plants in that year; of those, 278 had twenty employees or more and were thus substantial. Company and establishment counts have been declining over time. In 1992, 418 companies operated 576 establishments, 314 of those being large (20 employees or more). Consolidation is thus taking place gradually; in the process, establishments are being closed. Small establishments with fewer than twenty employees have been declining more rapidly than the larger ones. The presence of many small companies, however, implies the presence of a significant wholesale distribution system.

MATERIALS & SUPPLY CHAIN LOGISTICS

Milk is the only bulk ingredient required for cheese manufacturing. Milk production, however, is concentrated in 23 states; they produce 89 percent of all milk. California, Wisconsin, and New York account for 40 percent of all milk production. Not surprisingly, cheese production is heavily concentrated in Wisconsin, on the West Coast, and in the North East. A gallon of milk weighing 8.6 pounds produces about one pound of cheese. Logistics alone dictate that cheese is most economically produced near the occurrence of milk. The transportation costs of

one pound of finished product to a distant consumption market is much less than moving the milk.

In the production of processed cheese, the major raw material is natural cheese itself—which is then ground, melted, emulsified, and shaped. Processed cheese operations thus can be located more distantly from milk-producing regions. Processed cheese, requiring controlled atmospheres for optimal production, require the availability of nitrogen gas. Location of such facilities is somewhat influenced by the gas liquefaction plants where, typically, in the production of oxygen from air, nitrogen is also produced for industrial purposes. Oxygen production is normally closely located to centers of steel production, thus in heavy industrial corridors.

DISTRIBUTION CHANNEL

In 2005 some 2,400 dairy product wholesalers distributed dairy products valued at $39 billion. Some portion of that total was cheese, but Census Bureau data, at the wholesale merchant level, are not broken out by kind of product distributed so that precision is not possible—especially in the dairy category where many different types of distribution exist side by side. Two-tier distribution is common in situations where a large producer sells directly to a large grocery chain and the chain, in turn, sells directly to the consumer. The chain may perform a middleman function by buying for regional markets, holding the product in its own warehouses, and distributing the product to stores. A similar process is in place where chains buy private label merchandise; the manufacturer puts the product into packaging designed by the buyer and delivers directly to the chain. Two-tier distribution is also common for cheese products aimed at the fast food industry. The products—cheese slices for cheeseburgers, for instance—are produced under contract for the franchises and delivered directly by the producer. Variations on these patterns are common, and distributors are used, where producer-controlled deliveries are not justified by volume.

Major cooperatives play a large role in the distribution of products. The top dairy cooperatives—Dairy Farmers of America, Land O'Lakes, and Dairylea Cooperative Inc.—are major distributors as well as producers. A number of smaller cooperatives are also active in distribution, an example being Associated Milk Producers Inc. selling products produced by its members in the Upper Midwest. Commercial distributors will frequently have a history of production or be engaged in manufacturing although their principal activity is the distribution of goods on behalf of smaller dairies. Examples are Crystal Farms and Shamrock Foods. Crystal Farms has its own processing facility in Lake Mills, Wisconsin. The Arizona-based Shamrock Foods is a major dairy food distributor but also owns and operates Shamrock Farms.

KEY USERS

Virtually everyone eats cheese. Cheese contains very little milk sugar (lactose) and is therefore better tolerated even by the lactose-intolerant. Although some cheeses have a very sharp taste and a distinct odor, others are very mild and almost sweet. Connoisseurs of cheese may appreciate the sharp tones of strong Cheddar or the pungent aroma of Limburger. On the other end of the spectrum are the mild, soft cheeses offered to children by way of increasing their intake of calcium and protein. Among the most popular of these mild cheeses, marketed around the world and with a particular focus on children, are the Baby Bel and Mini Baby Bel cheeses, one of the La Vache Qui Rit (Laughing Cow) brand of cheeses produced by Fromageries Bel, S.A. Within this vast range of cheeses there is almost certainly one to match every taste.

ADJACENT MARKETS

Cheese is a source of protein in food. There are no direct substitutes for cheese, thus adjacent markets must be considered those that provide products high in protein content. Major sources of protein are animals or animal products, thus eggs, red meat, poultry, and fish—and certain categories of vegetables, principally legumes (peas, beans, and lentils).

In the past the public has consumed expensive foods when it could afford it—and animal proteins have always been more expensive than proteins harvested as plants. Early in the twentieth century, for instance, red meat consumption was limited by income. Late in the twentieth century it was limited by fashions. In the early decades of the century beef consumption was under 60 pounds per capita. Between 1964 and 1987, consumption increased sharply to more than 100 pounds per person—and then dropped again, and has continued to drop, reaching 93 pounds in 2005. In the early 1900s people could not afford to eat expensive meat. In the early years of the twenty-first century they chose not to, in order to avoid fats.

In more recent times, for instance, veal consumption dropped from 2.2 to 0.6 pounds per person between 1985 and 2005, pork consumption (despite the lean-pork movement) dropped in the same period from 66 to 64 pounds, lamb consumption from 1.6 to 1.2 pounds per capita, egg consumption from 216 eggs per person to 175. At the same time chicken consumption grew from 56 to 104 pounds per person, and turkey consumption increased from 12 to nearly 17 pounds per capita, and fish consumption from 9.7 to 11.5 pounds. Poultry and fish have been promoted as healthier sources of protein by health and dietary authorities. So have legumes. Legumes, however, have declined in the 1985 to 2005 period from 7 pounds to 6 pounds per person in consumption. The public appears to prefer meat to vegetables.

These data illustrate that in modern industrial markets, with extensive public communications and high levels of income, food products compete with one another based on rationales in which cost plays a relatively minor role. Cheese has done well in this environment and has grown in consumption.

RESEARCH & DEVELOPMENT

A significant waste product of cheese manufacturing is whey. It occurs as a thin liquid carrying whey proteins and residual amounts of lactose, casein, and butter fat. Whey proteins have come under intense study in the twenty-first century as the industry has been looking for ways to utilize whey. Considerable research is under way to identify precisely the function of whey proteins and to explore their potential in food and in pharmaceutical products. Alongside such rather basic R&D, producers in the industry—very much in parallel with producers in other branches of the dairy industry—have been very active in product development. Efforts are directed toward creating unique new snack products to exploit changes in lifestyle that favor casual eating.

CURRENT TRENDS

As the first decade of the twenty-first century is drawing to a close, the major trend in the cheese industry could be characterized as "steady on." This is a way of saying that the industry is on a path of steady growth established already in the closing decade of the twentieth century. The major, and continuing, uncertainties in the industry relate to milk pricing as the dairy industry continues to adjust to legislative manipulations of the national milk subsidy program initiated in 1996. Consumption of cheese across the board is growing; the public preference for natural (as contrasted to processed) cheese has been in place for more than a decade. The notable increase in share of Hispanic varieties of cheese is directly attributable to changes in the composition of the population—the Hispanic component having shown the most rapid growth between 1990 and 2005 (up 60 percent). Cheese was and continues to be an important ingredient in fast food; nothing indicates a cooling of the public's romance with pizza, burgers, nachos, and tacos. In the food industry generally, efforts continue to increase the share of food purchased in ready-to-eat or ready-to-microwave meals delivered in frozen or refrigerated form. Snacking continues to be a growing market. Cheese plays a role in all of these trends. At the institutional level, slow consolidation of the industry, already visible in the early 1990s, is still taking place, resulting in the elimination of very small producers. But this consolidation is influenced by the continuing dominance and effectiveness of large dairy cooperatives.

Another trend that first emerged midway through the twentieth century still continues strong. It is the production of dairy products aimed at the health-conscious. In the first decade of the 2000s a new aspect of this trend includes policies to avoid milk from herds treated with recombinant bovine somatotropin (rBST), a man-made growth hormone used to increase the milk production of herds. This growth hormone, produced by recombinant DNA techniques, came into use in 1994. The product, named POSILAC, is banned outside the United States for use in cattle, but milk obtained from cows injected with rBST has been found to pose no risk for people; foreign countries buy the milk but will not permit their own cows to be injected with the hormone.

Controversy continues to surround use of this productivity enhancer in the twenty-first century. Cheese makers are sensitive to this issue as they are to all dairy industry trends. The rise in popularity of foods labeled organic is part of the movement to use fewer chemicals in the production of our food, be those hormones used on livestock or industrially produced fertilizers used on crops.

TARGET MARKETS & SEGMENTATION

Cheese producers serve well-defined market segments. The two largest segments are institutional and consumer markets. The food industry itself represents the first major segment. These cheese buyers are companies that use cheese as an ingredient in the preparation of snacks, sauces, and ready-to-eat meals and companies that sell fast food.

The consumer market has different segments represented by three forms of cheese. Pre-sliced products intended for sandwich-making are predominantly processed cheese (Cheddar and Colby). Sliced natural cheeses have made their appearance as well and appear to be growing. The third form of consumer cheese itself includes three types of cheese. The block cheeses are intended to be sliced by the consumer; cream cheeses are intended to be spread by the consumer; and grated and shredded cheeses intended for use in cooking.

Although a major portion of all cheese consumed in the United States falls into two varieties, Cheddar and Mozzarella, cheese by its very nature is targeted at human taste buds so that every conceivable variety has its own segment of customers. In the wake of World War II, faced with the disorders left over from the global conflict,

Charles de Gaulle once famously said: "Only peril can bring the French together. One can't impose unity out of the blue on a country that has 265 different kinds of cheese." The same may be said of the world as a whole—with 2,000 varieties. But Charles de Gaulle—had he been thinking more about cheese than unity—might well have added to his comment another famous French phrase. He might have said: "But, as for cheese, *vive la différence.*"

RELATED ASSOCIATIONS & ORGANIZATIONS

American Dairy Association & Dairy Council, Inc., http://www.adadc.com

American Dairy Goat Association, http://adga.org/compare.htm

American Dairy Products Institute, http://www.adpi.org

California Cheese & Butter Association, http://www.cacheeseandbutter.org

International Dairy Foods Association (IDFA), http://www.idfa.org/about/index.cfm

New York State Cheese Manufacturers' Association, http://www.newyorkcheese.org

Wisconsin Cheese Makers Association, http://www.wischeesemakersassn.org

BIBLIOGRAPHY

Agriculture Fact Book 2001–2002. U.S. Department of Agriculture. March 2003.

Dairy Products: 2005 Summary. U.S. Department of Agriculture, National Agricultural Statistics Service. April 2006.

Darnay, Arsen J. and Joyce P. Simkin. *Manufacturing & Distribution USA* 4th ed. Thomson Gale, 2006, Volume 1, 927–929.

Frankhauser, David B. "Rennet for Making Cheese." University of Cincinnati Clermont College. 14 February 2005. Available from: <http://biology.clc.uc.edu/fankhauser/Cheese/Rennet/Rennet.html>.

"The Internet's Cheese Information Resource." CheeseNet. Available from <http://cheesenet.info/default.asp>.

Lazich, Robert S. *Market Share Reporter 2007.* Thomson Gale, 2007.

"Natural and Processed Cheese." *AP-42. Compilation of Air Pollution Emission Factors.* U.S. Environmental Protection Agency. January 1995.

Wattiaux, Michel A. "Milk Composition and Nutritional Value." Babcock Institute for International Dairy University of Wisconsin-Madison, Research and Development. Available from: <http://babcock.cals.wisc.edu/downloads/de/19.en.pdf>.

SEE ALSO *Milk & Butter*

Children's Apparel

—■—

INDUSTRIAL CODES

NAICS: 31–52 Cut and Sew Apparel Manufacturing, including the following six-digit industries: 31–5211, 31–5212, 31–5221, 31–5222, 31–5223, 31–5224, 31–5225, 31–5228, 31–5231, 31–5232, 31–5233, 31–5234, 31–5239, and 31–5291

SIC: 2311 Men's and Boy's Suits, Coats, and Overcoats Manufacturing, 2321 Men's and Boy's Shirts Manufacturing, 2322 Men's and Boy's Underwear and Sleepwear Manufacturing, 2325 Men's and Boy's Trousers and Slacks Manufacturing, 2326 Men's and Boy's Work Clothing Manufacturing, 2329 Men's and Boy's Clothing Manufacturing not elsewhere classified, 2341 Women's, Misses', Children's, and Infant's Underwear and Nightwear Manufacturing, 2361 Girl's, Children's, and Infant's Dresses, Blouses, and Shirts Manufacturing, 2369 Girl's, Children's, and Infants' Outerwear Manufacturing, and 2384 Robes and Dressing Gowns Manufacturing

NAICS-Based Product Codes: 31–52111 through 31–52119100, 31–5211B through 31–5211H100, 31–52121 through 31–52129100, 31–5212B through 31–5212J100, 31–52211 through 31–52215035, 31–52221 through 31–52227001, 31–52231 through 31–52233026, 31–52241 through 31–52243003, 31–52281 through 31–52283159, 31–52285 through 31–52285100, 31–52311 through 31–52319003, 31–52321 through 31–52321015, 31–52321 through 31–52323022, 31–52330 through 31–52330022, 31–52341 through 31–52347001, 31–52391 through 31–52399100, 31–52910 through 31–529102C03, and 31–52991 through 31–52995131

PRODUCT OVERVIEW

Throughout most of history children have either gone without clothing, climate permitting, or dressed in smaller versions of adult clothing. In the western world, children's movements and behavior were restricted by their clothing, a societal tendency described in "Centuries of Childhood," an exhibit at the Kent State University Museum. Infants were wrapped in swaddling clothes, which left them immobile, and toddlers were constrained by adult styles.

In 1762 Jean-Jacques Rousseau published his controversial novel, *Emile*, which championed childhood as a pure state to be cherished. His ideas created a demand for children's clothing that provided comfort and convenience. By the 1820s, however, the industrial revolution enabled the rising middle class to purchase ostentatious clothing for their children, and the youngsters again found themselves constrained, this time by crinolines, bustles, stays, long hair, and layers of petticoats.

Children continued to suffer with elaborate garments and high-heeled, narrow shoes until after World War I, when dress was simplified for all groups and ages. The growing ready-to-wear industry produced and marketed comfortable, active styles for children. With rising affluence and, in the late twentieth century, with numbers of two-income households increasing, families were willing to spend more on children's clothing.

By the early twenty-first century, the children's clothing industry served a market worth more than $30 billion annually. Some firms such as Carter's and Oshkosh mainly produced children's garments, while many adult clothing companies added children's lines. Teens, meanwhile, provided another lucrative market, pushing to

219

the forefront such labels as Pacific Sunwear, Hollister (the surf-oriented Abercrombie & Fitch brand), and American Eagle Outfitters.

Some style-conscious parents seemed to return to the dress-like-adults model for children, paying high prices for designer clothing such as $200 rhinestone-encrusted Seven for All Mankind flare jeans or Great China Wall hoodies priced from $395 to $595 each.

For most parents, however, price was an important consideration when buying children's clothing. This emphasis on value made the industry particularly susceptible to global market pressures. As with adult apparel manufacturers, most children's clothing manufacturers contracted with offshore firms whose employees were paid a fraction of what U.S. apparel workers would have been paid.

A series of treaties and quota systems tried to protect the domestic textile industry and to promote apparel manufacturing in neighboring countries such as Mexico, but World Trade Organization rules pushed for an end to quotas. Many observers feared that in 2008, when quotas on China were to be eliminated, the apparel industry in many developing countries would be destroyed.

Others argued that two important factors gave a substantial advantage to apparel firms located in the United States and neighboring countries. First, U.S. trade policy supported tariffs which added maximally 30 percent to the cost of Chinese products. Second, Wal-Mart pioneered a lean retailing model that was widely adopted throughout the industry. In this model, retailers replenished their shelves on a weekly basis to avoid overstocks and to keep current with customers' fashion demands. Transportation time and expense, therefore, created a second disadvantage for Chinese products. Whether these factors would be enough to protect the apparel industry in the Americas and the Caribbean would not be known until after 2008.

MARKET

The retail market for children's clothing—for children younger than 12 years of age—was worth $30.6 billion in the United States in 2005, according to market researcher Mintel International. This was a 13 percent increase in sales since 2002. The infant and toddler market was expected to reach $18.4 billion in 2010, according to Packaged Facts, a division of MarketResearch.com. This estimate is based partially on U.S. Census Bureau statistics that predict that the number of children under age 2 will increase 5 percent to 8.2 million by 2010, and the number of children aged 2 to 5 will increase 3.7 percent to 15.9 million. The number of women of child-bearing age is also projected to increase, to 60.5 million in 2010 and 62.7 million in 2020.

The U.S. teen population is expected to grow from 32.4 million in 2000 to 33.5 million in 2010. This group

has substantial discretionary income, averaging $1,500 per year at age 12 and increasing to nearly $4,000 per year at age 16 to 17, according to a report on the teen market prepared by the *Magazine Publishers of America*. In this report, clothing topped the lists of what teens planned to buy next and what they had bought last in 2003. The number of clothing stores catering to teens increased substantially.

According to the U.S. Census Bureau there were 6,558 children's and infants' clothing stores with sales of almost $7.1 billion and 24,539 family stores with sales of $63.9 billion in 2002. As is true of the entire clothing industry, most children's and teen's clothing is imported. The statistics in the Census Bureau's *2002 Economic Census* do not include a specific category for children's clothing. Instead the categories are as follows: Men's and Boy's Cut and Sew Apparel Manufacturing, Women's and Girls' Cut and Sew Apparel Manufacturing, and Infants' Cut and Sew Apparel Manufacturing. The shipped value of all cut and sew apparel manufacturing for men and boys was $11.6 billion; for women and girls, 18.8 billion; and for infants, $315.3 million, in 2002.

According to the U.S. International Trade Commission, the United States imported $70.9 billion of cut and sew apparel in 2006 and exported a fraction of that, $2.6 billion. Cut and sew apparel is produced by firms that buy the fabric and cut and sew garments from it, as opposed to those companies that both knit cloth and make clothing from the resulting material. Imports of clothing in 1998 totaled $48.7 billion and exports totaled $6.1 billion. For infants' apparel, imports totaled $2.27 billion in 2006, and exports totaled $28 million. In 1998, $1.4 billion was imported and $205 million was exported. Figure 57 presents an overview of the U.S. apparel industry, including accessories, from 1997 to 2006.

Back-to-school sales represent a substantial portion of the children's clothing market. The Census Bureau reports that $7.1 billion was spent in family clothing stores during August 2006, a month second only to December in clothing sales.

Total 2007 back-to-school shopping was projected to be $18.4 billion in a survey by BIGresearch, conducted for the National Retail Federation. According to this study, families with school-age children were expected to spend an average of $563.49, a 6.9 percent increase from $527.08 in 2006. Of that, $231.80 was expected to be spent on clothes and accessories, with the remainder being spent on footwear and electronics.

KEY PRODUCERS/MANUFACTURERS

Most of the apparel companies that serve men and women, also have children's apparel lines. In addition, there are a few major companies dedicated to children's

FIGURE 57

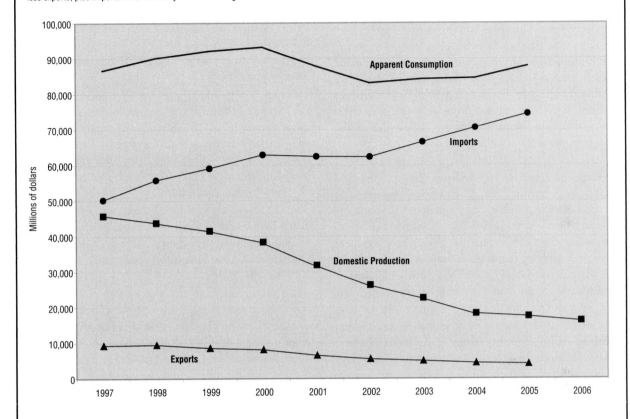

U.S. Apparel Industry, Production, Exports, Imports, and Apparent Consumption, 1997–2006

Shipments are presented in millions of dollars and include accessories as well as finished apparel. Apparent consumption is a calculation of domestic production, less exports, plus imports. Data were only available through 2005 for all measures other than domestic production.

Apparent Consumption

Imports

Domestic Production

Exports

Millions of dollars

SOURCE: Compiled by the staff with data from "Domestic Exports and Imports for Consumption of Merchandise by Selected NAICS Product Category: 1990 to 2000," *Statistical Abstract of the United States: 2001,* and "Domestic Expoerts and Imports for Consumption of Merchandise by Selected NAICS Product Category: 2000 to 2005," *Statistical Abstract of the United States: 2007,* and "Apparel: 2006," *Current Industrial Reports,* U.S. Department of Commerce, Bureau of the Census. January 2001, December 2006, and June 2007.

clothes. One of the best known of these is Carter's, a $1.3 billion marketer of Carter's and OshKosh branded apparel for babies and young children. The company reports on its Web site that it sells on average ten products for every child born in the United States. It markets its clothing as "adorable, comfortable, easy to care for, and very afford-able." Carter's has the largest share in the baby apparel market. OshKosh clothing is made for children ages 2 to 7 years.

Carter's. Carter's markets its products in department and specialty stores, as well as in its own chain of more than 180 retail stores. In 2003 the company launched the new brand, Child of Mine, which is sold exclusively at Wal-Mart. Carter's was founded in 1865 by William Carter, who knitted mittens in his kitchen. The Atlanta,

Georgia, company went public in 2003. In 2006 the company reported $1.34 billion in net sales.

Tween Brands, Inc. The company changed its name from Too, Inc. in July 2006 to reflect its focus on tweens—children from 8 to 18 years of age—as its core customers. Retail sales in 2006 were $883.7 million. Tween Brands became an independent company in 1999 after it sepa-rated from The Limited, the parent company of Victoria's Secret. The Limited established Limited Too independent stores and departments within its own Limited stores for sales to this age group.

Tween Brands sells its products at Limited Too and Justice stores. At the end of 2006 Limited Too had more than 722 stores in 46 states and Puerto Rico, with a few international franchise stores. These stores sell apparel, swimwear, sleepwear, underwear, footwear, lifestyle, and personal care items for active tween girls. The newer Justice chain focuses on moderately priced sportswear for tween girls. The first Justice store opened in 2004, and by the end of 2006 there were 159 Justice stores across the United States.

The company, however, had low sales in 2007, which led to disappointing per-share earnings and lower projections running through 2008. Company officials blamed delays in back-to-school dates and a shift in the Texas and Florida sales tax holiday, but Wall Street analysts suggested stiffer competition from discount and chain stores and fashion mistakes led to low sales in denim and casual bottoms in the volatile tween market.

The Gap, Inc. The Gap had revenues of $16 billion and 150,000 employees in 2006. The company was founded as a retail store in 1969 in San Francisco, California, by Doris and Don Fisher. In 2007 it had more than 3,100 stores selling the Gap, Banana Republic, Old Navy, and Piperlime brands of clothing, as well as the GapBody, GapKids, and babyGap brands. The Fisher family continues to control approximately one-third of the stock.

Gap is known for stocking its stores with its own brand of casual clothes for men, women, and children, featuring mainly T-shirts, jeans, and khakis. Banana Republic features urban chic clothing, while Old Navy is a budget merchandiser. All Gap products are private label, designed in-house, and manufactured by contract companies. The company states on its Web site that all factories selected to produce Gap products must meet the company's "Code of Vendor Conduct," with particular attention given to such issues as child labor and working conditions. Gap reported that the company terminated business with 23 factories, about 1.1 percent of the total, in 2006 for code violations.

Hanesbrands, Inc. Headquartered in Winston-Salem, North Carolina, Hanesbrands separated from Sara Lee Corporation in 2006. The company has approximately 50,000 employees and had $4.7 billion in net sales in fiscal 2005. Hanesbrands produces T-shirts, bras, panties, men's underwear, childrens' underwear, socks, hosiery, casual wear, and activewear. Company brands include Hanes, Champion, Playtex, Bali, L'eggs, Just My Size, Barely There, and Wonderbra. According to the company, its brands can be found in eight out of ten American households. In terms of sales, its T-shirts, fleece, socks, men's underwear, sheer hosiery, and childrens' underwear hold first place in the U.S. market. Its bras and panties are in second place.

Hanes Corporation is the outgrowth of two firms, each founded by a Hanes brother. J. Wesley Hanes established Shamrock Hills, a manufacturer of men's hosiery, in 1901. In 1902 Pleasant Hanes formed the P.H. Hanes Knitting Company and introduced men's two-piece underwear. Shamrock Hills changed its name to Hanes Hosiery Mill in 1910 and began to manufacture women's hosiery. The two companies merged in 1965 to form the Hanes Corporation. In 1971 Hanes Corporation acquired Bali Brassiere Company and Pine State Knitwear Company. The Just My Size brand for full-figured women was launched in 1984.

Levi Strauss & Co. Levi Strauss was founded in 1853 by Bavarian immigrant Levi Strauss, and shares of the company are still held privately by family members. Its stock is not traded in the United States, but Levi Strauss Japan, a company affiliate, is publicly traded in that country.

Levi Strauss, which has its headquarters in San Francisco, California, is the leading manufacturer of jeans and casual pants, with sales in 110 countries and more than 10,000 employees worldwide. Its brands include Levi, Docker, and Levi Strauss Signature. Net revenues for 2006 were $4.19 billion.

The company reports that it was the first global company to develop and implement a supplier code of conduct, and for 15 years it has monitored the employment practices of its contract suppliers. In 2006 it introduced a new program to help suppliers build management systems and capabilities that will enable them to meet these standards.

The VF Corporation. Headquartered in Greensboro, North Carolina, VF is one of the world's largest clothing companies, with more than 53,000 employees and annual revenues greater than $6 billion in 2005. It primarily manufactures jeans and sportswear. Its 40 brands include Wrangler, Lee, Riders, Rustler, North Face, Vans, Reef, Napapijri, Kipling, Nautica, John Varvatos, Jansport, Eastpak, Eagle Creek, Lee Sport, Majestic, and Red Kap.

VF Chairman and CEO Mackey J. McDonald told *Apparel* magazine in 2006 that the company was in the midst of transforming from a "category" apparel business to a "growing lifestyle brand" company, and that this change had resulted in three consecutive years of record earnings, with 2006 expected to be a fourth.

VF's history began with the Reading Glove and Mitten Manufacturing Company established in Pennsylvania in 1899. In 1919 the company began making undergarments and changed its name to Vanity Fair Mills. After acquiring the H.D. Lee Company in 1969, Vanity Fair Mills changed its name to VF Corporation. When Blue Bell Inc., with brands such as Wrangler and Jansport, was acquired, the VF Corporation became the largest publicly held apparel company.

Polo Ralph Lauren. Polo Ralph Lauren reported annual sales of $3.3 billion in 2005, 3.75 billion in 2006, and $4.3 billion in fiscal year 2007. The company began in 1967 with a collection of neckties by designer Ralph Lauren, who, as of 2007, continues to control most of the company's voting stock. In the twenty-first century the company designs and markets apparel, accessories, fragrances, and home furnishings, outsourcing production to a worldwide group of contract manufacturers. The company's brand names include Polo, Chaps, Lauren, and Club Monaco, which are sold at approximately 290 retail and outlet stores in the United States and at licensed stores around the world. Polo Ralph Lauren brand clothing is a favorite of many teens.

Fruit of the Loom. Headquartered in Bowling Green, Kentucky, Fruit of the Loom is a vertically integrated manufacturer of underwear and casual clothing. The company, which spins its own yarn, weaves or knits its cloth, and manufactures the finished clothing, was acquired in 2002 by Warren Buffet's Berkshire Hathaway. Fruit of the Loom is America's biggest seller of men's briefs, and it sells a variety of other underwear for men, women, boys, and girls. Additional products include T-shirts, activewear, casual wear, and children's clothing. Its brands include BVD, Munsingwear, and Gitano, as well as Fruit of the Loom.

The company was formed from two firms: B.B.&R. Knight brothers and Union Underwear. B.B.&R. Knight Brothers textile company was established in Rhode Island in the mid-nineteenth century. The brothers called their quality broadcloth "Fruit of the Loom," and that name was patented in 1871. As more women began buying ready-made clothing and linens, the retail market for cloth declined, and in 1928 the company began to license the brand Fruit of the Loom to clothing manufacturers. At about this time, Jacob Goldfarb established the Union Underwear Company. In the late 1930s he purchased a Fruit of the Loom license and began heavily promoting the name.

Union Underwear was acquired by the Philadelphia & Reading Corporation in 1955. The company had become the dominant producer of Fruit of the Loom products, and in 1961, Philadelphia & Reading acquired the Fruit of the Loom name to protect its use of the brand. The company acquired the BVD trademark in 1976 and began marketing this brand to upscale stores. A series of expansions and acquisitions followed. Until the 1990s Fruit of the Loom manufactured most of its products within the United States, but at that time it began moving its production out of the country. Fruit of the Loom now has more than 60 manufacturing and distribution companies around the world.

MATERIALS & SUPPLY CHAIN LOGISTICS

Material Inputs. Materials used by the apparel industry consist largely of fabric, both woven and knitted, as well as smaller quantities of items such as buttons, zippers, and elastic. While the apparel manufacturing industry in the United States has largely been supplanted by lower-cost off shore contract firms, the textile industry continues to be an important source of U.S. manufacturing employment. While an apparel manufacturing enterprise can be set up in a third-world country with little capital investment beyond a few sewing machines, textile mills require larger capital outlays, more advanced technology, and employees who understand how to use the technology.

A study published by Competitiveness Review in Winter 2006 and titled *Regional Trade Pacts and the Competitiveness of the U.S. Textile Industry*, reported that the industry from 1989 to 2001 was the third largest manufacturing industry in the United States, but that it was under severe pressure from foreign competition. Census figures show that in 2002 there were 3,932 textile mills in the United States, shipping $45.65 billion worth of products. The industry employed 269,064 workers.

Both textile and apparel manufacturing have been the subject of a series of contentious trade negotiations. The phasing out of the Multifibre Arrangement of 1974 put severe pressure on both industries by the end of the twentieth century. Pressure increased in 2002 when China joined the World Trade Organization and a series of safeguard quotas were due to expire in 2008.

The importing of textiles into the United States resulted in job losses, with as many as 100 plant closings in 2001 alone. The industry responded in two ways in the first decade of the twenty-first century. First, textile manufacturers achieved higher productivity with technology upgrades. Second, regional trade pacts led to realigned markets. Regional partnerships, encouraged by the trade agreements, cut transportation time and costs and built markets for U.S. textiles. Neighboring countries could ship apparel into the U.S. duty-free as long as the products were made with U.S.-produced fabric or yarn under the North American Free Trade Agreement (NAFTA), Central American Free Trade Agreement (CAFTA), and the Caribbean Basin Initiative (CBI). U.S. textile imports in real dollars doubled from $5 billion in 1989 to a little more than $10 billion in 2001. U.S. textile exports increased from $2.83 billion in 1989 to nearly $9 billion in 2001.

U.S. International Trade Commission data on textile mills do not correlate directly with the data used in the *Regional Trade Pacts and the Competitiveness of the U.S. Textile Industry*, nonetheless, they can be used to show that in the years from 2001 to 2006, the U.S. textile industry was holding its own. Textile imports increased from $6.34 billion in 2001 to $7.36 billion in 2006. Exports kept up with the rising value of imports, increasing from $7.37 billion in 2001 to $8.52 billion in 2006.

The effects of these changes on the apparel industry are complex. Protecting U.S. textile production would seem to drive apparel costs up, but supporters of the trade policies note that the agreements have allowed the industry to contract with companies in neighboring countries where labor costs are lower and then import the finished goods without paying any tariff. In addition, having garments made in nearby countries such as Mexico with lower transportation time and costs, not only saves money on shipping; the arrangement supported lower-cost, just-in-time inventory systems.

Other issues relate specifically to the material used in the production of children's clothing. Children's sleepwear sold in the United States must be made of flame-resistant material that self-extinguishes if a flame causes it to catch fire. In fact safety issues of all kinds receive particular attention when they involve children's clothing. In 2006, for example, Wal-Mart removed some brands of children's clothing that was made in China after they were found to contain a cancer-causing dye.

Supply Chains. Apparel has one of the most complex supply chains of any industry. Clothes are consumable products that are replenished regularly, but are dependent on fashion trends and seasons of the year. Apparel vendors must forecast trends and quantities weeks or months before the clothing is due on the retailers' shelves, factoring in manufacturing time. Forecasting trends and quantities was difficult when working with domestic factories, but in the twenty-first century, clothing companies often contract with a number of factories in several different countries. Communication difficulties are exacerbated by time zone differences. Forecasts now must take into account shipping and customs processing as well as manufacturing times.

Public attention on the working conditions in factories producing products under contract for large global firms rose during the late 1990s and the early years of the twenty-first century. Some companies have been severely criticized for the working conditions in factories run by the companies with whom they contract. Several companies have adopted codes of vendor conduct in an attempt to address this problem. In addition to vendor codes of conduct, companies are including other working condition clauses into the contracts with manufacturing subcontractors. Gap Inc., for example, is attempting to manage its supply chain so that factories are not forced to require excessive overtime to meet last-minute orders or changes.

Other complications can include bureaucratic chaos and/or inadequate infrastructure in a developing nation. India, for example, in the early 2000s was trying to develop a healthy manufacturing export industry, however, a substandard infrastructure system needed to be repaired. Some government officials stated an investment of $450 billion was needed between 2007 and 2012 to accomplish this. Only 2.4 million kilometers of India's roads were paved in 2007, and more than 1 million were unpaved. Drivers often had no rear-view mirrors and seat belts and communications systems were rare. Most rail lines, moreover, were old.

In 2007 India was spending 14 percent of its gross national product on its logistics system, compared to 8 percent in developed countries. These funds were raised by taxes, which drove up costs for Indian businesses.

Major carriers such as FedEx, UPS, and DHL began investing in India, thereby providing managerial expertise and allowing small businesses in the country to connect with the global economy. Services such as FedEx, however, concentrate on shipments of high-value goods with a critical need for speed to market, including high fashion apparel. When costs are critical, countries where transportation is fast and inexpensive have an advantage.

Still, it appears that major clothing companies will continue to consider India and other developing nations for apparel contracts even after quotas are removed from clothing made in China. Although Chinese products are often the least expensive, many buyers want to diversify their supply chains. In addition to the political risks of depending on China for all of their products, the buyers are also aware that terrorist attacks, hurricanes, earthquakes, and epidemics such as SARS can all disrupt supply lines.

DISTRIBUTION CHANNEL

The distribution systems used to get clothes from where they are made to the retail outlets from which they are sold has experienced many changes in recent decades. One of the largest U.S. shakeups of this distribution channel came in August 2005 when Federated Department Stores acquired the May Department Stores, which included Macy's and Bloomingdales. This and other consolidation deals increased competition in the apparel business because they resulted in fewer stores and less shelf space, while at the same time retailers were filling more of that space with store brands.

In 2007 clothing was being sold in department stores, specialty stores, discount stores, mass merchandisers, and warehouse clubs. The apparel makers also marketed products in their own retail stores and factory outlets, in catalogs, and on the Internet.

Cotton Incorporated, an industry group, conducted a Global Lifestyle Monitor survey in 2006, which included a survey of 4,000 U.S. consumers aged 15 to 54. The survey showed that U.S. consumers spend an average of $918 per year on apparel, buying 27 percent from department stores, 22 percent from chain stores, 19 percent from mass merchandisers, and 14 percent from specialty stores. When asked what was the most important factor when buying clothes, 77 percent said price.

In 2003 Chain Store Guide, a division of Lebhar-Friedman, conducted a study of 123 retailers operating 40,000 stores in 346 markets. The study concluded that discount apparel retailers had made dramatic gains in market share, accounting at that time for $70.2 billion of an estimated $182 billion apparel market. Wal-Mart held 24.6 percent of the discount market. Other major discounters were T.J. Maxx/Marshalls (10.1%), Target (8.4%), Old Navy (7.8%), Kmart (6.1%), Ross (3.8%), Charming Shops (3.2%), Burlington Coat Factory (2.7%), American Eagle Outfitters (1.9%), and Value City Department Stores (1.7%).

Harvard Business School conducted a study of the textile and apparel industries in 2000 to look at the potential of e-commerce. That study, supported by the Sloan Foundation, reported 1999 retail channels for an apparel and accessories business estimated at $179.8 billion as follows:

- Discounters, 20.5 percent
- Specialty stores, 22.4 percent
- Department Stores, 19.1 percent
- Major chains, 16.2 percent
- Off-price retailers, 6.5 percent
- Factory outlets, 3.7 percent
- Catalog, 9.6 percent
- Online, 0.6 percent
- Unreported, 1.4 percent

That study stated that the apparel industry could be segmented historically in several ways. When considering cost, a large segment of the industry competes on low cost, buying goods from distant suppliers and cutting costs by ordering and shipping in large lots. Such clothing was generally sold through mass merchants such as Kmart or Wal-Mart or at lower-end specialty stores and was a substantial part of the children's clothing market. Other firms chose higher costs for better quality and more fashionable

goods. These companies generally sold through department stores or high-end specialty stores.

KEY USERS

With the occasional exception of a baby on a blanket or a toddler who sheds garments on a summer day, all children wear clothing. Tweens and teens are often the ones who purchase the clothes that they wear, funded by part-time jobs, spending money from parents, or gift certificates from relatives and friends. Parents generally buy clothing for younger children, but many higher-priced items also come from relatives and friends. Parents also often receive generous quantities of clothing for newborn babies as gifts.

ADJACENT MARKETS

The fabric and apparel industries are closely tied, with U.S. government trade policy attempting to create markets for the domestic textile industry by allowing apparel imports from neighboring countries as long as the garments use U.S.-made thread or fabric. Wholesale fabric purchases by apparel manufacturers are also important to the textile industry.

A wide range of children's accessories can be considered adjacent to the children's apparel market because they compete for the clothing dollar. One indication of how these products compete can be seen by studying back-to-school sales figures. The National Retail Federation's (NRF) 2007 Consumer Intentions and Actions Back-to-School survey showed that families with school-age children expected to spend an average of $563.49 per child. Of that, $231.80 was intended for clothing and accessories, $108.42 for shoes, $94.02 for school supplies, and $129.24 for electronics or computer-related equipment.

From June 2006 to May 2007, children's footwear sales rose 9 percent over the preceding 12 months to $5.72 billion, according to the NPD Group, which conducted a study for the NRF. As is the case with apparel, far more children's footwear and accessories are manufactured abroad and imported than are manufactured in the United States. In the case of footwear, calculations by the U.S. Trade Commission show that in 2006 U.S. firms exported only $340 million worth of footwear, but the country imported $18.7 billion. In fact, the American Apparel and Footwear Association (AAFA) lobbied in 2007 to have tariffs removed, particularly on children's and lower- to moderately-priced shoes, saying the tariffs cost U.S. footwear companies $1.9 billion in 2006 to protect a "manufacturing industry that no longer exists."

Similar statistics are available for many other categories of accessories. In 2006 the United States imported

$18.7 billion of hosiery and socks and exported $340 million. In 2002 the shipments of 166 U.S. hosiery and sock mills were valued at $348 million. In the men's and boys' underwear and nightwear category, 2006 figures show that the United States imported $2.45 billion and exported $309 million. For women's and girls' lingerie, imports totaled nearly $5.37 billion, while exports were only $279 million.

RESEARCH & DEVELOPMENT

The Competitiveness Review's article concluded that "product innovation is a key to counteracting shrinking export markets…The recent use of nanotechnology, shuttleless looms, and robotics in textile manufacturing and the creation of 'smart' fabrics are examples of ways to counter global competition." To help the apparel and textile industries survive in a highly competitive global environment, research and development is conducted both by non-profit industry and education centers and by individual companies.

Cotton Inc., for example, is a non-profit industry group funded primarily by U.S. cotton producers. At its research center in Cary, North Carolina, it deals with all stages of cotton production. Its agricultural research has improved farming practices to enable the production of more cotton on less land. Other initiatives look at issues such as harvesting and ginning of cotton, fabric development, dyeing, and finishing. The organization monitors the global textile marketplace, advances in textile chemistry and processes, and consumer attitudes. Initiatives in 2006 included new programs to convince both Chinese customers and college-age Americans of the benefits of wearing cotton clothing. In response to concerns about global warming and sustainability, another new program attempted to educate the public about improvements in cotton growing that reduce its environmental impact.

Supply chain management is a central concern of the Textile Clothing Technology Corporation, or [TC]², also located in Cary, North Carolina. [TC]² is a not-for-profit organization that operates a demonstration center for leading edge technologies and a research facility for emerging technologies and business processes in the sewn goods and related soft products industries. At its demonstration center, sewn products are developed in a totally digital environment that often eliminates the need to make a physical sample and reduces cost, risk, and time when compared with the traditional product development process. Digital textile printing makes mass customization possible, bypassing the screen making process and enabling quick changes to color or design elements before printing. To enable custom fitting, the organization has also pioneered body scanning technology using white light.

The Harvard Center for Textile and Apparel Research, funded by the Alfred P. Sloan Foundation, conducts studies of how technology is transforming the way retailers plan and order merchandise as well as how manufacturers forecast demand, plan production, and manufacture and distribute their products.

Clemson Apparel Research is another university center that studies the textile and apparel industries. It describes itself on its Web site as a "premier national resource for high-performance textiles and related materials research and applications." In one program, the center developed BalancedFlow, which addresses supply chain issues with software that is designed to minimize the number of "days required to convert money invested in inventory into new money…[coming from consumers] at the end of the supply chain."

The importance of supply chain technology to VF Corporation was acknowledged in July 2006, when Mackey J. McDonald, chairman and CEO, told *Apparel* magazine that the company's supply chain organization and technology group played important roles in the company's future goals for growth. He said that VF had already implemented a common systems platform across several of its fastest-growing brands and that it planned to implement that common platform across the entire organization.

Radio-frequency Identification (RFID) tags on merchandise help manufacturers and retailers keep track of products during shipping and receiving and make possible automated replenishment of stock. Another way for manufacturers and retailers to track the needs and preferences of customers is through market research.

CURRENT TRENDS

Trends come and go in children's clothing at a rate that pressures retailers to constantly refresh their assortment and presentation. Teens, in particular, can keep abreast of the international style scene on Web sites such as Hypebeast and HighSnobiety. As for younger elementary school children, many want to dress like their older brothers and sisters.

Piper Jaffray, an investment firm, conducts a biannual "Taking Stock with Teens" survey. In the 2007 survey, teens were asked to rank apparel and footwear brands. Footwear companies Nike, Puma, Steve Madden, Adidas, Nine West, and Reebok all made the most trendy list. Polo Ralph Lauren rated highest among apparel brands, with Ralph Lauren Rugby particularly popular with younger buyers. Guess was next, boosted by its provocative, sexy black and white ad campaigns.

The next three apparel companies were board sports—surfing, skate boarding, snow boarding, and wake boarding—companies with surfer backgrounds. Quiksilver of Huntington Beach, California, markets boardshorts and wet suits, as well as snow gear such as jackets, goggles, and ski pants. In apparel, Roxy started with a swimwear line, then added sportswear and denim. It was the first surf apparel company to market a women's line. Volcom, another California company with appeal to the surfing, skateboarding, and snowboarding markets, sells a range of sportswear which the company says incorporates "distinctive combinations of fashion, functionality, and athletic performance."

The last apparel company on the list was Fossil, of Richardson, Texas, which began selling watches and now sells a full line of apparel and accessories. As teen girls went back to school in the fall of 2007, the favored jeans had a variety of leg shapes from wide to skinny, mainly in dark shades of blue. High waists were back, replacing low-rise in many cases. The look was basic and clean, largely without patches, embellishments, or holes. Jeans were matched with bright printed tops in plaids, polka dots, and stripes. Hoodies were seen as a must-have addition, not sloppy, oversized hoodies suitable for gym clothes, but form fitting garments in bright colors with graphics. Dresses were also popular: printed maxis, mini dresses, sundresses, sweater dresses, and babydolls. Boys, who tended to look to MTV and extreme athletes for styles, were wearing layered tees, distressed denim, and polo shirts.

Elementary school children were also asking for hoodies. Girls were favoring colorful prints and plaids, skinny or wide-leg jeans, and dresses. Boys were looking for straight, skinny or slim fit jeans with back-pocket details; patterned fleece jackets; layered tees; and long-length plaid shirts.

In another trend in early 2007, clothing manufacturers and retailers were concerned that climate change was beginning to change purchase patterns. An unusually warm January in 2007 cut into outerwear sales, but cold weather eventually made an appearance and helped stores dispose of their winter merchandise. Manufacturers responded by introducing more crossover styles as well as transitional coats and jackets with removable liners and outershirts and outervests designed for layering.

Environmental concerns fueled another trend: sustainable fabrics. The Organic Exchange, a nonprofit organization, reported in early 2007 that global sales of organic cotton products jumped from $245 million in 2001 to $583 million in 2005. The group projected that sales would reach $2.6 billion by the end of 2008. NPD reported in June 2007 that 18 percent of customers surveyed said they were interested in buying eco-friendly products compared to only 5 percent in 2000.

Sama Baby, for example, founded in 2006 in Jacksonville Beach, Florida, sells organic cotton clothing for children 0 to 2 years old. Its garments are produced in

India using certified organic cotton and earth-friendly dyes. Babies R Us has organic cotton items for sale on its Web site, and Wal-Mart has its own line of organic baby clothes under its George brand.

Even mainstream firms such as Levi Strauss have made use of the organic trend, introducing in 2006 a sustainable jeans line called Levi's Eco, with jeans made from organic cotton and waistband buttons from coconut shells. Reinforced stitching replaced metal rivets, and the hangtag was made from recycled board.

In 2007 Paxar launched a line of eco-labels to market to companies that make eco-friendly garments. Meanwhile, Patagonia, an outdoor apparel manufacturer that uses only organically grown cotton, received property tax abatements from Nevada for building a "green" distribution center with greatly reduced energy and water usage and improved storm water management.

The rapidly changing manufacturing environment spawned other trends. Global competition and pressure for lean manufacturing were on the minds of a group of industry executives who met with *WWD* magazine in 2006. Most reported that they would be consolidating production with fewer suppliers, focusing on those who were performing best. The goal would be to develop long-term collaborative relationships with these suppliers to meet the demands of retailers who wanted quality goods and who wanted to quickly bring in the hottest looks that would result in more full-price sales.

"We have to abandon some of our old habits," said Bob Zane, senior vice president, Liz Claiborne, Inc. "This is not the time for country-of-the-month sourcing. This is not the time for factory-of-the-season sourcing, and this is not the time for bargain hunting."

Zane was one observer who thought China would be in a dominant position after 2008. He said that with the emergence of China as a manufacturing power and the coming elimination of quotas on that country, the apparel business would soon look more like the footwear business. Chinese shoe factories are geographically close to material suppliers, employ tens of thousands of workers, including product development specialists, and have transparent pricing so buyers could know what each step of the process costs, he said.

TARGET MARKETS & SEGMENTATION

The clothing market for children 0 to 12 years of age was estimated at $30.6 billion by Chicago market research firm Mintel in 2006. In addition, the teen market was growing rapidly early in the first decade of the 2000s because the number of teens was increasing and these teens had substantial discretionary income. Manufacturers and retailers count on a number of strategies for targeting children and teens.

Much of the children's market is in the value category, with the average parent spending about $250 per year on a child's clothing in 2006, Mintel said. Many mothers want to buy clothing on sale, and large numbers of them shop at discount stores.

Vertis Communications, which tracks consumer behavior, stated in 2006 that 27 percent of adults with young children, 28 percent of adults with children aged 8 to 11, and 30 percent of adults with children aged 12 to 17 said "best value" was the most important consideration when selecting a store. Other considerations included "has clothes that fit me," "carries brand names I want," and "can get clothes for the entire family at one place." "Carries latest/trendy fashions" totaled 2 percent or less for the three age groups.

In the Mintel survey, mothers named Wal-Mart or Target as their preferred destination, with Old Navy next. GapKids, The Children's Place, Sears, JCPenney, and Kohl's also ranked in the top ten.

A small but growing group of parents, however, were buying high-priced clothing for their children. *Child* magazine was launched in the mid-1980s as the first parenthood magazine dedicated to fashion. By the early 2000s, there had been a growth in children's fashion, with celebrities such as Mary-Kate and Ashley Olsen and Hilary Duff designing children's clothing lines. Many women's apparel labels also began making children's clothing for mothers who wanted to dress their children in the brands they knew. Two-income families wanted to give their children the best, paying high prices even for premium pajamas.

Licensed apparel from popular movies sells well in children's clothes. Both Hasbro and Marvel introduced major apparel programs in 2007 for Transformers and Spider-Man 3. Products were available at mass to mid-tier retailers, including Wal-Mart, Kmart, Target, Sears, Kohl's, JCPenney, Federated, and Dillards, with prices ranging from $4 to $50.

In an effort to use blockbuster summer movies to reach the teen market for back-to-school shopping in 2007, JCPenney contracted with Screen Vision to produce 60-second movie theater spots. The ads promoted apparel items, including a new denim line, CP7.

Another factor supporting growth in the children's clothing market in the early 2000s was the growing racial and ethnic diversity of the U.S. population. Racial segments of the U.S. population that tend to have larger families, Hispanics, for example, are growing at a faster rate than are those with declining family size, Caucasians.

A rapidly growing market segment for men, women, and youths is plus-size clothing. The U.S. Center for Disease Control reported in 2002 that 16 percent of children 6 to 19 years of age were overweight. In 2007 just-style reported that while U.S. apparel sales growth had slowed to around 3 to 4 percent per year, the plus-size market was expected to grow at least 10 percent per year, reaching $62 billion in 2012. Deb Stores, which cater to junior women and are popular with teens, now offer plus-size departments in more than half of their stores.

Seasonal targeting is another way in which the apparel industry promotes its goods. As for most retail industries, Christmas is the biggest selling opportunity of the year, and this holds true for children's clothing even with a large percentage of their parents' gift dollars going for toys, games, and electronics. In October 2006, the NPD group conducted a survey that showed that 65 percent of customers were planning to buy clothing for Christmas gifts, and that 64 percent had done so in 2005. By mid-December 2006, the NRF reported that 47 percent of shoppers had bought clothing.

RELATED ASSOCIATIONS & ORGANIZATIONS

American Apparel and Footwear Association, http://www.apparelandfootwear.org

Clemson Apparel Research, http://car.clemson.edu

Cotton Inc. http://www.cottoninc.com

The Harvard Center for Textile and Apparel Research, http://www.hctar.org

National Retail Federation, http://www.nrf.com

[TC]² (Turning Research into Reality), http://www.tc2.com

BIBLIOGRAPHY

Anderson, Mae. "Market Spotlight: Back-to-School Shoes." Associated Press and Forbes.com. 6 August 2007. Available from <http://www.forbes.com/feeds/ap/2007/08/06/ap3991608.html>.

"Apparel: 2002," and "Apparel: 2006." *Current Industrial Reports.* U.S. Department of Commerce, Bureau of the Census. August 2003 and June 2007.

"Back-to-School Spending this Year to Top $18 Billion." National Retail Federation. Available from <http://www.nrf.com./modules.php?name=News&sp_id=342&op=printfriendly&txt=Nation>.

Baker, Stacy. "The Global Market for Discount Apparel." Just-style.com. June 2007.

"Centuries of Childhood." Kent State University Museum. Available from <http://dept.kent.edu/museum/exhibit/kids/kids.html>.

Chidoni, Loren. "From the Runway." Parents.com. Available from <http://www.parents.com/parents/printableStory.jsp?storyid=/templatedata>.

Clark, Evan and Ross Tucker. "Changes Sweeping Global Manufacturing." *WWD.* 14 March 2006, 8.

Data, Anusua, D.K. Malhotra, and Philip S. Russel. "Regional Trade Pacts and the Competitiveness of the U.S. Textile Industry." Competitiveness Review Winter 2006. 2007, 262.

"Estimates of Monthly Retail and Food Services Sales by Kind of Business: 2004." U.S. Department of Commerce, Bureau of the Census. Available from <http://www.census.gov/mrts/www/data/html/nsal04.html>.

"Estimates of Monthly Retail and Food Services Sales by Kind of Business: 2006." U.S. Department of Commerce, Bureau of the Census. Available from <http://www.census.gov/mrts/www/data/html/nsal06.html>.

Garbato, Debby. "Apparel Vendors Link a Global Chain to Retailers." *Retail Merchandiser.* May 2004, 34.

Gellers, Stan. "Weight Watchers: After a Winter Market by Record High and Low Temperatures, the Search is on for Transitional Outerwear." *Daily News Record.* 12 February 2007, 112.

Grannis, Kathy and Ellen Davis. "Online Clothing Sales Surpass Computers, According to Shop.Org/Forrester Research Study." National Retail Federation. Available from <http://www.nrf.com/modules.php?name=News&op=viewlive&sp_id=292>.

Grannis, Kathy and Scott Krugman. "Back-to-School Spending This Year to Top $18 Billion." National Retail Federation. 17 July 2007. Available from <http://www.nationalretailfederation.com>.

Hall, Cecily. "Coolness Factor: The Top 12 Footwear and Apparel Brands That Teens Consider to be the Trendiest." *WWD.* 7 June 2007, 17.

Hammond, Jan and Kristin Kohler. "E-Commerce in the Textile and Apparel Industries." Harvard Business School. Volume of BRIE Papers. Revised 19 December 2000.

Howell, Debbie. "Top Brands: A Comprehensive Nationally Projectable Consumer Survey of Brand Preference in 17 Product Categories." *DSN Retailing Today.* 24 October 2005, 38.

"Industry Statistics Sampler: NAICS 4481." 1997 *Economic Census.* U.S. Department of Commerce, Bureau of the Census. Available from <http://www.census.gov/epcd/ec97/industry/E4481.htm>.

"Industry Statistics Sampler: NAICS 4481." 2002 *Economic Census.* U.S. Census Bureau. U. S. Department of Commerce.

Available from <http://www.census.gov/econ/census02/data/industry/E31–33.htm>.

Kusterbeck, Staci. "The Apparel Top 50 Super Achievers Score High." *Apparel.* July 2006.

"The Looming Revolution—The Textile Industry." *The Economist.* 13 November 2004, 76.

McCormack, Karyn. "A Tough Break for Tween Brands." *Business Week.* 22 August 2007.

McNaughton, David. "Children's Clothier Carter Gains $61.3 Million from Initial Public Offering." *Atlanta Journal-Constitution.* 25 October 2003,

Malone, Robert. "Logistics: Bottlenecks to Growth." Forbes.com. Available from <http://www.forbes.com/2007/08/05/india-logistics-infrastructure-oped-cx_rm_0813logistics_print.html>.

Nolan, Kelly. "Intellifit: Dressing Room of the Future." *Retailing Today.* 4 June 2007, 9.

———. "Where Fashion Conscious Meets Bargain Hunter." *DSN Retailing Today.* 8 May 2006, 20.

O'Donnell, Jayne. "Competition Grows as Brands Lose Places to Hang Their Wares." *USA Today.* 29 August 2006, 1B.

"Purchasing Decisions of Back to School Shoppers Heavily Influenced by Ad Inserts." *Vertis Communications.* 28 August 2006. Available from <http://www.vertisinc.com/files/PressReleases/060828PR_VCF_Retail_BackToSchool.pdf>.

"Teen: Market Profile." Magazine Publishers of America. Available from <http://www.magazine.org/marketprofiles>.

"Textile Consumer." *Cotton Inc.* Fall 2006. Available from <http://www.cottoninc.com/TextileConsumer>.

"U.S. Interactive Tariff and Trade Database." U.S. International Trade Commission. Available from <http://dataweb.usitc.gov/scripts/user_set.asp>.

Van Dyke, Dawn. "AAFA, Members of Congress Introduce Affordable Footwear Act to Eliminate High Import Tariffs on Lower-Priced Shoes." American Apparel and Footwear Association. 12 June 2007.

Williams, Candy. "School Style." *Pittsburgh Tribune-Review.* 3 August 2007.

"W.S. Apparel Imports, 1997–2007." American Apparel & Footwear Association. Available from <http://www.apparelandfootwear.org/statistics/usimportsapparel0704.pdf>.

SEE ALSO *Athletic Shoes; Men's Apparel; Shoes, Non-Athletic; Women's Apparel*

Chocolate

———————

INDUSTRIAL CODES

NAICS: 31–1320 Chocolate and Confectionery Manufacturing from Cacao Beans, 31–1330 Confectionery Manufacturing from Purchased Chocolate

SIC: 2064 Candy and Other Confectionery Products and Chewing Gum, 2066 Chocolate and Cocoa Products

NAICS-Based Product Codes: 31–132040, 31–132071, 31–13207111, 31–132072, 31–13207221, 31–13207231, 31–13207241, 31–13207251, 31–132073, 31–13207360, 31–13207371, 31–13207381, 31–13297391, 31–13301, 31–133010, 31–13301000, 31–13301001, 31–13301004, 31–13301007, 31–13301015, 31–13301021, 31–13301028, 31–13302, 31–133020, and 31–13302000

PRODUCT OVERVIEW

Chocolate is by far the favorite flavor of Americans. The National Confectioners Association (NCA) reports that in one survey, 52 percent of Americans named chocolate as their first choice, with the next flavor choices, vanilla and berry, tied at only 12 percent each. As if especially designed to appeal to humans, cocoa butter has a melting point of 90 to 93 degrees Fahrenheit, just below human body temperature, which explains why chocolate melts in the mouth.

Historically, chocolate has been seen as an aphrodisiac, a hangover cure, an antiseptic, and a cure for burns, fevers, listlessness, rheumatism, and snakebites. Modern scientists now tell us that dark chocolate is a heart-healthy food. Chocolate candy's rich emotional appeal gives it a significant role in such major holidays as Valentine's Day, Easter, Halloween, and Christmas.

Chocolate candy is also big business. The U.S. Census Bureau reported that in 2005 chocolate manufacturing (wholesale) shipments totaled $12.13 billion. The NCA estimated that U.S. retail sales of chocolate candy grew 3.1 percent from 2005 to a 2006 total of $16.3 billion. (This figure has been adjusted to include imported chocolate and exclude candy exported by U.S. manufacturers.)

All chocolate begins with cocoa beans, which grow in pods on the cocoa tree. This tree grows only in tropical regions around the world, with the Ivory Coast in West Africa being the largest producer of cocoa beans. Humans have been enjoying chocolate for at least 2,500 years, with some civilizations elevating it to almost mystical status.

Most archeologists believe that the Olmecs (1500–400 BC), a tribe that populated what is now an area of Mexico and Central America, were the first to consume chocolate. The knowledge passed to the Mayans and Aztecs, who cultivated the beans, which they believed were man's inheritance from the god Quetzalcoatl. The cocoa beans were crushed and made into an unsweetened drink, but they were so important to the Aztec civilization that they were also used as currency.

The Mayans called cocoa the food of the gods, and the Aztecs simply called the beans "chocolatl," warm liquid. Montezuma, the great Aztec monarch, reportedly drank 50 or more servings of chocolate a day from a golden cup.

Christopher Columbus brought the first cocoa beans to Europe in 1502, but they were largely overlooked among the many intriguing objects he brought to the Spanish court. Hernando Cortes, on the other hand, recognized the commercial potential of cocoa beans, experimenting with chocolatl and adding cane sugar to make it more appealing to Spaniards. He established additional cocoa plantings in the Caribbean before returning to Europe. The drink found favor with the Spanish aristocracy, which kept it a secret for almost 100 years. Eventually, the knowledge of the chocolate beverage spread across Europe, with the drink becoming a favorite at the French and British courts. British King William II drank chocolate to help him recover from nights of gambling and drinking, and by the 1590s chocolate was in such favor at Hampton Court that a special kitchen was built just for preparing the beverage. By 1657 London had its first chocolate house. In 1753 Carl von Linnaeus, named the genus and species of the chocolate tree: *Theobroma cacao*, literally cacao, the food of the gods.

The invention of the steam engine made it possible to mechanize cocoa grinding, and the price dropped enough to make chocolate affordable to the general public. The invention of the cocoa press in 1828 made it possible to squeeze out about half of the cocoa butter, leaving a cake-like residue that could be processed into a fine powder. This lowered the price even further and improved the quality of the drink, giving it a smoother consistency.

In 1847 melted cocoa butter that had been combined with sugar and cocoa powder was first sold as eating chocolate. The candy, which was far smoother than the old grained chocolate, quickly became popular. A further improvement came in 1876, when it was discovered in Switzerland how to add milk to create milk chocolate. In the United States, Milton S. Hershey brought the first Hershey's milk chocolate bar to market in 1900. In 1923 the Mars company introduced its first candy bar, the Milky Way.

It takes about 400 cocoa beans to make one pound of chocolate, according to the NCA. Although there are a number of variations, the process is quite similar at most manufacturers. As the cocoa beans arrive, their variety and region of origin are carefully noted, because the flavor varies and the correct mixture of beans is needed for each product. The beans are cleaned, weighed, blended, and then roasted. After cooling, their thin, brittle shells are removed and blown away, leaving coarse pieces of chocolate known as *nibs*. Next the nibs are crushed between steel rollers or grinding stones. The nibs are about 53 percent cocoa butter, and the heat of the grinding process turns them to a liquid known as chocolate liquor.

If powdered cocoa is desired, hydraulic presses squeeze out much of the cocoa butter. The pressed cocoa cake is pulverized to make cocoa powder. For eating chocolate, cocoa butter is added to the liquid, enhancing the flavor and making the chocolate more fluid. Other ingredients such as sugar, vanilla, and milk are added, forming a paste. Conching—a form of machine kneading, and emulsifying—is done by beating the chocolate to break up sugar crystals, creating a smoother product. Careful cooling tempers the chocolate, which can then be poured into molds.

MARKET

The U.S. Census Bureau reports in "Confectionery: 2006," one in its series of *Current Industrial Reports* that American factories shipped $16.89 billion worth of confectionery products in 2006. This total is broken out into four categories, the largest of which (58.6%) was chocolate confections with 2006 industry shipments valued at $9.89 billion. Non-chocolate confectionery products had shipments of $4.73 billion, gums accounted for $1.77 billion, and an estimated half billion were shipped of unspecified confectionery products from firms too small to be included in the survey. By adding imports and subtracting exports, the Census Bureau estimated that consumption of chocolate candy in the U.S. market totaled $10.2 billion. Exports of all confectionery products totaled $511,286 and imports $781,747 in 2006.

The NCA reported that 2006 retail sales of all confectionery products totaled $28.2 billion, with $15.6 billion in retail sales of chocolate candy, $8.9 billion in non-chocolate candy sales, and $2.7 billion in retail sales of chewing gum.

The performance of premium and dark chocolate was particularly strong, according to Mintel International of Chicago. Mintel reported that sales of premium chocolate in food, drug, and mass merchandising stores (excluding Wal-Mart) increased 129 percent from 2001 to 2006, from $896 million to $2.05 billion. Dark chocolate sales increased 49 percent, from $1.26 billion in 2003 to $1.99 billion in 2006. Datamonitor, a market analysis firm, predicted that overall chocolate sales would grow 22 percent from 2005 to 2010, from $14 billion to $17.8 billion respectively. Dark chocolate sales, however, were predicted to grow 61 percent from $414 million in 2005 to $668 million in 2010.

The NCA reported that for the 52-week sales period ending April 22, 2007, chocolate candy sales had increased 4.4 percent, with chocolate boxes and bags up an impressive 9.3 percent, led by growth in the premium and dark segments. The NCA said 2007 sales through April 22 were supported by strong Valentine's Day and Easter seasons. These holidays, as well as Halloween and Christmas, are important for both the chocolate and non-chocolate candy industries.

The Chocolate Manufacturers Association (CMA) also cites the following figures as an indication of the size of the market it serves: 2005 exports of chocolate products (a broader category than chocolate candy) totaled more than $711 million to more than 160 countries around the world, and about 65,000 jobs in the United States were directly involved in the manufacture of these chocolate products.

The United States is the largest chocolate market in the world, according to statistics from the World Confectionery Report and Export Handbook and the NCA. These figures, only slightly different from the numbers cited above, put the 2005 U.S. retail sales market at $14.2 billion, 22 percent of the world total. The United Kingdom came next with $7.67 billion, or an 11.86 percent share; Germany at $7.23 billion, or 11.18 percent; France at $4.47 billion, or 6.92 percent; and Russia at $3.9 billion, or 6 percent.

KEY PRODUCERS/MANUFACTURERS

The largest chocolate candy maker in 2005 in terms of U.S. supermarket, drugstore, and mass merchandiser sales (excluding Wal-Mart) was Hershey with $1.9 billion in sales and a 43.5 percent market share. Masterfoods (renamed Mars in 2006) was second with $1 billion and a 22.89 percent share, and Nestlé third, with $368 million in sales and an 8.42 percent market share.

On a global scale, *Candy Industry* magazine published a list of the top 100 candy makers in January 2007, but these statistics are for all candy, not just chocolate. Cadbury Schweppes PLC led the list with $10.5 billion in sales, Nestlé was second with $9.7 billion, and Mars was third with $6.67 billion. Next came the Ferrero Group with $6.7 billion, Hershey Foods Corp. with almost $5 billion, and Wrigley with $4.7 billion.

Hershey Foods Corporation. Hershey's brands include Hershey's, Reese's, Hershey's Kisses, and Ice Breakers. Dark and premium chocolate brands include Hershey's Special Dark, Hershey's Extra Dark, and Cacao Reserve by Hershey's. In addition, Artisan Confections Company, a wholly owned subsidiary of Hershey, has acquired several premium chocolate makers and now markets Scharffen Berger, Joseph Schmidt, and Dagoba Organic chocolates. Hershey also has license agreements with Cadbury Schweppes affiliated companies to market and distribute York, Almond Joy, and Mounds worldwide as well as Cadbury and Caramello products in the United States. The company has an agreement with Societe des Produits Nestlé S.A. to market Kit Kat and Rolo products in the United States. On a dollar basis, Hershey's sales are about 75 percent chocolate and 25 percent non-chocolate products.

Milton S. Hershey, who owned a caramel manufacturing company, bought chocolate-making equipment at the 1893 Columbian Exposition in Chicago and began producing baking chocolate and cocoa as well as coatings for his caramels. Hershey built his own milk-processing plant and worked for three years to become the first American with a formula for manufacturing milk chocolate. The Hershey bar that resulted was an immediate success.

Hershey, located in Hershey, Pennsylvania, is the largest North American candy maker, and it employed more than 13,000 people worldwide in the middle of the first decade of the 2000s. After improving sales helped push up stock prices from 2001 to 2005, momentum slowed. The company's net sales fell 0.7 percent to $1.37 billion in the fourth quarter of 2006 from the year-earlier quarter, while net income dipped 10 percent to $153.6 million or 65 cents per share. Hershey responded in early 2007, announcing major restructuring that included cutting 1,500 jobs, eliminating one-third of its production lines, outsourcing cheaper items, and building a plant in Mexico.

In April 2007 Hershey announced a new strategic supply and innovation partnership with Barry Callebaut, the world's largest manufacturer of high-quality cocoa and industrial chocolate. The companies agreed to partner on a wide range of research and development activities. Callebaut also agreed to build and operate a facility to provide chocolate for Hershey's new plant in Mexico and to lease part of Hershey's Robinson, Illinois, plant and operate chocolate-making equipment there.

Mars, Inc. This company is an $18 billion, privately owned business operating in more than 65 countries. Its headquarters are in McLean, Virginia. Mars markets pet food, beverages, dinner products such as Uncle Ben's rice, and information services as well as candy. Its chocolate brands include M&M's, Mars, Milky Way, Snickers, and Twix.

Mars was founded in 1911 when Frank C. Mars opened a candy factory in Tacoma, Washington. He relocated to Minneapolis, Minnesota, in 1920 and introduced the Snickers and Milky Way bars. Mars and Hershey have fought each other over the years for leadership in the U.S. candy market. Mars acquired Dove International in 1986, and began marketing not only Dove candy and ice cream bars, but ice cream versions of 3 Musketeers, Milky Way, and Snickers as well.

In the candy market, however, Hershey regained first place in 1988 by acquiring the right to produce and market Mounds and Almond Joy bars. Mars responded in 1990 with the successful introduction of peanut butter M&M's, which cut into the market for Hershey's Reese's peanut butter cups.

Nestlé S.A. Nestlé, a Swiss company, is the largest food and beverage company in the world in terms of sales. It was founded by Henri Nestlé to market baby formula for infants who could not breastfeed. Today, Nestlé's 7,500 brands include coffee, bottled water, breakfast cereal, dairy products, ice cream, baby formula, and pet food as well as chocolate.

In 1875 Daniel Peter of Switzerland discovered how to combine Nestlé's condensed milk and cocoa to make milk chocolate. Peter, a neighbor of Henri Nestlé, founded a company that quickly became the world's leading maker of chocolate. As Nestlé's company grew, it also began making and selling chocolate, and in 1928 the company merged with Peter, Cailler, Kohler, and Chocolats Swiss, a company resulting from the merger of several Swiss chocolate makers including the firm founded by Peter. This merger added 13 chocolate plants in Europe, South America, and Australia.

In the United States, a subsidiary, Nestlé's Food Company Inc., was formed. The company's candy brands, including Nestlé's Crunch, Baby Ruth, and Butterfingers, are now produced by the snack division of Nestlé USA.

Cadbury Schweppes. Global leader Cadbury Schweppes resulted from a merger of the Cadbury candy company in Birmingham, England, and Schweppes beverage company in Geneva, Switzerland, in 1969. In early 2007 the company separated its beverage and confectionery units, dividing confectionery into four regional divisions: the Americas, Asia Pacific, Europe and a single unit focusing on Britain, Ireland, the Middle-East, and Africa.

Russell Stover. Another name worth noting in the list of chocolate makers is Russell Stover. In the United States, Russell Stover, which also owns the Whitman brand, dominates candy box sales. Russell Stover's 2006 sales totaled $93 million, with Whitman sales at $41 million.

MATERIALS & SUPPLY CHAIN LOGISTICS

Materials used by chocolate candy manufacturers include cocoa beans and products made from the beans such as chocolate liquor, chocolate cake and powder, chocolate coatings, and cocoa butter; sweeteners, including high-fructose corn syrup and cane and beet sugar; milk and milk products; nutmeats; fresh and dried fruits; and coconuts.

Hershey states on its Web site that cocoa is its most significant raw material, with other principal materials being, in order of magnitude, sugar, milk, peanuts, almonds and coconut. Chocolate makers globally use around two-fifths of the world's almonds and one-fifth of the world's peanuts. In the United States, chocolate manufacturers use about 3.5 million pounds of whole milk every day.

The supply chain for cocoa beans begins with small farmers in the Ivory Coast, Ghana, and other cocoa exporting countries. The farmer dries the beans, packs them, and delivers them to an exporting company, who inspects the cocoa and trucks it to a warehouse near a port. At the port, an independent grading agency inspects the beans before they are loaded onto ships. When the ship reaches its destination, it is unloaded and taken to a pier warehouse where it is sampled and inspected by the importer and declared to customs. Trucks then carry the cocoa to the manufacturer's facility.

The need for cocoa beans has drawn the industry into the contentious question of living and working conditions in the areas of West Africa where 70 percent of the beans are grown. In response to reports that child slaves were working on cocoa farms while living under inhumane conditions, particularly in the Ivory Coast, the World Cocoa Foundation (WCF) was formed in 2000. The goal is to promote "a sustainable cocoa economy through economic and social development and environmental conservation...." Its 60 candy company members include all of the major producers.

Working through this organization in the early 2000s, the cocoa industry, the United Nations, associations such as NCA and CMA, and local governments are together sponsoring programs to: (1) eliminate forced labor, (2) see that children are not harmed while working on family farms, (3) help small farmers get a better economic return on their produce, and (4) protect and enhance the environment in cocoa-growing regions. In one program, the NCA announced a partnership with UNICEF to provide greater access to education for children in the cocoa communities. Mars was participating in an initiative to provide vocational education in the Ivory Coast. Hershey was supporting another to provide teacher training in the cocoa-growing regions of Ghana and the Ivory Coast.

The WCF's Initiative for African Cocoa Communities (IACC) was formed in 2002 to bring together industry, farmer groups, non-governmental organizations (NGOs), development groups and others in an even stronger commitment of improving the lives of cocoa farmers and their families. Small cocoa farmers lose an average 30 percent of their crops to pests and disease, and at times they are faced with total crop failures. The IACC supports programs to educate farmers on more productive farming methods, as well as on better labor and environmental practices. The organization also deals with community issues such as HIV/AIDS and access to education. In another initiative, the United States adopted legislation setting up a protocol agreement that established a cocoa certification process that includes reports on the worst forms of child labor and forced labor and progress in reducing its incidence.

Ghana issued its first certification report in early 2007 as part of its own National Cocoa Child Labor Elimination Program. It stated that most children working in the country's cocoa industry are working on family farms and living with parents or a close relative. It added, however, that many children are still involved in hazardous tasks not appropriate for helping out on a family farm.

The large amount of sugar used by candy makers has involved the industry in trade policy. The NCA has for years argued that sugar price supports and trade restrictions put U.S. candy makers at a competitive disadvantage against manufacturers in other countries who can buy cheaper sugar on the global market. NCA says this is causing the loss of American manufacturing jobs, and it has lobbied in support of trade agreements such as the U.S. Central American-Dominican Republic Free Trade Agreement (CAFTA) and the North American Free Trade Agreement (NAFTA) that allow U.S. manufacturers to purchase some sugar from Central American growers at world market prices.

In 2006 the industry became involved in an effort by the Grocery Manufacturers Association to convince the U.S. Food and Drug Administration to revise food regulations to allow more flexibility. Of particular concern was a change that would allow manufacturers to substitute vegetable fat for cocoa butter and lower-cost milk substitutes for genuine milk products. Some chocolate companies such as Guittard Chocolate and See's Candies indignantly opposed the changes, while others, including Hershey saw no problem with them. It was predicted that it would be years before the FDA finally settled the issue.

Meanwhile a compromise within the European Union allows chocolate sold there to contain 5 percent non-cocoa vegetable fat so long as these nut butters come from an approved list of other tropical nuts. The presence of fats other than cocoa butter can be listed near the end of the list of ingredients rather than on the front of packages, as once had been suggested.

Another regulatory concern for chocolate makers arises when alcohol is used in candy fillings. In the United States, the use of alcohol in the fillings of some chocolate candies has subjected candy makers to different regulations in every state. Alabama, for example, prohibits the sale of candy containing "any vinous, malt, or spirituous liquor," while California permits the retail sale of any confectionery that contains less than 5 percent alcohol by weight.

DISTRIBUTION CHANNEL

Chocolate candy is sold through a varied and diverse network of distributors and retailers. The Hershey Company, for example, stated on its Web site that the company sells to grocery wholesalers, chain grocery stores, candy distributors, mass merchandisers, chain drug stores, companies that manage vending machines, wholesale clubs, convenience stores, concessionaires, and food distributors, for a total of more than 2 million retail outlets in North America. The product travels from manufacturing centers to a network of distribution centers and field warehouses located throughout the United States, Canada, and Mexico. The company generally uses contract carriers to deliver product from these warehouses to customers. Hershey's uses a real-time computing service, available 24 hours a day, seven days a week, to allow business customers to place and track orders.

Confectioner magazine reported in April 2005 on distribution channels for all candy. Although these figures included sales of non-chocolate candy, they provide a good indication of where chocolate candy is being sold. According to this report, 28.6 percent of candy was being sold in supermarkets, with 16.8 percent going directly from manufacturers to supermarkets' own warehouses or stores, 8.7 percent going indirectly to supermarkets by way of wholesale grocers and co-ops, and 3.1 percent going indirectly by way of candy and tobacco or specialty food distributors. The magazine said 7.8 percent of candy was going to wholesale clubs, although these clubs were only selling 3.9 percent of candy directly to consumers. The rest was being sold to small retail outlets or distributors and resold.

Confectioner listed a wide variety of independent retailers and alternate outlets, while noting that none was large enough to warrant individual designation on the chart. These outlets included beauty parlors, card and gift shops, video stores, college book shops, craft and hobby stores, hardware and home improvement outlets, liquor stores, toy shops, specialty stores, candy shops, and office supply and service outlets, as well as small independent outlets. Another group of sellers that did not warrant individual designations included manufacturing retailers (candy sold through company-owned stores), fund-raising operations, and government and military stores.

Basing its figures on input from Chicago-based Information Resources, Inc. (IRI), NCA's and CMA's monthly shipping reports, and the U.S. Department of Commerce, the NCA listed the following sales totals for various outlets from January through December, 2006: supermarkets, $4.3 billion; Wal-Mart, $3.2 billion; mass market stores other than Wal-Mart, $1.3 billion; convenience stores, $4.2 billion; drug stores, $2.4 billion; warehouse club stores, $2.1 billion; dollar stores $0.8 billion; vending machines, $1.2 billion; and bulk sales, $1.5 billion. A factor that must be considered in all warehousing and shipping of chocolate candy is its vulnerability to light, heat, humidity, and odor contamination.

One distribution issue that has been of concern to the industry is an October 2005 decision by the National Motor Freight Traffic Association, the group that maintains a freight classification system for the trucking industry, to change the classification for chocolate and other candy. The price of sending less-than-truckload shipments is negotiated based on this classification, which takes into consideration such transportability characteristics as density, stowability, ease of handling, and liability.

This group surveyed truckers and concluded that average density of candy shipments was less than was required for its classification of "65," which meant that a given number of pounds of candy was taking up more room in a truck. In 2005 the industry reclassified candy as "92.5," which resulted in higher shipping rates for the entire industry. The NCA appealed, and while it did not get candy back to its old classification, it obtained an agreement with the trucking industry on a sliding classification based on actual density. This meant that shippers of dense packages were charged less.

KEY USERS

Almost everyone eats chocolate candy. One small exception may be very young children. The CMA states on its Web site that older children are significantly more likely to prefer chocolate than younger children. With a large variety of reputedly healthier dark chocolates, sugar-free chocolates, and diet chocolate, even diabetics and those concerned about health or weight can and do eat chocolate candy.

ADJACENT MARKETS

Compared to other food products, candy and gum occupy a growing portion of the marketplace. IRI data for the year ending December 2006 show that in food, drug, and mass merchandising channels, candy and gum ranked third in volume of all food categories, behind carbonated beverages and milk. The confectionery category was growing, however, while carbonated beverage and milk sales were flat or declining. A more direct sales competitor, salty snacks came in fourth, and that category was also growing even faster than candy.

It can be argued that candy sales are increasing even more rapidly than shown in this survey, because dollar stores, which are increasingly stocking candy, and Wal-Mart are not included.

NCA figures also show that while candy and gum sales grew in 2006 in supermarkets, drugstores, and mass merchandisers (excluding Wal-Mart), snack food categories grew faster. Miscellaneous snacks grew 28 percent to

$220 million; snack and granola bars, 4.3 percent to $1.9 billion; coffee, 3 percent to $3 billion; bakery snacks, 3.5 percent to $860 million; salty snacks, 2.6 percent to $7.46 billion; and candy, 1.6 percent to $8 billion. Ice cream decreased 0.3 percent to $4.4 billion, and cookies, decreased 0.6 percent to $3.9 billion.

Within the U.S. candy category, IRI offered a breakdown of sales in 2006. U.S. chocolate candy sales totaled $4.58 million, non-chocolate candy totaled $2.3 billion, and gum, $1.1 billion.

Another way of looking at adjacent markets is to consider non-edible cocoa bean products such as cocoa butter and cocoa shell mulch. Cocoa butter is used in various skin-care products, while cocoa shell mulch is prized for its landscaping qualities, flexible use, and sweet aroma. Compared to candy, these cocoa bean-based products represent a very small market, and sales figures are not readily available.

RESEARCH & DEVELOPMENT

Those involved in the chocolate industry sponsor and conduct research in several areas. Pennsylvania Manufacturing Confectioners Association (PMCA), an international association of candy manufacturers, for example, awards one or two annual grants for graduate-level, non-proprietary university research in areas relevant to candy, manufacturers' problems, or industry issues, as well as one annual grant for research at Pennsylvania State University (Penn State).

Penn State has a long history of confectionery research, beginning in the 1920s. In addition to PMCA, support for Penn State candy research has come from a number of companies and organizations, including M&M/Mars, Hershey, CMA, the American Cocoa Research Institute, the American Association of Candy Technologists, and Dairy Management Inc. (on the use of fractionated milk fat in milk chocolate).

Research topics have varied widely, including work on the flavor and sensory quality of chocolate, the impact of cocoa butter in the diet, flavor-fade in peanuts, the continuous conching of chocolate, the influence of particle size and size distribution on the flow and sensory properties of chocolate, and the genetic transformation of the cocoa tree for improved disease resistance and seed quality.

Individual companies also conduct research, largely to develop and introduce new products. *Candy Industry* magazine reported on its Web site on a survey of 500 food and beverage processors by *New Products Magazine*. The magazine noted that while snack and candy makers as a whole devote 3 percent of their workforce to new product development, chocolate R&D accounts for less than 1 percent of the industry workforce.

CMA's Web site lists a number of 2006 and 2007 research publications on the health effects of chocolate. Flavanols are a class of chemicals found in raw cocoa beans, red wine, and green tea. Mars, Hershey, and Barry Callebaut are among the companies that have developed processes to retain most of the flavanols during processing, and they are marketing healthy chocolate with more flavanols.

Flavanols have been shown in some studies to relax blood vessels and increase blood flow, possibly lowering blood pressure and guarding against heart attacks. Other researchers have said flavanols may potentially balance the immune system, restrain brain aging, and even curb cancer cell production. Most health professionals warn, however, that chocolate is very high in calories and that unless portions are limited, it will contribute to obesity and associated heart risks.

CURRENT TRENDS

Premium chocolates and the overlapping category of dark chocolates were continuing to grow in popularity throughout the first decade of the 2000s. Mintel International reported that after 2002, when the premium chocolate category began to gain traction, it grew at a rate of about 14 percent a year, with consumers interested in dark chocolate both for rewarding themselves and because of the positive health news about dark chocolate. The importance of health as a motivation is illustrated by strong sales of non-premium dark chocolate such as Hershey's Special Dark bar.

Dark and premium chocolate cannot be completely combined into one category, however, because Mintel's survey showed that 43 percent of premium customers prefer dark chocolate while 42 percent favor milk chocolate. The health trend can also be seen in the strong growth of dietetic chocolate. Sales of low-sugar and no-sugar-added chocolate in U.S. food, drug, and mass merchandising stores (excluding Wal-Mart), according to ACNielsen, grew 478.2 percent between 2000 and 2005. Portion-control chocolate packaging was also gaining popularity as consumers looked for ways to control their weight. White chocolate was winning fans as well because it could boast that it contained no caffeine.

Premium no-sugar-added varieties appeal to diabetics as well as dieters. Guylian Chocolate of Englewood Cliffs, New Jersey, for example, established a premium no-sugar-added line, which uses Belgian chocolate and an extended conching process to give the bars a smooth texture and rich flavor. The bars are made with maltitol, a natural sugar substitute. Despite these trends, a great many chocolate buyers are not worrying about their weight. In fact, the sale of chocolate in bars, boxes or packages greater than 3.5 ounces has been increasing steadily.

The market for organic chocolate products was also expanding rapidly in the first decade of the twenty-first century. Dagoba Organic, sold by the Artisan Confections Company, a subsidiary of Hershey, and Theo Chocolate of Seattle, Washington, are two organic manufacturers who are gaining market share. Dagoba's new Apothecary line features a Cacao Elixir that is glycerin-based, sugar-free, and alcohol-free, and it offers botanical infusions made with cacao nibs. Theo Chocolate roasts its own Fair Trade-certified cocoa beans, noting that consumers are becoming increasingly aware of the child labor issues surrounding some cocoa beans.

As part of the gourmet chocolate trend, some observers predict that the most discerning consumers will search for cocoa from different continents, countries, or regions. More interest in ethnic chocolate is also expected, with dark chocolate being marketed with flavorings such as black or cayenne pepper, ancho chilies, lemon and lime, intense cinnamon, ginger, paprika, pine nuts, cloves, sea salt, rosemary, and even sun-dried tomatoes.

TARGET MARKETS & SEGMENTATION

Chocolate candy makers are not forgetting their core customers—women ages 25 to 40, who according to Mintel International, eat an average eight servings per month. New dark chocolate products aimed at this group are proliferating. Candy producers, however, are also looking for ways to appeal to other groups. Many of the new health-oriented varieties are particularly aimed at the aging Baby Boom generation (born between 1945 and the early 1960s).

Some new boxed candy collections are designed to attract younger, hipper buyers who have the money to treat themselves to an upscale assortment. One version is Target's new Choxie brand. Fanny May's new gourmet line, marketed under the Harry London brand, was presented in a sophisticated black-hinged box and featured richer, subtler flavors. In fact, new packaging is a popular way to reach targeted groups. Labels highlight the product's purported health benefits or other attributes. As seen above, portion control packaging, with individual packets containing 90 or 100 calories, has also been popular with consumers.

Retailers try to draw attention to new or gourmet products with various displays, including off-shelf displays in flower and greeting card sections. Store-within-a-store boutiques, positioned near such departments as the bakery and artisan breads section, can call attention to premium products. Giant Eagle stores in Pittsburgh, Pennsylvania, have even begun cross-merchandising dark chocolate in the wine department, touting red wine and chocolate as an appealing combination.

One group the candy industry wants to be careful about targeting is children. The NCA has announced guidelines on responsible confectionery marketing for its members. The organization has also joined the Children's Advertising Review Unit that seeks to ensure that advertising to young children is "truthful, accurate, and sensitive." In addition, the NCA joined the Ad Council's Coalition for Healthy Children to promote healthy lifestyles to children and parents, and has made materials available on nutrition and physical activity.

RELATED ASSOCIATIONS & ORGANIZATIONS

American Association of Candy Technologists (AACT), http://www.aactcandy.org

Chocolate Manufacturers Association, http://www.nca-cma.org

National Confectioners Association, http://www.candyusa.org

Pennsylvania Manufacturing Confectioners Association, PMCA: An International Association of Confectioners, http://www.pmca.com

Retail Confectioners International, http://www.retailconfectioners.org

World Cocoa Foundation, http://www.worldcocoafoundation.org

BIBLIOGRAPHY

"2005 Distribution Chart: Analysis of U.S. Confectionery and Gum Distribution and Sales." *Confectioner.* April 2005, 14A.

"ACNielsen Strategic Planner." ACNielsen. Available from <http://www.acnielsen.com/products/tools/strategicplanner>.

"Alcohol in Candy—Chart of State Regulations 2007." National Confectioners Association. Available from <http://www.ecandy.com/ecandyfiles/alcohol_confectionery_jan07.doc>.

Brubaker, Harold. "Candy-Maker Continues Its Reducing Regimen: In a Restructuring that Began in 2001, Hershey Will Lay Off 1,500, Cut Production Lines, and Outsource Cheaper Items." *Philadelphia Inquirer,* (Philadelphia, PA). 16 February, 2007.

"Candy Sales by Channel." *Confectioner.* November 2006, 12.

"Chocolate and Confectionery Manufacturing from Cacao Beans: 2002." *2002 Economic Census.* U.S. Department of Commerce, Bureau of the Census. December 2004.

"Cocoa, Chocolate, and Confectionery Research Group." University of Pennsylvania. Available at <http://foodscience.psu.edu/Research/Cocoa/cocoa3.html>.

"Confectionery: 2006." *Current Industrial Reports.* U.S. Department of Commerce, Bureau of the Census. June 2007.

"Confectionery Industry Review, 2006 Year-End." National Confectionery Association. Available from <http://www.ecandy.com/ecandyfiles/2006_Annual_Industry_Review_February_2007.ppt#387,1,Slide 1>.

Covino, Renee M. "Counting Candy: Sure There Are Industry Pressures, but the Confectionery Industry Numbers Tallied Up Nicely for 2006—with 2007 Already Breaking One Seasonal Record So Far." *Confectioner.* April 2007, 31.

"Dark Chocolate: Now That the 'Chocolate Bar' Has Been Raised Toward More Premium Varieties, Retailers Need to Strategize to Capture an Even Higher-Impulse Chocolate Market." *Confectioner.* March 2007, 28.

Darnay, Arsen J. and Joyce P. Simkin. *Manufacturing & Distribution USA,* 4th ed. Thomson Gale, 2006.

"Ghana Cocoa Certification Report to Guide Efforts to Help Children on Cocoa Farms." World Cocoa Foundation. Available from <http://www.worldcocoafoundation.org/commitements/ghana.asp>.

"Hershey and Barry Callebaut Announce Strategic Supply and Innovation Partnership." Hershey Company. 26 April 2007. Available from <http://www.hersheycompany.com/news/release.asp?releaseid=001642>.

"Industry Performance: Dept. of Commerce Annual Confectionery Report." National Confectioners Association. Available from <http://www.ecnady.com>.

"Investor Relations: FAQs." Hershey Company. Available from <http://www.thehersheycompany.com/ir/faq.asp#37>.

Kennedy, Maev. "Palace Serves Up History of Chocolate." *The Guardian,* (London, England). 14 April 2001, p 5.

Kolenc, Julie. "Chocolate Bar Sales Soar: Chocolate Bars Are Grabbing the Headlines as Consumers Discover New Varieties and React to Health Claims." *Professional Candy Buyer.* March-April 2007, 26.

Lazich, Robert S. *Market Share Reporter 2007.* Thomson Gale, 2006, Volume 1, 139–143.

Liebeck, Laura. "Sweet Ride: Candy's Gotten a Shot in the Arm from Popular Chocolate Segments Such as Dark, White, and Dietetic." *Progressive Grocer.* 1 March 2006, 58.

Marter, Marilynn. "A Turn for the Bitter in Chocolate Market: Sales of Dark Varieties Are Rising, Driven in Part by a Notion of Healthfulness." *The Philadelphia Inquirer,* (Philadelphia, PA). 26 October 2006.

Pacyniak, Bernie. "Don't Mess with the 'Gold Standard.'" *Candy Industry.* April 2007, 6.

Penn, Catherine. "Fast and Furious" *Candy Industry.* May 2007. Available from <http://www.candyindustry.com>.

Rehan, Kelly. "Delving Deeper into Dark: From Low-Sugar to Fair Trade, Organic to Dark, Chocolate Bar Manufacturers Expect to Turn Many of Last Year's Biggest Trends into Successful Mainstays in 2007." *Candy Industry.* January 2007, 40.

———. "From Farm to Pharma: Chocolate's String of Health Claims Ranges from Lowering Blood Pressure to Even Preventing Some Cancers." *Candy Industry.* April 2007, 12.

Roberts, Williams A. Jr. "Confounding Confections: With Gum and Mints Accounting for $3.3 Billion in Sales, Premium Chocolate Selling $1.5 Billion Worth Per Year and Sugar Confectionery Tallying $11.3 Billion Annually, There Is No Doubting the Vast Size of the Overall Confectionery Category." *Prepared Foods.* July 2005, 18.

SEE ALSO *Candy, Snack Foods*

Clocks

———————————■———————————

INDUSTRIAL CODES

NAICS: 33–4518 Watch, Clock, and Part Manufacturing

SIC: 3873 Watches and Clocks Manufacturing

NAICS-Based Product Codes: 33–45182, 33–45184, and 33–4518W

PRODUCT OVERVIEW

Among all instruments that measure time, clocks belong to the larger variety typically hung from walls, displayed in cases, or placed on the night table to wake us up when we need rousing. All watches, by contrast, are carried on the body, usually on the wrist, occasionally in a pocket or on a chain around the neck. Clocks are devices in which a mechanism is moved by a spring, by electric power running through a wire, coming from a battery or from solar energy, or by weights suspended from the mechanism by chains. Clocks of unusual design may also be moved by falling water, and sundials rely on the movement of the sun to cast a shadow on a semi-circular clock face. The mechanism moves coordinated pointers that indicate the current hour, minute, and optionally also the second, with time measured from zero hour presumed to take place at midnight. Two kinds of display are used, ignoring sundials: circular panels with numbers depicting the hours from 1 through 12, thus half of one day. In this scheme, the 12 serves both to mark noon and midnight, thus the beginning of periods as well as their termination. This standard clock face is called an analog indicator. The other kind is digital. It is a display which simply shows the time in numbers that change every minute, thus 3:15 A.M., for example.

There are many types of clocks. Alarm clocks, grandfather clocks (also known as longcases), hourglasses, mantel clocks, and sundials are common and well known. There are also clocks made for special markets such as high-precision chronometers and atomic clocks used in science. Clocks also serve symbolic uses. A biological clock refers to the effect of time and repetition on life processes. The Doomsday Clock is a symbolic clock face maintained since 1947 by the Board of Directors of the Bulletin of the Atomic Scientists at the University of Chicago. The clock is shown with the clock hands some number of minutes before midnight, with midnight symbolizing nuclear war. World events continually move the minute hand closer or further from the midnight hour.

Early Clocks. The ancient Sumerians and Egyptians relied on the stars to note the passage of time and to predict the seasons. The Egyptians observed that a group of stars (the Sirius or *Dog Star* constellation) rises next to the sun every 365 days. They also realized that there was a difference between the number of days in the year when measured by the position of the moon and sun. After making this observation, they were able to create the first 365-day calendar, dividing the seasons into four-month time spans. Later, they relied on sundials and water clocks to measure the passing of time; some hold that sundials and water clocks may have been used by even earlier civilizations. Sundials were used to mark the time by noting the shadow of the sun as it moved across the sky. Water clocks, first used around the fifteenth century BC, used the rise or fall of water in a tank to measure the passing of a certain amount of time. Water clocks could be a simple collection of bowls while some versions could be much more complex structures. Burning candles could also note the passing of time; candle clocks were used around the ninth century AD. Water clocks, candle clocks, sand clocks (hourglasses), and similar timekeeping methods were used around the globe.

These timepieces had various limitations. A sundial, for example, cannot tell time at night. A more reliable timepiece was needed. Some historians credit monks and religious institutions generally with leadership in this art. Monks and church members relied on clocks to schedule their services at consistent times. Religious organizations relied on bells and chimes to mark these times; the original meaning of the word clock was bell according to the *Oxford English Dictionary*. Bells and chimes were used outside monasteries and churches as well; they might be used to mark the beginning and end of a workday, for example, or the start of public meetings. Woodworkers, scientists, jewelers, and metalworkers improved upon these early timekeeping devices.

It is unclear when the first mechanical clocks actually appeared. Historians believe it was sometime in the late thirteenth century. The first clocks began to appear in public squares about this time. The clock of the Church of St. Eustorgio in Milan in 1309 is thought to have been the first clock in Europe. Mechanical clocks work with an escapement. The escapement is a balance wheel that mediates the transfer of the energy of the gravitational force acting on the weights attached to the clock's counting mechanism; it creates the clock's forward movement, or its ticking. The most common escapement was (and still is) the verge-and-foliot. In a typical verge-and-foliot escapement, the weighted rope unwinds from the barrel, turning the toothed escape wheel. Controlling the movement of the wheel is the verge, a vertical rod with pallets (hooks) at each end. When the wheel turns, the top pallet stops it and causes the foliot, with its regulating weights, to oscillate. This oscillation turns the verge and releases the top pallet. The wheel advances and the bottom pallet catches it. The process then repeats.

By the fifteenth century the first spring-driven clocks were invented. These clocks were smaller and lighter than their mechanical brothers; unfortunately, they were also just as inaccurate. The problem was the mainspring, which might make the clock run fast when wound tightly or slow when the clock had wound down. In a spring-driven clock, the spring is mounted on a balance wheel, which turns back and forth in sync with the spring's oscillations, simultaneously rocking the pallet from side to side. The pallet controls the turning of the gears connected to the

FIGURE 58

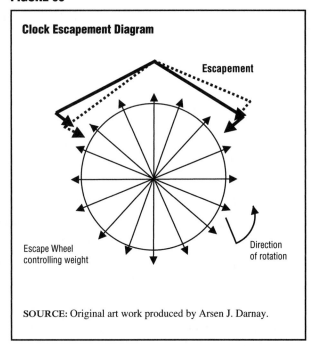

Clock Escapement Diagram

Escapement

Escape Wheel controlling weight

Direction of rotation

SOURCE: Original art work produced by Arsen J. Darnay.

clock's face and maintains a steady transfer of power from the mainspring to the clock's counting mechanism. More innovation would come in 1504 when Peter Henlein invented the first portable timepiece.

Galileo Galilei (1565–1642) was a famous astronomer, mathematician, and philosopher. One of the groundbreaking concepts with which he is credited involves uniform motion. In the late sixteenth century Galileo noticed that a pendulum took the same amount of time to swing completely through a wide arc as it did a small arc. In 1641, he set out to invent a device to keep a pendulum in motion; he was aware of the value such a device might have in keeping time. Galileo died before completing the device, but Dutch mathematician and astronomer Chris Huygens would use Galileo's theories to invent the first pendulum clock in 1656. Huygens' clock had an error of less than one minute per day, the first time such accuracy had been achieved. He would later refine the clock to be accurate within ten seconds. The pendulum clock was usually built with a case for the pendulum. Such clocks were referred to as longcase clocks, floor clocks, and even coffin clocks because of their narrow, long construction. Such clocks would not be known as grandfather clocks until 1875. A song that year by Henry Clay Work called *My Grandfather's Clock* is believed to have rechristened the longcase clock as grandfather clocks in the minds of millions of Americans. The song reportedly sold a million copies.

As in England so also in the United States the first clocks appeared in churches and in town squares. A church clock in Boston was constructed in 1650, and the first public clock in New York City was built in 1716. Clockmakers were already in business at that time but still few in number. Significant mass production of clocks, however, would not really begin in America until after 1800. The National Association of Watch and Clock Collectors notes that the main cost for manufacturers during this period was materials not labor. Clockmakers refined the miniaturization process of clocks and clock parts. Smaller clocks meant smaller costs and higher profits. Between 1840 and 1850 the average clock shrank in size approximately 30 percent. The small clocks, perfect for tables, shelves, and mantles, would be needed for new homes as the country's population expanded and moved into new territories. More than half a million brass shelf clocks were being manufactured by fifteen clockmakers in Connecticut alone in 1850. One in three homes in New England were thought to have a clock.

The country's economy soured after the Civil War. Many clockmakers who had rushed into the industry in the early nineteenth century were forced from the industry again. The market was left with just a handful of players, although these companies would dominate the industry for decades to come. The major players were the Ansonia Clock Company, the New Haven Clock Company, the Seth Thomas Clock Company, the Waterbury Clock Company, the William L. Gilbert Clock Company, and the E.N. Welch Manufacturing Company (eventually the Sessions Clock Company).

Clocks Become Precise. Until the mid-eighteenth century sailors could not calculate a ship's longitude (the angular distance east or west from the prime meridian of Greenwich, England). This created continual problems for the British, Dutch, and other seafaring countries. The British government in the Longitude Act of 1714 offered the equivalent of several million dollars in today's currency to anyone who could solve the problem. Astronomer and mathematician Gemma Frisius proposed a solution. A ship could record its longitude by recording the local time at the point of its departure and then compare the local time of the ship at noon, determined by measuring the highest point of the sun in the sky. The weight and pendulum-driven clocks at the time were incapable of the precision needed; also, such timekeepers could not be used at sea because of the ship's motion. Clock maker John Harrison invented the chronometer in 1735, an instrument capable of such precise measurement. Harrison built four different clocks, each more refined than its predecessor. In 1762 his last clock kept time to within two minutes' accuracy during a five-month voyage to Jamaica. A watch may be designated a chronometer only if it meets certain high timekeeping standards.

The clock industry continued to become more sophisticated. Clockmaker Levi Hutchins is credited with the first mechanical alarm clock in 1787. Hutchins, however, never patented the idea or had it mass-produced. French inventor Antoine Redier actually patented the first adjustable mechanical alarm clock in 1847. The first cuckoo clocks were produced in the Black Forest region of Germany in the middle of the eighteenth century. Cuckoo clocks are designed to look like a Bavarian home or chalet; the cuckoo appears from a small trap door and sounds its call as the clock is striking. Franz Anton Ketterer is credited with the invention of the clock, although historians have disputed this. Some historians also believe the first cuckoo clocks were built elsewhere in Germany, although the Black Forest region quickly became the center of production.

Siegmund Riefler's clock, in 1889, further advanced the pendulum clock with the invention of the nearly-free pendulum, accurate to a hundredth of a second a day. R.J. Rudd introduced a truly free pendulum clock in 1898. William Hamilton Shortt invented the clock that would take his name (the Shortt clock) in 1920. The clock has two pendulums, one that merely keeps time (the master) and the other to drive the time mechanism. The master pendulum is kept vibrating in a near vacuum. This clock is very accurate and was seen as being the pinnacle of timekeeping accuracy until the arrival of quartz clocks.

W.A. Marrison of Bell Laboratories built the first quartz clock in 1928. The room-sized clock was accurate to within one to two hundredths of a second per day. A quartz clock uses an electronic oscillator, which contains a crystal. An electronic field is applied to this crystal, which then vibrates. The vibration depends on the size of the crystal. The crystal then oscillates between 10,000 and 100,000 hertz (cycles per second). The high-frequency oscillation is converted to an alternating current, reduced to a frequency more convenient for time measurement, and then made to drive the motor of a synchronous clock or a digital display. Quartz clocks have no gears or escapements, which helps them maintain consistent frequency. Quartz clocks are very popular for several reasons. They keep time reliably. As well, the clock has fewer parts than a standard mechanical clock, which makes it inexpensive to mass-produce. Quartz watches would be produced in the 1960s. Seiko produced the first quartz wristwatch in 1969. Such watches sold well, and quickly threatened the dominant market share of mechanical watches.

Research on atomic clocks began in the 1930s, with the first clock built by the National Bureau of Standards in 1949. Atomic clocks are tuned to the frequency of the electromagnetic waves that are emitted or absorbed when certain atoms or molecules make the transition between two closely spaced, or hyperfine, energy states. Such clocks offer greater accuracy than other forms of clocks. They also are useful because the measured frequency is uniform across the same types of atoms. Thus, atomic clocks constructed and operated independently will measure the same time interval.

There are many types of atomic clocks although the most frequently used are cesium atomic, hydrogen maser, and the rubidium gas based. The first atomic clock in 1949 measured changes in ammonia molecules. The error between a pair of such clocks (the difference in indicated time if both were started at the same instant and later compared) was typically about one second in three thousand years. In 1955 the first cesium-beam clock (a device that uses as a reference the exact frequency of the microwave spectral line emitted by cesium atoms) was placed in operation at the National Physical Laboratory at Teddington, England. It is estimated that such a clock would gain or lose less than a second in three million years.

The standard in the United States is the NIST-F1 from the National Institute of Standards and Technology. This clock went into service in 1999 and should neither gain nor lose a second in 20 million years. A fountain atomic clock, the NIST F-1 consists of a 3-foot vertical tube inside a taller structure. It uses lasers to cool cesium atoms, forming a ball of atoms that lasers then toss into the air, creating a fountain effect. This allows the atoms to be observed for much longer than could be done with any previous clock. The NIST-F1 contributed to a group of clocks that establish Coordinated Universal Time, the official world time. Atomic clocks are used in a number of fields, including space navigation, global positioning services, and astronomy. The precision of the atomic clocks allowed the scientific community to define the second officially. In 1967, the Thirteenth General Conference on Weights and Measures formally defined a second as 9,192,631,770 vibrations of the cesium atom.

Daylight Savings Time. In the early 1900s, Europeans began to adopt summer time. They adjusted clocks by one hour every spring, moving the clock forward, and every autumn, moving the clock back by one hour. This became known as Daylight Savings Time in the United States. The plan was originally offered as an energy saving measure; more daylight meant less use of electric lights. It also no doubt played on public and personal concepts of productivity; the sun was still shining so there was still work to do. Ben Franklin is often erroneously credited with advocating Daylight Savings Time. In a humorous paper in 1784, Franklin urged the French to make maximum use of daylight hours to keep them from burning candles at night. He mentions nothing about resetting clocks, however. Daylight savings time was actually first proposed by William Willett in 1907. Daylight savings

time was seen as a way to conserve energy during World War I. The practice also found support among many in the business community in the United States although there was (and is) less support for it among the public. President Wilson signed daylight savings time into law in 1918. It was repealed after World War I. No national legislation mandated daylight savings time until 1966; before the Uniform Time Act of 1966, the matter was left to the states. In 1973 Congress enacted the Emergency Daylight Saving Time Energy Conservation Act in response to the Arab oil embargo. Daylight saving time was extended to eight months rather than the normal six. This was believed to save the country 150,000 barrels of oil per day during the winter months. Daylight savings time started three weeks early as part of the Energy Act of 2005.

MARKET

The Watch, Clock, and Parts Manufacturing industry in the United States shipped product valued at $641 million in 2005. Of that total the industry exported $256 million in goods but imported $3,039 million so that apparent consumption (production less exports plus imports) came in at $4,324 million or $4.3 billion. These results indicate that clocks and watches used in the United States are largely an import market—and increasingly so. Figure 59 provides a perspective on this subject for the period from 2000 to 2005.

The graphic shows apparent consumption, imports, domestic shipments, and exports as curves. Consumption grew at the anemic rate of 0.2 percent per year, imports increased at 1.7 percent, exports declined at the rate of 3.8 percent yearly, and domestic shipments slid at a rate of 7 percent per year in the 2000 to 2005 period. Domestic production shrank from a $1.2 billion to a $641 million industry in this five year period, nearly halving. An overlaid bar graph shows how net domestic production (production less exports) shrank from 19.6 percent of consumption at its peak in 2001 down to 8.9 percent.

Clocks versus Watches. Based on global estimates for the market developed by the Japan Watch & Clock Association (JWCA), world production of watches in 2006 stood at 1.19 billion units whereas clock production was 485 million units, showing that viewed in units, watches represent a larger industry. When measured in dollar value however, the clock industry worldwide is bigger than the watch industry because the average value of every clock sold is slightly more than four times the value of each watch sold. Using dollar valuations, translated from JWCA's reporting in Yen, the 2006 global market was 63 percent clocks and 37 percent watches—at the production level a $7 billion industry. An approximation of retail sales may be estimated by doubling that number. Clock

FIGURE 59

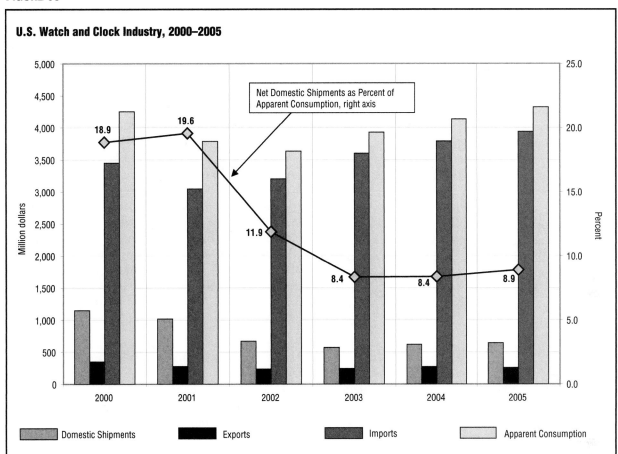

U.S. Watch and Clock Industry, 2000–2005

Net Domestic Shipments as Percent of Apparent Consumption, right axis

18.9 19.6 11.9 8.4 8.4 8.9

Domestic Shipments   Exports   Imports   Apparent Consumption

SOURCE: Compiled by the staff with data from "Table 1289. U.S. Exports and General Imports by Selected SITC Commodity Groups: 2002 to 2005." *Statistical Abstract of the United States: 2007* and earlier editions of the title, U.S. Department of Commerce, Bureau of the Census. December 2006.

shipments in units were growing (at 3.7% per year in the period 2000 to 2006) whereas watch shipments were declining (1.1% per year).

Details on the U.S. production, available only for 2002, indicated that clock production accounted for 79.5 percent of shipments in that year, watches for 17.8 percent, and categories not specified by kind for 2.7 percent. To get some measure of the total U.S. market in clocks, we can apply these proportions to 2005 domestic shipments and global distribution as reported by JWCA to imports. Using that method, we obtain an estimate showing that clocks represented $2.8 billion of U.S. apparent consumption in 2005 or 65 percent of the producer level market for watches, clocks, and timepiece parts.

Top exporters of clocks to the United States were China, dominating the export market with 82 percent of shipments of U.S. imports. Japan, Germany, and Taiwan each represented 3 percent of imports, and Hong Kong represents 2 percent.

Popular types of clocks include grandfather, wall-mounted, and mantle clocks. Grandfather clocks refer to devices that stand on the floor, long case clocks to those at least six feet tall. The average price of a grandfather clock is approximately $1,600. Mantel clocks are relatively small, moveable clocks traditionally placed on a shelf or mantel above the fireplace. The form is believed to have first appeared in France in the 1750s. Such clocks, which might retail for between $150 and $500 each, are highly ornate and are made from any combination of metal, porcelain, and wood. The vast majority of clocks are inexpensive electrical devices, many battery operated.

KEY PRODUCERS/MANUFACTURERS

Howard Miller Clock Company. This company was founded in 1926. Howard Miller was originally a division of office furniture manufacturer Herman Miller. Herman Miller's son Howard spun off the division in 1937. These two companies are now separate entities. Originally the company only made mantel clocks. Grandfather clocks did not become a part of the product line until the 1960s. In 1989 the company began making curio cabinets as well as clocks.

Bulova. A wholly-owned subsidiary of Loews Corporation, this company is known as a manufacturer of watches, clocks, and jewelry. Bulova manufactured the first clock radio in 1928 and the first electric clock in 1931. Bulova manufactures clocks for public places such as train stations, office buildings, and airports. It also was highly influential in early automobile clocks. In 1952 the company developed the Accutron, an electronic timepiece that promised accuracy to within two seconds per day. The timing mechanisms would be used by the National Aeronautics and Space Administration (NASA) in its programs. In 1959 Bulova was the first company to offer a one-year warranty on a clock radio. In 1968, it offered the first watches regulated by orbiting space satellites. In 2001 the company acquired the Wittenauer watch brand and in 2002 some assets from Heirloom, a grandfather clock maker.

Timex Corporation. Timex manufactures watches, clock radios, and alarm clocks. It was born out of the Waterbury Clock Company and its sister companies, Robert Ingersoll and the Waterford Watch Company. The company changed its name to the U.S. Time Company after World War II. The company introduced the Timex in 1950. The watch was immensely popular; more than one billion pieces have been sold since its introduction. Timex has approximately one-third of the global market and about the same in the United States, where it enjoys a 98 percent brand awareness among U.S. consumers. It was also one of the first U.S. watch firms to make and sell watches abroad (in the 1940s). In 1956 Timex released a series of highly memorable ads in which its watch was put through a series of tests to show its durability. The slogan "Timex Takes a Licking but Keeps on Ticking" is still remembered by many consumers. The company employs about 5,000 people and is owned by the Norwegian company Fred Olsen & Co.

Ridgeway Clocks. This compay began as The Gravely Furniture Company in 1926 and started producing grandfather clocks exclusively in 1960. This makes Ridgeway the oldest continuously produced grandfather clock brand in the United States. It is a division of Howard Miller Company, and is a producer of longcase clocks, mantle clocks, and wall clocks. Howard Miller acquired them in November 2004. The company claims several firsts, such as making the first grandfather clocks to chime "America the Beautiful," "God Bless America," and West-

minster chimes. The company's facilities are located in Ridgeway, Virginia.

Colibri Group. Known primarily for making luxury lighters, Colibri also manufactures jewelry and timepieces. Its subsidiaries include Colibri, Seth Thomas Clock Company, Kremantz Jewelry, and Dolan-Bullock. The company acquired Seth Thomas Clocks in 2001. Seth Thomas is one of the oldest American clockmakers. The Sligh Furniture Company was founded in 1880 by Charles Sligh. The company found success as a maker of office furniture, clocks, and cabinetry. It sold its clock designs to Bulova in 2005. Graphic designer Paul Rowen and industrial hardware manufacturer Les Mandelbaum started Umbra Inc. in 1985. The company manufactures stylish trash cans, dish towels, and other home furnishings. The company's clock line has received attention for its style and use of unusual materials in construction. Salton Inc. manufactures the popular Big Ben clock in addition to grills, lighting products, toasters, juice extractors, bread makers, griddles, waffle makers, and buffet ranges/hotplates.

MATERIALS & SUPPLY CHAIN LOGISTICS

Ignoring very large clocks produced for buildings and churches and assembled on site, as products, clocks extend from furniture-size objects to small devices a user can easily hold in his or her hand. Heavy grandfather clocks encased in solid mahogany represent one extreme, cheap alarm clocks the other. All of these devices have in common that they have high value in relation to weight so that transportation costs do not play a major role in their distribution. They are all precision instruments; the most valuable clocks are high-tech devices using advanced digital and quartz technologies. With most products purchased in the United States making a passage across the Pacific or the Atlantic oceans, logistical systems supporting the industry are global. Raw materials in no way limit the production of clocks.

DISTRIBUTION CHANNEL

Clocks are sold in specialized shops selling watches and jewelry, in electronics outlets, in furniture stores, in museum stores, over the Internet, in supermarkets, in department stores, in mass merchandising chains, and in drug stores—different channels carrying products at different price points. Approximately 58 percent of clocks were sold through mass merchandisers and warehouse clubs in the middle of the first decade of the twenty-first century. Specialty stores (jewelers, furniture stores) represented 22 percent of the market. Catalogs were the third largest category, taking 11 percent of sales. Department stores had 2 percent of sales. Supermarket participation was approximately 1 percent, and all other outlets accounted for 6 percent of distribution. Clocks typically move from producers through wholesalers servicing these quite different retail categories.

KEY USERS

An estimate made in 1990 by the Clock Manufacturers and Marketing Association put six clocks into the average American home. Based on data provided by the National Association of Home Builders, houses have increased in size over time. In 1973 new single-family homes had around 950 square feet; 2006 homes averaged 2,460 square feet—suggesting that more clocks are being used by the public.

Packaged Facts, a research firm, conducted a study on clock ownership in 2006. The company released its findings on the clock radio segment. It claimed that this segment only represents one-quarter of industry sales; however, the findings certainly offer some insight into the overall market. According to Packaged Facts, an estimated 115.6 million adults owned clock radios in 2003. They were most commonly owned by those 25 to 44 years of age, people who have attended college, work full-time, have household incomes between $75,000–200,000 and have children between the ages of two and nine.

ADJACENT MARKETS

As objects of luxury or conspicuous consumption expensive display clocks compete with other similarly prized home ornaments like art and sculpture. As devices that simply announce the time, clocks compete with appliances (like stoves and microwave ovens) that display the time continuously, even in the dark. In the form of clock radios, the time-telling function is essentially merged with a product probably purchased because it brings music and news. In that many clocks typically reach homes as gifts, the products compete with alternative items of decoration or utility, like china, glassware, and small appliances. Although cheap clocks are typically simply tossed when they finally fail to keep time, expensive devices depend on the presence of a watch/clock repair industry which still linger in more or less hidden corners of the economy. Jewelry stores typically maintain services for watch and clock repair and see some traffic in the repair of expensive clocks.

RESEARCH & DEVELOPMENT

Technology has helped timepieces refine their level of accuracy. The first clocks needed to be wound; later clocks were self-winding. The rise of mass production and miniaturization helped to usher in the pocket watch and wristwatch. Electric clocks proved more reliable than mechani-

cal clocks. The development of the microprocessor in the 1960s sparked the creation of LED (light emitting diode) and LCD (liquid crystal display) clocks and watches. Such clocks had digital faces (3:15 A.M., for example) rather than analog faces (the traditional 1–12 hours of a clock face). Deeper understanding of radio waves and atomic particles led to the creation of atomic clocks. Improvements in metallurgy and plastics have led to more durable clock escapements and components that resist rust, wear, and breaking.

Technology continues to drive the industry. Clock and watch manufacturers are now exploring new sources of power, such as sunlight and body heat. Some manufacturers are showing more interest in niche markets, such as clocks for the blind. American Innova offers an alarm clock for students that allows it to be preset for a different wake-up time for each day of the week.

CURRENT TRENDS

Market watchers have noticed a gradual decline in watch and clock sales. A significant reason is the proliferation of electronic devices that offer time displays: cellular phones, portable music players, pagers, and home and work computers. There are anecdotal reports of people deciding to stop wearing watches and relying on these devices as their primary timepieces. According to a Jewelry Consumer Opinion Council survey, those 18 to 24 years of age are those most likely to abandon their wristwatch in favor of their cell phone. Their interest in watches was not as timepieces but as fashion accessories. Their interest in watches was aimed largely at high-end luxury watches. Watchmakers may breathe a sigh of relief at the findings of this survey—fewer watches sold, but those sold are at higher prices. However, if young people never develop the habit of wearing a watch, it does offer a potentially gloomy prognosis for future watch industry growth.

The clock industry has been experiencing many of the same problems. People do use portable devices as their primary timepieces. At the same time what Packaged Facts calls the *casino effect* seems to be taking hold in some businesses. The casino effect describes the removal of clocks from public buildings and stores so consumers are unaware of how much time has passed (casinos have no visible clocks). It might, for example, persuade a book lover at the local Barnes & Noble or Borders to linger longer; more time in the store means the greater likelihood of the shopper spending money. Indiana's Bureau of Motor Vehicles removed clocks from 168 local license offices in 2005. The removal of clocks was to discourage clock-watching by employees and customers. The removal of clocks has been favored by other businesses to give chain stores a uniform appearance.

TARGET MARKETS & SEGMENTATION

The clock and watch industries remain poorly monitored and are in need of greater study. Industry players spent approximately $233 million in 2005 on advertising, according to Packaged Facts. Most of this spending was devoted to promoting watches. This spending has earned some wristwatch brands such as Timex, Seiko, Rolex, and Bulova, a high level of recognition. However, few clock makers have elevated their products to such a status. Those makers who have found this elevated status have done so largely because their brands have been successful in the wristwatch industry. Admittedly, high brand recognition in the industry may prove difficult as brand is a minor concern for the typical clock shopper. Fashion remains the primary impetus behind a clock choice.

Popular trends in 2007 according to clock makers include oversized and outdoor clocks, blonde woods, woven materials, and neutral colors. Infinity, Chaney, and Howard Miller are making oversized and outdoor clocks. Colibri Group has focused on oversized and theme clocks. Kirch Industrial makes high-end clocks made from steel and chrome. Clocks, finally, have also found new voices so that some announce the hour by bird song, fish sounds, dog barks, music, and even automotive honking.

RELATED ASSOCIATIONS & ORGANIZATIONS

American Watchmaker and Clockmaker Institute, http://www.awci.com

Antiquarian Horological Society, http://www.ahsoc.demon.co.uk

National Association of Watch and Clock Collectors, http://www.nawcc.org

BIBLIOGRAPHY

Bernard, Sharyn. "It's All in the Timing." *HFN.* 4 September 2006.

"Clocks and Watches." *Encarta.* 15 May 2007. Available from <http://escarta.msn.com.>.

"Grandfather Clock Refers to a Floor-Standing..." *Chicago Sun–Times.* 23 January 2005.

"More Than a Third of Adults Bought Watches in 2003." *Research Alert.* 19 March 2004.

Shuster, William George. "Upscale Clocks on the Rise." *Jewelers Circular Keystone.* March 1993.

"Trade in Goods and Services." Annual Revisions for 2005. U.S. Department of Commerce, Bureau of Economic Analysis. 9 June 2006. Available from <http://www.bea.gov/bea/di/home/trade.htm>.

"Value of Shipments for Product Classes: 2005 and Earlier Years." *Annual Survey of Manufactures.* U.S. Department of Commerce, Bureau of the Census. November 2006.

"Watch, Clock, and Part Manufacturing: 2002." *2002 Economic Census.* U.S. Department of Commerce, Bureau of the Census, Economics and Statistics Administration. December 2004.

SEE ALSO *Watches*

Coffee

———■———

INDUSTRIAL CODES

NAICS: 31–1920 Coffee and Tea Manufacturing

SIC: 2095 Roasted Coffee Manufacturing

NAICS-Based Product Codes: 31–19201 through 31–19204121

PRODUCT OVERVIEW

Coffee is an important world commodity. According to the International Coffee Organization, it is one of the most valuable primary products that is traded worldwide and for some exporting countries its traded value is second only to oil as a source of foreign exchange. Millions of people worldwide are employed in the coffee trade as growers, pickers, processors, millers, wholesalers, shippers, and roasters. Coffee has been an important commodity for at least 500 years. Stories about its origins and spread across the world are legendary.

History. One legend about the origin of coffee involves Kaldi the goat herder and takes place in Ethiopia on the east coast of Africa, across the Red Sea from the Arabian Peninsula. Kaldi discovered coffee after noticing that goats became so spirited after eating cherries from a particular tree they did not sleep at night. Kaldi reported this to the abbot of the local monastery. The abbot made a drink with the cherries and discovered that drinking it kept him and all the monks alert for the long hours of evening prayer. Support for this legend comes from the word coffee, allegedly derived from the region Kaffa in Ethiopia where

the goat herder Kaldi lived. Knowledge of the energizing effects of the cherries spread to the Arabian Peninsula.

By the 1400s coffee trees were cultivated in the area of Arabia known as Yemeni located directly east across the Red Sea from Ethiopia. By the 1500s, coffee tree cultivation spread northward throughout the Arab region giving Arabians a coffee monopoly.

By the 1600s European travelers returned from Arabia with beans to make the dark black beverage, sometimes called the wine of Araby. Even though coffee foes called it the bitter invention of Satan, coffee trade spread through Europe.

In the latter half of the 1600s, the Dutch usurped the Arabian coffee monopoly. They planted seedlings on the cluster of islands in the Indian Ocean between India and Australia in what is now Indonesia and was then a Dutch colony referred to as the Dutch East Indies. From there, coffee spread westward to the Caribbean.

One legend about the spread of coffee to the Caribbean island of Martinque north of the Venezuelan coast involves the French Naval Officer Gabriel Mathieu de Clieu. Serving in Martinique in 1720, de Clieu went to Paris on leave. He allegedly returned to Martinque with one coffee tree seedling. According to this legend, French Martinique stock spread southward across South America and westward into Central America.

According to other legends, the Dutch spread the coffee plant into South America and Central America from their colony Suriname in northeastern Brazil. The British are credited with introducing coffee into the legendary Blue Mountains of Jamaica in 1730. By 1825 the Blue

Mountain coffee from Jamaica was planted in the United States, in Hawaii.

Definition and Species. Coffee beans are the seeds of fruits which resemble cherries. The cherries have a red skin (the exocarp) when ripe. Inside the skin and beneath the pulp (the mesocarp), lie two beans, flat sides together, each surrounded by a parchment-like covering (the endocarp). When the fruit is ripe a thin, slimy layer of gelatinous mucilage surrounds the parchment. Underneath the parchment the beans are covered in another thinner membrane, the silver skin (the seed coat).

The two commonly traded coffee bean species are Arabica and Canephora, known as Robusta. Each require 60 inches of rainfall per year, and mild temperatures. For these reasons, coffee production is limited to the areas of the globe between the tropics of Cancer and Capricorn.

Arabica trees make up about 70 percent of worldwide coffee production. Arabica trees produce fine, mild, aromatic coffee beans that bring the highest prices. Varieties include the expensive Jamaican Blue Mountain. Arabica trees are costly to cultivate because they are often grown at high altitudes, 2,000 to 6,000 feet, where terrain tends to be steep and difficult to access. Arabica trees are also more disease-prone than their cousin, Robusta trees.

Robusta trees make up approximately 30 percent of worldwide coffee production. The tree is hearty, disease resistant, withstands warmer temperatures, and grows at lower altitudes than Arabica. Its taste is considered inferior; robusta beans have a distinctive bitter taste and approximately 50 to 60 percent more caffeine than Arabica. Robusta is primarily used in ground coffee blends and for instant coffees.

Taste Testing or Cupping. The legendary coffee beans pass through the hands of growers, pickers, dryers, processors, millers, wholesalers, exporters, shippers, and roasters. At every handoff, coffee beans are taste tested.

Taste testing is called cupping. Most handlers designate a special room for just the ritual of cupping. The taster—called the cupper—evaluates the beans for aroma, taste, and mouthfeel. To do so, the beans are roasted on the spot in a small laboratory roaster, ground, and infused in water boiled to a carefully controlled temperature.

The cupper noses the brew to experience its aroma. After letting the coffee rest for several minutes, the cupper breaks the crust by pushing aside the grounds at the top. After again nosing, the cupper slurps a spoonful with a quick inhalation. The objective is to spray the coffee evenly over the taste buds, assess it, then spit it out. Cuppers may sample beans from a variety of batches in a single day. An expert cupper can test hundreds of samples

of coffee a day and still detect subtle differences between aromas, tastes, and mouthfeel. Strong, highly recognizable notes are valued.

Because the aromatic notes of coffee beans are integral to the overall quality and experience of coffee, cuppers use precise aroma descriptors. For instance, animal-like is used to describe strong notes like wet fur, sweat, leather, or hides, and is not considered a negative. Some aromatic note descriptors double as tastes. Burnt/Smokey is both an aroma and taste descriptor associated with burning wood, and is used near the end of the journey when cuppers at roasters evaluate beans for blending or to indicate the degree of roasting necessary. Chocolate-like describes both the aroma and flavor of cocoa powder and chocolate, and is sometimes referred to as sweet. Caramel describes both the aroma and taste produced when sugar is caramelized without burning and is distinct from a burnt/smokey note.

Cereal/Malty/Toast-like is the descriptor for both aromas and tastes of uncooked or roasted grain (corn, barley, wheat, or malt extract), freshly baked bread, and freshly made toast. Earthy is used to describe a scent of fresh earth, wet soil, or humus associated with mold, and is considered undesirable. Floral describes an aromatic note of flowers like honeysuckle, jasmine, and dandelion found mainly in combination with fruity or green aromatic notes. Fruity/Citrus is a descriptor for both the aroma and taste of fruit associated with the scent of berries. The valuable perception of high acidity in coffee is correlated with the citrus characteristic. Grassy/Green/ Herbal describes scent notes of freshly mown lawn, fresh herbs or green beans, or unripe fruit.

Nutty describes the scent and flavor of fresh nuts, but not bitter almonds. Rancid/Rotten is a dual descriptor with rancid an indicator of fat oxidation in rancid nuts, and rotten an indicator of deteriorated vegetables or non-oily products. Spicy describes the aroma of sweet spices like cloves, cinnamon, and allspice, but not savory spices such as pepper, oregano, and Indian spices. Tobacco describes both the aroma and taste of tobacco but not burnt tobacco. Finally, winey is an overarching term used to describe the combined sensation of smell, taste, and mouthfeel experiences. It is generally perceived when a valuable strong, acidic or fruity note is found and does not apply to a sour or fermented flavor.

Cuppers utilize a broad array of precise aroma descriptors. Some describe both aroma and taste. The range of descriptors for tastes alone is smaller, consisting of five: acidity, bitterness, sweetness, saltiness, and sourness. In coffee, a taste of acidity is considered desirable, sharp, and pleasing. Coffee with too little acidity will taste flat and dull, coffee with moderate acidity is described as smooth, while coffee with high acidity is called lively.

Bitterness describes a taste characterized by the solution of caffeine, quinine, and certain alkaloids, and is considered desirable up to a certain level. It is affected by roasting and brewing. Detected on the back of the tongue, bitter is the characteristic note that brings together valuable dark chocolate and citrus fruit notes. Tastes for bitterness vary between countries. While Italians enjoy it, Americans are generally less keen for bitterness in coffee.

Sweetness describes tastes characterized by solutions of sucrose or fructose commonly associated with sweet aroma notes such as fruity, chocolate, and caramel, but also purity of taste or freedom from off-flavors of any type. Saltiness describes tastes characterized by a solution of sodium chloride or other salts. Sourness refers to excessively sharp, biting, and unpleasant flavors such as vinegar. The final factor cuppers evaluate coffee for is mouthfeel.

While cuppers use a broad array of aroma descriptors and a narrow well-defined set of five taste descriptors, only two descriptors for mouthfeel are used. Mouthfeel is characterized by either body or astringency. Body describes the physical properties of coffee. A strong but pleasantly full mouthfeel characteristic is preferred over thin mouthfeel. Astringency is characteristic of an after-taste consistent with a dry feeling in the mouth, and is undesirable in coffee. Mouthfeels, tastes, and aromas are assessed by cuppers at every stage as coffee beans change hands in the world coffee market.

MARKET

Coffee is crucial to the economies of many developing countries. For many of the least developed countries, coffee exports account for a substantial portion of foreign exchange earnings—in some cases over 80 percent. More than 50 countries grow and export coffee. Typical coffee growing countries, in order of export volume, are Nicaragua, El Salvador, Costa Rica, Ecuador, Tanzania, Kenya, Thailand, Rwanda, Burundi, Madagascar, Togo, Zambia, Panama, Bolivia, Venezuela, Zimbabwe, Nigeria, Haiti, Jamaica, Ghana, and Cuba, among about 30 others. Some are among the world's poorest countries.

Even though more than 50 countries export coffee and depend on it for economic livelihood, total production volume is dominated by only a few countries. The dominant coffee producing countries are Brazil, Vietnam, and Colombia, in that order.

Export Volume Worldwide. According to the International Coffee Organization in London, England, global coffee exports reached record levels in 2006. In that year, exports of all its member countries totaled 92 million 60-kilogram (kilo) bags. Figure 60 presents a year-on-year sequence of member countries' total export volume of 60-kilo bags, from 1977 to 2006. World coffee production almost doubled during this period. In 1977 world production was 48 million bags, and by 2006 it grew to 92 million bags. Figure 60 also shows how much export volume is controlled by both the top three and top 10 exporters.

Brazil is the top coffee exporter in the world. It exported 27 million 60 kilo bags in 2006, or almost one-third of the 91 million bags of total exports. Brazil towers over Vietnam and Columbia, the next two largest exporters. Vietnam and Columbia exported 14 million and 10 million bags in 2006, or about 50 percent and 60 percent less, respectively, than the behemoth Brazil with its vast plantations and mechanized farming systems.

The top three coffee producing nations exported 51 million bags in 2006, more than half of the 91 million total in the export marketplace. Since 2001 the top 3 exporters have controlled more than 50 percent of the world export market. For the 20-year period from 1980 until 2000, the top 3 firmly controlled 40 percent or more of the world commodity coffee market, dipping below 40 percent only twice, both times due to weather conditions in Brazil.

Market dominance by the top three exporters is growing due in part to increased production in Brazil, but also because of the rise of Vietnam. Vietnam is a new entrant into the world coffee commodity market and it advanced to the number two position in a matter of years. In 1981 Vietnam coffee exports were zero, while Brazil and Colombia dominated, exporting more than 13 and 9 million 60-kilo bags, respectively. From exports of zero, Vietnam raced to the top. From a production level of 2 million 60 kilo-bags in 1993, Vietnam tripled its exports to 6 million bags in 1997. By 2000 Vietnam's exports doubled again to close to 12 million 60 kilo-bags. In the seven years between 1993 and 2000, Vietnam went from being a minor player to being a world leader, replacing Colombia as the second largest exporter of coffee, a position that Colombia had held for a quarter century.

The top ten coffee producing nations of the world, in addition to Brazil, Vietnam, and Colombia, include, in order of export volumes: Indonesia, Peru, India, Uganda, Guatemala, Honduras, Ethiopia, and Mexico. Indonesia is the fourth largest player. Indonesia owns what the Dutch started in the 1600s, after obtaining its independence in 1949. In 2002 the top 10 crept up to control 80 percent of world export volume. For the five prior years, from 1997 to 2001, the top 10 exporting nations controlled 70 percent of the market, after hovering in the 60 percent range for almost 20 years.

World Import Percentages. Although about 50 countries export coffee, more than 155 countries import and drink the beverage. According to the International Coffee

Organization, imports totaled 119 million 60-kilo bags in 2005. Figure 60 depicts the top ten importers of coffee. The United States buys the most coffee. It imported 23 million 60-kilo bags in 2005, or close to 25 percent of world coffee exports.

In 2005 Germany purchased 17 million 60 kilo bags of coffee beans, or close to 20 percent of world coffee exports. Japan and Italy each imported around 7.5 million bags. France imported 6 million bags. Spain and Belgium/Luxemburg tied for imports at 4.4 million bags. Canada and the United Kingdom tied for imports of around 3.5 million bags. The Netherlands is the last country in the top ten importers at 3 million bags. People in the Russian Federation and Poland also enjoy consuming coffee and those countries imported close to 3 million bags each in

2005. The remaining 140 some countries purchased the remaining 14 percent of coffee.

U.S. Consumption Patterns. The USDA Economic Research Service has tracked per capita consumption of coffee and tea in gallons since 1911. Figure 61 depicts nearly a century worth of data on per capita consumption of coffee and nearly a half century of data on the consumption of carbonated soft drinks. Per capita coffee consumption since 1910 trended steadily upwards through the 1910s, 1920s, and 1930s, rising from 20 gallons per capita to more than 35 gallons per capita. In the 1940s, coffee was America's "hot drink." During that period, per capita consumption stayed steady at above 40 gallons and peaked at a record-setting all-time high of 48

FIGURE 60

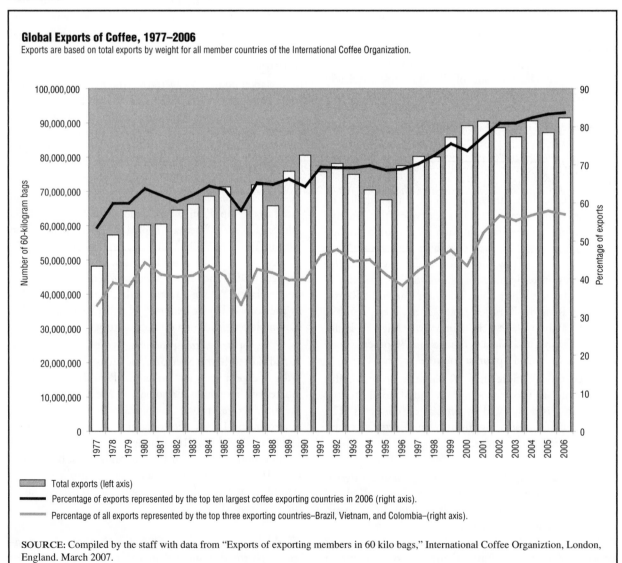

Global Exports of Coffee, 1977–2006
Exports are based on total exports by weight for all member countries of the International Coffee Organization.

Total exports (left axis)

Percentage of exports represented by the top ten largest coffee exporting countries in 2006 (right axis).

Percentage of all exports represented by the top three exporting countries–Brazil, Vietnam, and Colombia–(right axis).

SOURCE: Compiled by the staff with data from "Exports of exporting members in 60 kilo bags," International Coffee Organiztion, London, England. March 2007.

gallons in 1947. By 1950 coffee consumption started on a slow downward trajectory.

Coffee went cold in the 1970s. Cans of Folgers and Maxwell House lost supermarket shelf space to Coke and Pepsi. Folger's "Good to the last drop" was replaced by Coke's "It's the real thing." Coffee consumption plummeted. For a 20-year period from 1975 until 1995, coffee was a palliative for parents and "squares." Cool kids drank Coke and Pepsi. Pepsi reminded the young, "You've got a lot to live. Pepsi's got a lot to give." While per capita consumption of carbonated soft drinks grew from 30 to 50 gallons between 1975 and 1995, coffee consumption stayed below 30 gallons, reaching an all-time low of 20 gallons in 1994–1995.

Since 1995 U.S. coffee consumption has been percolating slowly upwards. Since the 1994–1995 low of 20 gallons per capita, coffee rose steadily until 2000 when it peaked—if such a low number can qualify as a peak—at 26.3. The peak of the new century is 50 percent lower than the 48 gallon per capita peak of the last century.

The slight upward trend since 1995 can be linked to the increase of away-from-home coffee sales. Changing consumption patterns driven by the reemergence of coffeehouses restructured the U.S. coffee marketplace. The coffee market is bifurcated into supermarket sales, which Folgers (Procter & Gamble) and Maxwell House (Kraft Foods, Inc.) have historically dominated, and away-from-home sales, with Starbucks dominating. Between 1996 and 2006 Starbucks grew to a $6 billion per year enterprise, creating copycat coffee shops across the county. Dunkin' Donuts changed its focus from donuts to coffee.

The number of U.S. coffee shops more than doubled in number in the first five years of the new century, growing from 12,600 to 21,400, according to market research firm Mintel International Group. This staggering growth equates to one coffeehouse for every 14,000 Americans. Mintel also reported that total U.S. coffeehouse sales increased from $3.2 billion to $8.4 billion in the same time period.

The total U.S. coffee market grew by more than 50 percent—from $19 billion to $29 billion—between 2002 and 2006 and is expected to grow by another $10 billion by 2011, according to independent market analyst Datamonitor statistics reported in the May 28, 2007 issue of *Nation's Restaurant News*. The majority of coffee is being consumed outside of the home. Datamonitor projected that "away-from-home purchases" will be 75 percent of total sales by the early 2010s. A large part of the rational for this high percentage projection is that coffee purchased away-from-home costs far more by volume than does coffee made at home.

U.S. Census Bureau Statistics. In 2002 the coffee and tea manufacturing industry reported total shipment values of $5.3 billion, according to the report titled "Coffee and Tea Manufacturing: 2002." Total coffee and tea product shipments declined almost 32 percent from $7.3 billion in 1997. Roasted coffee product shipments declined even more, from $5.6 billion in 1997 down to $3.6 billion in 2002, a total decrease of 35 percent. Since the 2002 U.S. Census Bureau report, the *Annual Survey of Manufactures* reported increases in coffee and tea manufacturing for 2003, 2004, and 2005, with increases over prior years of 5, 0, and 7 percent, respectively. Industry trade association statistics are unclear as to where the growth is. The National Coffee Association and the Specialty Coffee Association of America have different perspectives on growth in statistics they generate to track demand.

U.S. Coffee Demand. Coffee consumption has been on a downward trend that coincided with a sharp upward surge of carbonated soft drink consumption. The flip flop in consumption patterns is so drastic that even the National Coffee Association talks about its market gains relative to soft drink consumption. In its 2007 national coffee drinking survey the National Coffee Association reported that, for the first time since 1990, the percentage of U.S. adults who drink a daily cup of coffee exceeded those who drink a daily soft drink. Out of the nearly 3,000 adults sampled for the survey in January 2007, 57 percent said they drank coffee every day, up from 56 percent in 2006. For carbonated soft drinks, the results were 51 percent for 2007 versus 57 percent for 2006.

The National Coffee Association also reported that daily consumption of gourmet coffee beverages in 2007 fell to 14 percent from a 2006 high of 16 percent. The gourmet coffee subset known as espresso-based beverages was down 1 percent from its 2006 high of 6 percent. Statistics available from the Specialty Coffee Association of America (SCAA) support the opposite trend, reporting growth in the specialty coffee segment. How the terms *gourmet coffee* and *specialty coffee* are defined by different organizations has in impact on the statistics that result from a study of the subgroups.

The SCAA defines specialty coffee as coffee that has no defects and has a distinct flavor in the cup. Inherent in the definition is that specialty coffee does not include much Robusta.

The SCAA estimated the growing total market size with year-on-year sequence starting in 1999. The SCAA derives its growing numbers from five distinct distribution channels: coffeehouses with seating (15,500 locations), coffee kiosks without seating (3,500 locations), mobile coffee carts (2,900 locations), roasters on premise (1,900 locations) and supermarkets. From these five distinct

FIGURE 61

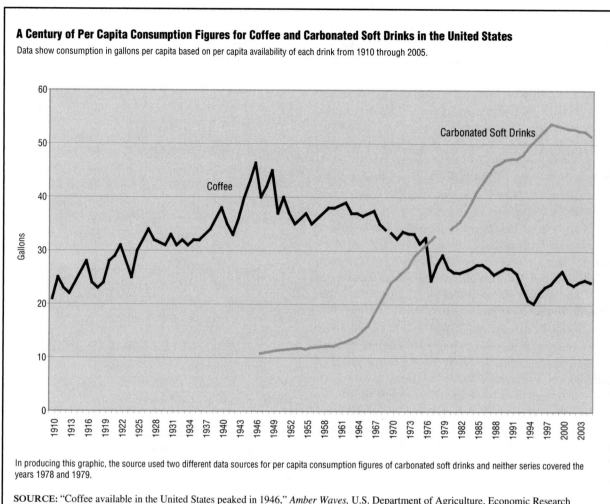

A Century of Per Capita Consumption Figures for Coffee and Carbonated Soft Drinks in the United States

Data show consumption in gallons per capita based on per capita availability of each drink from 1910 through 2005.

In producing this graphic, the source used two different data sources for per capita consumption figures of carbonated soft drinks and neither series covered the years 1978 and 1979.

SOURCE: "Coffee available in the United States peaked in 1946," *Amber Waves,* U.S. Department of Agriculture, Economic Research Service. June 2007.

channels, the SCAA proclaimed that specialty coffee grew from $7.5 billion in 1999 to $12 billion in 2006. Using these numbers, sales of specialty coffee were up 10 percent from 2004 to 2005.

Data from the U.S. Bureau of the Census reports, the *2002 Economic Census* and the *Annual Survey of Manufactures,* indicate a slight upward trend in the coffee and tea industry between 2003 and 2005, averaging growth of 6 percent per year. The National Coffee Association positions itself vis-a-vis the carbonated soft drink industry which it views as its main competitor. The SCAA sees coffee more on a par with wine, where growth involves a discerning consumer with a discriminating palate willing to pay more for a premium product, properly prepared. As discriminating consumers shift a greater share of coffee consumption to either small roasters offering a premium product on the retail side, or to away-from-home coffee

on the coffeehouse side, variations in the price of retail coffee have increased.

U.S. Retail Market Volume and Prices. Ground coffee purchases by volume at retail stores grew more slowly than did the U.S. population during the early years of the twenty-first century, according to the USDA Economic Research Service report from March 2007. The downward trend in supermarket coffee sales has been offset by increases in away-from-home coffee consumption.

The U.S. coffee total market retail value reported by Mintel International Group in October 2006 was estimated at $3.4 billion. The market total includes ground, instant, whole bean, and ready to drink coffee beverages like Frapuccino in a bottle. Ground coffee is the largest of the $3.4 billion, representing 70 percent. Ground coffee retail sales grew 30 percent over their 2004 levels, while whole bean sales grew 11 percent over the same period. Meanwhile, instant coffee sales lost ground. Because these total U.S. retail sales values do not include WalMart, Mintel takes great pain to estimate its effect on the total market. Mintel expects WalMart controls 15-17 percent of coffee sales and accordingly estimated its 2006 sales at around $650 million.

U.S. retail coffee prices per pound were also reported by Mintel International Group in October 2006. Prices were volatile for five the years between 2001 and 2006. The price started in 2001 at $3.25, dropping almost 10 percent to $3.01 in 2002, rising 10 percent to $3.19 in 2002 where it stayed steady in 2004. In 2005 the U.S. retail coffee price per pound increased 12 percent to $3.88. Worth noting is the fact that most specialty and premium coffee is sold at retail outlets in 12-ounce bags, not by the pound.

Java, Joe, Café, Caffe, Coffee. By any name, in any language, coffee is a popular beverage choice. Among all U.S. beverages regularly consumed, coffee ranks fourth in consumption, after carbonated soft drinks, fruit/vegetable juices, and milk. Coffee is not only one of America's favorite beverages, it is a great value. On average, each cup of coffee prepared at home costs less than a dime. From the choice of beverages consumed at home per eight fluid ounces, costs are as follows: coffee (5 cents), carbonated soft drinks (13 cents), milk (16 cents), bottled water (25 cents), beer (44 cents), orange juice (79 cents), and table wines ($1.30), according to the U.S. Bureau of Labor Statistics, Division of Consumer Prices.

KEY PRODUCERS/MANUFACTURERS

The U.S. coffee market is not a perfectly competitive market characterized by many small companies fighting for customers by lowering prices whenever possible. Kraft Foods, Inc. (with Maxwell House) and Procter & Gamble (with Folgers) have long dominated the market for coffee sold through supermarkets and grocery stores. Both are well-known and long-established manufacturers of consumer packaged goods. They each roast and grind coffee beans and package them in either vacuum packed bricks or cans for supermarkets. Neither of these coffee leaders operates coffee shops. Together they control more than 70 percent of supermarket ground coffee sales volume. From 2000 to 2004, market share volume was P&G (Folgers 38%) and Kraft (Maxwell House 33%). Folgers' market share increased to 42 percent in 2004, according to the USDA Economic Research Service March 2007 report.

Mass produced and vacuum packed coffee sold through supermarkets is not a premium product. The premium product is defined by fresh whole Arabica beans, sometimes flavored. Folgers and Maxwell House are not premium products primarily because during the roasting and grinding process, manufacturers presumably combine cheaper Robusta beans with a smaller amount of high quality Arabica beans.

Supermarket sales are one half of the newly bifurcated coffee industry. Since the mid 1990s, the increase of away-from-home coffee sales restructured the U.S. coffee marketplace. Starbucks dominates away-from-home sales in its coffee shops. Starbucks biggest competitor is Caribou Coffee Company. Along with the top two well-known and long-established players in supermarket sales, the newest entrants in the away-from-home sector—Starbucks and Caribou—are profiled. A few examples of small roasters, a growing segment in specialty coffee, are also listed.

Kraft Foods, Inc. Headquartered in Northfield, Illinois, Kraft Foods is the largest maker of food products for U.S. supermarket shelves. It produces Maxwell House and Yuban brands of ground coffee.

Kraft produces coffee at plants in Houston, Texas; Jacksonville, Florida; and San Leandro, California. The Jacksonville plant is the largest among those. Kraft coffee products take advantage of the lower priced Robusta beans, blending in a minimum of high quality Arabica beans to neutralize the harsher taste associated with Robusta beans. Since 1998 Kraft has distributed coffee for Starbucks' whole bean coffee including its Seattle's Best brand. The arrangement draws on Kraft's extensive network for selling, marketing and distributing packaged foods directly to supermarkets.

Kraft has roots as far back as 1767 in cheese, mustard, biscuits, crackers, and gelatin. James L. Kraft started out in 1903 as a cheese wholesaler in Chicago, Illinois. Eventually Kraft acquired Maxwell House Coffee. Maxwell House began in Nashville, Tennessee, when grocery wholesaler Joel Cheek created the company in 1892. Nashville's finest hotel, Maxwell House, served the coffee exclusively to its guests, giving the brand the prestige of the hotel, even though it was available at supermarkets and easily prepared at home.

A forerunner to General Foods Corporation acquired Maxwell House coffee in 1928. Maxwell House became the first nationally available brand of coffee specially ground for use in automatic drip coffee makers in 1976. Kraft remained independent until 1988 when it was acquired by Philip Morris, the tobacco company. Philip Morris had already purchased General Foods (in 1981). In 1989 the company formed Kraft General Foods from the two acquisitions and, in 1995, renamed this element Kraft Foods. Philip Morris, renamed Altria, spun off Kraft Foods in early 2007. Kraft had revenues of $34.4 billion in 2006 of which coffee beverages represented $4.8 billion.

The Maxwell House Web site boasts that, for more than 100 years Maxwell House has been exactly the same. Its slogan is still "Good to the last drop." It has a long way to go to catch up to changing coffee industry trends toward premium and whole bean coffee. While Maxwell House has expanded its line to include four categories of ground coffee: ground, flavored, convenience, and instant, it does not offer whole beans. Within its ground coffee offerings, nine choices are available, three each in mild, medium, and dark roasts. Maxwell House Master Blend, Maxwell House Slow Roast, and Maxwell House Decaffeinated are the mild choices. Original Maxwell House, Maxwell House Lite, and Maxwell House Original Decaffeinated are the medium roasts. Maxwell House French Roast, Maxwell House 100% Columbian, and Maxwell House French Roast Decaffeinated are the dark roasts. Each comes in either an 11.5-ounce or 34.5-ounce vacuum packed can. Flavor choices are limited at Maxwell House to two flavors in an 11.5-ounce can: Maxwell House Vanilla Flavored and Maxwell House Hazelnut Flavored. Convenience filter packs and singles are available, as well as instants.

Kraft also sells Yuban Original, Yuban Dark Roast, and Yuban Original Decaffeinated. Yuban is advertised as sustainable through its partnership with the Rainforest Alliance, which works toward sustainable agricultural and applies social and economic standards to coffee it certifies.

Procter & Gamble (P&G). Headquartered in Cincinnati, Ohio, P&G is the largest maker of household products for U.S. supermarket and superstore shelves. It produces Folgers and Millstone Signature brands of coffee. P&G roasts and packages most of its coffee in its New Orleans, Louisiana, plant. It operates a smaller roaster in Kansas City, Kansas.

While the P&G top selling Folgers brand remains a billion dollar business worldwide, even spokeswoman Tonia Hyatt has been forced into defending the brand. P&G's use of lower priced Robusta beans, blended with a minimum of high quality Arabica beans to neutralize the harsher taste of the Robusta beans has led to the brand being equated with old fashioned and non-discriminating tastes. "We have grown sales year after year," Hyatt proclaimed, "and I don't think we would have done that with poor quality coffee."

Folgers is available in four formats, ground, instant, pods, and the Café Coffees lines. Folgers ground line is broad. It includes Folgers Classic, Folgers Coffeehouse Series, Folgers Flavors, Folgers Simply Smooth, and Folgers Decaf. Instant can be in singles or crystals. The Coffeehouse Series includes Breakfast Blend, 100% Columbian, French Roast, Gourmet Supreme, and Gourmet Supreme

Decaf. The four Folgers Flavors are Chocolate Silk, Cinnamon Swirl, French Vanilla, and Hazelnut.

The Millstone Signature line reveals that P&G is slightly more in tune with trends than Kraft. It is 100 percent premium Arabica bean coffee offered in flavored and unflavored varieties. Millstone is available in bulk whole bean, prepackaged whole bean, and ground. Millstone offers online ordering of 1.5 ounce, 12 ounce, and 5 pound bags. Millstone Signature brand supports decaf, flavored, roast, and organic lines. Decaf has ten in its family: Decaf Bed & Breakfast Blend, Decaf Breakfast Blend, Decaf Caramel Truffle, Decaf Chocolate Velvet, Decaf Colombian, Decaf French Roast, Decaf French Vanilla, Decaf Hazelnut Cream, Decaf Irish Cream, Decaf Vanilla Nut Cream.

P&G is in tune with flavoring trends. Millstone Signature flavored utilizes flavors including caramel, chocolate, cinnamon, hazelnut, vanilla, peppermint, Irish Cream, Kahlúa, pumpkin, raspberries, almond, and maple. Millstone Signature roasts are available in light, medium, and dark. Light has a breakfast blend and a kona blend. Medium and dark both have the same six blends: 100% Columbian, Colombian Supremo, Foglifter, Maisonette Blend, Sumatra, and last but not least, Jeff Ruby's Steakhouse Blend. Organic supports five separate lines: Rainforest Reserve Rainforest Alliance Certified, Organic Mountain Moonlight Fair Trade Certified, Organic Deep Peruvian Forest Blend, Organic Nicaraguan Mountain Twilight Blend, and Organic Mayan Black Onyx Blend.

Contradicting its defense of Folgers somewhat, P&G went upscale in late 2006. It introduced Folgers Gourmet Selections. The new line includes seven flavors each in 12-ounce packages. Both ground and whole bean options are available in Vanilla Biscotti, Lively Colombian, Morning Cafe, Espresso Roast, Hazelnut Creme, Chocolate Truffle, and Caramel Drizzle. The new premium line is aimed at capturing some of the dollars specialty coffee makers such as coffeehouses, kiosks, and small roasters are getting by catering to the lucrative market of coffee connoisseurs.

The coffee industry is what some would call incestuous. Kraft has partnered since 1998 with its competitor Starbucks to distribute Starbrucks coffee and in 2007 P&G partnered with Dunkin Donuts. P&G will distribute Dunkin Donuts coffee packages to U.S. supermarkets and superstores. The Dunkin Donut range includes both whole bean and ground coffee in a variety of flavors and roasts.

Starbucks. Starbucks added the barista to the long list of coffee bean handlers that includes growers, pickers, dryers, processors, millers, wholesalers, exporters, shippers, and roasters. Starbucks was founded in 1971 in Seattle, Washington, by two teachers and a writer. Its legendary growth began only after entrepreneur Harold Schultz purchased the company to focus on the coffeehouse market rather than supermarket sales. Starbucks had 165 coffeehouses in the United States at the time of its 1992 initial public offering. Starbucks grew during the 2001 recession, announcing at the end of 2002 that, despite the weak economy, its profits were up 19 percent for the year. Starbucks has 39 presidents and senior vice presidents; 10 are women, or about 25 percent. It reached sales of $6.1 billion in 2006.

Instead of competing for a share of existing supermarket sales, Starbucks invented its own market. The strategy made Starbucks a tastemaker, according to a 2003 *New Yorker* article, which stated: "Starbucks changed not just what people drank but how they drank it. Instead of gulping down gas station swill on the fly, people learned to desire the experience of leisurely sipping a grande latte while eavesdropping on job interviews." The coffee colossus grew from 165 U.S. stores in 1992 to 9,814 stores in the United States by 2007. Starbucks had more than 145,000 employees at its nearly 10,000 stores throughout the country. Starbucks planned to open approximately 1,000 Company-operated locations and 700 licensed locations in 2007 and 2008. This phenomenal growth led Joseph Michelli to write an entire book about the Starbucks experience, while others wrote about Starbucks as a "third place"—what sociologists define as a necessary alternative to work and home. The success of the third place benefited by the rise of telecommuting and freelancing, and a move toward what Harvard historian Nancy Koehn called "affordable luxuries." Affordable luxuries allow American consumers to cultivate the aura of affluence at a relatively low price, if $4 represents a low price for a cup of coffee. Starbucks is so popular that a Starbucks Gossip Web site sprung up alongside an "I Hate Starbucks" Web site. Also popular is a Delocator Web site designed to guide the user toward independently owned cafés.

Starbucks purchases and roasts premium high quality whole Arabica coffee beans and sells them along with fresh, rich-brewed, Italian style espresso beverages at its coffeehouses. Starbucks also sells whole bean coffee at supermarkets through its licensing agreement with Kraft. Starbucks operates four roasting facilities located in Seattle and Kent, Washington, and in York, Pennsylvania. In April 2007, it announced the location of a fifth roasting plant near Columbia, South Carolina, scheduled to begin operating in early 2009.

In fiscal 2006 Starbucks bought almost 300 million pounds of coffee beans, or roughly 25 percent of all U.S. imports. Starbucks paid an average price of $1.42 per pound. About 10 percent or 32 million pounds of the total 300 million pounds was either shade grown coffee (2 million pounds), certified organic coffee (12 million pounds), or Fair Trade certified (18 million pounds). Starbucks is the largest purchaser, roaster, and distributor of Fair Trade certified coffee in the United States.

Caribou Coffee Company. The mythology surrounding Minneapolis, Minnesota, based Caribou Coffee is that in 1990, during an adventure through the Alaskan wilderness, its founders journeyed to the top of Sable Mountain. At the summit they were rewarded with a sensational view of boundless mountains, a clear blue sky, and a herd of caribou thundering through the valley. The "aha" moment and panoramic view led to Caribou Coffee. Its slogan involves variations on the phrase "Life is short. Make the most of it." The founders opened the first store in Minnesota in 1992. The company has a top leadership team of 15 individuals. Of these, six are women, or about half.

Caribou Coffee Company is the nation's second largest coffeehouse chain behind Starbucks. In April 2007 Caribou had 475 coffeehouses, plus 33 franchise locations, in 18 Midwest and Mid-Atlantic states, plus the District of Columbia. According to Mintel International Group, Caribou Coffee would need 12 years at its current expansion rate to catch up with Starbucks' current U.S. presence, while the number of new Starbucks stores opened in 2005 alone was greater than five times the entire Caribou Coffee franchise. In its battle with Starbucks for supermarket share, Caribou Coffee Co. made distribution deals with Shaw's Supermarkets, a unit of Supervalu, and Harris Teeter, a subsidiary of Ruddick Corp., to sell its packaged coffee beans in supermarkets in New England and the Southeast. In 2007 Caribou Coffee started a several-store pilot program with Bed Bath & Beyond to offer several varieties of its premium coffee beans at stores in six states. Caribou Coffee also sells to office coffee providers, airlines, hotels, sports and entertainment venues, college campuses, and other such commercial customers.

Caribou Coffee pledged that by 2008 at least 50 percent of its coffee beans will come from Rainforest Alliance certified farms. Caribou lines that currently bear the seal (and the percentage certified in each blend) include: Daybreak (50%), Colombia (100%), Guatemala El Socorro (100%), Caribou Blend (75%), Fireside Blend (30%), Espresso Blend (75%), French Roast Blend (75%), Reindeer Blend (30%), Perennial Blend (30%), and Amy's Blend (50%).

Others. Sara Lee Corporations sold its U.S. coffee retail brands (Chock Full o'Nuts, Hills Brothers, MJB, Chase & Sanborn; excepting Senseo) during 2005 to Segafredo

Zanetti Coffee Group of Bologna, Italy. In another industry sale, A&P sold Eight O'Clock in 2003 to Gryphon Investors, which subsequently operated as Eight O'Clock Coffee Company. In June 2006, Tata Coffee Ltd. of India acquired the brand.

Small Roasters. The SCAA sees a growth trend for small roasters who had shipments of $1.75 billion in 2006. These are typically small local retailers who roast on premises. They gained a competitive edge by roasting flavored beans at a time when mega-roasters of Maxwell House and Folgers were not providing the product. Examples include Our Coffee Barn in Spring Valley, Wisconsin; Prima Café in Oklahoma City, Oklahoma; and Baby's Coffee in Key West, Florida, which advertises itself as the southernmost coffee roaster in the United States. One small regional roaster called Dunn Bros. Coffee opened in St. Paul, Minnesota, in 1987. It has 100 locations in five states that roast beans on site. Small roasters can flavor coffee with basic equipment at an average of 25 to 50 cents per pound, and sell flavored coffees at a profitable margin.

Green Mountain Coffee Roasters of Waterbury, Vermont, is a rapidly growing regional roaster. It went public in 1993. Its partnership with Newman's Own Organic Coffee is a good indication of the growth in specialty coffee. In 2001 Green Mountain Coffee Roasters sales totaled $96 million. By 2006 sales more than doubled to $225 million. Green Mountain Coffee Roasters focuses on Fair Trade certified coffee; that segment grew 69 percent in 2006 over the prior year.

MATERIALS & SUPPLY CHAIN LOGISTICS

Besides coffee beans that have been grown, picked, processed, milled, wholesaled, exported, and shipped from abroad, U.S. manufacturers purchase a great deal of packaging materials. Materials consumed by coffee manufacturers are reported in a U.S. Census Bureau report titled "Coffee and Tea Manufacturing: 2002."

For the newly emerging and rapidly growing away-from-home segment, the paper cup is fundamental. For instance, Starbucks rarely spends on advertising and marketing. It lets its locations and cups do its marketing. The Starbucks cup is everywhere. Blogs show Kate Hudson, Sienna Miller, Nicole Ritchie, Renee Zellweger, and the ever present Mary-Kate and Ashley Olson twins holding Starbucks cups.

The 21,400 coffeehouses in the United States use a lot of cups each year. Projections vary. Starbucks and its partner, the nonprofit Environmental Defense, reported in 2006 that Starbucks uses almost 2 billion cups annually. Other sources report that Americans annually use more than 15 billion paper cups designed for hot beverages. That number is projected to grow to 23 billion by 2010.

At the coffee shop, coffee is the smallest portion of the beverage purchase price. For instance, one pound of beans makes about 40 cups of coffee. Assuming the beverage is made from premium coffee beans—the type roasters buy for $4 to $5 a pound—the value of the coffee is about a dime per cup. The cup and the lid can cost twice that much. Of course, the price of each coffee shop beverage has to cover real estate rents, salaries and benefits, taxes, marketing expenses, and other overhead costs.

At coffeehouses, the cup is as important as the coffee and costs more. For Starbucks, it is a marketing tool. The cup is so important to Starbucks that in 1996, the coffee colossus began a 10-year odyssey into finding the perfect cup. At that time Starbucks wanted to stop its practice of double cupping, the practice of providing a customer with two cups, one used to protect the hands for the heat generated by the coffee in the interior cup. As a replacement for the second cup, in 1997 Starbucks started using its 60 percent recycled corrugated paper insulating sleeve. In 1999 several Starbucks stores tested a two-layer cup with an outer layer made from 50 percent postconsumer fiber.

The coffee giant went back to the drawing board in 2001 and partnered with Solo Cup, headquartered in Highland Park, Illinois, with the goal of creating a more eco-friendly paper cup. A major obstacle was getting U.S. Food and Drug Administration (FDA) approval, since recycled content had never before been used in direct contact with food—especially not hot beverages. Mississippi River Corp. provided recycled pulp and paper mill MeadWestvaco (later called NewPage) developed new testing protocols. Solo Cup won FDA approval in 2005. The cups were rolled out in March 2006 at over 5,000 U.S. Starbucks locations. The new cup will conserve 5 million pounds of paper per year, or approximately 78,000 trees, according to Environmental Defense.

Focus on the cup continued. Green Mountain Coffee Roasters of Waterbury, Vermont, also introduced an eco-friendly disposable paper coffee cup. Called the ecotainer cup, it is the first hot paper beverage cup made from fully renewable resources. Unveiled in the summer of 2006, concurrent with the new Starbucks cup, the ecotainer cup is lined with a bio-plastic derived from corn, making it compostable under the proper conditions. Green Mountain Coffee Roasters and International Paper, its partner, were awarded a 2007 Sustainability Award by the Specialty Coffee Association of America. The cup's liner makes it unique. Conventional hot paper cups use a waterproof lining made from a petroleum-based plastic. By choosing to utilize a corn-based cup liner, Green Mountain Coffee Roasters will conserve the consumption of nearly a quarter of a million pounds of non-renewable petrochemical materials every year.

DISTRIBUTION CHANNEL

Coffee is an important world commodity. Its distribution channel is complicated. The legendary coffee beans pass through the hands of growers, pickers, dryers, processors, millers, wholesalers, exporters, shippers, roasters, and grocers, not to mention the many professional baristas now working at America's more than 21,000 coffee-houses. At every handoff, coffee beans are taste tested and a portion of every dollar spent on coffee is claimed. In May 2003 *Frontline World*, the Public Broadcasting Service program, reported that the profit is not equally shared between growers and harvesters in exporting countries and manufacturers and retailers in importing countries. *Frontline World* calculated the allocations of every U.S. retail dollar spent on coffee for growers, harvesters, shippers, roasters, and retailers.

Growing. Pruned short in cultivation, but capable of growing more than 30 feet high, a coffee tree is covered with dark-green, waxy leaves growing opposite each other in pairs. It takes 3 to 4 years for newly planted coffee trees to bear fruit. After the fragrant white flowers blossom, it takes one year for coffee cherries to grow. In Mexico, one of the smaller volume exporters, 90 percent of coffee farms are no bigger than 12.5 acres, the majority owned by local people. Growers receive 10 to 12 cents of every U.S. retail dollar spent on coffee.

Harvesting. Coffee cherries turn bright deep red when ripe. Coffee can be harvested in two ways. Strip picking involves harvesting the entire crop at once, by machine or by hand. Selectively picking involves harvesting only the ripe cherries by hand. This labor intensive method is used primarily to harvest the finer Arabica beans. A good picker averages 100 to 200 pounds of cherries a day, which will produce 20 to 40 pounds of coffee beans. Harvesters get between 10 and 12 cents of every U.S. retail dollar spent on coffee.

Processing. Processing must begin quickly to prevent spoilage. Coffee is processed using either a dry or wet method. Neither method is superior. Dry processing can take too long and / or muddle flavors or introduce bacterial and fungal taints. Wet processing can enhance valuable acidity at the expense of some of the desirable body. Even wet processed beans must eventually be dried. Drying affects the final quality of coffee beans. Over-dried beans are brittle and can be easily broken during the milling process. Under-dried beans are prone to rapid deterioration caused by fungi and bacteria.

The traditional dry method is used in countries where water resources are limited. Whole fresh cherries are spread on huge surfaces to sun dry. Cherries are raked and turned often to ensure even drying and to prevent mold and mildew. The process may take from several days to up to four weeks depending on weather. Almost all Robustas are dry processed. Mechanical coffee driers can be used. They tend to be faster, drying beans in less than two days. But cherries must be turned often to distribute hot air evenly or risk over-drying. Dry processed coffee beans fetch significant price premiums, due to a resurgent respect for old fashioned production processes. Dry processed coffee can result in fuller body, better mouthfeel, slightly lower acidity, and natural sweetness. Certain dry processed Sumatran coffees have become extremely expensive. Jamaican Blue Mountain sells for around $35 per pound.

Wet processing involves passing whole cherries through a pulping/stoning machine to separate the skin and pulp from the bean. Beans are then fermented in large tanks to break down the residual sticky mucilage adhering to the parchment surrounding the beans. Wet beans are 57 percent moisture and must be sun or mechanically dried. Wet processing is used for almost all Arabica beans, with the exception of those produced in Brazil. Fermentation adds quality to expensive Arabicas by increasing acidity, adding valuable complex notes. After processing, beans are milled.

Milling. Coffee is milled in three stages: hulling, polishing, and grading/sorting. For wet processed beans, machines remove the single remaining parchment layer. For dry processed beans, the entire three layer dried husk is removed. Polishing removes the silver parchment skin. While optional, polished beans are considered superior to unpolished ones. Grading/sorting involves a size scale of 10 to 20. A number 10 bean is the size of a 10/64 inch diameter hole, while a number 15 bean is 15/64 inch. Sorting removes beans deemed defective due to over-fermentation, insect damage, or bruising.

Exporting. Harvested, processed, and milled beans travel by ship in either jute or sisal bags loaded into shipping containers or bulk-shipped inside plastic-lined containers. Shippers share their cut with banks and insurance companies. Shippers need to move massive volume in order to make money. Shippers receive about 4 cents of every U.S. retail dollar spent on coffee.

Roasting. Because coffee beans travel by ship, most coffee is roasted near the shipping ports of New Orleans, New York City, San Francisco, and Miami. Most roasting machines maintain a temperature of about 550 degrees Fahrenheit. Beans are kept moving throughout the process to keep them from burning. When they reach an internal temperature of about 400 degrees, they begin to turn brown and the caffeol, or oil, locked inside the beans begins to emerge. This process, called pyrolysis, is at the heart of roasting. It produces the flavor and aroma of coffee. Roasted beans have a shelf life of about one week. Roasters receive the most of every retail dollar spent on coffee—about 70 cents.

Retailers. Packaged coffee is typically delivered directly to the warehouses of supermarkets and transportation costs are included in the price. Retailers receive 10 to 15 cents of every U.S. retail dollar spent on coffee.

KEY USERS

Adults drink coffee. In 2007 the National Coffee Association reported that 81 percent of American adults drink coffee sometimes, and that 64 percent of American adults drink coffee weekly. In 2003 the National Coffee Association reported that 82 percent of all coffee was consumed during morning hours, 4 percent was consumed at lunch, 5 percent in the afternoon, 4 percent at dinner, and 5 percent during the evening hours.

ADJACENT MARKETS

In order to brew fresh roasted coffee beans at home and extract from them the fullest flavor, consumers need grinders and coffee makers, either automatic drip coffee makers or espresso machines. Originally designed for the commercial market, residential espresso units became popular in the United States during the 1980s and 1990s. Manufacturers include Pasquini (units start at around $900) and Brugnetti ($1200). Nuova Simonelli is a premier Italian maker of espresso machines certified by the European Institute of Ergonomy.

Also adjacent to coffee are other beverages and in particular other beverages that contain caffeine. Many carbonated drinks fall into this category, such as Coke, Pepsi, and Mountain Dew. Figure 61 presents per capita consumption rates of coffee and carbonated drinks, showing the rise of carbonated drinks and the decline of coffee consumption.

RESEARCH & DEVELOPMENT

Research and development efforts have resulted in improvements in the wet processing methods used for coffee beans. Improvements decreased the quantity of water needed in the process and reduced the time needed to process while producing more nuanced flavors. For instance, traditional wet processing was invented around 1740 by the Dutch in the West Indies. It used large amounts of

water and involved several steps. In traditional wet processing, freshly picked cherries were separated by density in immense water tanks. Overripe and partially dry cherries have lower densities and float to the surface. Dense unripe and ripe cherries sink to the bottom. Denser cherries were pressed, again in water, between grated screens that removed the bulk of the cherry pulp. The de-pulped beans were typically then fermented in large tanks of water to remove the muscilage.

Besides requiring large amounts of water, the traditional method de-pulps dense cherries together, increasing the risk that unripe cherries get mixed in with the valuable ripe cherries. R&D resulted in an innovative semi-washed process that consumes 10 times less water than traditional systems. Instead of being done in water tanks, density separation is done by mechanical siphons such as those patented by Brazilian machinery maker Pinhalense. Mechanical siphon tanks separate cherries according to density with very little water consumption. Ecological wet processing removes the mucilage by friction instead of fermentation. Machines called mucilage removers use less water than traditional wate fermentation tanks.

The semi-washed system produces the same cup as natural coffee along with the advantage that there is no risk of unripe cherries. The semi-washed process is alleged to yield beans that are sweeter and somewhat less acidic than those that undergo a full washing process that includes the fermenting stage.

CURRENT TRENDS

Coffee beans go on a sort of journey that involves several numerous handlers. Each handler takes a cut of the consumer coffee dollar. Whether or not the cut is apportioned fairly is the question at the heart of fair trade coffee.

TransFair USA, in Oakland, California, is the only fair trade certifying label in America. The group, founded in 1998, created a "hipster" seal of approval for U.S. consumers concerned about trade inequities. Proponents say low prices that most companies pay to growers and pickers in economically disadvantaged countries cause widespread misery: poverty, unsafe work conditions, and forced child labor. TransFair has core requirements for importers, roasters, and growers.

Growers must be organized into cooperatives that agree to independent inspections and use sustainable methods of agriculture. In return, the growers were guaranteed a living wage of at least $1.26 per pound for their coffee (15 cents more if it were grown without pesticides). The TransFair USA Web site profiles about 30 coffee growing cooperatives, detailing how many members were in each cooperative, and how fair trade had helped meet local healthcare, education, and transportation needs for those in the community.

Fair trade coffee has grown since 1998 when 76 thousand pounds were imported. By 2002 imports grew to almost 10 million pounds and then doubled in the following year, reaching almost 20 million pounds in 2003. By 2006 imports of fair trade coffee had reached more than 60 million pounds. Fair trade coffee is available at roughly 20,000 retail outlets across the United States. Newman's Own Organics offers a fair trade line in cooperation with Green Mountain Coffee Roasters. Peru, Mexico, Nicaragua, Indonesia, and Ethiopia were the top five exporters to the U.S. of fair trade certified coffee in 2006.

TransFair USA had tracked the growth rate of both its traditional and organic fair trade certified coffee since 1999. It had also tracked estimated additional income to farmers as a result of fair trade totaling $92 million between 1999 and 2005. There was tremendous growth of organic fair trade coffee in 2006; it grew 94 percent over 2005.

Fair trade is becoming less a hipster seal of approval and more a part of mainstream commercial culture. Larger corporations offer some fair trade coffee. In April 2000 retail coffee giant Starbucks started to carry fair trade certified whole bean coffee. In October 2004 Burlington, Vermont, based Bruegger's announced it would serve fair trade certified coffee at its 242 locations. In November 2005 McDonald's announced that it would serve fair trade certified coffee in 658 of its restaurants in New England. Participating locations were to switch 100 percent of their coffee products to Fair Trade Certified organic coffee from Newman's Own Organics, roasted by Green Mountain Coffee Roasters.

New York, New York, based nonprofit Rain-forest Alliance works to conserve biodiversity and ensure sustainable livelihoods by transforming land use, business practices, and consumer behavior. Rainforest Alliance certified coffee purchases have doubled each year since 2003 (from seven million to 54.7 million pounds in 2006). All 1,000 U.S. Holiday Inn hotels carry Rainforst Alliance certified coffee, as do all Whole Foods in the United States and Canada.

When tastemakers succeed, they create competitors. The economist William Baumol estimates that innovators and their investors keep less than twenty percent of the economic benefits that their innovations create. The rest spills over. This is why there are thousands more independent coffeehouses today than when Starbucks started, and why McDonalds and Dunkin' Donuts sell premium coffee. Copying Starbucks has become common. Fancy restaurants, convenience stores, and fast food franchises have their own versions of Starbucks coffee. Consumers can find high quality coffee beverages at places such as Jack in the Box, Burger King, and 7-Eleven in addition to Dunkin' Donuts and McDonald's. Coffee beverages offered include espresso and espresso-based drinks such as cappuccino, latte or moccaccino; iced and cold coffee beverages; and flavored coffees and special blends.

Copycats are not only upgrading their offerings to more closely match Starbucks, they are changing their environments. For instance, McDonald's has redesigned some of its restaurants to include a "linger zone" where the youthful customers it hopes to attract can sink into cushy armchairs and sofas while using a wireless Internet connection to send instant messages and surf the Web. The Canton, Massachusetts, based chain of Dunkin' Donuts announced in October 2006 that it will remodel its 5,000 U.S. stores. The remodel is essentially to a coffeehouse environment with more comfortable seating. Dunkin' Donuts is shedding its pink and orange color scheme. Its enormous expansion strategy includes tripling U.S. locations to 12,000 by 2020. It will start in Cincinnati, Cleveland, Tampa, Charlotte, Atlanta, and Nashville and then move westward. Behind the remodel is the reality that coffee has surpassed donuts in popularity. Donuts account for 15 percent of sales, while coffee sales soared more than 40 percent from 2002 to 2006.

TARGET MARKETS & SEGMENTATION

According to the National Coffee Association, manufacturers target young people. It reported a major finding that coffee consumption among 18 to 24 year olds jumped six percentage points in a fourth, consecutive annual increase. The NCA's 2007 data reveal that consumption among the age group soared from 16 percent in 2004 to 37 percent in 2007.

According to the Specialty Coffee Association of America, manufacturers target daily gourmet coffee drinkers, who comprise 12 percent of the population. Daily gourmet coffee drinkers have a high average annual income—$68,400—and regularly purchase gourmet coffee outside of the home. Supermarkets/grocery stores are the preferred destination for gourmet coffee purchases (36 percent), followed by specialty coffee shops (26 percent), with 18 percent preferring convenience stores, according to the National Coffee Association

RELATED ASSOCIATIONS & ORGANIZATIONS

European Coffee Federation, http://www.ecf-coffee.org

Green Coffee Association, http://www.green-coffee-assoc.org

International Coffee Organization, http://www.ico.org

Mid-Atlantic Regional Roasters Group, http://www.marr
g.orgInternational

National Coffee Association of USA, Inc., http:
//www.ncausa.org

Rainforest Alliance, http://www.rainforest-alliance.org

Specialty Coffee Association of America, http:
//www.scaa.org

TransFair USA, http://www.transfairusa.org

BIBLIOGRAPHY

Anderson, Diane. "Evolution of the Eco-Cup." *Business 2.0.*
June 2006, 50.

Castle, Timothy J., and Joel Starr. "Green Coffee Processing."
Tea & Coffee Trade Journal. September 2006, 52.

Chambers, Kelley. "Increasingly Popular Fair Trade Coffee Available Near and Far." *Journal Record,* (Oklahoma City, OK).
14 March 2007.

"Chain Coffee Goes Retail." The Food Institute. 26 March 2007.

"Coffee '06 Mintel Research Overview." Mintel International
Group. October 2006.

"Coffee Surpasses Soft Drinks in Daily Market Penetration
among American Adults, Reversing 16-Year Pattern." National Coffee Association of U.S.A., Inc. Press Release, 3 March
2007. Available from <http://www.ncausa.org>.

"Dunkin' Focusing on Coffee as it Embarks on Major Expansion." *The Food Institute.* 16 October 2006, 1.

Ephraim Leibtag, Alice Nakamura, Emi Nakamura, and Dawit
Zerom. "Cost Pass-Through in the U.S. Coffee Industry."
U.S. Department of Agriculture, Economic Research Service.
March 2007, 28. Available from <http://www.ers.usda.gov/
publications/err38>.

"Glamorous Pouches Add Even More 'Perk' to Folgers' Gourmet." *Packaging Digest.* November 2006, 9.

"International Paper and Green Mountain Coffee Roasters
Receive Specialty Coffee Association of America's 2007
Sustainability Award." *Wall Street Journal Online.* Press Release,
21 May 2007.

"Investor Services Corporate Profile." Green Mountain Coffee
Roasters. 9 May 2007. Available from <http://www.greenmo
untaincoffee.com>.

"McDonald's New England Customers Will Drink Fair Trade
Coffee." Environmental News Service. 27 October 2005.
Available from <http://www.ens-newswire.com/ens/
oct2005/2005-10-27-09.asp#anchor3>.

"National Coffee Drinking Trends for 2007." *Tea & Coffee Trade
Journal.* May 2007, 130.

"New Post-Consumer Fiber Paper Cup." *Tea & Coffee Trade
Journal.* February 2007, 76.

Romeo, Peter. "The $6 Billion Gorilla: The Influence of the
Seattle-based Coffeehouse Giant is Evident in the Upgraded
Decor, Beverages and Business Practices of its Rivals."
Nation's Restaurant News. 29 January 2007, 12.

"Specialty Coffee Retail in the USA 2005," and "Specialty Coffee
Retail in the USA 2006." Specialty Coffee Association of
America. 2006 and 2007 respectively. Available from <http:
//www.scaa.org>.

"Starbucks Paper Project: Changing the Way Coffee is Served."
Environmental Defense Fund. 27 March 2006. Available
from <http://www.environmentaldefense.org/article.cfm?
contentID=791>.

Surowiecki, James. "The Tastemakers." *The New Yorker.*
13 January 2003.

Thorn, Bret. "Cool Coffee Creations Keep Guests Coming Back;
Java Consumption Projected to Stay Hot." *Nation's Restaurant
News.* 28 May 2007, 25.

SEE ALSO *Tea*

Comic Books

—————◆—————

INDUSTRIAL CODES

NAICS: 51–1120 Periodical Publishing

SIC: 2721 Periodical Publishing

NAICS-Based Product Codes: 51–112031511, 51–112031541, 51–112031571, 51–112031601, 51–112031631, and 51–112031661

PRODUCT OVERVIEW

A comic book is a periodical that contains a series of panels in the form of a narrative. A typical comic book is 32 pages long (10 of these pages being advertisements) and is printed on glossy paper; the average price in 2006 in the United States was $3.20. Comic books are typically in color and feature a single story. They tend to focus on one primary character or a particular series of them. The story and art in comic books have become more sophisticated as the medium has regained the popularity it had in its early years. The origin of the term comic books refers to the earliest comic books, which featured reprints of comic strips that had originally been printed in the newspaper.

The comic industry is organized around a number of historical periods known as eras or ages. The first comic book period was the Victorian Age, lasting from 1828 to 1883, according to the *Overstreet Price Guide*. The first comic strips begin to appear about this time. Word balloons were almost never used in Victorian cartoons. When they were, it would be for single panel cartoons. Narrative, not dialogue, drove a story. Swiss cartoonist Rudolphe Töpffer combined a narrative with multiple panels to produce the 40 page *The Adventures of Mr. Obadiah Oldbuck*. The story was published in Europe in 1837, and then published in the United States in September 1842. Töpffer's works are often called picture stories, but the sophisticated use of narrative and art is seen as the blueprint for the modern comic.

Lasting from 1883 to 1938 was the The Platinum Age. Familiar characters such as Dick Tracy and Little Orphan Annie made their first appearances in the late 1920s. Palmer Cox published a strip about a group of men known as the Brownies. Cox's Brownies are thought to be the first successful commercialization of cartoon characters. The elf-like creations were used to sell many products, including pianos, dolls, puzzles, and soap. A collection of their strips was printed in 1877. Cartoonist Richard Felton Outcault created the Yellow Kid, the first continuing comic strip character. When Outcault switched newspapers and took the Yellow Kid to the *New York World*, many readers made the jump along with him. A collection of Yellow Kid strips was published in 1896, the first time a collection of strips was published in pulp magazine form (other collections had cardboard covers, for example).

Comic strip characters were popular with both children and adults. It was becoming clear that comics wielded economic clout. They could help drive newspaper sales and could be used to sell products. At the turn of the century there were hundreds of comic strip syndicates with comic strips to sell and they were eager to license the stars of these strips to the highest bidder. The advertising market—largely a visually based medium like comics—was growing. Most of the licensed merchandise sales in the United States at the time came from comic book characters. Many of the Brownie and Yellow Kid based products sold had not been properly merchandised. Outcault secured the rights to his next character: Buster Brown. The famous face of the Brown Shoe Company, introduced at the 1904 World Fair, was the first fully licensed and syndicated comic strip.

The Golden Age of comics began in the early 1930s. As stated, repackaged comic collections appeared on newsstands from time to time. *Famous Funnies #1* was first published in July 1934. It is the first regularly published comic on newsstands and lasted for decades. The comic turned a profit within six months and by 1946 was selling half a million copies per issue. The repackaged collections were popular, but the comic syndicates charged high fees for the rights to their strips. Major Malcolm Wheeler-Nicholson decided it would cost less to simply hire writers and artists to produce new material. He assembled a staff of artists and writers eager for work in the midst of the Depression and published *New Fun Comics #1*, the first comic with all original material in February 1935.

Jerry Siegel and Joe Schuster got their start in comics on *New Fun Comics #6* in October 1935. The two men had been friends since their boyhood when they bonded over a mutual love of science fiction. They created a story called *The Reign of the Superman* in 1933, with Siegel writing and Schuster doing the art. The Superman was a bald, telepathic villain (who bears a passing resemblance to Superman's arch enemy Lex Luthor). The pair later reimagined their character as a hero. They decided to feature the character in the comic strip rather than comic book format. After several rejections, the Superman story was published in *Action Comics #1* in June 1938. Many cite this comic as a starting point for comic's Golden Age.

The Rise of Superheroes. The Superman story appeared with reprints of other comic strips. It featured a one page synopsis of Superman's voyage from his dying home planet to Earth. The complete Siegel and Schuster story would not actually be printed in its entirety until years later. The construction of the Superman myth was an ongoing process. The *Daily Planet* newspaper and Perry White would not appear until *Superman #4* and *Superman #7* in 1940. The first names of the Kents, the couple who find the infant Kal-El, changed several times in the 1940s. Kryptonite first appeared on the Superman radio show in 1943 and did not find its way into the comics until 1949.

Siegel and Schuster may have been inspired by a number of sources. They were certainly influenced by other heroes of the day such as Flash Gordon, Tarzan, and the Phantom. The first costumed hero made his initial appearance in February 1936. Indeed, some publishers initially turned down the story because of Superman's resemblance to a character known as The Gladiator. It is

also believed Siegel and Schuster might have at least been acquainted with Nietschze's Superman, a being of superior intellect and abilities that would be above modern conventions. Superman has been seen as a Christ-figure, the powerful being that descends from the heavens to help Man. As a being from a distant land, Superman has also been seen as a symbol of the thousands of immigrants that were transforming the American landscape in the late 1930s.

The comic sold nearly one million copies, and was followed by the first appearance of Batman in *Detective Comics #27* in May 1939. Artist Bob Kane is credited with creating Batman (originally known as the Bat-Man). Kane did indeed create Batman's basic look. But writer Bill Finger helped Kane refine the character's look, suggesting changes to Batman's cowl and cape as well as suggesting pointy ears on the costume to make the character more bat-like. Finger wrote the initial Batman stories, and was responsible for the brooding tone of the character, as well as the naming and characterization of both Bruce Wayne and Robin.

Dr. William Moulton Marston created Wonder Woman, the first real female superhero. She appeared for the first time in DC's *All-Star Comics #8* in 1941. In the origin story, Wonder Woman is an Amazon princess who falls in love with army officer Steve Trevor after he crashes onto Paradise Island. Wonder Woman joins Trevor in the outside world to help fight the Nazis. She is armed only with her magic lasso, which compels anyone trapped within it to tell the truth (clearly inspired by the lie detector machine Marston helped invent). Wonder Woman's first appearance was DC's second best-selling comic title of the day. At one point she had a readership of 10 million.

Batman, Wonder Woman, and Superman were soon joined by other DC heroes: The Flash, Green Lantern, and Aquaman. Timely Comics (soon to be Marvel Comics) published the first appearance of the Human Torch and the Submariner. Captain America first appeared in 1941. Other publishers offered their own superheroes. Readership grew during this period with the popularity of superhero comics. Lev Gleason's *Crime Does Not Pay* in 1946 ushered in an era of detective stories with a harder edge. Will Eisner also first published *The Spirit*, the story of a masked crime fighter, as an insert in comic pages. The Spirit would later make his way into regular comic book form. Writers and artists continually cite Eisner's storytelling techniques as inspiration in their own work. Joe Simon and Jack Kirby went after the older female audience with romance comics.

Comics found their way, along with chocolate and cigarettes, into the care packages sent to the soldiers that went off to fight in World War II. Average monthly sales, according to *Comic Book Superheroes Unmasked* increased from 25 million in 1943 to 100 million in 1953. By the early 1950s, according to the *Overstreet Comic Book Price Guide*, one in three periodicals sold in the United States was a comic book.

Comic Censorship. As comics soared in popularity, they drew the inevitable concern from parents and church groups. The medium had very little merit; the stories were simplistic and the art crude. A 1940 *Chicago Daily News* article characterized comic books as poisonous. FBI Director J. Edgar Hoover vilified them as well in 1947. There are reports of mass comic book burnings from the period. Dr. Fredric Wertham was a respected psychiatrist and director of several New York psychiatric hospitals. Much of his work involved how an individual's psychological development was shaped by his environment and social background, which, in the 1940s, was a topic that was just being seriously explored. In interviews with a number of juvenile delinquents, he discovered that all the young men were avid comic readers (as, of course, were most teenagers). Wertham published numerous articles against comic books and held symposiums on the evils of comics. In 1954 he published *The Seduction of the Innocent*, in which he discussed the negative influence of comic books on youth.

Wertham dismissed the entire comic industry in his book, using the most salacious examples from popular crime and horror comics. Wertham offered a number of assertions in his book. He suggested a veiled homosexual relationship between Batman and Robin. He characterized Wonder Woman as a poor role model for girls because she advocated power and self-reliance. And her costume made her too sexually suggestive. Superman was a fascist (Wertham would compare him to Lee Harvey Oswald after the Kennedy assassination). The book is full of Wertham's opinions; it is noticeably short on real data. Some chapter titles indicate Wertham's less than scholarly approach: "I Want to Be a Sex Maniac," "Bumps and Bulges," and "The Devil's Allies."

Wertham lobbied for government involvement. Comics were examined in 1954 during the Senate Judiciary Subcommittee on Juvenile Delinquency. The committee found that comic publishers did indeed need to clean up their act. Comic publishers had been watching this controversy unfold, of course, and took steps to police themselves rather than risk government involvement. Publishers formed the Comics Code Authority (CCA). The CCA is an independent board that evaluates each comic; comics bore this stamp of approval for decades to come.

The Comics Code created a set of guidelines for publishers. Criminals could not be presented in a sympathetic light. Police officers and other authority figures could not be presented in any way that challenged their authority. No police officers could die at the hands of a criminal. The word "crime" could not appear on the cover, nor could the words "horror" or "terror" appear in a comic's title. Other guidelines were issued on costume design and advertising. Wertham, it should be noted, never endorsed the code.

Comics and the 1960s. Some publishers ceased publication and others cancelled titles. Meanwhile a new generation of children came of age unfamiliar with comics. DC offered fresh takes on its characters and interest in comics was renewed. This period of the late 1950s marks the start of the Silver Age. DC Comics would have its popular heroes join up to form the Justice Society of America. The new superheroes proved to be very popular. According to legend, the publisher of DC Comics shared this news with the publisher of Atlas Comics (soon to be Marvel) during one of their regular golf games. Atlas Comic's then publisher Martin Goodman asked writer Stan Lee to come up with a team of his own. Lee agreed, and with his frequent collaborator Jack Kirby created the Fantastic Four. The comic was successful, and the company returned to some of its popular superheroes from the 1930s. Lee persuaded Goodman to change the company name to Marvel Comics after one of the company's early superhero comics. Marvel created many of the comic industry's most remembered heroes and villains during the 1960s: Daredevil, the Hulk, Ant Man, the Amazing Spider-Man, the Avengers, the X-Men, Doctor Doom, Galactus, Dr. Strange, and the Silver Surfer.

The major publishers followed the CCA during the 1960s, occasionally side-stepping it in clever ways the way Hollywood films got around the Hayes Code in the 1930s. Meanwhile, underground comic books, known as *comix* became popular in San Francisco in the 1960s. These comix (the x was used to distinguish these products from their mainstream brethren) were small press or self-published titles that took on the Vietnam War, rock music, the establishment, drug culture and, of course, sex. Companies such as Rip Off Press, Last Gasp, Kitchen Sink, and Weirdom Publications all operated without the Comics Code seal of approval. The movement got its inspiration in part from Harvey Kurtzman, publisher of *Mad* magazine and numerous satirical titles. *Zap Comix* is the most famous title from this period but a few other titles suggest just how far away underground writers were from mainstream comics and the Comics Code: *Amputee Love* (1975), *The New Adventures of Jesus* (1969) and *Slow Death* (1968). R. Crumb is perhaps the most famous writer/artist from this movement. His figure from the cartoon *Keep on Truckin* (a big-footed man caught in mid-strut) was a highly iconic figure from the 1970s. The figure and slogan appeared on everything from T-shirts to truck mud flaps. He went on to create album covers and Fritz the Cat, which was made into the first X-rated animated feature in 1972. As mainstream publishers began to challenge the Comics Authority with their own stories, the underground movement faded around 1980.

The 1970s. The Bronze Age begins in 1970. In the 1970s readership declined as comics had to compete with video games and other new forms of entertainment for children's attention. Marvel began to gain the upper hand over DC and started outselling DC in 1972. There are several reasons for this. One factor was cost. The two companies engaged in a price war, with DC titles selling for $0.25 and Marvel titles for $0.20 at the end of 1971. As well, the characters Marvel created in the 1960s (Fantastic Four, Spider-Man, the Hulk) were gaining devoted fans. Marvel also went after popular licensed properties, producing comics based on *Conan*, *Logan's Run*, *Micronauts*, and *Star Wars*. Taking its cues from *Dirty Harry* and other films from the period, Marvel introduced anti-heroes such as Ghost Rider, Wolverine, and the Punisher. Marvel was also publishing shocking stories and unconventional art. In 1973 readers of *Amazing Spider-Man* were shocked when the Green Goblin murdered Peter Parker's girlfriend Gwen Stacy. The murder of an innocent character was something readers had never seen before.

The industry began to introduce more minority characters during this period. Marvel introduced its first African American hero, Luke Cage, in 1971. Shang-Chi (clearly a nod to Bruce Lee and Kung-Fu) first appeared in 1973. DC introduced Black Lightning in 1977. At Archie Comics, Archie, Betty, and Veronica became friends with African-Americans Chuck Clayton, his girlfriend Nancy, Latino Frankie Valdez and his girlfriend Maria Rodriquez.

In the early 1970s, many people drew attention to rising drug use in the country. The U.S. Congress passed the Controlled Substance Act in 1970, seen as the basis for the modern war on drugs. In 1971 President Nixon referred to drug abuse as "public enemy number one in the United States." That same year, the United States Department of Health, Education and Welfare approached Stan Lee about writing a story about the dangers of drug use. Lee printed the story in *Amazing Spider-Man #96–98* (May–July 1971). The CCA refused to approve the story because of the drug references, but Marvel ran the issues without the CCA stamp. The story was well received, and the issues sold well without the CCA stamp. The CCA (fearing irrelevancy or finally getting the point) eased some of its policies later in 1971. Drug references were acceptable as long as they were shown in a negative way.

DC Comics was also exploring more socially relevant material. Denny O'Neill and Neal Adams had a memorable run on the *Green Lantern/Green Arrow* title in the early 1970s. The comic was near cancellation and the team took on the title with a sense of nothing to lose. The title took on the civil and social issues that were part of the overall culture of the day: racism, poverty, and the power of large corporations. Writer O'Neill has compared Green Lantern to the good cop who never questions authority; Green Arrow, in turn, a character reminiscent of Robin Hood, was a cross between that famous archer and the left-leaning activists of the day. The most memorable issues revealed the Green Arrow's ward Speedy to be addicted to heroin.

The 1980s. The Modern Age begins and carries on to the present day. The comic industry continued to go after more adult readers in the 1980s. Frank Miller took over Marvel's *Daredevil* comic in 1981, invigorating a minor character with dark stories and a naturalistic art style. Miller would offer a similar take on Batman in *The Dark Night Returns* in 1986. The story is set in the near future, with Batman pulled out of retirement to help fight crime in Gotham City, now overrun with crime and poverty. The dark, grisly tale is seen by many as the definitive take on the character, and was an inspiration for the first Batman film. Alan Miller released the limited series *Watchmen* the same year. The story details a group of past and present heroes who band together to solve the murder of one of their own. The story offers sharp characterizations, and also touches on the threats of nuclear war (a very real threat in this last decade of the Cold War) and Big Brother government. In 1989 the CCA modified the comics code. The new code allowed Marvel's Northstar to be the first openly gay superhero in 1992 (although Watchmen did have a few gay minor characters). A decade later Marvel developed its own rating system and now publishes without the CCA sticker (DC still does, however).

The growth of comic shops in the 1980s brought alternative titles to the attention of comics readers. Like comics from the underground period, alternative or independent comics are often written and drawn by one person. Harvey Pekar has published his autobiographical comic *American Splendor* since 1976, exploring the intimate details of his personal life and his job at a Veterans Administration (VA) hospital. Art Spiegelman's *Maus* told the story of Spiegelman's father's incarceration in a concentration camp. It won the Pulitzer Prize. Chris Ware's *Jimmy Corrigan: The Smartest Kid on Earth* tells the story of three generations of men in the Corrigan family and explores issues of loneliness and family dysfunction.

The 1990s and Beyond. After the stock market crash of 1987, newspapers and financial journals began to report on an investment opportunity that appeared to be outside the volatility of the market. Comic books were seen as one area that offered attractive investment possibilities. Some comic books that had originally sold for pennies were selling in the 1990s for thousands of dollars. Most of the investment advisors touting comic books were responsible enough to note that these comics were from the Golden Age in which few copies have survived. Nonetheless, not all investors were careful to look for Golden Age comics or first editions of the #1 issue of a new comic strip.

Speculators moved into comic book stores and started snapping up copies, convinced the comics would appreciate in value. The major publishers responded to the increased demand with large print runs. Marvel famously resorted to gimmicks to keep sales going, releasing issues with different covers (rare covers would make the comic more valuable). *X-Men #1 Vol. 2* is one of the most notorious examples of this trend; it was released with five different covers and had a print run of five million copies. Marvel and DC also flooded the market with action figures and similar merchandise. The industry pushed mylar bags and other products to protect a collector's comic book investment. DC then released its "Death of Superman" story (*Superman #75*) in 1992. The story generated national attention, and DC could not keep up with demand. Some comic stores had lines of customers; in Detroit, reportedly 175,000 copies of the comic were sold in one day. Many speculators expected to immediately sell the Superman comic at tidy profits. But the comic had a large print run; many copies existed, meaning the issues were worth only cover price. They soon found the other comics they had purchased could not be sold for high prices. The speculators left the market, and the comic industry collapsed. Sales fell from $850 million in 1993 to $265 million in 2000. The number of comic specialty stores (still a relatively new industry, it should be noted) fell from nearly 10,000 to 3,400 over the same period. Eleven of the twelve comic distributors shut their doors. Marvel filed for bankruptcy in 1996 after its too rapid expansion of Toy Biz, Panini, Fleer, SkyBox, and Heroes World Distributors. It continued to publish, however.

The success of the first X-Men film in 2000 helped stimulate interest in comics. Hollywood took notice of the film's box office, and recognized comics as good source material for films. Films starring Spider-Man, Batman, Superman and other comic characters followed the first X-Men film. The industry has seen sales increases in 2005 and 2006. The comic industry instituted a Free Comic Day each May to help lure young people to comic book stores.

MARKET

Comic books are sold through the direct market (comic stores) and mass market outlets. Comic book stores purchase most of their material for resale from Diamond Comics Distributors. These comic book stores typically do not have point of sale systems to track the sale of their products. What the local comic store purchases is non-returnable. In short, the industry tracks sales to the comic book store, not sales to the end user, and so this can make tracking actual comic sales difficult. Approximately 80 percent of comics are sold through the direct market. The balance of comics, graphic novels, and related materials are sold through retail chains. Bookscan tracks about three-quarters of this market. Historical data is even more difficult to track, and existing figures come from comic historians willing to assemble figures from old distributor reports. Clearly, figures in any study of the comic market are estimated, and any student of the industry may find contradictory figures in multiple sources.

According to a 1955 Senate report (part of the hearings on comic books and juvenile delinquency) comic book publishers were estimated to have been publishing 150 titles and generating annual revenues of $20 million in 1940. In 1950 there were 300 comic book titles being published with annual revenues of approximately $41 million. Average monthly circulation jumped from 17 million in 1940 to 68 million in 1953. In 1948 there were perhaps 400 comic publishers on the market.

The industry would slowly decline as it faced competition from television, movies, video games, and other entertainment. In the 1970s the industry struggled. Archie was the best–selling comic at this time, selling 500,000 copies. The growth of comic book stores in the 1980s helped save the industry. The speculator market also boosted the industry to amazing if unsustainable levels. In 1996 the industry saw monthly print runs of 120 million. After the market crash, the industry canceled titles and trimmed expenses. In 2006 a comic is considered successful with a print run of just over 100,000 copies. The *Amazing Spider-Man*, for example, which had a circulation of 234,290 monthly readers in 1995 had a circulation of only 112,564 readers in 2005. The overall comics market, including direct sales, newsstand sales, and trade paperback sales was worth between $475–550 million in 2006. The entire industry employed approximately 12,000 people, according to Matthew McCallister's *Comics and Ideology*. Figure 62 offers industry estimates of the annual sales in this industry in the late 1990s and early 2000s.

Comics are popular outside the United States as well. Manga is immensely popular in Japan, and the art form has spread to the United States and other countries. While manga literally means amusing drawings in Japanese, the

FIGURE 62

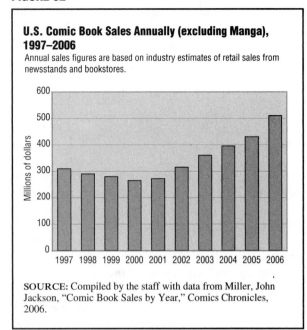

U.S. Comic Book Sales Annually (excluding Manga), 1997–2006

Annual sales figures are based on industry estimates of retail sales from newsstands and bookstores.

SOURCE: Compiled by the staff with data from Miller, John Jackson, "Comic Book Sales by Year," Comics Chronicles, 2006.

term is used to refer to all Japanese comics. The phenomenon began with the works of artist Tezuka Osamu, who created famous characters such as Astro Boy, Jungle Emperor Leo, and Princess Knight. He is also the artist behind the distinctive large eyes style of Japanese animation.

In 2004 total sales of manga in Japan reached ¥504.7 billion based on 1.38 billion publications. Manga represents 40 percent of book and magazine sales by one estimate. In the United States, sales of manga grew from $5 million in 2000 to nearly $200 million in 2005, according to Web site Icv2. In South Korea, comics and manga represent a quarter of all book sales, according to a *Business-Week* article. The comic market grew 12 percent in France in 2005, according to Syndicat National de l'Edition. The number of manga titles published in Europe jumped 45 percent to 754 in 2004, according to the Association des Journalistes et Critiques de Bandes Dessinées.

Comic Book Collecting. The first comic book conventions were held in the middle 1960s in Detroit, New York City, and San Diego. Comic book collecting as a legitimate industry took a major step forward in 2000. In 2000 the Comics Guaranty Corporation began to professionally grade comics, much like sports card and rare coin collectors get their collections professionally graded. Graders inspect the comic and then it is sealed in a hard plastic shell with the grade posted on it. A comic that goes through this process is described as being slabbed. With this process, a collector now knows the real value of the product he or she is purchasing. A graded book might rise in value from a few dollars to a few hundred.

KEY PRODUCERS/MANUFACTURERS

Two firms dominate the comic book publishing business in the United States. They are each profiled below, as are two other companies that are having a growing impact on the market in the 2000s. Figure 63 presents a picture of the market share picture of this industry by key players in 2006.

Marvel Comics. This publisher began as Timely Comics in 1939. Timely released *Marvel Comics #1* in October 1939, featuring the Human Torch and the Submariner, ruler of the undersea kingdom of Atlantis. The issue featuring these two heroes (still popular characters in the current Marvel universe) sold out the 80,000 copy print run. A second printing sold 800,000 copies. When the company returned to superhero comics in the 1960s, the company would rename itself after this best-selling title.

Stan Lee is the company's famous face, a writer and current Chairman Emeritus for Marvel. With his frequent collaborators Jack Kirby and Steve Ditko he created the Fantastic Four, the Hulk, Spider-Man, Iron Man, Thor, the X-Men, Daredevil, the Silver Surfer, Doctor Strange, and countless other supporting characters and villains. Lee wanted to bring a sense of realism and a psychological richness to comics. Science and atomic radiation figure

FIGURE 63

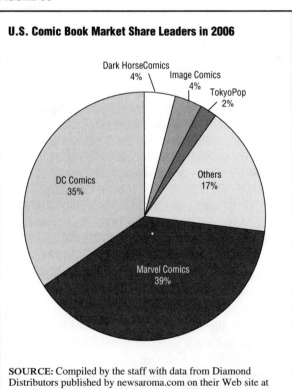

U.S. Comic Book Market Share Leaders in 2006

SOURCE: Compiled by the staff with data from Diamond Distributors published by newsaroma.com on their Web site at http://www.newsarama.com/marketreport/2006_year_end.html.

into the origin story of many characters; in the 1960s; the country had only just entered its Atomic Age, and was uncertain about the possible ramifications of this new power source.

The first issue of the Fantastic Four appeared in November 1961, featuring Mr. Fantastic, the Invisible Girl, the new Human Torch, and the Thing. The foursome received their powers after being bathed in cosmic rays during a space flight. There are some subtle politics in the story, for the rocket voyage was an effort to reach outer space before the Soviet Union (Yuri Gagarin was the first man to orbit the Earth in April 1961). Comic readers had never seen anything quite like this foursome. They were superheroes yet they had no secret identities. For several issues they had no costumes. Reed Richards (Mr. Fantastic) and Sue Storm (Invisible Girl) were in love and would marry; yet Ben Grimm (the Thing) had once been in love with her. The team bickers constantly, and often seems on the verge of disbanding.

Lee and Kirby created the Incredible Hulk, who first appeared in May 1962. The first issue details how Dr. Bruce Banner is struck by gamma radiation and then transformed into the Hulk. Certain story elements changed in early issues. The Hulk was first gray, then he was green. Banner initially changed into the Hulk each night; later, Banner's rage was the trigger to the transformation. The image of a man with a monster within is a classic motif from literature, and the comic has gone on to explore this theme in many ways.

The team turned to Norse mythology to develop Thor, who appeared in August 1962. That same month *Amazing Fantasy #15* was released, featuring the first appearance of Spider-Man. The series was about to be canceled, and this last issue was used as a showcase for the character initially dismissed by Lee's boss. Again, Lee broke standard comic conventions. Peter Parker was a teenager; teenagers were sidekicks, not the heroes on their own. Peter had a real life: dating problems, a sick aunt to care for, lost loved ones, and trouble paying rent. Spider-Man's own comic began in 1963 and the character is perhaps Marvel's most popular.

Iron Man first appeared in March 1963. The first X-Men comic appeared in September 1963. The X-Men are a team of mutants, which are people born with their super powers. Mutants are feared in society because of the abilities with which they were born. This idea of being born different and special (and to be feared for it) allows the comic to be seen as a veiled commentary on racism and homophobia.

DC Comics. This company began in 1935 as National Allied Publishing and was founded by Malcolm Wheeler-Nicholson, a former U.S. Army major and pulp magazine writer. Wheeler-Nicholson launched comic titles that featured original material instead of reprinted newspaper strips. One of the titles published was Detective Comics, which first appeared in 1937. By 1940 the letters DC began appearing on the company's comic covers. Readers would soon start referring to the company by these initials.

DC's characters are some of the most well-known in popular culture, including Superman, Batman, Robin, the Joker, Green Lantern, and Aquaman. DC created the first superhero team with the Justice Society of America; later, the Justice League of America was formed. The company controlled the superhero comic market in the 1950s and 1960s, then lost its number one position to Marvel. In the 1970s it competed with Marvel for leadership in the shrinking comic market. Both companies expanded too quickly; DC and Marvel published approximately 100 new titles between 1975 and 1978. DC launched 57 of these new titles. The expansion soon caught up with them and they canceled 31 titles in 1978. This event is known as the DC Implosion. The company restructured and recovered, and tried to tell more engaging, provocative stories. They went after new writers and artists, some of whom worked for Marvel.

DC publishes approximately 80 monthly comics and is highly involved in product merchandise. The company is owned by Warner Brothers. Other operations include comic imprints WildStorm and Vertigo, book imprint Paradox Press, and *MAD* magazine.

Dark Horse Comics. This company found success in part by the birth of the direct sales market in the 1980s. Some of the company's popular characters include *Concrete*, also known as Ron Lithgow, whose brain was transferred into his concrete body by aliens. The title has won several Eisner awards, one of the most prestigious awards in the comics industry. Dark Horse also publishes *Hellboy, Sin City, Star Wars* and *Buffy the Vampire Slayer*. Dark Horse, like Image, shares more of its profit with creators.

In 1992 Todd McFarlane, Jim Lee, and others were working on some of Marvel's top titles when they went of to start their own independent comic. The group created Image Comics. It is the most successful independent publishing venture to ever be launched from the comics industry. The small firm found success in part by new production techniques, such as digitizing its coloring process and raising page rates for creators. Marvel and DC would adopt a similar policy, paving the way for more creator owned titles. The company also found success by producing compelling titles, including McFarlane's Spawn. *Spawn #1* is the best selling independent comic as of early 2007. It has a cover date of May 1992 and sold 1.7 million copies.

Manga. Two firms are gaining market share in the United States by publishing Manga comics, the fastest growing type of comics in the world. Tokyo Pop is one, founded in 1996 and VizMedia is the other.

MATERIALS & SUPPLY CHAIN LOGISTICS

Comic book firms struggle to maximize profits. Some of this profit comes from a sensible approach to the manufacturing process. The production of a comic includes computers, inks, and adhesives. The major factor of course, is paper. It is unclear how much paper the industry consumes, but they have embraced some "green" elements in the production process.

Archie Comics was the first publisher to use soy inks and print on 100 percent-recycled paper in the 1990s. DC reports using soy ink on 90-95 percent of its titles. Dark Horse prints some of its line on recycled paper as well. Marvel does not make such production figures available.

Production materials are important but good comic books need good stories and good art. Good writing and good art are the keys to good comics. Comic books are often a collaborative effort, with a writer or writers working together to outline a story. Writers also collaborate closely with artists. The next step is the creation of a script. Comic scripts may be very specific, with descriptions about the number of panels per page and what should occur in each panel. Scripts from Marvel Comics tend to be much less structured, often not containing dialogue or complete breakdowns of panel by panel action. A lot of the pacing and panel content is left to the artists, also known as pencilers. The artist will then do some preliminary sketches to get a sense of how to best tell the story. Artists generally work on an 11x17 inch page. The best artists are those that can tell the story in a compelling way: a close-up on a character's face to find the story's emotional center, or an unusual angle to create dramatic tension. An artist typically spends six to seven hours on one page of the 22-page story he or she must produce.

The writer and penciler may collaborate on the sketches, refining the panel content or the flow of the story. An inker will then refine the drawings. The inker uses black ink to produce refined black outlines over the rough pencil lines. Inkers may also create shadows and black space to help create mood in a story. The letterer is responsible for the insertion of dialogue balloons. DC still has dialogue pasted onto the art boards; Marvel typically uses a computer for this process. Colorists colorize the black and white using Adobe Photoshop or some other digitized media; before the creation of such programs colorists used printing plates.

DISTRIBUTION CHANNEL

Until the 1970s, comics were sold through the newsstand market, which included drug stores, grocery stores, and toy stores. Magazine and book distributors could purchase comics directly from publishers on a fully returnable basis. Comic distribution changed in the 1970s with the creation of the direct market, or the local comic book shop.

Phil Seuling of East Coast Seagate and Bill and Steve Schanes of Pacific Comics approached comic stores about buying comics from them rather than periodical distribution firms. They would service comic stores directly, and offered financial deals not met by mainstream periodical distributors. Retailers could preorder comics and keep extra to sell in the store. New comics were offered at a 40 percent discount rather than the 20 percent offered by mainstream distributors. This allowed greater room for profit for the retailers.

The direct market owes some degree of debt to the underground comix movement, which had started only a few years before. Independent writers/artists such as Harvey Pekar and R. Crumb produced work explicitly for this direct market. The first 3,000 copies of *Teenage Mutant Ninja Turtles* in 1984 offered more validation of the market. Marvel and DC moved into this market with some trepidation. Marvel released *Dazzler #1* as a test case in 1981. Unavailable on newsstands, the comic sold 400,000 copies in the direct market. With the two major players on board, the direct market grew. There were 200 to 300 specialty comic shops in 1974; by 1980 this number had grown to 1,500 specialty stores. By 1993 there were approximately 9,000 such establishments in operation. After the market crash in the 1990s, fewer than 4,000 comic stores remain.

The direct market is credited with helping to save the comic industry in the seventies as readership sagged. It also proved to be a vital source of revenue for Marvel. The direct market was 6 percent of Marvel's gross sales in 1979, according to company and Mile High Comics estimates. By 1987 the figure was 70 percent.

Diamond Comics Distributors dominates the comic distribution market, with 85 percent of the global market in 2005. In the direct market, major comics publishers provide Diamond with a list of titles for which they solicit orders. These orders may not be returned, so the retailers must order wisely. Retailers place their orders with Diamond. The major publishers take the distributor orders and print enough copies off their comics or graphic novels to satisfy the orders. There may also be reorders in the rare case of a title selling out. Solicitations by the publishers begin three months before comics arrive in stores; retailers' orders are due roughly two months ahead of time. Comics are shipped to stores each Wednesday.

Melchior & Associates estimated that new comics represent 60 percent of market sales. There were approximately 2,500 specialty comics retailers. The top 700 stores placed most of the comic orders. There were another 1,000 to 1,500 stores that sold comics along with other merchandise (sports card and hobby stores, for example). These stores averaged sales of $70,000 to $100,000 annually.

The largest comics retail companies in North America have between six and eight stores with gross annual sales of between $2.2 and $3.5 million. No large chains exist for several reasons according to Melchior & Associates. Most storeowners lack the capital and business expertise for such an expansion. Problems with the industry include the typical store. Stores may be small, crowded and dimly lit. Some have argued that stores would benefit from becoming more user friendly: comfortable seating, and better lighting, and organization. Others feel this corporate approach might drive away the comic industry's major customers.

New comics can represent anywhere from 25 percent to 70 percent of a comic store's product mix, while graphic novels represent 5 to 15 percent. Games and other collectibles are also a part of the store's merchandise.

KEY USERS

Comic books are often seen as a product for children. But industry estimates show that the average comic book reader is a 28-year-old man who spends $1,300 to $1,500 per year on his hobby. There is a certain cycle at work here: writers tell sophisticated stories that appeal to adults. Adults keep buying comics, so publishers keep writing for them. Marvel has acknowledged the mistakes it made in the 1990s when it aged its heroes along with its readers. Spider-Man went from high school to college, got married and started a family—issues obviously beyond a young reader. Marvel estimates 70 percent of its readers are male, a figure about the same for DC. But 60 percent of Archie Comics readers are thought to be female.

Marvel and DC both started titles aimed directly at young readers. Marvel has found great success with its Ultimates line, which feature very contemporary takes on its classic characters (Spider-Man, for example, is back in high school). While the superhero genre is still dominated by male readers, girls and women do read them. In the manga category, readership may be as high as three-quarters female.

ADJACENT MARKETS

Comic characters have been licensed to sell products since the early days of comic books and strips. Marvel Entertainment was the sixth largest licensor of merchandise

in 2006, according to *Licensing Magazine*. Each year the magazine ranks the top licensors based on global sales of licensed merchandise. Marvel generated an estimated $4.8 billion through its movies and other products; Disney was number one with $23 billion. Marvel plans to move beyond simply licensing products to actual film production to generate even greater revenues. Marvel beat out Major League Baseball and the National Football League on the list. Warner Brothers, the owner of DC Comics, generated $6 billion and was number three on the list. However, this figure includes revenues generated from non-comic book properties owned by the company as well, such as Harry Potter.

RESEARCH & DEVELOPMENT

With the rise of the Internet and the increased availability of broadband connections, various analysts have promised us a paperless society. It was once broadly assumed that documents would one day all be sent online; people would read their newspapers online rather than over their cup of coffee. While we are far from paperless, technology has certainly changed the way we shop, listen to music, and gather news information.

Technology is also changing how comic books are distributed. Some comics are solely available online. Top Cow announced in early 2007 that many of its titles would be available online. Marvel and DC offer a few titles for viewing on their Web sites. Some comic fans are waiting for comics to be downloadable, as songs and movies are on iTunes. Industry leaders have many issues to work out, such as just what the comic might look like on an iPod and how to properly compensate creators. The key to online downloading is the demand for such an option. The first steps have already been taken in Japan. According to Japanese research firm Impress R&D, Japanese consumers spent $20 million to watch manga on their handsets in 2006.

CURRENT TRENDS

Since their creation in the 1930s, comic books have used war as a dramatic device. Indeed, Superman, Wonder Woman, and Captain America were off fighting Nazis before America entered World War II. Comic books were read by soldiers and used as instructional tools by the military. The comics from this period offer simple, moral visions of the world. There are two clear sides in a conflict—good guys and bad guys. Freedom is worth fighting for.

But as the conflicts in the world became more complex and in some cases ambiguous during the early 2000s, so too did comic book treatment of war. Comics began to face war and terrorism more directly. Spider-Man, who makes his home in New York City, visited the ruins of the

World Trade Center in a special issue with an all-black cover. Writer Sid Jacobson, creator of Richie Rich, created a graphic novel version of the *9/11 Report*. Jacobson describes his August 2006 work as graphic journalism. Archie Comics printed patriotic hymns and songs in the pages of its various titles.

Captain America, the symbol of American patriotism, made his comic debut in March 1941 (he was shown punching Hitler in the jaw on the cover of *Captain America #1*). The character has been popular since; more than 210 million Captain America comics have been sold worldwide. He played a major role in the *Civil War* miniseries Marvel launched at the end of 2006. In this miniseries, the government began requiring superheroes to register their services and secret identities. The registration polarized the superhero community, and the heroes found themselves falling into two camps. Iron Man led the group of heroes who favored the government regulation and control. Captain America led the group of heroes who opposed the government registration and control.

The story found much of its drama in Spider-Man, who initially sided with the pro-registration Iron Man. Iron Man persuaded Spider-Man that revealing his secret identity to the public would be a compelling way to persuade other masked heroes to register with the government. Spider-Man pulled off his mask and revealed himself to be Peter Parker at a press conference. Spider-Man would switch sides after he learned that Iron Man and the government were capturing heroes who refused to register and imprisoning them indefinitely in a secret prison.

The Civil War story played out as the United States neared the fourth anniversary of its invasion of Iraq. The parallels between the real life war and Marvel's story were clear. Many Americans fell into pro-war and anti-war groups as the war progressed, and the country could find no middle ground in the contentious debate. Indeed, the question posed on the cover of *Civil War #1—Whose Side Are You On?* could easily be asked by those who loved and loathed the Bush administration. The story touches on the need of the government to balance national security with the privacy of its citizens. The Superhero Registration Act feels very similar to the Patriot Act. The public debate in the story compares superheroes to weapons of mass destruction. The indefinite imprisonment of the superheroes deemed a threat by the government conjures up the image of Guantanamo, the Abu Ghrab scandal, America's secret prisons, and policies surrounding torture generally.

The grim story concludes with Captain America assassinated on his way to a courthouse to legally fight the registration act. The death of Captain America was covered in the media in March 2007, much as the death of Superman was reported on in 1992. Readers found rich symbolism in the story, in light of America's damaged reputation in the global community. Few characters in comics stay dead, however, and readers expect that Captain America will soon reappear.

TARGET MARKETS & SEGMENTATION

The comic book industry continues to try to gain more readers beyond its typical base of young men. Comic publishers are making a concerted effort to produce more titles to appeal to children. Many Baby Boomers remember learning to read with comics. Recognizing this, Marvel announced plans in early 2007 to release comics based on classic literature. As a visual medium comics remain useful teaching tools. Comics have been used in the military to educate soldiers on combat, personal health, and military policies. They have been also used as propaganda tools. UNICEF has used comics to teach children about land mines, AIDS, and sexual abuse. Cartoon and comic art have been shown in museum exhibitions. The first museum devoted solely to cartoon art opened in 1987 with an endowment from Peanuts creator Charles Schultz.

RELATED ASSOCIATIONS & ORGANIZATIONS

Comic Chronicles, http://www.comichron.com

DC Comics, http://www.dccomics.com

Marvel Comics, http://www.marvel.com

Museum of Comic and Cartoon Art, http://www.mocaany.org

National Association of Comic Art Educators, http://www.teachingcomics.org

BIBLIOGRAPHY

"The 1900s: a Century in Comics." Comics Buyers Guide. 12 December 2005. Available from <http://www.cgbxtra.com>.

"An Overview of the Direct Market in 2005." Comtrac.net. Available from <http://www.comtrac.net/cms/index.php?page=Articles_E_books>.

Condie, Stuart. "Report: Comic Book Hero Captain America Dies on the Page." *America's Intelligence Wire.* 7 March 2007.

Crawford, Philip Charles. "The Legacy of Wonder Woman." *School Library Journal.* March 2007.

Gustines, George Gene. "A Quirky Superhero of the Comics Trade." *New York Times.* 12 November 2006, 21.

Miller, Greg. "Owner Keeps Comics Coming." *Columbia Daily Tribune.* 3 June 2006.

Miller, John Jackson. "Comic Book Sales Figures by Year." The Comics Chronicles. Available from <http://www.comichron.com>.

Overstreet, Robert M. *Overstreet Comic Book Price Guide,*
    36th ed. House of Collectibles, Random House Publishing,
    May 2006.

Sanford, Jay Allen. "Two Men and Their Comic Books," *San
    Diego Reader.* 19 August 2004.

Stringer, Kortney. "Comics Bring in Ads." *Detroit Free Press.*
    19 June 2006.

Welsh, David. "Forget Manga. Here's Manwha." *BusinessWeek.*
    23 April 2007.

Wilensky, Dawn. "103 Leading Licensing Companies." *Licensing
    Magazine.* April 2007.

SEE ALSO *Magazines*

Construction Machinery

———————◆———————

INDUSTRIAL CODES

NAICS: 33–3120 Construction Machinery Manufacturing

SIC: 3531 Construction Machinery

NAICS-Based Product Codes: 33–31201, 33–31208, 33–3120Y, 33–31209 and 33–3120W

PRODUCT OVERVIEW

Tools used in construction include hand tools powered by human muscles alone such as hammers and shovels; hand-held power tools such as electric drills and saws; various supporting structures such as scaffolding, hoists, and ladders; and finally construction machinery. This last category of tooling extends from relatively small and light tools such as small-batch concrete or mortar mixers, jackhammers, and tampers—on up to heavy duty tractors, bulldozers, rollers, trenchers, and large drag lines—culminating in all types of cranes including giants used to support the raising of skyscrapers.

Structures are always built on solid ground, even those held up by submerged foundations. The preparation of the site itself is the first activity. The construction site must be cleared. It may be a forest or a preexisting building. The forest must go. Demolition cranes must destroy the other. Not surprisingly, snow plows are included under construction equipment, as are log skidders, log splitters, and chippers intended to reduce cleared trees, branches, and brush into manageable wood wastes. Dredging equipment for preparing underwater sites and demolition machines for destroying old structures are all part of the product mix.

Cleared surfaces are almost never level. Construction companies therefore deploy earth-moving equipment of all sizes. They use bulldozers and drag lines. Drag lines are cranes atop a mobile platform. The crane positions a bucket, its teeth into the soil. A cable, the drag line itself, is attached to the bucket horizontally. Winching it in causes the bucket to bite into the soil and fill up as it is dragged. Sometimes soil conditions are such that the builder must first secure the foundation to be built by sinking supports down to the bedrock. The machinery industry makes vertical earth augurs and posthole diggers to sink the necessary holes; it also supplies pile driving equipment to drive the poles down to the rock. Sometimes supports must be sunk into vertical rock faces. The builder thus buys or rents horizontal boring machines to accomplish this end. Alternatively the structure rests on foundations deep in the ground. The builder must excavate the holes first using trenchers, digger derricks, and similar equipment, including excavating drag lines. Excess dirt removed must be lifted into dump trucks. The powered shovels that fill the trucks, indeed the heavy-duty all-terrain trucks themselves, are yet other categories of construction equipment. Construction equipment is usually self-propelled, but some pieces travel to the site on their own wheels while others, of the tracked variety, arrive on trailers.

In the preparation of foundations, typically made of reinforced concrete, builders use concrete mixing machinery. Supporting these mixers, they use machinery to prepare aggregate before it is added to cement. Such equipment takes the form of rock crushers, slag mixers, sand mixers, and pulverizing and screening machinery.

Foundations are sometimes laid below grade in narrow strips which are first prepared by using powered ditching and trenching devices.

Once the foundations are in place and the structure begins to rise into the air, power cranes are used to lift all that is required to make the structure. In some applications, aerial platforms are hoisted up to serve as temporary working surfaces for the labor. Many types of cranes are used ranging from small, self-propelled units—distinguished by their crane-arm articulations and styles of movement—all the way up to the great tower cranes visible above high-rise structures. Tower cranes, incidentally, must themselves be assembled. They arrive at the construction site in large trailers. Once operational, they take the form of a capital T bolted to the foundation. The vertical bar of the T above its base is called the crane's *mast*. It comes in sections so that, as the building grows, the crane can grow with it. The crane's operators lift up the next section of the mast itself using the crane. The horizontal bar of the T is divided into a longer *jib*, which supports the load to be lifted, and a shorter *machinery arm*, which carries the lifting motor and a concrete counter-weight. A trolley runs along the bottom of the jib so that the lifting cable can be positioned near or far from the mast depending on the situation. A powered winch attached to the trolley actually does the lifting of the load.

Virtually all other work carried out in the raising of structures is done by laborers wielding hand-held power tools. To be sure, experimental robotic equipment for a variety of purposes (e.g., welding, roofing, plastering) has been under development in Japan and Europe since the late twentieth century, but thus far without commercial application.

Road and highway construction requires its own specialized tooling. Roads and highways are built up in layers. Very large rough stones or rubber constitute the bottom-most layer, well below the grade of the road in a deep trench. As the road surface gets higher, the material used is of ever finer size. Rock, gravel, and aggregate form successive layers finished using asphalt or reinforced concrete. Builders or suppliers employ hammermills for crushing or sizing rock and screening machines to separate stone or aggregate into different size classes. Aggregate spreaders are used to apply these surfaces. They are compacted by towed or self-propelled rollers. Asphalt mixing and application machinery rolls out the bituminous surface, asphalt sprayers to seal it. And, in road repair and maintenance, bituminous scrapers and planers are used to create an even surface, all loose material removed. Jackhammers prepare concrete surfaces for repair. Tampers and vibrators are needed to compact the material over which seals will be placed. Highway builders use automated machinery, running on temporary rails, to apply concrete to new highways and to finish its surfaces. In bridge building and maintenance specialized cranes are used. These are designed so that, resting on top of the bridge, they can be deployed to inspect the bridge's undersides or superstructures.

Machines developed for residential construction differ from their cousins used in commercial building erection, highway construction, pipeline development, underwater applications, and in underground construction (as in mining) primarily by size and power and special features required in different environments.

People in the industry usually group construction machinery into categories like earthmoving, lifting, bituminous/asphalt application, concrete and aggregate application, attachments and components, and then place those items that do not conveniently fit these categories into a "miscellaneous" bin. Attachments and components are separated in part because entire classes of producers specialize in making blades, buckets, rollers, and the like designed for attachment to power equipment that can accommodate different "working ends." Sometimes equipment is classed into light and heavy categories usually corresponding to residential and other construction applications. Some people, those at the Census Bureau, for instance, also define off-highway trucks as a special category. Figure 64 lists major equipment in this sector by type. Items in the table are drawn from the Census Bureau but arranged by categories commonly used in the construction machinery industry.

MARKET

Size and Trends. Shipments of construction machinery in the United States were valued at $30.08 billion in 2005. In 1997 the industry had total shipments of $21.7 billion; these declined to $19.1 billion by 2002 in a cycle of negative growth of 2.5 percent a year. Thereafter the industry bounced back and grew at a heady annual rate of 17.7 percent to its 2005 peak. Performance of this pattern typically indicates a cyclical industry, meaning that it responds to other cyclically growing/declining economic activities—in this case the activity of the construction industry generally. Construction machinery represents a very diverse mix of products. Many types of equipment are present at all construction sites but others are unique to the kind of building under way. Different rates of growth in types of construction affect the industry as a whole. Growth in housing stimulates small cranes, growth in office construction stimulates large tower cranes.

One way of tracking construction activity is by means of a statistical series published by the Census Bureau called *construction put in place*. During the 1997–2005 period, construction put in place grew at an annual rate

FIGURE 64

Construction Machinery by Major Categories

Earthmoving Equipment

Backhoes
Bulldozers
Ditchers and trenchers, self-propelled
Draglines, crawler
Drags, road construction and road maintenance equipment
Dredging machinery
Excavators (e.g., power shovels)
Graders, road
Land preparation machinery, construction
Plows, construction (e.g., excavating, grading)
Shovels, power
Surface mining machinery (except drilling)
Trenching machines

Lifting Equipment

Cranes, construction-type
Loaders, shovel
Shovel loaders

Bituminous Paving Equipment

Asphalt roofing construction machinery
Planers, bituminous
Rollers, road construction and maintenance machinery
Scrapers, construction-type

Concrete/Aggregate Machinery

Slag mixers, portable
Aggregate spreaders
Concrete finishing machinery
Concrete gunning equipment
Concrete mixing machinery, portable
Crushing machinery, portable
Crushing, pulverizing, and screening machinery, portable
Hammer mill machinery (i.e., rock and ore crushing machines), portable
Jack hammers
Mixers, concrete, portable
Mortar mixers, portable
Paving machinery
Rock crushing machinery, portable
Sand mixers
Scarifiers, road
Screening machinery, portable
Tampers, powered
Vibrators, concrete

Tractors, trucks

Construction-type tractors and attachments
Tractors and attachments, construction-type
Tractors, crawler
Trucks, off-highway

Attachments and Components

Blades for graders, scrapers, bulldozers, and snowplows
Bucket and scarifier teeth
Buckets, excavating (e.g., clamshell, drag scraper, dragline, shovel)
Grader attachments
Snow plow attachments (except lawn, garden-type)

Other Equipment

Augers (except mining-type)
Bits, rock drill, construction and surface mining-type
Rock drills, construction and surface mining-type
Cabs for construction machinery
Chippers, portable, commercial (e.g., brush, limb, log)
Extractors, piling
Highway line marking machinery

SOURCE: Compiled by the staff with data from the construction equipment lists used by the U.S. Bureau of the Census.

FIGURE 65

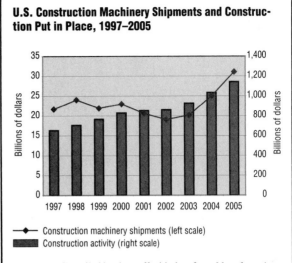

U.S. Construction Machinery Shipments and Construction Put in Place, 1997–2005

Construction machinery shipments (left scale)
Construction activity (right scale)

SOURCE: Compiled by the staff with data from *Manufacturing & Distribution USA,* 4th ed., Thomson Gale, Volume 2, 1034-1039, and "Value of Construction Put in Place Statistics," Manufacturing, Mining, and Construction Statistics. U.S. Department of Commerce, Bureau of the Census. 2006.

Reliable data on the construction machinery industry are only availalbe for the period beginning in 1997. In 1997 the Census Bureau introduced a new style of industrial coding. As a consequence data in the earlier period could no longer be compared to the 1997 and later periods—to the woe of all analysts. The historical continuity was broken. In the most recent eight-year period, for which comparable data are available, construction machinery shipments showed a pattern already well known. They follow trends in construction. But, one might say, they do so a little nervously. This is shown by a downturn in shipments in 1998 and 1999, despite healthy growth in construction put in place. The nervous behavior of construction firms in buying new equipment is illustrated more sharply by comparing year by year the actual rates of growth (rather than actual dollars of shipments) of machinery shipments on the one hand and construction put in place on the other. Such a comparison is shown in the figure.

We have no data for construction machinery shipments for 1996 directly comparable to data for 1997, hence the change from 1996 to 1997 cannot be shown. But in the years that follow, growth patterns in machinery shipments, compared with patterns in the growth of construction activity, suggest that equipment buyers, first of all, appear to anticipate softening of the construction markets. Beginning in 1999, the growth of construction activity slowed, 2000 growth was down from 1999 growth, 2001 from 2000, 2002 down further. Nonetheless, growth was present. But machinery buyers,

of 7.2 percent, from $653.4 billion to $1,143.7 billion. In comparison, the machinery sector grew 4.6 percent per year—more slowly than the activity that it was serving.

FIGURE 66

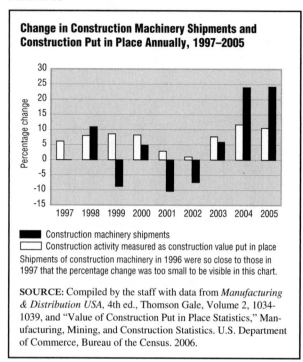

Change in Construction Machinery Shipments and Construction Put in Place Annually, 1997–2005

■ Construction machinery shipments
☐ Construction activity measured as construction value put in place
Shipments of construction machinery in 1996 were so close to those in 1997 that the percentage change was too small to be visible in this chart.

SOURCE: Compiled by the staff with data from *Manufacturing & Distribution USA*, 4th ed., Thomson Gale, Volume 2, 1034-1039, and "Value of Construction Put in Place Statistics," Manufacturing, Mining, and Construction Statistics. U.S. Department of Commerce, Bureau of the Census. 2006.

namely the construction industry itself, felt this slowing and responded by very weak purchases of construction machinery in 1999, 2001, and 2002—and also produced the weakest growth in machinery, in 2000, in this entire period. They became active buyers again as construction recovered its energetic growth in 2003 and then made up for delayed purchases in 2004 and 2005. The presence in this field of a large equipment rental sectors in a way aids and abets the construction industry's behavior. Builders don't have to commit to purchases. They can fill temporary needs by renting the equipment. A substantial used equipment market has the same consequence—companies have a lower-cost alternative to buying new equipment.

Sectors and Product Categories. In the 1997–2005 period, construction activity grew, albeit unevenly. Divided into private and public categories, private construction represents roughly 80 percent of the activity, and the bulk of that, approximately 70 percent, is residential construction. This very large category enjoyed the most rapid growth in the period; residential construction put in place increased at the rate of 10.5 percent per year. Home construction requires the lighter categories of equipment, unless it takes the form of apartment structures. The clearing of large residential housing tracts, of course, may require substantial earthmoving equipment, but in the lifting category smaller and usually self-propelled or towed equipment is used to lift timber and plywood to a one- and maximally a two-story height.

The private non-residential category grew at a very modest 1.3 percent per year by contrast. Within that category two sectors out-paced the total category. Office building construction advanced at a rate of 1.5 percent per year and commercial construction (e.g., warehousing, parking structures, stores, and similar facilities) saw 3.4 percent per year growth. All other sectors—utilities, manufacturing, pipelines, railroads, and others—grew at lower rates. Private construction, as a whole, thus favored the low end of construction machinery in this period.

Public construction, with 20 percent of total activity, does not have a dominant sector corresponding to the private sector's housing market. The two largest segments are educational construction and highway and street construction, each of these representing 27 percent of public construction expenditures equivalent to 6 percent of construction as a whole. In the 1997–2005 period, educational facility construction grew at 8.8 percent per year, the second highest growth in construction activity (behind housing), water supply systems (representing 1 percent of total construction) grew at 7 percent per year, and highway construction came fourth, with an annual growth of 5.5 percent.

Trends in the sector using construction machinery reflected trends in the sector supplying the equipment. In this eight-year period, tractor shovel loaders showed the strongest growth (10.8% per year); these machines are used in virtually all construction activities. Next were mixers and pavers (5% per year), reflecting both residential and highway construction. The next two categories were off-highway vehicles (3.8% per year) and graders (3.7% per year). Power cranes, draglines, and shovels lagged in growth (2.3% per year). Sluggish power crane production growth reflects the fact that large proportions of these products used in the United States are imports; slow shipments do not indicate slow demand. Two other major categories for which the Census Bureau reports detailed data, tractors, dozers, and skidders as a combined category and vehicle-mounted machinery of all kinds as another, both saw negative growth rates of 0.1 percent per year. The second of these categories, the mounted machinery, is most easily rented. Skidders, incidentally, had their origin in forestry for moving logs. Such equipment is used for land clearing.

How do these product categories relate to one another in absolute magnitude? Data for 2005, measured in shipments, show that the largest category was tractor shovel loaders (29.1% of shipments), power cranes, draglines, and shovels came next (16.5%) followed, in order, by graders (12.5%), tractors, dozers, and skidders (12%), mixers and pavers (8.4%), off-highway vehicles (7.8%), and vehicle-mounted machinery (2.4% of shipments). All other equipment and parts represented the remainder.

Exports and Imports. Construction machinery is both exported and imported in significant quantities. Precise data are not uniformly available for construction machinery across the board because the Census Bureau suppresses data as part of its policy to keep company information confidential. The dominance of Caterpillar, and its leading global share in the market (see below), in part explains data suppression. Despite this data problem, the Bureau's 2005 report on the subject, published as part of its *Current Industrial Reports*, still provides unambiguous data for approximately 36 percent of the total industry shipments. Based on those data, if true for the entire industry, it would appear that approximately 12 percent of total purchases of construction machinery by U.S. concerns came from abroad. Of $11.35 billion of shipments, $3.66 billion (32.3%) were exported. But $5.3 billion of the same kinds of products were imported, showing a trade deficit of $1.64 billion. The two largest import categories were hydraulically operated, tractor mounted excavators (imports equal to 114% of domestic shipments) and off-highway rear dump haulers (imports equal to 76% of domestic shipments).

Fundamental Driving Forces. The basic engine driving the market for construction machinery is gross fixed investment (GFI) expenditures. These data are similar to construction put in place—but signal commitments of capital rather than completion of building projects. GFI divides into residential housing, non-residential private investment in structures, and public investment. The latter may be further subdivided into state-and-local and federal expenditures. In 2005, the first three represented 51 percent, 22 percent, and 26 percent of gross investment nationally. And the public portion was 72 percent state-and-local and 28 percent federal. In the 1997–2005 period, residential gross investment grew at an annual rate of 10.4 percent, non-residential private investment in structures at 3.8 percent, public investment at 5.8, and all gross investment at 7.4 percent.

The construction machinery industry lagged this rate of growth, increasing at 4.6 percent a year in the 1997–2005 period. Total investment grew at an increased pace of 10.1 percent in the three-year period 2002 to 2005, and in that same period, the machinery industry grew much more rapidly at 17.7 percent. This lagging and then leading behavior of the machinery industry against the more ceremonious march of national investment decisions reflects both the changing mix of construction over time, another driving force, as well as the shifts in confidence within the highly cyclical construction industry.

KEY PRODUCERS/MANUFACTURERS

The leading producers of construction machinery in the United States, arranged here in order of their domestic market shares in this product category, are Caterpillar, Inc., Deere & Company, Ingersoll Rand Company, Case New Holland (CNH), and Komatsu Ltd. Caterpillar began as a tractor manufacturers, Deere and CNH had their origins in agricultural equipment, and Komatsu began in iron works in support of a mining company. Only Ingersoll Rand began as machinery builder. All of these leaders have their roots in the nineteenth century.

Caterpillar. This company began in 1890 as two separate companies, the first being Holt Manufacturing Co., headed by Benjamin Holt, the inventor of a steam tractor. Holt also first introduced the name Caterpillar as the brand name of one of his tractor lines. The other company, CL Best Tractor Company, was owned by another developer, of gas-driven tractors, Daniel Best. The two companies merged in 1925 to form Caterpillar. The company, with 2006 revenues of $41.5 billion, is the world's leading manufacturer of construction machinery, with 40 percent of its revenues derived from the category. Caterpillar is also dominant in mining machinery, forestry equipment, and power generation equipment. Like the other leaders in this industry, it is a global company rapidly expanding in Asia as the first decade of the twenty-first century closes. Caterpillar has approximately 23 percent of the U.S. and 20 percent of the world market.

John Deere. The founder of the Deere Company, invented the first self-scouring steel plow in 1837 and, in his early days, made them by hand in his own smithy. Deere's first self-propelled devices featured the Waterloo Boy tractors, introduced in 1918. The company became the top producer of wheeled tractors whereas Caterpillar was the leader in tracked tractors, and for some years in the past the two companies sold each other's products through their dealer chains. Deere, with revenues of $19.4 billion in 2005, realized 22 percent of its sales from construction equipment. The company's dominant business is still agricultural equipment, representing 54 percent of its sales. It also participates in forestry equipment and in lawn and garden products, including commercial equipment used in institutional and golf course maintenance. Deere's share of the U.S. construction machinery market is around 15 percent; its global share is around 4 percent.

Ingersoll Rand. This company had 2006 revenues of $11.5 billion of which about 12 percent came from construction machinery. Its three largest segments are climate control, compact vehicles, and security technologies. Ingersoll's market share in the United States was approximately 15 percent (4.4% worldwide). The company began in 1871 by introducing a rock drilling machine and continued, thereafter, as a leading producer of industrial machinery.

Case New Holland. This company is the fourth-ranked construction machinery producer in the United States with a 2005 market share of 13 percent (4.5% worldwide). Case New Holland (CNH) is itself a conglomerate of U.S., Canadian, and European companies. It began in the 1830s as the combination of Case, which made threshers, and International Harvester, which made reapers. Elements were added thereafter by acquisition. The Austrian Steyr and the Italian Fiat contributed tractor lines. Braud, of France, added a thresher line, the Canadian Flexicoil brought a line of seeders to the company. New Holland is also a producer of a variety of farm equipment. Early Case and International Harvester machines used Fordson tractors, a product of Ford Motors, the name itself derived from the fact that Henry Ford, having made his mark in automobiles, began to make tractors with his son and called the enterprise Fordson. CNH revenues from equipment in 2005 were $11.8 billion. Of that total 29 percent was derived from construction equipment, the bulk from the sale of agricultural machinery.

Komatsu Ltd. Last but not least is Komatsu Ltd., a Japanese firm that had a 7 percent share of the U.S. market in the early 2000s. Komatsu is the second largest construction machinery manufacturer in the world and has 10 percent share of the global market. Komatsu began in 1917, then called the Komatsu Iron Works. It was established by its parent, Takeuchi Mining Industry, a company that itself began in 1894. Later, as Komatsu grew in importance, it was spun off as an independent company. It established Komatsu American Corporation in 1970. This entity later created five manufacturing operations in North America, one located in Canada and four in the United States.

Other key players in the market, part of the top ten globally, are Terex, AB Volvo, Liebherr, Hitachi, and Metso.

Terex Corporation is an American company with 2005 revenues of $6.4 billion; it has a 6 percent share of the world market. The Swedish AB Volvo, the auto and truck producer, derives a portion of its $30 billion in annual sales from construction equipment and has a 5 percent share of the global market. The German Liebherr-Holding GmbH, with revenues of €5.3 billion, has 4.7 percent of the global market and is the world's leading producer of cranes, including tower cranes. Hitachi Construction Machinery Co., Ltd., is another major Japanese supplier of construction machinery and has a 4.2 percent share of the world market. Metso Minerals, a

Finish Company, is principally involved in mining machinery manufacturing but the share of its €2.17 billion revenues in 2006 came from construction machinery; the company enjoys a 3.1 percent share of the world market.

Smaller but important participants in the U.S. market are the following:

- Astec Industries, Inc.—$616 million, road equipment
- Bucyrus International—$738 million, primarily surface mining equipment
- Gehl Company—$478 million, loaders
- JLG Industries—$4.43 billion, aerial platforms, part of Oshkosh Truck
- Joy Global Inc.—$2.4 billion, primarily in mining machinery
- The Manitowoc Company, Inc.—$2.3 billion, 72 percent in cranes
- NACCO Industries—$3.2 billion, lift trucks
- Woods Equipment Company—attachments

The dollar values listed above are annual sales figures for 2005 or 2006 and, unless otherwise annotated do not necessarily represent sales to the construction machinery sector, although they include such sales.

MATERIALS & SUPPLY CHAIN LOGISTICS

Construction machinery falls into the heavy category of machinery production as a whole and is characterized by being a major consumer of iron and steel in all of its forms—plate, strip, and sheet; castings, stampings, and forgings; wire, springs, and rods. Leaders in the industry produce their own prime mover engines; others purchase these. The product category relies very heavily on fluid power for moving the working parts of its devices and manufacturers either make or buy hydraulic and pneumatic equipment. Wheeled devices require often quite massive tires that manufacturers either purchase from abroad or from the U.S. tire industry, centralized in Ohio. These fundamental inputs have caused the industry to be centered around concentrations of heavy industry, which tends to be located on or near the ocean or Great Lakes coasts or along major water ways. Substantial portions of production are exported, favoring the location of production facilities near water transport able to reach the open seas.

DISTRIBUTION CHANNEL

Construction machinery is sold using a two-tier distribution system in which producers use a network of dealers.

The dealers in turn sell to the ultimate customer. Parts for construction machinery, however, may move through a third tier of distributors located in major distribution hubs across the country. Distributors specialize in componentry, maintain large inventories, and supply the dealers in turn. Servicing dealers are the common form of distribution, meaning that the seller is prepared to repair the product at its own facilities—and in turn relies on either the manufacturers wholly-owned parts distribution center(s) or uses the services of a parts distributor. Dealers routinely participate in used equipment sales and also rent such equipment. A separate rental market exists as well, its operators purchasing new and used equipment to flesh out their range of offerings.

KEY USERS

Construction companies acting on behalf of buyers of construction services (builders, developers, highway departments; federal, state, and local agencies; industry, utilities) are the principal buyers of construction machinery. Public bodies engaged in construction (e.g., the Corps of Engineers) are included here. Rental houses are another, but much smaller market. Large corporations engaged in substantial building activity with in-house construction departments are a distant third. This category includes mining companies that, alongside mining machinery (often very similar to construction equipment), buy construction machinery for activities analogous to road construction that they undertake. Utilities, railroads, refiners, pipeline companies, and ports and port authorities (that engaged in dredging activities) are other similar buyers of machinery for use in-house. Cranes have their own unique markets. They may be designed to aid construction but are widely purchased by all kinds of institutional entities for doing lifting unrelated to construction.

ADJACENT MARKETS

As the history of participants in this product category indicates, products of a very similar nature are used in three other industries, two of these being, as it were, upstream from construction and one being downstream. The farm and agricultural industry in 2005 was a $20 billion industry selling very similar products for soil clearing, preparation, chemical spreading, and harvesting of products. One might say that the same basic machines are used by agriculture and construction, but while the farmer uses tractors to pull things, the builder uses tractors to push things. The lawn and garden industry ($8.3 billion in 2005) has similar equipment especially on the higher end of professional lawn and turf care. Both of these industries are very much involved in earth-moving or earth-preparation. Mining machinery is downstream from construction because it involves work deep underground or, in strip mining, major earth-moving activity over extensive terrain at a much greater scale. Mining machinery worth $10.7 billion was shipped in 2005. The largest of these industries is construction machinery with $31 billion in shipments.

Adjacent to construction equipment, in the sense that major growth in the sector would impact construction machinery negatively, is the production of manufactured homes. In 2005 this was a $7.3 billion industry entirely devoted to the manufacture of mobile homes. Throughout the twentieth and continuing into the twenty-first century, entrepreneurs and idealists have attempted to replace traditional construction methods in creating residential housing by factory-made housing placed on foundations, requiring only rapid assembly at the site. The mobile home has been the only commercialized expression of this vision. The mobile home industry has had a cyclic history. In 1989 it had shipments of $4.1 billion. It peaked in 1999 with shipments of $10.9 billion, declined again to $5.9 billion in 2003, rising slowly since. Attempts to go beyond the limited square footage of the trailer home has been attempted many times, never with commercial or lasting success. The general idea continues to be supported because substantial cost savings can be realized in mass producing homes. For this reason, efforts to achieve such an end continue, in the United States as well as in Japan and in Europe. Should they ever succeed, much of the volume of construction machinery, heavily dependent as it is on residential construction, would be transformed into industrial machine tools.

RESEARCH & DEVELOPMENT

Very substantial R&D expenditures characterize two of the leaders in this industry, Caterpillar and Deere, expending 3.2 percent and 3.1 percent respectively of their sales volume on R&D activity. These high rates are not matched by all of the participants. Much of the spending is directed toward product development aimed at improving or adding features that give the participants a marketing edge. Some of this expenditure is aimed at improving operator productivity in the field by using electronics to give the operator visual feedback on display consoles inside the operator's cab. Deere is a leader in this field. Such features are important in an industry where productivity rates can be quite low until operators acquire what might be called a sixth sense. In developing markets overseas, such features are important selling points. They permit owners to employ workers with less skill to operate expensive machinery. Other research, beyond feature development, is aimed at pollution control, particularly the control of diesel fuel emissions, which are tightening all over the world. In this industry, as in similar categories that massively employ fluid power, R&D is directed toward replacement of fluid-power actuation and feedback systems with electronic systems, promising greater speed at lower cost.

CURRENT TRENDS

As noted above, the driving force beneath this industry is gross investment in fixed structures. A very important trend is industrialization in parts of Asia, most notably in China and in India, in the Middle East, and in Latin America. Rapid growth in industrial production has not been matched by investment in infrastructure such as roads, ports, and office structures. As Prudential Equity Group put it, in a recent report titled *Capital Goods/Machinery*, "Despite continuing rapid economic expansion in most of the emerging markets, major underinvestment in local infrastructure has led to shortages, bottlenecks, and other infrastructure-related obstacles that the regional players and local governments are now eager to tackle. The reality is that major infrastructure upgrades have not been done since the 1970s in Eastern Europe and Russia, while countries in the Middle East as well as China and India are building their entire infrastructure from the ground up." Prudential's analysts predict an increase in constant-dollar investments in infrastructure globally from $900 billion in the 1995–2004 period to $3 trillion in the 2005–2014 period.

This tripling of global demand from the past to the next nine-year period is the major trend in the industry. It has produced major initiatives by the leading producers to participate actively in overseas markets by developing ties with local leaders in the major hubs of expansion.

TARGET MARKETS & SEGMENTATION

The large market segments in construction machinery are machinery used in residential construction—typically smaller in size—highway and road construction equipment, and lifting equipment of the larger variety targeted at the erection of multi-story building. Many categories of machinery are used in all types of construction.

As always in the case of industrial equipment purchased by companies rather than by individuals, active demand and the economy generally drive the markets rather than impulse or novelty. Industrial buyers are almost never subject to impulse purchases, instead, they carefully look at the bottom line, and—in highly mature industries like construction—innovation on a large scale is difficult to achieve. Targeting, by producers, is therefore achieved by lowering the buyer's acquisition costs, improving efficiency and thus lowering the buyer's operating costs. Producers segment their production based on funding trends. Not surprisingly, the leading association in this industry, the Association of Equipment Manufacturers, devotes substantial effort every year to the preparation of *outlook* publications aimed at helping producers see more clearly through the dark glass of the unknown future.

RELATED ASSOCIATIONS & ORGANIZATIONS

Associated Equipment Distributors, http://www.aednet.org

Association of Equipment Manufacturers, http://www.aem.org/Trends/index.asp

Canadian Association of Equipment Distributors, http://www.caed.org/home.asp

Construction Equipment Association (British), http://www.fmcec.org.uk

Japanese Construction Machinery Manufacturers Association, http://www.cema.or.jp/english/index.html

BIBLIOGRAPHY

AEM Industries Outlook 2007. Association of Equipment Manufacturers. November 2006.

Capital Goods/Machinery. Prudential Equity Group LLC. 22 September 2006.

"Construction Machinery: 2005." *Current Industrial Reports.* U.S. Department of Commerce, Bureau of the Census. July 2006.

Darnay, Arsen J. and Joyce P. Simkin. *Manufacturing & Distribution USA,* 4th ed. Thomson Gale, 2006, Volume 2, 1034–1039.

Industry Initiation–Machinery. Wachovia Capital Markets, LLC. 18 September 2006.

Lazich, Robert S. *Market Share Reporter 2007.* Thomson Gale, 2007, Volume 1, 389–396.

"Product Summary: 2002." *2002 Economic Census.* U.S. Department of Commerce, Bureau of the Census. March 2006.

SEE ALSO *Hand Tools, Machine Tools, Trucks*

Convenience Cleaning Tools

INDUSTRIAL CODES

NAICS: 33–9994 Broom, Brush, and Mop Manufacturing, 31–3230 Nonwoven Fabric Mills

SIC: 3991 Brooms and Brushes, 2392 Housefurnishings, not elsewhere classificed, 2297 Nonwoven Fabrics

NAICS-Based Product Codes: 33–99941 through 33–99941101, 33–99941311, and 33–99941316

PRODUCT OVERVIEW

Convenience cleaning tools are a relatively new product category emerging from the traditional manufacturing sector of Brooms, Brushes, and Mops. The brooms, brushes, and mops sector, referred to as stick goods, shipped more than $2 billion worth of products in 2005, according to the U.S. Census Bureau's *Annual Survey of Manufactures*. Convenience cleaning tools represent a growing portion of that total.

This essay will focus on convenience cleaning tools that contain disposable elements, including a number of specialized wipes. It will cover the home-cleaning market but not products produced for commercial and industrial cleaning.

Brushes were first made of natural materials such as hog bristles and horsehairs. Brooms generally began as birch and willow twigs and were home- and hand-made until the early 1800s. Some say that in 1797, Levi Dickenson, a farmer in Hadley, Massachusetts, made a broom for his wife from a variety of sorghum that he was growing for seeds, creating the first broom of this type. He and his sons began making and selling these new brooms, which were so popular that the sorghum became known as broomcorn straw. The Shakers soon began making brooms and are credited with inventing many improvements, including the flat broom. By about 1830, there were a number of one- and two-man broom shops in the United States, with about 60,000 brooms produced annually.

As the population moved west, it was discovered that broomcorn straw grew well in fertile Midwest fields, and the broom industry flourished with thousands of acres of broomcorn grown annually. Broom shops grew into broom factories, and the industry remains heavily concentrated in the Midwest even today. The introduction of new mass production techniques after World War II improved products and allowed for greater profitability. Plastics began to supplant broomcorn, with almost half of today's brooms now made of synthetic materials. Broomcorn brooms are considered superior by many because broomcorn stalks actually absorb dirt and dust, wear extremely well, and are moisture-resistant. Broomcorn brooms are the most expensive of the manufactured brooms.

Mops come in two major categories—wet mops, used primarily to clean kitchen and bathroom floors, and dry or dust mops. In 1893, Thomas W. Steward, an African American inventor, patented a mop made of yarn with a wringing mechanism. In 1950 Peter Vosbikian, a prolific inventor, developed the Automatic Sponge Mop, a sponge mop that used a lever and flat strip of metal to press against the wet mop and squeeze it dry. The success of this invention led to the formation of the Quickie Manufacturing Corp. Mops were improved in the 1940s and 1950s by new materials made available by the growing plastics industry.

In a revolutionary development in 1999, Procter & Gamble (P&G) introduced its Swiffer sweeper, a broom/dust mop. Disposable cloths, based on electrostatic technology that grabs and holds dirt, fit onto the head of the mop. Because the Swiffer captures dirt instead of pushing it around like a broom, the innovative new cleaning tool offered consumers better cleaning as well as convenience. As a result of Swiffer and the many similar products that followed, the convenience cleaning tools category soon became an important force in the industry. The success of the Swiffer proved that households were willing to pay for the convenience of disposable cleaning aids. A number of other disposable products and cleaning tools soon entered the market, including wet-mop systems, dusters, and bathroom cleaning devices.

MARKET

According to the *2002 Economic Census*, the value of U.S. shipments of brooms, brushes, and mops in 2002 was $2.2 billion with shipments originating in Ohio and Illinois accounting for 26.5 percent of all shipments. Household floor brooms were listed at $86.8 million; dry mops and dusters (including refills except for dusting cloths), $94.8 million; wet mops, except sponges (including refills), $97.8 million; and sponge mops (including refills), $106.7 million.

Convenience cleaning tools are only a portion of this total industry. The Census Bureau's figures do not provide enough detail to separate out the portion of total shipments attributable to these modern cleaning devices. In fact, a portion of the convenience cleaning tools sector as defined for this essay includes material wipes that have been presaturated with cleaning chemicals. These are not covered at all under the Census Bureau industry defined as brooms, brushes, and mops manufacturing, but are included as a very small segment of the huge category of nonwoven fabrics.

Because a great deal of money has been spent in recent years to advertise and promote the newest additions to convenience cleaning tools, interest in the sector has been high and the trade press has collected information with which educated assessments about the size of the market may be made. In a market report published by Citigroup Global Markets, Inc., cleaning tools are broken into three categories. Together, these categories were reported to have had retail sales in the United States of $951 million in 2004, with household sponges and clothes accounting for 47.2 percent, scouring pads accounting for 26.4 percent, and the brooms, mops, and wax applicators category accounting for 26.5 percent.

The fastest growing of these three categories is household sponges and cloths, which saw retail sales increase 34 percent between 2003 and 2004. The report states that

growth in this category was aided by such innovative convenience cleaning tools as Clorox's Toilet Wand and S.C. Johnson's Scrubbing Bubbles Fresh Brush. Neither of the other categories saw growth in retail sales between 2003 and 2004; in fact, the brooms, mops, and wax applicators category experienced a 4.3 percent decline in retails sales.

Business Week reported in 2005 that Swiffer had a 75 percent share of the quick-clean market. Also in 2005, P&G reported that 40 million households worldwide used a Swiffer, 12 million American households had a Swiffer WetJet, 22 million American households had a Swiffer duster, and 2 billion wet and dry Swiffer refill cloths had been sold since their introduction. Similarly, *HFN* reported in 2005 that P&G estimated that it controlled more than 70 percent of an $800 million quick-clean category. The company predicted growth of about 7 percent per year.

Although different surveys measured things differently, several of these figures give an indication of the strong market for convenience cleaning tools. Information Resources Inc. reported that for the year ending in August 2004, Swiffer held first place in sales in food, drug, and discount outlets (excluding Walmart) in the category of Floor Cleaners/Waxes/Wax Removers, with $26.8 million in sales. Clorox Ready Mop was in second place at $11.1 million in sales. More traditional products came far behind (Mop & Glo Triple Action, $9.3 million; Armstrong, $8.4 million, Future, $7.1 million.)

Other trade reports that look at wipes are helpful in understanding the market for convenience cleaning tools. Once dominated by baby wipes, the number of products in this category has been greatly expanded to include a variety of specialized wipes for personal and housekeeping use. Euromonitor International reported in March 2004 that wipes were experiencing double-digit growth and had evolved into a global market of more than $5 billion. The market research organization said household cleaning wipes, a small niche product only a few years before, had reached almost $2 billion in sales globally in 2003. Swiffer accounted for more than 20 percent of the 2003 sales of household wipes. Swiffer was closely followed by S.C. Johnson's Pledge brand, which features similar Grab-It products for household cleaning. The same report noted that largely due to the success of such products as the Swiffer WetJet and Pledge Grab-It Wet, floor wipes had grown to account for more than a third of global impregnated wipes sales in 2003.

In 2005 Packaged Facts, the publishing division of MarketResearch.com, was reported by *HFN* to say that the market for household wipes, including cleaning wipes and broom/mop wipes, was expanding at a compound annual rate of 8.3 percent, fueled almost entirely by household cleaner wipes such as Swiffer refills. The report said that P&G had 32 percent of the household wipes market.

Meanwhile, INDA, the Association of the Nonwoven Fabrics Industry, published a report in early 2006, which stated that in North America, double-digit growth had made the wipes market a $3.8 billion-a-year business, with $2.85 billion in the consumer sector and $0.87 billion in the industrial/institutional sectors. The consumer sector was reported to have grown 11.7 percent per year from 2000 to 2005, with annual growth in that sector projected to slow to 5.3 percent from 2005 to 2010. While baby wipes had historically been the largest consumer segment, household wipes had overtaken baby wipes with 45 percent of consumer wipes sales. This growth was attributed to the introduction of a number of new products, including floor-cleaning wipes, antibacterial wipes, polishing wipes, and others.

It can be seen from these figures that North American sales account for a very large portion of the household wipes segment and, it can be inferred, a similarly large piece of the convenience cleaning tools segment. This can be explained by the fact that the convenience products are marketed to households in developed countries where busier lifestyles, more single-person households, and growth in double-income families have created a pattern of more disposable income and greatly reduced levels of free time.

Euromonitor statistics show, however, that the Wipes segment is growing rapidly throughout the world. Western Europe's sales of dry electrostatic wipes were shown to have increased from $95.3 million in 1999 to $357 million in 2004, while sales of impregnated wet wipes jumped from $38.6 million in 1999 to $731.5 million in 2004. In the Asia-Pacific region during the same period, sales of dry wipes went from $160 million to $200 million and the sale of wet wipes grew by 44.2 percent, from $129 million to $186 million. Yet, local cultures can still have a major impact. When first introduced into Italy, Swiffer bombed so badly that P&G had to withdraw it. Italian women spend much more time cleaning than Americans do, according to the company's research, and they had little interest in products marketed as quick and convenient. There was far more appeal if products could be promoted as doing a better job and getting a home cleaner.

Convenience tools were conceived as a way to increase markets rather than to replace existing products, with families expected to buy a Swiffer as well as a traditional broom or mop. "We think that through innovation, we have injected growth in the category, rather than taking away from traditional stick goods," said Kelly Anchrum, spokeswoman for fabric and home care for P&G in a 2002 interview with *HFN*.

As the products gained in popularity, however, they did have a negative impact on stick goods, and products such as floor and furniture polishes took an even bigger hit. As noted earlier in the Citigroup report, the brooms, mops, and wax applicators category experienced a 4.3 percent decline in retails sales between 2003 and 2004. Also, Packaged Facts reported that sales of household cleaners dropped nearly 4 percent in 2004. To stay competitive, traditional mop and broom companies such as Butler, The Libman Co., and Casabella introduced disposable cleaning products that could be used independently or with the existing tools on the market such as Swiffer.

KEY PRODUCERS/MANUFACTURERS

P&G, which initiated the convenience cleaning tools market in the United States in 1999 with the introduction of Swiffer, remains the dominant player. P&G is a giant in a wide range of consumer goods and is the leading maker of household products in the United States. It markets its nearly 300 brands in more than 160 countries. In addition to the groundbreaking Swiffer line, P&G's product list also includes the Mr. Clean line, which has developed its own convenience cleaning products.

P&G did not originally invent the concept for Swiffer, a new type of electrostatic dust mop. A company called KAO was already marketing a similar product in Japan. The idea to produce a similar product came out of P&G's Corporate New Ventures Unit, an internal think tank, and rather than licensing the KAO mop (which was later licensed by S.C. Johnson and marketed as Pledge Grab-It), P&G decided it could do better by developing its own version.

When introduced, Swiffer took the market by storm, and by early 2007, the Swiffer group comprised a wide range of products. These included:

- Swiffer Dry, the original dust mop. The company said the cloths were extra thick with unique, V-shaped ridges that picked up more dirt, dust, and hair than traditional methods.

- Swiffer WetJet, a mop system designed to quickly remove everyday dirt from the floor. In 2006 P&G introduced on improvement, a scrubbing strip on the disposable cloth. Information Resources Inc. reported that for the year ending in August 2004, Swiffer products held first place in the floor cleaners/wax removers category in food, drug and discount outlets (excluding Wal-Mart Stores Inc.) with $26.8 million in sales (about 33 percent of the $82.1 million total for the category). This compared to $11.1 million for Clorox Ready Mop and $3.1 million for Pledge Grab-It Go Mop. The WetJet was introduced in 2001 at a price of $49.99, which was badly undercut by the Clorox ReadyMop, introduced in 2002 at a price of $24.99. The Swiffer WetJet price was quickly cut to match.

- Swiffer CarpetFlick, a non-powered lightweight sweeper for quick carpet cleanups between vacuuming. The CarpetFlick traps small bits onto a disposable adhesive cartridge. The company began developing this device in late 2003, recognizing that although it had revolutionized the market for cleaning hard floors, 75 percent of the floor space in U.S. homes was carpeted.

- Swiffer Sweep + Vac, which was promoted as picking up fine debris with the disposable dry cloth and larger objects with the vacuum, which is powered by a rechargeable battery. This product faced one of P&G's few bumps in the road in 2004, shortly after its introduction, when 175,000 of the machines were recalled because of reports of a potential fire hazard. P&G relaunched the device in May 2005.

- Swiffer Dusters, said by P&G to have the trapping power of Swiffer in a "fluffy, go-anywhere form."

- The Swiffer Max Cleaning System, with a sweeper and cloths twice as large as regular Swiffer for quicker cleaning.

- The Swiffer WetJet Power Mop, which was made available for multipurpose floor cleaning, wood floor cleaning, and antibacterial floor cleaning.

P&G's products under the Mr. Clean brand include the Mr. Clean MagicReach, designed for easier bathroom cleaning, and the Mr. Clean Magic Eraser products, which clean everything from walls to floors. P&G even introduced a Magic Eraser for auto wheels and tires. As early as 1999, P&G introduced Mr. Clean Antibacterial Wipe-Ups, premoistened cloths that cleaned and disinfected in one step. When the Magic Reach was launched in September 2003, it was a major hit, as was the Magic Eraser after its introduction. The Citigroup Report noted that in the household sponges and cloths category, P&G's leadership share expanded in 2004 to 51 percent, mainly owing to a 6.3-share-point increase for the Mr. Clean brand because of the Magic Eraser.

S.C. Johnson & Son, headquartered in Racine, Wisconsin, one of the largest family-owned and managed companies in the United States, manufactures a long list of home, personal care, and insect-control products, including the Pledge brand. Within weeks of the Swiffer launch, S.C. Johnson countered with the Pledge Grab-It, a very similar dust-mop system. Pledge marketing stressed the ability of Grab-it products to cut down dust, dirt, and hair, particularly for those with pets or allergies. In 2001, the company launched its Pledge Grab-It Wipes to compete with the Swiffer Wet, which had come onto the market earlier in the year. As of early 2007, the line included such products as Pledge Grab-it Floor Sweeper and dry cloths (unscented or in orange citrus scent), Vinegar Wet Floor Wipes, and Orange Wet Floor Wipes for vinyl, ceramic tile, linoleum, sealed wood, and laminated floors.

S.C. Johnson, already a strong force in the bathroom cleaning market with its Scrubbing Bubbles brand, entered the convenience tools market with the Scrubbing Bubbles Fresh Brush Toilet Cleaning System. In 2006, the company launched the Scrubbing Bubbles Automatic Shower Cleaner. Although that product does not quite meet the cleaning tools definition for this article because it has no disposable cloths, it is certainly another innovative tool to save consumers time and bother on an unpopular cleaning chore.

The Clorox Company, which enjoyed fiscal year 2006 revenues of $4.6 billion, has corporate headquarters in Oakland, California. It manufactures products in 25 countries and markets them in more than 100 countries. Clorox introduced all-purpose cleaning wipes in 1999, selling $66 million worth in the United States during the product's first year. Other manufacturers followed suit, but Clorox remained the leader with $123 million in sales in 2004.

Clorox entered the convenience cleaning tools market in 2002 with the ReadyMop, rolled out to compete with the Swiffer WetJet. The ReadyMop was promoted as capable of doing serious cleaning without the bucket, the dirty water, or the mop odor. *Consumer Reports* said the Swiffer mop had thicker pads that soaked up more liquids than the Clorox ReadyMop, that its handle was more comfortable to hold, that its cleaning pads were easier to install, and that it would cost less to use over the course of a year. Clorox, however, was said to be a pound lighter than the Swiffer and easier to maneuver.

Clorox also began marketing its own bathroom-cleaning systems, the Clorox BathWand, and the Clorox ToiletWand. The company promoted the ergonomic design of these products, which it said significantly reduced back strain.

There are private label versions of many of these products, as well as efforts by other cleaning companies to introduce convenience products, but P&G, S.C. Johnson, and Clorox remain the major players in the field. The Citigroup report gives an indication of the competitive success of the big three. Breaking down market share of individual product lines in various categories and subcategories, the report said that in 2004 in the Household Sponges and Cloths sector, Procter and Gamble had a 51 percent market share (Swiffer 42.9 percent and Mr. Clean 8.1 percent), S.C. Johnson, 12 percent (Scrubbing Bubbles Fresh Brush 6.6% and Pledge Grab-It 5.5%), and

Clorox, 9.8 percent (Toilet Wand 4.8 percent and Ready Mop 3.3 percent). In the larger Cleaning Tools Market, Procter & Gamble had 6.9 percent with Swiffer. In Floor Care Cleaners, Swiffer had 33.7 percent, S.C. Johnson's Pledge Grab-It, 16.1 percent, and the Clorox ReadyMop, 12.6 percent.

MATERIALS & SUPPLY CHAIN
LOGISTICS

The *Annual Survey of Manufactures*, published in all non census years by the Census Bureau, reported that in 2002 the broom, brush, and mop industry used $950 million worth of materials to produce $2.37 billion worth of products. The value of materials used increased to $1.01 billion in 2003, $1.07 billion in 2004, and $1.15 billion in 2005. The breakdown of these materials, found in the *2002 Economic Census* includes plastics products, metal products (shapes and forms), yarns and textiles, wood brush handles and backs, paperboard, plastics resins, and dressed hair (including bristle and horsehair).

Convenience cleaning tools require many of these same products. Additionally, they require large quantities of nonwoven fabrics for the disposable sheets they feature. In fact, the rapid growth of convenience cleaning tools has had a major impact on the production of nonwoven wipes. INDA estimated in 2005 that wipes, with $3.8 billion in sales, were consuming 2.9 billion square meters of nonwoven materials, almost triple the volume consumed a decade earlier. Spunlace and airlaid pulp accounted for almost three-quarters of that volume.

At one time, baby wipes were generally made by the airlaid process, in which a material such as wood pulp is bonded with resin and/or thermal plastic resins dispersed within the pulp. In 1973 DuPont introduced a spunlace process, also known as hydroentanglement because high-speed jets of water strike a web, causing the fibers to knot about one another without the use of a chemical bonding material. In the early 2000s spunlace material became popular for wipes because of its textile-like drape and softness, good mechanical and aesthetic properties, and good absorbency and wetting.

As the wipes market entered a period of double-digit growth, a number of nonwoven fabrics manufacturers increased their spunlace production capability, particularly in Europe where major manufacturers such as Orlandi, Sandler, Suominen Nonwovens, and BBA Fiberweb invested in the technology. In 2006 suppliers such as Spuntech, Jacob Holm, and Ahlstrom began to bring more spunlace capacity on-line in North America as well. Most commonly, vicose or rayon and polyester have served as the raw materials for spunlace, but the addition of cotton to the mix has become popular, particularly since the rising cost of petroleum products has made cotton

more price-competitive. The popularity of cotton is based on its purity, high absorbency rate, and durability. *Nonwovens Industry*, an on-line magazine, reported in October, 2006, however, that airlaid products were making a comeback, particularly for new applications and in products such as composite wipes that included both airlaid and sunlace materials. New processes and plants allowed airlaid manufacturers to offer significant cost savings.

DISTRIBUTION CHANNEL

Most convenience cleaning tools are produced by major corporations that manufacture a wide range of household goods. These manufacturers have fully developed distribution networks for moving their wide array of products from production plant to retail establishment. P&G, for example, sells its products in more than 160 countries, with regional distribution centers established to facilitate this worldwide commerce. Swiffer products are sold in North America, Latin America, Europe, the Middle East, Africa, and Asia. In the fiscal year that ended June 30, 2005, P&G's international sales in fabric and home care totaled $15.3 million. A worldwide network of distribution centers supports international sales. As of 2007 the newest such center was the Algerian and Moroccan unit which was built in 2003.

Similarly S.C. Johnson, with more than $6.5 billion in annual sales, sells products in more than 110 countries, conducts operations in more than 70 countries, and has manufacturing facilities in more than 20 countries. S.C. Johnson also relies on a network of modern regional distribution centers, having shifted in 1996 from a system in which most U.S. customers were served from the company's manufacturing plant in Racine, Wisconsin. In fiscal 2006, Clorox's Household Group North America reported $2.1 billion in sales. International sales totaled $0.6 billion of this with sales around the world.

In the business climate of the 1990s and the first decade of the 2000s, cost-cutting and increasing efficiency were important. P&G decided in 2006 to halve its 450 storage and shipping points around the world, in an effort to cut distribution costs and make better use of alternative fuels. After P&G acquired Gillette 2005, the two companies' distribution systems were combined to increase efficiency.

These companies also make use of the latest technology to control costs and assure that products are on the shelves when consumers look for them. One example can be seen in Lima, Ohio, where, in early 2007, P&G was building a 1-million-square-foot distribution center, due for completion in 2008. It was being equipped with an automated storage retrieval system employees call "The Rack," designed to handle 100,000 pallets at a time. In another example, the company worked with an in-house team and an outside partner, Data Ventures of Charlotte, North Carolina, to develop an item velocity monitor that greatly reduced the chance that P&G products would be out of stock on retail store shelves.

In another example of new distribution network innovations, Clorox installed Advance Order Picking software at each of its distribution centers to help plan and effectively load trucks. The software helps the company load trucks by drop location, with accurate axle weight information for each vehicle.

KEY USERS

Convenience cleaning tools are generally purchased and used by retail consumers who want to handle household cleaning chores as quickly and efficiently as possible and are willing to pay more to do so. Institutional sales are also an important part of the industry's sales mix. In many industries, new products are tested and used in institutional settings ahead of their penetration into the household market. With convenience cleaning tools, the opposite is true. New convenience cleaning tools tend to penetrate the household market before they are adapted for use in institutional settings.

ADJACENT MARKETS

When the Swiffer hit the market in 1999, it and the convenience cleaning tools that followed were seen as an opportunity for incremental rather than competitive growth. It was believed that most consumers who bought a Swiffer would still want to own a traditional mop and broom. This has to some extent proven true, with sales of brooms, brushes, and mops climbing from $2.26 billion in 1999, the year Swiffer was introduced, to $2.37 billion in 2005, according to the *Annual Survey of Manufactures*, although these figures do not show whether an individual item within the category such as "household floor brooms" gained or lost sales.

The market for household cleaners and polishes clea-dropped, however, as consumers used more impregnated wipes. According to Packaged Facts, the publishing division of MarketResearch.com, sales of household cleaners dropped nearly 4 percent in 2004, to $3.97 billion. Manufacturers who added convenience products such as impregnated wipes to their lines were still better off, however, because consumers were spending more on their total cleaning supplies purchases.

RESEARCH & DEVELOPMENT

Companies that invested in research and development were the ones to benefit from the increasing market for convenience cleaning tools. P&G is particularly known

for its heavy research spending. The company spent $1.94 billion on research and development in the fiscal year that ended June 30, 2005. In 2005, a company spokesperson said the company had gained market share 23 points in the last three years due to its innovations. In 2006 A.G. Lafley, P&G CEO, said at an analyst's meeting that innovation "is P&G's lifeblood." P&G not only develops new products and brands. The company maintains them through constant research with consumers, which led to improvements and re-releases of popular Swiffer products. These included a scrub strip that was added to WetJet disposable cloths as well as a second generation WetJet with an enhanced spraying mechanism and a fresh color—purple.

The other leading companies also emphasize research. The priority Clorox gives to innovative products can be read on the company's Web site, which has a special section for Innovations. In early 2007 the company listed seven new products, including the BathWand and the ToiletWand. Research on ergonomic design was cited in the section on the BathWand, while the ToiletWand blurb spoke of the revolutionary design, which was said to greatly outclean traditional toilet brushes.

Clorox also relied on research and development teams to take its traditional focus on health and wellness to a new level. Research teams worked with scientists to study germs and disinfection in various settings and then featured the results in promotion and advertising to convince parents the products were needed to protect families. One study, for example, looked at the use of Clorox products in day care centers, concluding that disinfecting helped stop the spread of germs.

CURRENT TRENDS

In the convenience cleaning tools market, one of the strongest trends is the continued introduction of product improvements and products that address new areas of cleaning as manufacturers try to expand their sales. Market analysts have noted the possibility of expanding into auto cleaning and pet cleanups. In fact, P&G has already entered the auto market with the 2004 introduction of Mr. Clean AutoDry, a car cleaning system that attaches to a hose. Each package comes complete with a quick hose connect system, AutoDry soap, an AutoDry filter, and a disposable wash mitt. Continued innovation is particularly important to the major manufacturers to combat the increasingly sophisticated private label products.

Convenience will continue to be the major emphasis of these products. Consumers want disposable, compact, time-saving, and efficient products, and manufacturers are eager to meet this need. The Scrubbing Bubbles shower cleaner is a good example of a product designed to tackle an old chore with little work on the part of the homeowner. Manufacturers also continue to stress ergonomics and simplicity and functionality in design. As noted, Clorox spends considerable promotional effort to convince homeowners that they will have fewer back problems using the BathWand.

The demand for kitchen and bathroom products is being fueled by the heavy investment homeowners are making in their homes, with many spending considerable sums to renovate kitchens and bathrooms. These homeowners are willing to spend money on products that help them to maintain these investments. Some analysts predict laundry and utility rooms will also be getting more attention, and manufacturers are introducing new products that simplify laundry and stain-removal.

Another trend is fragrance-driven innovation, seen in such products as S.C. Johnson's Pledge Vinegar Wet Floor Wipes, and Orange Wet Floor Wipes.

One concern for the industry, however, is the possibility of an environmental backlash by recycling-conscious consumers who are against disposable products. In response, manufacturers are introducing new wipes that are flushable and biodegradable. In another example, Oxi products—cleaners that use active oxygen—were first introduced as laundry detergents and stain removers. Oxi technology has now made its way into household cleaners. It is promoted both as a safe and effective alternative to chlorine bleach. Some smaller companies also emphasize products from botanically pure plant extracts. While environmental friendliness may not be the main selling point to a consumer, buyers often will choose the green product if price and effectiveness are comparable.

TARGET MARKETS & SEGMENTATION

As has been seen, the most basic target market for convenience cleaning tools is the large number of two-income or single-parent families with sufficient income to be willing to pay for convenience. The appeal of convenience is so strong that P&G was able to overcome commoditization and pricing pressures when it introduced Swiffer and then began expanding the line, solving problems consumers never knew they had. Those consumers could have used traditional mops and brooms, but the convenience of Swiffer, highlighted and emphasized in very expensive advertising campaigns, created a huge demand for the product.

The move into convenience products and the continuing introduction of new convenience devices has been based on various studies that identify potential customers. In 2002, for example, a study by the Soap and Detergent Association found that 25 percent of the Americans surveyed said that home cleanliness was a top priority for the coming year. In the same survey, however, 48 percent said they did not keep clean homes because they could not find the time to do so.

The introduction of convenience cleaning tools appealed to these busy consumers and transformed the way they think about household care. Such products as Swiffer dusters and Mr. Clean Magic Eraser represented completely new markets. S.C. Johnson's new shower cleaner actually introduces a new subsegment—preventative cleaners. Manufactures also follow the classic razor-and-blade model in marketing convenience products, making more money on refills than on handles. A purchaser of a traditional dust mop might have no further dealings with the manufacturer for months or years, but a new Swiffer owner can be relied on to regularly buy Swiffer refills.

Manufacturers have not relied solely on convenience, however, when targeting consumers. They advertise that their products will result in cleaner, safer homes. A study reported at the 2000 American Academy of Asthma, Allergy, and Immunology conference said Swiffer was more efficient in removing cat and dog allergens than other cleaning methods. The study, by the Institute of Respiratory Medicine at the University of Sydney, Australia, said Swiffer picked up more than 93 percent of cat and dog allergens from floors, in comparison to 16 percent by a standard broom and 13 percent by a dust mop.

Meanwhile, the Clorox public relations and research and development departments work together to target parents. As seen earlier, Clorox conducted studies of germs and disinfectants and used them to convince parents that Clorox could help protect their families. The company's health and wellness program includes television, print, and radio ads, a Web site for new mothers, and even Clorox educational kits for day care centers and elementary schools. Clorox, which had donated disaster relief money and products to the American Red Cross for some time, launched a new marketing partnership with the agency in July 2006 under the theme Dedicated to a Healthier World. Clorox is collaborating with the Red Cross on marketing, promotional events, and joint sponsorship of a NASCAR (National Association for Stock Car Auto Racing) race car.

Promotional efforts are central to the companies' efforts to convince consumers they need the new products. When P&G began shipping Swiffer WetJet in 2000, the launch was accompanied by heavy television and print ads and $1-off coupons. P&G estimated it was reaching 90 percent of its target households. P&G also agreed to be a presenting sponsor of 11 regional new-home shows, which in return featured the WetJet.

When the improved Swiffer WetJet came to market, P&G teamed with Buena Vista Home Entertainment on a sweepstakes tied to a special DVD edition of Disney's

Cinderella. Similarly, the introduction of Swiffer dusters was accompanied by a tie-in to a DVD release of the Jennifer Lopez film *Maid in Manhattan*. The dusters carried coupons good for $2 off of the DVD, while the video came with coupons good for $1 off of the dusters. In fact, the rollout of the duster also included a partnership with Tupperware Corp. to demonstrate Swiffer at the company's in-home parties, a "Swiffermobile" with testing equipment that could determine the age of dust, and "tell-a-friend" coupons in each box.

Other companies fought back. S.C. Johnson, for example, backed its 2002 introduction of the Pledge Grab-It mop with $8 million in marketing support.

RELATED ASSOCIATIONS & ORGANIZATIONS

American Brush Manufacturers Association (ABMA), http://www.abma.org

INDA, Association of the Nonwoven Fabrics Industry, http://www.inda.org

International Housewares Association for Retail and Wholesale Manufacturers and Buyers, http://www.housewares.org

Soap and Detergent Association, http://www.cleaning101.com

BIBLIOGRAPHY

Bernard, Sharyn. "Making a Clean Sweep; Thanks to Its Swiffer Line, Procter & Gamble Wipes the Floor with Competitors in the Huge Disposable Segment it Spawned." *HFN*. 1 August 2005, 26.

———. "More Stick Goods Companies Taking the Disposable Approach." *HFN*. 12 August 2002, 46.

"Broom, Brush, and Mop Manufacturing." *2002 Economic Census*. U.S. Department of Commerce, Bureau of the Census. Available from <http://www.census.gov/econ/census02/data/industry/E339994.htm>.

Curtis, Richard. "New Duster Cleans Up for Procter." *Business Courier Serving Cincinnati—Northern Kentucky*. 5 November 1999, 1.

Darnay, Arsen J., and Joyce P. Simkin. *Manufacturing & Distribution USA*, 4th ed. Thomson Gale, 2006, Volume 2, 1739-1743.

"How P&G Conquered Carpet." *Business Week Online*. 23 September 2005.

"INDA Releases Groundbreaking Report On North American Nonwoven Wipes Market." *INDA, Association of the Nonwoven Fabrics Industry*. Available from <http://www.inda.org/press/2006/WipesJan06.html>.

Lazich, Robert S. *Market Share Reporter 2007*. Thomson Gale, 2006, Volume 2, 556–557.

McIntyre, Karen Bitz. "Household Care: A Homerun for Wipes." *Nonwovens Industry*. February 2006. Available from <http://www.nonwovens-industry.com/portal/Wipes/articles.php>.

———. "Spunlace Market Report." *Nonwovens Industry*. Available from <http://www.nonwovens-industry.com/articles/2006/03/spunlace-market-report.php>.

"Mop Wars." *Consumer Reports*. July 2002, 10.

"No-Fuss, No-Muss Cleaning is the Aim." *MMR*. 4 October 2004, 15.

"North America: Wipes Put the Squeeze on Traditional Polishes." Euromonitor Press Release Archive. Available from <http://www.euromonitor.com/press.aspx>.

"P&G Analyst Meeting—Final." *Fair Disclosure Wire*. 14 December 2006.

"P&G's 'Swiffer' Proven to Remove Allergens." *Nonwovens Industry*. April 2000, 18.

Rausch, Tim. "Procter & Gamble Is Now Using Its 1 Million Square Foot Distribution Center." *Lima News*. 28 January 2007.

"Review of 2004 Trends in Category Market Share, Pricing, and Promotion." *U.S. Personal Care and Household Products Digest*. Citigroup/Smith Barney. 15 February 2005.

"Statistics for Industry Groups." *Annual Survey of Manufactures*. U.S. Department of Commerce, Bureau of the Census. Available from <http://www.census.gov/prod/2006pubs/am0531gs1.pdf>.

"Swept Away: New Technologies and Product Innovations Are Making a Clean Sweep in the Household-Cleaning Category." *Supermarket News*. 10 March 2003, 43.

"Wipes Make a Clean Sweep." Euromonitor Press Release Archive. Available from <http://www.euromonitor.com/press.aspx>.

Von Hoffman, Constantine. "Marketing Lifts P&G." *Adweek Online*. 31 October 2006.

Cookies & Crackers

———————●———————

INDUSTRIAL CODES

NAICS: 31–1811 Retail Bakeries, 31–1821 Cookie and Cracker Manufacturing

SIC: 2052 Cookies and Crackers

NAICS-Based Product Codes: 31–18110141, 31–182121, 31–182122, 31–182123, 31–18212Y, 31–182141, 31–182142, 31–182143, 31–18214Y, and 31–1821W

PRODUCT OVERVIEW

If we combine all products made from flour, cookies and crackers represent the second largest product category based on dollar sales (17.7%), second only to the output of retail and commercial bakeries (43.6%) and edging out the third largest category, breakfast cereals (17.6%), by just a smidge. Other large groupings in order are frozen products (8.2%), flours and doughs (6.2%), and pastas and tortillas (3.2% each). Cookies and crackers are quite different products, to be sure, different as sugar is different from salt. In this case sugar beats salt in that the industry produces roughly six cookies for every four crackers.

Crackers. Cookies and crackers are both unleavened, baked, wheat products. The biscuit appears to go back to Roman times because its name is derived from the Latin word *biscoctum* meaning twice-cooked. We still use that word when we *concoct* something, thus cook things together. From biscoctum we have the Italian *biscotto*, its plural being *biscotti*, and finally the English biscuit. The same product, in German, is *zwieback*, meaning the same thing. These products were baked once, sliced, and baked again to remove all possible moisture. The dried goods had very long life and were used as food in war and in travel by land or sea. The old naval staple, hardtack, was a simple biscuit made of flour, water, and salt. The *hard* speaks for itself; *tack* was old English for coarse, inferior food. The value of twice-baked goods came from the high carbohydrate and protein content of wheat, the most nutritious food grain grown. It provided all necessary nutrients needed by sailors except Vitamin C, one reason why scurvy was feared by the seafarers of old. Crackers are the modern descendents of biscotti, and the fact that these products are now principally consumed as snacks tells us something about the improvement of food supplies in modern days. The soldiers during the Civil War would have been surprised by this transformation. An old song, cited on G.H. Bent's Web site—a company that baked hardtack for those soldiers in the nineteenth century—recalls the soldiers' gloomy attitude toward the product:

> There's a hungry, thirsty soldier
> Who wears his life away,
> With torn clothes, whose better days are o'er
> He is sighing now for whiskey
> And, with throat as dry as hay,
> Sings, "Hard crackers, hard crackers, Come again no more."

The American term for biscuits—crackers—originated in 1801 with Josiah Bent, the founder of G.H. Bent. Bent began baking biscuits for merchants bound on transatlantic voyages. During baking the biscuits made a crackling sound which Bent introduced as a trade name for his product, the coldwater cracker.

Cookies. By contrast with the utilitarian cracker, cookies have always represented delicacies and, in culinary usage, are classed with cakes. Wheat comes in hard and soft varieties. Hard wheats have a high gluten content and therefore high protein content: gluten is made of two amino acids, the building blocks of protein. Gluten produces a tough dough structure which acts to confine the carbon dioxide produced by yeast added to dough and thus causes the dough to rise. The best bread is made of wheat with a high gluten content. Cakes are relatively light, soft, and easy to slice. Soft wheats with low gluten content are ideal for cakes, generally, and are used in cookies as well. When the cookie crumbles, it does so because the wheat is soft. Soft wheats are grown in humid climates with high temperatures both by day and night, thus east of the Mississippi and in the Pacific Northwest. Cookies are made of flour, water, sugar, eggs, and some kind of fatty substance, usually but not necessarily butter. One kind of cookie, shortbread, got its name because *short* once also meant *crumbly.* The name is said to come from the Dutch word *koekje* meaning little cake. The German word *keks* has a sound very close to *cakes*—both words signaling the lineal descent of cookies from their larger predecessor. Some believe that cookies originated when cooks put in bits of cake dough to test whether or not their ovens were hot enough, and, munching the test results, discovered a new treat.

From earliest times cookie bakers added spices, seeds, berries, raisins, and crumpled nuts to the dough to give their products extra flavors. One of the great cookie favorites of all time, the chocolate chip cookie, came about, one might say almost inevitably. The creator of the category was Ruth Wakefield who, with her husband, owned and operated the Toll House Inn near Whitman, Massachusetts. While baking cookies one day, she crumpled up a bar of semi-sweet Nestlé chocolate and added it to the dough. She thought the chocolate would melt, but it did not. The delightful result pleased everyone. Later Wakefield sold her recipe to Nestlé in exchange for all the chocolate she would ever need—and the company introduced semi-sweet Toll House chocolate morsels as a product.

Industrial Origins. The industry originated with bakeries initially making other products, principally bread, with biscuits baked for the travel trade as a sideline. These origins in the retail bakery sector are still visible in the modern industrial distribution of cookie production at least. The retail bakery industry continues to make cookies; cookies represent just a little under 5 percent of that industry's output. Retail bakeries, however, account for less than 1.3 percent of all cookies and crackers produced.

The first company producing biscuits principally for sailing vessels originated in 1792. This was the Pearson & Sons Bakery in Massachusetts. The company made biscuits called pilot bread. As noted above, the Josiah Bent Bakery began operations some years later, also specializing in crackers. These two companies were later merged, along with six other bakeries, to form the core of what later became the country's largest cracker and cookie producer, Nabisco.

Product Categories. In both the crackers and cookies categories, products have proliferated into many varieties. Some feel for leading types of products is provided by the U.S. Census Bureau which provides product-level breakdowns of the industry in those years for which it conducts full surveys. Figure 67 presents data from the *2002 Economic Census,* showing industry category shares based on shipment values in that year.

In the crackers category the "all other" line accounts for more than half of total volume, each subcategory, however, too small for detailed treatment by the Census Bureau. Names of varieties of crackers have tended to be brand names in the past, thus even the word cracker began in that fashion. Saltines were once a Nabisco name, but the company lost its trademark when the term became a widely used generic designation. Graham crackers are named after their inventor, the Rev. Sylvester Graham. These products, made of coarse-ground wheat, have a high fiber content and were originally intended as a health

FIGURE 67

U.S. Cookie and Cracker Shipments by Type of Product, 2002

Percentages are presented based on shipment value for the product relative to total industry shipments.

Products	Percent of Total Category
Crackers	
Saltines	20
Cracker sandwiches	13
Graham crackers	9
Cracker meal crumbs	4
All other	54
Cookies	
Sandwiches	21
Chocolate chip	11
Toasties/ice cream sadwich wafers	9
Creme-filled cookies	6
Oatmeal cookies	5
Ice cream cones and cups	4
Marshmellow cookies	1
All other	43

SOURCE: Compiled by the staff with data from *2002 Economic Census,* U.S. Department of Commerce, Bureau of the Census. December 2004.

food. Ritz crackers, a Nabisco brand, is another favorite similar to saltines. Matzo bread is classified as a cracker and is an ethnic specialty. Cream crackers, a thicker variety of matzo is yet another type; many different varieties compete for the market and are classified as flavored, sprayed, sponge-type, and so on. Cracker sandwiches consist of two layers with a filling, usually cheese, in the center.

Cookie sandwiches match their cracker cousins in appearance but have a sugar filling in the center. They are the largest single category and Nabisco's Oreos is also the leading cookie brand. Unlike crackers, which are typically consumed with dips or spreads and in which salt and cheese are the principal flavoring ingredients, cookies are consumed without auxiliaries but with a liquid to enhance the eating experience (milk, coffee, tea, and cocoa). More flavors and fillers are used. Sugar, cream, chocolate, and marshmallows are highlighted by the Census data, but nuts and fruits are common ingredients. Fig products, Nabisco's Fig Newtons being prominent, are included under the All Other category. For industrial purposes, ice cream cones, cups, and wafers have been added to the cookie category. These products are eaten with ice cream. The Census Bureau has also added one of the latest innovations in sweet snacking to the cookies and crackers industry, the toastie—while, somewhat inconsistently, carrying other similar snack products, like breakfast bars, under breakfast cereals.

MARKET

Measuring the market for the cookies and crackers category is best accomplished at the production level, thus looking at industrial shipments as reported by the Census Bureau. As these products move from producers into the wholesale and retail channels, the category is lost to view because its identity is hidden in aggregations with other grocery and supermarket products. Market research firms produce occasional snap shots of the industry at the retail level, usually reported by leading brands. These data are almost always partial and exclude results from Wal-Mart. Wal-Mart does not report its over-the-counter retail sales except in aggregates in its annual and quarterly reports.

Industry Shipments. Shipment data for 2005 indicate that the cookies and crackers industry had sales of $11.4 billion. Sales in 1997 were $9.2 billion and in 2002 $10.1 billion. The growth rate between 1997 and 2002 was 2.0 percent per year, between 1997 and 2005 2.7 percent. Industry sales in the later period had picked up, but early indications, derived from private market research data, suggest another weakening of sales in 2006 despite a flurry of new product introductions by producers.

More detailed data, thus at the product level, were available only for 1997 and 2002. These data indicated

that the cracker category was advancing at a barely visible 0.2 percent per year whereas the larger cookie category was growing at the rate of 1.5 percent per year in that earlier period. Stronger growth in the later years (2002–2005) may have been due to an upsurge in sales of the snack food industry which, as they say in political jargon, had coattails. Crackers are not included in the snack food category for purposes of industrial reporting, but the contexts in which crackers are usually consumed *are* the snacking context.

A broad look at our category's performance is provided in Figure 68. In this graphic shipments of all food products, snack foods, and the cookies and crackers industry are shown as an index, with the index set at 100 for the year 1997 for all three categories. Changes from that base year are then displayed for successive years. This analysis shows that over the 1997–2005 period, the All Food category outperformed cookies and crackers by a small margin, although not every year, showing a 27 index point gain over 1997 in 2005 versus a 24 point gain for cookies and crackers. During this same period, the snack food category, which is principally nuts and chips, gained 54 index points, the strongest growth taking place in the 2002–2005 period. The bottom line here, at least for this eight-year period, is that cookies and crackers have grown at about the same rate as food as a whole. The industry is strongly associated with impulse-eating but has not benefited much from the growth of the snack food industry. Put into annual growth rates, food advanced at 3.0 percent per year, cookies and crackers at 2.7 percent, and snack foods at 5.5 percent per year in this period.

Product Detail. Within the cracker category, the Census Bureau shows only details for 2002; no numbers are available for 1997 to serve for comparison. Within the cookie category, some feel for trends is provided for at least the early part of this period. Among named products in the sweet category, the ice cream cones and cups exhibited the strongest growth, advancing at the rate of 7.9 percent per year; the second product with strong growth was cookie sandwiches, growing at 6.6 percent. Ice cream consumption—along with snack food consumption—has defied a growing resistance to fattening foods; the cream itself has thus given the cookie category a lift by needing cones. Among products showing the worst performance have been the two richest categories. Crème-filled cookies declined at the rate of 0.1 percent per year and chocolate-chip cookies lost sales at a surprisingly high rate of 7.7 percent per year.

Factors Influencing Growth. The food industry as a whole has been undergoing a transformation that began in the 1980s and has yet to run its course. The essence of this

FIGURE 68

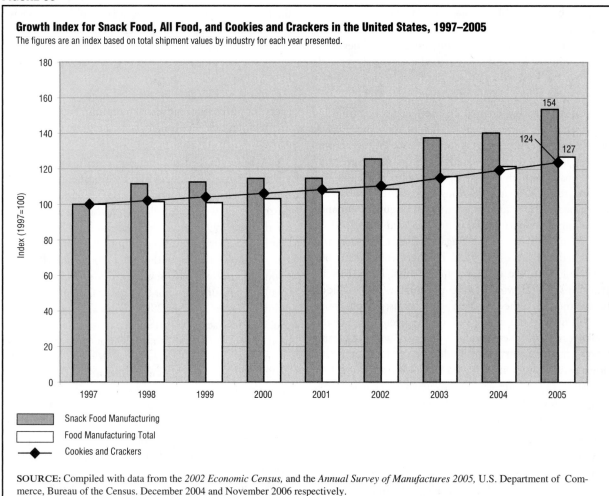

Growth Index for Snack Food, All Food, and Cookies and Crackers in the United States, 1997–2005

The figures are an index based on total shipment values by industry for each year presented.

SOURCE: Compiled with data from the *2002 Economic Census,* and the *Annual Survey of Manufactures 2005,* U.S. Department of Commerce, Bureau of the Census. December 2004 and November 2006 respectively.

crisis—as it is sometimes labeled in the industry itself—is the public's concern with health. That concern, however, is not uniformly shared by the public. The consequence is what might be viewed as at best ambiguous and at worst schizophrenic consumer behavior, at least as statistics resolve the picture. What many people, and certainly most health authorities, view as foods to be consumed with great care, if at all, are growing (e.g., butter, ice cream, cheese, and snacks) while other categories are struggling. Cookies and crackers belong in the embattled group. The specifics that bear on this transformation will be discussed more fully under Current Trends below.

KEY PRODUCERS/MANUFACTURERS

Kraft, Inc. This company is the largest producer of cookies and crackers in the United States with some twenty-eight brands of cookies, twenty-three brands of crackers, and one brand of ice cream cone. Kraft illustrates

the vast process of acquisitions and divestitures that has characterized the food industry in the twentieth century. The company began as a cheese producer in 1916 but has grown by acquisition into a diversified food company. Its role in the cookies and crackers industry is anchored to its ownership of Nabisco, itself a company that grew by the assembly of many scores of bakeries before it emerged as National Biscuit Company. Standard Brands Inc. merged with Nabisco in 1981. Philip Morris acquired Kraft in 1988. When Philip Morris acquired Standard Brands, Nabisco became part of Kraft in 2000. Philip Morris renamed itself Altria. In 2007 it spun off Kraft, Inc. Kraft became an independent corporation once again. Kraft's sales in 2006 were $34.4 billion. Of that total its Snacks and Cereals components, which is principally crackers and cookies, represented $6.4 billion of that total, including North American sales, thus the United States and Canada. Kraft's share of the cookie market was around 37 percent, its share of crackers 47 percent, making it the top producer in both categories.

Kellogg Company. Ranked second overall in sales in this industry and the top producer of breakfast cereals, Kellogg had sales in 2006 of $10.6 billion of which $3.3 billion were in what Kellogg labels snacks foods, principally cookies and crackers. Kellogg entered this business in 2000 with the acquisition of Keebler Foods Company. Keebler began operations in 1853 and, until both were acquired, competed as an independent company with Nabisco, both companies arising in the nineteenth century. Keebler offers sixteen brands of cookies, eleven brands of crackers, and also sells ice cream cones. Keebler's brands include Sunshine, owned by Sunshine Biscuits Company until the latter was merged with Keebler in 1996. The company's share of the U.S. cookie market is 11 percent and of the cracker market 24.2 percent. Keebler is also the leading producer of a special kind of cookie most people buy to help young girls raise money—the Girl Scout cookie. Kellogg is also the owner of another leading cookie producer, Murray Biscuit Company, which has a 5 percent share of the cookie market.

Campbell Soup Company. This company is third ranked in cookies, with 8 percent and fourth in crackers, with 9 percent of the market. Its sales in 2006 were $7.3 billion, of which the company's baking and snacking segment accounted for $1.7 billion. Campbell Soup's participation, as that of the other two leaders, is through the activities of its wholly-owned Pepperidge Farm company, best known for its bread, but Pepperidge Farm also ranks high in cookies with four major brands and in crackers with two. The company was founded in 1937 by Margaret Rudkin who created a high quality natural bread. One of her sons suffered from allergies; the new bread was originally created for the boy, but friends and neighbors loved it too. Rudkin lived on a Connecticut farm at that time. Sour gum trees, pepperidge trees, grew there. The trees obliged the budding entrepreneur by lending their name to her successful startup. Pepperidge Farm operated as an independent corporation until it was acquired by Campbell Soup in 1960.

McKee Corporation. This family-owned and privately held company, originating in the 1930s, is best known by American consumers for its Little Debbie brand of cookies. McKee, with a 6 percent market share in cookies, also makes a line of cereals and cakes. The company is based in Collegedale, Tennessee, and reports sales of around $1.1 billion.

The companies presented above accounted for more than two-thirds of all cookie sales and for 80 percent of cracker sales in 2005. This industry, however, is quite extensive, with participation by some 322 companies operating 378 establishments. In addition to such domestic participation, cookies are also imported, with Danish butter cookies in fancy round tins playing a role, however minor, in our Christmas celebrations every year.

MATERIALS & SUPPLY CHAIN LOGISTICS

Among materials consumed by the industry as measured in dollars and reported in federal statistics, the largest clearly identified category is packaging materials (aluminum foil, plastic, coated and uncoated paper, bags, paperboard, and even glass containers). Wheat flour and other wheat products are next, followed by white and brown sugar, fats and oils, and chocolate. Production is concentrated on where the population is concentrated, thus on the West Coast, a diagonal line extending roughly from Missouri to New York State, in Texas, and in Georgia and Florida, suggesting that raw materials flow to the population centers, are converted into cookies and crackers near their markets, and travel short distances to retail outlets. Illinois, Georgia, and California are the three largest producing states.

DISTRIBUTION CHANNEL

Based on reporting by such magazines as *Grocer*, *Grocery Headquarters*, and *Progressive Grocer*, which rely on market research reporting by such firms as ACNielsen and Information Resources Inc., roughly three quarters of all sales take place in the grocery channel, including superstores in that category. The remainder is distributed between drug stores, dollar stores, warehouse clubs, and miscellaneous outlets that include stores run by filling stations. The grocery channels rely on wholesale distribution centers. Large chains own their own distribution or operate using a mix of wholly-owned and third party wholesalers. Smaller grocers make use of independent wholesalers. In this as in other grocery product categories producers make use of routes, operated by themselves or by third parties, to keep distribution centers adequately stocked.

KEY USERS

Being basic food products, everyone eats cookies and crackers. Under pressure to respond to growing health concerns, the industry has developed products aimed even at those who must avoid sugar or people on diets. Some people lean more toward salt than sugar, and vice versa, but the category offers products for either taste.

ADJACENT MARKETS

A somewhat artificial division has been created by the historical development of crackers and cookies on the one hand and nuts and chips on the other. The last two are assigned to the snacking category by industrial reporting services like the Census; crackers are very often purchased ahead of parties and dinners and are viewed as appetizers, cookies as desserts. Appetizers are sometimes referred to as hors d'oeuvres, a French phrase meaning *outside of work,* thus they are snacks or break time foods, not specifically *food ahead of eating.* At the same time, in commercial categorization used by virtually all of the major companies, cookies and crackers are classed as snack foods. All this by way of saying that snack foods are the most obvious adjacent markets to cookies and crackers—but, alas, in many definitions, they *are* snack foods.

Health-conscious people who see too much snacking as unhealthy suggest that the snacking urge be satisfied by eating fresh vegetables like carrots and radishes or fruits like apples, pears, oranges, grapefruit, or berries. Health drinks based on dairy products like yogurt are suggested as good replacements for foods too rich in processed sugar and in carbohydrates. In one sense formal diet regimes represent anti-snacking and are therefore genuine adjacent (in the sense of alternative) markets.

RESEARCH & DEVELOPMENT

Producers in this industry are putting substantial R&D dollars into making new products that taste as good as ever but have very little of the substances that actually produce pleasing taste sensations: sugar, salt, and fat. Milk whey is a byproduct of cheese-making very rich in complex chemistry. Whey is much studied to isolate from it chemicals that can enhance the taste of cookies while permitting the removal of sugar. Use of spices permits eliminating some of the salt that gives crackers their desirable tang. Saturated fats occur in dairy products, red meat, chocolate, and coconuts. The term saturated derives from the fact that such fats have high levels of hydrogen. Trans fats also have high hydrogen content, but the element is introduced artificially to raise the boiling points of vegetable oils and to retard their spoilage. Replacing fats from such sources with fats derived from olives, canola, nuts, avocadoes, corn, cottonseed, safflower, soybeans, and fish can render cookies or crackers more heart-friendly. Saturated fats build up in the arteries and result in heart disease. Sugar intake is associated with diabetes. High levels of salt consumption lead to high blood pressure, but hypertension is also associated with unbalanced levels of potassium, magnesium, and calcium in the blood. Ensuring ideal minerals content in foods is part of the industry's attempt to fight off the assault on snacks by the critics of modern food consumption.

Considerable development effort is also expended on much more mundane commercial efforts such as producing portion-packaged snacks sold on the basis of calorie content—the 100-calorie packaging movement. Products in these packages are often miniaturized. More objects, by sheer count, are in the package than heretofore, although each object is smaller than in bulk packages. This is intended to satisfy the consumer's psychological perceptions. Reduced-size packaging is priced higher than bulk packaging if product weight is used as the measure, but the consumer accepts this pricing as a service for helping him or her exercise constraint. Developing miniaturized products, their packaging, and the refining of marketing messages also falls under the same budget categories as taking out sugar, salt, or trans fats.

CURRENT TRENDS

Two contradictory trends influence the industry. One is widespread concern with the deteriorating public health, in part blamed on food content and on consumption habits. The other is the embrace of snacking as a way of eating if not a way of life. The industrial response to these self-canceling movements has been to segment the market into slices and to put out products pleasing both sides and shades of behavior in between.

The health concern centers on the increasing incidence of the milder but ultimately very destructive Type II diabetes, the type which is not treated with insulin injections. Similar concern accompanies the growing incidence of heart disease. These conditions are associated with a population getting excessively heavy. Obesity is associated with lack of exercise and consumption of too much and the wrong kinds of food.

The Centers for Disease Control and Prevention (CDC) began measuring the population's weight status in 1963. The National Health and Nutrition Examination Survey (NHANES), produced by CDC, reports the results. The second of these surveys, for the period 1976–1980, reported that 47.1 percent of the adult population was overweight or obese; those judged obese were 15 out of 100 adults. NHANES 2003–2004 showed an increase from 47.1 to 66.2 percent, with 32.9 percent being obese. The last figure certainly raised eyebrows across the nation—and looking at crowds confirmed what CDC was measuring scientifically. That survey also noted the substantial increase in children who were overweight. Of those aged 2–5 13.9 percent had too much weight, 8.9 percent higher than in the 1976–1980 period. Among those in older segments total rates and increases over the earlier period were even higher. CDC also tracks how much people exercise and concludes that less than half (48%) of the public pursues sports or other energetic activities at levels recommended by health authorities.

People are also simply eating too much. The U.S. Department of Agriculture (USDA) tracks food consumption through its Center for Nutrition Policy and Promotion. Data from the Center indicate that food intake between 1970 and 2004 increased substantially, from 3,200 to 3,900 kilocalories per capitum, up 22 percent. Consumption of carbohydrates increased by 26 percent, protein intake least, by 23, and fat consumption most, by 30 percent. A high proportion of the carbohydrate consumption in the U.S. diet comes from sugar.

Underlying some of these changes have been shifts in lifestyle—the new perpetual busyness. The Families and Work Institute (FWI) reported in 2002 that dual-earner couples worked 91 hours, up from 81 in 1977. In the same period households headed by a single parent increased from 9 to 17.6 million. All households increased 47 percent between 1977 and 2002, single parent households by 100 percent. According to USDA's Economic Research Service 36 percent of household food budgets were spent on eating out in 1975, 48 percent in 2005. Parents have less time to prepare meals. Meals are being replaced by snacking or eating out.

Public reactions to the consequences are creating the pressures on the food industry to modify foods. People snack more, exercise less, and reach for ready-to-eat bars or bags of food to cope with social stress. The food industry, the aim of which is to maximize profits, targets its products to each market and every niche within it.

TARGET MARKETS & SEGMENTATION

The foregoing discussion of trends points at basic marketing and targeting practices in this industry. Products are aimed to attract those concerned with health but still wishing to indulge, those aiming to diet, but not too fiercely, and those who do not care. Producers for this market formulate products that have low- or no-salt in the cracker categories, products that avoid saturated fats, cookies with low- or no-sugar content, and portion-packaged products aimed at those wishing to control their snacking urges while still at the check-out counter. At the same time, items rich in taste continue to be offered in traditional bulk packaging. The ultimate choice is in the consumer's hands.

RELATED ASSOCIATIONS & ORGANIZATIONS

American Bakers Association, http://www.americanbakers.org

American Institute of Baking, https://www.aibonline.org

Independent Bakers Association, http://www.mindspring.com/~independentbaker

International Dairy-Deli-Bakery Association, http://www.iddba.org/default.htm.

Retail Bakers of America, http://www.rbanet.com

Salt Institute, http://www.saltinstitute.org/4.html

Snack Food Association, http://www.sfa.org

BIBLIOGRAPHY

"Bent's Hardtack Cracker." G.H. Bent Company. Available from <http://www.bentscookiefactory.com/hardtack.htm>.

"Carbohydrate Counter." Carbohydrate-counter.org. Available from <http://www.carbohydrate-counter.org/cereal/search.php?cat=Wheat&fg=2000>.

Darnay, Arsen J. and Joyce P. Simkin. *Manufacturing & Distribution USA*, 4th ed. Thomson Gale, 2006.

"Food CPI, Prices and Expenditures." U.S. Department of Agriculture, Economic Research Services. Available from <http://www.ers.usda.gov/Briefing/CPIFoodAndExpenditures/Data/table1.htm>.

"General Summary, 1997 Economic Census." U.S. Department of Commerce, Bureau of the Census. June 2001.

"The History of Bent's." G.H. Bent Company. Available from <http://www.bentscookiefactory.com/history.htm>.

Lazich, Robert S. *Market Share Reporter 2007*. Thomson Gale, 2007.

"National and State Population Estimates." U.S. Department of Commerce, Bureau of the Census. 22 December 2006. Available from <http://www.census.gov/popest/states/NST-ann-est.html>.

The National Study of the Changing Workforce. Families and Work Institute. 2002.

Peters, Jeremy W. "Fewer Bites. Fewer Calories. Lot More Profit." *The New York Times*. 7 July 2007.

"Prevalence of Overweight and Obesity Among Adults: United States 2003–2004." U.S. Department of Health and Human Services, National Center for Health Statistics. Available from <http://www.cdc.gov/nchs/products/pubs/pubd/hestats/overweight/overwght_adult_03.htm#Table%201>.

"Product Summary: 2002." U.S. Department of Commerce. Bureau of the Census. March 2006.

Stradley, Linda. "History of Cookies." What's Cooking America. Available from <http://whatscookingamerica.net/History/CookieHistory.htm>.

"Toll House® Cookies." The Great Idea Finder. Available from <http://www.ideafinder.com/history/inventions/tollhouse.htm>.

"Wheat." Walton Feed, Inc. Available from <http://waltonfeed.com/self/wheat2.html>.

"Whole Grains: High in Nutrition and fiber yet low in fat." MayoClininc.com. Available from <http://www.mayoclinic.com/health/whole-grains/NU00204>.

SEE ALSO *Bakery Products, Snack Foods*

Copiers

—■—

INDUSTRIAL CODES

NAICS: 33–3315 Photographic and Photocopying Equipment Manufacturing

SIC: 3861 Photographic Equipment and Supplies

NAICS-Based Product Codes: 33–33153

PRODUCT OVERVIEW

Laborious, Awkward Beginnings. Before the first copying machine was introduced by 3M Corporation in 1950—the Thermo-Fax™ machine—documents were routinely duplicated as their originals were typed. Two to three thin sheets of so-called onion-skin or tissue paper, each faced with carbon paper, were placed behind the original. The keystrokes transferred the letter image to the copies by impact. To this day some people ask for a cc of a document, meaning a carbon copy.

When more copies were desired than could be reasonably produced in this way, people typed text on special ditto masters. The masters consisted of two overlaid sheets. The upper sheet was used to type on, the backing sheet was covered by a layer of tinted wax, usually colored blue. Typing on the master caused wax from the second sheet to adhere to the reverse side of the front sheet, the wax mirroring the shape of the letters typed. The front sheet, with its reverse marked with wax, was then placed on a printing drum face down exposing the wax images on its back. As the drum turned, an alcoholic solution dissolved just enough of the wax images so that the master, contacted with another sheet—the copy—would transfer the image to the other sheet. Dittoing was limited because the wax wore off after a while. The alcoholic solution could also soften both master and copy and cause the machine to jam.

A more robust method of duplication was by stencil and was known as mimeographing. Stencils were made of thin, waxed paper—more wax than paper but visibly fibrous. Typewriters used without a ribbon could cut a stencil, as the phraseology of the time had it. Held up to light, a stencil looked like a photonegative, the print bright. Stencils had stiff backing sheets to permit their easy handling in typewriters. The stencil itself, with the backing sheet removed, was laid over a perforated printing drum, the inside of the drum generously rubbed down with ink. The drum was then turned, mechanically or by hand, to print the stenciled image to sheets of paper fed mechanically. Many more and sharper copies could be printed in this way, the drum re-inked from time to time. The process could go on until the stencil itself tore.

In organizations of some size, especially those producing a large amount of paper for internal distribution, small offset presses were used in-house to print documents and forms. Images to be printed were made by photographic methods using sensitized metal plates or specially manufactured stiff paper plates on which text was typed directly using electric typewriters. In the new century this technology is used principally by printing companies.

Thermofaxing involved specially manufactured paper sensitive to infrared energy. The original was laid on top of the sensitized sheet and inserted into a machine that applied heat to the original. In all those places where the original had carbon, or print, the carbon itself heated up

297

and passed this heat on to the sensitized sheet. White areas did not heat up. The copy reproduced the image by darkening under the impulse of heat. The method was widely used because it was very handy, simple, and did not include solvents or inks. But nobody much liked the product. Thermofax paper had a tendency to curl, to grow dark, and to become brittle with time. Costs of the process were low, around 1 cent per copy or less, and the equipment was not expensive either. Thermofaxing was a half-step in early morning semi-darkness between labor intensive and/or messy methods of duplication and the brilliant sunrise that came with the invention of xerography.

The Birth of Copying. The discovery of electrophotography, as the art was initially called, is an extraordinary story of the persistence of one individual and of the deafness of large, rich institutions to the knock of opportunity. Chester Carlson, a patent attorney trained in physics, found the need to copy patent applications tiresome and tedious. The job involved not only retyping applications but re-doing their intricate drawings as well. He thought that some method of easy copying would be very beneficial and set to work in his spare time to see what approach might work.

As the story is told by David Owen in *Copies in Seconds*, a published paper by Paul Selenyi, a Hungarian physicist, who had experimented with inscribing images using a stream of ions, reminded Carlson of the much earlier work of the German professor, George Christopher Lichtenberg, who had used very fine powder to make visible patterns on charged pieces of amber in 1777. These two ideas eventually led to Xeroxing.

In the crucial experiment that proved the concept, Carlson used a flat piece of zinc thinly coated with sulfur as a receptor of his image. The coated plate, rubbed in a dark room with a bit of fur, would take on a charge of static electricity. Carlson was using a piece of glass as the document to copy. On this glass he had written, in black India ink, the phrase *10.-22.-38 ASTORIA*, the date of the experiment and the name where it was taking place, namely in Astoria, Brooklyn, New York. Still in darkness Carlson laid this glass on the charged zinc-sulfur plate and then exposed the two items to very bright light. Removing the document, Carlson saw nothing on the plate, but, expecting the image to be there, he next sprinkled lycopodium powder over it. This substance consists of the yellow spores of club moss, the same extremely fine powder Selenyi had also used in his ionic printing trials. Blowing on the plate to remove the powder, he saw the image of his message perfectly reproduced in yellow powder on the plate after the excess was easily blown away. The powerful light he had directed at the plate had caused all areas to

lose their charge except those shielded by the dark India ink on his document; the light itself had removed the static charge on the plate everywhere else. The areas still charged held on to the powder very tightly. Carlson next laid a sheet of wax paper over this image and applied a hot iron to pick up the powder from the plate. The powder adhered to the heated wax. The waxed paper had become a perfect copy of the message on the document. The first Xerox copy had been made. It can still be viewed at the Smithsonian Museum in Washington, D.C.

It took Carlson seven years of effort to find backers for the serious development necessary to turn this discovery into a copying machine—and then it took another fifteen years before the first effective copier came on the market. His invention protected by patents, Carlson contacted twenty companies trying to get them interested. He found no takers. In 1944 Battelle Memorial Institute, a not-for-profit research organization, agreed to help Carlson develop the technology. Battelle enlisted the help of Haloid Company, a small producer of photographic paper, to bring the product to market. Haloid introduced the first clumsy Model A in 1949. The breakthrough came in 1959 when the 914 copier, fully automated, went on sale and became an immediate success. Haloid changed its name to Xerox in 1961. The name itself, trademarked in 1948, was fashioned by joining the Greek word *xeros* meaning dry to the Greek, *graphein*, meaning to write.

In the commercial model Carlson's zinc plate is a drum whose surface can be charged. The document is laid print-side down on glass. Powerful light is directed at the document and reflected down on the drum as it turns. A precise reflection is ensured by using mirrors. Where light reaches the drum, thus where the document is white, the charges on the drum's surface are neutralized. The printed or other dark areas of the document remain charged on the drum. The drum, as it turns, passes a source of toner which corresponds to the yellow lycopodium spores. The toner is oppositely charged and adheres selectively to the drum. As the drum continues to turn, it touches the sheet that will hold the copy. The sheet itself is also given a charge but greater than that of the drum so that the toner transfers to the paper, also by electrostatic attraction. A strong source of heat is then applied to the paper to cause the toner to melt and fuse into the fibers of the sheet.

Digital Copying. Toner for copiers is made of tiny, spherical carbon and polymer particles in size between 5 and 7 microns. A micron is one millionth of a meter and thus is extremely tiny. Printer resolution is defined in dots per inch (dpi). A good quality printer will have a 600 dpi resolution meaning that a square inch will hold 600 dots—any one of which may be black or white (or some other color and white). A standard 8.5 x 11 inch sheet of paper at 600 dpi will accommodate 56,100 dots. In preparing documents on a computer, the machine can readily divide the page into many thousands of sequential dot-lines, known as rasters, and send these lines to a printer dot by dot and line by line. Similarly, a high-resolution scanner can look at a document raster by raster and capture what it sees as 56,100 separate bits of data, storing them in computer memory as 0s or as 1s depending on whether the tiny areas are white or dark. These modern capabilities in combination have produced both laser printers and digital copiers.

Laser printing was invented by Gary Starkweather in 1969; he was an optician working for Xerox. It occurred to him that he could user lasers, capable of producing very tiny beams of light, to inscribe a page on a xerographic drum. He could send light for every dot he wanted to be white and simply withhold the light from dots he wanted to be black. Spots refused the light would attract toner. All other areas would be neutral. A computer that had already divided a document in its memory into a matrix of dots could send the signals to the laser sequentially. Digital copying developed from laser printing. A digital copier first scans a document into a copier's memory and then laser-prints that digital image to paper by means of drum and toner.

Since the development of digital copying, ordinary copiers are called analog, which reflect light from the entire page to be copied or digital, which send light in dot-sized increments. The advantage of digital copying is that in making many copies less energy is used; less light is used. The scanned image can also be faxed, permitting functions to be combined. The digital copier can thus be used as a fax machine, a scanner, and as a printer. Such devices are often referred to as multi-function printers or MFPs. In the first decade of the twenty-first century digital printers were rapidly displacing analog machines. Furthermore the two categories, copiers and printers, were converging into a common imaging technology.

Categories, Segments, and Pricing. In addition to the digital/analog divide, copiers are classed as personal and business copiers, monochrome and color, single- or multi-function, and are divided into segments by speed, the speed expressed as pages per minute or ppm. There are seven speed segments with the following ppm ratings:

- PC – 1 to 10 ppm
- Segment 1 – 11 to 20 ppm
- Segment 2 – 21 to 30 ppm
- Segment 3 – 31 to 40 ppm
- Segment 4 – 41 to 69 ppm
- Segment 5 – 70 to 90 ppm
- Segment 6 – 91 ppm or higher

The first segment is referred to a PC instead of by number and stands for personal copier. Resolutions associated with digital copying begin at 300 dpi and reach 2,400 dpi at the highest end delivering quality equivalent to professionally printed, glossy magazines.

Personal copiers cost around $300–500; inkjet copiers can be cheaper. Inkjets work like digital copiers but, instead of toner, tiny bubbles of ink are released for each dot in place of toner, essentially boiled to a burst. Digital copiers for business start around $1,500 per unit for 15 ppm production rates increasing in prices as speed advances. High-end copying systems will cost around $40,000, according to *Business Week*'s Buying Guide, but mass copying systems, used in printing businesses and major corporate centers, producing at rates of 100 ppm and greater and monthly volumes of 600,000 to 800,000 pages, will cost $100,000 and more. Color systems cost 20 to 30 percent more than comparable black and white printers. Special, optional features can raise the price of acquisition further. Among these are networking features and automatic document feeders (ADFs) capable, for instance, of copying bound documents without human assistance. Copying requires toner. Toner is purchased separately and, in the industry, represents a larger market than the hardware itself for copier companies with a sufficiently large product placement in the market. A reasonable ratio, based on separate reporting by corporations suggests that for every dollar spent on hardware, an average of $1.35 is spent on toner.

MARKET

Data on domestic production of copiers from the U.S. Census Bureau are available only for the year 2002. In that year domestic manufacturers shipped goods valued at $615.2 million. Data collected for the *1997 Economic Census* five years earlier suppressed all data. The Census Bureau does that when revealing shipment data would disclose the sales of one or two producers. The value in 1997 was greater by inference because the total industry category of which copiers are a part (Photographic and Photocopying Equipment) had total shipments more than four times greater in 1997 than 2002—$4.5 billion in 1997 versus $1.1 billion in 2002. As part of the domestic manufacturing economy, this sector was in sharp decline during the period 2000 to 2006 with most of domestic demand satisfied by imports. Exactly the same patterns also characterize the computer and peripherals categories.

With Census Bureau reporting covering only a tiny fraction of total activity, market research organizations have filled the vacuum by aggregating data on the industry and publishing it for stiff fees to subscribers only at costs of thousands of dollars per report. Two organizations prominently follow this industry: InfoTrends, Inc., and

FIGURE 69

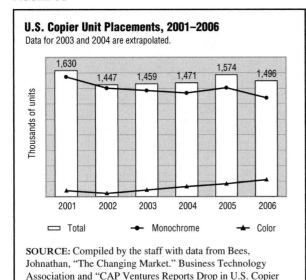

U.S. Copier Unit Placements, 2001–2006
Data for 2003 and 2004 are extrapolated.

SOURCE: Compiled by the staff with data from Bees, Johnathan, "The Changing Market." Business Technology Association and "CAP Ventures Reports Drop in U.S. Copier Market," InfoTrends, Inc. 2007 and 2003 respectively.

Lyra Research, Inc. These companies occasionally release data on unit placements but do not release dollar valuations. Dollar estimates on the market occasionally appear. Thus M2 Communications Ltd., put the market in 2003 at $24 billion, a number also echoed by *Business Week* and mentioned by Hewlett-Packard spokespeople.

The market in units from 2001 through 2006 is shown in Figure 69, the numbers shown based on unit placements reported by InfoTrends. The data show an essentially flat market during this period. From the peak in 2001 to 2006 unit placements declined at an annual rate of 1.7 percent, monochrome copiers at the rate of 3.8 percent, but color copiers showed very healthy growth of 22.4 percent per year.

Although dollar valuations are difficult to get, a general feel can be obtained by looking at user group subdivisions provided by InfoTrends for 2002. In that year personal copiers represented 12.1 percent and business copiers 87.9 percent of units. If we assume an average price of $500 per unit for personal copiers, a mid-level average of $5,000 per unit for 85 percent of the market, and a $100,000 per system expenditure for 2.75 percent of top end buyers, the calculation produces an average unit price of $7,073. This would suggest that the $24 billion market in 2003 was put together from hardware valued at $10.3 billion and toner supplies of $13.7 billion. Applying the same ratios to 2006 unit placements, the market in that year would have been $24.9 billion, $10.6 in hardware and $14.3 billion in toner. To be sure, these are very approximate numbers but provide an interesting assessment of the hardware to toner relationship. These values compare to data for computer printers, based on much better statistical information available from the government's

Current Industrial Reports showing hardware sales in 2005 of $8.7 billion and toner/ink cartridge sales of minimally $11.8 billion.

KEY PRODUCERS/MANUFACTURERS

Participants in the U.S. market are Brother, Canon, Hewlett-Packard, Hitachi, Konica Minolta, Kyocera, Muratec, NEC Corporation, Océ, Olivetti, Matsushita, Ricoh, Samsung, Sharp, Toshiba, and Xerox. Of these companies eleven are Japanese, one is Korean, one is Dutch, one Italian, and two are U.S. corporations.

Brother Industries, Ltd. This Japanese company began as a sewing machine manufacturer. Brother later added typewriters. That experience helped it enter the computer printer market. The company entered the copier market by way of laser printers and hence by the digital copying route.

Canon Inc. This company is the leading copier manufacturer in the world with U.S. market share of approximately 37 percent. The company is well known as a camera producer as well and came into the market by developing its own copier technology. Canon, in effect, invented copying for a second time by developing an alternative technology, known as New Process and abbreviated as NP. It introduced the product in 1970. Notably, the NP series did not violate any of the Xerox patents. Canon initially under-priced Xerox in the business market and then introduced desk-top copiers aimed at the small office and home market. Canon's leadership opened the market widely to competition. Canon's sales in 2006 were ¥4,157 billion, or $33.8 billion.

Hewlett-Packard (HP). This company is the world's top-ranked producer of laser printers. The strong convergence between printers and copiers led HP to announce its entry into the copier business in 2003.

Hitachi Ltd. This company is a major producer of electrical equipment, originating in 1910 as an electrical repair shop. Hitachi became an electronics company and, by this route, also began offering a wide range of electronic products, including printers and copiers made by Hitachi Koki Imaging Solutions, Inc. Hitachi also responded to the shift from analog to digital copying by establishing a major reseller support activity in the United States in 2000.

Konica Minolta Holdings, Inc. (KMH). KMH is the result of the merger of Konica and Minolta in 2003. Both companies began in the camera field. KMH subsequently sold off its camera business but remains an important participant in the printing and copying industries.

Kyocera Corporation. Kyocera is a diversified global corporation selling telephones, cameras, solar generating systems, wireless network systems—and printers and copiers. Kyocera's origins were in ceramics. The company operates under the Copystar name in the United States.

Muratec America, Inc. This company is the subsidiary of Murata Machinery, Ltd. Muratec illustrates trends in technological convergence in another way. The company began operations in the United States as a seller of fax machines and later built its comprehensive product line of copiers around this expertise.

NEC Corporation. NEC is a global electrical/electronics company with substantial U.S. based manufacturing operations. The company was originally called Nippon Electric Company but changed its name in English to NEC in 1983. The company's products extend across the entire spectrum of computing, including imagining systems.

Océ N.V. This is a Dutch company which, in the United States, is the inheritor of the copying business originally developed by Pitney Bowes. That company spun off its imaging activities in 2001 to an independent entity known as Imagistics. Océ acquired Imagistics in 2003.

Olivetti SpA. This Italian typewriter company offers a full line of printers and copiers, including multi-function devices incorporating faxing. The company is most active in the United States through Royal Consumer Information Products, Inc., the one-time Royal Typewriter Company. Royal sells an extensive line of toners for printers and copiers, with products offered for all major brands.

Matsushita Electric Industrial Company, Ltd. This company is better known in the United States as Panasonic and operates as Panasonic Corporation of North America. The company offers a complete line of office systems ranging from color scanners to complete document management systems at the high end.

Ricoh Company Ltd. Ricoh is one of the top four copier producers in the world. Like Canon, Konica, and Minolta, Ricoh began as a camera company. Ricoh introduced its first copier in the 1950s and the first digital fax machine in the early 1970s. In addition to growing its business from within, Ricoh has also expanded by acquisitions. In the United States it sells under the Lanier and Savin names. It also owns the German firm Gestetner. Ricoh's sales in 2006 were $16.8 billion. Three-quarters of Ricoh's revenues are associated with its imaging products, thus printers and copiers.

Samsung Electronics. Samsung is a global conglomerate based in South Korea. It grew from a company trading in agricultural products and foods into an electronics giant. In the United States Samsung sells mobile phones, TV sets, computers and peripherals, home appliances, semiconductors, telephone systems, and professional equipment. The company's copier lines are part of its computer/peripherals business.

Sharp Electronics Corporation. This company is ranked second overall in the copier business with market share in the United States of approximately 13 percent. Sharp's 2006 revenues were ¥2,797 billion ($24.1 billion). The company's copier activities are part of its Information Equipment segment, 15 percent of its sales, thus $3.6 billion. That segment, however, includes computers, monitors, printers, calculators, and other related product categories. Sharp's largest segment is Audio-Visual and Communications Equipment, thus television sets and DVD players, accounting for 39 percent of its sales. The company also sells electronics components.

Toshiba Corporation. Toshiba celebrated its 125th birthday in 2000; the company began as a manufacturer of electrical products and components and evolved into a global electronics company. Toshiba divides its product offerings into four groupings: color copiers and copiers for small, medium, and large work groups. The company offers products across the line in the electronics category as a whole, including laptops, printers, and hard drives.

Xerox Corporation. This pioneer in the product category is the third-ranking producer with a market share of 10 percent in the United States. From the introduction of the 914 copier in 1959 until 1970 Xerox dominated copying because its invention was well protected by 600 patents. Canon's ability to circumvent these patents in 1970 heralded a more competitive era. That competition was further stimulated by a 1975 ruling by the U.S. Federal Trade Commission requiring Xerox to freely license its technology of dry copying to others. The company thus lost market share steadily as, above all, Japanese companies entered the market. Xerox, however, remains the dominant provider of high-end systems across the world. Company sales in 2006 were $15.9 billion.

MATERIALS & SUPPLY CHAIN LOGISTICS

The flow of components which finally results in a finished copier is very much the same as the flow of components across the entire electronics industry. In effect this means that components are predominantly manufactured in Asia with some production in Europe as well. Most assembly and testing operations are also conducted outside the United States in areas with relatively low labor costs. In the latter years of the twentieth century manufacturing first began to move away from industrialized countries toward countries in which labor coast were lower.

This movement continues in the twenty-first century and much of manufacturing and assembly work in moving from the developed Asian economies like Japan's and South Korea's to China and elsewhere. The inherent characteristics of copiers, as of other electronic products, favor this logistical dispersal of functions: componentry is light in weight but high in value thus able to bear freight charges. Most products reaching U.S. customers arrive in the United States at sea ports.

DISTRIBUTION CHANNEL

Distribution of personal copiers takes place through office supply chains, mass merchandisers, electronics retailers, and computer stores. Most of the volume, however, passes through business-to-business equipment dealers. Such dealers often specialize in a single major brand of copier or carry two brands that do not directly compete for the same market. Dealers are often active in computer distribution, systems integration, and service activities as well. The dealer networks, thus, have comprehensive technical capabilities to serve larger operations effectively. They typically service equipment after installation and may sell supplies.

KEY USERS

Copiers are communications tools in that they imply distribution of the same information to many. Copiers are purchased by businesses or institutions and by individuals operating in similar contexts, thus the self-employed, the researcher, the freelance writer or analyst. Most high-end systems serve institutional or corporate bodies for internal communications. Some systems, however, are purchased by professional printing organizations to be used side-by-side with offset and other printing operations for high-volume jobs.

ADJACENT MARKETS

For those who use single-function analog copiers adjacent markets are printers, fax machines, and scanners. With multi-function copiers—more often referred to as multi-function printers (MFPs)—scanning, printing, faxing, and copying are built-in features requiring only a certain amount of employee training in their use. MFPs, to the uninitiated, may at first appear daunting. Operating such machines, however, rapidly becomes habitual, even if the user must first key in a password.

In that modern copiers have scanning capabilities, thus the ability to digitize the image of document pages, an alternative market is optical character recognition (OCR) software. A scanned image can be viewed by the user but cannot be manipulated. OCR software is designed to analyze graphic images line by line and to recognize the shapes of letters and other symbols. These are then stored sequentially as numbers as defined by the American Standard Code for Information Interchange (ASCII). In other words, the images in black and white become characters that can be edited as text.

Dealers in this field participate in a substantial services market in which networking computers, MFPs, and Internet connections is job one. Networking of computers, printing from remote stations, balancing the load between multiple MFPs, and similar systems-integration work is not considered part of the copier business but at least in part creates a demand for it.

Alternatives to the copier technology are offset and letterpress printing services. These traditional technologies, however, are actually being eroded by copiers rather than offering genuine alternatives to it.

RESEARCH & DEVELOPMENT

Major thrusts in R&D in this industry are driven by the universally-perceived transition from analog to digital copying. Opportunities in this area are offered by miniaturization of large business systems, particularly the use of laser printer. MFPs based on inkjet technology are available fairly widely at desktop size. The future of copying is multi-colored. Research is devoted to making color-lasers better by improving toners and reducing the time it takes to put down four colors instead of one, thus increasing copying speeds. Resolution is important and is still receiving development dollars with the industry foreseeing the ultimate replacement of offset printing with very high performance color copiers.

CURRENT TRENDS

Perhaps the most visible trend in this industry is the fusion of copying and printing. It is already virtually impossible to tease apart the two segments when looking at the reports of corporations engaged in selling both kinds of machines. Hewlett-Packard's entry into copiers as late as 2003 is an indicator of trends; copier producers have made the move in the other direction; as they adopted digital copying, they also entered the printer markets. In the developed world both industries are mature and are growing slowly. The ability to cluster functions such as scanning, faxing, copying, and printing gives producers opportunities to tempt its customers to upgrade.

The second major trend is from black and white to color copying. As shown earlier, color copiers are growing in the United States and elsewhere; monochrome machines are losing ground. The process is relatively slow, explained by Lyra Research, Inc., because corporations, while adding color capabilities here and there, are not willing to upgrade up-and-running systems until these must be replaced.

TARGET MARKETS & SEGMENTATION

Copier products, by their very nature, are produced beginning from design to final shipment with a particular user in mind, that user defined by his or her throughput requirements, thus maximum page production in a month or, for home users, in a minute. As already noted earlier, the industry uses a PC (personal copier) and six other segments, each defined by output per minute. Personal copiers form a distinct segment moving through conventional retail channels. Choice of machine to purchase is negotiated in the stores as buyers decide what kind of speed will suit them. In the business channel where specialized dealers represent the interface between producers and users, machine or system selection takes place in the course of surveying requirements. Marketing approaches are typically based on promoting technical features and cost efficiencies achievable by the user.

RELATED ASSOCIATIONS & ORGANIZATIONS

Association for Computing Machinery, http://www.acm.org

Business Technology Association, http://www.bta.org/i4a/pages/Index.cfm?pageID=2031

Computer & Communications Industry Association, http://www.ccianet.org

Consumer Electronics Association, http://www.ce.org

Electronic Industries Alliance, http://www.eia.org

IEEE Computer Society, http://www.computer.org/portal/site/ieeecs/index.jsp

National Electrical Manufacturers Association, http://www.nema.org

Portable Computer and Communications Association, http://www.pcca.org

BIBLIOGRAPHY

Avery, Susan. "Double-digit Cost Savings Possible with New MFPs." *Purchasing.* 15 August 2002.

Bees, Johnathan. "The Changing Market." Business Technology Associates. Available from <http://bta.org/i4a/pages/Index.cfm?pageID=2109>.

"CAP Ventures Reports Drop in U.S. Copier Market." InfoTrends, Inc. 3 June 2003. Available from <http://www.capv.com/public/Content/Press/2003/06.03.2003.b.html>.

"Color Copier Buying Guide." *Business Week.* Available from < http://businessweek.buyerzone.com/office_equipment/copiers-color/printable_bg.html.#pg8>.

"Computers and Peripheral Equipment: 2005." *Current Industrial Reports.* U.S. Department of Commerce, Bureau of the Census. January 2007.

Conner, Stephanie. "Convergence is Setting the Tone Today for Office Copier." *Boston Business Journal.* 8 June 2001.

Darnay, Arsen J. and Joyce P. Simkin. *Manufacturing & Distribution USA,* 4th ed. Thomson Gale, 2006, Volume 2, 1088–1092.

"House and Store Brands – Disrupting the Imaging Supplies Industry." InfoTrends, Inc. 13 June 2007. Available from <http://www.capv.com/home/Multiclient/housebrands.html>.

"HP Unveils Solutions to Transform Copier Market, Optimize Imaging and Printing for Businesses." Hewlett-Packard. Press Release, 18 November 2003.

Lazich, Robert S. *Market Share Reporter 2007.* Thomson Gale, 2007, Volume 2, 534–535.

"New Products." *Quick Print Products.* May 2001.

Owen, David. *Copies in Seconds.* Simon & Schuster, Inc. 2004. Reprinted in *Engineering & Science,* No. 3. 2005.

"The Story of Xerography." Xerox Corporation. 9 August 1999. Available from <http://a1851.g.akamaitech.net/f/1851/2996/24h/cacheB.xerox.com/downloads/usa/en/s/Storyofxerography.pdf&g;.

SEE ALSO *Copy & Printer Paper, Optical Goods, Printers*

Copy & Printer Paper

―――――――●―――――――

INDUSTRIAL CODES

NAICS: 32–2121 Paper (except Newsprint) Mills, 32–2233 Stationary Product Manufacturing

SIC: 2621 Pulp and Paper Mills and Manufacturers, 2678 Stationary, Tablets, and Related Products

NAICS-Based Product Codes: 32–21211 through 32–21213491 and 32–22331 through 32–2233691

PRODUCT OVERVIEW

Early surfaces used to convey words or images included clay tablets, palm leaves, snake skins, and tortoise shells. The inner bark of the papyrus plant and the skin of animals also provided a support used to proclaim the laws of the land, carry the sacred word, and declare love. Or war. The more common papyrus plant and animal parchment are not in the strictest sense true papers. But they were widely used as such.

The papyrus plant was plentiful in the marshlands of the Nile River Valley. Egyptians cleaned its bark, wove strips evenly side by side, and pressed them to produce a flat writing surface. Animal skins were plentiful in Europe. Early Europeans cleaned and stretched animal skins to create parchment. Sheepskin was preferred. Parchment paper was used in Europe for more than one thousand years.

True paper is made from pulped plant fiber mixed with water, screened, drained, and dried. The paper from pulp method is 2,000 years old. It is generally attributed to Ts'ai Lun, a private councilor in the court of Emperor Ho Ti in southern China around 110 AD. Ts'ai Lun beat hemp, cotton rags, and mulberry trees into a watery pulp. Thinly spread pulp was left on the frame to dry, resulting in a flat writing surface.

News of paper made from pulp spread to the west a thousand years later, but Europeans preferred parchment. Knowledge—and the writing used to convey it—was in the purview of the Church. Only within the Church did living conditions allow monks the time to carefully copy manuscripts using calligraphy. Around 800 AD, monks used parchment for the 680 page Book of Kells, a religious manuscript with intricate calligraphy surrounded by magnificent decoration. Parchment was deemed the only fit support to carry sacred words and images. Paper from pulp was for pagans. As a result, its use spread slowly through Europe.

Italians became the first adopters of paper from pulp. By 1250 Italians were European leaders in paper from pulp and dominated the European paper market for 100 years. The French began producing paper from pulp around 1350. Germany contributed to the spread of paper from pulp another 100 years later when, around 1450, a German named Johann Gutenberg invented the letterpress.

Letterpress printing involves a reusable cast metal alphabet that can be arranged into paragraphs, locked in place, inked, and pressed onto paper. Gutenberg's first large-scale project was, not surprisingly, religious. For two years he assembled the moveable type to print 200 copies of a 2-volume bible that sold in Frankfurt in 1455. Roughly fifty Gutenberg bibles survive in spectacular condition due to the consistently high quality of paper from pulp used.

Gutenberg is commonly accepted to be one of the inventors that revolutionized the written word and eventually the entire world. Within 50 years, 2,500 Gutenberg presses were in use in Europe. The availability of the letterpress for printing replaced the need to carefully copy manuscripts using calligraphy. Printed material became more available. Literacy increased. The printing press took knowledge—and the printing used to convey it—out of the hands of the church. It also established a direct relationship between papermakers and printing press owners. Press operators needed a consistent supply of high quality paper, and were the main purchasers of paper. Individual businesses and homes purchased relatively little paper.

For the 300 years that followed Gutenberg's 1450 revolutionary invention, the production of paper made from cotton rags boiled and pounded into pulp was an art practiced by individuals. European papermakers modified the oriental method of drying pulp on the frame. A 1568 woodcut by Jost Amman is the earliest known image that depicts the process of making paper from pulp. The woodcut shows an individual craftsman working over a large, round, wooden, waist high vat. He is pulling a screen surrounded by a wooden frame through the pulp to form one sheet of paper. The screened frame, or mould, captures the pulp and lets water drain, allows the craftsman to express the sheet off the mould, and then return to the vat to make another sheet. The woodcut also shows that early paper producers needed a paper press to squeeze water from sheets to hasten drying time, and an apprentice helper. The vat, the mould, the press, and the apprentice allowed the paper artisan to make an unlimited number of sheets from a single wooden frame.

In the early years of the 1700s, a shortage of cotton rags affected papermaking. The cotton rag shortage lasted for about 150 years. The critical shortage contributed to a view of paper as rare, expensive, and valuable. It was not a common household product to be tossed into the trash. An anonymous poem from the 1769 Boston Newsletter demonstrates the importance of cotton rags for papermaking through rhyming:

Rags are as beauties which concealed lie, but when in paper, how it charms the eye! Pray save your rags, new beauties to discover, For of paper, truly, everyone's a lover. By the pen and press such knowledge is displayed As wouldn't exist if paper were not made.

In 1850 a German solved the cotton rag shortage problem when he devised a new paper from pulp method that used a readily available renewable resource: the trees in Germany's forests. Wood was chipped, beaten with water into a pulp, screened, drained, and dried. Two years later an Englishman perfected the use of woodpulp by adding chemicals that hastened the wood pulping process.

Since then, chemists, scientists, and artists have tweaked the pulping process. Wood for paper is pulped through a combination of thermo, mechanical, and chemical processes. Mass production of paper emerged from Germany and America because trees were plentiful. The dominant definition of modern paper became paper from woodpulp. By the turn of the 1900s, modern paper from pulped wood began to overtake paper from pulped cotton.

Modern paper is made from the cellulose in wood fiber. All plants contain cellulose so any plant can be used to make paper including hemp, straw, or marsh plants like cattails. Good pulp has long cellulose fibers to facilitate interlocking. Sheets are strong because cellulose is hydrophilic. When cellulose pulp is screened and drained, the deposited layer of water-loving cellulose fibers dries into strong sheets because the water affinity is so great it pulls the fibers tightly together during the drying process. Pressing helps.

Copy and printer paper. Generically known as multi-purpose paper, copy and printer paper is used in laser and digital copiers and printers to generate words and images. Affordable copy and printing machines became widely available in the 1980s, coincidental to the invention of the computer and the Internet, which together are seen as comparable to Gutenberg's revolutionary letterpress because they also changed how knowledge was disseminated. Affordable printers let consumers—not the church and not printing press owners—create documents with the words and images of their choice. More people used more copy and printer paper in more places, including businesses and homes.

Copy and printer paper is usually white. The standard size is 8½ by 11 inches. This is referred to as letter size. Copy and printer paper is also available in 8½ by 14 inches, a size known as legal. Less popular but available is 11 by 17 inches, a size commonly called tabloid. The paper used in copy and printer equipment is a high grade product, but not considered premium paper. Paper grade is determined by weight, brightness, opacity, and texture.

Paper weight. The numbers that refer to a paper's weight are based on the weight of a 500-sheet ream of 17 by 22 inch paper. Standard copy and printer paper weight is 20 pounds, pegged to the weight of a 500-sheet ream of 17 by 22 inch paper. These sheets are converted by cutting into quarters to produce four 500-sheet reams of letter size sheets. Therefore, one 500-sheet ream of standard 20 pound paper weighs five pounds.

Brightness rating. The numbers used to rate the brightness of paper range from 80 to 113. Higher numbers represent brighter paper. Brightness is a measurement of how well light is reflected from the paper. High brightness is valued because it gives good contrast between the ink and paper. Paper mills control whiteness by adding minerals and dyes. Newspaper appears yellow next to the bright white of copy paper.

Opacity. Opacity is a measure of transparency. High opacity paper is not easily penetrated by light. Low opacity paper allows light to pass through, making it translucent. Opacity is influenced by weight and brightness. Standard 20 pound copy and printer paper is not translucent, but it is also not good for two-sided printing because light passes through, letting ink show from the other side.

Texture. The texture of paper is a factor of fineness and coarseness. Standard 20 pound copy and printer paper is valued for its smooth surface. Laser copiers and printers require smooth paper because they use heat and toner to print words and images. Digital printers jet water-based ink off the print head and onto the paper. They benefit from textured paper that facilitates drying and reduces bleeding.

Paper is available in a wide range of weights, brightness, opacity, and textures. Sales of the equipment that uses copy and printing paper are increasing. Predictions are for paper sales to continue to increase.

MARKET

Put the call for a paperless society on hold. Even though in the early 1990s, visionaries promised a paperless society due to the availability of the Internet and electronic communication, North Americans still churn out more than 1.2 trillion sheets of copy and printer paper every year. According to *Consumer Reports*, the availability of affordable printers explains why Americans now use twice as much copy and printer paper as they did in 1985. According to a 2005 article in *Geographical*, at the height of the dotcom boom between 1995 and 2000, the world's largest manufacturer of copy and printing paper (Canada) almost doubled its exports. *Geographical* estimates that e-mail caused copy and printer paper use to increase by 40 percent during that era.

At the paper mill and in the paper industry, copy and printer paper is called uncoated freesheet. Uncoated freesheet product shipments were $8.1 billion in 2002, part of the $41.2 billion per year paper industry, according to a U.S. Census Bureau report titled "Paper, Except Newsprint Mills, Manufacturing: 2002." This $8.1 billion is consumed by commercial printers, large and small businesses, and individual consumers with home offices. Some estimate that approximately $4.0 billion reaches the small business and home office market.

Uncoated freesheet is shipped from the mill in three major categories. In order of market size they are: (1) bond freesheet, (2) other writing paper, and (3) cover and text papers. Each represents approximately one-third of the uncoated freesheet industry as can be seen in Figure 70.

Bond freesheet had a product shipment value of $2.9 billion in 2002, down from $3.5 billion in 1997 or a decline of 16 percent. Bond is a term historically used to refer to 100 percent cotton content paper that has been generalized to mean premium paper. Bond freesheet is shipped from the mill in two formats: bond writing paper or form bond paper in rolls. Approximately three-quarters is bond writing paper, and one-quarter is form bond paper in rolls. Both showed decline between 1997 and 2002. The decline in form bond paper in rolls was notable. That class was down 35 percent. Fewer preprinted forms are made from premium quality bond paper. Forms are more often made available on the Internet, in an electronic format ready to be printed by the end user.

The category "other writing paper" represented just over one-third of the market, 34.4 percent in 2002. Since 1997, mills have been making more of this category of paper and it is growing in importance to the industry. In 1997, other writing paper accounted for 18 percent of freesheet product shipments; by 2002 it had grown to represent 30 percent. Other writing paper had a product ship-

FIGURE 70

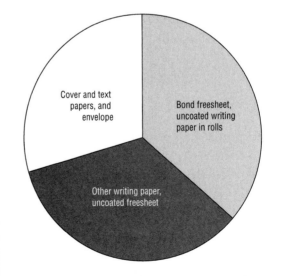

U.S. Copy Paper Market in 2002 by Type of Paper

Total manufacturer product shipments of these three types of paper in 2002 totaled $8.1 billion. The total U.S. paper manufacturing industry had industry shipments of $142.7 billion in the same year.

Cover and text papers, and envelope

Bond freesheet, uncoated writing paper in rolls

Other writing paper, uncoated freesheet

SOURCE: Compiled by the staff with data from "Paper, Except Newsprint Mills, Manufacturing: 2002," *2002 Economic Census*, U.S. Department of Commerce, Bureau of the Census. January 2005

ment value of $2.8 billion in 2002, up from $2.1 billion in 1997, an increase of 32 percent. Other writing paper is shipped in four formats. The major format is paper for communication or copying, comprising approximately 90 percent of shipments. Paper for communication or copying had a product shipment value of $2.1 billion in 2002, up from $788 million in 1997, an increase of 156 percent. The other three classes are: (1) technical and reproduction paper, (2) writing tablets, and (3) ledger, onionskin, and wedding paper.

Cover and text paper is the category that makes up the final third of this sector of the market. Production of this category is declining. It includes envelope paper and stock for coating. Cover and text paper had a product shipment value of $2.4 billion in 2002, down from $3.3 billion in 1997, a decline of 28 percent.

"I have seen the future and it is 8½ by 11," Chris Harrold of Mohawk Paper Mills, Inc., in Cohoes, New York, joked with colleagues. For 500 years, paper makers focused on large printers who fueled the demand for paper because, like Gutenberg, they had the printing presses upon which to use it. Harrold's comment was a reference to the need for the paper industry to forge a closer relationship with small businesses and individual consumers with home offices, a market that had for 500 years purchased relatively little paper. Demand is beginning to be driven by a different type of user in the computer age—the user of small, affordable copiers and printers, as evidenced by the 156 percent increase in paper for communication or copying, an uncoated freesheet category characterized primarily by reams of 8½ by 11 inch letter size sheets.

Only one uncoated freesheet product category is growing: paper for communication or copying. It grew by 30 percent annually between 1997 and 2002. According to Conservatree, a nonprofit organization within the paper and environmental fields, copy and printer papers have higher profit margins than other types of papers. According to *Print Week*, copy and printer papers are the only segment of the industry that has seen double-digit growth in recent years. The driving force behind this growth in freesheet paper shipments is laser and digital copying and printing equipment used by small businesses and individual consumers in small home offices. This high margin, high growth category is predicted to continue to expand.

According to the July 2006 edition of *Pulp & Paper*, the trend is toward more home use of copy paper. *Purchasing Magazine* reported in around the same time that paper mills in the U.S. had instituted five price increases on a quarter-to-quarter basis for copy and printing paper in 2005 and 2006. Producers of uncoated freesheet paper pushed for higher prices to offset their rising costs for fibers, energy, and chemicals.

After slowing in 2000 through 2003, the industry as a whole began to see recovery in 2004. That year, the Pulp & Paper Products Council in Canada reported that copy and printing paper shipments in North America, by weight, rose 4.7 percent to 28.1 million tons.

Copy and printing paper products are classic nondurable consumer goods. Nondurable goods are purchased for immediate or almost immediate consumption and have a life span ranging from minutes to three years. Nondurable goods are destroyed by their use so consumers need to repeatedly replenish their supply throughout the year. For instance, the average U.S. office worker uses 10,000 sheets of copy and printer paper per year. Since paper is sold in reams of 500 sheets, that equates to 20 reams per year. One ream of International Paper's Hammermill brand Copy Plus 20 pound paper is 2 inches thick. In one tall stack, 10,000 sheets of standard copy paper weighs 100 pounds.

The nondurable consumer goods market is characterized by a large variety of affordable products to tempt consumers. For example, one ream of Domtar Copy 20 pound paper is $2.99. Because of the combination of high margins and double-digit growth, copy and printer paper products are available in a wide range of weights, brightness, opacity, and textures. The explosion of offerings includes specialty laser paper, inkjet paper, and photo paper. The combination of a large variety of affordable products results in price competition among manufacturers to gain market share. Brand loyalty for copy and printer paper is seen as generally low because consumers view it as an undifferentiated commodity and buy on price.

KEY PRODUCERS/MANUFACTURERS

The top North American manufacturers of copy and printer paper control approximately 70 percent of the market. According to *Market Share Reporter 2007,* the top three uncoated freesheet manufacturers in North America are Domtar Corporation (34%), International Paper (25%), and Boise Cascade, LLC (10%).

Boise Cascade, LLC. With its paper headquarters in Boise, Idaho, Boise Cascade, LLC is the third largest manufacturer of uncoated freesheet paper in North America. It employs close to 10,000 workers and owns four paper mills located in International Falls, Minnesota; St. Helens, Oregon; Wallula, Washington; and Jackson, Alabama. A recycling plant in Jackson, Alabama, de-inks 300 tons per day of used copy and printer paper for use in recycled-content papers. Boise Cascade, LLC, was formed in 2004 when Madison Dearborn Partners purchased the paper and wood products (and building materials distribution) businesses from Boise Cascade Corporation, which is now known as OfficeMax, Inc. Since the 2004 change of ownership, one key strategy of Boise Cascade, LLC has been to focus on copy and printer paper used in businesses and homes. This reflects a new orientation toward the consumer and away from commercial printers. In 2007, for the fourth consecutive year, Boise Cascade will conduct online voting to allow business and home consumers to determine the design for its limited edition holiday themed carton for its Boise X-9 multi-use copy paper.

Boise-branded copy and printer papers are two: X-9 (available in 16, 20, and 24 pound weights, 92 brightness, letter, legal, tabloid sizes) and X-9 PLUS (20 pound weight, 96 brightness, letter, legal, tabloid sizes, designed as a high-end paper for offices with multiple machines). Boise-branded laser paper lines are called Everyday Laser, Presentation Laser, Enhanced Color Copier Paper, Glossy Color Laser, and Cover. Boise-branded inkjet paper is called Presentation Inkjet. Boise branded recycled copy and printer paper lines are made with a minimum 30 percent post-consumer fiber and have names evocative of nature: Aspen, Aspen 50, and Aspen 100. Colored papers are sold in a variety of weights under Boise-branded lines MP Colors and MP Brites.

Domtar Corporation. Headquartered in Montreal, Canada, with its operational headquarters in Fort Mill, South Carolina the mammoth Domtar Corporation employs nearly 14,000 people in North America. It is the largest papermaker in North America, controlling one-third of the U.S. production of uncoated freesheet paper. Since its acquisition of Weyerhaeuser Company in March 2007, it is the second largest uncoated freesheet producer in the world. The newly merged Domtar operates 16 uncoated freesheet mills in North America with a total annual capacity of 5.2 million tons of paper. Paper comprises most of Domtar's sales (77%). The rest are from its distribution unit (13%) and lumber (9%) from five sawmills. It focuses on designers, commercial printers, publishers, and converters. EarthChoice is its newest brand consisting mostly of premium papers. In an attempt at greenwashing (appearing environmentally-friendly while doing just the opposite), Domtar re-labeled old standards Nekoosa Bond, Nekoosa Linen, and Nekoosa 25% Cotton under the EarthChoice umbrella and introduced EarthChoice Office Paper, a Forest Stewardship Council-certified multi-function paper. The multi-platinum recording group Barenaked Ladies released their 2006 album with CD covers printed on Domtar Earth Choice paper. Domtar also manufacturers copy and printer papers for the home and the office.

Papers for the home include the Domtar-branded lines Recycled Copy, Copy, Hots, Laser, Pastel, Premium Inkjet, and Premium Laser. Paper products for the office include the Domtar-branded lines Copy, Colors, Microprint Coated Laser, Microprint Color Copy, Microprint Digital Publishing Text, Microprint Ink Jet, Microprint Laser, and Multi-System Ultra (designed for all office machines).

International Paper. Headquartered in Stamford, Connecticut, International Paper is the second largest manufacturer of uncoated, freesheet paper in North America. It is a global paper company with manufacturing operations in the United States, Europe, Latin America, and Asia. With annual sales in the $22 billion range, the company employs approximately 60,000 people worldwide. In the United States, International Paper operates 18 mills. International Paper's products are focused on professional printers and converters as well as the home and office users. In addition, International Paper operates an extensive private label program. International Paper's Hammermill brand paper products were featured on "The Office," a television comedy that takes a painfully funny look at the "Dunder Mifflin" paper supply company in Scranton, Pennsylvania. In a September 2006 episode of the Emmy-winning comedy series, Hammermill was featured when the starring character, Michael (Steve Carell) attended a paper industry trade show.

International Paper makes Hammermill brand paper designed for business and home use in product categories including copy paper and multi-purpose paper. It sells specialty inkjet and laser printer paper. It also has color copier paper, cover paper, and card stock. International Paper is detailed in the Current Trends section of this essay by way of demonstrating the explosion of offerings in this high-margin, high-growth category of uncoated freesheet paper.

MATERIALS & SUPPLY CHAIN LOGISTICS

The Census Bureau reported that in 2002 the U.S. paper manufacturing industry used $15.1 billion worth of materials to produce $41.2 billion worth of products. While the Census Bureau reports on product shipments by category, it does not break down the value of material inputs buy these same product categories and therefore, one is left with only industry-wide figures for materials consumed. The value of all materials used in manufacturing paper in the United States declined by 14 percent between 1997 and 2002, from $17.6 billion to $15.1 billion. Industry-wide shipments declined as well, but by a smaller 10.9 percent. The majority of the materials used to produce paper are organic and inorganic chemicals; woodpulp purchased from paper mills; and pulpwood bolts and logs.

Copy and printer paper products require the same inputs as all other paper. To make the light, white, and bright sheets of paper needed for copy and printer paper products, pure white pulp is needed. Higher grade pulp is commonly made by a combination of thermal, mechanical, and chemical pulping processes. Thermal steam is used to soften wood particles before mechanically pulping them. For high-grade paper, the wood particles are chemically treated before entering the pulper. As a result, the pulp is very white and bright, properties suited to copy and printer paper manufacture.

Industry-wide spending for organic and inorganic chemicals decreased 20 percent between 1997 and 2002, from $3.9 billion to $3.1 billion. The industry decreased its spending in every chemical class purchased, save for chalk. Chalk is an inorganic chemical also known as calcium carbonate used in papermaking as a filler and a pigment. Spending for chalk increased 6 percent over the period from $262 million to $278 million. Because opacity is valued, calcium carbonate is added to pulp to increase opacity by reducing transparency and thus the amount that ink can be seen through the paper.

The top three chemicals purchased, in order of 2002 expenditures, are starch, clay, and chalk, inorganic chemicals used primarily to improve paper strength. Starch is used at the wet end of the papermaking process to improve strength and at the dry end as a sizing. Clay is added during the pulping stage to fill in pores of wood fiber to create strong paper. Among all chemicals needed, the industry decreased spending the most for chlorine, caustic soda, and titanium dioxide.

Chlorine purchases decreased 73 percent from $82 million in 1997 to $23 million in 2002. Chlorine is used primarily to make pulp white. Caustic soda purchases decreased 54 percent from $248 million in 1997 to $113 million in 2002. Caustic soda is known by scientists as sodium hydroxide and by laypeople as lye; it is used to hasten pulp cooking and to make pulp white by removing impurities known as lignin. Titanium dioxide is a lustrous, lightweight, pure white pigment used to boost whiteness. Industry-wide purchases of titanium dioxide decreased 39 percent from $366 million in 1997 to $200 million in 2002.

All pulp—whether thermo, mechanical, or chemical—is yellowish because wood fiber is yellow. Color is controlled by adding dye to correct the yellowish tint. For example, violet dye absorbs green and red reflected light, reducing yellowness. Because whiteness is highly valued, paper mills also use fluorescent whitening agents. Also known as optical brightening agents, they absorb ultraviolet light and reradiate it as blue light. This explains why the high-grade paper appears bluish. For example, Boise

Cascade, LLC, emphasizes that its Boise branded X-9 and X-9 PLUS feature an enhanced blue-white shade.

Industry-wide spending for purchased woodpulp fell by almost one-third between 1997 and 2002, from $4.1 billion to $2.8 billion. Manufacturers purchase woodpulp from two sources. The largest expenditure is for woodpulp obtained at the market rate from other paper mills. The other source is woodpulp produced at affiliated mills. Expenditures for woodpulp purchased at the market rate decreased 50 percent, from $3.3 billion in 1997 to $1.7 billion by 2002. Expenditures for woodpulp produced at affiliated mills increased 30 percent from $890 million to $1.1 billion. This was, in part, the result of industry consolidation as well as the movement of manufacturing off shore.

Industry-wide spending for pulpwood bolts and logs decreased 11 percent between 1997 and 2002, from $3.0 billion to $2.6 billion. Pulpwood bolts and logs are generally classified as either softwood or hardwood. Softwoods are integral in the making of high-grade paper products. In order of expenditure value, softwoods are classified as southern pine; chips, slabs, cores, and other mill residues; softwoods such as Douglas Fir and Jack Pine; spruce and true fir; and hemlock. The expenditure for softwoods was $1.5 billion in 2002. Spending in all softwood categories decreased, save for one. Douglas Fir and Jack Pine spending more than doubled, from $123 million to $259 million, an increase of 111 percent. Softwoods are known by laypeople as conifers. Conifers like spruce, fir, pine, balsam, and hemlock are preferred because they are fast growing, plentiful, easy to grind into pulp, and produce long fibers. Long fibers are favored for chemical pulp that needs to be bleached to high brightness and whiteness, qualities especially valued in the production of copy and printer paper products.

Hardwood pulpwood bolts and logs, along with chips, slabs, cores, and other mill residues, are also consumed. Industry-wide spending for hardwood products decreased 20 percent between 1997 and 2002, from $1.5 billion to $1.2 billion.

Paper mills use packaging materials as another raw material, after chemicals, woodpulp, and pulpwood bolts and logs. Packaging materials are used to prepare packages and cartons for letter, legal, and tabloid size reams of copy and printer paper for shipment through the distribution channel. Industry-wide spending for all packing material was $1.2 billion in 2002. The types of packaging products used include paperboard containers, boxes, and corrugated paperboard; packaging paper and plastics film; and glues and adhesives.

DISTRIBUTION CHANNEL

The distribution channel for copy and printer paper is characterized by vertical integration. The top three North American manufacturers of uncoated freesheet are each vertically integrated. Boise Cascade, LLC has 30 distribution locations in North America. Domtar has more than 80 paper distribution facilities in North America. International Paper owns xpedx, which has more than 250 distribution branches located primarily in the United States. The distribution channel is changing, with less emphasis on large commercial printers—which had been the focus of the industry for more than 500 years—and a new focus on individual businesses and homes.

As a result, the distribution channel for copy and printer paper is broadening. It is becoming characterized by a large number of smaller buyers. *Multichannel Merchant* reported in March 2007 that office supply catalogs are no longer sent only to large businesses, but also to small businesses and homes. The availability of affordable printers created a significant increase in the number of active buyers. According to New York-based list brokerage services firm ParadyszMatera, the names of 8.9 million 12-month buyers from office supply merchants were available for rent or exchange during the fourth quarter of 2006, up nearly 18 percent from the fourth quarter of 2005. The available active buyers list is only a portion of potential buyers within the distribution channel for copy and printer paper. For instance, the largest office supply merchant, Staples, does not sell its client lists.

Staples tops the distribution channel for copy and printer paper, which is dominated by office products superstores. *Office Products International* reported in 2005 that the vast majority of small and medium-sized businesses buy their copy and printer paper from the top three office products superstores, in this order: Staples, Office Depot, and OfficeMax. In a study conducted by Lyra Research among paper buyers from businesses of up to 99 employees, 75 percent of respondents purchased paper from the top three. Other distribution channel outlets included the mass market retail/warehouse club channel (Costco, Sam's Club, Wal-Mart), Boise Office Solutions, xpedx, and Corporate Express.

The distribution channel for copy and printer paper is characterized by purchase incentives. Office supply companies typically offer either free shipping or a gift with purchase. Both were equally popular, with each offered by 36 percent of catalogs received by ParadyszMatera. A typical incentive is "buy two cases of copy paper, get the third free." From 2004 to 2006, 47 percent of office supply catalogs received by ParadyszMatera offered a gift with purchase as an incentive to buy. Common types of gifts offered were bags (such as Staples' business attaché with

laptop holder) and electronics (such as Amsterdam Printing & Litho's MP3 player).

KEY USERS

Key users of large amounts of copy and printer paper tend to be businesses such as law firms, health care providers, manufacturers, accounting firms, civic/nonprofit groups, retailers/wholesalers, marketing/advertising agencies, and architects. Office workers use more than non-office workers. Key users of lesser amounts of copy and printer paper are individual consumers who have inkjet or laser printers.

Schoolchildren are another audience for copy and printer paper. Copy and printer paper is used in preschool as a means to communicate with parents. Teachers continue to funnel paper to schoolchildren with assignments and notes to parents. Many schoolchildren go to college. During college, they have both the time and the need to search the Internet. Printed material drives traffic to cyberspace and, once there, the information available drives people to print. This reinforces what will be for many a lifelong connection between fiber space and cyberspace.

ADJACENT MARKETS

All products made with wood are a potential adjacent market to the market for copy paper as they are alternative uses for the raw material needed to produce paper. As the demand for wood used in construction increases, the pressure on the commodity price is increased. This, in turn, influences the raw materials cost for paper mills and pushes the price of paper up.

Printers, copiers, and the inks and toners required to operate them are all adjacent markets to the market for copy paper. The toner used in most printers and copiers is stored in disposable cartridges that need to be replenished frequently. The toner business is an fact a lucrative one and some low-end printers are sold at a break-even point, or even at a loss, in order to generate a stream of ongoing toner replacement sales for that machine into the future. Dealing with the increasing quantity of printed paper being generated by printers and copiers in homes and businesses leads to another market adjacent to copy and printer paper, the market for paper shredders. Because users of copy and printer paper may worry about privacy and the increased threat of identify theft, the paper shredder is becoming a device seen commonly in offices of all sizes, from small home offices to large corporate centers.

Successful people appear to use less copy and printer paper because they more efficiently use advanced digital technology. Bill Gates explained, in an article he wrote in 2006 for *Fortune*, that he does not use much copy and printer paper in his office. "On my desk I have three screens, synchronized to form a single desktop. I can drag items from one screen to the next. Once you have that large display area, you'll never go back, because it has a direct impact on productivity. The screen on the left has my list of e-mails. On the center screen is usually the specific e-mail I'm reading and responding to. And my browser is on the right-hand screen. This setup gives me the ability to glance and see what new has come in while I'm working on something, and to bring up a link that's related to an e-mail and look at it while the e-mail is still in front of me."

RESEARCH & DEVELOPMENT

Paper scientists continually tweak the blends of chemicals, woodpulp, pulpwood bolts and logs, and machine specifications to perfect office paper products. Researchers study the effects of different combinations of paper weight, brightness, opacity, and texture.

R&D within the industry resulted in erasable paper. Scientists at Xerox's Research Centre of Canada invented a way to print temporary documents. Images lasted for one day. The technology could lead to a significant reduction in paper use. Researchers developed compounds (similar to inks) that change color upon absorption of a certain wavelength of light and gradually fade to nothing in 16 to 24 hours. The paper can be reused.

R&D efforts also resulted in antibacterial paper. Domtar introduced such paper in January 2007. New York-based Gould Paper is the first wholesaler to market this Domtar product. Treated with a silver compound, the paper also guards against odors and the growth of fungus, mold, and mildew. Lab tests demonstrated a 99 percent reduction of methicillin resistant Staphylococcus aureus, a type of staph infection resistant to certain antibiotics. Domtar is targeting health care providers and government offices as the natural customer base for these antibacterial paper products.

CURRENT TRENDS

The combination of high margins and double-digit growth rates have stimulated the spread of copy and printer paper products available in a wide range of weights, brightness, opacity, and textures. The explosion of offerings includes specialty laser paper, inkjet paper, and photo paper.

In the coming years, even more combinations of paper weight, brightness, opacity, and texture will be available. One trend that is seen now and expected to continue is the use of color in printing. Papers designed to take color require greater weight with higher brightness ratings so colors will really pop, and higher opacity so inks from the other side will not show through. Notable is the wide range of specially-designed copy and printing paper; it is now available in gloss levels in the range glossy, semi-gloss, semi-matte, and matte. Glossy paper is commonly used for printing photographs. The demand for paper suitable for color is on the rise.

International Paper's Hammermill brand paper products are a good example of the explosion of copy and printer paper offerings. Besides specialty laser paper, inkjet paper, and photo paper, Hammermill-branded copy paper includes Tidal MP (20 pound weight, 92 brightness, value-priced, multi-purpose paper in letter and tabloid size) and Copy Plus (20 pound weight, 92 brightness, workhouse paper in letter, legal, and tabloid size).

International Paper's Hammermill-branded multi-purpose paper includes Fore MP (20 and 24 pound weights, 96 brightness, letter, legal, and tabloid sizes as well as 3-hole punched, designed for newsletters, reports, manuals), Multipurpose (20 pound weight, 96 brightness, letter size only, designed to run efficiently through all office machines), Premium Multipurpose (24 pound weight, 96 brightness, letter size only, a heavy paper), and OfficeOne Business Gloss (32 pound weight, 96 brightness, letter size only, coated on both sides for laser and digital images).

International Paper's specialty copy and printer paper includes Hammermill Inkjet (24 pound weight, 96 brightness, letter size only, uncoated but with a special surface to enhance ink absorption and maximize drying), Hammermill Laser (24 pound weight, 96 brightness, letter size only, smooth surface designed for heat of laser process), and Hammermill Laser Print (24, 28, and 32 pound weight, 96 brightness, letter, legal, and tabloid sizes and 3-hole punch, a premium paper designed for both monochrome and color printing), Hammermill Color Laser Paper (26 pound weight, 98 brightness, letter size only, designed for 2-sided color printing), and Hammermill Color Laser Gloss Paper (32 pound weight, 90 brightness, letter size only, coated on both sides for 2-sided color printing).

International Paper also makes Hammermill-branded Color Copy Paper (28 and 32 pound weights, 98 brightness, letter, legal, tabloid, ledger size), Color Copy Cover (60, 80, and 100 pound weights in letter, tabloid and 18 x 12 inches), and Color Copy Gloss (32 pound weight, 90 brightness, in letter tabloid and 18 x 12 inches, designed for high speed color laser printers, not suitable for inkjet printers).

Other products include Hammermill-branded color Copy Gloss Cover (80 pound weight), Cover (67 pound), and Card stock (110 pound). International Paper also makes Hammermill Pastels in 20 pound weight, with a 30 percent recycled content, available in pastels such as blue, canary, cream, golden rod, gray, green, lilac, pink, salmon, and turquoise. Hammermill Fore MP also comes in 16 pastel colors (20 and 24 pound weight, 30 percent recycled, in letter, legal, and ledger size). Available colors are blue, buff, canary, cherry, cream, goldenrod, gray, green, ivory, orchid, lilac, peach, pink, salmon, tan, and turquoise.

TARGET MARKETS & SEGMENTATION

One important and influential target market for paper makers are graphic designers. Printed material has to compete with all the other media, including electronic media. Graphic designers influence decisions regarding the purchase of paper as well as other materials. They can encourage the use of higher-quality premium paper at heavier weights and more unique textures.

Paper makers also target those who are concerned about environmental issues. Environmental ethics remain important, since paper is a perennial problem for environmentalists. Some mills have become certified to ISO 4001 production standards, which take into account energy consumption and emissions, as well as recycled properties. Most mills provide products that sport a green certification. Four different certification levels are offered by the Forest Stewardship Council (FSC). The basic certification is FSC. It means the FSC-certified fiber content in the paper, even though virgin, comes from operations that comply with the Forest Stewardship Council's sustainable forestry practices.

The FSC Pure Material seal denotes paper made with 100 percent FSC material from an FSC-certified forest. The paper product with the Pure Material seal has been sold and/or processed by an FSC chain-of-custody certified company. The FSC Mixed Pulp seal denotes paper made from any combination of FSC-certified forests, company-controlled sources, and/or recycled material. Company-controlled sources are controlled in accordance to FSC standards. They exclude illegally harvested timber, wood from forests where high conservation values are threatened, and/or practices that violate civil and traditional rights, or wood from areas that have been converted from natural forest to plantations. The FSC Recycled seal denotes products that contain 100 percent post-consumer waste material.

RELATED ASSOCIATIONS & ORGANIZATIONS

American Forest and Paper Association, http://www.afandpa.org
Conservatree, http://www.conservatree.com
National Paper Trade Association Alliance, http://www.gonpta.com
National Resources Defense Council, http://www.nrdc.org
Wisconsin Paper Council, http://www.wipapercouncil.org

BIBLIOGRAPHY

"Better Shred than Read." *OfficeSolutions.* May-June 2006, 14.

"Boise Holds Carton Design Contest." *Official Board Markets.* 5 August 2006, 6.

Chiger, Sherry. "Office-supplies Buyers? We Got That." *Multichannel Merchant.* 1 March 2007.

"The ePaper Revolution." *Geographical.* August 2005, 36.

Gates, Bill. "How I Work." *Fortune.* 17 April 2006, 45.

Genn, Adina. "Great Neck-based SilverCo Creates Anti-disease Paper, Targets Germs." *Long Island Business News.* 26 January 2007.

"Hammermill Paper Featured on 'The Office'." *American Printer.* 6 October 2006.

Lazich, Robert S. *Market Share Reporter 2007.* Thomson Gale, 2007, Volume 1, 218.

Main, John. "Less Office, More Home Use (of Copy Papers)." *Pulp & Paper.* July 2006, 28.

"OP Superstores Grab Paper Sales." *Office Products International.* May 2005, 23.

"Paper, Except Newsprint Mills, Manufacturing: 2002." *2002 Economic Census.* U.S. Department of Commerce, Bureau of the Census. Available from <http://www.census.gov/econ/census02/>.

"Pulp Nonfiction (Printer and Copier Papers)." *Consumer Reports.* October 2003, 6.

Rooks, Alan. "Printing and Writing Papers: Half Full or Half Empty Glass?" *Solutions—for People, Processes and Paper.* April 2005, 42.

Roberts, Lauretta. "Green Impressions." *Design Week.* 12 October 2006, 19.

Schwarcz, Joe. "Writing the Book—On Paper." *Canadian Chemical News.* April 2004, 8.

Stundza, Tom. "High Demand, Tight Supply will Boost Transaction Prices." *Purchasing.* 15 June 2006, 42.

"Technical Tutorial: The Definition of 'White' Paper." *Print Week.* 6 July 2006, 46.

"Xerox Develops Erasable Paper." *Print Week.* 7 December 2006, 8.

SEE ALSO *Copiers, Premium Paper, Printers, Recycled Paper*

Cosmeceuticals

———◼———

INDUSTRIAL CODES

NAICS: 32–5620 Toilet Preparation Manufacturing

SIC: 2844 Perfumes, Cosmetics, and Other Toilet Preparations Manufacturing

NAICS-Based Product Codes: 32–56207281, 32–56207291, 32–5620D121, 32–5620G1, 32–5620G111, 32–5620G121, 32–5620G131, 32–5620G3 through 32–5620G331, 32–5620G351, and 32–5620G3B1

PRODUCT OVERVIEW

A new class of products blurs the line between skin care and medicine. One element that unites this new class is that its products contain active ingredients that manufacturers claim perform some special anti-aging function upon the skin. The American Academy of Dermatology (AAD) calls these products *cosmeceuticals*, because they are formulated to be more than mere color cosmetics. Cosmeceuticals, the AAD proclaims, improve skin functioning and prevent premature aging. The most common active ingredient is an alpha hydroxy acid (AHA) known as glycolic acid. Its active function is to increase the rate of skin exfoliation; quicker cell turnover is said to make skin smoother and softer, give it a more even tone, and reduce age spots. Another common active ingredient is a beta hydroxy acid (BHA) known as salicylic acid.

The term cosmeceutical is apt because it reflects the blurred line between products classed as cosmetics (nontherapeutic) and products classed as drugs (therapeutic).

The word implies that a product sold as a nontherapeutic cosmetic has some therapeutic or drug-like benefit.

The U.S. Food and Drug Administration, however, does not recognize the term cosmeceutical. The FDA oversees implementation of the Food, Drug, and Cosmetic Act (Act). That Act defines drugs as products with therapeutic benefits that are subject to FDA review and approval, and defines cosmetics as articles other than soap applied for cleansing, beautifying, promoting attractiveness, or altering the appearance. According to the FDA Center for Food Safety and Applied Nutrition, "a product can be a drug, a cosmetic, or a combination of both, but the term cosmeceutical has no meaning under the law." Examples of products that the FDA recognizes as being a combination of both—and thus subject to FDA oversight—are sunscreens, fluoride toothpastes, hormone creams, anti-perspirants, and anti-dandruff shampoos.

The $28 billion per year industry that makes and markets products for cleansing, beautifying, promoting attractiveness, or altering the appearance is commonly known as the cosmetics and toiletries industry. Cosmetics and toiletries—with the exception of those few listed above—are not subject to FDA approval. Their ingredients are not FDA approved. Manufacturers may use any ingredient or raw material (except for color additives and a few prohibited substances) to make products without government review or approval.

Companies are not required to substantiate anti-aging performance claims or conduct anti-aging safety testing. This lack of governmental oversight resulted in this new class of products being characterized as cosmePSEUDO-cals in a 2005 Harvard publication. Labeling regulations, however, do apply. The Fair Packaging and Labeling Act requires an ingredient declaration on all cosmetics and toiletries products, whether they are anti-aging or not. Ingredients must be listed in order of quantity.

Whatever they are called, this new class of products is considered to consist of classic nondurable consumer goods. Nondurable goods are purchased for immediate or almost immediate consumption and have a life span ranging from minutes to three years. Nondurable goods are destroyed by their use so consumers need to repeatedly replenish their supply throughout the year. Generally this equates to a large variety of products to tempt consumers. Nondurable goods are often products that are used every day, such as cosmetics, hair care including shampoo and conditioner, and creams and lotions.

This new class of products is embodied by the entire L'Oréal Paris Age Perfect line. Its 2007 ads featured actress Diane Keaton in print and TV supporting the line of face cosmetics and face creams. *Procalcium* is the name of the line's anti-sagging and anti-fragility moisturizer. It re-densifies and hydrates skin. In 2004 and 2005 L'Oréal spent between $10 to $20 million on U.S. advertising for its Age Perfect product line, according to ACNielsen Monitor-Plus. Money is spent because the market for these products is growing and profitable, whatever they are called, or however they are classified.

MARKET

While beauty may be subjectively held in the eye of the beholder, the entire $28 billion per year U.S. cosmetics and toiletries market is objectively growing. *Cosmetics International* reported in January 2007 that demand for anti-aging products will fuel new product introductions and reformulations of existing products through 2010. The market research company Packaged Facts predicted double-digit growth of this new class of products with active ingredients that make anti-aging claims. According to the U.S. Census Bureau report "Toilet Preparation Manufacturing: 2002," the three major cosmetics and toiletries product divisions, from largest to smallest, are: cosmetics; hair care products; and creams and lotions, as depicted in Figure 71.

The Census Bureau classification system highlights growth of product classes within each of the major divisions that are potentially part of this new class of cosmeceuticals. However, defining the size of the anti-aging market is complicated. Each of the three major divisions depicted in the pie chart had product classes that experienced double-digit, triple-digit, and even quadruple-digit growth due in part to products that have been reformulated to incorporate active ingredients their makers claim fight aging.

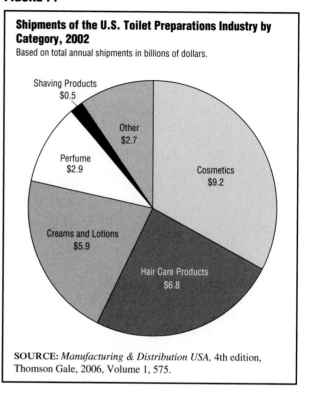

FIGURE 71

Shipments of the U.S. Toilet Preparations Industry by Category, 2002

Based on total annual shipments in billions of dollars.

Shaving Products $0.5
Other $2.7
Perfume $2.9
Cosmetics $9.2
Creams and Lotions $5.9
Hair Care Products $6.8

SOURCE: *Manufacturing & Distribution USA*, 4th edition, Thomson Gale, 2006, Volume 1, 575.

Anti-aging products initially introduced were designed to target wrinkles and were primarily in the creams and lotions sector. Anti-aging products quickly extended to cosmetics and hair care, notable expansions to even larger industry sectors with more opportunity for growth and profits.

Successful advertising contributed to this trans-industry expansion. For instance, Olay (made by Procter & Gamble) used the seven signs of aging concept to support the introduction of new products designed to target more than just wrinkles. The seven signs of aging are: fine lines and wrinkles, rough skin texture, uneven skin tone, skin dullness, visible pores, blotches and age spots, and skin dryness. Thereafter, Olay's advertising instructed consumers that aging also causes skin dullness, discoloration, and brown spots—not just wrinkles. Products with active ingredients said to help brighten those areas can be purchased. Aging also causes hair dullness, fading, and thinning. Not surprisingly, in January 2007 newly reformulated hair care products with active ingredients said to address those problems became widely available.

Because active ingredients are added to many products across all three major divisions of the cosmetics and toiletries industry in order to create new anti-aging formulations, it is difficult to accurately track the growth of these products and measure the size of the market. The products covered in this essay had shipments of $10.8

FIGURE 72

U.S. Shipments of Creams & Lotions and Related Products by Specific Product Category, 1997 and 2002

The product classes covered here are a subset of those in the Census Bureau's creams, lotions and oils division of the Toilet Preparations Manufacturing industry. Figures are in thousands of dollars unless otherwise specified.

Product Class	1997 Shipments	Percent of Total	2002 Shipments	Percent of Total	Percentage Change 1997 to 2002
Cleansing Creams	378,976	12.9	2,059,669	37.2	443.5
Premoistened Towelettes	647,433	22.1	1,112,713	20.1	71.9
Hand Lotion	678,038	23.1	911,291	16.4	34.4
Moisturizing Creams	884,453	30.1	805,213	14.5	-9.0
Body Lotion, excluding bath lotions	255,851	8.7	458,174	8.3	79.1
Facial Scrubs and Masks	58,599	2.0	118,991	2.1	103.1
Cosmetic Oils	30,752	1.0	75,304	1.4	144.9
Total	2,934,102		5,541,355		88.9

SOURCE: "Product Summary Tables: 2002," *2002 Economic Census*, U.S. Department of Commerce, Bureau of the Census, March 2006.

billion in the United States during 2002, the last year for which detailed shipment data are available at the product level. Figure 72 demonstrates potential product classes affected, at least in part, by active ingredients that manufacturers claim perform some special anti-aging function. One caveat relates to the Census classification system used as a basis for this analysis of cosmeceutical products. Two Olay lines—Regenerist and Total Effects—announced in January 2007 that each had moisturizers available with a "touch of sunless tanner for a sun-kissed glow and younger looking skin." These newly reformulated products could be classed as moisturizers, self-tanners, active anti-aging products, or even all three. These hard to classify products contributed to double-digit, triple-digit, and even quadruple-digit growth depicted in Figure 72.

Lip cosmetics experienced triple-digit growth. Shipments of lip cosmetics increased 157 percent between 1997 and 2002 from $1.0 billion to $2.7 billion, in part due to the introduction of reformulated products that helped the class exceed double-digit growth predictions. Growth occurred among lip stick, lip gloss, lip liners, and lip creams. Because lip skin is vulnerable to damage from aging, lipstick reformulations became more robust—functioning both as a color cosmetic and as a lip plumper that fights aging by creating a youthful pout.

According to Mintel's Global New Products Database (GNPD), 431 new lip plumpers were launched into the market in 2004. The pillar of skin care, Clinique (made by Estée Lauder) has a best selling lip cream called All About Lips. Its active ingredient is the common beta hydroxy acid salicylic acid. It exfoliates lips, minimizes fine lines, and attracts moisture to prevent lip color applied afterward from creeping into fine lines around lips.

Face powder and foundation creams were flat. In the five-year period between 1997 and 2002, shipment value remained steady, staying close to $925 million. This class had a tremendous amount of anti-aging product introductions, yet remained flat because it is a highly competitive area. Manufacturers tend to focus research and development on face powder and foundation creams because they not only have high margins but, for most women, these face products are a high loyalty category. America's Research Group reported that women have a penchant for buying brand name cosmetics; 35 percent reported they are very loyal to a face cosmetic brand. Once a woman commits to a face powder or foundation cream brand, she rarely wants to change. Makers count on this committed consumer for sales and profits.

Face powder shipment value grew 22 percent between 1997 and 2002 from $327 million to $400 million. Double-digit growth was due, in part, to the introduction of anti-aging product formulations. One example is Cover Girl (made by Procter & Gamble) Advanced Radiance Age-Defying Pressed Powder with skin brightening ingredients. Using it evens out skin tone to deal with signs of aging including rough skin texture, uneven skin tone, skin dullness, visible pores, and age spots. Supermodel Christie Brinkley, age 52, is the spokesmodel. Cover Girl Smoothers Pressed Powder is formulated with special light-reflecting ingredients (more than likely nano-sized titanium dioxide pigments) that help minimize the appearance of fine lines.

Foundation creams dropped 12 percent in the same period, from a shipment value of $597 million down to $527 million. New formulations include Bobbi Brown Luminous Moisturizing Foundation (an Estée Lauder brand). It fills in fine lines to effectively mask signs of aging. Its active ingredients eschew the commonly used AHAs and BHAs and focus instead on an all-natural blend of hexapeptides and milk thistle extract, alleged to be both collagen-boosting and skin-firming. It retails for $45 for 1 ounce. In October 2006 Bobbi Brown Luminous Moisturizing Foundation won the *More Magazine* Best Of Beauty award. More Magazine is dedicated to women over the age of 40 and has a growing circulation of affluent, educated, and accomplished women.

Cleansing creams experienced triple-digit growth. Growth of 444 percent is due in part to anti-aging products with active ingredients. The facial cleansing regimen now includes toners. As part of its two-step cleansing regimen, Biomedic/LaRoche-Posay (owned by L'Oréal) offers Conditioning Gel toner. It both tones skin and promotes healthy cell renewal by exfoliating the surface layer of skin. More important for fighting aging, its continued use can reduce hyper-pigmentation caused by age spots. Active ingredients include hydroquinone (a skin lightener), salicylic acid, and glycolic acid in a 5.6 percent solution. It is available only at a doctor's office and retails for $29 for 6 ounces.

Premoistened towelettes experienced double-digit growth. Growth of 72 percent between 1997 and 2002 from $647 million to $1.1 billion was due in large part to the introduction of new cleansing cloths, wet cloths, and self-foaming discs, such as Olay Total Effects Age Defying Cleansing Cloths and Age Defying Wet Cleansing Cloths. Double-digit growth was due, in part, to an ever-increasing array of anti-aging products. One example is the Olay Regenerist Eye Derma-Pod. It comes in single-use silver packets that include the eye cream and the sponge to use to massage the cream around the eyes. The sponge exfoliates, while the cream de-puffs and plumps.

Hand and body lotions experienced double-digit growth. Growth of 47 percent between 1997 and 2002 can be traced in part to the introduction of anti-aging products. A recent introduction from Dove Pro·Age promises luminous skin. Multi-functional Beauty Body Lotion is a moisturizer, an exfoliator, and a sunscreen. Its active ingredients include AHA (5% solution) to exfoliate to reduce age spots and wrinkles; glycerin (13% solution) to hydrate; olive oil (1% solution) to add nutrients; and SPF 5 to protect skin from sun damage. America's Research Group reported that women have a penchant for buying brand name cosmetics and beauty products, and that 30 percent reported they are very loyal to a hand and body lotion brand. Competition among manufacturers for this committed consumer who must repeatedly replenish her supply of body lotion is great.

Facial scrubs and masks experienced triple-digit growth. Growth of 103 percent between 1997 and 2002 was due in part to the introduction of anti-aging active products. Anti-aging facial scrubs benefit skin that is dull, rough, sallow, mottled, lined, and unevenly toned due to sun damage. Facial scrubs and masks increase exfoliation to remove damaged epidermal cells to reveal newer and healthier skin beneath. Peels not only retexturize skin via exfoliation but also rejuvenate skin by stimulating cell renewal. Neutrogena Advanced Solutions Facial Peel claims to deliver the results of a professional level 20 percent glycolic facial peel. Its active ingredients involve the naturally derived CelluZyme technology, which works with skin's pH to stimulate surface cell renewal, diminish fine lines, and restore vibrancy. It is modeled on the professional level 20 percent glycolic facial peel introduced by BioMedic/LaRoche-Possay in 1991. A 1.4 ounce container is sold for $11.99 and up.

Cosmetic oils experienced triple-digit growth. Growth of 144 percent between 1997 and 2002 was due in part to the introduction of anti-aging active products. Cosmetic oil serums target fine lines, wrinkles, discoloration, and loss of elasticity due to collagen breakdown, slower cell-turnover, and dryness. SkinCeuticals (owned by L'Oréal) offers a line of Corrective Products that repair, lighten, and exfoliate to restore a youthful appearance. Its Serum 20 AOX+ is an antioxidant alleged to neutralize the free radicals that contribute to cellular aging and cause skin cancer. It also reduces fine lines and wrinkles and stimulates collagen synthesis. The active ingredients in Serum 20 AOX+ are a 20 percent solution of L-ascorbic acid (a fancy name for Vitamin C) and ferulic acid (an antioxidant found in seeds and leaves of plants). It uses a Duke-patented technology that allows the skin to be as absorbent as possible. Once absorbed, the product cannot be washed or rubbed off, remaining effective for a minimum of 72 hours. It retails for approximately $120 for 1 ounce.

Shampoos experienced triple-digit growth. The growth of 324 percent can be traced in part due to reformulated products to deal with the effects of aging. The most recent example of anti-aging hair care is the Dove Pro·Age line launched in February 2007. Because hair can become thinner with age, Dove introduced shampoos that deliver fullness and protect against brittleness and breakage. Makers tend to focus on hair care products, because for most women it provokes high customer loyalty. America's Research Group reported that 35 percent of women reported that they are very loyal to their hair care brand.

Hair rinses experienced quadruple-digit growth. The growth of 2,120 percent can be traced in part to innovative products that are anti-aging. Because silicone coats the hair strand, newly reformulated conditioners and hair rinses form protective films on hair that provide shine. For example, John Frieda (made by Kao) has a Luminous Color Glaze that uses silicone hybrids to create a trademarked color-illuminating technology that boosts shine and adds a hint of color. These types of rinses are anti-aging because shine and color are affiliated with youth, as opposed to dullness and graying which is associated with aging.

KEY PRODUCERS/MANUFACTURERS

Key producers run the gamut from high-end prestige makers like The Estée Lauder Company to low-end mass marketers Procter & Gamble and Unilever. In between sit Johnson & Johnson and L'Oréal. Key producers are profiled in alphabetical order.

The Estée Lauder Company. Estée Lauder began operations in 1946, primarily as a skin care company. Its portfolio consists of thousands of product across 26 brands, including Clinique, Bobbi Brown, and Origins. The Clinique franchise is a pillar of cosmetics and toiletries industry, so big it overshadows rival beauty brands. Clinique's Repairwear line includes Intensive Eye Cream with antioxidants and peptides, Day SPF 15 Intensive Cream, Intensive Night Cream, Deep Wrinkle Concentrate for Face and Eyes, and Extra Help Serum. The latter retails for $56 for one ounce. Bobbi Brown's anti-aging products entirely eschew AHAs or BHAs, relying instead on natural ingredients. Bobbi Brown consumers recognize that AHAs and BHAs cause exfoliation, which thins the skin, and thin skin is a sign of aging.

Two bestselling Bobbi Brown anti-aging products are Intensive Skin Supplement ($60 for 1 ounce) and Overnight Cream ($60 for 1.7 ounce), designed to work in unison. Active ingredients in the former include white birch extract, vitamins A, C, and E, green tea extract, and grape and mulberry extract. The latter includes protein activator milk thistle extract, cat's claw extract, almond, apricot, and avocado oil.

Estée Lauder's Origins Collection utilizes the alleged antioxidant properties of mushrooms in its Mega-Mushroom Face Serum and Face Cream. This line does not promise to erase wrinkles; it promotes healthy skin. Its marketers believe the desire of consumers for the look of vitality will eventually replace the desire to look youthful.

Johnson & Johnson. Headquartered in Racine, Wisconsin, Johnson & Johnson is one of the largest family-owned and family-managed companies in the United States. It manufactures and markets thousands of products in hundreds of categories, all related in some way to health and cleanliness. J&J formed in the 1880s as a pioneer of

ready-to-use surgical dressings that provided antiseptic wound treatment. It is still in the pharmaceutical industry, and is able to utilize its pharmaceutical R&D to make anti-aging products. J&J grew in part by acquiring Neutrogena in 1994. It also makes Aveeno products, among others. Neutrogena is not only a top seller, it is also one of the most expensive mass market product lines. Some consumers equate price with quality.

As the pharmaceutical company that developed Retin-A and Renova, proven anti-wrinkle products that won FDA approval in 1996, J&J is an anti-aging industry leader. It not only sells anti-aging products; it also has products that are age-reversing. These include the Neutrogena Advanced Solutions Facial Peel earlier mentioned, Neutrogena Advanced Solutions Skin Transforming Complex, and Neutrogena Advanced Solutions Nightly Renewal Cream. Active ingredients in the latter are retinol, AHAs, and dimethylaminoethanol (DMAE). It works overnight with skin's natural, nightly repair cycle.

The L'Oréal Group. L'Oréal owns numerous companies. Two it gained through acquisitions specialize in anti-aging and are undisputed leaders in product advancements. These are BioMedic/LaRoche-Posey and SkinCeuticals. For L'Oréal, brand acquisition lets it penetrate new markets and reposition a well-respected brand to gain international audiences. BioMedic/LaRoche-Posey is the generally acknowledged grandfather of the serious cosmeceutical industry. La Roche-Posay originated as a pharmaceutical laboratory in France and BioMedic was founded in 1990 in Phoenix, Arizona. Its products were introduced to the American Society of Plastic and Reconstructive Surgeons in 1991, and became an overnight sensation.

In 1989 L'Oréal Group purchased BioMedic/LaRoche-Posey, which maintained its own formulation laboratory. Now more than 25,000 board-certified plastic surgeons and dermatologists endorse its MicroPeel product, available only through a physician. The line includes 18 products to address fine line and wrinkles, along with body products for those with very dry skin and facial products for dehydrated skin. In 2005 L'Oréal purchased SkinCeuticals.

Formed in 1997 in Dallas, Texas, SkinCeuticals was a relative latecomer to the serious cosmeceutical industry. It quickly differentiated itself by publishing clinical studies in respected medical journals usually reserved for pharmaceutical research. SkinCeuticals products contain high levels of active ingredients not generally available at mass merchandisers and are three time as expensive as BioMedic/LaRoche-Posey. The line includes products to prevent and correct aging, along with products to cleanse, tone, and moisturize. SkinCeuticals introduced the concept of products targeted at younger women who want to prevent aging instead of curing it later.

Procter & Gamble. Cincinnati-based Procter & Gamble has been a cosmetics and toiletries industry leader since its establishment in 1837. P&G has more than 100 brands available in nearly 130 countries. It has grown through acquisitions of brands such as Olay, Cover Girl, and most recently Doctor's Dermatologic Formula (DDF).

In 1985 P&G bought Oil of Olay. Olay began in the home of a chemist in the 1950s who created Oil of Olay Beauty Fluid, a face moisturizer that resembled fluids found in young skin. Within five years, it was phenomenally successful. In 2000 P&G shortened the brand name to Olay. It is a worldwide brand with over $1 billion in global sales. The Olay line has been expanded to include Olay Definity and Olay Regenerist. Definity features cleansing, moisturizing, and protecting products designed to work together to repair past skin damage and give luminous skin. Regenerist is the newest line; so far it consists of 15 products including a toner that reduces age spots, a lip treatment that diminishes vertical lines, an eye lifting serum that gives complete turnaround in 24 days, a cleanser that speeds the skin's natural regeneration process, and a night recovery moisturizing treatment that gives a mini-lift every night.

In January 2007 P&G acquired DDF. DDF was a pioneering brand in 1991 when it created one of the first dermatology-based skin care lines. The DDF line is sold only in department stores and spas. "The addition of DDF to our portfolio provides us with the opportunity to reach new consumers in new distribution channels," said the President of P&G Global Skincare regarding the 2007 acquisition.

Unilever. Unilever is a behemoth. It is an international manufacturer of products in various industry sectors like food, home care, and personal care. In 2003 Unilever executed its largest ever North American launch, spending $110 million to promote a new line of Dove hair care products. It banked on consumer loyalty to its Dove cream bar, launched 50 years earlier and number one in North America—consistently capturing close to 50 percent of the bar soap market.

In January and February 2007 Unilever once again spent hundreds of millions on an even larger launch of Dove Pro·Age. Unilever turned anti-aging on its head with the tagline "This isn't anti-age, this is pro-age." In January 2007 Unilever kicked off the line's national advertising campaign. Advertisements boldly queried, "Too old to be in an anti-aging ad?" Images of four regular women (one black, one pudgy, one slender, and one grey-haired) are presented in sequence. The 2007 product rollout is

aimed directly at women aged 45 and over, the fastest growing demographic group. The product offerings include body lotion, body cream oil, hand cream, neck and chest beauty serum, body wash, antiperspirant, shampoo, and conditioner.

As it did in 2003 with Dove hair care, Unilever is banking on the loyalty of the female consumer to the well-loved Dove brand name. Unilever doubled its U.S. business in the five years prior to 2006 and reports $1 billion yearly in U.S. retail brand sales.

MATERIALS & SUPPLY CHAIN LOGISTICS

This industry spends 70 percent on packaging, and 30 percent on ingredients that go into the product. Toiletries are frequently sold with extensive packaging to differentiate them from competitors' products.

The $28 billion per year toiletries industry as a whole spent $8.0 billion in 2002 on the total cost of materials needed to make all kinds of toiletries. Of this, $5.6 billion was spent on packaging and other materials, and $2.4 billion was spent on ingredients used in the production of the products themselves.

The primary categories of ingredients needed to make toiletries of all kinds according to data reported by the Census Bureau are, from largest to smallest in terms of industry-wide spending, as follows:

- Perfume oil mixtures and blends, essential oils (natural), and perfume materials (synthetic organic)
- Other synthetic organic chemicals
- Bulk surface active agents (surfactants)

The class of products known as cosmeceuticals is based entirely upon active ingredients alleged to perform some special function upon the skin or hair. The most common active ingredients are AHAs and BHAs. New active ingredients are frequently rolled out. Peptide copper complex repairs skin tissue while marine carotenoid astaxanthin is an antioxide that reverses photo damage. Important botanical active ingredients include mushrooms, milk thistle, green tea, soy, chamomile, aloe, pomegranate, and avocado.

Active ingredients have been heavily marketed by makers with almost evangelical promotions. Evangelicism often elicits skepticism. One Miami dermatologist told *Body and Soul* in November 2006, "A product's scent, packaging, price, or celebrity endorsement makes no difference to your wrinkles. It's the right active ingredients that do the work." The dermatologist beseeched consumers to read labels and make sure active ingredients are among the first three ingredients listed, since The Fair

Packaging and Labeling Act requires ingredients to be listed in order of quantity.

The burden is on the consumer to conduct an intensive scrutiny of labels. A closer scrutiny of five commonly used active ingredients follows.

Alpha Hydroxyacid Acids (AHAs). AHAs come from nature. They include malic acid from apples and citric acid from citrus fruit. The most commonly used AHA is glycolic acid, from sugar cane. It is generally regarded to be among the most effective active ingredient. It is proven to improve skin tone and texture, reduce skin discoloration and age spots, and lessen fine wrinkles.

Glycolic acid is the ingredient to look for when scrutinizing ingredients listings. It is effective because it has a small molecule which allows it to penetrate skin. Glycolic acid helps dissolve the substance that holds the keratinized skin cells together to increase cell exfoliation and eventual renewal. Because glycolic acid exfoliates, it thins the stratum corneum. This results in a healthy new glow: the dead cells of the stratum corneum no longer obscure the living cells of the epidermis beneath, so skin is tangibly radiant. With continued use, the epidermis eventually thickens, further restoring radiance since young skin is thicker than old skin.

With continued use over years, glycolic acid encourages skin to manufacture new collagen, elastin, and hyaluronic acid, and to renew itself. Skin is less dull, rough, sallow, mottled, lined, and unevenly toned due to years of sun damage and aging.

The pH level of a product is an independent factor that influences its effectiveness. A drop of 2 in pH level correlates to an increase of 30 percent in cell turnover rate. Typically, brands sold in stores have a pH of 6, while brands sold in doctors offices have a pH of 4. The doctor-dose will perform 30 percent better based solely on its slightly more acidic nature.

Beta Hydroxy acids (BHAs). BHAs are commonly shown on ingredients lists as salicylic acid. They have a larger molecule than AHAs. The larger molecule keeps BHAs on the skin surface allowing it to better exfoliate. The larger molecule size of salicylic acid produces less irritation than AHAs so it is a good choice for those with sensitive skin.

Algae Peptides. Quite the rage, a peptide is a combination of amino acids; amino acids are protein units of structure essential to human metabolism. These small molecules can penetrate the epidermis to reach the skin layer (the dermis) beneath. Algae peptides produce rapid visible changes, such as tightening and the appearance of firming. When applied to the skin, such peptides are picked up by binding sites on the cell of the epidermis as well as the dermis. This increases the ingredients' range of activities, which includes cell turnover and collagen synthesis. Algae peptides can enhance hydration, smoothness, elasticity, and fullness of skin.

Antioxidants. Antioxidants include alpha lipoic acid, beta-carotene, coenzyme Q10, green tea, idebenone, panthenol, and topical vitamins A, C, and E. Oxidation is a binary process that involves the union of oxygen with a radical. Free radicals are produced in cells as byproducts of normal biochemical processes, particularly the metabolism of sugars and fats. Once produced, free radicals can damage almost any biological structure they come into contact with. Free radicals contribute to cellular aging and increase the risk of skin cancer due to changes in the cellular structure.

By neutralizing free radicals, antioxidants may prevent both cellular aging and skin cancer. Vitamin C is perhaps the most effective and most common. However, it also oxides immediately when exposed to air, rendering it ineffective.

Retinoids. Retinoids are the only active ingredient that have established scientific backing. They have been tested and approved by the FDA. Retinoids are vitamin A derivatives that act as agents for anti-oxidizing, anti-pigmenting, and exfoliating. They also increase production of collagen and hyaluronic acid, which is how they help firm the skin.

The pharmaceutical company J&J developed Retin-A and Renova, the products that won FDA approval in 1996. Prescription retinoic acids include Avage, Differin, Retin-A, and Renova. Non-prescription retinol is available in a slew of products. Neutrogena offers an anti-wrinkle cream with retinol in its Healthy Skin line. But it does not disclose the percentage of retinol in the product. Lack of full disclosure is what makes some consumers prefer the rapidly growing doctor-dosing distribution channel.

DISTRIBUTION CHANNEL

Cosmeceutical products are distributed in three distinct channels. Within the industry these channels co-exist and are often referred to as *prestige, masstige,* and *doctor-dosing*. Prestige products are classified based primarily on the location where they are distributed, such as major department stores and specialty boutiques, and are generally higher-priced. Lower-priced products are known as masstige because they are distributed via the mass market. Doctor-dosing involves products made with a higher dosage of active ingredients and sold primarily at medical doctors offices.

The department store distribution channel for prestige products is a giant. This channel consistently represents approximately 40 percent of sales of cosmeceuticals in North America. This channel underwent a shakeup when Federated Department Stores acquired May Department Store Company. At the time the deal was finalized in August 2005, Federated owned 458 stores nationwide, mostly known as Bloomingdale's and Macy's, while May operated 491 department stores nationwide under the names Lord & Taylor and Marshall Field's, among others. In September 2006, the May Department Store nameplates were formally changed to Macy's. All prestige companies—or 40 percent of the industry—had to deal with department store consolidations. An estimated 75 stores closed.

The mass market distribution channel is broadening in part because store loyalty is less prominent with masstige products than for prestige products. The broadened distribution channel for lower-priced masstige products includes a throng of outlets: drugs stores such as CVS, Rite Aid, and Walgreens; food stores including Whole Foods and health food stores; mass merchandisers including KMart, Target, and Wal-Mart; and nontraditional retailers including all Internet sites. The broadened channel also includes warehouse club stores such as Costco and Sam's Club.

The doctor-dosing distribution channel is growing. It is a distribution channel of choice for prestige manufacturers because of the intimate environment within the doctor's office (and sometimes spas and specialty stores). This distinctive channel distributes products made with a higher dosage of active ingredients. Some are twice the strength of active ingredients sold in stores.

Higher dosing is based on the assumption that purchase and use decisions will be influenced by doctors with intimate knowledge of patients' skincare needs and better placed to recommend appropriate products and ensure they will be properly used. Doctors can steer patients toward prescription products when appropriate, and steer patients away from dangerous product combinations. Internet Wire reported on U.S. demand for cosmeceuticals in January 2007 and suggested this channel of professional products will experience the fastest growth.

Manufacturers such as BioMedic/LaRoche-Posey and SkinCeuticals have sales representatives that go door-to-door to doctors' offices to convince them to stock products. When medical doctors agree to distribute the product, it is a very valuable product endorsement. This channel is more likely to use ingredients proven effective. To convince highly educated medical doctors who are generally skeptical of efficacy claims, makers in this channel use clinical studies published in respected medial journals usually reserved for pharmaceutical research.

Some manufacturers prefer the doctor-dosing channel because it provides geographic diversity and statistics

show that sales increase when professionals are present to influence choices. The doctor-dosing distribution channel helps certain makers maintain their prestige reputation. For instance, SkinCeuticals lost some of its prestige after it was purchased by L'Oréal because L'Oréal failed to control the distribution channel and allowed SkinCeuticals products to be sold on almost any Internet site. Doctors then tended to disassociate with the line.

A popular product available only through the doctor-dosing distribution channel is the BioMedic/LaRoche-Posay MicroPeel 30 Plus. It costs approximately $95 and can only be performed under the direct supervision of a medical doctor who will also warn and educate the patient about the increased need for sunscreen and sun avoidance after using the product. In-office peels are more effective than store products because under the auspices of a medical doctor, peels with a pH of 1 or 2 are applied. The result is a dramatic improvement that lessens unevenness and roughness, and lightens darkened pigment (known as age spots). The active ingredient in MicroPeel Plus 30 is a 30 percent solution of salicylic acid. Such a high concentration of salicylic acid, administered at such a low pH level, sloughs off dead cells to stimulate the skin's natural renewal processes. For 7 to 10 days afterwards, skin continues the MicroPeel Plus 30 renewal process, shedding its dead outer layer and stimulating cell renewal.

KEY USERS

Women of all ages purchase anti-aging products with active ingredients. Women are the primary users of anti-aging products according to a 2004 study conducted jointly by the National Consumers League and Harris Interactive. A reported 90 million Americans had used procedures or products in an effort to reduce the appearance of aging. Reportedly, 76 percent of women use these products compared to 18 percent of men.

Some women spend more money per ounce on anti-aging products with active ingredients than on the cost of gold, which is $600 per ounce. Cosmeceuticals are used for a variety of reasons including to improve the self, to attract the opposite sex, and to hide the effects of aging. Spending on cosmeceuticals may provide insight into the desires of the women who buy them. The real product being purchased is hope for a bright and luminous future.

ADJACENT MARKETS

Some believe that one good alternative to high priced anti-aging products is a healthy attitude about aging. According to one British Web site, the real seven signs of aging can include any of the following: emotional maturity, wisdom, self-confidence, self-esteem, owning a house, sense of perspective on life, sexual confidence, career development, and financial security. According to Citibank's

March 24, 2007 advertisement in the *New York Times*, women really can get rid of wrinkles and worry lines, when they learn about saving money and making financial investments in their future. According to rationalists, no readily available cosmeceutical rivals the benefit of being born with good genes, and leveraging those genes through a lifelong program of adequate rest, balanced diet, mental stimulation, and exercise. According to Coco Chanel, "A woman's unhappiness is to rely on her youth. Youth must be replaced with mystery."

Sunscreens. The sunscreen market both affects and is affected by cosmeceuticals. Any product reformulated with an SPF value could be considered an anti-aging product. Any product reformulated with AHAs and BHAs that temporarily thin skin via exfoliation should consistently be paired with a pure sunscreen product with a high SPF rating. The FDA recently published scientific evidence from studies it sponsored that suggest the use of AHAs and BHAs increases the risk of sunburn. As a result, consumers who use products with AHAs and BHAs must religiously use sunscreen products, and be encouraged to consistently avoid the sun. It is estimated that 80 percent of aging is caused by the sun's rays. Consumers purchase anti-aging products to protect against aging. Their use is oddly contradictory because they may cause age spots and wrinkles to occur more rapidly, if sunscreen and sun avoidance is not implemented properly.

Dermatologists, plastic surgeons and medical doctors. More and more consumers view dermatologists as having the skills and tools needed to reverse or halt the aging process. Consumers increasingly choose cosmetic procedures such as lasers to remove age spots. If a consumer wants absolutely to reverse the effects of aging, her medical doctor can prescribe drugs with proven ingredients. Well-known examples of prescription anti-aging products are Retin-A and Renova, products developed by Johnson & Johnson that won FDA approval in 1996. A Boston dermatologist reminded consumers that there is no such thing as a miracle in a jar, explaining that, "If it were a miracle, its use would flop over to a drug."

If a consumer wants absolutely to reverse the effects of aging, plastic surgery firms skin and removes wrinkles. It might be cumulatively less expensive to save the money from potential expenditures on products and apply the total amount to services that are guaranteed to firm skin and remove wrinkles. For instance, a consumer could potentially spend approximately $100 per month on prestige anti-aging products. Over ten years, the $12,000 of cumulative spending on these products would easily have been enough to pay for a surgical face-lift.

RESEARCH & DEVELOPMENT

The quest for sustained youth and beauty that sells cosmetics is age-old, although the ingredients used to achieve the desired outcome change. Typical R&D budgets increase approximately 1 to 2 percent per year. R&D budgets related to anti-aging products are predicted to increase 10 percent per year through 2010. One focus of R&D will be to meet consumer demand for multi-functional products. R&D will involve product reformulations to create combination products that can be moisturizers, self-tanners, active anti-aging products, sunscreen, or even all four. Other R&D will focus on ingredients.

Ingredients are rapidly changing and manufacturers are making even more efficacy claims. This increases the threat of FDA oversight, especially regarding labeling. AHAs and BHA have now been shown with FDA sponsored research to contribute to the increased risk of sunburn. Future R&D may well focus on replacements for AHAs and BHAs.

Although the FDA does not require warnings, it notified manufactures in 2005 that it would start enforcing a labeling requirement for "Warning—the safety of this product has not been determined." This was necessary because of the explosion of new products and new ingredients that had not been proven safe primarily because they could potentially be used in dangerous combinations or could contribute to the increased risk of sun sensitivity. Any additional labeling requirement is estimated to potentially cost the industry $35 million in the one year that would be required to implement it. Makers might find replacements for AHAs and BHAs, or choose to forego their use. Another approach might be to educate consumers about increased risks.

Future R&D may result in increased consumer education programs. Consumers who use products with AHAs and BHAs that exfoliate and temporarily thin the skin must use sunscreen. Sun avoidance is recommended. To deal with the fact that AHAs and BHAs are known to thin the skin, and that thin skin is a sign of aging, future R&D may result in products specifically formulated for different skin types. As the anti-aging field evolves, it will emerge that no one type of product or type of ingredient is appropriate for all skin types. Thin-skinned women usually have luminously thin skin when they are young. Women with thin skin and women with dry skin should avoid products with AHAs and BHAs that encourage exfoliation. Thick-skinned women tend to have more problems because thick skin is proven to be sluggish; skin like that benefits by an injection of AHAs and BHAs in any format to encourage the exfoliation and skin renewal process that follows.

CURRENT TRENDS

In an October 2006 article titled "Cosmeceuticals to go Mass by 2008," *Cosmetics International* reported predictions that 40 percent of cosmetics and toiletries will use active ingredients to make some kind of efficacy claim by 2008. The rapid expansion of anti-aging ingredients into toiletries products represents a fundamental shift that transcends the entire $28 billion year industry.

Industry-wide emphasis on ingredients has created a consumer who is learning to be vigilant about reading the fine print. Some brands like SkinCeuticals have an entire glossy color catalog that explains the special function of each ingredient. Some brands—especially those within the doctor-dosing channel—offer more transparency about ingredients. They go beyond the letter of the Fair Packaging and Labeling Act to list the percentage of each ingredient, its solution and pH levels, and to explain its alleged function. Biomedic/La-Roche Posay evidences this trend towards more transparency in the doctor-dosing channel, and Dove Pro·Age evidences this trend in the mass market distribution channel.

In the future, the mass market distribution channel will begin to operate more closely on the doctor-dosing distribution channel model. The mass market will begin to substantiate efficacy claims with clinical studies published in respected medical journals usually reserved for pharmaceutical research. The mass market will take on more responsibility for warning and educating consumers about the increased need for sunscreen and sun avoidance.

Bobbi Brown may be ahead of the curve on one trend. The line entirely eschews all common anti-aging ingredients. The brand is marketed as AHA and BHA free. It relies on all natural ingredients like grape extract, white birch, and whey protein. Bobbi Brown may be presciently foreshadowing the future, since new science sponsored by the FDA suggests AHAs and BHAs increase the risk of sun damage, a key factor that contributes to aging. If the FDA follows through on its labeling threat, Bobbi Brown will save money by avoiding having to relabel even one product.

Manufacturers have introduced glycolic peels, and microdermabrasion kits that claim to offer gentler, at home versions of clinical procedures. Manufacturers can follow the classic razor-and-blade model in marketing these products, making more money on refills than on the initial kit.

Estée Lauder is also taking a new approach with the launch of its Perfectionist Power Correcting Patch. The new product is said to target deep eye lines and wrinkles and offers a glimpse into the future trends. It emits a tiny micro-current of energy (1.5 volts) that allegedly allows the formula to penetrate more effectively into the epidermis. The inspiration for the patch came from the medical community's use of an electrical current to deliver medication into body tissue.

TARGET MARKETS & SEGMENTATION

Manufacturers hope for double-digit, triple-digit, and even quadruple-digit growth in certain product classes. Manufacturers target growing product classes and growing demographic segments with money to spend. Manufacturers know that creams and lotions are the most lucrative cosmetics and toiletries division, and that many female consumers are loyal to much-loved brands. America's Research Group reported that women have a penchant for buying brand name cosmetics: 35 percent are very loyal to a face cosmetic brand, 35 percent are very loyal to a hair care brand, and 30 percent are very loyal to a hand and body lotion brand. Manufacturers target the high margin nondurable goods that loyal consumers need to repeatedly replenish their supply of throughout the year. Makers count on this committed consumer for sales and profits.

Significant opportunity involves the 78 million U.S. Baby Boomer generation (those born between 1945 and the early 1960s). Many are not happy about reaching middle age and want to retain a youthful look. Many retired with a large nest egg and expect to be active, look young, and maintain healthy for well into old age. They can afford products they hope will either mask, stave off, or reverse aging.

This large segment of consumers with disposable income is an important target market for cosmeceuticals. The aging population in the developed world puts a premium on youth. There is apparently no limit to what customers will pay to keep the years from showing. The targeted segment is women who can afford to buy anti-aging products and are willing to do anything to fight against the appearance of aging.

Many women devote income to products that promise to make them look and feel younger than their age. Reportedly, 76 percent of women use these products compared to 18 percent of men. Women over 40 represent the market. Revlon recently kicked off its biggest launch in more than a decade with Vital Radiance, aimed specifically at women age 50 and older. ACNielsen estimated that female heads of households over age 45 account for nearly 70 percent of mass retail cosmetics purchases. The tagline "This isn't anti-age, this is pro-age" was used in February 2007 to support the Dove Pro·Age product rollout. The phrase is aimed directly at women aged 45 and over.

RELATED ASSOCIATIONS & ORGANIZATIONS

American Academy of Dermatology, http://www.aad.org

Cosmetic, Toiletry and Fragrance Association, http://www.ctfa.org

The European Cosmetic Toiletry and Perfumery Association, http://www.colipa.com

Research Institute for Fragrance Materials, http://www.rifm.org

Synthetic Organic Chemicals Manufacturers Association, http://www.socma.com

BIBLIOGRAPHY

"AHAs and UV Sensitivity: Results of New FDA-Sponsored Studies." U.S. Food and Drug Administration, Center for Food Safety and Applied Nutrition. Available from <http://www.cfsan.fda.gov~dms/coshauv.html>.

"Alpha Hydroxy Acids in Cosmetics." U.S. Food and Drug Administration, Center for Food Safety and Applied Nutrition. March 2002. Available from <http://www.cfsan.fda.gov~dms/cos-aha.html>.

Beatty, Sally. "Hot at the Mall: Skin Care Products from Physicians; Cosmeceutical Creams Top Anti-aging Market." *The Wall Street Journal*. 14 November 2003.

"Beta Hydroxy Acids in Cosmetics." U.S. Food and Drug Administration, Center for Food Safety and Applied Nutrition. January 2006. Available from <http://www.cfsan.fda.gov~dma/cos-bha.html>.

Brown, Laurel. "Cosmeceuticals or CosmePSEUDOcals: Examining the FDA's Under-sight of Celebrity Dermatologists in the Cosmeceuticals Industry." Harvard University, Legal Education Document Archive. Available from <http://leda.law.harvard/leda/data.722/brown.05html>.

Ciraldo, Loretta. "6 Weeks to Sensational Skin." *Rodale*. 2006.

"Cosmeceuticals." U.S. Food and Drug Administration, Center for Food Safety and Applied Nutrition. Available from <http://www.cfsan.fda.gov~dms/cos-217.html>.

"Cosmeceuticals to Go Mass by 2008." *Cosmetics International*. 6 October 2006, 6.

Edgar, Molly. "Pro Age takes Dove's Real Beauty to Next Level." *WWD*. 12 January 2007, 5.

"Facing the Future: Consumers are Increasingly Comfortable with Advanced Technologies When it Comes to Saving Face." *Soap Perfumery & Cosmetics*. June 2006, 26.

Gustke, Constance. "Loyal Customers: Top 10 Consumer Product Categories that Generate the Most Brand Loyalty." *WWD*. 28 September 2006, 18.

"Is It a Cosmetic, a Drug, or Both? (or Is It Soap?)" U.S. Food and Drug Administration, Center for Food Safety and Applied Nutrition. Available from <http://www.cfsan.fda.gov~dms/cos-217.html>.

"Labeling for Topically Applied Cosmetic Products Containing Alpha Hydroxy Acids as Ingredients." U.S. Food and Drug Administration, Center for Food Safety and Applied Nutrition. January 2005. Available from <http://www.cfsan.fda.gov>.

"L'Oréal, Keaton Seek Perfection" *AdWeek*. 12 May 2006. Available from <http://www.adweek.com/aw/national/article_display.jsp?vnu_content_id=1002501756>.

"Luminous Moisturizing Foundation." Bobbi Brown. Available from <http://www.bobbibrowncosmetics.com/templates/products./sp_shaded>.

Mathews, Imogene. "Consumer Comfort in Cosmeceuticals: Mass and Prestige Fight the Signs of Aging." *Global Cosmetics Industry*. March 2005, 44.

"The New Face of Beautiful Aging: Product Appeal has Moved Beyond Basics Making 40-plus Celebrities the Hottest Faces for Ad Campaigns." *Global Cosmetics Industry*. January 2007, 43.

"OTC: Glamour for All Ages." *Chemist & Druggist*. 17 February 2007, 4.

"Over 50 is No Longer Such a Bad Place to Be." *MMR*. 8 May 2006, 97.

Prior, Molly. "Cosmeceuticals Reach the Mainstream." *WWD*. 15 December 2006, 6.

Public Resource Center. American Academy of Dermatology. Available from <http://www.aad.org/public/Publications/pamphlets/Cosmetics.htm>.

"The Seven Signs of Aging." The F Word: Contemporary UK Feminism. Available from <http://www.thefword.org.uk/features/2001/06/the_signs_of_ageing>.

"Toilet Preparation Manufacturing: 2002." *2002 Economic Census*. U.S. Department of Commerce, Bureau of the Census. Available from <http://www.census.gov/econ/census02/>.

"US Skin Care Set to Grow to $7 Billion by 2010." *Cosmetics International*. 12 January 2007, 4.

SEE ALSO *Cosmetics, Creams & Lotions, Sun Care Products*

Cosmetics

————■————

INDUSTRIAL CODES

NAICS: 32–5620 Toilet Preparation Manufacturing

SIC: 2844 Perfumes, Cosmetics, and Other Toilet Preparations Manufacturing

NAICS-Based Product Codes: 32–5620D111, 32–5620D121, 32–5620D141, 32–5620D241, 32–5620G111 through 32–5620G131, 32–5620G311 through 32–5620G331, 32–5620G351, and 32–5620G381 through 32–5620G3A1

PRODUCT OVERVIEW

Color cosmetics fall into five well-defined categories, depending on where they are used on the body—most often the female body: lips, eyes, face, nails, and cheeks. Because applying color cosmetics is part of what the French used to refer to as a woman's *toilette*, color cosmetics became known as toiletries. The French word toilette refers both to the room where the toilet is, and activities generally conducted in that room. As a result, the expression "to do one's toilette" emerged. This includes applying makeup, or applying color cosmetics.

The use of color cosmetics was historically associated with women of suspect morals. The advent of the cinema changed the perception of color cosmetics. Beautiful actresses appeared on the screen adorned with lush lips accentuated by red lipstick, color-shaded eyes, and thick black mascara. The emergence of the chain, or dime, store in the 1920s made color cosmetics available at the mass

market level. After World War II, a number of cosmetic manufacturers emerged. For instance, Estée Lauder began operations in 1946 and with its Clinique brand became a pillar of prestige color cosmetics.

The major types of products associated with doing one's toilette are defined in a U.S. Census Bureau report titled "Toilet Preparation Manufacturing: 2002." The five major toilette product divisions, in order from largest to smallest based on 2002 market size, are:

1. Cosmetics
2. Hair preparations
3. Creams, lotions, and oils
4. Perfumes
5. Shaving preparations

Figure 73 presents the size of each of these five major product divisions based on 2002 data. This essay focuses on a subset of the cosmetics division of the toilet preparation manufacturing industry.

MARKET

The market for all toilet preparation products in the United States was valued at almost $28 billion in 2002 according to the Census Bureau. Cosmetics accounted for $9.2 billion or 33 percent of the industry. Cosmetics are the largest and most profitable single division of toiletries.

The color cosmetics discussed in this essay are a subset of all cosmetics as defined by the Census Bureau. The subset covers five well-defined product classes, according to where they are used on the body: lips, eyes, face, nails, and

FIGURE 73

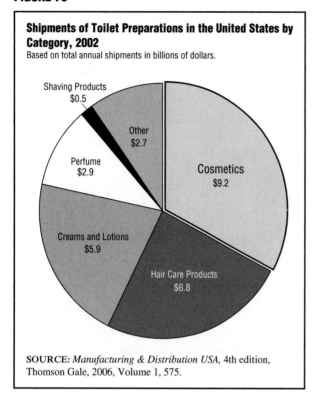

Shipments of Toilet Preparations in the United States by Category, 2002
Based on total annual shipments in billions of dollars.

Shaving Products $0.5

Other $2.7

Perfume $2.9

Cosmetics $9.2

Creams and Lotions $5.9

Hair Care Products $6.8

SOURCE: *Manufacturing & Distribution USA*, 4th edition, Thomson Gale, 2006, Volume 1, 575.

cheeks. Figure 74 lists the products in the color cosmetic categopry which had shipments in 2002 of $5.1 billion.

Lips. Manufacturers' shipments of lip cosmetics—the largest product class within color cosmetics—almost tripled in size during the five-year period between 1997 and 2002, exploding in value from $1.0 billion to $2.7 billion, a growth of 157 percent. Lip cosmetics are a growing product class. In 1997 lip cosmetics represented 27 percent of the value of color cosmetics shipments. By 2002 lip cosmetics represented more than half (53%) of the total $5.1 billion color cosmetics shipment value.

This triple-digit growth may be seen by some as evidence of the 2001 recession because of an alleged economic indicator known as the lipstick effect. During a recession, consumers supposedly tend to spend more money on small, comforting items such as lipstick rather than large luxury items such as cars.

The triple-digit growth in lipsticks may also be evidence of innovations in lip care product formulations that resulted in a multi-functional product. Because lip skin is vulnerable to damage from sun, wind, and aging, reformulated lipsticks have been made to function as both color cosmetics and lip protection. Lipstick, lip glosses, and lip liners do double- or even triple-duty, functioning as colorants, sunscreens, moisturizers, and even plumpers. Other functions besides coloring include benefits such as gloss, long-lasting wear, and transfer resistance.

L'Oréal Paris, for instance, offers nine different lip cosmetic formulations: sheer gloss, moisturizing, zero-transfer, zero-transfer gloss, 8-hour, 8-hour liquid, shimmering 8-hour, plumping, and volumizing. Zero-transfer refers to zero-transfer when kissing. Even that old standby Chap Stick has been reformulated to be a skin protector with SPF 15.

Face. Manufacturers' shipments of face powder and foundation creams—the second largest product class within color cosmetics—were flat during the five-year period between 1997 and 2002, staying close to $925 million. This category combines two product classes: face powders and foundation creams. Face powders grew 22 percent in the five-year period between 1997 and 2002, from a total shipment value of $327 million to $400 million. Foundation creams dropped 12 percent in the same period, from a shipment value of $597 million down to $527 million. Even with the 12 percent drop in foundation creams, women still prefer cream foundation over face powder.

Overall, foundation's drop in shipments combined with powder's gain meant that face powder is becoming almost as popular as cream foundation. Face powder and foundation creams together remain a large and competitive color cosmetic product class. Makers tend to focus on face powder and foundation creams because for most women, face products remain a high loyalty category. America's Research Group reported that women have a penchant for buying brand name cosmetics; 35.2 percent reported that they were very loyal to a face cosmetic brand.

An article in *Chain Drug Review* summarized dollar sales of face powder and foundation creams for the year ending in January 2006 using data from Information Resources, Inc., for purchasing at drugstores, discounters, supercenters, supermarkets, warehouse clubs, and dollar stores. Such stores represent the mass market only, which is less than 50 percent of the total market and is commonly called masstige to characterize it as separate from the upscale prestige market. Prestige color cosmetic sales represent approximately 40 percent of the market. The top five face brands, with dollar sales and percent change over the prior 12-month period, were:

- Revlon Age Defying foundation—$34.8 million, up 88.9 percent

- Cover Girl Clean powder—$26.0 million, down 11 percent

- Cover Girl Clean powder—$21.40 million, down 12 percent

- L'Oréal True Match foundation—$19.6 million, down 13 percent

FIGURE 74

U.S. Shipments of Cosmetics by Product Class in 1997 and 2002
Figures are in thousands of dollars unless otherwise specified.

Product Class	1997 Shipments	Percent of Total	2002 Shipments	Percent of Total	Percentage Change 1997 to 2002
Lipsticks, Lip Glosses, and Lip Pencils	1,047,770	27.1	2,696,865	52.8	157.4
Face Powder and Foundation Creams	924,324	23.9	926,803	18.2	0.3
Eye Cosmetics	962,873	24.9	869,321	17.0	-9.7
Nail Enamels, Polishes, and Polish Removers	513,192	13.3	418,139	8.2	-18.5
Blushers	416,707	10.8	192,738	3.8	-53.7
Total	3,864,866		5,103,866		32.1

SOURCE: "Product Summary Tables: 2002." *Economic Census 2002*, U.S. Department of Commerce, Bureau of the Census, March 2006.

• Maybelline Dream Matte foundation—$18.6 million, up 53 percent

Eyes. Manufacturers' shipments of eye cosmetics—the third largest product class within color cosmetics—declined 10 percent during the five-year period 1997 to 2002, from $963 million down to $869 million. Eye cosmetics generally consist of mascara, eye shadows, and eye liner. Part of the explanation behind the decline in eye makeup shipments is the more natural look that has been popular in Hollywood, with simple eyes and a focus on kissable lips.

Cosmetics manufacturers have not, however, lost interest in eye cosmetics. New products are regularly introduced, especially mascaras and eye liners. According to *Chain Drug Review* the following five masstige eye brands led U.S. sales in drugstores, discount stores, supercenters, supermarkets, warehouse clubs, and dollar stores for the year ending in January 2006. Those sales are listed as well as the percentage change over the prior 12-month period for all five leading products:

• Maybelline Great Lash mascara—$48.5 million, up 5 percent

• L'Oréal Voluminous mascara—$31.4 million, up 8 percent

• Maybelline XXL mascara—$28.3 million, up 153 percent

• Cover Girl Eye Enhancers mascara—$26.9 million, down 4 percent

• Revlon ColorStay eye liner—$22.7 million, up 15 percent

Of interest is the fact that bestselling brands are often produced by the same company so that manufacturers compete against themselves within some categories. L'Oréal, for example, owns both Maybelline and L'Oréal, and both brands have best selling mascara.

Nails. Nail polishes, enamels, and polish removers are the fourth largest color cosmetics class. Shipments of nail cosmetics declined 13 percent in the United States between 1997 and 2002, from a total of $513 million to $417 million. The use of nail polish appears to be on the decline. Long, brightly colored nails have been eclipsed by a natural look. Perhaps working women are too busy to apply color to their nails.

Cheeks. Blushers, the smallest color cosmetics classes, declined 54 percent in the five-year period between 1997 and 2002, from a U.S. shipment value of $417 million down to $193 million. This decline is likely the result of the 22 percent increase in face powder products that can be used to provide contour to the face but are not categorized as blushers.

The color cosmetics subset of the toilet products manufacturing industry grew 32 percent between 1997 and 2002, from industry shipments of $3.9 billion to $5.1 billion. Lip products, the largest group of color cosmetic products, saw the greatest growth while other product groups like those for cheeks and nails declined. Most manufacturers of color cosmetics have a full line of products, including items from each of the product categories. The toilet products manufacturing industry as a whole grew 23 percent in the five-year period from 1997 to 2002, from $23 billion in 1997 to $28 billion in 2002.

KEY PRODUCERS/MANUFACTURERS

Within the color cosmetics industry, some manufacturers have only top-drawer brands, and want to keep it that way. For instance, Chanel stays away from the masstige market fearing that association with the mass market will hurt, not help, sales. Other manufacturers such as L'Oréal have an impressive inventory of brands sold side-by-side at mass merchandisers such as Kmart and Target so that they have brands that compete with one another for sales.

A report by Citigroup Global Markets highlights the three top mass market color cosmetic makers who remained neck and neck over the period 2001 and 2004. In 2004 L'Oréal had a color cosmetic market share of 34 percent; Procter & Gamble (P&G) had a color cosmetic market share of 21 percent; and Revlon had a color cosmetic market share of 21 percent. Leading manufacturers are profiled in alphabetical order with note of popular brands.

Estée Lauder. The Estée Lauder portfolio consists of thousands of products across 26 brands, including Clinique. The Clinique franchise is a pillar of color cosmetics, so big it overshadows rival beauty brands. Lauder reported in August 2006 that Estée Lauder and Clinique are its fastest-growing prestige color cosmetic brands, as measured by same-store sales. Lauder also sells M-A-C and Bobbi Brown. It is committed to color cosmetics, where cream foundation tends to have the best margins. For most women, foundation is a high loyalty category. Once a person commits to Clinique or Bobbi Brown foundation, she or he rarely wants to change. Bobbi Brown planned to open 30 more U.S. doors in 2007. Lauder has 40,000 beauty consultants that helped make Clinique—and recently Bobbi Brown—pillars of color cosmetics. It renewed investment in training and educating its beauty consultants for one reason: consumers buy more under the auspices of a knowledgeable professional.

L'Oréal. Headquartered on Fifth Avenue in Manhattan, L'Oréal is the most successful mass market color cosmetic maker in the United States. L'Oréal is the U.S. color cosmetics leader with 34 percent of the mass market share. Its two strongest color cosmetics brands are Mabelline NY (taking 19% of mass color cosmetics market share) and L'Oréal Paris (with 15% of mass color cosmetics market share). L'Oréal grew both organically with its existing brands and through acquisitions such as Maybelline New York in 1996. L'Oréal is in a unique position in the color cosmetics market, selling both high-end products at department stores and low-end products at drugs stores and chain stores. Its Lancombe product line is more prestigious than Clinique and a longstanding color cosmetics brand.

Procter & Gamble. Cincinnati-based Procter & Gamble has been a giant in a wide range of consumer goods since its establishment in 1837. It produces nearly 300 brands in five segments: personal & beauty, house & home, health & wellness, baby & family, and pet nutrition & care. *ICIS Chemical Business Americas* reported in December 2006 that P&G is the second largest color cosmetics maker with 20.5 percent of the mass color cosmetics market. Its two strongest masstige color cosmetics brands are Cover Girl (taking 18% of market) and Max Factor (taking 3% of the market).

Revlon. According to *ICIS Chemical Business Americas*, Revlon moved up to capture 20.7 percent of the color cosmetics mass market in 2006 (with the Revlon brand taking 15% and the Almay brand taking 5%). Revlon was founded in 1932, by the Revson brothers and a chemist who contributed the "L" to the Revlon name. Starting with a single product—a nail enamel that used pigments instead of dyes—Revlon offered women a rich-looking, opaque nail enamel in a wide variety of shades. Revlon market has segments include cosmetics, skin care, fragrance, and personal care. Revlon's products are offered at affordable prices and it sponsors events that raise awareness and money for women's health issues. In the 1990s Revlon revitalized its color cosmetics business. It introduced the first transfer resistant lipcolor which led to a full ColorStay Collection of transfer-resistant color cosmetic products. Revlon's color cosmetic brands include Revlon, ColorStay, New Complexion, Revlon Age Defying, Super Lustrous, Almay, and Ultima II.

Other companies with important roles within the industry include Del Laboratories, Inc., Johnson & Johnson, Noxell Corporation, Shiseido Cosmetics Ltd., and Unilever. Avon and Mary Kay are two companies with a direct sales distribution model for color cosmetics.

MATERIALS & SUPPLY CHAIN LOGISTICS

The materials used by manufacturers in the production of color cosmetics consist of the ingredients necessary to produce the cosmetics as well as the materials needed to package them. Color cosmetics packaging plays an important part in marketing makeup products. Packaging provides color cosmetics with a means of product differentiation.

As a whole, the toiletries industry spent $8.0 billion in 2002 on the total cost of materials needed to make all kinds of toiletries, including color cosmetics. Of this, $5.6 billion was spent on packaging and other materials, and $2.4 billion was spent on ingredients. In other words, this industry spends 70 percent on packaging, and only 30 percent on buying ingredients that go into the cosmetics themselves. Industry-wide expenditures for plastic containers including jars, tubes, tubs, and bottles were $1.2 billion in 2002, more than on any other single ingredient class.

Setting the issue of plastics aside, this section examines that portion of the $2.4 billion industry-wide purchases of ingredients that end up on the lips, eyes, face, nails, and cheeks. The review is based on Census data for the entire toilet products industry because information about materials consumed is provided only at the industry-wide level, not at the individual product class level. The primary categories of ingredients needed to make toiletries of all kinds are, from largest to smallest in terms of industry-wide spending, as follows:

- Perfume oil mixtures and blends, essential oils (natural), and perfume materials (synthetic organic)—representing 41 percent of ingredient spending

- Synthetic organic chemicals—representing 30 percent of ingredient spending

- Bulk surface active agents (surfactants)—representing 10 percent of ingredient spending

Perfumes are used in the production process to impart a pleasant aroma to both the product and the packaging. Together, perfume ingredients purchased for their pleasant aroma accounted for almost 40 percent—or $935 million—of total industry ingredient spending in 2002. When purchasing aromas, more than 50 percent of the $935 million—$522 million—went toward perfume oil mixtures and blends. The cost to purchase essential oils (natural) decreased 7 percent between 1997 and 2002, dipping from $174 million down to $161 million.

The cost for purchasing perfume materials (synthetic organic), however, almost quadrupled in the five-year period that ended in 2002, ballooning from $67 million to $252 million, an increase of 278 percent. Synthetic organic perfume materials are created primarily from chemical compounds obtained during petroleum distillation, a process which separates petroleum into fractions according to their boiling temperatures. The advantages of synthetics are substantial. The almost quadrupling of their use shows how rapidly they became pervasive as ingredients. For manufacturers, they make economic sense because they do not depend on plant material and plant harvest from year to year. The result is stable prices, consistent supply, and more than 2,000 odor profiles to chose from (instead of only 200 plant-derived profiles).

Synthetic organic chemicals represent 30 percent of the cost of the ingredients needed to make color cosmetics. They are the largest single class of ingredients consumed by the toiletries industry (not including the materials used to make the packaging). Synthetic organic chemicals are generally derived from petroleum products during its separation into fractions according to boiling ranges. In color cosmetics, organic chemicals are most important as preservatives integral to making products with a long shelf life. Consumers expect color cosmetics to last years. Examples are the parabens. Parabens include methyl, propyl, ethyl, and butyl. They provide broad spectrum anti-microbial protection to toiletries products. The U.S. toiletries industry almost tripled its use of synthetic organic chemicals between 1997 and 2002, mushrooming

from purchases valued at $243 million to $721 million, an increase of 197 percent.

Surfactants are used to adjust the surface tension of a liquid or cream liquid to assist emulsifying in color cosmetics such as lip sticks and glosses, cream foundation, cream eye shadows, and cream blushes. Wetting agents that help products spread, surfactants are integral to making creamy color cosmetics that go on evenly and smoothly. The U.S. toiletries industry decreased spending on surfactants by 26 percent in the five-year period between 1997 and 2002, from $330 million to $243 billion.

DISTRIBUTION CHANNEL

Color cosmetic products exist side-by-side as both high-end products and low-end products. Product lines on either end of this spectrum are distributed through a distinct channel. Within the industry these channels co-exist because they serve different consumer segments. They are often referred to as the prestige channel and the masstige channel.

Prestige products are classified primarily based on price and are distributed via the department store channel. Masstige products are primarily lower-priced products distributed via a mass market distribution channel.

The department store distribution channel is a classic and large channel for prestige products. Prestige brand names were built through their exclusive availability in high-end department stores and this channel represents 40 percent of North American sales of cosmetics. This channel underwent a shakeup when Federated Department Stores acquired May Department Store Company. At the time the deal was finalized in August 2005, Federated owned 458 stores nationwide, mostly known as Bloomingdale's and Macy's, while May operated 491 department stores nationwide under the names Lord & Taylor and Marshall Field's, among others. In September 2006, the May Department Store nameplates were formally changed to Macy's. All prestige companies, or 40 percent of the color cosmetics industry, had to deal with department store consolidations. In effect, the Federated May consolidation was a merger of the two biggest distribution channels for prestige color cosmetics. An estimated 75 stores closed. Even with the department store distribution channel shake up, department stores account for 40 percent of sales within the color cosmetics industry.

The department store distribution channel for prestige products involves delivering the product via a company-trained beauty consultant. For instance, Estée Lauder has 40,000 beauty consultants that helped make Clinique—and later Bobbi Brown—a pillar of color cosmetics. Lauder renewed investment in training and educating its beauty consultants for one reason: consumers buy more under the auspices of a knowledgeable professional. Respondents to a recent Bobbi Brown survey indicated that a lesson with a makeup artist was one of the strongest influences on spending. It will remain so.

The department store channel offers makers a controlled environment, which they know generally equates to higher sales. Consumers also prefer the prestige channel since mistakes made in masstige product purchases can be cumulatively expensive. Many women may have a dozen or more products at home that were never used because colors were too bright, too gold, too orange, or all wrong. It is not uncommon for a woman to have a drawer that includes a Max Factor eye shadow (triple), a Revlon wet-dry shadow in lavender (single), and Oil of Olay midnight red lipstick—all used only once.

The mass market distribution channel is broadening in part because store loyalty is less prominent with masstige products than for prestige products. The mass market distribution channel grew to include a large variety of places to buy products. Lower-price products are sold through a throng of outlets: drugs stores such as CVS, Rite Aid, and Walgreens; food stores including Whole Foods and health food stores; mass merchandisers including KMart, Target, and Wal-Mart; and nontraditional Internet retailers. The broadened masstige channel includes warehouse club stores like Costco and Sam's Club. Popular Web sites that allow for online sales include cosmeticamerica.com and sephora.com.

In 2005 Lauder's internet operations contributed more than 30 percent to its U.S. sales surge. On that model, Lauder expanded e-commerce internationally by launching sites for Estée Lauder, Clinique, and M-A-C in the United Kingdom. Lauder has 17 single brand marketing sites, 11 with e-commerce capabilities. Because color cosmetics are fashion-trend focused, often just as a consumer finds a product color she prefers, it is discontinued and replaced with a trendier color. This has led to Web sites for discontinued color cosmetics. One example is cosmeticsandmore.com. The Urban Decay brand also has a website with one section for discontinued items called RIP. It explains that because its customers are cool and up on what is hip, it regularly introduces new colors and sells discontinued items on its Web site.

The mass market distribution channel is expanding. Cosmetics manufacturers actively courted a broader distribution channel using technologies including cable TV, cell phones, and podcasts. For instance, QVC, a direct sales cable network, is no longer perceived to be a sales channel for exclusively down market items. Philosophy—a prestige product sold at Macy's, Bloomingdale's, and Nordstrom—partnered with QVC eight years ago and is now the home shopping network's top beauty brand, edging out Bare Escentuals. QVC accounted for 48

percent of Philosophy's net wholesale volume of $120 to $150 million in 2006.

Bare Escentuals went to QVC in 1997. Bare Escentuals still relies on infomercials—not generally associated with the prestige air that matches its price points—to educate consumers about its crushed minerals, patents pending, and organic ingredients, making it easier for the consumer to part with $60 for 0.15 ounce of Rare Minerals Skin Revival Treatment, meant to be worn while sleeping for luminous skin. Even the pillar of prestige department store sales, Clinique, moved beyond TV and glossy magazines to health clubs, Weight Watchers magazine, and college campuses in order to reach a diverse audience.

KEY USERS

Users of color cosmetics are predominantly women of all ages. The Food and Drug Administration (FDA) helps identify key users of color cosmetics by referring to the generally accepted uses of cosmetics, including color. The FDA defines cosmetics as "articles other than soap that are applied to the human body for cleansing, beautifying, promoting attractiveness, or altering the appearance." From this classic definition, it can be assumed that the key user is primarily a female with an interest in beautifying herself and promoting attractiveness; thus, use of color cosmetics may increase during activities involving dating, a process which often involves attractions.

The products purchased are dependent upon the financial circumstances of the consumers who buy them. Products purchased are also dependent upon the talent of the consumers who buy them. Some women do not need to purchase prestige products in order to utilize the services of a trained beauty consultant because they inherently know what works. Women with large, well-spaced eyes quickly learn to use eyeliner to highlight and draw attention to them. Women with narrow, closely-spaced eyes tend to stay away from eyeliner in order to highlight something else, perhaps luscious lips. Women learn early what their best feature is: lips, face, eyes, cheeks, or nails, and purchase products that work best for them. Barbra Streisand, for example, is well known not only for her voice, but for her striking nose and long manicured nails. She uses color cosmetic products to draw attention to one and away from the other.

ADJACENT MARKETS

The apparel industry is a market that is strongly adjacent to color cosmetics. The use of color cosmetics is strongly influenced by fashion trends. Changes in fashion impact sales of color cosmetics. For the most part, shifts in fashion tend to increase the sales of one type of cosmetic, say darkly colored lip sticks, while dampening demand for another such as bright blue eye shadow. Since most cosmetics manufacturers produce an impressive inventory of color cosmetics, they tend to be prepared for fashion shifts. The cosmetics industry also directly affects fashion trends and is, therefore, well-suited to plan ahead in order to take full advantage of sales opportunities that grow out of shifting fashions.

Knowledge, too, can impact the use of color cosmetics and is therefore an adjacent market. Knowledge about skin care products in particular influences cosmetics usage since skin care products are closely adjacent to the cosmetics industry. Books, tapes, and DVDs are available for the consumer to learn skin care and color cosmetics beauty lessons in her own living room. Ostensibly, this knowledge will save money and time.

The consumer who has made purchasing mistakes that became cumulatively expensive might well benefit from such books and tapes. Recent examples include Laura Mercier: *The Flawless Face*, a VHS tape that demonstrates expert applications of makeup and shares insider tricks. Another is *Your Makeup: Simple Steps to Amazing Looks*, a DVD by celebrity makeup artists who reveal their secrets.

Bobbi Brown has four best selling books. Her first book, *Bobbi Brown Beauty: The Ultimate Beauty Resource* (HarperStyle 1997) was a beauty guide for women of all ages. Her next book *Bobbi Brown Teenage Beauty* (Cliff Street Books 2000) was a New York Times Best Seller. *Bobbie Brown Beauty Evolution* (Harper Resource 2002) was for women of every age. *Bobbi Brown Living Beauty* (Springboard Press 2007) redefines beauty for women over 40. The books are successful at creating increased interest in the entire Bobbi Brown franchise.

RESEARCH & DEVELOPMENT

Color cosmetics are fashion-trend focused. Research and development activities must be ongoing because most products stay on the shelf less than four years. New products must be developed to meet the constant demand for a cycle of renewal. Color cosmetics R&D tends to focus on where the market growth is robust. In 2006 Lauder's research centered on developing advances in lip glosses, foundations, and mascaras. Not surprisingly, these segments correlate to robust product classes. Lip cosmetics, for example, grew at the rate of 157 percent in the five-year period between 1997 and 2002. The color cosmetics industry is competitive, and constantly changing due to scientific advancements in the ingredients used to make the products.

One such scientific advancement is related to pigments, which supply the color in color cosmetic products. R&D resulted in an innovation called borosilicate pigment technology that produces multi-dimensional pigments that make distinctive color-shifting effects available in color cosmetics. Borosilicate flakes are conducive to pearlescent pigment and can be fine-tuned for both transparency and sparkle. The flake pigment allows the unaided eye to discern colors particle-by-particle, which provides the multi-color effect. L'Oréal was among the first to introduce products utilizing this new innovation and signed Hollywood starlet Scarlett Johansson to promote its new line of color-changing cosmetics. The brand is known as HIP because it makes uses of special high intensity pigments. HIP was launched at U.S. chain, drug, and mass market outlets in February 2007. It is a comprehensive range of color cosmetics, including products for lips, face, cheeks, and eyes.

CURRENT TRENDS

In 2000 L'Oréal established its Institute for Ethnic Hair & Skin Research in Chicago, Illinois—the first and supposedly only research facility operated by a beauty company to conduct basic scientific research on the unique properties of the skin of people of African descent. In 2006 Lauder reinvigorated Prescriptives, a line designed around the customized color cosmetics concept. The Prescriptives line is uniquely positioned to appeal to women of diverse ethnic backgrounds. These are indications of trends in color cosmetics. Makers are formulating products to meet the needs of minorities including African American women and women of Hispanic origin.

Another trend involves the ever-expanding mass market distribution channel. Recognizing the effects of consolidation of the department store industry, and the trend toward more women buying from mass merchandisers, Lauder subsidiary BeautyBank created three new color cosmetics and skin care brands in 2004 for Kohl's, a discount chain with 750 stores in the United States. Lauder will retain ownership of the three brands. This represented Lauder's first foray into the low-end or mass market. One brand called Grassroots rolled out in 2006. Another is American Beauty for which actress Ashley Judd is the spokesmodel. All are available on Kohl's Web site.

TARGET MARKETS & SEGMENTATION

Cosmetics manufacturers target market segments that provide significant growth opportunities, such as lip cosmetics with its triple-digit growth or cream foundations with their highly loyal consumers. In color cosmetics, women over the age of 50 are a targeted segment along with ethnic groups with unique needs that have not been fully met.

Women over the age of 50 represent a market that has been traditionally underserved by the cosmetics industry, a fact that has been changing during the first years of the twenty-first century. ACNielsen estimates that female heads of households who are over age 45 account for nearly 70 percent of mass market color cosmetics purchases.

L'Oréal launched a cosmetics collection for women in their 50s and 60s with actress Diane Keaton as the spokesmodel. Revlon kicked off its biggest launch in more than a decade—Vital Radiance. Vital Radiance is built on the premise that women who are 50 years old or older need to update their choice of color cosmetics as well as their application techniques to reflect the fact that their skin is changing—lines are appearing, color is diminishing, and lashes and eyebrows are becoming more sparse.

Ethnic and racial groups are also segments of the market that have been traditionally underserved by the cosmetics industry. These segments of the market are receiving increasing attention from cosmetics manufacturers. The fastest growing segments of the U.S. population by race and ethnicity are, in order of growth, Hispanics, Asian Americans, and African Americans. Together, these three segments of the population represented 24 percent of the population in 1990 and by 2005 they accounted for 31.5 percent of the population. This trend in demographics paints a picture of a population whose dominant skin tone is darkening. Color cosmetics for ethnic skin types are a product group designed for a growing market.

One problem that cosmetics manufacturers have encountered in their pursuit of new products for the ethnic market is the fact that chemists who are formulating new products often have little knowledge of the ethnic demographic. Upcoming formulations for black skin are predicted to include natural shea butters, plus cocoa and peanut oils because black skin is frequently dry and needs to be moisturized. Procter & Gamble's Cover Girl brand is targeting this market in glossy magazine advertisements and television commercials that feature beautiful African American singers Queen Latifah and Beyonce as spokesmodels.

Mag Beauty Puig Fragrance and Personal Care Division NA maintains that paying attention to the buying preferences of Hispanic women helps them distribute Hispanic-oriented cosmetics under the Maja and Heno de Pravia brands, both strong sellers in many Latin American countries. Brand name recognition can drive sales that will open up the Hispanic market. Mag Beauty advertises its two brands heavily in Spanish language magazines.

The unmet needs of women over 50 and of ethnic minorities will drive future color cosmetics R&D. Successfully selling to these groups will contribute to the broadening of the distribution channel for color cosmetics.

RELATED ASSOCIATIONS & ORGANIZATIONS

Cosmetic, Toiletry and Fragrance Association, http://www.ctfa.org

The European Cosmetic Toiletry and Perfumery Association, http://www.colipa.com/site/index.cfm?SID=15588

Research Institute for Fragrance Materials, http://www.rifm.org

Synthetic Organic Chemicals Manufacturers Association, http://www.socma.com

BIBLIOGRAPHY

"Brief History of Beauty and Hygiene Products." Duke University. Available from <http://scriptorium.lib.duke.edu/adaccess/cosmetics-history.html>.

"Chanel Makeup." Chanel. Available from <http://www.chanel.com/mprod>.

"Color Cosmetics Glow." *ICIS Chemical Business Americas.* 18 December 2006, 1.

Darnay, Arsen J. and Joyce P. Simkin. *Manufacturing & Distribution USA,* 4th ed. Thomson Gale, 2006, Volume 1, 574–578.

"Facts for Features: Black History Month." U.S. Department of Commerce, Bureau of the Census. 5 December 2006.

Guste, Constance. "Loyal Customers: Top 10 Consumer Brand Categories that Generate the Most Brand Loyalty." *WWD.* 28 September 2006, 18.

"Innovations in Lipstick Technology." *Cosmetics & Toiletries Newsletter.* 9 September 2003, Volume 118, Number 9.

Lazich, Robert S. *Market Share Reporter 2007.* Thomson Gale, 2007, Volume 1, 295–297.

"L'Oréal Paris Skincare." L'Oréal Paris. Available from <http://www.lorealparisusa.com/cosmetics/lin-up>.

"A Market that Requires Attention." *MMR.* 24 April 2006, 1.

Monks, Richard. "Salon Patrons, Retail Shoppers Aren't Mutually Exclusive." *Chain Drug Review.* 11 September 2006, 69.

"Over 50 is No Longer Such a Bad Place to Be." *MMR.* 8 May 2006, 97.

"P&G Product Site Map." Procter & Gamble. Available from <http://www.pg.com/product_card/brand overview>.

Pitman, Simon. "Englehard Unveils Special Effects Technology for Color Cosmetics." Cosmetics Design.com. Available from <http://www.cosmeticsdesign.com/news/printNewsBis.asp?id-65902>.

Schaefer, Katie. "The Road to More Effective Ethnic Skin Care." *Cosmetics & Toiletries Magazine.* February 2007, 88.

"Table 14. Resident Population, by Race, Hispanic Origin, and Age: 2000 to 2005." *Statistical Abstract of the United States: 2007."* December 2006, 15.

"Table 16. Resident Population by Race, Hispanic Origin Status, and Age—Projection: 2010 and 2015." *Statistical Abstract of the United States: 2007."* December 2006, 18.

"Toilet Preparation Manufacturing: 2002." *2002 Economic Census.* U.S. Department of Commerce, Bureau of the Census. Available from <http://www.census.gov/econ/census02/>.

"U.S. Personal Care and Household Products Digest." Citigroup Global Group and Smith Barney. 15 February 2005.

Walker, Rob. "Earth Cover: How a Cosmetics Company Replaced Romance with the Glow of Rationality." *New York Times Magazine.* 21 January 2007.

SEE ALSO *Cosmeceuticals, Sun Care Products*

Countertops

INDUSTRIAL CODES

NAICS: 32–6199 Plastics Product Manufacturing not elsewhere classified, 32–7991 Cut Stone and Stone Product Manufacturing, 33–7110 Wood Kitchen Cabinet and Countertop Manufacturing, 33–7215 Showcase, Partition, Shelving, and Locker Manufacturing, 32–7122 Ceramic Wall and Floor Tile Manufacturing

SIC: 2434 Wood Kitchen Cabinets, 2541 Wood Partitions and Fixtures, 2542 Partitions and Fixtures, Except Wood, 3089 Plastics Products, not elsewhere classified, 3253 Ceramic Wall and Floor Tile, 3281 Cut Stone and Stone Products

NAICS-Based Product Codes: 32–61998, 32–6199A, 32–6199W, 32–79911, 32–79914, 32–79917, 32–7991W, 33–71101, 33–71104, 33–7110B, 33–7110F, 33–71220, 33–7215A, 33–7215E

PRODUCT OVERVIEW

The word counter designates any raised surface supporting work to be accomplished, be that the exchange of goods at a store, transactions to be accomplished at a bank or at an airline ticket counter, food preparation to be done in a kitchen, or personal care to be accomplished in the bathroom. The counter acquired its name in the early days of open markets where money changers set up operations and counted out coins and bills. Counters almost always take the form of a cabinet useable also for storing goods, supplies, or implements beneath the countertop. The countertop refers to the topmost layer of the cabinet usu-

ally sold separately from the cabinet itself because it has a specially formed layer of material chosen for its attractiveness or functional characteristics.

Countertops come in six distinct categories: laminates, solid synthetics, stone, ceramics, wood, or steel. The counter itself is usually framed in wood or metal and is finished in the same material or, in cases where architectural features are important, are faced in stone. Six different materials are used for countertops, each with particular strengths and weaknesses.

Laminates. The structural base of laminated countertops is a solid wooden board to the top and sides of which a covering made typically of plastics has been laminated to make a composite object. The laminating material may also be made of strong kraft or printed sheets sealed by protective layers of synthetics. By far the most common laminate, however, is a sheeting of solid-colored plastic material. Laminated countertops are the dominant product in the market for all applications, domestic as well as commercial. They are cost-effective and run approximately $10 to $40 per square foot of surface.

Solid Surface. The term refers to countertops made of a single synthetic material formed by the mixing a mineral compound with polyester and/or acrylic resins. A popular mineral is alumina tri-hydrate, ATH. ATH is derived from bauxite ores and represents 45–70 percent of the content of the material and supplies the structural strength of the countertop whereas the polyester or acrylic lends the object sheen, water-, and stain-resistance. Solid surface counters are popular because they are seamless and offer uniform appearance. They are flexible and easy to shape during installation. Solid surface countertops have some disadvantages though, as they are vulnerable to high heat and scratch relatively easily—although the scratches may be buffed out. They run from $70 to $150 per square foot.

Stone. Stone counters come in two versions: those made of natural and those of engineered, or modified, stone. Natural stone is an attractive but expensive material more often encountered in flooring and on walls than on counters, but advancing efficiency in quarrying has reduced the cost of natural stone so that it is appearing in high-end counters as well. Siliceous and calcareous stone varieties are used. Siliceous stones include granite, quartz-based stone, serpentine, slate, and soapstone. Calcareous stones include limestone, marble, onyx, and travertine. These stones vary by color, thickness, texture, strength, and in the overall aesthetics of their appearance. Granite is the most durable but must be resealed periodically. Stone is porous. Marble is attractive but less durable and more easily chipped. Soapstone requires less maintenance than most of the other stones; it was the stone used as countertops in early New England settlements. A limitation of stone counters is that the user must refrain from cleaning them with abrasive powders or devices. Natural stone countertops range in cost between $70 and $100 per square foot in the usual installation. In more luxurious projects the stone may cost the buyer as much as $300 per square foot.

Engineered stone is a manufactured product in which a selected type of stone is combined with a pigmented polymer resin to produce a surface similar to solid surface (all-synthetic) countertops but with a more natural look. Stone is usually reduced to particles of various sizes and then slurried with polymer, the slurry used to form the countertops in molds. Before letting the mixture set, the producer must mechanically free all air from the slurry and must also pressurize it so that stresses that may be present as a result of particle alignments do not later cause the countertop to crack. Engineered stone offers some advantages over natural stone; it is more scratch-resistant, and the natural porosity of stone is countered by the presence of the polymer; the surface is thus water-resistant. Some consumers like such surfaces because they offer a consistent color often lacking in natural stone. Engineered stone countertops, however, cannot be seamed together invisibly.

Ceramic Tiles. Ceramic tiles are made from pressed clays with a matte finish or a glaze produced by the use of metallic oxides and colorants to pre-stain the ceramic before baking. Individual tiles are heat-resistant, seamless, and have the natural attractiveness of all ceramics if well

made. The disadvantage of countertops formed of tiles is that tiles may be chipped or cracked accidentally if the user drops heavy objects; grout lines at the interface of tiles may become stained, and the stains may prove difficult to remove. The cost per square foot is affected by various factors. Handmade tiles may drive up the price. At the low end costs will run between $10 and $20 per square food.

Wood. Wood and butcher block counters are typically made of glued layers and pieces of hardwood maple. Other types of wood might be used as trim or as accents. Most counters are one to six inches thick. Part of the appeal of wood is that as a natural material it gives a room a warm feeling. But the material has its drawbacks: knife marks show, the wood needs to be resealed periodically, and butcher block wears out. Wood may cost $50 to $70 per linear foot.

Stainless steel. Stainless steel is steel alloyed with chromium; the chromium helps the surface avoid oxidation, or rusting. Stainless steel is easy to clean and is attractive, but produces an industrial sort of ambiance. Noise problems can be minimized if a layer of plywood is installed beneath the counters. Stainless steel shows tracks or activity and leaves fingerprints showing if it is touched after cleaning. Apart from expensive natural stone surfaces, it is the most expensive counter on the market running in the range of $85 to $100.

The six categories briefly characterized above leave a wide range of other specialized solutions for the discriminating customer who wishes to install something unique. Such a buyer might opt for lavastone which is non-porous and thus delivers a natural stone-counter that will not stain, comes in many desirable colors, and offers takeoff-points for conversation, for example, about volcanic eruptions and such. Those inclined to show off wealth may opt for semi-precious stones set in binder materials that will show off the mineral, yet provide a smooth surface. Paper-based countertops have been used for some time in the commercial sector. Glass counters can offer rich textures and colors. The surface is non-porous and easy to clean. Modern glass products are able to withstand the rigors of most kitchens, bathrooms, and store counters where such tops are typically used—jewelry stores, optical departments, and many other department stores where allowing customers to look through the glass at the product is desired.

MARKET

Estimating the Market. Estimating the total market for countertops is difficult because data from the U.S. Bureau of the Census are available in detail only for a limited range of countertops—for kitchen and bathroom cabinets

FIGURE 75

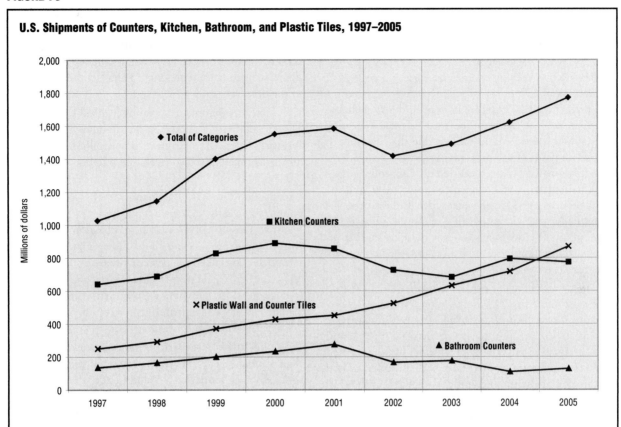

U.S. Shipments of Counters, Kitchen, Bathroom, and Plastic Tiles, 1997–2005

SOURCE: Compiled by the staff with data from "Value of Product Shipments: 2001," and "Value of Product Shipments: 2005," *Annual Survey of Manufactures*, U.S. Department of Commerce, Bureau of the Census. January 2003 and November 2006.

and for plastics tiles used for walls, floors, and countertops. Data for stone, metal, ceramic, or glass countertops are hidden from view. In the stone products industry, for instance, industry shipments for granite, limestone, and marble are provided, but categories are too broad for proper statistical analysis. Stainless steel shipments data do not break out countertops. The glass products industry provides a single shipment number for architectural glass and automotive glass in combination. The architectural glass mentioned is not windows; windows are reported separately.

The U.S. Census Bureau reports shipment data for countertops in three categories—kitchen and bathroom counters and plastics tiles used for wall and counter coverings. Shipment trends from 1997 to 2005 are shown in Figure 75. It is well to note that these are shipment data and not the value of the product as installed in a kitchen or a bathroom. This subset of the total market produced shipments in 1997 of $1.0 billion rising to $1.8 billion in 2005, growing at the rate of 7.1 percent per year. Most of that heady growth, however, came from plastic tiles, increasing at a rate of 16.9 percent per year. The category

includes wall and ceiling tiling as well as counters though counters are a small percent of the total. Plastics-laminated kitchen countertops grew at the more modest rate of 2.4 percent and bathroom countertops actually exhibited a decline at the rate of 0.8 percent.

An overview of the market can be seen in market research data provided by Freedonia Group Inc., a research firm, as published in *Wood & Wood Products* magazine. These data show kitchen and bathroom countertop demand expressed in millions of square feet. Demand in 1997, as reported by Freedonia, was 353 million square feet rising to 467 million square feet by 2007, representing annual growth of 2.8 percent. In 2007 laminated countertops represented 59.3 percent of the square footage demanded. Other categories defined by type, in order of size, were solid surface countertops (11.2%), natural stone (7.1%), and engineered stone (5.4%). The second largest category was actually All Other, including plastics and ceramic tiles, metal, wood, and glass in combination. In order of overall growth, natural stone led with growth of 10.2 percent per year, engineered stone with 9.5 percent, solid surface with 2.8 percent, and the two largest categories, laminates and All Other, experienced demand growth at 1.7 percent per year each. The growth rate for engineered stone, a recently introduced product, reflects annual change from 2002 forward. Figure 76 presents materials shares, based on square footage demand for 2007.

Freedonia's uses an average cost per square foot of countertops of $25 at the production level, suggesting a total market in 2007 of $11.7 billion for the kitchen and bathroom cabinet market. The retail market, reflecting installation charges, would well double that amount. If the middle point of the price ranges shown above are assumed, laminated products would cost $25, solid surface $110, natural stone $85, engineered stone $100, and the All Other category $40 per square foot. If these numbers are applied a market size in 2007 of $21.2 billion is obtained at retail. How much larger is the total market including countertops in stores, banks, and in a great variety of service activities?

To get a feel for this expanded market, the following may be assumed. In 2007 construction activity, as reported by the Census Bureau in dollars is divided into private residential (46.3%), private non-residential (29.4%), and public sector construction (24.1%). If all the sectors used counters at the same rate as residential construction, the total market at the production level would be $25.3 billion. This approach, however, would overstate the market. Large components of nonresidential and public construction are marginal or non-users of counters—office buildings, utilities, and road construction. A closer look reveals

that 29 percent of these other sectors are in the countertop market. These include commercial construction on the private side and educational construction on the public side. If the numbers are readjusted to include in total construction only sectors that are installing counters, residential construction represents 75 and all other 25 percent of the base.

This now permits us to produce a ball-park estimate of total countertop market. At the production level, starting with Freedonia's estimate, $15.6 billion is obtained and at the retail, installed level, $28.3 billion.

Trends in Materials. Laminates are the most popular countertop thanks to low cost and superior performance features. The category, however, has been under pressure from more expensive and more showy materials. Laminates have lost nearly 7 percent of their share between 1997 and 2007. They represented 66.1 percent of square footage in 1997 and 59.3 percent ten years later. The leading makers of laminate include Wilsonart, Formica, Nevamar, Pioneer, and Laminart.

Solid surface countertops have grown more rapidly than laminates but had 11.2 percent of the market (in square footage) both in 1997 and in 2007. Leading makers of solid surfacing include DuPont Corian, Formica, Wilsonart, and Avonite.

Consumer preferences have shifted to stone surfaces because of the luxury and style that such materials offer. Granite and slate were once quite expensive. Low cost providers from Asia and other regions have helped to drive costs down. Demand for natural stone was 12.5 million square feet in 1997 and 33.1 million squares in 2007. Engineered stone is a relatively new material. Its primary advantage is that it has the attractiveness of stone but is less porous and more heat-resistant. Stone has achieved the greatest gains in market share, natural stone a 3.5 percent gain in the ten-year period, engineered stone a 9.5 percent gain since 2002. Leading makers of engineered stone include Zodiaq, Clanfield, CaesarStone, and Silestone.

KEY PRODUCERS/MANUFACTURERS

DuPont Corian. The DuPont company's origins go back to the early nineteenth century. Founded by E.I. DuPont, the company made only one product from 1802 to 1840: gunpowder. DuPont expanded into other arenas and soon set up mining and manufacturing facilities outside of its home on the Brandywine River in Delaware. The company would go on to be a leader in high explosives, paint, dye, ammonia, and cellophane manufacturing. In the 1960s the company offered the public some of its greatest inventions: Lycra (1962), Kevlar (1965) and Tyvek building wrap (1966).

FIGURE 76

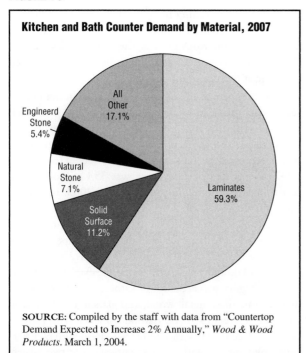

Kitchen and Bath Counter Demand by Material, 2007

All Other 17.1%

Engineerd Stone 5.4%

Natural Stone 7.1%

Solid Surface 11.2%

Laminates 59.3%

SOURCE: Compiled by the staff with data from "Countertop Demand Expected to Increase 2% Annually," *Wood & Wood Products.* March 1, 2004.

The 1960s also marked the development of polymethyl methacrylate as a synthetic type of marble surfacing suitable for kitchen countertops. In 1967 DuPont introduced the material as Corian in three colors: cameo white, dawn beige, and olive mist. The material was an improvement on existing nonporous surfaces because it was resistant to chipping, cracking, and staining; scratches and scorches could be scoured or sanded away. By the mid-1970s, Corian was also being used as a surfacing material in bathrooms.

According to DuPont, market research conducted in the early 1990s revealed that the public regarded Corian as a luxury item. DuPont set out to reduce the cost of the material, increase the number of colors, and to promote the product's durability and attractiveness. DuPont substituted a continuous single-sheet production method for the earlier batch process and substantially cut production costs. The company also increased available colors from eighteen in 1992 to eighty-one in 2000. DuPont has also successfully marketed its product outside the residential sector. Corian is now used in hospital counters and in furniture construction.

Formica Corp. Daniel J. O'Conor and Herbert A. Faber of Westinghouse invented formica in 1912. The material was originally conceived as an electronic insulator and a substitutive for mica, a mineral frequently used as insulation in electrical equipment. The two men named the product using its intended function (*for-mica*). The product was discovered in 1913.

The growing electronics market in the years after World War I prompted the company to move into other markets. Improvements in manufacturing processes allowed Formica to produce sheets that imitated marble or wood surfaces. This material was soon a hit with the public. According to the *Modern Design Dictionary*, in the years after World War II, approximately 2 million of the 6 million homes built in the United States had Formica countertops.

Formica was bought and sold five times during the late 1980s and in the 1990s. By the end of the decade, the company was heavily leveraged and could not pay its debts. One measure of how far the company had fallen was the loss of its market share. In the late 1980s Formica had 50 percent of the market for decorative laminates. Fifteen years later, the company's share was at 25 percent of the market. Wilsonart International, meanwhile, had seen its fortunes go in the opposite direction. Over the same period, according to a March 1999 *Business Week* article, its market share in this sector increased from 25 to 50 percent. In 2002 the company declared bankruptcy.

Formica is headquartered in Cincinnati and employs more than 3,800 people. It has 14 manufacturing plants in North America, Europe, and Asia. In May 2007 the company was purchased by New Zealand-based Fletcher Building Ltd.

Dal-Tile. Dal-Tile is the largest maker and distributor of ceramic tiles. The company operates ten manufacturing facilities in the United States and in Mexico and employs more than 10,000 people. Dal-Tile began operations in 1947; it established its first glazed wall tile manufacturing facility and corporate headquarters in Dallas, Texas. Since that time, operations have greatly expanded and now incorporate the manufacture of glazed and unglazed floor tile, glazed and unglazed ceramic mosaics, and unglazed quarry tile. In 2002 the company was acquired by Mohawk Industries, the second largest maker of carpet in the United States. In that year Dal-Tile reported a 26 percent share of the overall ceramic tile market. Its annual production capacity was 527 million square feet. Top brands include Daltile and American Olean. In 2006 its parent corporation, Mohawk, reported Dal-Tiles sales at $1.9 billion.

Wilsonart International. Ralph Wilson founded the Ralph Wilson Plastics company in 1956. By the 1970s the company had established fifteen warehouses across the country to help distribute its laminate product to customers. By the 1980s the company began diversifying its product line and offering new colors and patterns. The company became one of the leading laminate distributors in the world. Its brands include many popular names in countertops and flooring including Estate Plus, DuoLink, Oasis, Roca, Earthstone, and Gibraltar. Wilsonart also sells adhesives, cleaning, and trim products. The company reported 3,402 employees and revenues of $742 million for the fiscal year ended December 2006. It is a subsidiary of Illinois Tool Works.

MATERIALS & SUPPLY CHAIN LOGISTICS

Stone. Dimension stone is defined as natural rock material quarried for the purpose of obtaining blocks or slabs that meet specifications as to size (width, length, and thickness) and shape. Granite, sandstone, marble, slate, and other types of natural stones are considered dimension stone. Such stone is a popular choice for countertops. Part of the reason this sector has seen strong interest is that there are new supplies. New supplies of stone are available from China, India, Africa and other regions not available ten years earlier.

According to the U.S. Geological Survey, there were 100 companies in this industry in the United States. They operated 136 quarries in thirty-five states—Wisconsin, Vermont, and Georgia were most likely to be the location of these quarries. Stone production has been increasing since 1991 when production was placed at 1.16 million tons. In 2006 production was 1.53 million tons valued at $275 million. Limestone represented 38 percent of mine output in 2006, followed by granite (27%), marble (14%), sandstone (13%), slate (1%) and other types (7%). Imported stone is highly popular. An estimated 2.5 million metric tons were consumed in 2006, worth $2.7 billion.

Plastics and Ceramics. There are a number of ways to process plastics, including film casting, rotational molding, extrusion, and lamination. Laminate wall and counter coverings move in the market in the same production channel as a number of other products that are used every day, including wastepaper baskets, flowerpots, certain kinds of door hardware, and various parts of shoes. This industry saw shipments of more than $16 billion and generated employment of 63,400 people in 2005. Manufacturers should continue to see steady growth rates as plastics continue to be integrated into numerous consumer products. As for ceramics, more of the floor and wall tile industry is being satisfied through imports. According to Freedonia one-quarter of the ceramic market came from imports in the early 1990s. By 2001 three quarters of the ceramic tile supply was imported, largely from Italy and Spain, countries long considered expert manufacturers in the trade.

DISTRIBUTION CHANNEL

The Kitchen Cabinet Manufacturers Association is a trade association for cabinet and countertop makers. It lists nine points of purchase for consumers: builders, remodelers, retail showrooms, architects/designers, wholesale distributors, home centers (e.g. Home Depot or Lowe's), direct-to-consumer channels, lumber yards, and multi-family building projects. Among companies that claim more than $25 million in sales annually by outlet break down as follows: retail dealers represented the largest outlet of kitchen cabinets in the market, claiming 44 percent of sales in 2006. Home Centers continued to lose market share, falling from 20 percent in 2005 to 16.6 percent in 2006. Builders' position in the market slipped somewhat during this period from 17 percent in 2005 to 15.3 percent in 2006 owing in large degree to the housing slump. The market share held by distributors was up from 2005 to 2006 from 15 percent to 22 percent, respectively.

According to an estimate by Lowe's, building material stores and home centers represented 27 percent of cabinet and countertop sales in 2005. This was down only slightly from the 28 percent market share they held in 1999. Cabinets and countertops were also the fastest

growing product at Lowe's stores. From 1999 to 2004 sales saw annual growth of 14 percent.

Customer Control. One growing trend in the distribution of kitchen cabinets and countertops is known as customer-controlled ordering. The use of sophisticated Web sites by cabinet and countertop manufacturers allows customers to log onto the site, make selections, and place orders themselves directly with the company.

Once only used by designers kitchen and bathroom planning software is becoming more available than ever as either downloadable software from a Web site or purchased at a store for home use. Such software allows customers to plan their own kitchens without the necessity or cost of a designer. Some kitchen planning software enables a direct interface with the cabinet company's inventory and ordering systems through which purchases can be made without having to deal with a salesperson. Planning software is also available increasingly for in-store use in conjunction with a representative at dealers, distributors, and home centers.

KEY USERS

According to Freedonia the remodeling segment accounted for 70 percent of all countertop sales in 2004. Bathroom and kitchen remodeling has long been seen as offering excellent returns on investment for the homeowner.

The Census Bureau estimates that $228 billion in total home renovation was performed in 2006. The renovation market continued to be driven in the new century by numerous factors, including a growing housing stock, the age of existing homes, and the increased wealth of homeowners, particularly the top fifth of the population. It will also be driven by consumers' desire for increased comfort and luxury. The total housing stock exceeded 126 million in 2006. The median age of these homes was thirty-two years in 2005, up from twenty-three in 1985. In other words, homes on the market were getting older and more likely to be in need of updating. Buyers of existing homes are likely to do some remodeling in order to make the home more their own.

Approximately 15.4 million U.S. households (14% of all households) remodeled a bathroom and 10.6 million remodeled a kitchen in 2005. According to data from Mediamark Research, reprinted in the *Statistical Abstract of the United States*, another 1.9 million households added a bathroom and another 5.8 million installed new countertops. Some homeowners do the work themselves while others contract out the work. The budget for such work varies as well. For example, of the 5.8 million households purchasing new countertops, 1.5 million households spent less than $1,000. Another 1.3 million spent

$1,000–$2,999, and 1.2 million spent more than $3,000. Approximately $21 billion was spent on bath and kitchen remodeling in 2005.

The National Kitchen and Bath Association offered a slightly different analysis on remodeling done in the United States. The association noted that of the 10.3 billion baths remodeled in 2006, 5.5 million were master baths, 3.4 million were other full baths, and 1.3 million were powder rooms. A total of 5.2 million new bathrooms were planned; approximately half were full baths, one-third were master baths, and the remainder were powder rooms. A total of 9.2 million kitchens were remodeled.

ADJACENT MARKETS

The countertop industry is seen as something of a tag-along to the kitchen cabinet industry writ large. Both industries are ultimately separate with different players and material needs and uses. There is some overlap, however. Both are subject to changes in the housing market and are shaped by changing consumer tastes. There is also some overlap from a manufacturing perspective while some lumber companies that mill cabinetry will also mill the wood for countertops. Some companies assemble both cabinetry and countertops. There is also some overlap between laminate countertop makers and the laminate flooring industry.

Kitchen Cabinets. The kitchen cabinet industry saw 127 consecutive months of growth at the end of 2006. This impressive performance ended in early 2007 when sales in January fell 12.7 percent. There were a number of reasons for the decline, including rising interest rates, record-high home costs, and a rising inventory of new homes. Most analysts see this downturn as a necessary economic correction. The industry is seen as strong and is seen as responding quickly to changes in consumer tastes, readily investing in new technology, and embracing new production strategies. Market research firm Freedonia Group Inc. estimated this market to be worth $16 billion in 2008. Trade journal *Kitchen and Bath Business* places the value of the market slightly lower at $14 billion.

Other Related Industries. The same companies that produce kitchen cabinets also produce bathroom vanities and related cabinetry. According to the July 2007 edition of *Kitchen and Bath Business*, of the more than $10 billion in sales by manufacturers, 86 percent was generated by cabinets and just over 10 percent by the sale of bathroom vanities. The remaining market share was spread among countertops, millwork, and other products.

RESEARCH & DEVELOPMENT

The most important area of research in this industry is the effect of countertop production on the environment. Concerns include environmental damage through the production of raw materials, energy used in the manufacturing process and in transporting goods, and the use and disposal of harmful chemicals.

Manufacturing and Materials Consumption. Many materials used in countertops (granite, quartz, slate) come through mining. This process notoriously consumes a great deal of energy and can cause harm to the land and surrounding ecosystems. Cement is a major ingredient of concrete; concrete countertops are a hot trend in high-end homes. However, the production of cement is the third largest creator of carbon dioxide, a gas linked to global climate change. Natural stone is a preferred countertop material for some people. It is not recyclable but is natural. Some countertop materials also consume considerable energy. Ceramic tiles must be fired twice in kilns. Stainless steel is another desirable counter material; its production consumes considerable energy but steel can be easily recycled.

Transportation. Shipping is an environmental issue because of the fuel burned by trucks and ships transporting goods. Environmental organizations have urged consumers to buy food and goods produced locally to reduce the levels of carbon monoxide and other harmful gases released into the atmosphere from the burned fuel. This energy consumption is an issue for countertop makers as well. Freedonia notes that net imports of ceramic tiles increased 13 percent from 1999 to 2004. Environmental organizations have urged consumers shopping for stone counters to seek out stone that might be taken from nearby quarries. Stones from local quarries would require less fuel for transportation and be more environmentally friendly.

Chemicals. Chemicals are also a concern for countertop makers. Laminates and paper-based counters are manufactured using particleboard and VOC (volatile organic compound) adhesives. VOC adhesives contain chemicals that affect the environment and worker health and safety. Formaldehyde and other harmful chemicals may be used in the manufacturing process as well.

Some companies have taken steps to make their counters more environmentally friendly. PaperStone and Richlite, for example, manufacture paper-based counters using pulp from sustainably managed forests. Durat manufactures solid surface countertops that are composed of 50 percent recycled plastics. Unfortunately, some materials are simply not environmentally friendly. For example, laminate counters are inexpensive and easy to clean but are not recyclable; an old counter will simply end up in a landfill.

CURRENT TRENDS

The countertop generally follows overall trends in housing. Aside from the temporary dead stop that this industry was experiencing in the latter years of the first decade of the twenty-first century because of the subprime lending implosion—lender practices to write mortgages for borrowers with low-to-no qualifications—the trends in housing were bigger and better. According to the Census Bureau the average new home in the first decade of the new century was 2,434 square feet, up from 1,660 in 1973. Kitchens have long been seen as the center of activity in a home; they have expanded in size with total square footage to accommodate the homeowner's taste for high-end, commercial grade appliances and for considerable storage. The average size of a kitchen in 2005 was 285 square feet, up from just 90 square feet in 1950. With larger kitchens, has come increased demand for more surface finished in more expensive materials.

Some homeowners prefer separate stations for food preparation and cleaning; one in four kitchens had more than one sink in 2006, according to the National Association of Home Builders (NAHB). For some people the kitchen is even more important than the living and dining areas. Thirty-four percent of Americans favored a home in which kitchens were larger than average even if it meant smaller-than-average living quarters, according to the NAHB. According to the Census, 24 percent of single family homes had three or more bathrooms in 2004 while only 12 percent did in 1987. Another recent trend affecting countertops is separate facilities for men and women in the master suite. This typically means his and her vanities and dressing areas but may include separate toilets and shower stalls as well.

TARGET MARKETS & SEGMENTATION

Kitchen and bathroom improvements are among the more popular home remodeling projects, and countertops are a key aspect of many of these projects. A number of factors go into the selection of a countertop—cost, texture, durability, appearance, and availability. New colors and textures will help drive the market. In value terms, sales of kitchen countertops will continue to outpace those of bath countertops through 2009, a reflection of ongoing interest in larger kitchens as well as the shift in focus from viewing the kitchen as a work area to a social space.

RELATED ASSOCIATIONS & ORGANIZATIONS

Kitchen Cabinet Manufacturers Association, http://www.kcma.org

National Association of Home Builders, http://www.nahb.org

National Kitchen and Bath Association, http://www.nkba.org

BIBLIOGRAPHY

Ahluwalia, Gopal. *Home of the Future International Builders Show.* National Association of Home Builders. 12 January 2006.

Christie, Les. "Honey, I Stretched the House Again." *CNNMoney.com.* 25 July 2006. Available from <http://money.cnn.com>.

"Countertop Demand Expected to Increase 2% Annually." *Wood & Wood Products.* Available from < http://www.allbusiness.com/wood-product-manufacturing/sawmills-wood-preservation/774777-1.html>.

Freedonia Focus on Ceramic Tile. Freedonia Group Inc. September 2002.

"How Formica Got Burned By its Buyouts." *Business Week.* 22 March 1999.

"Kitchen & Bath Countertops." Freedonia Group Inc. MarketResearch.com. 1 October 2005. Available from <http://www.marketresearch.com/product/display.asp?productid=1187213&g=1>.

Mineral Commodity Summaries 2007. U.S. Department of the Interior, U.S. Geological Survey. 2007.

Pennock, Alex et. al. "Choose the Best Countertop Material for Your Home." Green Home Guide, Inc. 18 April 2006. Available from <http://www.greenhomeguide.com>.

"Soft Landing," *Kitchen & Bath Design News.* January 2006.

"Table 969: Home Remodeling—Work Done and Amount Spent 2005." *Statistical Abstract of the United States: 2007.* U.S. Department of Commerce, Bureau of the Census. December 2006.

"Trends in Countertops for Kitchens and Baths." *Solid Surface.* July-August 2005.

SEE ALSO *Kitchen Cabinets*

Creams & Lotions

INDUSTRIAL CODES

NAICS: 32–5620 Toilet Preparation Manufacturing

SIC: 2844 Perfumes, Cosmetics, and Other Toilet Preparations Manufacturing

NAICS-Based Product Codes: 32–5620D through 32–5620D271

PRODUCT OVERVIEW

Creams and lotions are, along with soap, among the oldest products used by humans on the skin. Early humans—with skin exposed to wind, sun, and extreme temperatures—likely used oils and fats to relieve the pain associated with dry, burned, and chapped skin. Human skin is the body's largest organ and a complex biological system known as the integumentary system. As the interface between a body and the environment, skin serves many functions. It provides a protective layer around the body, keeps moisture in and toxin out, regulates body temperature, and produces vitamin D, which is essential for the growth and strength of human bones. While humans have long experimented with products used on the skin's surface, the dermis, creams and lotions have changed.

Gone are the days when the use of bar soap followed by the application of face cream was considered appropriate skin care. Creams and lotions have an expanded role in a well-rounded skin care regimen. Creams, used as facial cleansers and moisturizers, and lotions, used as hand and body lotions, are large and robust product classes within the Toilet Preparations Manufacturing industry as defined by the U.S. Census Bureau. Most are marketed with claims that their use transforms the skin and prevents or delays signs of aging.

Facial cleansing creams are available in formulations including foam, fizz, cream, milk, and gel. Toners, disposable cleansing cloths, facial scrubs and masks, and even exfoliation are considered part of a well-rounded skin care regimen. Creams and lotions oils are available in multi-function formulations, so that one product can provide cleansing, moisturizing, self-tanning, and sun screening benefits to the user. This makes it difficult to separate creams and lotions into discrete classes. While the standard Dove and classic Pond's brand names still exist, they each now lead a cavalry of multi-functional products within their respective lines.

The U.S. Food and Drug Administration (FDA) refers to skin care products as cosmetics. The U.S. Census Bureau refers to skin care products like creams and lotions broadly as toiletries and reports on them every five years in the *Economic Census* and during off years in the *Annual Survey of Manufactures*. Because both governmental agencies use generic terms to refer to a broad class of products, the skin care market is referred to, generally, as the cosmetics and toiletries industry. The FDA does not regulate cosmetics and toiletries, and critics like the Campaign for Safe Cosmetics claim that consequently it is a marketplace where rule of law barely exists. Others insist it is a highly self-regulated and well-functioning market.

Manufacturers may use any ingredient or raw material (except for color additives and a few prohibited substances) to make cosmetic and toiletries products without

FIGURE 77

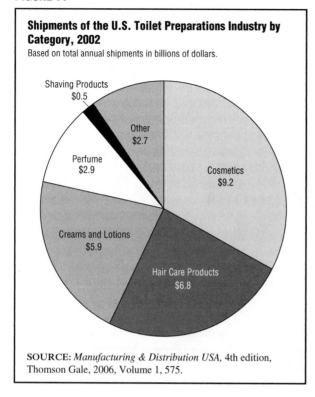

Shipments of the U.S. Toilet Preparations Industry by Category, 2002

Based on total annual shipments in billions of dollars.

- Shaving Products $0.5
- Other $2.7
- Perfume $2.9
- Cosmetics $9.2
- Creams and Lotions $5.9
- Hair Care Products $6.8

SOURCE: *Manufacturing & Distribution USA,* 4th edition, Thomson Gale, 2006, Volume 1, 575.

ranging from minutes to three years. Nondurable goods are destroyed by their use so consumers need to repeatedly replenish their supply throughout the year. Generally this equates to a large variety of affordable products for consumers to choose from in the market.

MARKET

Creams and lotions are part of the $28 billion per year U.S. cosmetics and toiletries market. The U.S. Census Bureau refers to such skin care products as toiletries and reported the value of all shipments of them in one of the *2002 Economic Census* series of reports titled "Toilet Preparation Manufacturing: 2002." The major toilet preparation divisions are presented in Figure 77 and are, from largest to smallest: (1) Cosmetics; (2) Hair preparations; (3) Creams, lotions, and oils; (4) Perfumes; (5) Shaving preparations, and (6) Other toilet preparations. The creams, lotions, and oils accounted for 21 percent or $5.9 billion of the total $28 toiletries industry in 2002.

The creams and lotions discussed in this essay are a subset of the Census Bureau's third toiletries division, creams, lotions, and oils. Product classes discussed are listed in Figure 78 with total U.S. shipments for 1997 and 2002.

government oversight, review, or approval. Companies are not required to substantiate performance claims or conduct safety testing. Labeling regulations, however, do apply. The Fair Packaging and Labeling Act requires an ingredient declaration on cosmetics and toiletries products. Ingredients must be listed in descending order of quantity.

Creams and lotions are classic nondurable consumer goods. Nondurable goods are purchased for immediate or almost immediate consumption and have a life span

Cleansing Creams. Shipments of cleansing creams mushroomed between 1997 and 2002, rising from $379 million to $2.0 billion, a growth of 428 percent. This represents an 85 percent annual growth rate over the five-year period, striking growth by any industry yardstick and particularly within the mature cosmetics and toiletries industry. This growth demonstrates that consistently over the five-year period, more U.S. women and men concluded that the unique nature of the skin on the face requires a special product to properly cleanse it. Cleansing creams

FIGURE 78

U.S. Shipments of Creams & Lotions and Related Products by Specific Product Category, 1997 and 2002

The product classes covered here are a subset of those in the Census Bureau's creams, lotions and oils division of the Toilet Preparations Manufacturing industry. Figures are in thousands of dollars unless otherwise specified.

Product Class	1997 Shipments	Percent of Total	2002 Shipments	Percent of Total	Percentage Change 1997 to 2002
Cleansing Creams	378,976	12.9	2,059,669	37.2	443.5
Premoistened Towelettes	647,433	22.1	1,112,713	20.1	71.9
Hand Lotion	678,038	23.1	911,291	16.4	34.4
Moisturizing Creams	884,453	30.1	805,213	14.5	-9.0
Body Lotion, excluding bath lotions	255,851	8.7	458,174	8.3	79.1
Facial Scrubs and Masks	58,599	2.0	118,991	2.1	103.1
Cosmetic Oils	30,752	1.0	75,304	1.4	144.9
Total	2,934,102		5,541,355		88.9

SOURCE: "Product Summary Tables: 2002," *2002 Economic Census,* U.S. Department of Commerce, Bureau of the Census, March 2006.

became the leader within the creams and lotions major division, leapfrogging over moisturizing creams to become most popular.

According to *Market Share Reporter 2007*, top facial cleansers are: Olay, Pond's, Cetaphil, and Neutrogena. Olay, the number one best seller, is manufactured by Procter & Gamble. An examination of Olay's facial cleansing products shows a substantial collection including Hydrate & Cleanse Micro-Bead Cleansing Serum, Hydrate & Cleanse Antioxidant Lathering Face Wash, Hydrate & Cleanse Night Nourishing Cream Cleaner, Dual Action Cleanser & Toner, Deep Cleansing Face Wash, Moisture Rich Cream Cleanser, and Clarify Foaming Cleanser. That covers only Olay cleansing creams. Cleansing with Olay can also be accomplished with various non-woven fiber cloths, often referred to as premoistened towelettes.

Premoistened Towelettes. Shipments of premoistened towelettes grew at a healthy pace between 1997 and 2002, rising from $647 million to $1.1 billion, a growth of 70 percent. Such double-digit growth is striking by most industry yardsticks. More Americans are turning to towelettes premoistened with creams, lotions, and oils as part of their skin care routine. For instance, facial cleansing can be accomplished with such Olay towelettes as Daily Facials Clarity Lathering Cloths, Daily Facials Moisture Balancing Self-Foaming Discs, Daily Facials Lathering Cleansing Cloths-Hydrating for Normal to Dry Skin, and Daily Facials Night Cleansing Cloths with soothing lavender and chamomile.

Hand Lotions. Between 1997 and 2002 hand lotions product shipments grew 34 percent, from $678 million to $911 million. According to *Market Share Reporter 2007*, the top hand and body creams in 2005 were Vaseline Intensive Care, Aveeno, Jergen's Natural Glow, Olay Body Quench, Eucerin, Cetaphil, and Nivea Body.

A visit to a major metropolitan mass merchandiser such as Target produced a substantial list of products that highlights why Vaseline Intensive Care is a leader in hand lotions. The Vaseline Intensive Care brand includes a broad variety of ancillary products such as: Vaseline Intensive Care Cocoa Butter Deep Conditioning Extra Rich Cream and Cocoa Butter Hydrating Lotion; Vaseline Intensive Care Aloe Cool & Fresh Light Moisturizing Lotion and Aloe Cool & Fresh Body Lotion; Vaseline Intensive Care Total Moisture; Vaseline Intensive Care Daily Skin Shield with SPF 15; Vaseline Intensive Care Healthy Body Glow with a Touch of Self-Tanner; Vaseline Intensive Care Intensive Rescue Moisture Locking Lotion; Vaseline Intensive Care Renewal Age Redefining Body Lotion with AHAs; and Vaseline Intensive Care Intensive Rescue Heal and Repair Balm and Intensive Rescue Healing Foot Cream.

Moisturizing Creams. The only creams and lotions product class that declined between 1997 and 2002 was moisturizing creams. Shipments of these products in the United States dropped 9.7 percent between 1997 and 2002, from $884 million to $806 million. As a result, moisturizing creams lost their historic first place spot due to the vigorous growth of cleansing creams and premoistened towelettes. Many creams and lotions have been reformulated to perform simultaneous functions such as cleansing, moisturizing, self-tanning, and sunscreening. Consequently, it may be assumed that this decline in the shipments of moisturizing creams is not a sign of reduced use of moisturizers but rather a sign that other products are now being used to fulfill the moisturizing function. For instance, Dove Energy Glow can be purchased as either a face moisturizer or a body lotion.

According to *Market Share Reporter 2007*, the top facial moisturizing creams in 2006 were Olay, Neutrogena, and Pond's. An examination of the classic Pond's Cold Cream that moisturizes as it cleans might explain the decline in shipment value of the moisturizing creams class. Pond's classic Cold Cream that removes make-up while it also moisturizes has a whole family of classic products to support. Additions to the classic brand include Dry Skin Cream, Deep Cleanser & Makeup Remover with cucumber in a lighter lotion format, Clean Sweep cleansing & make-up removing towelettes that clean down to the pores, and Exfoliating Clean Sweep cucumber cleansing towelettes that gently exfoliate while removing make-up. Pond's also sells Dramatic Results age-defying towelettes that fight visible signs of aging while cleansing.

Body Lotions. Shipments of body lotions grew 79 percent between 1997 and 2002, rising from $256 million to $458 million. According to *Market Share Reporter 2007*, the top hand and body creams in 2006 were: Vaseline Intensive Care, Aveeno, Jergen's Natural Glow, Olay Body Quench, Eucerin, Cetaphil, Nivea Body, and Jergen's.

Aveeno, the number two best seller after Vaseline Intensive Care, is made by Johnson & Johnson. Top selling Aveeno body lotions encompass six lines. Daily Moisturizing utilizes natural colloidal oatmeal, natural oils, and emollients; it is available in both a lotion and a creamy moisturizing oil. Stress Relief Lotion is natural colloidal oatmeal combined with lavender and essential chamomile and ylang-ylang oils. Continuous Radiance is a self-tanner line that utilizes soy and color enhancers gradually deepening skin color. Positively Radiant body lotions use soy to even out skin tone/texture and comes in formulations including a moisturizer with SPF 30, a moisturizer with

SPF 15, a daily moisturizer, and an anti-wrinkle cream. Positively Smooth body lotions are formulated with skin-silkening soy extract and skin conditioners to minimize the appearance of unwanted hair.

Facial Scrubs and Masks. Shipments of these products more than doubled between 1997 and 2002, rising from $59 million to $119 million. This robust product class includes products such as Olay Daily Facials Intensives Clay Mask with lanolin clay and marine extract. Contributing to the growth in this category were products such as Pond's Clear Solutions strips designed to instantly unclog pores and remove unwanted blackheads.

Cosmetic Oils. Cosmetic oils grew at an annual rate of 28 percent between 1997 and 2002, rising from $31 million to $75 million in shipment value. One of the long time leaders in the product class is Johnson & Johnson's Baby Oil, a product recommended in the January 2007 edition of *In Style Magazine* as the best eye makeup remover on the market.

The market for creams and lotions is a mature and healthy market. The more robust product classes grew at rates as high as 85 percent annually. As creams and lotions become ever more multi-functional, it has become harder to classify them as simply cleansing creams or moisturizing creams, or even to determine whether a product was a hand lotion, a body lotion, a self-tanner, an anti-aging product, or possibly all four. Manufacturers of these multi-functional products are discussed in next section.

KEY PRODUCERS/MANUFACTURERS

Branding plays a key role in marketing toiletries and penetrating the more robust product classes. Manufacturers tend to invest heavily in well-established brands. The connection between a brand and the company that owns that brand is not always obvious to customers. Makers of top selling cream and lotions brands are profiled in alphabetical order.

Biersdorf. In 1911 Beiersdorf develop a skin cream based on one of the first-ever water-in-oil emulsifiers and named it Nivea, from the Latin word *nivius*, meaning snow-white. The Nivea brand grew to encompass not only facial care, but hair care, shaving, bathing/shower, body care, baby care, and sun care products all sold in 150 countries.

Biersdorf markets five lines under its famous name: Nivea Cream, Nivea Body, Nivea Soft, Nivea for Men, and Nivea Visage, which focuses on face care only. The Nivea Visage line includes Moisturizing Toner, Makeup Remover, Sun Kissed Facial Moisturizer, Anti-wrinkle and Firming Cream, All Around Protection Cream, and

Q10 Advanced Wrinkle Reducer available in a day cream, a night cream, and an eye cream.

Nivea for Men features Revitalizing Eye Creme Q10 with coenzyme Q10, a natural component of the skin that allows the cream to be quickly absorbed to instantly revitalize eyes; it reduces dark circles, under-eye puffiness, and fine lines.

According to *Market Share Reporter 2007*, Beiersdorf captured 14 percent of the 2005 U.S. body lotion market and 8 percent of the hand lotion market. Recent innovations in its Nivea Body line include Smooth Sensation, a daily lotion for dry skin that delivers intense moisture in a light, fast absorbing formula enriched with gingko extract, shea butter and vitamin E.

S. C. Johnson & Johnson (J&J). Headquartered in Racine, Wisconsin, Johnson & Johnson is one of the largest U.S. family-owned and family-managed companies. It manufactures and markets thousands of products in hundreds of categories, all related in some way to health and cleanliness. J&J was formed in the mid-1880s when it pioneered ready-to-use surgical dressings that applied the pharmaceutical theory of antiseptic wound treatment. It is still in the pharmaceutical industry, and is able to utilize its pharmaceutical R&D in making creams, lotions, and oils.

According to *Market Share Reporter 2007*, Johnson & Johnson consistently captures 28 percent of the U.S. creams, lotions, and oils mass market. It makes Aveeno, Lubriderm, and Neutrogena products, among others. In 2006 J&J introduced a natural products infant care line called Johnson's Soothing Naturals, its biggest product launch in a decade. The line includes a lotion, a cream, and a balm, all touted as using the healing power of pure vitamin E, special minerals to regulate "cell water balance," and olive leaf extract.

Neutrogena makes Body Lotion, Body Oil, Firming Body Moisturizer with Active Copper, Norwegian Formula Body Emulsion, Norwegian Formula Body Moisture, Norwegian Formula Fast Absorbing Hand Cream, Norwegian Formula Foot Cream, Norwegian Formula Hand Cream, Relaxing Overnight Body Cream, and Summer Glow Daily Moisturizers SPF 20.

Procter & Gamble. Cincinnati-based Procter & Gamble was established in 1837. P&G is a giant in a wide range of consumer goods and is the leading U.S. maker of consumer products. It markets its nearly 300 brands in more than 160 countries.

P&G makes products in five main segments that it labels as: Personal & Beauty, House & Home, Health & Wellness, Baby & Family, and Pet Nutrition & Care. According to *Market Share Reporter 2007*, P&G consistently captures 31 percent of the U.S. creams, lotions, and oils market. Two of its popular Personal & Beauty products are Olay and Noxema.

Banking on Noxzema's 102 year-old name, P&G spent an estimated $15 to $25 million on advertising push to update Noxzema's attitude and look. A French illustrator based in Germany created sexy yet cheeky characters to appeal to potential consumers. P&G told *WWD* in June 2006 that Noxema had a 7.7 percent dollar share in face cleansing and admitted that while they had been chasing the technology trends, their re-emphasis on Noxema helped get lost consumers back. While the heart of Noxema's business is still its tub of tingling cleansing cream, a bevy of products has been launched over the past five years, including the Triple Clean acne line and a face-cleansing line with a citrus fragrance, which is lighter than the traditional eucalyptus, camphor, and menthol blended scent that is the original Noxema cream.

Unilever. Unilever is an international manufacturer of leading brands in food, home care, and personal care. Unilever's personal care brands include Vaseline Intensive Care, Pond's, and Dove. Vaseline Intensive Care and Pond's were highlighted in the market section.

Unilever created two ancillary lines for its Pond's classic product. One is called Cleanse and Purify and consists of six additional products for exfoliating, toning, and dermabrasion. The other is a Pond's restore, protect and age-defy line.

Dove's newest line is New Dove Body Care. It uses a unique 24-hour Nutri-Serum to add moisture and essential nutrients to skin and includes six products. New Dove Energy Glow Daily Moisturizer with subtle self-tanners; New Dove Intensive Nourishing Lotion to nourish skin with essential nutrients; New Dove Sensitive Skin Lotion unscented and dye-free for sensitive skin; New Dove Cool Moisture Lotion with green tea and cucumber extracts; New Dove Regenerating Night Lotion to fuel skin's nightly renewal process with honey and nourishing shea butter micro-pearls; and New Dove Intensive Firming Lotion to make skin firmer and smoother after two weeks with a blend of collagen and seaweed extract.

In addition to the four firms profiled, other large manufacturers of creams and lotions worldwide include Kao Corporation, headquartered in Tokyo, Japan; Shiseido Company, Ltd., also headquartered in Tokyo, Japan; and L'Oréal USA headquartered in New York, New York.

MATERIALS & SUPPLY CHAIN LOGISTICS

The materials used by manufacturers in the production of creams and lotions consist of the ingredients necessary to produce the product themselves as well as the materials used to package those products. The packaging used with toiletries serves two roles, to contain the product through its distribution and to help create the products' look, important in helping a product stand out from the crowd on store shelves.

Packaging is so important that the industry as a whole spends more than half of its expenditures for materials on packaging materials. The toilet preparations manufacturing industry in 2002 spent a total of $8.0 billion on materials of all sorts. Of this total, $2.4 billion was spent on ingredients for making the toiletry products and $5.6 billion was spent on packaging materials—containers made of plastic, metal and glass, paper and paperboard, plastic, and the like.

In terms of the ingredients that actually end up in the products sold, the primary categories—according to data reported by the U.S. Census Bureau—are, from largest to smallest in terms of industry-wide spending, the following:

1. Perfume oil mixtures and blends, essential oils (natural), and perfume materials (synthetic organic)

2. Other synthetic organic chemicals

3. Bulk surface active agents (surfactants)

Perfumes are used to impart a pleasant aroma to both the product and the packaging. Together, perfume ingredients purchased for their pleasant aroma accounted for almost 40 percent or $935 million of the annual $2.4 billion industry-wide ingredient cost in 2002.

When purchasing aromas, more than 50 percent of the $935 million—$522 million—went toward perfume oil mixtures and blends. The cost to purchase essential oils (natural) decreased 7 percent between 1997 and 2002, dipping from $174 million to $161 million. Natural essential oils are expensive, so if possible they are replaced in formulations by perfume materials (synthetic organic).

The cost for purchasing perfume materials (synthetic organic) almost quadrupled in the five-year period that ended in 2002, ballooning from $67 million to $252 million. Synthetic organic perfume materials are created primarily from chemical compounds obtained during petroleum distillation, a process that separates petroleum into fractions according to its boiling temperature. Synthetics both mimic fragrances found in nature, and provide fragrances not found in nature. The quadrupling of spending in this class was driven by products such as the expanded Noxema line scented with eucalyptus, camphor, and menthol and Olay Daily Facials Night Cleansing Cloths scented with lavender and chamomile. The advantages of synthetics are tremendous. Synthetic organic perfume materials have enlarged the fragrance library, resulting in more than 2,000 odor profiles to choose from (instead of only 200 plant-derived profiles).

Synthetic organic chemicals are generally derived from petroleum during its separation into fractions according to boiling ranges. In creams and lotions, organic chemicals are most important as preservatives integral for making high performance, non-toxic products with a long shelf life. Consumers expect creams, lotions, and oils to last a long time, perhaps up to three years. Examples are the parabens. Parabens include methyl, propyl, ethyl, and butyl—all provide broad spectrum antimicrobial protection to toiletries products. The U.S. toiletries industry almost tripled its use of synthetic organic chemicals between 1997 and 2002, from purchases valued at $243 million to $721 million, an increase of 197 percent.

Surfactants are used to adjust the surface tension of a cream, lotion, or oil. They are wetting agents that help cleansing and moisturizing creams, and hand and body lotions, spread smoothly and evenly onto the face and skin. They are integral to making toiletries. The U.S. toiletries industry decreased spending on surfactants by 26 percent in the five-year period between 1997 and 2002, from $330 million to $243 billion.

DISTRIBUTION CHANNEL

Creams, lotions, and oils are distributed through two distinct channels: the department store and the mass market. Within the industry, these channels are often referred to as prestige and masstige. Prestige products are classified based on the location where they are sold, primarily through the department store distribution channel. Lower-price products are called masstige to differentiate them from the higher-end prestige products and because they are sold using a mass market distribution channel.

The department store distribution channel is a classic distribution channel for prestige products. Prestige brand names were built on the department store distribution channel, which represents 40 percent of the North American market for prestige products. The department store distribution channel involves delivering prestige creams, lotions, and skin oils via a gift with purchase. This involves the teaser of a free gift with purchase to entice the customer to buy.

The gift with purchase model was invented by Estée Lauder and became an almost instant classic. It became a national sensation when Clinique promoted it heavily as part of its twice-yearly Bonus Time to encourage consumer spending. The gift with purchase has power over the consumer and can determine the outcome of a transaction.

The mass market distribution channel is growing as the number of stores and outlets expanded during the end of the twentieth and beginning of the twenty-first centu-

ries. This growing distribution channel for lower-priced products includes drugs stores such as CVS, Rite Aid, and Walgreens; food stores; mass merchandisers such as Target, Kmart, and Wal-Mart; and nontraditional retailers such as e-commerce Web sites. This channel also includes warehouse club stores BJ's, Costco, and Sam's Club.

The mass market channel has begun to copy the classic department store gift with purchase concept. Lubriderm Daily Moisture recently offered a variation on the gift with purchase by shrink-wrapping a free 3.3 ounce Lubriderm Daily Moisture with Sea Kelp Extract to the product. A free gift with purchase helps capture attention in the mass market channel where a wide range of products are sold side by side on the shelf. Eye catching and innovative packaging is used to capture the attention of buyers looking at a wall of creams and lotions.

KEY USERS

Adult women are the primary users of creams, lotions, and oils. Some creams, lotions, and oils are specially formulated for men and for children. For instance, Nivea has a line dedicated to men called Nivea for Men. Johnson's introduced an all natural line dedicated to baby care. Women are the primary purchasers of these products as well, although the products are designed for men or children. Consequently, the marketing appeals used to sell creams and lotions are almost always geared to attract the attention of women.

ADJACENT MARKETS

Markets adjacent to creams and lotions are moisturizing and sun protection products, self-tanning creams and sprays, cosmetics, hair care products, and apparel. Apparel is used in part to protect the skin from the elements, as are creams and lotions. Multi-functional products that may be a hand lotion, a body lotion, a self-tanner, an anti-aging product, or possible all four are adjacent to more traditional creams and lotions. Moisturizers double as sun blocks, foundation creams also exfoliate, and hair sprays contain sun protection.

The large variety of multi-functional creams, lotions, and oils product choices causes some consumers to choose to buy less. In January 2007, *The America's Intelligence Wire* reported on a Manhattan dermatologist who advocates skin-care minimalism. Most people, according to the dermatologist, need just two products: a gentle cleanser and a good sunscreen. These are available at almost any drugstore or grocery store. In the same article, a clinical professor of dermatology at Tulane University Health Sciences Center in New Orleans said, "A $200 cream may have better perfume or packaging, but as far as it moisturizing your skin better than a $10 cream, it probably won't."

Books about how to care for the skin are also an adjacent market to creams, lotions, and oils. Books explain what individual chemical ingredients do, and give tips on how to buy just the ingredients needed, and not pay more for packaging and celebrity endorsements. One example is *Six Weeks to Sensational Skin,* published by Rodale in 2006. It recommends reading product labels and does not make product recommendations, letting the consumer decide which ingredients are needed based on a listing of ingredients and their functions. Similar books include the 2005 fifth edition of *A Consumer's Dictionary of Cosmetic Ingredients* and the 2003 sixth edition of *Don't go to the Cosmetics Counter Without Me* by Paula Begoun.

RESEARCH & DEVELOPMENT

Manufacturers of creams and lotions tend to be secretive about their R&D endeavors. Generally R&D focuses on reducing material costs and creating products for the robust product classes like cleansing creams that grew 428 percent and premoistened towelettes that grew 70 percent between 1997 and 2002. Much R&D focused on "chasing the technology trends," explained P&G in a *WWD* article in June 2006.

Technology trends tended to result in products that were multi-functional. For instance, innovations in silicone waxes allow melting points to be formulated and then modified. When incorporated into emulsions these silicone waxes lower the surface tension of skin care oils and greatly improve their spreading properties. As a result, new silicones waxes allow oils to be utilized in more pleasing formulations. For instance, Dove introduced a line of moisturizing products that utilizes this type of breakthrough technology to combine oil with cream to form a rich, creamy consistency that gives the benefit of oil but in an indulgent way that is not greasy.

Research and development often results in products containing natural ingredients. Natural implies that ingredients used in the creams, lotions, and oils are extracted directly from plants as opposed to being produced synthetically. Labeling implies that products containing natural ingredients are good for the skin. For instance, J.R. Watkins began in 1868 by incorporating natural ingredients into creams, lotions, and oils. Its more recent Watkins Apothecary line includes shea butters and lotions to build on its approach of combining nature with science. Many Watkins products contain an exclusive blend of six botanical essences designed to nourish all skin types: hypericure, cornflower, linden, matricaria, calendula, and chamomile. Its natural ingredient-based skin care products include body oils, dry oil mists, and hand salves. Dry oil mists are available in Aloe & Green Tea, Citrus & Chamomile, Lavender, Vanilla, and Herbal

Extract. Shea butters are available in Aloe & Green Tea, Lavendar, Mango, and Vanilla.

Research and development efforts focused on the benefits of creams and lotions made from organic plants resulted in more than 127 products sold under the Dr. Hauschka brand name. Dr. Hauschka began 40 years ago as a holistic beauty company and is owned by WALA Heilmittel of Germany. Dr. Hauschka entered the U.S. market in 1972 and developed a cult following in Hollywood. Dr. Hauschka has a network of certified biodynamic, or sustainable, gardens to grow the organic plants it uses in its exclusive formulations. Its products are sold in 1,200 stores, mostly specialty boutiques, health food stores, and spas. One three-item kit includes its best-selling products—cleansing cream, facial toner, and rose day cream.

CURRENT TRENDS

Cleansing creams—the product class that grew 428 percent over 5 years—have expanded the definition of the cleansing regimen. More facial cleansing creams than ever are available in formulations including foam, fizz, cream, milk, and gel. Exfoliating and dermabrasion have become standard components of the skin care regimen. Creams, lotions, and oils are available in multi-function formulations. One product can cleanse, moisturize, apply a self-tanning chemical, and protect the skin from sun damage.

The trend toward multi-functional product formulations has produced a tendency on the part of the consumer to pay more attention to labeling. This scrutiny of ingredient listings makes it more obvious that the FDA does not require cosmetics and toiletries products and their ingredients to undergo approval before they are sold to the public. Manufacturers may use any ingredient or raw material (except for color additives and a few prohibited substances) to make cosmetic and toiletries products without government oversight, review, or approval. Companies are not required to substantiate performance claims or conduct safety testing. Labeling regulations, however, do apply.

Long lists of chemicals on facial cleansers and body lotions can arouse suspicions, and organizations like The Campaign for Safe Cosmetics emerge in an environment of suspicion about the ingredients used on some toiletry products. The Campaign for Safe Cosmetics claims that the chemicals used in cosmetics and toiletries, even in small amounts, may be cumulatively causing damage to humans and their environment. Their campaign calls for more transparency regarding the chemical found in products.

The cosmetics and toiletries industry is sensitive to the image that it operates in an uncontrolled market where rule of law barely exists. It counters this image with well-established, self-regulation programs. Manufacturers are interested in self-regulation by way of avoiding the imposition of government regulation. The most well-known of industry-sponsored self-regulation is the Cosmetic Ingredient Review (CIR), sponsored by the Cosmetic Toiletry and Fragrance Association. CIR is conducted by a panel of experts who evaluate cosmetic ingredients for safety and publish detailed reviews of the resulting safety data. A finding of safety by the CIR provides a degree of confidence that the ingredient can safely be used in cosmetics and toiletries.

Even if the entire cosmetics and toiletries industry is, as it insists, working under a well-established program of self-regulation, organizations such as The Campaign for Safe Cosmetics are still working for more transparency about chemical ingredients used in the manufacturing process. The goal of The Campaign for Safe Cosmetics is to have the toiletries industry phase out the use of certain chemicals linked to cancer and birth defects. In February 2007, the Environmental Working Group published a report titled *SkinDeep* that called attention to potential cancer causing ingredients in cosmetics and toiletries. A precedent for continent-wide and industry-wide transparency is the Registration, Evaluation, Authorization and Restriction of Chemicals (REACH) program in place in the European Union, which went into effect in 2007. The *New York Times* reported on February 15, 2007, in an article titled "Looking at the Bottle and What's in It," that momentum is building for greater oversight of chemicals used in everyday products, including cosmetics and toiletries.

TARGET MARKETS & SEGMENTATION

The market for creams and lotions tends to be segmented based on gender, age, and skin type. In addition, for all product classes, less expensive, mass market products are offered as well as more prestigious, higher priced creams and lotions.

Women are the largest target market for creams and lotions. They use more creams and lotions than do men and women are influential in the purchasing decisions for products designed for men and babies. Manufacturers of creams and lotions market products to inspire brand loyalty from women early in life. A woman who has used Oil of Olay from an early age will be more likely to transition to one of Oil of Olay's anti-aging products as she gets older than a woman who has never tried the Oil of Olay line of products.

Aging skin is targeted with products that are designed for it. The Baby Boom generation (those born between 1945 and the early 1960s) is a prime market for products designed to increase factors like moisture, elasticity, and luminescence. If products do not promise to increase those factors, they promise to decrease signs of sun damage and aging.

Manufacturers target men with products designed especially for them. These products frequently contain more masculine scents and are marketed with a more masculine tone.

Manufacturers target people with particular skin conditions with specially designed creams and lotions. One such manufacturer is Galderma Laboratories, maker of Cetaphil Gentle Skin Cleanser.

RELATED ASSOCIATIONS & ORGANIZATIONS

The Campaign for Safe Cosmetics, http://www.safecosmetics.org

Cosmetic, Toiletry and Fragrance Association, http://www.ctfa.org

The European Cosmetic Toiletry and Perfumery Association, http://www.colipa.com

Research Institute for Fragrance Materials, http://www.rifm.org

Synthetic Organic Chemicals Manufacturers Association, http://www.socma.com

BIBLIOGRAPHY

"About Us." The Campaign for Safe Cosmetics. Available from <http://www.safecosmetics.org/about>.

"Baby Care's Nature Makes Section a Frequent Destination." *MMR*. 13 November 2006, 26.

Darnay, Arsen J., and Joyce P. Simkin. *Manufacturing & Distribution USA,* 4th ed. Thomson Gale, 2006, Volume 1, 565–568

Edgar, Michelle. "Dr. Hauschka Celebrates 40 Years with Remedies." *WWD*. 19 January 2007, 6.

"Facing the Future: Consumers are Increasingly Comfortable with Advanced Technologies When it Comes to Saving Face." *Soap Perfumery & Cosmetics*. June 2006, 26.

Lazich, Robert S. *Market Share Reporter 2007*. Thomson Gale, 2007, 317–319.

Nagle, Andrea. "Noxzema Develops a Personality." *WWD*. 23 June 2006, 6.

"The New EU Chemicals Legislation—REACH." European Commission—Enterprise & Industry. Available from <http://www.ec/enterprise/reach/index_en.htm>.

"Our Gardens and Growers." Dr. Hauschka Skin Care. Available from <http://www.drhauschka.com/about/our-gardens-growers>.

Singer, Natasha. "Looking at the Bottle and What's in It." *New York Times*. 15 February 2007.

"Skin Deep." Environmental Working Group. Available from <http://www.ewg.org/reports/skindeep>.

"The Skinny on Skin: It Might Make you Smile but the Efficacy of That Cosmetic is Doubtful." *The America's Intelligence Wire.* 5 January 2007.

"Toilet Preparation Manufacturing: 2002." *2002 Economic Census.* U.S. Department of Commerce, Bureau of the Census. December 2005.

"Watkins, Inc." *MMR.* 12 February 2007, 41.

SEE ALSO *Cosmeceuticals, Cosmetics, Sun Care Products*

Dishwashers

INDUSTRIAL CODES

NAICS: 33–5228 Other Major Household Appliances

SIC: 3639 Household Appliances, not elsewhere classified

NAICS-Based Product Codes: 33–52285, 33–522851, 33–52285110, 33–52285111, 33–52285113, and 33–52285197

PRODUCT OVERVIEW

A dishwasher is a cleaning appliance that pushes multiple jets of water through spray arms onto neatly arranged, soiled dishes causing food, grease and grime to fall away, leaving dishes and utensils clean and ready to reuse. The development of the modern dishwasher has enabled people tasked with cleaning a kitchen after a meal to accomplish that task with ease. The device has one primary goal, washing dishes, glassware, pots and pans with higher temperature water and with specialized cleaning cycles that are more effective than manual methods of cleaning. The convenience factor alone has made dishwashers a standard fixture in the kitchens of the industrialized world in the twenty-first century.

Early History. The earliest patent noted in America for a dishwasher was in 1850 issued to Joel Houghton, who built a wooden machine with a hand-turned wheel that splashed water on dishes. The functionality of this machine was minimal. Later, in 1865, L.A. Alexander obtained a patent for a device that used a hand crank and gearing to spin dishes through the dishwater, again lacking in functionality and feasibility. Through the late nineteenth century other dishwashers were designed but, like the early machines for washing clothes, they were large contraptions that used steam and large supplies of heated water to soak many dishes at one time. In some models, the dishes were held on cradles that rocked through the water; others had paddles that sloshed water around the dishes or circular racks that held the dishes and rotated to circulate them through the water. There were machines designed with an assortment of propellers or plungers that drove water over dishes in an effort to clean them. As the designs for such machines evolved, all manner of systems were used to move the water over dishes usually held in a stationary racking system.

Josephine Cochrane is credited with designing the first hand-operated mechanical dishwasher. In a paper titled "Inventing the Dishwasher," John H. Leinhard tells the story of Josephine, who was born Josephine Garis in 1839. She married merchant and politician William Cochran and lived as a socialite in Illinois. Josephine Cochran was so bothered by the damage to her seventeenth century china that she began to do the washing of her own fine china. Frustrated with this mundane and laborious task, she set out to design a machine that would wash dishes. In the late nineteenth century, Cochrane established the Garis-Cochran Dish-Washing Machine Company to promote and sell the results of her invention, a hand-operated mechanical dishwasher. Cochran's big break came in 1893 when her machines were used in the vast kitchens at Chicago's World Fair. Nonetheless, for the mass market, this early dishwasher and others like it were ahead of their time. The machines required large quantities of hot water not available in most homes at the time—and the machines were bulky. Their only practical applications were in large institutional settings. Cochran's company grew through the years and later became a part of KitchenAid. In 1949 the first KitchenAid dishwasher based on Cochran's design was brought to market.

By the mid-1950s special low-sudsing dish-washing soaps were developed especially for dishwashers increasing their cleansing abilities. While the dishwasher remains one of the home appliances with the lowest penetration rate of all large household appliances, it has become a standard in many U.S. households.

Types of Machines. Dishwashers come in a large range of sizes and offer an extensive range of features. They fall into two basic categories: Portable Dishwashers (standard, countertop, and convertible models) and Built-in, Under-the-counter Dishwashers, (Built-ins, Dish drawers, and Fully Integrated models).

Portable dishwashers offer home owners the option of in-home dishwashers without the concern of permanent placement. Portable dishwashers are self-contained units that can be transported form one location to another. The intake feed is not permanently attached to a water source. When is use, a portable dishwasher is hooked up to a kitchen faucet using an adaptor. The water for the faucet is carried to the dishwasher through an intake tube and after cleaning is expelled through the output valve into the sink drain for disposal. A portable dishwasher is useful in kitchens short on space. The downside of this appliance is that, while in operation, the faucet it is hooked up denying water for other uses.

The current trend for portable dishwashers has seen the advent of the portable countertop dishwasher. These newly designed units offer the convenience of washing up to four place settings on the countertop while using less water than would be used during a manual hand washing cycle. The units are designed to fit under most wall cabinets with a total height of 17 inches and are created with a quick-connect faucet adaptor making hookups reasonably simple. The ease and simplicity of these devices outweighs their limitations in size for use in apartments, small galley style kitchens, break rooms, and mobile homes.

Convertible dishwashers bridge the gap between standard portable dishwashers and built-in dishwashers leaving the owner with the possibility of flexible or permanent placement. The convertible dishwasher, like the portable dishwasher, is rolled to the sink for operation. A tap adapter must be used to connect two hoses to the kitchen sink. One tap is for filling the dishwasher with water and the other is for removing waste from the chamber. The electrical cords and hoses connected to the dishwasher store neatly behind the unit. The convertible units are converted to built-in units by removing the exterior wall and rollers. The idea that a dishwasher can be portable today and built-in tomorrow, keep these units popular in many applications like apartments, older homes, and mobile homes.

Built-in dishwashers are permanently installed under a 36 inch high countertop, industry standard height, leaving only the front door and control panel exposed. The doors of the dishwashers are available in many finishes to coordinate with design trends in the kitchen industry. A standard dishwasher is 34 inches high, 24 inches wide, and 24 inches deep. Standard European models are slightly smaller but fit the same rough opening as models produced for the North American market. Built-in dishwashers are hard-wired for electrical purposes and are permanently installed on the hot water line with release into the drain tee or food-waste disposal line.

Drawer dishwashers are available under several brands depending on geographic location; Fisher & Paykel, Kenmore, KitchenAid, and Bauknecht are providers of such units in the United States. These dishwasher drawers are based on the concept of the filing cabinet with each dishwasher having two fully independent cabinets. They are marketed as either an individual drawer for small living spaces or as a double drawer for roomier kitchens.

Features. The interior of a dishwasher, also known as the tub, can be comprised of plastic or stainless steel, the later being the current trend. Older model or lower-end dishwasher interior finishes are formulated of baked enamel on steel and are prone to erosion and chipping. The process to fix such chips in the baked enamel requires the finish to be cleaned of all dirt and corrosion and then patched with a special compound or a good quality two-part epoxy, leaving the interior of the dishwasher spotty at best. Stainless steel dishwashers do not have these corrosion problems. The stainless tubs also resist hard water staining, provide better sound dampening, and preserve heat to dry dishes faster. A stainless steel tub comes at a premium.

The addition of food waste disposals in dishwashing units is another trend in this industry. North American dishwashers in the mid- to high-end range claim these units can eliminate large pieces of food waste from the dishwasher by sending them through a disposal system similar to that of a garbage disposal. The benefit of such a system is that it makes a pre-rinsing of dishes unnecessary thereby saving both time and labor. KitchenAid offers devices with this feature in its high-end line of machines.

Modern dishwashers provide a control panel that includes digital displays on the front panel. The displays use computer chips and light emitting diodes (LEDs) to signal the status of the machine at any time. Products with built-in food elimination systems may also have turbidity sensors that determine the number of rinses needed to obtain the perfect wash.

Energy consumption is a key concern in the manufacture of environmentally friendly appliances. Water consumption or the amount of water required to run the programmed cycles varies by brand and type of cycle. On average North American dishwashers use 8 gallons of water per normal operational cycle. European dishwashers utilize 4 to 6 gallons of water per normal operational cycle. Energy Star-rated dishwashers beat the minimum federal standards for energy consumption. Products with such a rating benefit the environment and cost less to operate since they use less energy.

MARKET

An analysis of the market for dishwashers begins with a look at the industry referred to by the U.S. Census Bureau as Other Major Household Appliance Manufacturing. Approximately 40 to 45 percent of industry shipments, the range depending on the year, is represented by dishwashers. The industry as a whole covers water heaters, both gas and electric; dishwashers; food waste disposers; trash compactors; and floor waxing and polishing machines. This industry as a whole, in the United States, saw steady growth over the period 1997 to 2005 with industry shipments growing 32 percent over this period or 3.5 percent per year. Industry shipments in 1997 were $3.23 billion and in 2005 they stood at $4.26 billion. Census Bureau data do not provide shipment data for dishwashers specifically.

Within that industry is a sector in which dishwashers are a major part, referred to by the Census Bureau as Household Appliances and Parts not elsewhere classified. Product shipments in this industry saw similar growth rates over the period 1997 to 2005, growing from $1.59 billion in 1997 to $2.13 billion in 2005.

A more detailed, product-level view is provided by the Census Bureau's *Current Industrial Reports* series which reports data on domestic shipments, imports, and exports of dishwashers as such. The available data series permits us to look back to 2001 with the most recent data provided being for the year 2006. In 2001 dishwasher shipments stood at $1.39 billion, increasing to $1.95 billion by 2006, producing a growth rate of 7 percent annually, double that of the total industry of which dishwashers are a part. In that the other category is dominated by water heaters, this growth rate indicates that the performance of dishwashers is shaped by special household initiatives rather than merely mirroring new residential construction being put in place. The pattern of sales over the 1997 to 2006 period is shown in Figure 79. Data for the period from 1997 to 2001 are estimates based on the broader category, Household Appliances and Parts not elsewhere classified.

FIGURE 79

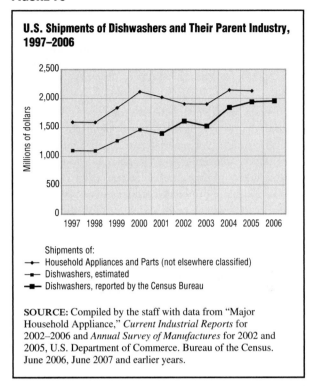

U.S. Shipments of Dishwashers and Their Parent Industry, 1997–2006

Shipments of:
- Household Appliances and Parts (not elsewhere classified)
- Dishwashers, estimated
- Dishwashers, reported by the Census Bureau

SOURCE: Compiled by the staff with data from "Major Household Appliance," *Current Industrial Reports* for 2002–2006 and *Annual Survey of Manufactures* for 2002 and 2005, U.S. Department of Commerce. Bureau of the Census. June 2006, June 2007 and earlier years.

Imports and exports play a rather minor role in this industry, particularly when compared with other household appliances. Dishwasher exports at the level of $108.6 million in 2002 were matched by imports of $77 million, giving the United States a trade surplus that year. By 2006 imports had increased to $163 million, exports stood at $155.2 million, producing a very small trade deficit of 7.8 million.

The penetration rate of dishwashers in U.S. households, equal to the percentage of households that have at least one dishwasher, stood at 60.5 percent in 2005. This means that nearly 40 percent of households did not have a dishwasher, representing a potentially large market for new dishwashers. However, more than half of dishwashers sold in the United States annually are replacements. The life expectancy of an average dishwasher built in the early twenty-first century stood at eleven years, although some low-end machines were only expected to last for three years. *Appliance Magazine,* in a 2006 overview of the appliance industry, projected that replacement sales of dishwashers in 2007 would represent 4.8 million units.

KEY PRODUCERS/MANUFACTURERS

Seven manufacturers of dishwashing machines are competing for the consumer's purchasing dollar. The top two U.S. manufacturers are Whirlpool and General Electric. Leading foreign manufacturers are Electrolux, BSH Bosh und Siemens Hausgeräte GmbH, LG Electronics, and Miele.

Whirlpool Corporation. Whirlpool, headquartered in Benton Harbor, Michigan, has brand names recognized by anyone who ever separated dark colors from light. Whirlpool is the number one U.S. home appliance maker and second worldwide, after Sweden's AB Electrolux. In addition to Whirlpool branded products the company sells products it makes under brand names including KitchenAid, Bauknecht, Roper, and Magic Chef. Approximately 14 percent of Whirlpool's 2006 sales came from its Kenmore products, produced under contract for retailer Sears Roebuck and Company. With the purchase in 2006 of Maytag, the company had secured this leading position for the foreseeable future.

Whirlpool has annual sales in the range of $19 billion and more than 80,000 employees worldwide. In addition to ovens, stoves and ranges, this company manufactures washers and dryers, refrigerators, freezers, dishwashers, trash compactors, room air conditioners, and microwaves ovens. It has a strong international presence and is a leader in the home appliance field worldwide.

Electrolux. Founded at the beginning of the twentieth century as Aktiebolaget Electrolux, this Swedish firm was first involved with floor-cleaning devices and later expanded into larger household appliances and commercial-grade kitchen machinery. In 2005 the company's worldwide net sales were $16 billion, and it employed more than 57,000 people in 150 countries. Nearly half of Electrolux's net sales are made in Europe with an additional 40 percent originating in North America. Latin America is its next largest market, and represented 5 percent of the company's net sales in 2005.

Fisher & Paykel Appliances. This company is a major appliance manufacturer located in Auckland, New Zealand. In November 2001 Fisher & Paykel Industries Limited was separated into two independent publicly listed companies, Fisher & Paykel Appliances Holdings Limited and Fisher & Paykel Healthcare Corporation Limited. In October 2004 the company acquired Dynamic Cooking Systems, Inc. (DCS). DCS, a leading U.S. manufacturer and distributor of premium cooking appliances, was acquired for US$33 million in a debt free state. This acquisition allowed Fisher & Paykel the ability to leverage market presence while maintaining its higher quality of engineering. The DCS product line has been seen as complementing the existing Fisher & Paykel product line.

In June of 2006 Fisher & Paykel acquired Elba, the Italian cookware business from De Longhi. The purchase price for the acquisition of Elba was US$ 98 million. Elba, based near Treviso, Italy, manufactures and distributes cookware products including freestanding stoves, built-in ovens and cooktops. Elba exports to more than fifty-four countries.

BSH Bosch und Siemens Hausgeräte GmbH. Headquartered in Munich, Germany, this company is a 50/50 joint venture between Robert Bosch and Siemens. BSH is one of Europe's largest appliance manufacturers. BSH's major appliances include dishwashers, ovens, microwaves, washers and dryers, refrigerators, freezers, and vacuum cleaners. It also makes small appliances, such as coffeemakers and hair dryers, as well as motors and pumps. The company's primary brands are Bosch and Siemens, but it also produces its own brands, including Balay, Coldex, Constructa, Continental, Gaggenau, Lynx, Neff, Pitsos, Profílo, Thermador, Viva, and Ufesa.

General Electric Corporation. General Electric (GE) is associated with most the major modern technological markets. The company produces an enormous range of products including aircraft engines; locomotives and other transportation equipment; kitchen and laundry appliances; lighting; electric distribution and control equipment; generators and turbines, and medical imaging equipment. GE is also one of the largest financial services companies in the United States, offering commercial finance, consumer finance, and equipment financing. To round things out, the company also owns the NBC television network. Kitchen appliances were a very small part of GE's total revenues of $163 billion in 2006.

LG Electronics Inc. (LG). This large company is based in Seoul, South Korea. LG makes many of the sorts of things that have excited game show contestants for decades. The company's seventy-five subsidiaries worldwide design and manufacture display and media products (TVs, VCRs, plasma display panels), home appliances (refrigerators, microwaves, air conditioners), and telecommunications devices (wireless phones, handsets, switchboards). LG owns Zenith Electronics and launched a flat-panel display joint venture with Philips Electronics (LG Philips Displays). After Asia, LG generates most of its revenue from sales in North America; the company established a North American headquarters in 2004. Founded in 1958 as Goldstar, the firm is a member of South Korea's LG Group.

Miele. Headquartered in Gütersloh, Germany, Miele is a manufacturer of household appliances, commercial and business equipment, and fitted kitchens. It was founded in 1899 by Carl Miele and Reinhard Zinkann and has always been a family-owned and family-run company. In additional to large household appliances Miele produces vacuum cleaners, commercial wet cleaning machines, lab glassware washers, dental disinfectors, and medical equipment washers. Miele was presented with an award in Munich on February 7, 2007, for being the most successful company in Germany that year. In the category Best Company, Miele beat the 2006 winner, Google, which came in second in 2007, ahead of Porsche which ranked third. Miele had fiscal year 2006 revenues of €2.54 billion (US$3.26 billion).

MATERIALS & SUPPLY CHAIN
LOGISTICS

The production of any heavy, durable good requires a significant capital investment. Making dishwashers is no exception. Plastics, metals and other heavy materials must be transported to the manufacturing site and fabricated. Forging, injection molding, welding, baking and insulating are all factors in the design and development of a dishwasher.

Steel and plastic are the major materials formed into dishwashers. The basic structure is made primarily of a steel frame and a steel door panel. Sheets of stainless steel are purchased and fabricated into the required shapes by the factory. The door and the wrap-around cabinet for standalone models are purchased as coiled sheet steel. Other components are typically bought from vendors. The racks that hold the dishes are made either of steel or plastic. Steel is delivered to the factory as coiled wire. Once formed, the tines are coated with a plastic arriving in a powdered form, typically polyvinyl chloride (PVC) or nylon.

The inner box or tub that holds the racks and spray arms is a single piece injection molded in the plant. This does not include the piece lining the inside of the door. The injection molded component is formed of calcium-reinforced polypropylene. This polymer is used because of its exceptional strength and good performance in the presence of chemicals, water, and heat. Additional parts, including the cutlery basket, detergent containers, wash tower, and spray arms are also injection molded. Motors, pumps, and electric controls are typically sourced from the outside unless the company is also a major producer, as is the case with General Electric.

The componentry and materials used to make dishwashers are also commonly used in the production of a wide array of industrial and consumer products and are normally available in major manufacturing centers. No unusual logistical problems, therefore, influence the industry. Large manufacturing centers coincide with population centers, thus products rarely travel far to reach their consumers.

DISTRIBUTION CHANNEL

Producers in this industry sell through wholesale distributors who, in turn, supply retailers. The presence in the market of large mass merchandisers like Sears, Roebuck and Company; Home Depot; Lowe's; Wal-Mart; Costco; and others—companies that maintain their wholly-owned distribution centers—means that producers sell directly to the wholesale arm of a group of retailers. Other retailers purchase from independent wholesale distributors.

Built-in dishwashers, the dominant category, require professional installation. The plumber or electrical contractor who installs such machines is a part of the distribution channel and is required to make the machine functional. The installer may be employed by the retailer, by the builder of new homes, or by the contractor doing a kitchen remodeling job for the homeowner.

In recent years, rapid technological advancements and the globalization of economic activities have resulted in fierce competition. The concept of electronic/virtual distribution has taken a front seat both in the literal sense that the showroom is virtual and the soft sense that a portion of the sales activity itself takes place on the Internet. Consumers can evaluate a product by comparing competing models, feature to feature, by clicking around on a screen. Consumers can also buy products directly on the web. In direct purchasing from a large online seller, the consumer may be eliminating one level of the distribution and saving some money.

KEY USERS

Dishwashers are used in commercial, institutional, and residential environments. Commercial and institutional users are functionally identical. Restaurants are a typical representative of the former and a hospital or a nursing home of the latter. Commercial and institutional buyers often require dishwashers of larger capacity engineered for heavy throughput and continuous use. Some institutional buyers deploy batteries of the same dishwashers used in the home. Residential users may live in houses or condominiums with space enough to accommodate built-in dishwashers or in apartments where space limitations can be overcome by using smaller portable dishwashing machines.

ADJACENT MARKETS

Is it more efficient to wash dishes by hand? Scientists in Germany at the University of Bonn studied this issue and found that the dishwasher uses only half the energy, one-sixth of the water, and less soap than hand-washing the same number of dishes, each batch soiled to the same extent. The study also showed that dishwashers delivered a higher level of cleanliness. The study concluded that dishwashers manufactured after 1994 used an average of 10 gallons of water per cycle, while older machines use up to 15 gallons. Newer machines also used less energy.

John Morril of the American Council for an Energy-Efficient Economy advises that dishwashers are an efficient appliance if consumers comply with two simple criteria: "Run a dishwasher only when it's full, and don't rinse your dishes before putting them into the dishwasher." Morril also recommends not using the drying cycle as the water used in most dishwashers is hot enough to evaporate quickly if the door is left open after the wash and rinse cycles are complete.

The trajectory of the housing market has a strong bearing on the market for all household appliances such as dishwashers. Since the average house has six major appliances, the number of new houses built annually has an important impact on sales of all major household appliances. Housing itself, thus, becomes an adjacent market.

Although only approximately one quarter of the sales of dishwashers each year in the United States are directly associated with new housing stock, the strength of the housing market also has an important impact on sales associated with upgrading appliances. Since 1990 a kitchen remodeling project has been among the most popular ways to spruce up a home before trying to sell it.

The housing market in the United States soared to record levels during the period 2002 through 2005, buoyed by low interest rates and flexible lending practices. Housing starts as well as sales of existing housing stock have both set records during this period. Housing starts, as measured by building permits issued, grew every year from 2000 to 2005. Sales of existing homes peaked in 2005, according to the National Association of Realtors, reaching just under 6 million homes sold. The pattern of strong growth in the housing market came to an end in 2006 when sales began to slow and the inventory of housing stock on the market grew.

The food service industry is another market adjacent to dishwashers. The more people eat outside the home, the less wear-and-tear they cause to their dishwashers and the longer life those dishwashers are likely to have. The food industry, in turn, will buy more dishwashers. The tendency of Americans to eat out has been rising for several decades. Between 1986 and 1996, total food service sales in the United States grew 69 percent to $286 billion. Growth continued at a brisk pace and by 2005 had reached $476 billion according to the U.S. Department of Agriculture. The indirect impact that this has had on appliance sales is difficult to quantify but all cultural influences on food preparation do have some bearing on the fluctuating demand for kitchen appliances.

RESEARCH & DEVELOPMENT

Research and development focus in this industry has been to improve energy efficiency. The motivation in this direction has been provided, in part, by initiatives undertaken by the U.S. Department of Energy (DOE) in the department's energy conservation moves. DOE has funded research, some of it in cooperation with manufacturers. Curiously this general movement is a reversal of an earlier thrust in industry. Throughout the 1950s and 1960s industry took cost out of products by using less material, thus reducing insulation and increasing energy use to compensate for insulation losses. Since the late 1970s the industry has been working with materials able to insulate products better to bring down energy consumption. In the dishwasher industry, energy for pumping water, for heating water above temperatures delivered by water heaters, and for drying dishes are the chief focus of conservation. The industry has been successful in lowering consumption, achieving energy savings of approximately 30 percent between 1990 and 2005 when comparing models from those two periods.

Development of new materials is another continuing thrust in the industry aimed at keeping dishwashers cleaner, thus easier to maintain, and to produce better results for the ultimate user. An example is the development of stronger plastics with superior detergent resistance. Water conservation is a focus of research as is the modification of detergents so that less detergent is used; less of it therefore reaches wastewater, while the detergent itself delivers superior cleaning power. Such detergents, to be sure, cost more.

CURRENT TRENDS

In the later years of the first decade of the twenty-first century, perhaps the most striking development directly affecting dishwasher purchases has been the virtual freeze-up of residential construction activity. Inventories of unsold homes were on the rise and new housing starts were plummeting. The consequent reduction in demand for new equipment for new construction has been a serious obstacle for all household appliance makers. This phenomenon is known as the sub-prime lending crisis and arose from reckless mortgage lending practices by one layer of the financial industry. Mortgage lenders had written a large number of contracts with home buyers in the early 2000s, thus with buyers only marginally or not at all qualified to service their loans. These poorly-secured and therefore sub-prime mortgages, along with others, have been sold and used as assets by their secondary buyers to leverage yet other loans. Widespread defaults by sub-prime borrowers created a crisis in credit in 2007. The ultimate resolution of the issue remained murky at the time this essay went to press. Observers in 2007 foresee at least two to three years of sluggish housing markets ahead.

Favorable technical advances in the product category, discussed in the Research & Development section, represent good news for the product category. Upward spiraling trends in energy costs, the costs of plastics, and the availability of clean water suggest that products that wasted such resources would suffer in the future. Dishwashers were well designed to adapt to these trends.

TARGET MARKETS & SEGMENTATION

The target markets for dishwashers are all households large enough to accommodate fully functional kitchens. The dishwasher market is paced by growth in disposable income. The income of the upper fifth of households has increased at a much higher rate than the income of all households. This growth was likely responsible for rapid growth in dishwasher shipments. The industry's promotional activity is directed at the high income consumer indirectly, thus at: (1) homebuilders, developers, and architects who make decisions on which array of new appliances to install in new construction and, (2) at contractors and kitchen-retailers who sell into the kitchen-remodeling business.

RELATED ASSOCIATIONS & ORGANIZATIONS

The Alliance to Save Energy, http:///www.ase.org

The American Council for an Energy-Efficient Economy, http://www.aceee.org

The Appliance Standards Awareness Project (ASAP), http://www.standardsasap.org

Association of Home Appliance Manufacturers (AHAM), http://www.aham.org

Building Owners and Managers Association, http://www.boma.org

North American Retail Dealers Association, http:///www.narda.com

BIBLIOGRAPHY

"2006 Appliance Industry Outlook." U.S. Department of Commerce, Office of Health and Consumer Goods. Available from <http://www.ita.doc.gov/td/ocg/outlook06_appliances.pdf>.

"Business History Project, Fisher & Paykel." The University of Auckland. Available from <http://www.businesshistory,aukland.ac.nz/fisher_paykel/key_events.html>.

Bonnema, Lisa, "Expanding Its Reach." *Appliance Magazine.* April 2004, B2.

Darnay, Arsen J., and Joyce P. Simkin. *Manufacturing & Distribution USA,* 4th ed. Thomson Gale, 2006, Volume 2, 1418–1421.

Davies, Scott. "Fisher & Paykel Acquires Italian Cooking Company Elba." Press Release. Fisher & Paykel Appliances Limited. 15 June 2006. Available from <http://www.fisherpaykel.com/press/>.

"Demand for Major Appliance in China to Grow." *Appliance Magazine.* 15 November 2006.

Fenster, J.M., "The Women Who Invented the Dishwasher." *Invention & Technology.* Fall 1999, 54–61.

"Household Appliances, Not Elsewhere Classified." *Encyclopedia of American Industries,* 4th ed. Thomson Gale, 2005, Volume 1, 1038–1042.

Le Blanc, Jenny. "Expansion in the U.S." *Appliance Magazine.* February 1997, B11.

Leinhard, John. H. "Inventing the Dishwasher." *Engines of our Ingenuity.* University of Houston. Available from < http://www.uh.edu/engines/epi1476.htm>.

"Major Household Appliances: 2006." *Current Industrial Reports.* U.S. Department of Commerce, Bureau of the Census. July 2006.

Otto, Reinhard, Ruminy and Herbert Mrotzek. "Assessment of the Environmental Impact or Household Appliances." *Appliance Magazine.* April 2006.

"Plastics Wash Out Old Design" *Design News.* 18 May 1998.

Roggema, Paul. "Merloni Expanding in Central and Eastern Europe." *Appliance Magazine.* February 2004.

"The Share-of-Market Picture for 2005." *Appliance Magazine.* September 2006.

"World Major House Appliances Demand to Reach 367 Million Unites in 2007." *Appliance Magazine.* 30 January 2004.

"World Major Household Appliances — Market Research, Market Share, Market Size, Sales, Demand Forecast, Market Leaders, Company Profiles, Industry Trends." Freedonia. January 2006. Available from < http://www.freedoniagroup.com/World-Major-Household-Appliances.html>.

SEE ALSO *Microwave Ovens, Ovens & Stoves, Refrigerators & Freezers, Washers & Dryers*

Disposable Diapers

———◆———

NAICS: 32–2121 Paper (except Newsprint) Mills, 32–2291 Sanitary Paper Product Manufacturing

SIC: 2621 Pulp and Paper Mills and Manufacturers, 2676 Sanitary Paper Products

NAICS-Based Product Codes: 32–2121L through 32–2121L131 and 32–22913 through 32–22913131

PRODUCT OVERVIEW

Early diapers for infants were formed from animal skin and wool. For thousands of years people have dealt in various ways with baby urine and feces, some more successful than others. Swaddling clothes used to wrap young babies helped. Another solution was early toilet training.

Around 800 AD cotton was introduced in southern Europe. Its use spread slowly into Western Europe and North America. By the 1300s Mediterranean farmers cultivated cotton and shipped the fiber to the Netherlands for spinning and weaving into cloth. Textiles were valuable fibers that were re-used between generations of families. Cotton rags were used over and over again for baby urine and feces.

Innovations in the late 1700s such as the cotton gin and the water-powered spinning machine made manufactured cotton cloth more available. Even after its widespread use, cotton cloth for diapers had limitations. Problems included leaking, discomfort, and cleanup. Early diapers were fastened with pins that had no clasps to keep them from poking the delicate skin of infants. The

modern pin with the safety clasp was invented in 1849. Cotton diapers held in place with diaper pins were de rigueur from the 1850s until the 1950s, when the consumer market began to grow after World War II.

Disposable diapers were developed simultaneously in Europe and the United States during the decades of the 1930s through the early 1950s. Entrepreneurs worked to overcome the limitations of the cotton diaper held in place with diaper pins. A Swedish firm developed a 2-piece diaper: a disposable wad of shredded paper pulp covered with gauze was inserted into reusable plastic pants. In the 1940s an American housewife and a British mother each developed 2-piece models. In 1949 Johnson & Johnson introduced CHUX disposable diaper. It was a 1-piece product with shredded paper wadding between a plastic back and a tissue lining. In 1950 the Swedes introduced rolls of shredded paper wadding covered with mesh that could be cut and fit into reusable plastic pants.

During the 1950s and the 1960s, disposable diapers were primarily 2-piece products that depended on reusable plastic pants. They were manufactured by various makers in France, Germany, Belgium, and Italy in Western Europe and, in North America, by Scott Paper and International Paper. Disposable diapers were a premium product intended for niche markets due to their expense relative to cotton diapers. Cotton continued to prevail.

Procter & Gamble developed its Pampers disposable diaper after it acquired Charmin Paper Company and began research into new products that used the tissue paper Charmin made at its Wisconsin paper mill.

Procter & Gamble test launched Pampers in Peoria, Illinois, in 1961. The test diaper was a rectangle. In between its plastic backing and rayon lining were multiple layers of tissue paper. The diaper was held in place with diaper pins. It featured what was known as a Z fold; inner edges were pleated to provide better fit around upper legs. In 1968 Kimberly-Clark acquired Kimbies, a 1-piece disposable diaper with fluff pulp superior to both the paper wadding in CHUX and the tissue paper in Pampers. In 1969 Procter & Gamble rolled out Pampers nationwide with a heavy advertising campaign.

During the 1960s, two inventions emerged—super absorbent polymers and nonwoven fabric—that eventually contributed to the creation of the modern diaper. Superabsorbent polymers were developed simultaneously at Dow Chemical and Johnson & Johnson.

Superabsorbent polymers are pepper-like flakes that absorb up to 300 times their weight in liquid. When superabsorbent polymers became widely available in the 1980s, they replaced shredded paper wadding.

Nonwoven material was first used by a Swedish company in a rectangular diaper pad. When nonwovens became widely available in the 1990s, they were incorporated into the diaper manufacturing process. Together these two inventions reduced manufacturing costs and helped create an industry-wide dynamic of producing improved products over many decades with few price increases.

By the mid-1970s Johnson & Johnson, Kimberly-Clark, and Procter & Gamble each had 1-piece disposable diapers in the national market. The 1-piece dislodged the 2-piece model and the private label market developed. In 1972 Procter & Gamble upgraded Pampers, adding adhesive tape to replace safety pins and switching from multiple layers of tissue paper to fluff pulp. Scott Paper test marketed gender-specific Raggedy Ann and Andy diapers, but quickly withdrew from the market to focus on private label diapers. In 1973 Procter & Gamble introduced Pampers to Western Europe.

In 1976 Procter & Gamble test marketed Luvs as an affordable brand. Its improvements included a fitted shape, elastic leg openings to help prevent leaks, and improved fastening tape. In 1978 Kimberly-Clark replaced Kimbies with its Huggies brand. Huggies also had a fitted shape and elastic leg cuffs. During the late 1970s to make more absorbent diapers, manufacturers began adding more fluff pulp to diapers. By the early 1980s disposable diapers were bulky, thick, and wide in the crotch.

During the late 1970s and early 1980s, Western European and North American consumers began to substitute disposable diapers for cotton diapers. Consumers switched because the advantage (ease of use) overwhelmed the disadvantage (higher cost).

Due in part to the freedoms associated with the late 1960s, disposable diapers were valued because they liberated parents from time consuming diaper chores of soaking, rinsing, washing, drying, and folding.

Disposable diapers are one of the six innovations that contributed to the liberation of women, according to the European Disposables and Nonwovens Association. Disposable diapers are listed along with voting, driving, equal pay, maternity leave, and the birth control pill as factors that contributed to female freedom.

In the 1980s advances in polymers made possible important improvements in disposable diaper design and performance. Johnson & Johnson stopped marketing disposable diapers under its own brand names in 1981 and refocused its manufacturing toward production of private label diapers, just as Scott Paper had earlier. In 1984 diaper makers exploited superabsorbent polymers. In 1985 resealable tape was used for the first time. Within one year, Kimberly-Clark's Huggies and Procter & Gamble's Pampers each reduced bulk by 50 percent by using superabsorbent polymers. The original bulky shredded paper diaper held 275 milliliters of liquid, or one cup. A diaper made with superabsorbent polymer held 500 milliliters, almost twice as much fluid. Another benefit of superabsorbent polymers is that they do not easily release the absorbed fluids under pressure of a toddler's bottom. Improved products were introduced to consumers without cost increases.

During the 1990s both of the top manufacturers of disposable diapers once again exploited the benefits of superabsorbent polymers to introduce ultra-thin models that were a further 30 percent thinner. Velcro fasteners replaced resealable tape. Nonwoven fabrics were exploited to create cloth-like backings that were better than plastic. New models featured waste dam leakage barriers and stretch breathable side panels. Private label brands grew as they incorporated many of these innovations. They helped hold prices down. Diapers got thinner yet better while costs stayed low. The once niche product was accessible even to low income consumers.

MARKET

Disposable diapers are part of the larger sanitary paper products industry. According to data gathered and published by the U.S. Census Bureau in its *Annual Survey of Manufactures*, this industry as a whole had shipments of $8.5 billion in 2005, down from a level of $9.5 billion in 2002. During the period 1997 to 2002 industry shipments for the sanitary paper products manufacturing industry grew at an annual pace of 3.7 percent, from $7.8 to $9.5 billion. In the years since the 2002 Economic Census, shipments declined by $1 billion.

In order to analyze the disposable diaper portion of this larger industry, one is limited to data that are published in the U.S. Economic Census years because it is only

FIGURE 80

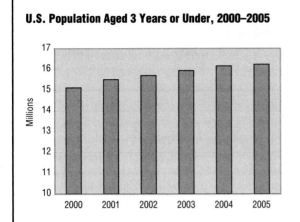

U.S. Population Aged 3 Years or Under, 2000–2005

SOURCE: Compiled by the staff with data from "Residential Population by Race, Hispanic Origin, and Single Years," *Statistical Abstract of the United States: 2007*, and previous editions of the same, U.S. Department of Commerce, Bureau of the Census. December 2006 and earlier years.

in these years that data at the product level are gathered. In 2002 disposable diapers—a Census Bureau category which includes not only children's diapers but feminine hygiene and adult incontinence products as well—made up 53.4 percent of the sanitary paper product manufacturing industry in the United States. The slight increase in the number of children aged three years and under in the United States after 2002 gives reason to believe that shipments of disposable diaper likely grew in this period as well. Figure 80 shows the growth in population aged three years or younger for the period 2000 through 2005.

In 2005 an estimated $8 billion worth of children's disposable diapers were sold in the United States, at the retail level, according to a report from Freedonia Group, a Cleveland, Ohio, market research firm. Freedonia predicted that U.S. disposable diaper sales would increase at an annual pace of 1.4 percent through 2010 to become a $9.1 billion year retail market.

Freedonia's predictions may not pan out. *Supermarket News* reported that disposable diaper sales declined nearly 6 percent to $943 million in food stores for the 52 weeks ending February 19, 2006. Figures were based on data provided by Information Resources, Inc., that do not include Wal-Mart, club, or dollar stores. *Supermarket News* tallied the top disposable diaper brands to emphasize that private label brands hold the number three position in food stores. The top three disposable diaper brands were Huggies ($198 million), Pampers Baby-Dry ($167 million), and private label ($147 million).

The emergence of the national premium market controlled by Kimberly-Clark with Huggies and Procter & Gamble with Pampers created a private label market

characterized by value-priced products. Competitive pressures between the two markets keep prices low. Between 1997 and 2004, according to European Disposable and Nonwovens Association (EDANA), real prices of disposable diapers in Europe declined by 20 percent. According to *Nonwovens Industry*, in 1990 the U.S. price of a standard disposable diaper was 22 cents. Almost 15 years later, even with countless improvements, a standard disposable diaper was approximately the same price.

Parents can save $200 per year in diaper costs if they switch to private label brands, according to *Consumer Reports*. The best private label brands cost approximatley 20 cents per diaper. The national premium brands cost approximately 30 cents per diaper. Of the 12 disposable diapers *Consumer Reports* ranked, the top five were premium products. The remaining seven were private label brands from stores like A&P, Albertson's, Kmart, Kroger, Target, and Wal-Mart. Private label brands cost approximately 30 percent less than premium brands. *Consumer Reports'* ranking of the top ten brands, with cost per diaper in 2004, reads as follows:

1. Pampers Custom Fit Cruisers ($0.30 per diaper)

2. Huggies Supreme ($0.32 per diaper)

3. Pampers Baby-Dry ($0.28 per diaper)

4. Huggies Ultratrim ($0.29 per diaper)

5. Luvs Ultra Leakquards ($0.23 per diaper)

6. Baby Basics Ultra Leakage Protection (Albertson's) (costs $0.21 per diaper)

7. America's Choice Ultra Thin Stretch (A&P) (costs $0.22 per diaper)

8. Ultra Comforts (Kroger) ($0.21 per diaper)

9. White Cloud (Wal-Mart) ($0.21 per diaper)

10. Simply Dry (Stop & Shop) ($0.19 per diaper)

Tension between the premium and value market segments was exhibited during the summer of 2005. Kimberly-Clark raised diaper prices 4.6 percent to match a Procter & Gamble price increase and to respond to a 7 percent increase taken in January 2005 by private label manufacturers. By December 2005 Procter & Gamble rolled back its price increase. The December 2005 price rollback was 2.8 percent on Pampers and 3.8 percent on Luvs. Procter & Gamble kept plans to raise prices within its super premium Pampers Baby Stages of Development line by 5.4 percent in April 2006. This move formalized the existence of a new three-tiered market consisting of premium and value market segments, complemented by a super premium segment where higher prices are more acceptable.

Nonwovens Industry remarked in December 2005 that low prices in the diaper market are due in part to the dominance of Wal-Mart and other big box mass retailers. They demand low prices and, considering Wal-Mart is responsible for approximately 60 percent of U.S. diaper sales, diaper manufacturers must meet demands. For instance, big box retailer Costco stopped carrying Procter & Gamble's Pampers Baby Stages of Development line at the majority of its stores during the summer of 2005. Rivals Kimberly-Clark and private label brands replaced the super premium line.

Disposable diapers are classic nondurable consumer goods. Nondurable goods are purchased for immediate or almost immediate consumption and have a life span ranging from minutes to three years. Nondurable goods are destroyed by their use so consumers need to repeatedly replenish their supply throughout the year. For instance, the average U.S. baby uses approximately 4,000 diapers prior to toilet training. Newborn babies may use 12 diapers per day or 84 per week for the first few months. Of course, toddlers use fewer diapers than newborns, perhaps six per day, or 42 per week. Babies in between the two extremes of newborn and toddler probably use 60 diapers per week. The average U.S. baby wears diapers until age three. Parents spend between $1,500 and $2,500 on diapers. Depending on baby's developmental stage, parents spend $50 to $70 per month.

The nondurable consumer goods market is characterized by a large variety of affordable products to tempt consumers. The best diapers prevent leaks, fit well, fasten securely, and are affordable. Brand loyalty for even the best disposable diaper is seen as generally low because consumers view them as an undifferentiated commodity and buy on price. After decades of tremendous product improvements and few price increases, consumers have come to expect low prices. Key producers of these ultra-thin, absorbent, and well-fitting disposable diapers are next.

KEY PRODUCERS/MANUFACTURERS

The top two North American manufacturers of disposable diapers consistently controlled 85 percent of the market during the first years of the twenty-first century. According to *Market Share Reporter 2007*, the Kimberly-Clark and Procter & Gamble split was 49 percent to 35 percent, in favor of Procter & Gamble, for the year ending June 2005.

Kimberly-Clark took the lead in 2006. The gap between the two key producers narrowed to one percent. Kimberly-Clark controlled 43 percent to Procter & Gamble's 42 percent. Whatever the split is from one week to another, the one thing which held firm during the early

2000s was the fact that between then, the two leaders controlled 85 percent of the market.

Irving Personal Care. This company is a relatively recent entrant into sanitary products manufacturing. It is a part of the J.D. Irving Ltd. family of companies based in Dieppe, Canada, which owns lumber and paper businesses and employs around 8,400 in North America.

Irving Personal Care began in 1988 with the purchase of a paper mill in New Brunswick, Canada. In 1990 it constructed a second plant there to produce private label products. Due to the intervention of the U.S. Department of Justice (DOJ)—which required Kimberly-Clark to divest its Scotties brand when it acquired Scott Paper Company in 1996—Irving acquired a paper plant in Fort Edward, New York. The DOJ decision was based upon the recognition that entry into the sanitary products market is difficult, requiring a significant investment in plant equipment and brand building, and that a new entrant was necessary to restore the competition lost when Kimberly-Clark and Scott Paper merged.

In 2001 Irving acquired a fourth mill in Toronto, Canada. After spending $19 million in 2003 to add a production line to its New York plant, Irving announced the first major new brand entry into the disposable diaper market in 25 years. Irving's Little Tikes branded premium disposable diapers rolled out to 1,000 U.S. locations in late 2005 and to 1,900 Eckerd and Brooks drugstores in early 2006.

Procter & Gamble (P&G). Established in 1837, Cincinnati-based Procter & Gamble has been a leader in the disposable diaper market since it introduced Pampers in 1961. P&G is a behemoth in consumer goods. It markets its nearly 300 brands in more than 160 countries.

Procter & Gamble launched its super premium Baby Stages of Development line in 2002. In 2006 P&G improved the line's absorbency and told investors the line represents more than half of Pamper's U.S. sales. The line includes Pampers Swaddlers for newborns and Custom-Fit Cruisers for crawlers. Pampers Baby-Dry diapers have koala fit grips, Sesame Street designs, a breathable cloth-like cover, and a bigger waistband with wider grips to make fastening easier.

In 2005 Pampers Active Fit was voted Product of the Year for Baby Care and revolutionized the sector by increasing elasticity at the sides and making the waistband 20 percent wider at the front. *Advertising Age* reported that as of August 2006, Procter & Gamble's most recent innovation to its basic Pampers Baby-Dry line was Caterpillar Flex, offering better fit, fastening, and flexibility. Procter & Gamble also improved leakage protection for its value-priced Luvs brand, billed as more leak-proof than the higher-priced Huggies products.

Kimberly-Clark. Established in 1872 in Neenah, Wisconsin, and headquartered in Dallas, Texas, Kimberly-Clark employed more than 55,000 people in 37 countries in 2006. Kimberly-Clark markets its many products in more than 150 countries. The personal care giant entered the disposable diapers sector in 1968 through the acquisition of Kimbies. It launched Huggies in 1978 to replace Kimbies.

Huggies Supreme was launched in 1994, creating the super premium segment. Huggies Supreme was named America's favorite diaper by *American Baby* in 2005. Kimberly-Clark introduced Huggies Supreme Gentle Care and Huggies Supreme Natural Fit diapers to replace Huggies Supreme during the fall of 2006. Supreme Gentle Care diapers are for the youngest disposable diaper wearers. They feature a cottony nonwoven liner called Cuddleweave that is extra gentle for younger babies and features better umbilical-cord cutouts for newborns. Supreme Natural Fit diapers are for older disposable diaper wearers. They feature an even thinner (10%) and more flexible hourglass shape called Hugflex with flexible sides that stretch to increase baby's mobility. Improved printing technology gives the diapers contemporary graphics for a more underwear-like look. The 2006 rollout was Kimberly-Clark's biggest launch in 12 years, since its 1994 introduction of Huggies Supreme.

Private Label Brands. Top rated private label brands are Baby Basics Ultra Leakage Protection (Albertson's), America's Choice Ultra Thin Stretch (A&P), and Ultra Comforts (Kroger), according to *Consumer Reports*. Associated Hygienic Products manufactures private label disposable diapers including Ultra Comforts for Kroger.

In 2004 Kroger recognized Associated Hygienic Products as one of its outstanding corporate brand vendors of the year. In 2005 Associated Hygienic Products launched Accordion-Stretch on its private label diapers for more stretch and better fit. Associated Hygienic's private label products feature its patented dry-lock acquisition layer, uni-cuff leak barriers, and wide-stretch fastening system.

Tyco Healthcare manufacturers Wal-Mart's White Cloud private label brand of diapers. Baby Time diapers (Wegmans) feature a Baby Snoopy design and a flex-fit system touted as wider and more flexible to move with baby to prevent leakage. Cuddle Ups (Brookshire) features the usual superabsorbent polymers, enhanced elastic leg gathers, and stronger fasteners, but is priced 15 percent less than Huggies and Pampers.

MATERIALS & SUPPLY CHAIN LOGISTICS

Modern diapers are layered to redistribute urine from a soft nonwoven fiber liner to an absorbent core of fluff pulp and superabsorbent polymers protected by a plastic back. In many brands of diapers the leak-proof poly film plastic back has been replaced by a nonwoven and film composite.

According to EDANA, the average baby diaper is comprised of 43 percent fluff pulp, 27 percent superabsorbent polymer, 22 percent polypropylene/ethylene, 3 percent adhesives, and 1 percent elastics.

Fluff pulp is essential. Its availability influences diaper manufacturing costs. Leading suppliers Rayonier, Koch Cellulose (Georgia-Pacific), and Buckeye Technologies rely on long-term relationships with makers in order to achieve economies of scale. Superabsorbent polymer is also essential. The bulk of global superabsorbent polymer production is swallowed up by the sanitary products manufacturing industry, *ICIS Chemical Business Americas* reported in February 2007. The use of elastics has increased as manufacturers honed fit. Once found only in the leg cuff, elasticized material is now found throughout the diaper in waistbands, on side panels, and in closure systems. For example, Pampers Baby Fresh was launched in July 2006 with highly elastic fasteners, known in the industry as ears.

Improving quality while keeping manufacturing costs low has been the paradoxical challenge of disposable diaper manufacturers for almost five decades. Intense competition, pricing pressures, and market maturity create an unwillingness among producers to increase prices. The Census Bureau reported that in 2002 the sanitary paper products manufacturing industry used $3.1 billion worth of materials to produce $8.6 billion worth of products. The value of materials used remained static between 1997 and 2002, hovering right around $3.1 billion. During the same period, product shipments grew 15 percent from $6.5 billion to $8.6 billion.

The main types of materials used to produce diapers and other sanitary paper products are paper, woodpulp (also known as fluff pulp), and nonwoven fabrics. Industry-wide spending for paper decreased 12 percent between 1997 and 2002, from $779 million to $689 million. Industry-wide spending for woodpulp increased between 1997 and 2002, from $195 million to $503 million, an increase of 158 percent. Industry-wide spending for nonwoven fabrics decreased 43 percent between 1997 and 2002, from $472 million to $271 million.

In January 2007 *Nonwovens Industry* shed some light on the disposable diaper industry's unique ability to grow by 15 percent and continually introduce improved products, while keeping cost expenditures on the raw materials consumed in production at a static level. Because the absorbent core is the most expensive part of the disposable diaper, makers adjust the ratio of woodpulp to superabsorbent polymers to improve absorbency while saving costs.

Other cost savings that emerged as part of the challenge to make diapers better yet cheaper has been to reduce the amount of materials. For example, fitted hourglass shaped diapers use less materials, especially less nonwoven fabrics. Cost savings related to nonwoven fabrics can be further explained by trends away from poly-laminated backings. Also, more companies buy poly films and nonwovens separately and put them together during diaper production. Kimberly-Clark in Neenah, Wisconsin, makes all the Huggies in the Midwest. It also makes its own nonwoven fabrics.

After using paper, woodpulp, and nonwoven fabrics to make diapers, manufacturers use packaging materials to prepare cartons of disposable diapers for shipment through the distribution channel. U.S. industry-wide spending for packing material was around $400 million in 2002. The types of packaging products used (in order of expenditures) were paperboard containers, boxes, and corrugated paperboard; packaging paper and plastics film; and glues and adhesives.

DISTRIBUTION CHANNEL

Disposable diapers are distributed in corrugated paperboard boxes that contain packages of diapers wrapped in plastics film. Disposable diapers are available at a throng of outlets including:

- Drugs stores such as Brooks, CVS, Eckerd, Rite Aid, and Walgreens

- Food stores including A&P, Albertsons, Brookshire, Kroger, and Wegmens

- Health food stores including Whole Foods

- Convenience stores like 7-11 and Tom Thumb

- Gas stations such as Holiday and Super America

- Mass merchandisers including Kmart, Target, and Wal-Mart

- Nontraditional retailers and warehouse club stores like Costco and Sam's

- E-commerce Web sites that offer online shopping for diapers

Within these various distribution channels, the national premium disposable diaper brands are sometimes sold as loss leaders. Retailers do so because of the financial importance of diaper buyers, who buy more on average than other shoppers. Diaper shoppers can be counted on to need lots of other items, so stores lure them with price reductions.

Within the category, private label brands are used to grow margins, which can be near zero. For instance, the Cuddle Ups line covers 13 stock keeping units including convenient, jumbo, and mega packs, and provides an example of why these private label brands are important to the channel. Cuddle Ups are profitable. Its margins are 15 to 20 percent. Cuddle Ups' margins reflect the critical role of private label brands in the distribution channel. Because national brand margins are low, retailers need the higher margins on private label brands to maintain gross-margin integrity within the disposable diaper category.

Within the distribution channel, disposable diapers are promoted in hospitals. The use of in-hospital promotions is an important tool in reaching moms early in their diaper decision making process. Procter & Gamble has long dominated the in-hospital sampling distribution channel.

The disposable diaper distribution channel involves advertising on an in-hospital network known as The Newborn Channel. It provides educational programming in 1,840 hospitals' maternity wards, reaching around 3.4 million new mothers in the United States each year. An estimated 82 percent of new mothers are exposed to the Newborn Channel. The network is jointly operated by NBC Universal, iVillage, and GE Healthcare. On the network, Kimberly-Clark had category exclusivity for its Huggies Supreme 2006 launch.

Within the distribution channel, size matters. The old disposable diapers filled with paper created a transportation problem. By trimming the size of a diaper by more than half, more diapers fit on a truck. Diaper manufacturers and distributors were able to cut transportation expenses that contributed to keeping costs low for consumers.

Size also effects the way diapers are sold. They take up a lot of space, comprising one of the largest product categories in food, drug, and discount stores. Store shelves are divided into increments of four feet, so diapers, for example, might be presented as a 20-foot set. Bulky diaper packages of the 1980s took up much of that length, contributing to high out-of-stock rates. For a fast-moving, bulky item like diapers, restocking problems prevailed. By making smaller diapers and by extension smaller packages, diaper makers insured their products would be on the shelves a greater percent of the time.

The first U.S. Web site that provided home delivery of brand name diapers at wholesale prices with free shipping opened in 2005. Known as 1-800-Diapers, it delivers Pampers, Huggies, and Luvs at prices below Target, Wal-Mart, and other discounters. Because it carries much bigger wholesale boxes, 1-800-Diapers' price per diaper is low.

KEY USERS

The range of key users covers the single parent, the working parent, the cost conscious parent, and the parents of multiple children. Key users are parents of young children and child care providers. The key user of disposable diapers can be characterized as a shopper in a hurry. The harried consumer may value one-stop-shopping in a superstore where she can get everything she needs. Alternately, she may value a small local store where she can get in, get the diapers, and get out. The vast majority of parents choose disposable diapers over cloth diapers, although reliable figures as to what percentage that is are not available.

An estimated 10 percent of households experiment with cloth diapers. An active Internet community supports parents who choose cloth diapers. Useful Web sites are BorntoLove.com with information on laundering, costs, and suppliers and DiaperPin.com which sponsors a parent forum.

ADJACENT MARKETS

The most obvious product adjacent to disposable diapers is cloth diapers and diaper services which deliver the same to parents. This market is very small, only a fraction of the total U.S. linen laundering market. According to the Census Bureau, the linen laundering service industry had receipts of $3.5 billion in 2002. Most of these receipts were for services provided to restaurants, hotels, hospitals, and institutions that provide uniforms for their workforce.

Other products whose markets are adjacent to the market for disposable diapers include baby wipes, bibs, formula, food, and toys such as rattles. Diaper bags too can be considered adjacent to disposable diapers. They are needed to carry not only diapers, but wipes, bottles, snacks, and toys. One househusband was quoted in an *Nonwoven Industries* article saying, "I do not want to carry a girly diaper bag. I want something cool and funky."

Baby and toddler toiletries are a growing adjacent market. They are typically differentiated through ergonomically designed packaging. They are an important adjacent market for the top two key producers.

Kimberly-Clark forayed into baby toiletries in 2004 when it extended its Huggies brand into the baby bath and body market. Leading its array of toiletries is Huggies Liquid Powder. It goes on as a liquid and dries to a pow-

der. The Huggies bath and body line comprises more than 20 stock keeping units including shampoo, baby lotion, diaper rash cream, disposable wash mitts, and variations of Kimberly-Clark's established baby wipes. New Huggies products include a proprietary shea butter moisturizing formula in baby lotion and baby wash and extra thick and soft disposable washcloths and toddler mitts using proprietary nonwoven fabric.

Procter & Gamble expanded into toddler toiletries in 2005 with its Pampers Kandoo Toddler Care line. Pampers Kandoo Flushable Wipes are sold in a colored, easy-to-open, pop-up tub that dispenses one wipe at a time. Lightly moistened for gentle, easy cleaning, they are made to fit small hands and come in two scents: fresh splash and jungle fruits. Pampers Kandoo Foaming Handsoap, also in fresh splash and jungle fruits, has a wide pump top easy for little hands to press. Procter & Gamble's established Pampers Baby Wipes in a pop-up tub are available in lavender, scented natural aloe, unscented natural aloe, and sensitive chamomile. Its Pampers Sensitive Wipes target newborns with sensitive skin.

RESEARCH & DEVELOPMENT

Early research and development resulted in a changed disposable diaper. The form of the product changed through the incorporation of superabsorbents to make products thinner, and the use of nonwovens instead of plastic backings to make products softer. Later research and development decreased manufacturers' reliance on superabsorbent polymers, since supply is tight with the bulk of production swallowed up by the sanitary products industry.

Superabsorbent polymers producer Tredegar's makes products called AquiDry, AquiDry Lite, and AquiSoft that help makers reduce superabsorbent polymer use by as much as 25 percent without compromising performance. Superabsorbent producer Lysac Technologies introduced Actofil for baby diapers. It enhances the diffusion of superabsorbents in diapers so manufacturers can reduce superabsorbent polymer use by up to 20 percent.

Recent research and development decreased manufacturers' consumption of polyethylene plastic resin. Thinner-gauge nonwovens have reduced costs for manufacturers. *Chemistry and Industry* September 2006 reported that research at Kimberly-Clark resulted in biodegradable but breathable poly film. It is a mix of a biodegradable polyester with a calcium carbonate (clay) filler. The compound produces a film thinner than a human hair that is waterproof, breathable, and biodegradable when composted.

The latest research and development resulted in treated diaper liners, known as rash guards or skin wellness liners, that further protect baby's bottom from urine and feces. Typical treatments are a three part combination of emollients, viscosity enhancers, and botanical active ingredients. Emollients and viscosity enhancers can be mixtures of petrolatum, vegetable-based oils, mineral oils, lanolin, glycerol esters, alkoxylated carboxylic acids, alkoxylated alcohols, and fatty alcohols. Botanical active ingredients include aloe, echinacea, willow herb, and chamomile, green, black, oolong, and Chinese teas.

CURRENT TRENDS

Successful innovations that started as trends include absorbent cores, elastic leg bands, superabsorbent polymers, resealable tape fasteners, elastic waistbands, Velcro fasteners, breathable backing from nonwoven fibers, graphic designs, and newborn diapers notched for the umbilical cord. Children may be less fussy because they are more comfortable with a diaper that does not sag, drip, or fits poorly with pins at the hips.

Disposable diaper innovations have been so successful, in fact, that parents keep children in disposable diapers until they are approximately three years old on average. In the past century, the average age of toilet training crept up from 1.5 years to sometimes even beyond the age of three. This change is attributed to the ease of using disposable diapers.

The current trend—common in a mature industry like disposable diapers—is to expand the range of product sizes and types. In the early days of the industry, disposable diapers were sold as either small, medium, or large sizes. Most of the innovations focused on technical and functional issues related to absorbency. The expanded range of sizes and types is a more consumer-relevant focus on baby development. Makers link new introductions to the disposable diaper wearers' developmental stages. The expanded range of shapes and features are focused for newborn babies, crawlers, toddlers, walkers, and toilet trainers. Different versions correspond not only to the wearers' size and weight, but to their developmental stage, frequency of urination or bowel movement, and where the child is with controlling those functions.

A secondary trend is disposable diapers that allege an ecologically sound alternative to the major national brands. The Environmental Protection Agency estimates that yearly more than 3 million tons of disposable diapers arrive in landfills in America. Seventh Generation and gDiapers are part of the reason the disposable diaper category is one of the fastest-growing sectors in the natural products arena.

Seventh Generation, based in Burlington, Vermont, got into disposable diapers in 2004. Since then, its category involvement has grown over 300 percent. Seventh Generation's selling point is that its diapers are 100 percent chlorine-free. Seventh Generation's chlorine-free diapers have a brownish tint, different from the blindingly white premium national brands.

gDiapers, based in Portland, Oregon, sells a flushable and compostable 2-piece disposal diaper. gDiapers include a washable, reusable outer pant and a flushable liner made of biodegradable, all-natural fiber. The interior uses elemental, chlorine-free, tree-farmed fluff pulp, and tiny sodium polyacrylate crystals to absorb wetness.

TARGET MARKETS & SEGMENTATION

Three disposable diaper segments emerged due in large part to the industry's tendency toward continuous product improvements. Coexisting segments are value, premium, and super premium.

The value segment incorporates many of the innovations that emerged first in the national premium brands. The value segment targets the price-conscious consumer. Value-priced private label disposable diapers cost 15 to 20 percent less than premium diapers.

The premium segment is controlled by the top two manufacturers. Prices are low in the premium segment, due to tension between it and the value segment. Makers target the super premium purchaser, and niche markets like disposable training pants and disposable swimming pants. Training pants are estimated to cost $0.80 each. Swimming pants are estimated at $0.85 each.

The main strategy for growth in a mature U.S. market where the population of babies is growing only approximately 1 percent annually and price increases are hard to come by has been to introduce super premium products. Kimberly-Clark and Procter & Gamble one-up each other with a better diaper and try to persuade consumers to trade up from the base Huggies UltraTrim or Pampers.

Super premium products can be characterized by higher absorbency such as extra capacity for nighttime use, skin-care benefits such as rash guards and skin wellness liners, and superior fit specially designed for crawlers or toddlers. Pampers, for example, markets Cruisers for walkers with super stretchy leg cuffs and sides to better prevent leaks as the child moves around. However, these improvements make a soiled diaper more comfortable for the infant or child, which may result in fewer diaper changes each day.

Known for offering solid performance and enhanced fit, super premium is the one segment where margins are still high. For instance, the Huggies Supreme Gentle Care and Huggies Supreme Natural Fit introduced in fall of 2006 cost 14 to 20 percent more than the premium Huggies brand, according to an August 2006 article on the battle for the bottom line in *Advertising Age*. Super premium—also known as top tier supreme—is where most of the category growth is, accounting for half of the $4 billion North American disposable diaper business.

The secondary strategy for growth in a mature U.S. market where consumption of disposable baby diapers is threatened by a decreasing birth rate is to target niche markets. Kimberly-Clark and Procter & Gamble each have disposable training pants and swimming pants.

In 1989 Kimberly-Clark introduced Huggies Pull Ups training pants with gender-specific models, the first ever disposable training pant to facilitate toilet training. In 2003 Kimberly-Clark launched Huggies Convertibles, which could be either pulled up like training pants or put on like a diaper. This dual functionality solved one of the problems that parents had with Pull Ups, namely the need to remove a toddlers pants and shoes in order to change them.

Other diaper manufacturers offer disposable training pants, but Kimberly-Clark holds the market lead. For instance, in 2003, Procter & Gamble's introduction of Easy Ups training pants gave Pampers a 19.5 percent share of the training pants segment, where it had previously not competed.

In 2006 *Nonwovens Industry* reported that Procter & Gamble launched a new Pampers swim pant and incorporated specialized absorbency for boys and girls on its Easy Ups training pants. Irving Personal Care Little Tikes entered the training pants category in 2006 with the debut of Snug'n Snoozzz Overnight Training Pants.

Kimberly-Clark targets this segment that can be counted to spend almost $1,000 per year on PBS Kids Sprout, a 24-hour preschool network and Web site. A for-profit channel created by Comcast Corp. and PBS, Kids Sprout programming includes *Sesame Street, Bob the Builder, Barney & Friends*, and *Teletubbies*. The video-on-demand element allows Kimberly-Clark to use vignettes to drive moms and moms-to-be to its Huggies Baby Network, Pull-Ups.com, and two new Web sites called Huggies Happy, Healthy Pregnancy and Huggies Happy Baby.

RELATED ASSOCIATIONS & ORGANIZATIONS

Absorbent Hygiene Products Manufacturers Association (AHPMA), http://www.ahpma.co.uk

European Disposables and Nonwovens Association (EDANA), http://www.edana.org/index

Real Diaper Association, http://www.realdiaperassociation.org/clothdiapering_inthenews.php

BIBLIOGRAPHY

"Absorbent Articles with Non-aqueous Compositions Containing Botanicals." *Soap Perfumery & Cosmetics*. March 2006, 55.

Angrisani, Carol. "Bottoms Up: When it Comes to Quality and Promotions, Private-Label Diapers Have Come a Long Way, Baby." *Supermarket News*. 17 April 2006, 43.

"Bath/Body Items Make Huggies Brand More Interesting." *Chain Drug Review*. 20 December 2004, 15.

Bitz McIntrye, Karen. "Hygiene Component Suppliers Provide Form, Fit and Function: Pricing Pressures, Demands for Innovation Continue to Define Industry." *Nonwovens Industry*. December 2005, 28.

Blanchfield, Lindsey. "Superabsorbent Polymers Soak up Potential." *ICIS Chemical Business Americas*. 26 February 2007.

"Branding News: Design Choice—Pampers Active Fit." *Marketing*. 16 March 2005, 9.

"Diaper Portal Offers Discounts." *Nonwovens Industry*. July 2005, 20.

"Diapers: Wegmans Food Markets Inc. is Providing Baby Time Diapers." *Private Label Buyer*. October 2005, 84.

"Disposable Diapers: Time to Change Brands?" *Consumer Reports*. March 2004, 34.

Dyer, Davis. "Seven Decades of Disposable Diapers: A Record of Continuous Innovation and Expanding Benefit." European Disposable and Nonwoven Association. August 2005 Available from <http://www.edana.org/index.cfm>.

Ebenkamp, Becky. "Little Tikes Toying with 'Super' Diapers." *Brandweek*. 28 November 2005, 6.

Gladwell, Malcolm. "The Disposable Diaper and the Meaning of Progress." *The New Yorker*. 26 November 2001.

"Greener Nappies." *Chemistry and Industry*. 4 September 2006, 17.

"K-C Revamps Supreme Diaper Line." *Nonwovens Industry*. October 2006, 14.

"Kimberly-Clark Corp—Private Particulars." *Private Label Buyer*. May 2005, 14.

"Kimberly-Clark Puts $20 Million Toward Diaper Debut." *Promo*. 20 September 2006.

"Making Room for Baby—Diapers Private Label Category Review." *Private Label Buyer*. December 2006, 36.

Neff, Jack. "The Battle for the Bottom Line: P&G, K-C Push Innovations in $5B Diaper Category." *Advertising Age*. 28 August 2006, 3.

Odell, Patricia. "Pampers Out, Huggies in at Some Costco's: Report." *Promo*. 16 June 2005.

"P&G Revamps Diaper Pricing." *Nonwovens Industry*. January 2006, 12.

"Product Lines: 2002." *2002 Economic Census*. U.S. Department of Commerce, Bureau of the Census. November 2005.

Richer, Carlos. "Competition Afoot in the Diaper Market." *Nonwovens Industry*. January 2007, 32.

"Sanitary Paper Products Manufacturing: 2002." *2002 Economic Census*. U.S. Department of Commerce, Bureau of the Census. Available from <http://www.census.gov/econ/census02/>.

Thedinger, Bart. "Demographic Trends & Offshore Competition Moderate Growth of Hygiene Films." *Plastics Technology*. October 2003, 68.

"Two Huggies Supreme Gentle Care Diapers." *MMR*. 18 September 2006, 13.

"Uniqueness Plus Price Sells." *MMR*. 14 November 2005, 34.

Wilcox, Tyler. "Help Parents Clean Up with Eco-friendly Diapers." *Natural Foods Merchandiser*. October 2006, 48.

Williamson, Richard. "K-C First to Sponsor PBS Kids Sprout." *ADWEEK Southwest*. 28 September 2005.

Distilled Spirits

—————●—————

INDUSTRIAL CODES

NAICS: 31–2140 Distilleries

SIC: 2085 Distilled and Blended Liquors

NAICS-Based Product Codes: 31–21401 through 31–214049B1

PRODUCT OVERVIEW

Spirits are beverages that contain purified ethyl alcohol—purified and potable ethanol. They are made from a fermented mash of grains, fruits, or vegetables. Such beverages are low in sugar and contain at least 35 percent alcohol by volume. Examples of distilled spirits include brandy, gin, tequila, vodka, and whiskey. Such drinks are referred to by a number of terms, including distilled beverages, liquor, hard liquor, mixed drinks, and alcoholic beverages. Distilled beverages do not include wine or beer.

All forms of alcoholic beverages—beers, wines, and liquors—are based on fermentation, the breakdown of carbohydrates into alcohol. Yeast is the catalyst in the process. Liquor production involves the extra step of distillation. Distillation involves the boiling off of alcoholic vapors from the fermented mash through intense heat and collecting them as condensed liquid. Alcoholic beverages are made from carbohydrates, in essence, sugars. The most common sugars used to distill liquor come from grapes, sugarcane, molasses, corn, rye, barley, wheat, and potatoes. A variety of additional products are used in the distillation process to flavor the resulting alcoholic beverage.

A still is needed for distillation. Stills consist of three parts: a vessel, in which the substance to be distilled is heated; a condenser, in which the vapor is liquefied; and a receiver, in which the product is collected. Most commercial producers use stills made from copper or stainless steel with copper interiors. The copper removes sulfur-based compounds from the liquid. Copper piping also reduces the level of copper in the wastewater by-product, which is often used in animal feed at large distilleries. Copper is also used in stills because it is an excellent conductor of heat and resists corrosion. Nonetheless, some corrosion is unavoidable, and stills typically undergo repairs every eight years.

Commercial distilleries generally use a column still, often called a continuous still or coffey still. Robert Stein invented the first such still in 1826; however, Aeneas Coffey improved upon Stein's design. Column stills have two columns or chambers. The columns are designed with a series of graduated levels through which the liquid passes as it is heated. The vapors rise through the levels and become more concentrated as they progress upward. Cooling begins in the higher levels of the second column.

Column stills are a more efficient means of production than pot stills. A pot still is a single chamber still in which heat is applied directly to the pot. Alcohol has a boiling point of 173 degrees Fahrenheit, lower than that of water, which reaches its boiling point at 212 degrees Fahrenheit. Consequently, when the liquid is heated, the alcoholic portion boils off as vapors and is separated from the initial liquid. The vapor is sent through a condensing coil and in the process is cooled and becomes liquid again. The first distillation results in a liquid with an alcohol strength of 25 to 35 percent by volume. The alcoholic liquid is then concentrated further in a second distillation which brings its alcohol level to approximately 70 percent. The measure of alcohol by volume should not be confused with another measure of alcoholic level used for spirits: proof. Proof is a measurement of the alcohol content in a beverage. Each degree of proof equals a half percent of alcohol, so a beverage labeled as 100 proof contains 50 percent alcohol by volume.

The alcohol level of distilled spirits is much higher than that of beer or wine. Beer averages 2 to 8 percent alcohol content by volume and wine averages from 8 to 14 percent. The alcohol levels in both beer and wine are limited to a maximum of 15 percent by volume, beyond which yeast is adversely affected and cannot ferment. Liquors achieve a higher percent of alcohol by volume because of the distillation process, which concentrates the alcohol and separates it from the original liquid.

The Manufacturing Process. The production of alcoholic beverages is complex. Grains or vegetables high in carbohydrates are placed into an automatic mash tub. The tub is fitted with agitators to break down the fibers of the raw material. The mash is heated to a boiling point to eliminate the growth of harmful bacteria. The mash is then poured into stainless steel vats. Yeast is added and the fermentation process begins. The sugar in the mash is converted into ethyl alcohol.

The liquid, known at this point in the process as wash, is transferred into a column still. The wash is heated in the analyzing column. The wash enters the top of the analyzing column and steam is injected at the bottom of the column. The vapor travels through a number of perforated plates and then travels through a tub connecting the two columns. The second column is called the rectifying column, and it holds the condenser. This is where the liquid is cooled and condensed. The liquid is distilled a second time. Various factors influence the level of the final alcohol content of distilled liquor: temperature, water, and any flavoring substances that have been added, also known as cogeners. If the liquid is more than 95 percent alcohol it will be flavorless because the cogeners have been boiled away.

The production process varies based on the distiller and the liquor being produced. Some distillers use a column still while others use pot stills. Some ingredients require an extra production step. For example, Scotch whiskey is produced with barley and barley must go through a several week process of germination before the fermentation process can begin. This makes the starches in the barley more soluble. Most whiskey is distilled twice, although single malt whiskies are only distilled once. Some vodka is distilled as many as six times. The wood of the cask in which a liquor is stored can affect the color and taste of that product. The place of origin can affect taste also, with malt whiskies made near the coast having a briny taste. The length of the aging process is another element that varies in the production of different spirits.

A Brief History of Distillation and Liquor. Fermented beverages and crude forms of distillation were known in Babylonia, ancient China, and other early cultures. Alcoholic beverages were a regular part of life, playing vital roles in celebrations, oaths of allegiances, religious ceremonies, and trips into battle. Beer and wine were also known to early societies.

The Arabs and Persians refined the distillation process. Arab Alchemist Jabir ibn Hayyan is credited with inventing early stills, known as alembics. An alembic is a still consisting of two chambers, called retorts, connected by a tub. These stills were typically glassware with long, angled necks. *Al-ambiq* is the Arabic word for still, which ultimately comes from the Greek word *ambix*, or cup. Early alembics were designed to cool the distillate, making the collection of the alcohol more efficient. These early alchemists began to understand the connection between alcohol and its possible use in medicine and science. Tenth century Persian physician Rhazes and Arab surgeon Abulcais, for example, wrote of using alcohol as anaesthetic and as a solvent in drugs.

The origin of the word alcohol is not known with certainty. Some sources credit alchemist Paracelsus with inventing the word. Historians believe the word to be Arabic in origin, for al- is a common Arabic prefix. The root word most likely comes from the original *kohl*, which was a fine powder produced by grinding and used by women to paint the eyes. Arab alchemists, however, used the word to refer to any powder or substance produced in a number of ways. The use of the word alcohol to refer to a powder produced by heating a substance into a gaseous state and then cooling it was first used in 1543, according to the *American Heritage Dictionary*. The word alcohol was used to refer to any substance that had gone through the distillation process by 1672.

Because of the association between alcohol and the sciences, alcohol was believed to have medicinal properties. Alcoholic beverages have been used to treat colds, fevers, frostbite, snakebites, and many other ailments. By the Middle Ages alcoholic beverages were known as the "water of life"—in the Gaelic *uisge beatha* or in Latin *aqua vitae*.

By the twelfth century, Irish whisky, German brandy, and French cognac were found on the European continent. Catholic monks performed much of the early distilling and winemaking in Europe. As they set out to establish new monasteries they took their distilling practices with them. Irish clergy, for example, are believed to have introduced the distillation process to nearby Scotland somewhere between the twelfth and fourteenth centuries. As the process became better known it was improved upon. The earliest distillery in Scotland dates from 1494. This pattern took place on the European continent as well, and stills in rural homes became common.

Naval exploration, war, and trade would further stimulate the development of distilleries. Christopher Columbus brought sugarcane from the Canary Islands to the Caribbean. The sugarcane, used in rum production, prospered in the region's warm climate. The local plantations began manufacturing rum in large quantities. The Caribbean region was popular with ships from Colonial America as part of the growing slave trade. The ships brought the drink back to New England; a rum production industry was established in this region by the 1660s. British soldiers may have first encountered gin in Holland in the late 1580s. Protestant William of Orange of England promoted the production of distilled beverages—grain spirits as they were known at the time—after restricting the importation of wine from Catholic countries in the late seventeenth century. As the British Empire began its expansion in the eighteenth century, it brought gin and other grain spirits along with it.

During the exploration of the American continent, European settlers poured into the middle of the country and began to plant corn, wheat, and other grain products. However, the transportation of such goods was difficult. Without the benefit of roads and rail, it could cost more to ship the grain to commercial centers than the grain was actually worth. Farmers began to distill their excess corn into whisky. This was one way to generate income. The distilling of the corn into whisky made it portable and easier to sell in surrounding areas. Kentucky's first commercial distillery began in 1789.

In 1791 Secretary of the Treasury Alexander Hamilton prompted Congress to levy a tax on liquor. A major reason for this tax was to help pay down the national debt. In 1794 farmers rebelled against the tax in what has come to be called the Whiskey Rebellion. The tax was later repealed, and the Whiskey Rebellion has been given some credit for the development of the whisky industry in Kentucky and Tennessee.

The term moonshine was first used in the late eighteenth century to refer to illegally produced whisky. The term comes from the fact that whisky smuggling usually took place late at night and away from the eyes of law enforcement—under the shine of the moon. Moonshine soon came to refer to all forms of liquor produced in the home or produced where it is illegal to do so. The government often tried to collect excise taxes on this illegal liquor. The brutality of the encounters between moon-shiners and law enforcement helped fuel a temperance movement in the United States.

The United States passed the Eighteenth Amendment on January 16, 1920. This constitutional amendment prohibited the sale, manufacture, and transport of alcohol; an exception was made for medicinal products. In 1933 the Twenty-first Amendment to the U.S. Constitution repealed the nationwide prohibition. The amendment gave individual states the right to restrict or ban the sale of alcohol, leading to a patchwork of state and local laws designed to regulate the sale and consumption of alcohol. Many states continued to enforce prohibition after the Twenty-first Amendment was passed. In 1966 Mississippi became the last state to repeal prohibition. The passage of the Twenty-first Amendment also paved the way for federal and state authorities to tax wine, spirits, and beer, which generate billions of tax dollars annually for local, state, and federal authorities. Other countries also went through periods of Prohibition, including the Soviet Union (1914–1925), Iceland (1915–1922) and Norway (1916–1927).

The home production of wine was permitted after Prohibition ended. In 1978 President Jimmy Carter signed legislation permitting the home brewing of beer. This legislation has been given some credit for the rise in microbreweries and small craft brewers that was to follow. The production of liquor in the home, however, is not legal in the United States without a license. Periodic attempts to legalize spirit production for personal use have failed. A major reason is that alcohol is one of the most heavily taxed of consumer goods; approximately one-third of the retail price of a bottle of liquor goes to state and federal authorities.

Bourbon. Bourbon is an American form of whiskey named for Bourbon County, Kentucky. Most bourbon is distilled there, but bourbon may be manufactured anywhere in the United States where it is legal to do so. Elijah Craig, a Baptist minister, is credited with inventing bourbon. Fifty-one percent of the grains used in bourbon production must be corn, although most distillers use 70 percent corn. Bourbon must also be distilled to no more than 160 proof, and aged in new charred oak barrels for at least two years. After aging, it is diluted with water and bottled. Bourbon must be put into the barrels at no more than 125 proof. Bottling proof for whiskey must be at least 80 proof.

Gin. Gin is believed to have been first produced in Holland in the early seventeenth century. The first step in gin manufacturing is to distill the neutral spirit alcohol. Gin makers seldom make their own neutral spirit. They typically purchase it from a distiller who manufactures neutral spirits for a variety of uses. The second step involves distilling the spirit again, this time adding various botanicals for flavorings. Gin usually requires juniper berries, but other flavorings may be added, such as citrus peel, ginger, or caraway seeds. Distilled water is added to bring the alcohol content to 80 to 95 proof. Dry gin is the most popular type of gin. It is produced in a column still. Dutch gin, however, is produced in a pot still. Other types of gin include Plymouth Gin, Old Tom gin, and Golden gin. Contrary to popular belief, sloe gin (flavored with fresh sloe berries) is actually a liqueur, not a gin.

Liqueurs. Distilled beverages with added flavorings and a relatively high sugar content such as Grand Marnier, Frangelico, and American style schnapps are generally referred to as liqueurs. They are flavored with fruits, herbs, spices, flowers, seeds, roots, plants, and barks. They appeared by the thirteenth century and may have been invented by the Dutch. As with wine and other liquor, monks were the major distillers of such beverages in Europe. In fact, some types of liqueurs still carry names that originated with the monks who once distilled them, Chartreuse liqueur, for example.

Rum. Rum is produced from fermented molasses or sugar cane. Most countries within the Caribbean get their molasses from Brazil. There are three main categories of rum: Cuban, Jamaican, and Dutch East Indian. There are also various grades of rum: light rum, gold rum, spiced rum, dark rum, flavored rum, overproof rum, and premium rum. There is no preferred distillation method for making rum. Some producers use pot stills while others use column stills, but most rum is aged in bourbon flasks. Because of the warm climate of the Caribbean region, rum ages more quickly than scotch or cognac, which are made in cooler climates.

Spanish speaking countries such as Cuba and Panama typically produce light rums with a clean taste. English speaking countries produce darker rums with heavier tastes. Jamaica is considered by many to be the best maker of rum. French speaking countries such as Haiti and Martinique produce their rum from sugarcane as opposed to molasses and they typically grow their own sugarcane.

Tequila. Tequila is made by fermenting and distilling the sweet sap of the agave plant. It may be colorless or be a pale gold color. Tequila is made in and around the small town of Tequila, in Mexico's Jalisco province. In order to be classified as tequila, the liquor must be produced from blue agave plants grown in a precisely delineated area covering five Mexican states: Guanajuato, Jalisco, Michoacan, Mayarit, and Tamaulipas. Mexican law states that tequila must be made with at least 51 percent blue agave; the remaining 49 percent of the input material is usually sugarcane. Tequilas labeled 100 percent Blue Agave are considered the best. Tequila has 40 to 50 percent alcohol by weight. There are four categories of tequila: blanco, joven abocado, reposado, and añjo. Tequila is mixed with lime juice and orange liqueur to make the margarita cocktail.

Vodka. Vodka is both odorless and colorless. Its exact origins are unknown, but it likely had its start in Poland or Russia. The word vodka comes from the Slavic word *voda* (or woda), meaning water. The drink was first made from distilled potatoes and then from corn. However, most vodkas are made from cereal grain products such as wheat. Distillers may own fields and produce the grain themselves or purchase it from suppliers.

Vodka has a particularly high alcohol content. Vodka can have a proof as high as 145, but water is added to bring the proof down to a range between 80 and 100. Vodka is often mixed with other beverages and is a main ingredient in such popular drinks as Screwdrivers and Bloody Marys.

Whiskey. This liquor, which is spelled either whiskey or whisky, is a shortened form of *usquebaugh*, a word the English took from the Irish Gaelic language meaning water of life: *uisge* means water; *baugh*, or *beada*, means of life. Whiskey is a broad category of distilled spirits, covering a number of types of spirit. Each type has a unique production method that varies by amount of water used, ingredients, type of oak cask in which the whiskey is stored, and the manner of distillation. These methods have an effect on color and flavor.

Straight whiskey includes bourbon, Tennessee whiskey, and rye whiskey. It is made from at least 51 percent of a particular grain, must not exceed 160 proof (80 percent alcohol), and must be aged in oak barrels for 2 years. Blended whiskey is a combination of two or more 100 proof straight whiskeys blended with neutral spirits, grain spirits, or light whiskeys. The malt in malt whiskey is allowed to germinate to a certain extent and then distilled two to three times. Light whiskey has been distilled to a high alcohol level, typically more than 160 proof, and then diluted with water to a greater extent than harder whiskeys. It gets its distinctive character from being stored in charred oak containers. Such whiskies are generally used for blending. Single-malt whiskey is made only from malted barley and from a single distillation.

In order for a whiskey to be considered Scotch whisky it must conform to standards laid out under Scottish law. It must be produced with Scottish water and barley. Its alcohol strength must not exceed 94.8 percent by volume; a higher level would compromise its flavor. It must be aged for at least three years in Scotland, although most are aged 5 to 10 years. After aging, Scotch whisky may be bottled elsewhere.

MARKET

The alcoholic beverage market consists of wine, beer, and liquor. According to the U.S. Census Bureau, U.S. manufacturers shipped $20.7 billion worth of beer in 2005, $10.4 billion worth of wine, and $5.3 billion worth of distilled beverages.

The U.S. distilleries industry saw its shipments generally increase in the 1990s and early 2000s. Figure 81 presents Census Bureau shipment data for the industry from 1992 through 2005. While there was steady growth during most of the 1990s, the pace of growth increased sharply in the first decade of the 2000s. With an exception in 2002, which saw reduced shipments, the industry saw steady growth in the period from 2000 to 2005 during which it grew a total of 27 percent, equal to a healthy 5.4 percent annual rate of growth.

Some of this increase can be attributed to demographics. The population of those in their thirties and forties increased during this decade. Many people in these age groups consume alcohol and are more likely than younger consumers to be able to afford liquor. Younger consumers tend to select beer as a preferred alcoholic beverage when making purchasing decisions.

Besides growth in the population of key age groups segments, the United States was experiencing a strong economy during the 1990s. This period is often called the technology or Internet boom. The stock market was strong. Consumers could afford to purchase liquor,

FIGURE 82

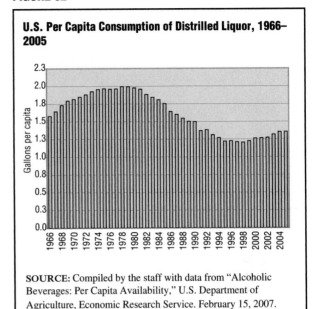

SOURCE: Compiled by the staff with data from "Alcoholic Beverages: Per Capita Availability," U.S. Department of Agriculture, Economic Research Service. February 15, 2007.

which is more expensive than beer and wine. As well, the industry was more visible. In 1996 the industry lifted the self-imposed ban on advertising that had been in place since 1936 for radio and 1948 for television. The ending of this ban generated some protests by the public. Former President Clinton urged the liquor industry to reconsider and some members of Congress considered enacting legislation to make the ban on such advertising formal law. Some states moved to ban such advertising on their local airwaves.

The liquor industry voluntarily agreed to certain standards. For example, the two major alcohol industry trade groups, the Distilled Spirits Council of the United States and the Beer Council, agreed only to advertise on broadcasts and in publications in which 70 percent of the audience is above the legal drinking age and advertisements must encourage responsible drinking habits.

While the major networks shun liquor advertising, cable channels and satellite radio are more receptive to the alcoholic beverage industries advertising dollars. The industry made a major push into the cable market and according to the Center on Alcohol Marketing and Youth at Georgetown University, the number of liquor commercials on cable networks increased from 645 in 2001 to more than 37,000 in 2004.

Shipments of alcoholic beverages by U.S. brewers and distillers slipped in the period from 2001 to 2002. A recession in early 2001 caused a decline in discretionary purchases, and high priced liquor sales dropped. The terrorist attacks against the United States on September 11, 2001, further depressed the sale of high priced liquor as

FIGURE 81

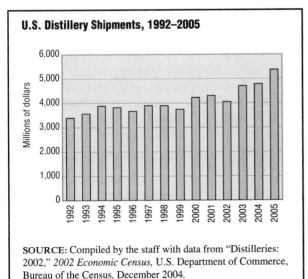

SOURCE: Compiled by the staff with data from "Distilleries: 2002," *2002 Economic Census,* U.S. Department of Commerce, Bureau of the Census. December 2004.

many bars and restaurants saw sales decline as consumers stayed close to home. From 2002 through 2005, the value of shipments grew steadily.

According to the Census Bureau there were 45 firms in the United States devoted to distilling alcoholic beverages in 1998. These establishments provided employment for 9,900 people. The industry saw a noticeable increase in number of participating firms in 2001, when the number of establishments increased to 75. In 2003 and 2004 the number of participating firms fell slightly to 65.

The major players in this industry are both large and small. In 2002, 35 percent of U.S. establishments in this industry had 1 to 4 employees, 16 percent had 20 to 49 employees, 23 percent had 100 to 249 employees, and 26 percent had 250 or more employees.

Consumption Trends. U.S. per capita consumption of alcoholic beverages declined in the 1980s and 1990s. Figure 82 presents per capita consumption data showing a decline from a high of 28.8 gallons per capita in 1981 to a low of 24.7 gallons per capita in 1995. The consumption of distilled beverages during the period fell from 2 gallons per person per year to 1.2 gallons. It rose again thereafter to reach 1.4 gallons per capita in 2005.

Faced with declining consumer demand, the major distilleries began to consolidate. The top twelve liquor firms in the United States in 1984 controlled 67 percent of liquor sales. In 1988 they had increased their share of the market to 72 percent. Guinness plc and Louis Vuitton Moet Hennessey, a distiller of cognac and champagne,

formed a joint marketing agreement in 1987. Guinness also purchased its distributor, Schenley Industries, that same year. Grand Metropolitan acquired distributor Liggett Group in 1980. Grand Metropolitan acquired Heublein Inc., which markets more than 100 brands of spirits, wines, and beer.

Retail Sales. For a long period in the liquor industry, the dark spirits such as whiskey, scotch, and bourbon were the most popular types of liquor. According to industry tracker Impact Databank, the top three selling brands of spirits in the 1960s and 1970s were Seagram's 7 Crown American blended whiskey, Seagram's VO Canadian whisky, and Canadian Club whisky. But during the following decades, Americans began searching for lighter liquors and became interested in vodka, gin, and other white spirits.

Figure 83 presents U.S. consumption of hard liquors by type and shows that vodka is the most popular form of distilled spirit in the United States. Vodka's share of total consumption was 23.4 percent in 1995; by 2005 it had an estimated market share of 27.1 percent. Rum also increased over this period as a percent of liquor consumption by type, representing 12.8 percent of consumption by 2005, up 45 percent from 1995 share. Meanwhile straight liquor (bourbon and whiskey) fell from 10.1 percent to 8.3 percent of consumption.

In 2006 a total of 176.6 million 9-liter cases of distilled spirits were sold at the retail level in the United States. Sales were up 3.7 percent from 170.2 million in 2005. Bacardi rum was the best selling brand, selling 9 million 9-liter cases. Smirnoff vodka was second with 8.5 million cases. Captain Morgan was third with 5.5 million cases.

Liquor has also reclaimed market share in the overall alcoholic beverage category. In the 1970s liquor represented 44 percent of all alcoholic beverages consumed. As alcohol consumption declined, liquor's share of overall consumption fell to 29 percent in 1995. The improved sales and visibility of the liquor industry in the early twenty-first century have helped liquor reclaim share of the retail alcohol market. According to the Distilled Spirits Council of America and Adams Beverage Group, in 2006 the liquor industry accounted for 32.8 percent of gross alcoholic beverage sales. That same year, beer accounted for 50.7 percent and wine accounted for the remaining 16.6 percent.

Global Sales. Diageo plc, based in the United Kingdom, Suntory Ltd, based in Japan, and Allied Domecq, also based in the United Kingdom, were the top producers of liquor in 2004 based on dollar sales. According to estimates by market research group Euromonitor retail sales of liquor worldwide totaled $254 billion in 2004.

FIGURE 83

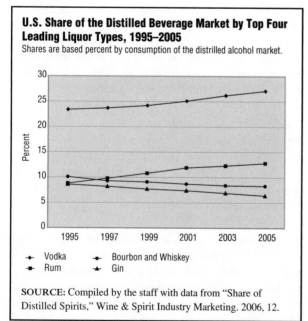

U.S. Share of the Distilled Beverage Market by Top Four Leading Liquor Types, 1995–2005

Shares are based percent by consumption of the distrilled alcohol market.

Vodka · Bourbon and Whiskey
Rum · Gin

SOURCE: Compiled by the staff with data from "Share of Distilled Spirits," Wine & Spirit Industry Marketing. 2006, 12.

This was expected to increase to $263 billion in 2005. Whiskey was the best-selling liquor worldwide. In 2005 it was expected to generate $61.7 billion in retail sales, or 23.4 percent of industry sales. India is the top market for whiskey. Irish whisky has had a presence in the Indian market since the nineteenth century when the British set out to establish its presence in that part of the world. The industry tracking firm just-drinks.com expected Asia to represent 60 percent of spirits consumption worldwide during the period from 2006 to 2008.

KEY PRODUCERS/MANUFACTURERS

Diageo plc. Diageo was formed in 1997 by the merger of Guinness plc and Grand Metropolitan plc. The company produces many of the best-known brands in the market: Johnnie Walker Scotch whiskey, J&B Scotch whiskey, Smirnoff vodka, Popov vodka, Baileys Irish Cream, Captain Morgan rum, and Gordon's gin. It also distributes Jose Cuervo tequila. It was a leading company in the United States, the United Kingdom, Ireland, Russia, Brazil, India, Korea and Australia. According to estimates by Deutsche Bank and Euromonitor the company had 22.1 percent of the U.S. spirits market in 2006.

Fortune Brands Inc. Fortune Brands is a holding company that acts as a distributor of distilled spirits, golf equipment, and numerous products for the home, including Moen faucets, MasterBrand cabinets, Master Lock padlocks, and Therma-Tru doors. Its popular liquor brands include Jim Beam, DeKuyper, Knob Creek, and Maker's Mark. Fortune Brands is located in Deerfield, Illinois. It employed 33,000 people and earned revenues of $8.7 billion for the fiscal year ended December 2006. The company had a 12 percent share of the U.S. spirits market in 2006.

Bacardi & Company Limited. Bacardi is another major company in the industry. The company's origins can be traced to 1862 when Don Facundo Bacardi Massó purchased his first distillery in Cuba. The company expanded into Spain, Mexico, Puerto Rico, and the United States at the turn of the twentieth century. Bacardi produces more than 200 different brands including: Bombay Sapphire gin, Martini & Rossi vermouth, Dewar's Scotch whisky, DiSaronno Amaretto, Grey Goose vodka, Blue Agave tequila, and B&B and Benedictine liqueurs. The company had 11.1 percent of the U.S. spirits market in 2006.

Constellation Brands. The company was founded by Marvin Sands in 1945. The company's spirit brands include Black Velvet Reserve, Effen vodka, Caravella, 99 Schnapps, and Paul Masson. The company markets more than 250 alcohol brands in nearly 150 countries. It markets itself as the largest wine producer by volume.

The company reported sales of $5.2 billion for the fiscal year ended February 2007. It operates more than 50 production facilities and has 9,200 employees worldwide. The company had a 10.2 percent share of the U.S. spirits market in 2006.

MATERIALS & SUPPLY CHAIN LOGISTICS

In 2002 U.S. distilleries spent $1.3 billion on material inputs to their operations. They spent $70.5 million on wooden casks, barrels, and similar types of products. Whisky is aged in new or used charred barrels for several years before it is shipped to market. Producers purchased $22.7 million in grain such as corn, rye, and barley. Approximately $155 million was spent on neutral spirits used in the production of vodka, gin, and other liquor. Approximately $256 million was spent on glass and plastic containers. Other expenditures included spending on grapes, flavorings, malt, and packaging materials.

Liquor production involves water, yeast, and grain products. The corn industry received increased attention from the media in the early twenty-first century because of the possibility of using corn in the production of ethanol fuel, an alternative to petroleum products of which the United States is heavily dependent on imports. According to the Department of Agriculture, corn production increased from 8.9 billion bushels in 2002 to 11.1 billion in 2005. The value of the corn increased as well, from $20.8 billion in 2002 to $24.4 billion in 2003 and then fell to $22.1 billion in 2005. In 2006 the value of corn production increased noticeably to $33.8 billion. The increased interest in corn is reflected in the increased harvest between 2006 and 2007. In 2006 an estimated 10.6 billion bushels were produced; by 2007 this figure was expected to climb to 13.3 billion. Iowa was the top state for corn production in 2006.

Barley is used primarily in whiskey production. According to the Department of Agriculture, the value of barley production fell from 2003 to 2006. In 2003 total production was valued at $755.1 million; in 2006 production was estimated to be worth $497.5 million. As with corn, estimates for 2007 show an increase in barley production. Total harvest area increased from 2.9 million acres in 2006 to 3.5 million acres in 2007. Total production was estimated to climb from 180 million bushels to 223.4 million bushels. Idaho was the top state for barley production in 2006.

Rye is another grain used in liquor production. According to the Department of Agriculture, the value of production fell from $26.5 million in 2004 to $23.5 million in 2006. Production of rye was expected to increase in 2007. Oklahoma was the top state for rye production in 2006.

Water is an important material used by distillers. The amount of water and its overall character—the amount and type of minerals in the water—will influence the taste of liquors. The responsible use of water by distillers is an important issue in the twenty-first century. Many distillers are implementing sustainable production practices as they deal with rising prices for water and increased pressure on the national aquifers.

DISTRIBUTION CHANNEL

Liquor is sold in the United States through a three-tier system of producer-wholesaler-retailer. This means distilled spirits (like beer and wine) move from the producer to a designated distributor at the state level and then on to a legally licensed retail establishment, restaurant, or bar. This chain is highly regulated. Distillers, vintners, and brewers are collectively known as the first tier; wholesalers and distributors are the second tier; and the retailer, restaurant, and bar are characterized as the third tier.

In 2006, the largest company based on sales was Southern Wine & Spirits of America Inc. Other major distributors include Glazer's Distributors, National Distributing Company, and Young's Market Company. Most distributors are privately owned. Approximately half of industry revenue comes from the sale of beer, 30 percent from liquor, and 20 percent from wine. Distributors tend to specialize in either beer or a combined list of wine and liquor, but some handle all three.

According to the Census Bureau's *Annual Survey of Manufactures*, there were an estimated 1,800 wine and liquor distributors in the United States in 2004. In 1987 there were 1,900 establishments in this industry—the highest number of establishments reported between 1984 and 2004. The distribution industry, like the liquor industry, saw a number of consolidations take place in the 1990s as large national distributors acquired the small, local companies in the market. Many state laws prohibit alcoholic beverage manufacturers from owning retailers, but producers are allowed to own distributors, and some distributors are allowed to own retailers.

In 2004 the liquor distribution industry employed more than 57,000 people. Sales in 2004 were estimated to be $39.3 billion. Sales per establishment were approximately $21.7 million. The wine and liquor distribution industry is fragmented largely because each state has different laws and regulations surrounding the distribution of alcohol. In 2006 the top 50 wine and liquor distributors held more than 70 percent of the U.S. market.

Retail Sales. In 1978 many states deregulated liquor prices that had previously been set by the government. Deregulation allowed supermarkets and convenience stores to enter the business of selling alcoholic beverages. Some feared that the corner retail liquor store—generally a small, independent operation—would become extinct. While the industry did experience some economic hardships in the 1990s, by the start of the twenty-first century the industry saw its fortunes reverse. The number of beer, wine, and liquor stores increased from 24,830 in 1998 to 26,037 in 2004, according to the Census Bureau. Employment increased from 129,129 to 142,294. Retail sales in beer, wine and liquor stores were approximately $13.4 billion in 2004. The average beer, wine, and liquor store generated $400,000 in sales and employed approximately four people. California, Colorado, Connecticut, Florida, Illinois, Massachusetts, Michigan, New Jersey, New York, Ohio, Pennsylvania, and Texas represented nearly two-thirds of the market based on value.

Wal-Mart has also helped shape the retail liquor market. The retail giant represents a significant portion of retail sales of many consumer products, from cookies to diapers. Much of their success comes from the fact that Wal-Mart purchases directly from manufacturers. This allows them to offer products at low prices. While the company had always had small amounts of alcohol available on its shelves, in 2003 the company quietly started a push into the liquor market. This was perceived as a risky move for them. The company does not permit alcohol at its headquarters and is very proud of its family friendly image. In 2005, it teamed up with liquor firm Diageo to develop new merchandising and products. In addition to new products Wal-Mart would stock some of Diageo's best-known brands, such as Johnnie Walker Scotch and Smirnoff vodka.

Wal-Mart sold $1 billion worth of alcohol in 2004, representing a small portion of the $285 billion it generated in the 2004 fiscal year. The company's new alcohol policies were met with some resistance. However, by 2005 the company appeared to be staking its claim in the market. It adopted a common liquor store policy of establishing an outlet just across the county line of dry counties (dry counties do not permit alcohol sales). Glazer's Wholesale Distributors, the third largest alcohol distributor in the United States, invested in new management software in 2005. The software—which manages sales and inventory and helps Glazer restock shelves more efficiently—is more compatible with Wal-Mart's systems.

Dry Regions. There are 415 dry counties in United States. An estimated 18 million people live in such dry counties. Alcohol sales are not permitted in these regions, although many enjoy a lucrative trade in illegal liquor. From 2002 to 2006 grocery stores, real estate developers, restaurant groups, and Wal-Mart have spent $15 million on local campaigns to legalize the sale of alcoholic beverages in stores and restaurants where their sale is currently prohib-

FIGURE 84

Trends in Alcoholic Beverage Consumption by Age Group and Type of Beverage, Select Years, 1992–2005

Figures are percents and show preferred drink by type and age group.
The total by age group and year do not equal 100 percent due to rounding.

Age Group/ Type of Beverage	1992/ 1994	1997/ 1999	2004/ 2005
Age 21 to 29 years			
Beer	71	56	48
Wine	14	20	16
Spirits	13	22	32
Age 30 to 49			
Beer	48	47	40
Wine	31	32	37
Spirits	17	17	21
Age 50 years plus			
Beer	28	30	30
Wine	37	46	45
Spirits	30	16	20

SOURCE: Eadie, Graeme and Nick Bevan, "Trends in Drink Consumption by Age Group," *2005 US Spirits Review*, Deutsche Bank. September 15, 2006. Based on data from *Adams Liquor Handbook 2005*.

ited. The reversal of such policies has caused controversy; however, many residents of these areas have been persuaded by arguments regarding the economic development and increased tax revenues that would be generated by the legal sale of alcoholic beverages.

KEY USERS

According to the *National Survey on Drug Use and Health 2004*, approximately half of Americans over 12 years of age claimed to be current consumers of alcohol.

The *Adams Liquor Handbook 2005* offers some insight into drinking preferences by age group. As can be seen in Figure 84, liquor became the preferred alcoholic beverage for those aged 20 to 49 years during the 1990s and early 2000s. The table presents these alcoholic beverage preferences by age group and type of alcoholic beverage for three time periods. The growing rate of consumption by a young segment of the population is important to the distilled spirits industry. While gaining new customers in the younger demographic, liquor appears to be on the decline with older drinkers. Those aged fifty and older moved away from liquor in favor of both beer and wine.

Liquor and Cultural Trends. Alcohol consumption increased in the 1990s. With a strong economy consumer confidence was high and spending on luxuries grew. Men started smoking cigars in greater numbers, for example. A

brief cigar craze occurred and cigar imports reached an all time high point of 417.8 million units in 1997. The circulation of *Cigar Aficionado* magazine rose from 141,000 in 1994 to nearly 400,000 in 1996. In 1998 the television show *Sex and the City* began its run on cable television. The show featured four young female friends living in New York City and one of the activities that they were depicted doing frequently was drinking cocktails during their nights out on the town. Social observers of the time used the term Cocktail Nation to describe the sudden increase in cigar clubs and martini bars.

Many men favor dark spirits such as whiskey or scotch. Women tend to favor white spirits such as gin or vodka. While many older drinkers have established drinking habits, younger drinkers are seen as being more willing to experiment. Liquor companies actively market to younger drinkers in an attempt to take advantage of their perceived willingness to try new things. Liquor companies have also introduced low calorie and flavored products, which they market primarily to women.

Binge and Moderate Drinking. Research continues on the effects of drinking on consumer health and behavior. Binge drinking, defined as five drinks or more in one sitting, continues to be a problem in the United States. The consumption of excessive alcohol plays a role in car accidents, violence, and other traumatic injuries. Approximately 64,000 deaths in 2000 were attributed to heavy drinking according to the Center for Disease Control and Prevention (CDC). A CDC report released in 2007 showed that 15 percent of the U.S. population were binge drinkers. Approximately three-quarters favored beer, 17 percent favored liquor, and 9 percent favored wine. Teenage binge drinkers were much more likely to binge on liquor. The CDC theorized that adult binge drinkers favor wine or beer because it is less expensive than liquor. Teenagers, who by definition are drinking illegally, are not hampered by considerations of cost since they are obtaining the alcoholic beverages they binge drink from others, often their parents' liquor cabinet.

Unlike binge drinking, moderate drinking offers some health benefits. For example, the Harvard School of Public Health tracked 11,711 men diagnosed with hypertension from 1986 through 2002. The study, published in January 2007 in the *Annals of Internal Medicine*, found that men who consumed 15 to 29.9 grams of alcohol per day—the equivalent of one to two drinks—reduced their risk for a fatal heart attack by 30 percent. Nonetheless, excessive drinking, defined as more than 3 drinks per day, was not encouraged by the researchers.

ADJACENT MARKETS

Wine. The development of wine occurred in the region of Mesopotamia and around the Caspian Sea. The early civilizations made important discoveries about pruning, irrigation, soil, and vine harvesting to ensure the best grape production possible. Manufacturer shipments of wine were worth $10.4 billion in 2005. California dominates the industry in the United States, making 90 percent of the wine produced in the country. Exports of California wine increased from $641 million and 282 million liters in 2003 to $672 million and 382 million liters in 2005. However, major wineries are found in other states, and small wineries are beginning to proliferate throughout the United States. According to WineAmerica, the national association of American wineries, there were wineries in all 50 states in 2006. Approximately 45 percent of the nation's 4,280 wineries were in California, Washington, Oregon, New York, Virginia, Texas, Pennsylvania, and Michigan, with each of these states having more than 100 wineries.

Beginning in 1970, wine sales in the United States surged. During the 1980s sales were flat or falling. Beginning in 1994 sales began to climb steadily again and continued to grow in early 2007. This growth was attributable to a number of factors: new reports stating that moderate wine drinking is healthy, increased and improved marketing campaigns, and those in their twenties and early thirties adopted wine drinking at a younger age than earlier generations.

Beer. Manufacturers shipped $20.7 billion worth of beer in 2005. A total of 1,367 brewers operated in the United States in 2005. The Beer Institute characterized twenty-one of these brewers as traditional brewers and the remaining 1,346 as specialty brewers. The traditional brewers are large, industry leaders by volume, while the specialty brewers include microbreweries and brewpubs, most of which produce specialty and craft beers. The large number of specialty brewers highlights the dramatic changes that have taken place in the U.S. beer brewing market during the last quarter of the twentieth century. In 1960 there were only 175 brewers in the United States.

The U.S. craft brewing renaissance has had a significant affect on the market. Craft brewing is the fastest growing segment of the alcoholic beverage industry. While 2004 proved to be a difficult year for the industry as a whole, the craft beer segment grew at a rate of 7 percent. Moreover, craft brewers have been able to grow despite the proliferation of importers during the 2000 to 2004 period.

RESEARCH & DEVELOPMENT

Manufacturers are following the work of craft distillers and putting an emphasis on fresh ingredients, new flavors, and unique production methods. The Empire Winery and Distillery produces the award winning V6 vodka brand, which is distilled six times in copper pots from malted rye and distilled water. The vodka is then filtered through oak and maple hardwood charcoal until it is essentially pure. The Silver Creek Distillery filters its vodka five times through charcoal and lava rock garnet. Bendistillery manufactures vodka with hazelnuts and espresso beans. Some luxury vodka brands are manufactured using arctic or deep-sea water. Ocean Vodka, based on the island of Maui, produces vodka with water taken from 3,000 feet below the Pacific Ocean.

In 2006 and 2007 consumers began to demand products that were produced locally; such products are preferred for two reasons. First, they are environmentally friendly since less energy is required to move them to market. Second, they are beneficial to the local economy. Distillers have responded to this increased demand for local products. Silver Creek Distillery, based in Idaho, for example, manufactures its quadruple-distilled potato vodka entirely from Idaho Russet Burbank potatoes.

R&D efforts have also been expended on liquor packaging. At the end of 2006 Allied Glass Containers celebrated what it claimed was an industry first, the manufacturing of a tall, elegant 700 milliliter (ml), spirit bottle weighing just 295 grams (g). Produced at its Knottingley, England, facility, the container is the culmination of a project that started in 1999, when the company's working weight for a 700ml bottle was 435g. Allied reported that the lowest average weight achieved for this size bottle is approximately 340g.

CURRENT TRENDS

Healthier Drinks. In a battle against their expanding waistlines, many Americans embraced healthier lifestyles in the 1990s and early 2000s. Food and beverage marketers capitalized on this trend by releasing more foods grown organically, as well as products low in sugar, in carbohydrates, and in trans fats. The food industry produced healthier versions of cookies, potato chips, bread, ice cream, and beer.

For the health conscious, the cocktail was just empty calories, potentially high in sugar and high in carbohydrates. In 2007 the liquor industry promoted drinks with more nutritious additives: pomegranates, green tea, berries, and fresh juices.

Premium Categories. The spirits industry constantly looks for ways to innovate. From 2004 to 2005 the industry introduced 283 new types of spirits. Offering premium and super premium brands are another way the industry stimulates sales. The market for high-end spirits came into being in 1997, when the Sidney Frank Importing Company launched the Grey Goose brand. This brand was intended to create a sense of luxury, a brand that could unseat vodka market leader Absolut. In 2005 economy class vodka brands, those priced at less than $10 per bottle, witnessed a sales decline, while premium brands, costing between $15 and $30 per bottle, grew by 7 percent. Super premium varieties, those costing in excess of $30 per bottle, saw growth of 12 percent between 2004 and 2005. Other spirits are seeing strong sales in their premium segments as well. Sales of premium tequila, those costing $40 or more, increased more than 20 percent from 2002 to 2006.

TARGET MARKETS & SEGMENTATION

The spirits market in the United States performed well in the first decade of the twenty-first century. Markets being targeted are those associated with premium products and flavored liquors. Young customers are the targets for many of these flavored beverages and in particular, young women. Liquor companies have targeted drinkers between 21 and 29 years of age and the results of this targeted marketing are tangible. Between the early 1990s and the middle of the first decade of the twenty-first century, the alcoholic drink preferences of those aged 21 to 29 shifted strongly in favor of liquors at the expense of beer of and wine. In the period from 1992 to 1994, these drinkers preferred beer to liquor at a rate of 5.4 to 1. Twelve years later, in the period from 2004 to 2005, their preference was still in favor of beer but at a substantially reduced rate of 1.5 to 1.

A growing Hispanic population in the United States offered liquor companies another growing market to target with products designed to appeal to the Hispanic taste sensibility. Distillers are also producing and marketing products that appeal to the wealthy and health conscious Baby Boom market.

RELATED ASSOCIATIONS & ORGANIZATIONS

American Distilling Institute, http://www.distilling.com
Association of Canadian Distillers, http://www.canadian distillers.com

Distilled Spirits Council of the United States, http://www.discus.org
Kentucky Distillers Association, http://www.kybourbon.com

BIBLIOGRAPHY

"Adult Binge Drinkers Prefer Beer, While Teens Opt for Liquor." CNN. 8 August 2007. Available from <http://www.cnn.com/2007/HEALTH/08/07/binge.drinking.ap/index.html>.

Bakalar, Nicholas. "Vital Signs: Heart Health, Study Links Alcohol Level to Lower Risk of Coronary." *New York Times.* 9 January 2007.

Ball, Deborah and Ann Zimmerman. "Hard Stuff." *Wall Street Journal.* 17 August 2005.

Barrionuevo, Alexi. "At $300 a Bottle Do You Have to Tip? The Allure of Expensive Liquors in Pretty Packages." *New York Times.* 16 December 2006.

"A Billion Dollar Gamble in Whisky." *Time.* 12 April 1971.

"Distilleries: 2002." *2002 Economic Census.* U.S. Department of Commerce, Bureau of the Census. December 2004, 1–8.

"The Global Spirits Market." Just-drinks.com. March 2006, 19. Availalble from <http://www.just-drinks.com/store/product. aspx?id=52174&lk=rotw_arch>.

Greenberg, Marc. "Beverage Industry. A Cup Half Full: Gulping the U.S. Profit Pool." Deutsche Bank. 17 May 2007, 14.

Hanson, David J. Ph.D. "History of Alcohol and Drinking Around the World." D.H. Handson. Available from <http://www2.potsdam.edu/hansondj/controversies/1114796842.html>.

Levy, Clifford J. "Drink, Don't Drink. Drink, Don't Drink." *New York Times.* 9 October 2005.

Nagle, James. "Vodka Gains on Bourbon as Favorite Liquor in U.S." *New York Times.* 13 January 1975.

Plotkin, Robert. "Vodka Booming." *Beverage Dynamics.* May-June 2007, 12.

Seelye, Katherine Q. "Trickle of Television Ads Releases Torrent of Regulatory Uncertainty." *New York Times.* 12 January 1997.

Spiegler, Marc. "The Cocktail Nation." *American Demographics.* 1 July 1998.

"Top 20 Distilled Spirits Brands 2005–2006." *Wine Handbook Annual 2007.* Adams Beverage Group. 2007, 187.

Warner, Melanie. "With Business Leading a Push, Liquor Comes to Dry Bible Belt." *New York Times.* 12 August 2006.

Williams, Alex. "Alcohol Goes on a Health Kick." *New York Times.* 15 July 2007.

SEE ALSO *Beer, Wine*

DVD Media

———————•———————

INDUSTRIAL CODES

NAICS: 33–4613 Magnetic and Optical Recording Media Manufacturing

SIC: 3695 Magnetic and Optical Recording Media Manufacturing

NAICS-Based Product Codes: 33–461301 through 33–46130213, and 33–46130615

PRODUCT OVERVIEW

A DVD is a disc measuring 4.72 inches in diameter that is used to store digitized information. The DVD was introduced in the 1990s as an optical storage medium and, ultimately, as a replacement for compact discs (CD). The abbreviation DVD is used to stand for both Digital Video Disc and Digital Versatile Disc. Those who use the latter term argue that the abbreviation should stand for Digital Versatile Disc because of the media's many non-video applications.

The design of the DVD is an advancement on the design of the earlier CD. These two types of optical storage media share the same dimensions and are created in much the same way. An optical disc has millions of bumps or pits, depending on which side of the disc from which they are viewed, arranged in one long line that spirals from the center of the disc out. These bumps are arranged into tracks. Each track is separated by 1.5 microns (millionth of a meter) of space. The areas between the bumps are known as lands. The lands and bumps represent the zeros and ones of digital information. A laser in the CD or DVD player deciphers the digital information on the disc. The laser light reflects off the bumps differently than it does off the lands. Sensors in the disc player detect these differences and are thus able to read the information stored on the disc.

The primary difference between the CD and the DVD has to do with their different storage capacities. By reducing the size of the bumps and lands on the surface of an optical storage disc more data can be placed on the same physical surface area. As laser and material technology improved, the bumps and lands on the CD were reduced in size and the DVD was developed. On a CD a bump measures 0.83 microns long and 125 nanometers (billionths of a meter) high. On a DVD the bump length is 0.4 microns allowing for a much greater density of information on a single disc. A single sided DVD holds approximately seven times the capacity of a CD. A double-sided, double-layer version can store 24 times as much information as a standard CD. As the DVD format became popular in the late 1990s, it generated numerous new and competing formats for how data are stored on this high density, optical storage media.

The Creation of the DVD Format. The DVD was the result of competing technologies in the early 1990s. Philips and Sony backed the MultiMedia Compact Disc (MMCD) data formatting technology. Toshiba, Time-Warner, Matsushita Electric, Pioneer, Thomson, and JVC supported the Super Density Disc data formatting technology. In order for disc drives or players to be able to read from a CD or DVD, the information on the disc must be organized by a standard known to the disc drive. In a sense, they must speak the same language.

FIGURE 85

Types of Recordable and Rewritable DVDs

Type	Full Name	Frequency of Writing Capacity	Storage Capacities Single Sided SL	DL	Year Developed	Supporting Industry Group
DVD-R	DVD Recordable	Writable a single time	4.7GB,	8.5GB	1997	DVD Forum
DVD-RAM	DVD Random Access Memory	Rewritable 100,000 times	4.7GB,	9.4GB	1999	DVD Forum
DVD-RW	DVD Rewritable	Rewritable 1,000 times	4.7GB,	8.5GB	1999	DVD Forum
DVD+RW	DVD Rewritable	Rewritable 1,000 times	4.7GB,	8.5GB	1997	DVD+RW Alliance
DVD+R	DVD Recordable	Writable a single time	4.7GB,	8.5GB	2002	DVD+RW Alliance

SL stands for single layer
DL stands for double layer
GB stands for gigabyte or a thousand million bytes

SOURCE: Compiled by the staff with data from both the DVD Forum Web site and the DVD+RW Alliance Web site, Available from http://www.dvdforum.com and http://www.dvdrw.com. 2007.

Thus, they must share a data formatting technology. The president of IBM, Lou Gertsner, brought these two technology camps together, hoping to avoid a protracted battle like the one that occurred during the introduction of the video cassette recorder (VCR) between supporters of the Betamax format and supports of the VHS format. Philips and Sony abandoned the MMCD and the companies agreed upon a new standardized technology. The DVD specification Version 1.5 was announced in 1995 and was finalized a year later. In May 1997 the first DVDs and DVD players appeared on the market.

That same year the DVD Forum was created. The DVD Forum is a group of consumer electronics and computer equipment manufacturers that work to standardize DVD formats. They are a trade organization and do not set official policy. The DVD+RW Alliance was also formed in 1997. It includes leading electronics manufacturers such as Hewlett-Packard, Dell, Microsoft, Yamaha, and Sony. These industry leaders disagreed with the guidelines established by the DVD Forum. They developed the DVD+R (recordable) and DVD+RW (rewritable) formats to be more compatible with existing consumer DVD players, which the Forum's DVD-RAM (random access memory) format was not. The DVD Forum has not endorsed the rival formats. The DVD+R and DVD+RW formats have become quite popular with consumers in the first decade of the 2000s. They offer faster writing, better internal linking, and support for drag-and-drop desktop file manipulation, but they are also more expensive than the DVD-RAM technology.

DVD Capacities. Any discussion of DVD media can be confusing because of the various disc capacities, the misuse of terms, and the competing formats. There are four different sizes categories of DVDs: DVD-5, DVD-9, DVD-10, and DVD-18. They are distinguished by their data storage capacity.

- DVD-5 (Single Sided Single Layered) holds approximately 4.37 gigabytes (GB) of data.

- DVD-9 (Single Sided Dual Layered or Single Sided Double Layered) holds approximately 7.9 GB of data.

- DVD-10 (Double Sided Single Layered) holds approximately 8.75 GB of data.

- DVD-18 (Double Sided Dual Layered or Double Sided Double Layered) holds approximately 15.9 GB of data.

Movies and music that are distributed on DVDs can not be written to. They are not sold as storage media. DVDs sold for use as storage media can be written to. Figure 85 presents a summary of recordable and rewritable DVDs by type. Each is discussed separately in more detail.

Writable DVD Formats. The DVD-R is the format supported by the DVD Forum. The DVD+R is the format supported by the DVD+RW Alliance. These names are pronounced in several different ways. DVD-R, for example, is pronounced as DVDR, DVD Dash R, or DVD Minus R. This recordable format may be written to only once and is best suited for the archiving of computer data or for the distribution of data or video. These discs hold up to 4 hours of DVD quality video or 16 hours of VHS quality video. Subtle differences exist between the DVD+R and DVD-R formats; the systems for tracking, error management, and speed control are more accurate at high speeds in the DVD+R than in the DVD-R.

DVD+RW and DVD-RW discs can be written to up to 1,000 times. Both formats have storage capacity of approximately 4.7GB, which is equivalent to the storage capacity of 6 CDs. This capacity allows 8 hours of video to be stored on a disc. The rewritable DVD format is popular because it can be erased and written to with ease. This makes it ideal for daily backups of computer data. It is also used in camcorders and personal video recorders. Backers of the DVD+RW such as Sony and Phillips claim this format is more compatible with consumer DVD players.

The DVD-RAM format is the most complex and expensive of the formats. It is also the least popular. The disc has defined tracks that allow the machines that read them to locate the exact track to write or to erase. This function makes the disc work like a computer hard drive. The format can be written to repeatedly and is seen as the most reliable format because of its built-in safeguards to manage errors and defects. It is well suited to data backups and archiving. DVD-RAM is more popular in consumer devices such as camcorders and set-top boxes than in computers. DVD-RAM can be written to up to 100,000 times before it is no longer usable.

Commercially produced DVDs may be played on any standard DVD player or DVD drive because of the manner in which such DVDs are produced. Commercially, mass produced DVDs used to distribute movies, music, and software, are made in a stamping process that is different from the writing process used to write onto rewritable or recordable DVDs. Audio and video content that is written to a disc on a personal computer may not be compatible with all DVD players. There is no guaran-

tee that a DVD made on a computer or other electronic device will play on all standard DVD players.

As a relatively new technology, standards were still evolving for this medium in the latter half of the first decade of the twenty-first century and uncertainty in its use was expected to continue for the foreseeable future. As hardware and software formats are agreed upon by all parties involved—those who make the DVDs themselves, those who make the materials to be placed on the DVDs, and those who make the devices used to read or play the DVDs—the complexity that characterized this storage medium in the latter years of the first decade of the twenty-first century will diminish.

MARKET

The number of producers of recording media in the United States is declining. According to U.S. Census Bureau figures, there were 173 establishments involved in the manufacturing of magnetic and optical media in 2002, down from 257 establishments in 1997. Most establishments were based in California, a state with numerous software and technology firms. Employment fell sharply between 1998 and 2004. U.S. employment in the recording and optical media industry fell from 20,476 workers in 1998 to 5,754 workers in 2004. Employment in this industry has followed the basic trend seen in the United States in other electronics industries. First, much of the production capacity has been moved to other countries. Second, technological improvements have increased productivity and this increased productivity has reduced the demand for labor.

As can be seen in Figure 86, the value of U.S. manufacturing shipments of magnetic and optical recording media have been declining since the mid-1990s. In 1997 shipments were valued at $5.9 billion. By 2005 shipments had fallen to $1.6 billion. In addition to blank discs, these establishments manufacture blank data tapes, reels, cassettes, and cartridges.

Retail Sales Figures. Recordable DVDs generated sales of $50.7 million at supermarkets, drug stores, and discount stores for the year ending December 3, 2006, excluding sales at Wal-Mart. This figure is approximately half of the $101.2 million in sales generated by blank CD-ROMs for the same period, but sales suggest the rising popularity of the DVD media format. According to *MMR* magazine, sales fell 2.4 percent for CD-ROMs between 2005 and 2006 while sales of recordable DVDs increased 38 percent. For the 2005 to 2006 sales period, floppy disc sales fell 42 percent to $3.7 million.

Global Demand. According to the Japan Recording Media Association (JRMA), global demand for DVD media

FIGURE 86

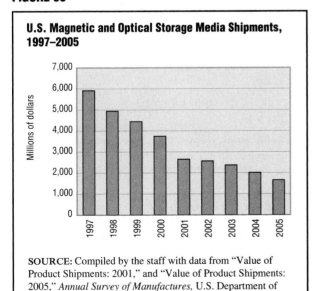

U.S. Magnetic and Optical Storage Media Shipments, 1997–2005

SOURCE: Compiled by the staff with data from "Value of Product Shipments: 2001," and "Value of Product Shipments: 2005," *Annual Survey of Manufactures*, U.S. Department of Commerce, Bureau of the Census. June 2004 and November 2006.

products was strong in the middle of the first decade of the 2000s. Demand for recordable DVDs was anticipated to grow 21 percent and reach 5.7 billion units in 2007. This growth was anticipated to continue, reaching 6.9 billion DVDs in 2009. Growth was anticipated for rewritable DVDs as well. The JRMA forecasted rewritable DVD demand to climb from 495 million units in 2007 to 541 million units in 2009. This demand was forecasted based on expectations of continued growth in the shipment of personal computers and DVD recorders.

According to estimates from industry analysts Santa Clara Consulting, and Understanding & Solutions, Verbatim was the leading brand of recordable DVD products worldwide in 2005. It had a 16.9 percent market share that year as seen in Figure 87. The TDK brand had 14 percent of the market; Memorex had 10.7 percent; Sony, 8.2 percent; and Imation, 6.5 percent.

KEY PRODUCERS/MANUFACTURERS

Imation. Imation was formed in 1995 from parts of the 3M Corporation. The 3M Corporation had stagnant revenues for some time and the move was seen as a way to stimulate growth. The Imation name comes from the words imaging, information, and imagination. In 2006 it acquired Memorex and in 2007 it acquired TDK. These acquisitions solidified the company's place in the optical media and flash storage markets and gave Imation commanding market shares across the various DVD media categories. Imation had 2,070 employees in 2007 and reported revenues of $1.5 billion for the fiscal year ended December 2006.

Memorex. Memorex is the leading producer of recordable DVD media in the United States. Memorex had 28 percent of the U.S. retail market through the first quarter of 2006 based on dollar sales, according to The NPD Group. Memorex also manufactures other recording media. Established in 1961, the company first drew attention with its shattered glass ad campaign in 1972. In these ads, jazz singer Ella Fitzgerald would sing a note that shattered a wine glass while being recorded with Memorex audiotape. The tape was replayed and Fitzgerald's recorded voice again shattered the glass, indicating the crystal-clear quality offered by Memorex tape. The ad asked the question "Is it live or is it Memorex?" The company still uses this slogan and has found success through its brand recognition and strong marketing. The Memorex brand is marketed in more than 25 countries and is on the shelves of 21 of the top 25 consumer retailers. In recent years, the company was sold several times. As of early 2007 the company was a division of Imation, which manufactures a variety of computer data storage products. Memorex reported revenues of $430 million in 2005.

FIGURE 87

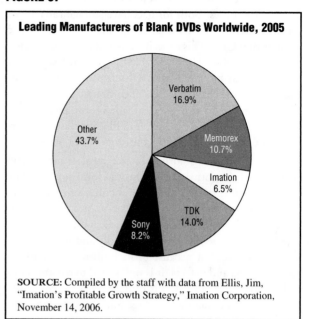

Leading Manufacturers of Blank DVDs Worldwide, 2005

Verbatim 16.9%
Memorex 10.7%
Imation 6.5%
TDK 14.0%
Sony 8.2%
Other 43.7%

SOURCE: Compiled by the staff with data from Ellis, Jim, "Imation's Profitable Growth Strategy," Imation Corporation, November 14, 2006.

TDK. TDK, founded in 1935, is based in Japan. TDK employed nearly 54,000 people and saw revenues of $6.8 billion for the fiscal year ended March 2006. While its name is associated with audio tape and similar media, the company was founded to commercialize a magnetic material known as ferrite. Ferrite is used in the production of a variety of electronic components, such as high-frequency devices, capacitors, magnets, and transformers. It is from these products that TDK derives the bulk of its income.

Verbatim. Verbatim was founded in 1969 and is a division of the Mitsubishi Chemical Corporation of Japan. It earned revenues of $130 million for the fiscal year ended December 2005. Verbatim was the leader in the global market in 2005 in a number of media categories, including DVD-R, DVD+R, CD-R, and CD-RW.

MATERIALS & SUPPLY CHAIN LOGISTICS

DVD production varies depending on the format. If a disc is single-sided, then it is composed of a recording side and a dummy side. A double-sided disc will use two recording sides.

The disc is manufactured through an injection molding process. It consists of three layers: a polycarbonate plastic substrate; a thin layer of aluminum or some similar reflective material which serves as the surface off of which the laser operates; and an acrylic layer that provides a protective seal for the bumps, lands, and aluminum layer below. A label is often painted on top using screen-printing methods and is often thought of as a fourth layer. Some

DVDs consist of multiple readable layers and it is through the use of these multiple layers that manufacturers are able to produce the four different types of DVDs with varying capacities: DVD-5, DVD-9, DVD-10, and DVD-18.

The primary materials used in the manufacture of DVDs are polycarbonate plastic, acrylic, silver, silver alloy, silicon dioxide, germanium, and zinc sulfide.

The manufacturing process for the DVD-R and DVD+R formats, those intended for use as recordable storage media, differs slightly from the process used to make DVDs intended for use as a distribution media for music, movies, and software. The DVD-R, for example, has code prewritten onto it in what is known as the Control Data Zone of its lead-in area. This area of the disc surface is reserved for instructional codes. It is in this area that code designed for preventing illegal copying resides. Such computer codes are intended to prevent the direct copying of prerecorded DVD-Video discs encrypted with the Content Scrambling System (CSS).

DISTRIBUTION CHANNEL

Most companies that produce DVDs are involved in the production of many types of digital recording media. Both Memorex and TDK, for example, make a variety of recording products and have well established distribution networks in place through which they can send new products. The distribution channels used by DVD manufacturers were established first for the blank CD. As the technology advanced, and its applications grew, the number of outlets through which DVDs were made available also grew.

Blank recording media can be purchased in almost any type of general retail outlet. They are found in office supply stores, electronics stores, movie rental stores, music stores, many grocery stores, drug stores, dollar stores, and even in some gas station and convenience stores. The Big Box retailers such as Costco, Sam's Club, and Wal-Mart, also carry digital recording media and usually offer a full array of blank DVDs. Such stores have lower prices because they purchase directly from the manufacturer.

Office supply stores were once the primary retail outlet for digital recording media. Office supply stores such as Office Depot and Staples represented a significant share of office product sales in the early twenty-first century. According to *Market Share Reporter 2007*, office supply stores and warehouse clubs, as well as other discount chains, represented approximately 30 percent of the overall office supply market in the United States in 2005. Grocery stores, the Internet, independent dealers, contract specialists, and other chain stores held the rest of the market.

CD and DVD media were among the general merchandise categories showing the strongest growth in supermarkets in 2002. Maxell's blank DVD-R was among the top 50 best selling brands of general merchandise at supermarkets in 2007, according to *Chain Drug Review*. The brand's sales were up more than 52 percent; among the top 50 brands its sales increases were second only to Duracell Coppertop batteries.

KEY USERS

Businesses first used DVDs, drives, and recorders for computer storage purposes, but consumers soon saw the value of these products as well. Archiving and data storage were the most popular reasons consumers gave for wanting a DVD recorder, according to a 2004 study by International Data Corp. Market researcher In-Stat predicted the worldwide DVD recorder market will grow from approximately 20 million units in 2006 to approximately 38 million units in 2008. According to In-Stat, the market will be affected by a continued fall in DVD prices, improved features in DVD recorders, and consumers' understanding of the benefits of DVD optical media and recording devices.

Consumer electronics such as camcorders, recorders, and cameras are increasingly becoming disc-based. The disc-based technology allows consumers to store and organize photos and provides high-quality video output. Digital video recorders (DVRs), also known as Personal Video Recorders (PVRs) have become popular ways to record television programming. Previously consumers had to use a VCR and videotape. Programs recorded with DVRs or PVRs are saved to a rewritable hard disk, instead of a videotape. Consumers prefer these digital devices because live broadcasts can be instantly replayed and commercials can be skipped. Also, saved programs can be organized by title and content. The industry started in 1999 with TiVo and ReplayTV. Users purchased the stand-alone devices and then paid a subscription fee; however, consumers did not really embrace the technology until it was integrated into cable and satellite TV set-top boxes.

DVR firms estimate that there were 7.4 million households with DVRs in the United States in early 2005, or 6.5 percent of all U.S. households. Estimates from analyst eMarketer suggest that there will be more than 47 million households with DVRs by 2009. The estimate may seem high, but cable and satellite television penetration is increasing in the United States. The Cable and Telecommunications Association for Marketing claims that cable households with a DVR have more than doubled in 2006 from 2005. Thirty percent of digital cable households have a DVR compared to 22 percent of satellite households. Three-quarters of these digital cable households received their DVR from their cable provider.

ADJACENT MARKETS

Declining Formats. DVD media is part of the larger market of recording media. Demand for many of these formats has reached its peak and is declining. According to the Japan Recording Media Association, CD-R/CD-RW demand will fall from 6.5 billion units in 2007 to 5.5 billion by 2009. Audio CD-R hit peak demand in 2004. Demand is expected to fall from 258 million units in 2007 to 225 million units in 2009. The older data storage formats, audiocassettes and tapes, have been on the decline for years and are expected to decline further as users recognize the limitations of such media and upgrade their storage hardware. Global demand for blank audiocassettes was anticipated to fall from 228 million units in 2007 to 137 million units by the end of the first decade of the twenty-first century. Demand for full-sized blank video-cassettes and floppy diskettes were forecasted to drop by 20 percent per year through the first decade of the 2000s.

Interest in Electronic Devices. As consumers become more interested in computers and entertainment, the use of DVD media will become more prevalent. Many businesses use DVDs as reliable means for archiving data. Consumers are being encouraged to perform similar backups on their home computers. Santa Clara Consulting Group forecasts the worldwide installed base of DVD writers will climb to more than 180 million by the end of 2005.

DVD players and camcorders continue to become common in American homes. According to statistics from *Appliance* magazine more than half of U.S. households (55%) owned a camcorder in 2005, up from 32 percent in 1998. Seventy percent of households owned a DVD player in 2005; only 2 percent did in 1998. Approximately 65 percent owned a computer; 44 percent reported having one in 1998. The combination of falling prices and attractive new features will likely make these devices irresistible to consumers.

RESEARCH & DEVELOPMENT

DVDs have been on the market for only about a decade, yet next-generation formats are already appearing. The high-definition industry gained a foothold in the American electronics market in 2006, which translated into increased sales of high-definition televisions, DVD players, and discs. This industry is still in its infancy, and neither the public nor electronics makers have embraced the Blu-ray or the HD format, the two formats competing to be the standard for the next generation of high definition recording.

In the early 2000s the Chinese government has been trying to find a technology to compete with the DVD in order to reduce its licensing fees and royalty payments

FIGURE 88

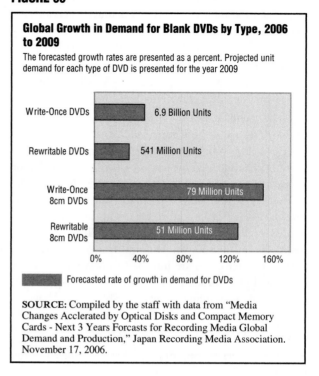

Global Growth in Demand for Blank DVDs by Type, 2006 to 2009

The forecasted growth rates are presented as a percent. Projected unit demand for each type of DVD is presented for the year 2009

- Write-Once DVDs: 6.9 Billion Units
- Rewritable DVDs: 541 Million Units
- Write-Once 8cm DVDs: 79 Million Units
- Rewritable 8cm DVDs: 51 Million Units

Forecasted rate of growth in demand for DVDs

SOURCE: Compiled by the staff with data from "Media Changes Acclerated by Optical Disks and Compact Memory Cards - Next 3 Years Forcasts for Recording Media Global Demand and Production," Japan Recording Media Association. November 17, 2006.

to foreign countries that use DVD technology. The development of the Enhanced Versatile Disc (EVD) was announced in China in November 2003. The EVD uses converters developed by On2 Technologies. Development stalled after legal disputes developed between the consortium of companies developing this technology and On2 Technologies. The replacement for the EVD is the Advanced Video Standard (AVS). AVS was approved as a national standard in China in March 2006; however, it has been adopted by only a handful of Chinese companies.

The Forward Versatile Disc (FVD) was developed in Taiwan and is a less expensive competitor of the high-definition disc. It is manufactured like a regular DVD and uses a red laser, but track width has been shortened slightly to allow the disc to have 5.4 GB of storage per layer as opposed to the standard 4.7 GB.

The Holographic Versatile Disc Alliance (HVD Alliance) was formed in 2005. Its technologies combine single beam holographic storage and DVD technologies to provide cartridge capacities starting at 500 GB. The HVD Alliance has not yet brought any products to market.

Another new development is the protein-coated disc. In 2006, Professor V. Renugopalakrishnana and his colleagues at Harvard Medical School coated a disc with light-sensitive proteins called bacteriorhodopsin. These proteins enter into an intermediary state when exposed to light. This intermediary state acted as a sort of binary system. Renugopalakrishnana and his colleagues were able to alter the DNA of the protein and prolong this state for several years. This disc could hold up to 50 Terabytes (50,000 GB). It was unclear when (if ever) this disc would be available.

CURRENT TRENDS

The DVD media market was strong enough to support the plethora of formats that continued to co-exist at the end of the first decade of the 2000s. There was no clear leader in the field based on format and manufacturers supported either the *Plus* or the *Minus* technologies. Manufacturers have split again, this time in support for the next generation of DVDs.

Next-generation DVDs offer greater storage capacity than their predecessors. There are two competing formats. The first is the Blu-ray Disc developed by the Blu-ray Disc Association (BDA). The BDA includes electronics, personal computer, and media manufacturers such as Apple, Dell, Hitachi, Hewlett-Packard, JVC, LG, Mitsubishi, Panasonic, Pioneer, Philips, Samsung, Sharp, Sony, TDK, and Thomson. Most recordable media uses a red laser to read and write data. The Blu-ray, however, uses a blue laser. It is this distinction which gives Blu-ray its name. There are two types or recordable Blu-ray discs. The BD-R disc can be written to once, while BD-RE can be erased and reused multiple times. A single layer Blu-ray disc has a 25 GB capacity while a dual layer has 50 GB. The dual layer can store approximately 9 hours of high-definition film. A single layer disc has a read mechanism of approximately 36 megabits per second (Mbit/S). A double layered disc's read mechanism has a speed of approximately 72 Mbit/S.

The HD-DVD (High-Definition) format is Blu-Ray's competitor. In November 2003, the DVD Forum voted to support the format as the high-definition replacement to the original DVD. Like Blu-ray, the HD-DVD disc has its own prominent supporters, including Microsoft, Intel, NEC, Sanyo, and Toshiba. There are two forms of recordable HD discs. The HD DVD-R is the writeable disc version of HD DVD, and has a single-layer capacity of 15 GB. HD DVD-R has slower write speeds than the competing BD-R format and lower storage capacity. The HD-DVD disc has about the same read mechanism speed as a Blu-ray disc.

Most analysts feel that consumers are waiting to see which format will win the battle before making an expensive investment in media equipment. In June 2006, the HD-DVD format appeared to be the winner, taking 70 percent of DVD video sales while Blu-ray had the remaining 30 percent. By January 2007 the two companies saw their positions reversed. Blu-ray reported a 67 percent market share in that month while the HD-DVD format claimed the remaining 33 percent.

TARGET MARKETS & SEGMENTATION

Businesses and individual consumers are both target markets for DVD media. The media is a reliable method of storing and transferring data. Electronics and software companies have turned to this product as a format on which to release software and video games that require large storage capacities. The DVD can also store photographs, which stimulated the digital photography market. As a sign of how important data storage is to consumers, computers are routinely built with CD or DVD drives capable of writing to recordable and rewritable CDs or DVDs. DVDs can be purchased in a wide variety of retail outlets as well as through e-commerce Web sites.

RELATED ASSOCIATIONS & ORGANIZATIONS

Blu-Ray Disc Association, http://www.blu-ray.com

DVD Forum, http://www.dvdforum.org

DVD+RW Alliance, http://www.dvdrw.com

International Recording Media Association, http://www.recordingmedia.org

Optical Storage Technology Association, http://www.osta.org

Video Software Dealers Association, http://www.vsda.org

BIBLIOGRAPHY

Blass, Evan. "Protein Coated Discs Could Enable 50TB Capacities." *Engadget.* July 2006.

Chan, Joshua. "DVD Players Surpass VCRs in U.S. Households." *International Business Times.* 19 December 2006.

"Change a Constant in Blank Media." *MMR.* 8 January 2007, 89.

Lazich, Robert S. *Market Share Reporter 2007.* Thomson Gale, 2007, Volume 2, 626.

"Magnetic and Optical Recording Media Manufacturing: 2002." *2002 Economic Census.* U.S. Department of Commerce, Bureau of the Census. December 2004.

Macklin, Ben. "The Rise of the DVRs." *IMedia Connection.* 12 August 2005.

"Media Changes Accelerated by Optical Disks and Compact Memory Cards—Next 3 Years Forecasts for Recording Media Global Demand and Production." Japan Recording Media Association. 17 November 2006. Available from <http://www.jria.org/english.html>.

"The Role of Standards in Accelerating Markets." Consumer Electronics Association. 27 September 2005. Available from <http://www.ce.org>.

"Top 50 General Merchandise Brands." *Chain Drug Review.* 21 May 2007, 37.

"Verbatim Global Recordable CD and DVD Media Sales." *E Media Live.* 19 March 2006.

SEE ALSO *Movies on DVD, Personal Computers*

Ethanol

———■———

INDUSTRIAL CODES

NAICS: 32–5193 Ethyl Alcohol Manufacturing

SIC: 2869 Industrial Organic Chemicals, not elsewhere classified

NAICS-Based Product Codes: 32–51930111 and 32–519303

PRODUCT OVERVIEW

Ethanol is ethyl alcohol, also referred to as common alcohol, the intoxicating component of alcoholic beverages. Structurally it has two carbon atoms as its spine. The first carbon atom (C) is bonded to three atoms of hydrogen (H), the second carbon atom is bonded to two atoms of hydrogen plus a hydroxide (OH). A hydroxide is a single oxygen-hydrogen pair. The chemical description, reflecting these components, is rendered as C_2H_5OH or as CH_3CH_2OH. In chemistry the front part of the chemical, thus without the hydroxide, is known as an ethyl group. The terms used for this chemical reflect the presence of that group, hence names like ethyl alcohol, ethyl hydroxide, or ethanol. Ethanol is colorless, occurs as a liquid, is soluble in water, and is highly flammable, making it an excellent fuel. In the presence of excess oxygen its combustion will produce water and carbon dioxide, the gas that we breathe out. Ethanol's most commonly used cousin is methanol alcohol, wood alcohol, commonly known as rubbing alcohol. Methanol lacks the second carbon and its two hydrogen atoms. Methyl alcohol is toxic to humans if ingested but competes with ethanol as a gasoline modifier.

A Glance at History. Those who think that the use of ethanol as fuel is a recent innovation may be surprised to learn that the first engines made used ethanol as their fuel, and that ethanol was available for such experimentation because it had already been used as a fuel to power indoor lamps. Nicholas Otto, the German inventor of the 4-cycle internal combustion engine (1861) used ethanol for fuel—as did the father of the American auto industry, Henry Ford (in 1896), to fuel his first vehicle, the quadricycle). Ford designed the Model-T (1908) to run on ethanol, gasoline, or a mixture of the two.

Spurred by demand during World War I (1917–1918), ethanol production was at 50 million gallons per year as the war ended. In the 1920s Standard Oil began to add ethanol to gasoline to increase octane levels and to reduce engine knock. Ethanol use increased until the end of World War II (1945). It was used as an additive to gasoline, the product used in the Midwest and referred to as gasohol. War time demand caused its production to spike in various uses. Between 1945 and 1978, however, very low petroleum prices essentially brought fuel-ethanol production to a halt.

Interest began to revive for environmental reasons. Ethanol could be used to replace lead used in gasoline as an anti-knock agent. The energy crisis in 1979 put ethanol back on the front burner. Amoco Oil Company, soon followed by others (Ashland, Chevron, Beacon, and Texaco), began to sell gasoline-ethanol blends. In 1980 Congress passed the Energy Security Act in which incentives for ethanol production appeared. Other actions followed, including tariffs against relatively cheap Brazilian ethanol (made from cane) and subsidies for ethanol. In 1983 these subsidies had reached 50 cents per gallon. Interest in ethanol, since the 1980s, has grown or waned depending on the price of oil.

The international conflicts of the new century have caused oil prices to rise and uncertainties concerning future supplies have fueled interest in this substitute for oil. In the latter part of the first decade of the twenty-first century ethanol had once more become a prominent issue in planning the United State's energy future. This interest was unlikely to wane again soon as world supplies of petroleum were beginning to peak, with a gradual decline, thereafter, reasonably certain.

Sources and Production. Ethanol is produced by the fermentation of sugar contained in grains, sugar cane, or sugar beets—anything with adequate sugar content can be used to make alcohol. All carbohydrates in plants are sugars, starch being a very common form of it, used by biological systems to store glucose for future use. Grains are made up of three major components: the outer bran, the inner mass called endosperm, and the innermost germ which holds the genetic code. The endosperm or kernel is largely carbohydrate or starch, approximately 66 percent. It is structurally very similar to hydrocarbons, but with the hydroxide (OH) added.

In the United States the principal raw material for ethanol is corn. Based on data from the Economic Research Service of the U.S. Department of Agriculture (USDA), approximately 18 percent of all corn was used for ethanol in 2006. Other than the corn used in the production of ethanol, the largest use of corn in that year was for animal feed (50.8%), exports (19.1%), corn syrup (4.4%), and all other uses (7.3%). All other uses included starch, sweeteners, alcohol other than ethanol, cereals, corn sold for direct consumption, and seed. A portion of the corn converted to fuel alcohol, approximately 34 percent of inputs by weight, ends up as animal feed. This is the residue after the starch-rich endosperm has been separated and fermented. Viewed from a broad perspective, people mainly consume corn as meat, as dairy food produced by cattle fed the corn, as corn sugar and sweeteners, and as a fuel additive. Very little corn reaches us as corn. When it does, it generally comes to us as breakfast cereals, tortillas, and as corn chips in a bag.

Ethanol is made by wet or dry milling of corn. Dry milling is rapidly becoming the standard production method because it is simpler and cheaper than wet milling. The wet milling process is fundamentally the same as that used for making food grade corn products. It consists of cooking the corn in water laced with sulfuric acid; this causes the thin hull around each kernel to separate. The product is ground and screened to separate the three major grain components. The germ is used to extract corn oil, the bran to produce proteinaceous feeds, and the starchy kernel is fermented and made into alcohol. The residues are also sold as feed. The fermentation process begins by adding enzymes that convert starch to dextrose. Ammonia is added to feed yeast used in the actual fermentation process. The fermented liquid, similar to beer, is passed through a distillation tower to cause the alcohol to separate from the water. Ethanol boils at 173° Fahrenheit (F), water at 212° F, making this process straight-forward. The distilled ethanol still contains water and must be dried. Producers use molecular sieves to accomplish this last task. The sieves are towers filled with hard and very porous materials, of which the pores are uniform, tiny, and powerful enough to absorb molecules of small size. The sieves fill with the tiny molecules; the alcohol, which is more massive, passes through. Sieves are then restored by using heat to remove the water. The alcohol is denatured, meaning rendered unfit for human consumption, by adding a small amount of gasoline. Ethanol is then ready to ship. The wet residue left over by distillation is centrifuged to remove the water; the solids are sold as animal feeds. Wet milling produces four kinds of outputs: oil, protein feed, carbohydrate residues, and ethanol; thus requiring sales of these products to be competitive.

Dry milling emerged as a simplification of the wet milling process. The corn is simply ground into a meal. Mixed with water it produces a mash. The mash is brought to a high temperature to kill off bacteria and after cooling, is fermented with yeast. The rest of the process is the same as in wet milling. The residual solids, however, contain germ and bran as well as residual carbohydrates. The wet residue is handled somewhat differently. Solids are separated first and the remaining liquor is rich in sugar. It is concentrated further into a syrup known as condensed distillers soluble (CDS). CDS is then mixed with the solids producing dried distillers grains with solubles, known as DDGS or simply as DDG. DDG is sold as a substitute for soybean feed but has a lower protein content (27% versus soybean feed's 49%). New ethanol plants are predominantly dry mills. They are technically simpler, require less capital, and products only two things, ethanol and DDG.

Ethanol as a Fuel or Additive. Data on the energy content of ethanol are somewhat variable, ranging from a low of 76,000 British Thermal Units (Btu) per gallon to a high of 83,000 Btu. (A British Thermal Unit is equivalent to the energy required to heat one pound of water by 1 degree Fahrenheit.) Using information provided by the Energy Information Administration (EIA), part of the U.S. Department of Energy, ethanol has 76,330 Btu compared with the 116,090 Btu of regular unleaded gasoline. Ethanol's lower heating value, which translates into less actual work done by the fuel (to use the terminology of physics), means that to get the same mileage using ethanol that we get using gasoline, we have to burn 1.52 gallons of ethanol. This result comes about because ethanol's carbon content is 52.2 percent by weight (gasoline's carbon ranges from 85% to 88%). Ethanol and gasoline have similar hydrogen content (ethanol 13.7% and gasoline between 12% and 15%). A substantial part of ethanol (34.7% by weight) is oxygen. There is, consequently, less fuel to burn in the alcohol, but ethanol carries some of the oxygen needed for combustion in the fuel itself. For this reason it burns cleaner than gasoline.

Since 1995, under authority passed by Congress in that year, the Environmental Protection Agency (EPA) mandated the use of reformulated gasoline (RFG) in urban areas with poor air quality. RFG is required to contain at least 2 percent oxygen, supplied by an oxygenating agent. It represents approximately 31 percent of all gas sold. In the latter years of the first decade of the 2000s most of this oxygen was provided by methyl tertiary-butyl-ether (MTBE), a product based on methanol and manufactured by the petroleum industry. Its presence in gasoline reduces carbon monoxide in exhausts. In 1999 environmental sampling began to show that MTBE, which has carcinogenic characteristics at sufficient concentrations, was showing up at high levels in wells and ground water. Announcements of these findings by the EPA caused a stir of legislative activity in Congress, its aim to eliminate this type of oxygenation agent. Legislative activity spurred rapid expansion of ethanol capacity. Ethanol is the logical MTBE replacement and was already in use for about 15 percent of RFG shipped.

In 2001 when George H. W. Bush took office, the policy on MTBE changed as pressure to eliminate this oxygenating agent declined. State legislatures filled in where the federal government had stepped back from legislating the replacement of MTBE. Seventeen states participated in this movement, banning the MTBE using a number of different phase out timelines. In response the petroleum industry, at its own initiative, began a process of replacing MTBE voluntarily even in areas where it was not banned. The industrial deadline for eliminating MTBE was informally set for 2006. When that year came, very little MTBE remained in routine use.

Approximately 31 percent of gasoline is sold as RFG—in 2005 the total was nearly 43.5 billion gallons. To achieve a 2 percent oxygen content in such a quantity of gasoline, a straightforward way to get there is to produce gasoline with at least 7 to 10 percent ethanol content. MBTE replacement, therefore, represented a potential new ethanol market of approximately 3 to 4

billion gallons—at the high end a doubling of the 3.9 billion gallons already produced in 2005 and sold as blended gasoline. Demand for ethanol was high and growing in the late years of the first decade of the twenty-first century and imports were growing.

Ethanol cannot be shipped in pipelines economically because it has a great affinity for water and water is often present in pipelines. In taking up the water, the ethanol or gas-ethanol blend loses energy and incorporates particles of rust and dirt. For this reason producers ship the alcohol in rail cars, trucks, or barges and blend it with gasoline at distribution centers for truck-based delivery to filling stations. Summer heat also causes ethanol to volatilize more easily than gasoline. Special blending efforts, involving extra processing and additives, are necessary to prevent this in formulations where a fixed oxygen content is mandated by law at all times, not just on cool days. Another way to achieve this end is to use more than the minimum amount of ethanol to reach the 2 percent oxygen-by-weight level.

In addition to its most recent deployment as an oxygenation agent, ethanol is sold blended into gasoline and designated by the letter E followed by a number. E10 is a blend containing 10 percent; E15, 15 percent alcohol, and so on. E10 is widely used in the United States. The highest ethanol-content gasoline available in the country is E85. E100 is used in Brazil and contains approximately 4 percent water.

MARKET

The last published Economic Census, for 2002, reported shipments of ethyl alcohol was $2.79 billion. Of this total fuel ethanol was $2.17 billion (77.7%). Wet milling plants produced the bulk, $1.35 billion, dry milling plants the rest, $810 million. The EIA reported ethanol production to have been 2.13 billion gallons in 2002; thus shipments were worth $1.02 per gallon to the wholesaler. The year was not representative however because prices were at record lows. Anticipating rapid transition from MTBE to ethanol as a oxygenating agent, the industry had added too much capacity. The recession that began in early 2001, and reduced travel due to the 9/11 terrorist attack, influenced total gasoline consumption. Prices of ethanol crashed from levels in 2001 of $1.35 at the low and $1.80 per gallon at the high end.

By 2006 production had increased to 4.86 billion gallons. As reported by the Chicago Board of Trade, ethanol briefly reached historic highs in excess of $4 per gallon in June 2006, although prices dropped again later that year. For most of the year they were between $2.00 and $2.70 per gallon. If we assume an average price of $2.35 that year, ignoring the peak at $4 and the valley at $1.70 per gallon, the market in 2006 was approximately $11.4 billion. Between 2002 and 2006 production increased at the rate of 22.9 percent per year.

Figure 89 illustrates energy production from three types of renewable energy technologies: alcohol, solar, and wind power. The data are shown in quadrillions (1,000 trillions) of Btus, a measure used by government to track energy production from all sources using a common denominator. A single quad of energy is equivalent to 250 million barrels of gasoline and 380 million barrels of ethanol. These equivalencies reflect the differences in energy content delivered by the two kinds of fuel; therefore the chart is indexed by fractions of a quad of energy.

Although the slope of the curves shown does not make this obvious, wind power showed the most rapid growth from 1989 to 2004. It increased from 0.022 quads to 0.143 quads, a 6.5-fold increase and an annual growth rate of 13.3 percent. Ethanol was second. It grew from 0.071 quads to 0.296 quads, a 4-fold increase and a growth rate of 10 percent per year. Solar power was flat; growing at less than 1 percent (0.9%) annually. These new forms of energy represented inconsequential portions of total power in 2004. All three technologies together produced 0.5 percent of all power, thus 0.5 quads out of a total of 100.4 quadrillion Btus consumed in the United States. Of that total, petroleum-based energy represented 86.2 quadrillion Btus.

Issues and Controversies. The ethanol fuel industry is unusual and beset with controversy. Ethanol represents the only major initiative in the United States to turn a food crop into transportation fuel and has passionate opponents and promoters. Opponents include environmentalists, some energy economists, and free market supporters. Proponents are, generally speaking, agricultural interests and those seeking energy independence, both in the private and the public sector. Two issues divide these camps.

The first is ethanol's energy balance. Opponents wonder if ethanol actually produces a net surplus of energy, or if it consumes more energy than it yields. Competing studies produce contrary results. The second issue is subsidy. Ethanol is subsidized at fairly high levels by the federal and by selected state governments. Promoters favor subsidies, arguing that such subsidies help to speed up the development of an industry that has the potential of reducing the United States' energy-independence. Opponents question subsidies to an industry already up and running profitably in the nineteenth century. They argue further that if ethanol is viable as an alternate to petroleum based energy, that rising oil prices will naturally favor it without a subsidy—if it has a positive energy balance. If not, increasing fossil fuel costs will simply make ethanol

FIGURE 89

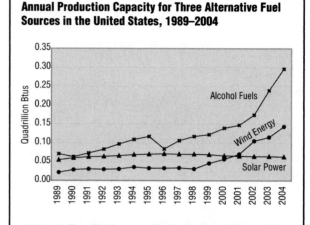

Annual Production Capacity for Three Alternative Fuel Sources in the United States, 1989–2004

SOURCE: Compiled by the staff with data from "Historical Renewable Energy Consumption by Energy Use Sector and Energy Source, 2000–2004," U.S. Department of Energy, Energy Information Administration.

uneconomical to produce. Subsidies and energy balance are thus closely linked issues.

Apart from these major issues, others are rooted in broader philosophical considerations. Environmentalists favor conservation and drastically reduced energy consumption to cope with diminishing oil supplies. Food crops should feed people, not move people about in cars. They fear that massive corn-based alcohol production will lead to irreversible soil erosion. They see solar and wind power as the technologies of choice. Free marketeers favor market solutions to the energy challenge. They see ethanol production as government intervention. Many promote using nuclear power to create hydrogen fuel by electrolytic splitting of water. Nuclear power is hampered by excessive government regulation. A closer look at the two major issues follows.

Energy Balance. Energy balance is based on the results of a measure of energy returned by (on) energy invested, abbreviated EROEI. This equation is also referred to as energy return on investment (EROI). The basic question becomes, is any particular fuel worthwhile if its production uses more energy than the fuel generated?

Even in the Stone Age, people had to expend some effort to get heat; they had to gather the firewood. No fuel used is free. The general rule is that substances with the highest energy density or Btu value, have the highest EROI. Modern agriculture is very energy-intensive: it requires fertilizers, pesticides, herbicides, machinery, and fuel to make it work; heaters to dry the grains; cars and trucks to ferry labor to and from the fields; trucks and trains to transport the crops; metal and cement to hold up the silos that store them—each process requires the expenditure of energy. To make ethanol, additional energy is required, both in the agricultural processes and in manufacturing the vessels and grinders used at the plants to convert corn into fuel.

EROI is usually rendered as a ratio to one, one indicating energy in, the other number the yield produced. Where the product is energy itself, a ratio of 1:1 means that the energy you made took as much as you consumed, not a worthwhile endeavor. Data published by Cutler J. Cleveland of Boston University in *Energy* shows that crude oil had an EROI of 100 in 1930, but this ratio had declined to 20 by 2000; depleted oil fields required more energy to get the crude out by that year. Similarly, coal had an EROI of 100 in 1950 but only 80 in 2000. Cleveland's calculations indicated current EROI for gasoline as ranging between 6 and 10, a way of saying that energy equivalent to a gallon had to be expended to get six gallons minimally and ten gallons maximally. His calculations produced a negative ratio for corn ethanol, thus less than 1, suggesting that it required more energy to make than it delivered. If energy is equivalent to money, a product with a negative EROI requires subsidy.

Two prominent academicians who have studied ethanol in depth are David Pimentel at Cornell and Tad Patzek at University of California, Berkley. Both have reached the same conclusions as Cutler Cleveland. Opposing views are held by Hosein Shapouri, a leading analyst of this issue at the U.S. Department of Agriculture, Michael S. Graboski of the Colorado School of Mines, John McClelland of the National Corn Growers Association, and Michael Wang of the U.S. Department of Energy. Many others also participate in what was an ongoing debate in the first decade of the 2000s.

Those concluding that ethanol manufacturing has a negative EROI have tended to include more elements in their analysis, thus energy consumption associated with making the machinery and the equipment used as well as fuels used in farming or production. In a 2003 study, for example, Pimentel concluded that it took 99,119 Btu to make one gallon of ethanol with 77,000 Btu, a deficit of 22,119. In a rebuttal, Graboski and McClelland concluded that it took 58,942 Btu to produce ethanol at 76,000 Btu, a surplus of 17,058. Pimentel included energy used in manufacturing machinery and equipment for the agricultural and production activities and assigned all energy consumed to the ethanol output. Graboski and McClelland excluded energy associated with machinery and equipment and gave a credit of 12,351 Btu to account for by-product shipments of the hypothetical plant studied. They also assumed a higher corn yield per acre based on a 9-state study and a higher output of ethanol per bushel of corn (2.68 versus Pimentel's 2.5 gallons).

In the early twenty-first century the debate was just beginning. Both studies cited above produce an EROI very close to 1—Pimentel's results yielding 0.8 and Graboski/McClelland's 1.3—both derived by dividing output by energy used. In comparison with the EROI of gasoline, which is between 6 and 10, or even wind power's EROI of 4 to 5, ethanol's result is nothing to write home about. EROI is not an absolute economic indicator but is an indicator of energy sufficiency. All systems close to an EROI of 1 indicate very high expense. An EROI below 1 means that no net energy can be produced for use outside the system. If ethanol turns out to be such a technology, it will always depend on some other form of energy to sustain it. It may, however, even then, be a viable way to produce liquid fuel for transportation.

Subsidies and Costs. Since 1979 federal subsidies for ethanol have been available in the form of partial exemptions for the excise tax and income tax credits. Excise taxes are imposed on the sale of certain goods, including fuels. The U.S. General Accounting Office (GAO) conducted a study of such incentives for the 1980–2000 period. The GAO reported the result in constant year 2000 dollars. Based on that study the industry benefited by receiving $11.7 billion in subsidies. The same study also noted that the petroleum industry received, in the 1968–2000 period, subsidies equivalent to $149.6 billion. Promoters of ethanol point to the more massive subsidies for petroleum as justification. Both are fuels, both get subsidies.

A closer examination of the GAO data is required to understand that the ethanol subsidies are substantially larger than those available for gasoline and related petroleum products because much less ethanol is produced than gasoline. GAO data for the 1989–2000 period are used here because consistent ethanol energy data are available from 1989 forward. From 1989 to 2000, total subsidies available for gas were 0.3 cents per gallon. For ethanol the actual disbursement was 54 cents per gallon, nearly 200 times higher. Subsidies were continued beyond 2000 as well and in 2004, Congress enacted legislation establishing the Volumetric Ethanol Excise Tax Credit (VEETC). It provides a 51 cent per gallon subsidy for ethanol through 2010. The VEETC is collected by the blenders of ethanol, thus the gasoline producers, not the producers of ethanol.

In addition to the federal subsidies, 19 states provided tax incentives for alcohol fuels. All told 41 states had some kind of incentive program related to ethanol in the form of inducements to convert to alternative fuel vehicles or to purchase them; mandating biofuel use in state-owned fleets; and incentives offered distributors and retailers.

Ethanol costs more to make than gasoline. Pimentel presented detailed data on cost of production indicating $1.48 per gallon. To produce sufficient ethanol to equal the Btu content of gasoline would thus cost $2.25 (1.48 x 1.52, 1.52 being the Btu difference between ethanol and gasoline). According to EIA data, the average wholesale price of gasoline from 1989 to 2005 was 82 cents per gallon, with a low of 53 cents in 1998 and a high of $1.68 in 2005; the differences reflect the changing price of crude. Figure 90 illustrates the differential between ethanol and gas prices over the 2003 to 2007 period, expressed in price per million Btus using Ethanol 85 as the ethanol grade (85% ethanol, 15% gasoline). The cost differentials in this period favored gasoline. It cost $3.40 less in 2003 at the low end and $4.90 less per million Btu in 2004. Calculating the effect of the subsidy on ethanol makes the cost difference less alarming.

E85 is 85 percent ethanol. Therefore 850,000 Btus of one million come from pure ethanol. This number, divided by 76,300 Btu per gallon (the energy content of ethanol discussed above), produces 11.14 gallons of ethanol. Multiplying the 11.14 gallons by 51 cents (the prevailing subsidy means that a million Btu of E85 received $5.68 in subsidy each year. The subsidy was more than sufficient to erase E85's higher price, it provided additional funds used to make the fuel more competitive. Without the subsidy ethanol would still be purchased as a replacement for MTBE in order to oxygenate gasoline in high-pollution urban areas. To achieve greater sales, however, alcohol may very well depend on the presence of the government subsidy.

Ethanol as an Alternative to Gasoline. Ignoring the marginal energy balance of ethanol, its subsidy, and its inherent reliance on petroleum fuels to produce its fertil-

izers and to power its agricultural machinery, the question arises: Can ethanol entirely replace gasoline as the primary transportation fuel in the United States? The answer, not likely.

Assuming a high yield of corn per acre (140 bushels), one acre will yield 375 gallons of ethanol, equivalent to 247 gallons of gasoline in energy content, thus 5.9 barrels per acre of gas-equivalent fuel. Gasoline consumption in 2005 stood at 3.34 billion barrels. If all of this fuel had been replaced by ethanol, it would have required 566 million acres planted with corn. According to USDA's Natural Resources and Conservation Service, all cropland in the United States was 368 million acres in 2002, down from 420 million in 1982 and 381 million in 1992. The answer to the question posed above is therefore *no*. Current acreage devoted to corn for all purposes was approximately 80 million acres in 2005. If all of this acreage had been used to produce ethanol, it would yield 470 million barrels of gasoline-equivalent fuel—14 percent of U.S. consumption, up from current levels of around 1.6 percent. Even that usage would have cut into food production. Using all corn for ethanol, however, does indicate an upper boundary. At best, ethanol can displace only 14 percent of gasoline usage if the nation gives up most food, but not all animal feed, uses of corn.

To illustrate boundaries, we may examine the goals set by President George W. Bush in his 2007 State of the Union address. The President called for cutting gasoline consumption by 20 percent by 2017 to be achieved in part by producing 35 billion gallons of alternative fuels. The president's object would translate into using 93 million acres of corn for ethanol, well in excess of all corn acreage planted in 2005. A production of 35 billion gallons would replace approximately 16 percent of gasoline usage in 2005 (much less in 2017 unless natural growth in consumption was artificially constrained). An additional 4 percent would have to come from more fuel-efficient vehicles, electric cars, or straight-forward curtailment of travel.

The ethanol industry is very complex. The industry could see growth stimulated by mandated use of alcohol for fuel oxygenation specifically driven by the need to replace MTBE in gasoline production because it pollutes water. Ethanol production is also supported by a generous subsidy without which it would probably be used strictly as an oxygenation agent. Rising crude oil prices causes gasoline to be more expensive. This will favor using ethanol even though ethanol costs will also rise because its production demands substantial amount of fossil fuel. Natural limits to growth, however, are set by shrinking crop acreage and an unfavorable energy balance. Ethanol has the characteristics of an industry strongly supported by national and state government policies. The industry

FIGURE 90

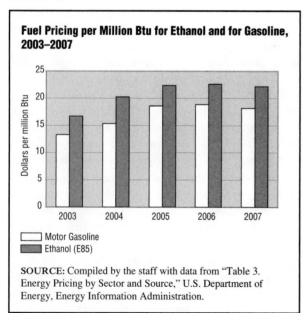

Fuel Pricing per Million Btu for Ethanol and for Gasoline, 2003–2007

SOURCE: Compiled by the staff with data from "Table 3. Energy Pricing by Sector and Source," U.S. Department of Energy, Energy Information Administration.

rests on environmental regulations, agricultural support, and national security considerations, not an innate market force, favorable economics, or popular demand.

KEY PRODUCERS/MANUFACTURERS

Based on data provided by the Renewable Fuels Association, 74 companies participated in ethanol manufacturing in 2004. They operated 87 plants. Of these plants 77 used corn as the input raw material. Nine plants produced ethanol from such products as cheese whey, wheat starch, brewery wastes, and sugar-starch combinations.

The major producers are Archer Daniels Midland (ADM), Cargill Corporation, Aventine Renewable Energy, VeraSun Energy, Pacific Ethanol, Inc., AE Staley Manufacturing, U.S. BioEnergy Corporation, and Hawkeye Holdings LLC.

In the middle of the first decade of the twenty-first century then producing sector was in substantial flux. Capacity was being added rapidly in response to the change-over from MTBE to ethanol oxidants in gasoline and the major publicity provided by the Bush Administration's plans to transform the nation's use of fuels. In efforts to obtain capital for expansion, a number of companies had gone public in 2006, including VeraSun, Aventine, and Hawkeye. Some of these, and others, were adding to their sales by developing relationships with smaller producers to act as their exclusive distributors, thus boosting their own volume but limiting their risks. Aventine for example, had production capacity of 696 million gallons in 2006 but also sold an additional 493 million gallons as a distributor on behalf of other producers. Cargill had capacity of 230 million gallons but expected to be able to distribute 750 million gallons all told. Archer Daniels Midland (ADM), which at one time had enjoyed a 60 percent share of the market continued to be a dominant factor, but with its market share reduced to 24–25 percent. ADM was also boosting capacity. ADM and Cargill are major, diversified corporations with low exposure, to the uncertainties associated with ethanol. AE Staley is also diversified and is part of Tate & Lyle, PLC, a global company based in the United Kingdom. Ethanol plays a major role in the operations of the other leaders.

Archer Daniels Midland. Based in Decatur, Illinois, ADM had sales in 2006 of $36.6 billion divided among oil seed operations (32.5%), corn (13.3%), agricultural services (42.2%), and all other activities including food and feed ingredients (12%). ADM's ethanol operations were part of its corn activities producing 7.4 percent of total ADM sales.

Cargill Corporation. This company is privately held and, for that reason, is not as well known as ADM. Cargill,

however, is one of the largest companies in the United States and is the largest privately held firm, with sales of $75.2 billion in 2006. The company is a global trader in agricultural and food products, provides pharmaceutical, offers financing products, and produces industrial goods from agricultural commodities. Of these ethanol is one product, estimated to represent a little over 2 percent of Cargill's business, a sufficient level of participation to make Cargill one of the top three participants in the business based on its distribution, not on its production activities. Cargill is based in Minneapolis, Minnesota.

Aventine Renewable Energy. Aventine ranks with Cargill in second or third place. The company has its headquarters in Perkin, Illinois, and had sales in 2006 of $1.59 billion, largely from ethanol manufacturing and its by-products. of ethanol manufacture. Aventine operates its own plant and is partial owner of Nebraska Energy LLC, jointly owned with the Nebraska Energy Cooperative. The company also sells the products of ten other producers under partnership agreements. The partners are located in Illinois, Kansas, Minnesota, Iowa, South Dakota, and Wisconsin—the very heartland of ethanol production. The company went public in 2006.

VeraSun Energy. Located in Brookings, South Dakota, VeraSun Energy was a $558 million company in 2006. It operated three plants producing 226 million gallons of ethanol. VeraSun was also building three others, two in Iowa and one in South Dakota. Pacific Ethanol, Inc., located in Sacramento, California, is the leading producer of ethanol in the western United States. The company operated four plants with a capacity of 101 million gallons. Pacific had sales in 2005 of $226 million.

AE Staley Manufacturing. Located in Decatur, Illinois, AE Staley is part of Tate & Lyle, PLC. The company is best known for its participation in corn sweetener and other food ingredients. The company's total sales are estimated at approximately $1 billion—of this total ethanol is merely a fraction. In 2004 the company had installed capacity to produce 65 million gallons at a plant in Loudon, Tennessee.

US Bio Energy Corporation. This company is headquartered in St. Paul, Minnesota. US Bio Energy Corporation was operating three plants in 2007 with total capacity of 250 million gallons. It was engaged in building five other facilities expected to lift its capacity to 650 million gallons. The company reported sales of $125 million for 2006.

Hawkeye Renewables. This company, operating as Hawkeye Holdings LLC, went public in 2006. The company reported capacity of 215 million gallons and sales of $89 million for that year.

MATERIALS & SUPPLY CHAIN LOGISTICS

Corn prices have a strong bearing on the cost of ethanol, not on its price. Gasoline prices are set by crude oil prices and by demand. Ethanol prices are pegged to gasoline's and to the level of subsidy that fuel-alcohol enjoys. Ethanol producers have little leverage in the market, and some producers make this point bluntly in their communications with stockholders when discussing risk. If corn prices shoot up, producers cannot pass on this increased cost to the oil companies. Gas prices, however, were moving upward as a consequence of international uncertainties arising from unrest in the Middle East, and of shrinking reserves.

Corn prices are set by demand and by factors that still influence agriculture: soil quality, the seasons, and the weather, not least the winter's length and precipitation. The later the planting, the lower the yield; but if the farmer plants too early, frost can hurt the crop. Other factors are the cost of fuels and fertilizers, both influenced by oil prices. In the early 2000s corn prices were rising in response to the prospect of selling substantially more corn for ethanol. Farmers were also planting more corn, thus increasing supplies.

Most ethanol production in the United States is done in the Midwest, near the source of its agricultural raw material. Four states have more than 10 plants. They are Iowa (16), Minnesota (14), and South Dakota and Nebraska (11 each). Illinois, Kansas, and Wisconsin, with 7, 6, and 5 plants respectively, make the second tier. All told these states have 83 percent of production plants, and also the dominant share of corn production. The region has three major population centers—Minneapolis/St. Paul in Minnesota, Kansas City, Kansas on either side of the Missouri/Kansas border, and Chicago in Illinois. To service most of the population across the country, however, especially urban areas where RFG is mandated, the product must move substantial distances. Ethanol cannot be transported through pipeline so truck, rail, and barge transportation is used to get it to blending or distribution locations. Such movement typically involves the intermediation of distributors.

DISTRIBUTION CHANNEL

Ethanol is delivered under direct contractual arrangements between large producers and the principal buyers of ethanol—the major gasoline producers. Producers also act as distributors for others, aggregating production from multiple plants. Independent distributors are also active. All parties involved seek as much predictability as possible

aiming to obtain long term contracts that guarantee sales in the future at adequate prices matching their projected costs without leaving too much on the table, especially in a market where gas prices are generally rising. Ethanol buyers have the balance of power. There are many ethanol suppliers. Alcohol is but a fraction of total fuels reaching the consumer at the pump. Mandated product use, such as the use of oxygenation agents, depend on legislative action which may be rapidly reversed. The same holds for subsidies. These factors influence the details and timing factors integrated into distribution contracts.

In the context of distribution, and in relation to contracts and pricing particularly, the substantial risks of ethanol become visible. The cautious investor is provided the straight talk in producers' filings with the Securities and Exchange Commission, such as 10-K reports. In the middle of the first decade of the 2000s, and by contrast, exaggeration and hyperbole frequently accompany discussions of ethanol as a solution to the U.S. dependence on imported fossil fuel.

KEY USERS

The ultimate users of ethanol are people at the pump, but the public at large is not actively participating in buying ethanol. The demand is rooted in institutional mandates rather than in consumer demand. A small segment of the public, in fact, views ethanol as an inferior fuel, a view that ethanol producers feel obliged to correct. Other key users are the petroleum companies. To satisfy public mandates for oxygenated gasoline, they buy ethanol because a more technically and economically superior product, MTBE, is in process of being banned for potential health risks. Gasoline refiners buy ethanol because, with the excise tax credit they can obtain a product at lower cost than gasoline.

ADJACENT MARKETS

When ethanol is viewed as a renewable source of energy, adjacent markets are other renewable technologies, such as biomass, hydroelectric, geothermal, wind, and solar power. Biomass represents 47 percent of renewable energy, the bulk of which is provided by burning wastes—primarily industrial and agricultural wastes combusted to produce heat. Ethanol is classed with biomass but is just 4.8 percent of all renewable energy. Hydroelectric represents 45 percent of total renewable energy and more than 80 percent of the renewable energy used for electrical generation. Wind energy, 2.3 percent of renewable energy, is the most rapidly growing category. Wind energy is also generously subsidized but has a much more favorable energy balance.

If ethanol is viewed as corn, its adjacent markets in the 2000s were also its principal markets: animal feeds, sugar and sweeteners, exports, and human food products.

RESEARCH & DEVELOPMENT

R&D related to this market actually encompasses activities far beyond ethanol. Research extends into the use of ethanol, methods of blending ethanol, control of ethanol's volatility in hot weather, and its potential replacement in oxygenation applications by another hydrocarbon are all under study in the context of gasoline production. Within the ethanol industry itself, process improvement is an important research and development goal. That effort, in turn, is supported by agricultural research to improve corn cultivation at lower rates of fertilizer, herbicide, and pesticide use. R&D efforts also extend into territories far removed from fuels use such as genetic modification of corn varieties. At the point of ethanol utilization, development work and engineering adaptations by the automobile industry round out the rather complex R&D picture.

An important area of research is the replacement of corn itself with agricultural and wood waste as the principal source of sugar. Switch grass, common on the North American prairies, is an often-touted candidate crop as a source for ethanol to be used to keep our transportation systems running as petroleum based fuel become scarce. Switch grass is, however, somewhat limited because it requires more acreage than corn to produce equivalent quantities of ethanol and therefore requires capital investments nearly 3.5 times greater for the same output, based on the present state of the technology. Similar problems limit other raw materials.

CURRENT TRENDS

The major trend in the first decade of the twenty-first century was expansion. Ethanol had been singled out by President Bush as an important candidate to reduce domestic use of gasoline. The switch from methyl-tertiary-butyl-ether (MTBE) to ethanol was well underway. Consolidation among producers had not yet begun but was in prospect once it became clear that this new emphasis on ethanol was here to stay. In an industry very much subject to disruption, changes in the price of crude oil and therefore gasoline prices, a strong undercurrent of anxiety also registered in the reports filed by producers with the SEC. Quite conceivably a settling of international disputes, resolution of the conflict in Iraq, and tensions with Iran could produce at least a temporary fall of fuel prices. Such a price drop, even if relatively temporary, could expose ethanol producers to difficulties—too much capacity and falling prices.

TARGET MARKETS & SEGMENTATION

Ethanol has two established markets and one still in process of development. One of the established markets is as a replacement for MTBE. The other established market is as a supplement fuel used as an additive to gasoline. The MTBE and blending markets are of roughly the same size, once all MTBE has been removed. The industry's long term objective, and the market that is still under development, is increasing use of ethanol as pure ethanol for transportation fuel—something already a reality in Brazil. This last objective is naturally limited by available crop land and by ethanol's marginal energy balance. That balance, however, may improve over time.

RELATED ASSOCIATIONS & ORGANIZATIONS

American Coalition for Ethanol, http://www.ethanol.org

American Corn Growers Association, http://www.acga.org/renewable_energy/default.htm

Corn Refiners Association, http://www.corn.org/web/ethanol.htm

National Corn Growers Association, http://www.ncga.com

Renewable Fuels Association, http://www.ethanolrfa.org

BIBLIOGRAPHY

"CBOT® Ethanol Futures Contract." Chicago Board of Trade. March 2007.

Cleveland, Cutler J. "Net Energy from the Extraction of Oil and Gas in the United States." *Energy* 2005, 30.

"Ethanol Timeline." U.S. Department of Energy, Energy Information Administration. Available from <http://www.eia.doe.gov/kids/history/timelines/ethanol.html>.

"Full Text of 2007 State of the Union Speech." MSNBC. 23 January 2007. Available from <http://www.msnbc.msn.com/id/16672456/>.

Graboski, Michael S. and John McClelland. "A Rebuttal to 'Ethanol Fuels: Energy, Economics and Environmental Impacts.'" National Corn Growers Association. Available from <http://www.ncga.com/ethanol/main/index.asp>.

Griscom Little, Amanda. "Mikey Likes It." *Grist.* 9 December 2004. Available from <http://www.grist.org/news/muck/2004/12/09/little-johanns/>.

"How Ethanol is Made." Renewable Fuels Association. Available from <http://www.ethanolrfa.org/resource/made/>.

Patzek, Tad W. "Thermodynamics of the Corn-Ethanol Biofuel Cycle." *Critical Reviews in Plant Sciences.* CRC Journals, Taylor & Francis. 2004, 519-567.

Pimentel, David. "Ethanol Fuel: Energy Balance, Economics, and Environmental Impacts are Negative." *Natural Resources Research.* June 2003. Available from <http://www.ethanol-gec.org/netenergy/neypimentel.pdf>.

"Prices." *Ethanol & Biodiesel News.* 16 April 2007.

Shapouri, Hosein. "The 2001 Net Energy Balance of Corn-Ethanol." U.S. Department of Agriculture. Available from <http://www.ethanol-gec.org/netenergy/NEYShapouri.htm>.

"Synergy in Energy: Ethanol Industry Outlook 2004." Renewable Fuels Association. February 2004.

"Tax Incentives for Petroleum and Ethanol Fluids." U.S. General Accounting Office (GAO). GAO/RCED-00-301R, 25 September 2000.

Whims, John. "Pipeline Considerations for Ethanol." Kansas City University, Department of Agricultural Economics. August 2002.

SEE ALSO *Gasoline*

Fertilizers

———■———

INDUSTRIAL CODES

NAICS: 32–5311 Nitrogenous Fertilizer Manufacturing, 32–5312 Phosphatic Fertilizer Manufacturing, 32–5314 Fertilizer (Mixing Only) Manufacturing

SIC: 2873 Nitrogenous Fertilizers, 2874 Phosphatic Fertilizers, 2875 Fertilizers, Mixing Only

NAICS-Based Product Codes: 32–53111, 32–53114, 32–53117, 32–51321, 32–51324, 32–51327, 32–531401, and 32–531402

PRODUCT OVERVIEW

The major fertilizers used are based on nitrogen, phosphorus, and potassium. Potassium is often referred to as *kalium* in Latin and is abbreviated *K*. Producers of fertilizers routinely put the *NPK* content of fertilizers on the packaging. Gardeners and farmers know just which combinations of these three vital ingredients they need at which time of the season.

Nitrogen. Nitrogen is 78 percent of the earth's atmosphere. The element, however, occurs in paired atoms tightly bonded to each other and thus are not reactive chemically. Very small amounts of nitrogen come down in rain, produced by lightning. Living things are built of proteins of amino acids. Every amino acid has an amino group in which nitrogen holds hydrogen atoms and an acid group made up of carbon bonded to oxygen atoms. Without nitrogen to help build it, no life can exist. Nitrogen-capture from the atmosphere, consequently, is a basic cycle in nature. It is efficiently accomplished by bacteria living in symbiotic relationship with legumes and other families of plants. In biology nitrogen is most commonly present as ammonia (NH_3) and as ammonium (NH_4) compounds. Bacteria producing these groups are rewarded by their hosts with payments of sugar. These are rhizobial bacteria, named after the word for *root* in Greek; they are most concentrated in root systems. When plants die some of the nitrogen remains present in the decaying vegetation and some is converted back into atmospheric nitrogen by other, so-called denitrifying, bacteria. These activities together constitute the nitrogen cycle.

Humanity has traditionally employed three methods of supplying nitrogen to the soil. One has been to plow back plant wastes that still hold ammonia. Another has been to use animal wastes as manures. Wastes are rich in nitrogen (ammonium hydroxide, a liquid, as decaying protein, and as urea, the last made in the liver and expelled in urine). Human wastes were also used to fertilize certain crops. The third approach used has been crop rotation. Planting crops like beans, peas, peanuts, lentils, and soybeans—or clover or alfalfa—renews soil fertility because these legumes (formally the *Fabaceae*) are most strongly associated with rhizobial bacteria that naturally fix nitrogen. Traditional approaches continue in use in parts of the world and even in industrialized countries by those engaged in organic gardening and agriculture. The dominant modern approach is to use manufactured nitrogen products, which require substantial amounts of energy to make.

The industrial method of nitrogen fixation was invented in the first decade of the twentieth century (1908–1910) and is known as the Haber-Bosch process. It is easy, conceptually, to grasp what happens. Methane is one atom of carbon surrounded by four atoms of hydrogen. Ammonia is one atom of nitrogen surrounded by three atoms of hydrogen. If we pluck the carbon out of methane and replace it with a nitrogen, the job is done. We will have one hydrogen left over or, if we react the product further, we can keep methane's fourth hydrogen as well to make ammonium compounds such as the common solid fertilizer ammonium nitrate. In actual fact the Haber process and its variants require multiple sequences in which methane is processed into a synthesis gas containing the right proportion of nitrogen and hydrogen. The actual synthesis of ammonia is the last of several stages. Each stage is facilitated by catalysts and heat. Temperatures ranging between 572 and 1022 degrees Fahrenheit are needed—as well as relatively high pressures which also require energy. Furthermore, and most relevantly, the raw material of nitrogen fertilizer is itself humanity's cleanest high-energy fuel, natural gas. The carbon taken out of methane is released as carbon dioxide.

Ammonia is a gas that liquefies under pressure or when cooled. It can be used in that form if injected below the soil grade to minimize its loss to the atmosphere. It dissolves in water readily to form ammonium hydroxide and can be sprayed on crops. It also comes in solid forms of nitrate fertilizers or is compounded with phosphate.

Phosphorus. Like nitrogen, phosphorus is a crucial element of proteins. A phosphate group is present in every component of deoxyribonucleic acid (DNA) and in ribonucleic acid (RNA), the genetic code and its messenger/transfer agent. Neither life nor its hereditary transmission is possible without the P in NPK. Phosphorus is a widely occurring element. It is very reactive and is not encountered in elemental form. It occurs as phosphate rock, made up of phosphate and calcium. The word phosphate is the collective designation of a family of compounds derived from phosphoric acid (H_3PO_4). Phosphorus-rich deposits have formed, and continue to form, at the bottom of oceans, the phosphorus thought to originate from once-living creatures. The largest deposits of phosphate rock in the United States (in Florida and North Carolina) were at one time at the bottom of oceans, revealed by marine fossils embedded in the rock. Other phosphorus-rich rock deposits, however, have no such tell-tale signs, with the consequence that geologists still continue to study the origins of phosphorus in nature.

Producers mine and crush phosphate rock to reduce it to fine size. Screening removes waste to the extent possible. The aim of processing is to concentrate the active element as phosphorus pentoxide (P_2O_5), achieved by processing the beneficiated rock with 93 percent sulfuric acid. Phosphate content is increased; gypsum (calcium

sulfate) is a byproduct. Phosphorus comes to crops in the form of superphosphate, a dilute solution containing 20 percent phosphate, or as triple-superphosphate with 40 to 50 percent phosphate. It can be enriched by compounding the phosphorus with ammonia into ammonium phosphate; this fertilizer may be applied as a solid. All manner of combinations of phosphorus with other nutrients are common.

Potassium. This element is fundamentally involved in cellular regulation, cell structure modification, and in the operation of the nervous system. The concentration of potassium ions inside living cells continuously changes to maintain electrical equilibrium or to concentrate energy potential. Leaves of plants can orient themselves toward the sun and our muscles can tense or relax only because potassium mediates such actions. The element is present in all living entities.

Potassium is very active chemically and occurs in combination with chlorine, oxygen, sulfur, and magnesium. The name is derived from the English *potash*, a word literally derived from ash found at the bottom of a pot in which wood had been burned. This ash, dissolved in water, becomes rich in potassium carbonate (K_2CO_3). The word is somewhat loosely applied to various compounds of potassium, thus to potassium chloride, potassium oxide, potassium hydroxide, and potassium sulfate.

Two major ores of potash are sylvinite (potassium chloride) and langbeinite (potassium magnesium sulfate). The ores have formed by the gradual evaporation of ocean waters in which they had been dissolved long ago and are closely linked to ordinary salt deposits. Most domestic potassium is obtained from three mines in New Mexico (77% of U.S. production), the rest from three operations in Utah and one in Michigan.

Potash is either mined as rock and then processed to concentrate the potassium or is obtained as a brine by pumping water into inaccessible underground deposits. Whether as rock or as brine, the raw input has a high content of clay, insoluble minerals, and ordinary salt. Wet-processing by flotation and screening separate tailings (clay, salt, and fines) from potassium-rich fractions. The latter are dried and further processed into desirable grain sizes for distribution. Large areas of land are associated with potash processing whether for waste disposal or for solar drying of brines, which is also practiced.

Major product categories are muriate of potash (potassium chloride), a dry crystal, potassium sulfate, also dry, applied directly or sold for blending into mixed fertilizers, potassium nitrate used in blends, potassium hydroxide used in liquid fertilizers, and sulfate of potash magnesia, also called Sul-Po-Mag or K-Mag. This last product, a dry substance, is made from langbeinite ores that naturally occur with magnesium.

Minor Minerals. Also important for supporting plant life are minerals usually referred to as secondary fertilizers in the industry. They are more readily available in soil but are incorporated in the major nutrients as additional helpers. These are calcium, magnesium, and sulfur. Calcium is part of triple superphosphate, magnesium in K-Mag, and sulfur is introduced in ammonium sulfate and in potassium fertilizers.

Natural and Manufactured Fertilizers. In looking at the history of modern fertilizer utilization, it is important to keep in mind that fertilizers in the modern sense almost always refer to manufactured products based either on natural gas or mined minerals that require substantial inputs of energy. Within the nitrogenous fertilizer category, for instance, only 1.5 percent of output in 2002 was fertilizer of organic origin, thus activated sewage sludge, processed tankage, and other organic wastes. The modern

industry emerged early in the twentieth century. Fertilization by manuring, crop rotation, or letting the soil rest were extensions of natural processes. Intensive fertilization lifted crop yields dramatically. Nitrogen fixation ushered in the Green Revolution in underdeveloped parts of the world. Rapid population growth was one consequence. More food has been available to sustain more people. Life spans also lengthened. The negative consequences of such interventions, although known for a long time, began emerging into full view in the last decades of the century past. These include soil erosion, water pollution, and questions regarding the sustainability of such practices in the long run.

MARKET

The Market in Dollars. In 2005 the three major industries under which the Census Bureau classifies fertilizer production had shipments of $13.75 billion. Of this total nitrogenous fertilizers represented 35.4 percent, phosphatic fertilizers 37.8 percent, and the fertilizer mixing industry 26.9 percent, with shipments, respectively, of $4.86,

FIGURE 91

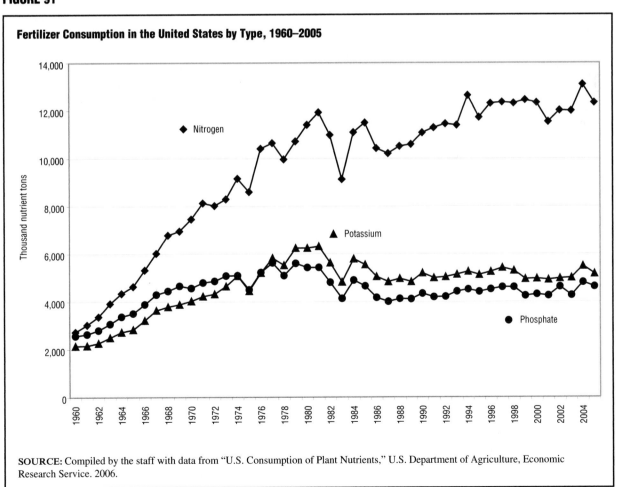

Fertilizer Consumption in the United States by Type, 1960–2005

SOURCE: Compiled by the staff with data from "U.S. Consumption of Plant Nutrients," U.S. Department of Agriculture, Economic Research Service. 2006.

$5.20, and $3.69 billion. In the Census Bureau classification potassium as a nutrient does not have a unique industrial categorization; potash derivates are present as inputs in the other three industries. Potassium is incorporated into nitrogenous and phosphatic products and is also used in the production of mixed fertilizers. Pricing and nutrient tonnage data provided by the Economic Research Service of the U.S. Department of Agriculture (USDA) indicate that potassium-based products represented approximately $1.27 billion of the $13.75 billion total market in 2005. Organic fertilizers represented an estimated volume of $73 million in 2005, thus barely visible.

In 1997 the market for fertilizers was $12.76 billion, indicating annual compounded growth of 0.9 percent from 1997 to 2005. The component industries, however, had quite divergent rates of growth. Nitrogenous fertilizers increased from a 1997 base of $3.96 to $4.86 billion, showing the most rapid growth, 2.6 percent per year. Phosphatic fertilizers declined from a 1997 level of $5.47 to $5.20 billion in 2005, eroding at the rate of 0.7 percent per year. The fertilizer mixing industry represented a middle ground, growing at 1.3 percent yearly from $3.32 to $3.69 billion.

The growth in the nitrogenous fertilizer sector during this period was due almost entirely to increasing prices rather than growing consumption. All nitrogen-based fertilizers had annual price increases above 3 percent, with the widely used anhydrous ammonia topping the list with a 4 percent annual price increase in that industry. Phosphate shipments declined despite much lower price increases of 1.9 percent per year for superphosphate and 1.4 percent for diammonium phosphate. This category, however, had a small quantitative increase in product shipments measured in weight. Potassium chloride had the highest level of price increases, 6.1 percent per year. It also declined in tonnage shipped between 1997 and 2005.

The Market in Tonnage. Figure 91 shows the dramatic increase in nitrogen-based fertilizer consumption going back to 1960. In that year 2.7 million tons of nitrogen nutrient were consumed by U.S. agriculture. By 2005 tonnage consumed had increased to 12.3 million tons, a 4.5-fold increase. In that same period, phosphate tonnage consumed increased from 2.57 to 4.64 million tons (a more modest 1.8-fold increase). Potassium tonnage grew from 2.15 to 5.17 million tons, a 2.4-fold increase.

Nitrogenous fertilizers have the most direct relationship to yield per acre, indirectly illustrated by the dramatic increase in nitrogen usage in agriculture as shown in the graphic. The largest single crop produced in the United States is corn (maize). Corn is typically grown in monoculture, the same acreage planted with corn year after year,

a type of farming that tends to exhaust the soil so that fertilizer use is vital to maintain productivity. In recent decades, however, it has become clear that adding fractionally more fertilizer does not produce an equivalent, linear, increase in yields. Plants are unable to use all of the fertilizer and it runs off into rivers and pollutes groundwater sources. In the more recent period, for which Census Bureau data in dollars are available, nitrogen consumption has been largely flat. This is also visible in Figure 91. Consumption of nitrogen fertilizers actually decreased fractionally between 1997 and 2005, from 12.35 to 12.34 million tons between those years—although 2004 was a record year of consumption with 13.1 million tons reaching croplands across the country. Phosphates had a small increase in tonnage during this period (25,800 tons), potassium a significant tonnage loss (251,600 tons).

Another view of the market is provided by looking at apparent tonnage of fertilizers applied per acre of cropland. According to the National Resources Conservation Service (NRCS), an element of the USDA, cropland in the United States is decreasing. The decrease is largely attributable to use of land for development to accommodate a growing population and the services it needs. In 1982 NRCS classified 420 million acres as cropland, in 1992 381 million, and in 2002 368 million. Using these data as benchmarks, we can calculate the average weight of fertilizer applied per acre on average in these years. The results are shown in Figure 92.

It is worth noting that in these years nitrogen allocated to the acres of cropland increased from 52 to 60 to 65 pounds per acre. Thus, even in the period after 2002,

FIGURE 92

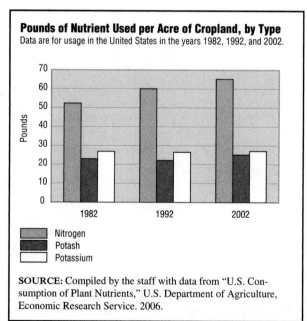

Pounds of Nutrient Used per Acre of Cropland, by Type
Data are for usage in the United States in the years 1982, 1992, and 2002.

Nitrogen
Potash
Potassium

SOURCE: Compiled by the staff with data from "U.S. Consumption of Plant Nutrients," U.S. Department of Agriculture, Economic Research Service. 2006.

when nitrogen fertilizer tonnage remained essentially flat, more nitrogen was still applied per acre if cropland continued to decrease at the rates indicated by the NRCS for earlier years. Application rates for phosphates (rendered as potash in the graphic) went up somewhat. In the potassium category, application rates were essentially unchanged. The application rates shown here are, to be sure, gross averages, derived by dividing total tonnage of nutrients by total acres of cropland. Actual rates of application are significantly higher and are noted in the Key Users section.

Factors Influencing the Market. Modern agriculture, much as everything that characterizes industrial civilization, is dependent on ample supplies of fossil energy for cultivation, for fertilizing the soil, and for killing off competing herbs or insects.

The extent of energy use may be illustrated by an analysis produced by David Pimentel of Cornell University published in *Natural Resources Research* in June 2003. Pimentel's analysis focused on corn production, the nation's largest grain crop, in the context of ethanol production. The growing of corn, Pimentel found, requires energy equivalent to 13.7 million British thermal units (Btus) per acre. The corn produced for this expenditure of energy (around 7,660 pounds per acre) has a caloric content equivalent to 50 million Btus. Thus, at the point of harvesting, the energy balance is favorable, a ratio of 1 unit yielding 3.65 units of energy. To be sure, the energy efficiency diminishes thereafter because the crop requires substantial additional processing and transportation before it appears as food on the table. One frequently encounters estimates stating that a single calorie delivered by the modern food industry requires ten calories of energy as input. In the field itself, however, the balance is positive.

In Pimentel's analysis, of energy consumed in growing corn the largest proportions are accounted for by fertilizers, 37.1 percent of the total. Other major categories are herbicides and pesticides (2.7%) and fuels used to generate electricity used, to move farm machinery, and to transport fuels, seeds, and chemicals to the parcel of land (20.7%). Energy consumption directly related to modern agriculture accounts for minimally 60 percent of all energy used, not counting energy required to manufacture the farm machinery itself or that used in irrigation. The single largest energy-consumptive component in agriculture is nitrogen fertilizer, accounting for 32 percent of all energy used.

Based on this analysis, which is representative of studies carried out by others, the fertilizer industry is tied to future trends in energy generally and to the availability of natural gas particularly. The tie-in—fuels and agriculture—has become even more obvious and complex since the Bush Administration's announcement in the presi-

dent's 2007 State of the Union address, which set a goal of replacing 20 percent of the nation's fossil fuel use by 2017 by substituting ethanol for gasoline. Corn-production, highly dependent on fossil fuels, is thus viewed at the highest levels as a viable replacement for fossil fuels. Not everyone agrees. Richard Manning, a critic of modern agriculture, wrote an article for *Harper's* in 2004 titled "The Oil We Eat." Manning's chief point was that in modern agriculture humanity transforms oil into food. The new national policy thus suggests another kind of title for an as yet unwritten article, "The Food We Pump." Time and experience will resolve the evident contradictions inherent in this national initiative. In the short term, however, substantially higher rates of corn production will certainly require expanded uses of nitrogen and other fertilizers— either on abandoned acreage reclaimed for agriculture or taken from other crops to grow more corn.

In the absence of extraordinary departures, such as a major shift to ethanol as fuel represents, the fertilizer industry appears either to have reached or to be in the process of reaching natural limits to its growth. With the exception of phosphate fertilizers, which have shown a slight tonnage advance in the 1997 to 2005 period, the other major nutrients have exhibited flat growth as measured in tonnage of nutrients consumed. In large part such results are due to low marginal additions to yield by marginal additions of fertilizer, in part due to water pollution concerns, not least ground-water contamination with nitrogen, in part due to shrinking acreage.

Total acreage in croplands had been declining. At the same time, organic farm acreage, according to data provided by the USDA's Economic Research Service, has exhibited explosive rates of growth. Certified organic farm acreage increased from 638,500 acres in 1995 to 1.72 million acres in 2005, growing at an annual rate of 10.4 percent, representing a three-fold increase in this 10-year period. Total organic acreage, to be sure, remains less than 1 percent of total cropland, but the growth rate at least indirectly indicates response to consumer demand which, itself, is motivated by health concerns. Of the total organic acreage, 33 percent is devoted to gains, with wheat and corn representing the largest components in that order. Beans of all kinds represented 9 percent, and fruits and vegetables 6 percent each of total organic acreage in 2005.

KEY PRODUCERS/MANUFACTURERS

As reported by the Census Bureau, some 116 companies participated in nitrogenous fertilizer and 31 in phosphatic fertilizer manufacturing in 2002. In essence these companies, aside from a handful of potash mining companies reported under mining statistics, were the key producers in the industry. Those engaged in potash mining were also active in phosphate and nitrogen production. A much larger number of companies, 373, were involved in fertilizer mixing operations, taking their inputs from primary producers. Unlike other mature industries, especially in the chemicals sector, fertilizers are not dominated by three or four major companies with the rest of the participants a distant second. Instead, production capacity is widely distributed, especially in nitrogen. Geographically limited potash and phosphate mining have also limited participants. But these companies compete on a global scale with mining companies overseas exploiting indigenous deposits.

In 2001 most of the domestic nitrogen capacity was owned by 19 companies of which the leading producer, the Kansas City-based cooperative, Farmland Industries, had a 16 percent capacity share. Since that time the U.S. nitrogen sector experienced both consolidation and contraction. Natural gas prices moved up in response to demand and later as a consequence of supply curtailments caused by the 2005 hurricane season which ended with two storms, Katrina and Rita, devastating the U.S. Gulf Coast. Long before that time Farmland Industries were bankrupt due to cost squeezes (2002); its elements were sold. Koch Industries, Inc. purchased Farmland's nitrogen business and thus moved into the top ranks of nitrogen producers. Mississippi Chemical Company, a diversified fertilizer producer, filed for bankruptcy in 2003. Price pressures continued to affect the U.S. nitrogen sector well into the first decade of the 2000s so that, in 2006, according to data published by PotashCorp, approximately 30 percent of domestic capacity was largely unused and ammonia was brought in by producers from their overseas-based operations where gas prices were low.

Among the leaders in fertilizers are Terra Industries, Inc., Koch Industries, Inc., CF Industries, Inc., The Mosaic Company, Agrium Inc., and PotashCorp. The last two are Canadian corporations but with U.S.-based nitrogen facilities.

Terra Industries, headquartered in Sioux City, Iowa, is most likely the leading producer of nitrogen-based fertilizers in the United States, reporting production of 4.6 million tons of anhydrous ammonia, 4.3 million tons of urea ammonium nitrate, and 1.8 million tons of ammonium nitrate. The company had sales of $1.8 billion in 2006.

Koch Industries has its headquarters in Wichita, Kansas. After its acquisition of Farmland's nitrogen operations, it had in excess of 3.5 million tons of capacity. Koch is a highly diversified and privately held organization active in food product and fertilizer sales in 60 counties. Koch has 80,000 employees.

CF Industries (CFI) reported sales of $1.9 billion in 2006. The company, operating for many decades as a cooperative, went public in 2005. CFI is located in Deerfield, Illinois.

Agrium, based in Calgary, Alberta, reported sales of $4.2 billion in 2006. The company is a diversified fertilizer producer and also one of the leading retailers of agricultural products to farmers. The company is thus forward-deployed into distribution, the retail portion of its business dominating the production of fertilizers.

The Mosaic Company is what might be called an instant Fortune 500. The company came into being in 2004 when Cargill Corporation's Cargill Nutrition, active in nitrogen fertilizers, came to be combined with IMC Global to form a $5.3 billion (2006) fertilizer company. IMC Global has been a leader in phosphate mining since the 1940s, exploiting its New Mexico mineral holdings by operating its Carlsbad production plant. With IMC now a part of it, Mosaic is the nation's largest phosphate fertilizer producer.

PotashCorp is also known as Potash Corporation of Saskatchewan (PCS). The company had sales in 2006 of $3.7 billion in U.S. dollars. CFI, Agrium, and PCS are all significantly involved in phosphate and potash fertilizer production in addition to producing nitrogen. PCS, indeed, is the leading producer of potash in North America and holds a commanding market share in that product in the world as well. With headquarters in Saskatoon, Saskatchewan, the company owns the largest potash rock deposit in the world. PotashCorp's most profitable ammonia production is centered in Trinidad. Substantial natural gas deposits were discovered there but not exploited until the 1970s and 1980s. Low cost gas available in Trinidad has been one source of ammonia imported into the United States by PCS and others to displace domestic capacity.

Whereas the companies highlighted above are well known in the industry, the public at large is not likely to recognize many of the names, if any. The name most likely to spell fertilizer to the public is Scotts, formally The Scotts Company or The Scotts Miracle-Gro Company (after Scotts merged with Miracle-Gro). Scotts is the unquestioned leader, with fiscal year 2005 sales of $2.3 billion, in the consumer lawn and garden industry.

MATERIALS & SUPPLY CHAIN LOGISTICS

Potash is geographically concentrated in just a few locations across the world. The largest reserves are in Canada (4,400 million metric tons), in Russia (1,800 million), and in Belarus (750 million). U.S. reserves stand at approximately 90 million metric tons. The tonnages cited represent the weight of potassium carbonate (K_2O) in the rock deposits. Metric tons are 2,205 pounds. More than three-quarters of domestic potash production comes from

New Mexico. In 2006, however, more than 80 percent of domestic consumption was imported from Canada.

Phosphate rock is mined in 40 countries across the world. In 2005, 46.1 million metric tons of phosphorus pentoxide equivalent was mined across globe, the United States being the largest producer (10.5 million metric tons), China the second largest (9.1 million), and Morocco the third (8.3 million). Despite high production rates domestically, the United States is a major importer of phosphate rock; the country is a major exporter of potassium fertilizers.

As of 2005, total world reserves of natural gas stood at 6,359 trillion cubic feet (cf). U.S. reserves were 204 trillion cf or 0.3 percent of the world total. In 2004, the last year for which data were available at time of writing, world consumption of natural gas was 99.7 trillion cf. The United States represented 22.5 percent of that consumption (22.4 trillion cf). In the period from 1980 to 2004, the U.S. growth in consumption was a modest 0.5 percent per year, but world consumption rose at a much more rapid rate of 2.68 percent yearly. If we assume gas consumption worldwide at current rates without any growth at all, gas reserves will be exhausted in just under 64 years. If the 2.68 percent growth rate continues, gas runs out in 47 years. If growth accelerates, gas will run out sooner.

If the United States relies entirely on its own reserves (which it does not at present), gas will be mined out in nine years requiring either import of ammonia fertilizers or of natural gas. The movement of natural gas requires considerable energy because the gas must be shipped as liquefied natural gas, or LNG, in special tankers. The gas is held in liquid form at approximately -261 degrees Fahrenheit and under pressure (approximately 3.63 pounds per square inch). Providing the cryogenic temperatures, pressures, and expensive insulated shipping adds to the cost of the fuel/raw material. Rising natural gas prices in the 2000 to 2006 period had already caused cutbacks in domestic production and increases in ammonia imports.

Unless the rosiest expectations of techno- and natural-resources optimists turn out to be right—that large new fields of gas will be discovered as needed—the fertilizer industry is likely to undergo dramatic changes in the early third of the twenty-first century, principally because its most important raw material will be exhausted. Very expensive nitrogen fertilizers will affect, in the first instance, the nation's and the world's grain products. Yields will diminish as organic wastes or crop rotation will have to be substituted to return fertility to the soil.

DISTRIBUTION CHANNEL

Fertilizers reach their ultimate users—farmers—by means of some 7,400 merchant farm supply establishments that had approximately $50 billion in sales in 2007 based on data reported in *Manufacturing & Distribution USA* (*MDUSA*). These organizations also supply the lawn service industry. The relatively small fraction of total tonnage that reaches homeowners is sold in lawn and garden outlets, hardware stores, and by mass merchandisers. The dollar volume includes herbicides, insecticides, seed, and other supplies along with fertilizers.

Farmers' cooperatives play a very important role in distribution. Among 75 leading distributors listed in *MDUSA*, with total sales of $27.6 billion, 26 were cooperatives. Many cooperatives as well as profit-making wholesaler/retailers in this industry are also directly involved in manufacturing at least some of the products they sell in a pattern similar in other branches of what might be called the agricultural-industrial complex.

KEY USERS

When looking at key users of the products of this industry, it is important to distinguish between customers of finished fertilizer and customers who buy chemicals from fertilizer manufacturers. For instance, 78 percent of ammonia production is converted to fertilizers, but 22 percent is sold for industrial uses. Fertilizer manufacturers cultivate industrial sales; these are less affected by increases in price of the underlying natural gas. A portion of phosphate manufacturing is sold as animal feed supplements. Buyers of supplements are less influenced in their buying decisions by price because the supplements represent very small proportions of total feeds but deliver well-known benefits.

The principal users of fertilizers are farmers. Which fertilizers they use and how intensively will depend on the crop or crops they plant. Using four major crops as an example (corn, wheat, soybeans, and cotton) data from the USDA indicate that highest use of nitrogen is associated with corn, with 97 to 98 percent of acres treated. Wheat is second with 80 to 88 percent of acres treated. Cotton is third with 79 to 86 percent of acres receiving nitrogen. Soybean, a crop that by its very character is a natural nitrogen fixer, has the lowest rates of application, 13 to 18 percent of soybean acreage.

Corn farmers are the chief users of fertilizers. They applied the most nitrogen per acre (127–136 pounds), the most phosphate (56–59 pounds), and the most potash (80–84 pounds). Cotton farmers applied the second highest rates of nitrogen (84–100 pounds per acre), soybean farmers the second highest rates of phosphate (47–50 pounds) and of potash (76–88 pounds). Wheat farmers generally occupied a middle ground in these highly fertilized crop varieties.

The role of the ordinary household in fertilizer usage is somewhat difficult to determine because Census data do not provide product breakdowns detailed enough to discern the ultimate destination of major types of fertilizers. At the same time, commercial reporting tends to lump together all lawn and garden chemicals and uses retail pricing as its basis—difficult to relate back to production data. What little information is available indicates that the homeowner consumes somewhere between 13 and 14 percent of the industry's total production.

ADJACENT MARKETS

The modern fertilizer industry represents a major break with traditional farming. It deploys technology and fossil energy to lift soil fertility by artificial means. In effect markets adjacent to fertilizers lie in two directions—in the distant past and in a possible, if problematic, genetic future.

The distant past suggests a return to natural and organic ways of fertilizing soil without the use of nitrogen derived from hydrocarbon sources. Organic farming is expanding quite rapidly and in so doing is using ancient farming methods in new and modern ways. This approach has an enthusiastic but still small public following. It is more labor intensive and consequently produces food that costs more. The costs of modern fertilizers, however, are also rising and may rise dramatically in the future, thus leveling the playing field.

A futuristic solution to avoid fossil-rich fertilizers would lie in modifying plants genetically so that, for example, they attract and hold symbiotic nitrogen-fixing bacteria. Public reaction to genetically modified foods, however, has been mixed. The dangers of "messing with Mother Nature" are becoming more and more visible. Examples in the early 2000s were puzzling die-offs of pollinating bees, mad cow disease, and large dead zones in the waters of the Gulf of Mexico. European resistance to genetically modified foods was spreading to the United States where, for instance, resistance to the use of bovine growth hormone, produced by recombinant DNA modification, was voluntarily shunned by major elements of the dairy industry as a result of adverse public reactions.

Looming shortages of fossil fuels, already signaled by spiking gasoline and very high natural gas prices in the United States, combined with rapidly shrinking supplies of gas, strongly suggest that adjacent markets may become primary markets in the future for ensuring fertile soils and ample food. If all goes well the transition will be smooth and gradual. The fertilizer industry, however, is sure to be transformed in the process.

RESEARCH & DEVELOPMENT

Fertilizer usage is supported by an extraordinarily diverse research activity based on agricultural schools present in every state and in part supported by funds made available by the U.S. Department of Agriculture. USDA's research funding, for all categories, of which fertilizers are a small slice, was running around $1.7 billion annually in the first decade of the 2000s. Research is also supported by the Environmental Protection Agency in the context of ground water pollution. Most projects are focused on specific issues related to crops, or varieties within them. A significant number of the studies being done are aimed at discovering optimal application rates over time to curtail overuse of nutrients and their loss to runoff.

CURRENT TRENDS

As the first decade of the new century was drawing to a close, a handful of developments were pointing, if somewhat vaguely or contradictorily, to the future shape of the fertilizer industry. The most significant of these was rising costs of energy, more specifically of natural gas. In the 1960s, based on data from the Energy Information Agency, well-head prices for natural gas averaged 16 cents for 1,000 cubic feet. In the 1970s prices had risen to 50 cents, in the 1980s to $2.08, in the 1990s they averaged $1.92, and then, in the early 2000s, gas had risen to an average (2000–2006) of $4.96. The price in 2006 was $6.42 per 1000 cf. This run-up in the price of the raw material for ammonia was causing the shuttering of numerous U.S. ammonia plants and growing imports of nitrogen fertilizers from regions with lower gas prices. Total consumption was also flattening out.

Acreage devoted to organic farming was growing at a surprisingly strong rate in the 1995 to 2005 period, 10.4 percent per year overall, 14.9 percent in corn, 11.2 percent in wheat, and 13.4 percent per year in oats. The growth in organic farming might be seen as the early and barely visible greening of a new order, yet to emerge fully into view. Organic agriculture is the old order reborn in a new shape, likely to take a stronger hold as its unfavorable economics are made ever more viable by rising energy costs.

If rising energy prices portend the weakening of industrial agriculture and organic farming the nascent emergence of a response, the national emphasis on ethanol fuels is difficult to fit into the picture. Ethanol is made from corn, the most intensively fertilizer-dependent crop. It would appear that as nitrogen fertilizers grow ever more expensive because of diminishing fossil fuels, fuels based on corn can only be a temporary solution to the problem of transportation energy. Hence the policy promoting ethanol fuels is clearly a short-term fix rather than a sustainable strategy.

Emerging near-term problems in the nitrogen sector will not necessarily impact the entire fertilizer sector in a uniform manner. Rising ammonia costs will tend naturally to cause more farmers to rotate crops. Leguminous crops require proportionally more phosphorus and potassium fertilizers while leaving soil richer in residual nitrogen. Minerals-based fertilizers, therefore, will benefit from pressures created by shrinking natural gas supplies.

TARGET MARKETS & SEGMENTATION

Fertilizers are formulated for specific crops, seasons of application, and specific soil conditions. Product purchasing is closely tied to scientific essaying by farmers who buy their products based on soil analysis and their own experience. The industry routinely produces basic nutrient mixes with the most commonly employed proportions of NPK in the product. Segmentation is therefore largely dictated by the requirements of the targeted plant species, regions, and climates. At one extreme of targeting, refrigerated tanker trucks arrive at the field to dispense liquid ammonia by underground injection. At the other, the home gardener will uncap a tiny bottle of a rose nutrient solution and apply two or three drops to a single plant. Between these two extremes exist very large segments each served by specialized formulations.

RELATED ASSOCIATIONS & ORGANIZATIONS

The Fertilizer Institute, http://www.tfi.org/factsandstats/fertilizer.cfm

International Fertilizer Industry Association, http://www.fertilizer.org/ifa

Minerals Information Institute, http://www.mii.org

BIBLIOGRAPHY

"2006 Overview of PotashCorp and Its Industry." Corporate Brochure. Potash Corporation. Undated.

Darnay, Arsen J. and Joyce P. Simkin. *Manufacturing & Distribution USA,* 4th ed. Thomson Gale, 2006.

Helikson, Helen. "The Energy and Economics of Fertilizers." *Energy Efficiency & Environmental News.* September 1991.

Kirk, David. *Biology Today.* Random House, 1972.

Kostick, Dennis S. "Potash." *Mineral Commodity Summaries.* U.S. Department of the Interior, U.S. Geological Survey. January 2007.

Manning, Richard. "The Oil We Eat." *Harper's.* February 2004.

"Modern Agriculture Footprints." *Redefining Progress.* 2005. Available from <http://www.redefiningprogress.org/pdf_files/Modern_Agriculture_Footprints.pdf>.

"National Resources Inventory, 2002 Annual NRI." U.S. Department of Agriculture, Natural Resources Conservation Service. April 2004. Available from <http://www.nrcs.usda.gov/technical/land/nri02/landuse.pdf>.

"Nutrient Management." U.S. Department of Agriculture, Economic Research Service. Available from <http://www.ers.usda.gove/publications/arei/eib16/chapter4/4.4/>.

"Organic Production." U.S. Department of Agriculture, Economic Research Service. Available from <http://www.ers.usda.gov/Data/Organic/>.

"Phosphate Rock." *2005 Minerals Yearbook.* U.S. Department of the Interior, U.S. Geological Survey. August 2006.

Pimentel, David. "Ethanol Fuel: Energy Balance, Economics, and Environmental Impacts are Negative." *Natural Resources Research.* June 2003. Available from <http://www.ethanol-gec.org/netenergy/neypimentel.pdf>.

"Table 1.2.–11. Description of Fertilizer Materials." *Agronomy Guide 2007–2008.* Pennsylvania State University. Available from <http://agguide.agronomy.psu.edu/cm/sec2/table1-2-11.cfm>.

"U.S. Natural Gas Wellhead Price." U.S. Department of Energy, Energy Information Administration. 30 April 2007. Available from <http://tonto.eia.doe.gov/dnav/ng/hist/n9190us3A.htm>.

"World Proved Reserves of Oil and Natural Gas." U.S. Department of Energy, Energy Information Administration. 9 January 2007. Available from <http://www.eia.doe.gov/emeu/international/reserves.html>.

Firearms

INDUSTRIAL CODES

NAICS: 33–2994 Small Arms Manufacturing

SIC: 3484 Small Arms Manufacturing

NAICS-Based Product Codes: 33–29941 through 33–29943546

PRODUCT OVERVIEW

The manufacture, sale, and ownership of firearms have played an integral role in the development and culture of the United States since its inception. From the establishment of the first national armory at Springfield, Massachusetts, in 1794 to the technological advances of the nineteenth century and the often controversial federal oversight of the twentieth and twenty-first centuries, the gun industry has helped shape a national consciousness that is, for good or ill, distinctly American. Once a necessity in the acquisition of food and for general protection, guns became a topic of political debate in the late twentieth century. The U.S. firearms industry found itself in an uncertain regulatory environment in the early twenty-first century.

This essay will confine itself to a discussion of the firearms commonly known as small arms, as opposed to those used for military purposes. Small arms is a term first seen in English in the early eighteenth century and denotes those weapons designed for personal, generally hand-held, use. The category includes such weaponry as rifles, shotguns, revolvers, pistols, assault rifles, and submachine guns.

The very early history of guns, including the invention of gunpowder, is somewhat murky. Gunpowder or black powder, as it became known, is variously alleged to have originated in China, Arabia, Germany, or yet someplace else entirely. The fog begins to lift, however, by the fourteenth century, when the initial references to cannons were made in Italy and England. Handgun references appeared by the middle of that century, and the ensuing evolution of small arms is fairly well documented.

Advances in the ignition systems of firearms were crucial to the development of modern guns. Beginning with muzzleloaders, which were fired by applying a lighted match or wick to gunpowder and a projectile that had been loaded into the muzzle end of the weapon, firing mechanisms on small arms gradually became safer and more reliable. In the matchlock systems developed in the early fifteenth century, for example, the lighted wick was no longer in a person's fingers. It was now in the mechanism of the gun. The wheel lock of the next century did away with the need for a match at all, replacing it with a steel and iron pyrite interaction that created a spark to light the powder. Flint and steel combined to create the necessary spark in the flintlock ignition of the late seventeenth century, and the percussion lock, or caplock, of the early nineteenth century was the forerunner of contemporary self-contained ammunition.

By the nineteenth century, the United States had become a hotbed of weaponry innovation, especially in an area of the Connecticut River Valley known as Gun Valley. Among the many notable names of that era and a sampling of their inventions were: Samuel Colt (inventor of the revolver), Richard J. Gatling (first machine gun), Sir Hiram Stevens Maxim (semiautomatic rifle and fully automatic machine gun), and John M. Browning (semi-automatic pistol, gas-operated machine gun, and the Browning Automatic Rifle, or BAR). It was a heady time in which fortunes were made, technological ground was broken, and the subject of small arms engendered very little discussion at all. Not incidentally, it was also a time of fewer people, more available land, hunting as a means of procuring sustenance, and the U.S. Civil War fresh on the collective mind of the populace.

The world had changed by the onset of the twenty-first century. An increasingly urbanized society became divided on the need or desirability of small arms in the hands of its citizens. One faction saw personal gun ownership as a fundamental entitlement guaranteed by the Second Amendment of the U.S. Constitution. Hunting, competitive shooting, and self-protection were considered perfectly legitimate pursuits. The other side blamed the relative accessibility of small arms for such societal ills as crime and suicide. It also voiced concern over the proliferation of such weapons worldwide. The small arms industry was caught in the middle.

The 1990s were particularly volatile for the gun industry. Increased government regulation, such as 1994's Brady Handgun Violence Prevention Act, had a mixed impact as consumers stocked up on weapons before new measures went into effect. The unprecedented initiation of litigation against manufacturers by major cities, starting with New Orleans in 1998, was another immense challenge as the prospective costs of defending such lawsuits forced manufacturers to explore new ways to prosper in the coming years. Some companies, for instance, filed for bankruptcy protection, while others entered into settlement negotiations. Still others broadened their product offerings, branching into such ancillary markets as specialty clothing and sporting goods. Perhaps most notably, however, the adversity prompted the famously competitive industry to begin to band together as a group.

The small arms industry is in the durable goods sector, with products that do not quickly wear out and are unlikely to need replacement in a consumer's lifetime. Despite two factors that are causing uncertainty for the industry—concerns presented by a changing customer base and the social debate about imposing greater gun control regulation—it is in little fear of extinction. The twenty-first century saw sales, particularly of handguns, on the rise once again. Manufacturers continue to investigate new marketing avenues, from cutting-edge technologies to foreign-made weapons to shooting accessories to the Internet. Cultivating or reviving non-traditional customer bases, such as women and youth, are also in play. Market growth is an ongoing worry, but hunters still hunt, police officers still police, and competitive shooters still compete. Small arms are what each of these groups use.

MARKET

It is notoriously difficult to pinpoint the numbers of U.S. citizens who own guns or, accordingly, the number of weapons within a given household. One reason for this is that such statistics are typically gathered by survey and people have many motivations for giving false information. By combining survey statistics provided by the National Rifle Association (NRA) and *Reason Magazine*, the percentage of private gun owners appears to be stable at between 39 and 49 percent. Another way to track private firearm ownership is to rely on the National Instant Background Check System (NICS) that was initiated with the passage of the Brady Act of 1994. The background check is required for gun purchase or permits. The NICS numbers, as cited by *Shooting Industry*, indicate that gun ownership is flourishing. In May of 1999, for instance, the number of NICS background checks performed was 576,272. In May of 2007, that number was 803,051. The rise is even more notable when one considers that the May 2007 numbers are closer to those normally associated with the peak buying season in the autumn. Neither a survey nor the NICS is a perfect system, as they do not account for firearms already in the household or those that were illegally obtained. Nonetheless, one can accurately glean some ownership trends from such data.

Excise taxes, calculated as a percentage of wholesale receipts, are yet another way to assess activity within the small arms industry. According to *Shooting Industry*, excise taxes demonstrated a 5.6 percent increase in sales for all firearms from 2005 to 2006. Handguns did particularly well, posting a 21.94 percent increase in 2006 over the previous year. Research and Markets data, as cited in a *Business Wire* article, show the industry's overall 2006 revenue at approximately $2.15 billion with a gross profit of nearly 36 percent.

A vital point about the small arms industry, however, is the ongoing influence of the global marketplace and foreign trade. For many years, U.S. firearm exports have been eclipsed by the number of imports. *Shooting Industry* pointed out that in 2003, for example, the value of gun exports was $42 million and the value of imports was $380 million, creating a trade deficit of a whopping 84 percent. While that was not necessarily a negative situation for such industry participants as importers, distributors, and dealers, it certainly posed a problem for U.S. manufacturers. Nor, given the popularity of foreign-made weapons, was the trend apt to reverse itself anytime soon. Manufacturers addressed the challenge in several ways, including cutting costs, making better guns, and adding foreign-made firearms to their own product offerings. The Remington Arms Company jump-started matters in 2004 with its introduction of a line of Russian-made shotguns. It went on to add others, and competing companies began to follow suit. Results of the manufacturer's efforts were quickly realized—U.S. exports rose 18.66 percent in 2005 and imports fell 2.2 percent. The top three exporting companies in that year were Remington, Smith & Wesson, and Sturm, Ruger and Company. The primary importing countries were Brazil, Austria, and Italy.

The political climate also has a tremendous impact on the small arms industry. An example of this is the dramatic reduction in the number of licensed gun dealers through the end of the twentieth century. According to the *Christian Science Monitor,* there were approximately 245,000 licensed dealers in the United States in 1996. By 2006 that number had dropped nearly 80 percent. The decrease was clearly attributable to tightened government firearm regulations that had been put in place during the Democratic administration of President Bill Clinton. The Brady Act, and new zoning and reporting requirements for dealers were among those changes. Gun control advocates heralded the downsizing as a victory against crime, while the pro-gun side maintained that the measures had simply driven out individuals who had received licenses in order to buy guns at wholesale prices. The crux of the matter does not lie in which position is correct, however. Instead, the point is that the decrease did take place and politics played an integral role in that process.

Finally, a discussion of the small arms market must include mention of two distinct factions within its domestic confines—private citizens and law enforcement. No numbers were available to express the exact percentage of buyers within these markets, but there is little question that one may be at least partially offsetting the other. Particularly in the wake of the September 11, 2001, terrorist attacks against the United States, the government budget for small arms is expanding, according to the *New York Times,* as quoted in the *International Herald Tribune.* However, government agencies are also investing more money into guns and homeland security. It follows that police forces, especially in major cities, are doing so as well. Thus, it may be that any increases in small arms sales owe more to the law enforcement market than to the civilian market.

KEY PRODUCERS/MANUFACTURERS

According to the U.S. Census Bureau's *Current Industrial Report* series, there were 177 small arms manufacturing companies in the United States in 2002. Most were located in Texas, Connecticut, Pennsylvania, and Massachusetts, with California and Wisconsin close behind. As is the case with many mature industries, the small arms business had undergone its share of upheaval throughout the years. The venerable Winchester factory, for example, closed its doors in March of 2006. Colt's Manufacturing had seen various owners and undergone bankruptcy pro-

ceedings. Others, such as Springfield Armory, no longer bore any real relationship to their storied pasts. Nonetheless, familiar brands remained among the top manufacturers of 2005. Those included Marlin; Mossberg; Remington; Smith & Wesson; and Ruger.

Marlin Firearms Company. Founded by John Mahlon Marlin in 1870 in New Haven, Connecticut, Marlin makes rifles. Such famous characters as sharpshooter Annie Oakley and cowboy movie actor Tom Mix were Marlin fans. The company was bought by a syndicate and became the Marlin Rockwell Corporation in 1915, but came into a new family legacy in 1924 when it was bought by Frank Kenna for just $100 (and a large mortgage). The Kennas still owned Marlin in 2007. In 2000 the company acquired H&R 1871, the biggest manufacturer of single-shot shotguns and rifles in the world. Marlin was the third largest maker of rifles in 2005.

O.F. Mossberg & Sons. The second largest producer of shotguns in 2005, Mossberg was founded in 1919 by Oliver F. Mossberg. It began making .22 rifles in 1922 and introduced its shotgun line in 1957. Since that time, shotguns have been a primary focus of the company. Mossberg is headquartered in North Haven, Connecticut.

Remington Arms Company. Begun in upstate New York in 1816 by Eliphalet Remington, the strictly long-gun company is now located in Madison, North Carolina. It claims to be one of the oldest continuously operating firearm manufacturers in the United States, as well as the sole U.S. producer of both guns and ammunition. It has been the top U.S. manufacturer of firearms for five consecutive years as of 2005.

Smith & Wesson. Perhaps most famous for introducing the .44 magnum popularized by actor Clint Eastwood in the movie *Dirty Harry,* Smith & Wesson's past has been more checkered than most. It was founded, after a failed initial partnership by Horace Smith and Daniel B. Wesson in Springfield, Massachusetts, in 1856 and was run by Wesson's descendents until 1967. A long history of innovation and profit was notably marred in 2004 when the company's then-chairman was discovered to be a convicted felon and accounting irregularities were investigated. By 2005, however, it was back on top as the leading U.S. manufacturer of handguns, and in 2007 it implemented a turnaround strategy of product expansion and acquisition (Thompson/Center Arms). As a result profits soared 41 percent.

Sturm, Ruger & Company. The number two producer of both handguns and rifles in 2005, William Batterman

Ruger opened for business in 1949 after working as a gun designer for the original Springfield Armory. Although a relative latecomer to the small arms business, Ruger quickly established a reputation as an industry leader with an enviable balance sheet. The company makes rifles, shotguns, pistols, and revolvers, and is the only leading U.S. small arms manufacturer that is publicly held.

Springfield Armory, Inc.. The original Springfield Armory was the first national armory in the United States. It was in operation until 1968. The name was adopted by Robert Reese in 1974, when he founded a firearms manufacturing company in Geneseo, Illinois. Springfield ranked third among the leading handgun producers of 2005.

Others. The Austrian-based semiautomatic pistol maker Glock bears special mention as the producer of this small arm of choice for many—particularly police officers. Other notable manufacturers include Savage, Kimber, Beretta, Beemiller, and Bushmaster.

MATERIALS & SUPPLY CHAIN LOGISTICS

The basic materials required for small arms manufacturing are fairly straightforward. They include iron, steel, copper, and aluminum, as well as various plastic products and fastening devices such as bolts, nuts, and screws. At least as important as the raw materials necessary to production are the technical design and, often, the artistry of the weapon. Many guns, for instance, are elaborately engraved to be aesthetically pleasing. Others rely on trademark innovations to set themselves apart. Accuracy, reliability, and safety are further important considerations, depending on the nature of the weapon (target shooting vs. self-protection, for example). In short, the small arms business depends less on the manufacturing materials involved than the expertise behind it.

Guns can be, and are, manufactured all over the world. In this respect, especially with regard to imports, cost may come more in alignment with expertise. Just as with domestic manufacturers, however, this possibility depends largely on the maker—a Glock, for example, would hardly be considered a cheap import. Thus, while costs also come into play when looking at supply chain logistics, where the weapon is made ultimately matters very little.

DISTRIBUTION CHANNEL

Small arms manufacturers generally get their products to consumers through retail outlets. Those outlets primarily consist of gun dealers, sporting goods stores, and broader-based, Big Box retailers such as Wal-Mart. Retailers

must possess a federal license. Most major manufacturers also maintain websites with information that includes firearm offerings, dealer/retailer locations, and accessories that can be purchased directly from the maker online. Accessories may include gun stocks and barrels, collectibles, and apparel.

A subset of the gun dealer faction involves the highly-charged retail category known as gun shows. Gun shows are temporary exhibitions held in such public spaces as shopping malls, hotels, or stadiums. In addition to weapons, gun parts, ammunition, knives, collectibles, and gun literature are often found for sale. The controversy surrounding gun shows stems from the NICS provision of the Brady Act, which only applies to those in the business of selling firearms and, thus, those who hold a license. Unlicensed vendors have no obligation to conduct the NICS background check required of licensees. The underlying assumption in this loophole is that private citizens have a right to sell or trade weapons from their own collections, much as they sell their cars without going through a car dealership. The concern, however, is that potential buyers who would be denied if a background check were conducted can bypass that problem by patronizing an unlicensed dealer. Debate about the issue continues, as do gun shows.

KEY USERS

Small arms purchases can loosely be divided into five categories—hunting, competitive and target shooting, self-protection, law enforcement, and collecting. Although the ranks of hunters have been decreasing over the years, public support for hunting appears to be on the rise. The number of hunters dropped 4 percent from 2001 to 2007, while the number of those who disapprove of the sport has fallen from 22 percent in 1995 to 16 percent in 2007. States, anxious to maintain or increase hunting license revenues in order to keep up conservation efforts, are banding together with gun groups to reverse the declining trend. Prime among such campaigns are those aimed at attracting young people and families to the sport.

Competitive and target shooting also have their aficionados. The NRA alone sanctions approximately 10,000 shooting tournaments per year and conducts over 50 national championships. The National Shooting Sports Foundation (NSSF) estimates that there are approximately 17 million active target shooters in the United States. Those involved in sporting clays grew by 8.4 percent from 1998 to 2005.

It is difficult to estimate how many gun owners purchase weapons for self-defense. *Reason Magazine,* however, cited Gallup Polls of 1999 and 2000 that placed the percentage of private owners who have firearms for protection against crime at 65 percent.

A major outlet for small arms is law enforcement. Once the domain of the revolver, Colt and Smith & Wesson had the market tied up for many years. Ruger and others eventually joined the fray, but the law enforcement market drastically changed when greater criminal firepower prompted departments to switch to 9-millimeter handguns in the 1980s. This brought in overseas competition, bringing Glock to the forefront. By the turn of the century, Glock enjoyed a market dominance of over 70 percent.

Collectors are another important group of gun owners. Many collectors appreciate weaponry primarily for its artistic or investment value. As with many of these users, this objective in gun ownership may overlap with other interests. The collector may also enjoy target shooting, for instance. Or a police officer might hunt in his spare time. Thus, it is not uncommon for a gun-owning household to have more than one small arm.

ADJACENT MARKETS

The small arms industry supports an array of other markets. Prime among these is ammunition. Federal data showed 110 ammunition manufacturing companies operating in the United States as of 2002. It was over a billion dollar per year industry at the time, and supported nearly 7,000 employees.

Other adjacent markets range from specific firearm accessories, including scopes, holsters, gun racks, and cleaning equipment, to corollary hobby and professional supplies for reloading and gunsmithing, to broader sporting equipment and accompaniments such as binoculars, camping gear, pocket knives, and GPS systems. Additional markets include everything from eye/ear protection, decoys, and clothing to targets, shooting rests, and duck calls.

Another, less quantifiable influence, is the crime rate and/or other perceived threat to the U.S. citizenry. For the reasons cited above, these factors are hard to translate into reliable statistics. A good example, nonetheless, is the surge in weapons sales after the attacks of September 11, 2001. Tangible or not, self-defense can be a motivator among the gun-buying public.

RESEARCH & DEVELOPMENT

Much of the recent research and development surrounding the small arms industry focuses on safety and crime control. It should be noted that not all of this investigation is being undertaken by the manufacturers themselves, but by such independent institutions as the New Jersey Institute of Technology (NJIT) and the National Institute of Justice (NIJ).

One of these investigations is the quest for a so-called *smart gun* that would only function in the hands of an authorized user. Colt, Smith & Wesson, and the foregoing independent institutions are among those who have explored the possibilities of such a firearm. The idea behind the development of a smart gun is to prevent gun deaths in such situations as children playing with firearms or police officers who have their own weapons turned against them by suspects or criminals. The various technologies looked into have included biometrics (reading unique body signatures) and radio frequency devices. A prototype was introduced at the NJIT in 2004, but commercial availability remained in the future as technology was fine-tuned and political debate continued over the viability and necessity of safe guns.

Another intriguing and controversial development was a technology known as *micro-stamping*, which would stamp a firearm's make, model, and serial number onto shell casings every time the gun was fired. Applicable to semiautomatic weapons, as revolvers retain their casings in their chambers, California's state assembly passed a bill in 2007 that would require such technology on all semiautomatic pistols sold in the state beginning in 2010. The premise in this case was to provide a further means of evidence gathering for law enforcement. Predictably, pro-gun advocates objected and gun-control fans cheered. It was not made clear how such technology would be useful in the case of illegal, unregistered weapons.

CURRENT TRENDS

Today's trends in small arms manufacturing, from new research and development efforts to the incorporation of foreign-made weapons into domestic product lines, are largely fueled by the saturation level within the industry and the consequent quest for fresh markets. Although a fairly stable market continues to exist, the maturity of the industry and the inherent nature of firearms combine to offer manufacturers a continuing challenge in finding new customers. For instance, given the shelf life of most small arms, a gun owner often has no real need to replace an existing weapon. He or she may covet the latest thing, but, unlike, say, the instance of a broken refrigerator, there is no necessity to buy a new one. Developing smart guns and cultivating young shooting enthusiasts are but two avenues toward that goal.

TARGET MARKETS & SEGMENTATION

While markets and segmentation with the small arms industry have been largely covered in the forgoing discussion, the role of women should be given special attention. According the NSSF, 16 percent, or over 3 million, of all active firearm hunters in 2005 were female. That same year, 23 percent, or 5 million, target shooters were women. These figures alone make women an attractive target market for gun makers, but there is another consideration that gives them even more allure—the youth market. That is, as the industry attempts to maintain its longevity by reeling in another generation of avid consumers, there are hardly better champions it could have than the women of that next generation. By involving more women and, by hopeful extension, families, in small arms pursuits, gun makers can see a saturated market base become filled with potential.

RELATED ASSOCIATIONS & ORGANIZATIONS

Brady Campaign to Prevent Gun Violence, http://www.bradycampaign.org

National Rifle Association (NRA), http://www.nra.org

National Shooting Sports Foundation, http://www.nssf.org

United States Fish & Wildlife Service, http://www.fws.gov

United States Practical Shooting Associations, http://www.uspsa.org

Violence Policy Center, http://www.vpc.org

BIBLIOGRAPHY

Ayoob, Massad. "Guns 50th Police: It's Been a Helluva Ride the Last Half Century, with an Almost Complete and Diametric Reversal of the Traditional Paradigm." *Guns Magazine.* January 2005. Available from <http://findarticles.com/p/articles/mi_m0BQY/is_1_51/ai_n7581229>.

Clayton, Mark. "Hunters as Endangered Species? A Bid to Rebuild Ranks." *Christian Science Monitor.* 27 September 2005. Available from <http://www.csmonitor.com/2005/0927/p01s02-ussc.htm>.

"The False Hope of the 'Smart' Gun." Violence Policy Center. Available from <http://www.vpc.org/fact_sht/smartgun.htm>.

"Gun Ownership: The Numbers." *Reason Magazine.* May 2001. Available from <http://www.reason.com/news/show/28021.html>.

"Gun Shows: Arms Bazaars for Terrorists and Criminals." Brady Campaign to Prevent Gun Violence. Available from <http://www.bradycampaign.org/facts/faqs/?page=second>.

"Guns and Ammo: History." Dyer Laboratories, Inc. Available from <http://www.dyerlabs.com/guns_and_ammo/history.html>.

"Guns, Gun Ownership, & RTC at All-Time Highs, Less 'Gun Control,' and Violent Crime at 30-Year Low." National Riffle Association, Institute for Legislative Action. Available from <http://www.nraila.org/Issues/FactSheets/Read>.

"History of Firearms." Today's Hunter in South Carolina. Available from <http://www.hunter-ed.com/sc/course/ch2_history_of_firearms.htm>.

Marks, Alexandra. "Why Gun Dealers Have Dwindled." *Christian Science Monitor.* 14 March 2006.

Moyer, Ben. "Hunting: Number of Hunters is Dropping, But Not Public Support for Those Who Hunt." *Pittsburgh Post-Gazette.* 1 July 2007.

"National Survey of Fishing, Hunting, and Wildlife-Associated Recreation." U.S. Department of the Interior, Fish & Wildlife Service. Available from <http://federalasst.fws.gov/surveys/surveys.html>.

"Shots Fired at Bayonne Range Prove Smart Gun Technology Works." Press Release. New Jersey Institute of Technology. 16 December 2004.

"Small Arms." *Encyclopedia of American Industries.* Thomson Gale, 2006.

"Small Arms Ammunition Manufacturing: 2002." *Current Industrial Reports.* U.S. Department of Commerce, Bureau of the Census. January 2005.

"Small Arms Manufacturing: 2002." *Current Industrial Reports.* U.S. Department of Commerce, Bureau of the Census. January 2005.

"The Small Arms Manufacturing Industry's Revenue for the Year 2006 Was Approximately $2,150,000,000." *Business Wire.* 11 April 2007. Available from <http://www.allbusiness.com/services/business-services/4317925-1.html>.

Thurman, Russ. "Business Hits Robust Level: An Energized Industry Enjoys Brisk Sales." *Shooting Industry.* July 2007.

———. "It Ain't Your Grandfather's Gun Business: Intense Government Scrutiny, Relentless Anti-gun Assaults, Increased Imports and An Erratic Economy—They've All Changed the U.S. Firearm Business!" *Shooting Industry.* July 2004.

Wayne, Leslie. "Gun Maker Banks on Pentagon." *New York Times.* 11 April 2006.

Yi, Matthew. "Assembly OKs Micro-stamp on Some Guns." *San Francisco Chronicle.* 30 May 2007.

Fitness Equipment

INDUSTRIAL CODES

NAICS: 33–9920 Sporting Goods Manufacturing

SIC: 3949 Sporting and Athletic Goods Manufacturing

NAICS-Based Product Codes: 33–99207131, 33–99207141, and 33–99207198

PRODUCT OVERVIEW

Fitness equipment is a subset of the larger sporting goods sector. Fitness equipment includes items such as treadmills, free weights, weight machines, and elliptical trainers.

A treadmill is a device that consists of a moving belt on which a person walks or runs while remaining in one place. Treadmills were used as early as 1875 in the agricultural sector when animals were put on treadmills to help power butter churns or threshing machines. Cardiologist Robert Bruce and Wayne Quinton developed the first treadmill for medical use in 1952. Bruce later sold the design to Stairmaster, now known as Nautilus, who began to market the treadmill as a fitness device. Some of the earliest manufacturers include Tunturi, Aerobics, Inc., and Woodway. The earliest models consisted of a motor, belt, and a deck. In the 1980s and 1990s more advanced models included inclines, programmable workouts, and more durable decks. They could also transfer data from a PalmPilot.

Free weights such as dumbbells are not attached to any sort of machine. Dumbbells typically come in pairs and consist of a bar with weights attached to either end.

Barbells, similar in construction to dumbbells, have adjustable weights. Because free weights are not attached to a machine, they require extra muscle exertion to stabilize each movement. As a result a weight trainer can gain greater muscle size and strength. However, it is more difficult to train specific muscles using free weights.

A weight-training machine uses a weighted pulley system to vary resistance. Some may offer a wide range of exercises while others may be designed to address a particular muscle group. A cable machine consists of a rectangular steel frame approximately 3 meters wide and 2 meters high, with a weight stack at each end. The cables that connect the handles to the weight stacks run through adjustable pulleys. A Smith machine consists of a barbell that sits on steel runners; the user may only move the weight up or down. A series of slots allow the barbell to be secured into place at any time, unlike a regular barbell. Other weight-training machines include butterfly machines, used for strengthening the chest muscles; lateral pull down machines, used for strengthening the muscles of the back, forearm, and the biceps; and leg presses, used for strengthening the leg muscles.

An elliptical trainer, also known as a cross trainer, is a stationary exercise machine used to simulate walking or running while remaining in place. Such devices first became popular in the 1990s, and are driven by leg power. The user also grips shoulder level handles and uses them in a push-pull motion to provide a secondary source of aerobic activity. The handles travel in a somewhat elliptical pattern, hence the name of the device. The basic features of elliptical machines include the drive system, the resistance system, and the stride length. Rear drives are considered superior to front end drives; the movement of their parts is much smoother. Inexpensive models have a resistance level that must be set manually and a preset stride length. Higher end models have automatic versions of these features.

Early Attitudes about Physical Fitness. Physical fitness was an important part of early civilizations. People gathered food, built shelter, and ensured the safety of the local village. While evidence of physical training exists in many ancient cultures, perhaps the Greeks have been the most influential to Western civilization. Herodicus (ca. 480 BC), a physician and athlete, strongly advocated proper diet and physical training. Hippocrates (460–377 BC), best known for his Oath of Medical Ethics, was regarded as the greatest physician of his day, advocating proper exercise and the importance of hygiene. Claudius Galenus or Galen (131–201 AD) was a physician whose writings on the anatomy were highly influential to Western cultures.

The concepts of proper diet and exercise remained highly influential throughout the eighteenth and nineteenth centuries. The first physical education classes began in schools in the nineteenth century. The first department of physical education at an American college was established at Amherst College in 1860. By 1950 there were over 400 colleges and universities in the United States offering a major course of study in the field of physical education.

In 1953 a controversial study was released called *Muscular Fitness and Children*. The study revealed that approximately 57 percent of children in the United States failed to achieve "minimum standards for health" compared to only 8.3 percent of European children. This study was released, it should be noted, as the United States was increasingly concerned about the rise of Communism and the Soviet Union. Young men needed to be fit for military combat, but young men of the United States, according to the study, appeared to be falling behind. In June 1956 President Eisenhower created the President's Council of Youth Fitness. President John F. Kennedy would later change the organization's name to the President's Council on Fitness to suggest that all Americans, not just young people, should be physically fit.

Modern Fitness. Some health clubs existed during the nineteenth century, although they were primarily private organizations. The Young Men's Christian Association (YMCA) was founded in 1844 in London, England. The first YMCA facility in the United States opened in Boston in 1851. The YMCA was, and still is, concerned with "the improvement of the spiritual, mental, social, and physical condition of young men." Eventually, this concern would expand to all people regardless of race, religion, or nationality. Early exercise classes included exercise drills,

the precursor to aerobics. Members used wooden dumbbells and heavy medicine balls to maintain physical fitness. Swimming pools and bowling alleys were popular during the 1880s. Many sports that people currently play for fitness and recreation were invented at the YMCA: volleyball, racquetball, basketball, and football. In the twentieth century, the popularity of the YMCA waxed and waned through times of peace and times of war.

The fitness equipment market became popular in the 1950s and 1960s. In 1957 Universal Gym Equipment released the first multi-station, weight-training machine. The company, owned by Harold Zinkin, a former Mr. California, produced a machine with a bench, racked weights, and pulleys that allowed an athlete to perform a number of exercises safely. This machine is credited with introducing the concept of circuit weight training (moving from one exercise to another) to the fitness world. During the 1960s the growing computer and microprocessor industry helped Dr. Keene Dimick invent the first computerized stationary bike in 1968.

The 1970s saw a major fitness movement in the United States. Gas shortages prompted some people to take up bike riding; bike sales hit a decade high of 15 million bicycles in 1973. The sport of bodybuilding grew in popularity after the release of the documentary movie *Pumping Iron* in 1977, featuring Arnold Schwarzenegger. The first neighborhood gyms began to appear during this decade as well. Dr. Kenneth H. Cooper coined the term *aerobics* in a book of that title published in 1968. Jane Fonda and Richard Simmons helped bring this fitness activity into the mainstream in the 1980s. By one estimate, the number of aerobic participants in the United States grew from an estimated 6 million in 1978 to 22 million in 1987.

By the 1980s people weren't just exercising at the gym. Home gyms became more popular as people purchased exercise bikes and treadmills for home use. Jane Fonda's workout books and videos were influential in helping people to think about working out at home and not just at the gym.

The Next Wave of Equipment. Jerry Wilson created his Soloflex machine in 1978. Previous weight lifting machines from Nautilus or the Universal Gym were simply too heavy and cumbersome for home use. Wilson designed a system that used weight straps made from heavy-duty rubber, and was also small enough to fit into any home gym. Wilson advertised the Soloflex on infomercials and with a number of ads starring model Scott Madsen. The device became popular and 30,000 machines were sold in 1984. Tessema Dosho Shifferaw invented the Bowflex exercise machine in 1979. While the Soloflex used a series of rubber straps in its weight system, Bowflex used a power rod system that offered users increasing levels of weight resistance. Bowflex of America was formed in 1986. The company began a series of direct marketing campaigns that helped propel sales. The Stairmaster was another popular fitness machine on the market in the 1980s.

MARKET

According to the Sporting Goods Manufacturers Association (SGMA), manufacturers shipped $771 million in fitness equipment to the home market in 1988. The market grew quickly during the following decades. Manufacturer shipments totaled $990 million in 1990, $2.9 billion in 2000, and $3.5 billion in 2006. New products helped drive industry sales as the first elliptical machines and cross country machines arrived on the market during the 1990s. Elliptical machines became popular quickly because they put less stress on the joints than stair climbers. Meanwhile older types of fitness equipment saw dropping sales. According to the SGMA, consumers purchased 10.4 million stair climbers in 1989; by 1997 only 300,000 were sold. Stationary bike sales fell from 3.1 million to 900,000 over the same period. However, treadmills remained with sales increasing from 800,000 to 3.1 million during this period. Abdominal machines became a hot trend in the 1990s; numerous fitness experts and their celebrity endorsers appeared on infomercials talking about the importance of this body part. Abdominal exercise machines were the fastest-growing product category in 1996 with sales jumping 200 percent, from $75 million in 1995 to $225 million in 1996. In 2006 the total fitness equipment market was valued at $4.7 billion, 76 percent of which went to the home market and the remainder going to clubs and institutions.

Health club membership grew from 17.3 million in 1987 to 32.8 million in 2000, according to the International Health, Racquet and Sports Club Association. By January 2007 that number reached 42.7 million. Attendance increased as well; club members went to the gym an average of 72 days per year in 1987. This figure was up to 92 days in 2005. The number of members who attended the gym more than 100 days per year increased from 5.3 million to 17.6 million over this same time period.

Total exercise equipment sales were $4.7 billion in 2006, according to the SGMA, up from 3.7 billion in 2000. Treadmill sales steadily grew from 2000 to 2005 before declining in 2006 as can be seen in Figure 93. Home gyms and free weights gained steadily throughout this time period. Elliptical machines had steadily increasing sales from 2002 to 2005 as well and then sales jumped dramatically in 2006, more than tripling from $201 million in 2005 to $725 million the following year.

The global fitness equipment market was valued at approximately $7.5 billion in 2005, according to estimates from the sporting goods company Amer Sports.

FIGURE 93

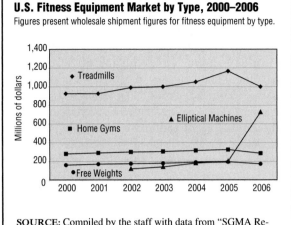

U.S. Fitness Equipment Market by Type, 2000–2006

Figures present wholesale shipment figures for fitness equipment by type.

SOURCE: Compiled by the staff with data from "SGMA Recreation Market Report," editions 2001, 2004, and 2005, and "Manufacturers Sales by Category Report," Sporting Goods Manufacturers Association. 2002, 2004, 2006, and 2007 respectively.

The North American commercial and consumer markets accounted for 35 percent of this total. ICON Fitness was the leading producer of fitness equipment worldwide with sales of $852 million in 2005, followed by Nautilus with sales of $617 million. Life Fitness was the third largest producer with sales of $593 million. Technogym and Precor rounded out the top five with sales of $333 million and $311 million, respectively.

FIGURE 94

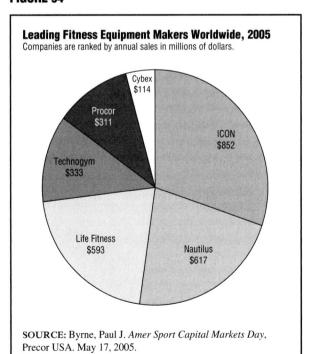

Leading Fitness Equipment Makers Worldwide, 2005

Companies are ranked by annual sales in millions of dollars.

SOURCE: Byrne, Paul J. *Amer Sport Capital Markets Day*, Precor USA. May 17, 2005.

KEY PRODUCERS/MANUFACTURERS

ICON Health & Fitness, Inc. Scott Watterson and Gary Stevenson founded Weslo, Inc. in 1977. In the beginning their company imported kitchenware, tableware, and marble products. The company's entry into the health and fitness industry began with the manufacturing of trampolines. The company then expanded their business into treadmills, exercise bikes, rowing machines, and home gyms. The company leaders decided to focus solely on fitness equipment and sold off its Weslo business in 1988. Welso was then acquired by investment bank Bain Capital, who incorporated ICON Fitness in November of 1994 from several entities: Weslo, ProForm Fitness, Legend Products, Inc., and American Physical Therapy Inc.

ICON Fitness then acquired the HealthRider in 1996 and NordicTrack in 1999. In 2005 ICON produced nearly 4 million treadmills, incline trainers, elliptical machines, stationary bikes, home gyms, weight benches, yoga and Pilates equipment, and other fitness accessories. ICON has facilities located in China, Europe, Canada, and the United States. Its popular brands include ProForm, HealthRider, Reebok, Image, Weider, JumpKing, and NordicTrack. It employed nearly 5,000 people in 2006.

Nautilus Inc. The company's origin began with the creation of Bow Flex of America Inc. in 1986 and found success with its Bowflex home workout machine which was patented in 1987. The company entered into a partnership with Schwinn Cycling and Fitness. Through this partnership the company released the Schwinn Bowflex. The partnership lasted until Schwinn declared bankruptcy in 1993. The company continued to refine its products and had success in the direct marketing industry. By the late 1990s they acquired the Nautilus, Stairmaster, and Schwinn exercise brands. The company manufactures and markets fitness equipment for commercial and home use under the Bowflex, Nautilus, Stairmaster, Schwinn, and Trimline brand names. Its Bowflex equipment is marketed to home users through television ads, the Internet, and direct marketing efforts. They have also moved into the fitness apparel market with its acquisition of Pearl Izumi. The company employs approximately 1,500 people.

Life Fitness. Keene Dimick, Ph.D., a chemist and inventor, developed the Lifecycle exercise bike, the first computerized fitness equipment. Ray Wilson and Augie Nieto purchased the rights to the device for $50,000, and introduced the Lifecycle 2000 exercise bike for commercial use. Life Fitness competes primarily in the premium segment with approximately 80 percent of its sales in the commercial market and the remaining 20 percent in the home market, and employs approximately 600 people.

Technogym. Technogym was founded in Seattle, Washington in 1996. The company manufactures products primarily for gyms and corporate wellness centers. It has equipment in 35,000 fitness centers and 20,000 homes worldwide.

Precor. Precor was founded in 1981 in Woodinville, Washington. Precor was sold to Finnish firm Amer Sports in 2002, which counts sporting goods brands Wilson and Salomon among its holdings. Precor receives two-thirds of its sales from the club/institutional segment. The company has 733 employees.

Cybex. This company was founded in 1947 and is based in Medway, Massachusetts. Cybex offers over 150 different models of equipment to the commercial and consumer markets. Cybex commercial products are sold to health clubs, hotels, pro sports teams, and other commercial users primarily through Cybex direct sales; consumer products are sold through specialty fitness dealers. Internationally Cybex products are sold through distributors in 87 countries across the globe and a direct sales force in the United Kingdom. The company has two manufacturing facilities and a workforce of 560 employees.

MATERIALS & SUPPLY CHAIN LOGISTICS

Fitness equipment is primarily made of aluminum, carbon steel, or some other metal alloy. Such metals are preferred because they are both light and durable. The manufacturing of an exercise bike, for example, begins with the construction of the bike frame from one of these metals. The frame is then powder coated and painted to create a durable, attractive finish. The structural elements require metal, plastic, or rubber and are manufactured by injection molding, roll wrapping, lay-up, or similar processes. The various parts are then attached: drive belts, wheels, cranks, pedal straps, handrails, and flywheels. Epoxy adhesives or metal brackets and bolts may be used for this step. Control panels with liquid crystal displays are also added. Decals indicating the name of the company or brand are then added.

The basic parts of a treadmill are the rollers, the belt, the deck, and the motor to turn the belt. The construction of the rollers is of particular importance for they effect wear on the bearings and belt. High-end equipment has more powerful motors and a more resilient track to provide better cushioning for the runner.

Free weights are made from iron, steel, or hard rubber. The bar is made from iron, kevlar, or a metal alloy that would ensure durability and the product's straightness. In high-end home gyms the materials used can be expensive: laser-cut stainless steel dumbbells, hand-made weight pulleys, and leather upholstered bench press seats. Yoga rooms might feature sandstone coverings, onyx decking, sea grass, radiant heating built into the floor, and expensive cabinetry.

Supply Chain Issues. The competitive fitness equipment market has prompted companies to make various improvements. Companies once used the Internet merely to address issues such as the shipping status or availability of its products. Companies now use the Internet to collaborate on new product designs, and to monitor customer opinions and market trends. Some companies have introduced policies to streamline production efforts and reduce waste, planning out new shipment routes to reduce costs, for example. Supply chain software helps manufacturers manage their inventories and orders more efficiently. Software installations have also smoothed out the overall production process. For example, Life Fitness had to continually run tests to ensure that their products were meeting quality standards. Software installed in July 2007 now alerts production managers through email, Blackberries, and similar devices only if a problem has been discovered.

DISTRIBUTION CHANNEL

Fitness equipment makers market their products through a number of methods, including direct marketing, sporting goods stores, specialty dealers, mail order, and online sites. As stated, the home market has become the largest sector for fitness equipment firms. Health clubs are the next largest category, many of which are independently owned. Because of this many health club owners are bypassing specialty dealers and making purchases and service agreements directly with the manufacturer. The hospitality industry is another market for fitness equipment makers. Many hotels began adding workout facilities in the 1990s as an added amenity for their customers. Gyms and saunas can be an expensive but ultimately lucrative business for hotels. Hotel revenue from health clubs grew 4.07 percent from 1999 to 2005, according to PKF Hospitality Research. Room revenue grew 1.42 percent and revenue from golf fell 1.27 percent over the same period.

There are an estimated 8,500 senior living centers and apartment complexes in the United States, which is a growing market for fitness firms. The medical community has repeatedly stated the importance of exercise in maintaining good health, particularly for seniors. A survey by Vital Health found that nearly two-thirds of those in senior living facilities used their wellness centers (centers that offer many health services in addition to fitness opportunities) at least three times a week.

Other popular markets include country clubs, academic institutions, and hospitals. Some businesses have

entered into partnerships with nearby gyms or installed their own. The idea behind this is that an employee in good health will ultimately cut down on the health care expenses of the company. Some airports have also installed workout facilities.

Sales in Retail Channels. The home market has become the largest sector of the fitness equipment market. According to a July 2004 survey by the National Sporting Goods Association, department stores gained considerable market share in the industry between 1998 and 2003. Their share of treadmill sales, for example, increased from 38 percent to 54.1 percent during this period. Discount stores saw their share of unit sales drop from 13.7 percent to 7.3 percent. Specialty fitness stores lost market share as well, dropping from 6.3 percent to 6.2 percent. Sporting goods stores saw a modest drop in sales as well; their market share dropped from 15.1 percent to 15.0 percent over this period.

Department stores grabbed market share in the multi-purpose home gym market as its unit share increased from 15.6 percent to 24.0 percent from 1998 to 2003. Discount store unit sales dropped from 20.8 percent in 1998 to 7.3 percent over the same period. Specialty fitness stores increased from 4.2 percent to 5.2 percent. Sporting good stores saw their share jump from 15.6 percent in 1998 to 29.4 percent in 2003.

Department stores and sporting good stores saw similar market share gains in other sectors of the fitness equipment market. Sears seemed to have identified the fitness equipment trend early on, they were among the first retailers to stock elliptical trainers. They also marketed their fitness equipment well. Prominent displays featured the higher end equipment; these displays allowed customers to test the equipment as well. Customers were also comfortable making their first fitness equipment purchase in a familiar department store. Specialty retailing was still a developing industry in the 1990s. Shopping at a new specialty store (there were only 18 NordicTrack Fitness at Home stores open by the end of 1992) was intimidating to some customers.

Many fitness products aree sold through infomercials. Infomercials are commercials that might run as long as a regular television show and are devoted to promoting a particular product or service. Marketdata Enterprises, a research firm, estimated that they generated $2.6 billion in sales in 2004; just under half of the products sold (48%) were of the self-improvement category (fitness, diet books, etc.). The Bowflex is a major advertiser in this market. Other leading products have been the Air Climber, the Gazelle Free Style Elite, and the Slendertone Ab Belt. By one estimate one in 60 of these advertisements is believed to turn a profit. Other distribution channels have proven popular for fitness equipment makers. Mail order purchases represents approximately 10 percent of fitness equipment sales. Many manufacturers have online sites through which they sell their products. There are also third-party vendors that sell these products, often at a discount.

KEY USERS

Approximately 70 percent of health club members were between 18 and 54 years of age in 2005. Women outnumber men at health clubs. A 2005 estimate by the International Health, Racquet & Sports Club Association estimated that women represented 57 percent of health club memberships. A major reason for this division may be the growing popularity of women-only fitness clubs.

According to the National Sporting Goods Association (NSGA), millions of Americans participate in fitness programs. In May 2007 the NSGA released a list of the most popular fitness activities. They defined core participants as those who participated in the activity at least 50 times per year, or at least once per week. Data from 2000 also show the types of activities pursued by men and women. It shows that men and women seem to favor resistance training nearly equally. Other activities are favored by one sex over the other. Figure 95 presents some of the participation figures by gender studied by the NSGA in its 2001 study.

Fitness Equipment Purchases. According to a 2001 study by the National Sporting Goods Association 32 percent of fitness equipment purchases were made by those 35–44 years of age. Those 45–64 years of age made approximately 29 percent of purchases followed by those 25–34 years of age with 22 percent of purchases. People under 24 years of age made 8 percent of purchases. Seven percent of purchases were made by seniors. Women made 52 percent of the purchases. Men were more likely to purchase home gyms and weight sets. Women were more likely to purchase elliptical trainers. Households with incomes of at least $100,000 per year represent roughly 14 percent of U.S. households. They purchased 54.4 percent of elliptical trainers, 30.6 percent of treadmills, and 25.2 percent of multi-purpose home gyms in 2001.

ADJACENT MARKETS

There were 29,069 health clubs in the United States in 2006, according to the International Health, Racquet & Sports Club Association. This figure has risen steadily since 1998, when there were 13,000 clubs. Membership in 2006 was 41.3 million; another 23.6 million people visited health clubs as non-members, meaning they pay a fee to the club or institution for the use of their facilities. The Denver, Colorado, region had the greatest percent-

FIGURE 95

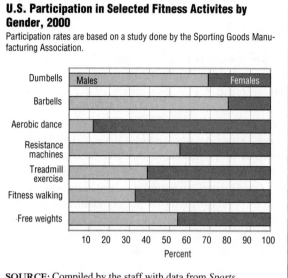

U.S. Participation in Selected Fitness Activites by Gender, 2000

Participation rates are based on a study done by the Sporting Goods Manufacturing Association.

SOURCE: Compiled by the staff with data from *Sports Participation in America, 2001*, Sporting Goods Manufacturers Association. 2002.

age of its population belonging to a health club; a quarter of the population in that metro area (25.1%) belonged to some sort of workout facility. Columbus, Ohio, was a close second with 24.9 percent of its population belonging to a health club. The industry saw revenues of $17.6 billion in 2006. The health club market is fragmented. The leader based on revenues is 24 Hour Fitness, with revenues of $1.1 billion in 2005 and a 7.1 percent market share. One of the trends in the industry is to appeal to target markets. For example, some gyms such as Curves target women. Miracles Fitness and Club 50 are aimed at men and women over 50 years of age.

RESEARCH & DEVELOPMENT

As technology advances new features are added to fitness equipment. New equipment tracks weight lifting efforts and measures progress. New commercial treadmills offer hand and chest sensors to monitor heart rates more accurately. Some vary the incline automatically to simulate running in real world conditions. Some treadmills come with entertainment centers with liquid crystal displays. Newer treadmills include better shock absorbers and minimize noise from the treadmill motor and belt. Some equipment is aimed at seniors and their special needs. For examples, the Resistance Chair Exercise System from Fitter International Inc., is a home or light clinical exercise system designed for the mature adult market. The chair is designed for a variety of rehabilitation applications, from chest and shoulder mobility to balance and lower-body stabilization.

CURRENT TRENDS

New exercise machinery and equipment comes onto the market regularly although only a fraction of those new devices gain a staying power. Trends in the last decade have favored equipment and small machines that can be easily used in the home and reasonably easily moved for storage. Travelers, too, have encouraged a trend in the equipping of hospitality venues.

Hotel guests increasingly demand the same workout experiences on the road as they do at home. Hotels have responded. Hilton, Doubletree, and Embassy Suites hotels in North America added new equipment in 2006 and 2007. Hyatt spent $8 million on fitness equipment in 2006, and offers in-room yoga mats, on-call personal trainers, and workout videos as part of its Stay Fit program. W Hotels started selling Puma athletic gear; it offers guests Puma bikes, route maps, and iPods loaded with running guides.

TARGET MARKETS & SEGMENTATION

Baby Boomers are credited with helping to kick off the fitness movement of the 1970s. Exercising with equipment and weekly trips to the health club may have been initially perceived to be just for the wealthy. To be sure, a gym membership and workout equipment require a considerable expenditure, but in the twenty-first century fitness was increasingly seen as a necessary part of life, not just a luxury. As the nature of work has changed—shifting from an agricultural to a manufacturing economy and then from manufacturing to a service-oriented economy—more U.S. workers are employed in occupations that require little physical activity. Consequently, it has become necessary for many to schedule physical activity into their daily or weekly routines in order to maintain a level of activity sufficient to support health. Health clubs have responded to these changes by designing flexible membership programs and offering a wide range of fitness programs—a something for everyone philosophy.

RELATED ASSOCIATIONS & ORGANIZATIONS

50-Plus Fitness Association, http://www.50plus.org

Aerobics and Fitness Association of America, http://www.afaa.com

International Health, Racquet & Sports Club Association, http://www.ihrsa.org

National Association of Health and Fitness, http://www.physicalfitness.org

Sporting Goods Manufacturers Association, http://www.sgma.com

BIBLIOGRAPHY

"About the Industry." International Health, Racquet & Sports Club Association. Available from <http://cms.ihrsa.org/IHRSA/viewPage.cfm?pageId=149>.

Adams, Mike. "U.S. Weight Loss Market Worth $46.3 Billion in 2004." *Newstarget.com.* 30 March 2005.

"Back Exercises: Machine Lateral Pull-Down." Full Fitness.net. Available from <http://www.fullfitness.net/routines/machine_lateral_pulldown.html>.

Byrne, Paul J. *Amer Sport Capital Markets Day.* Precor USA. 17 May 2005. Available from <http://www.amersports.com/presentations/en_US/Amer_2005_report_cm_2_en.pdf>.

"Core Participants: the Focus of SGMA's New Sports Participation Study." Sporting Goods Manufacturing Association. 2002. Available from <http://www.sgma.com>.

"Health Club Member Attendance." International Health, Racquet & Sports Club Association. Available from <http://cms.ihrsa.org/IHRSA/viewPage.cfm?pageId=615>.

Higgins, Michelle. "Hotels Tone Up Their Treadmills." *New York Times.* 26 November 2006.

Kratzman, Val Arthur. *US Fitness Industry.* Tekes National Technology Agency. Available from <http://www.tekes.fi/julkaisut/US_Fitness.pdf>.

Macmillan, Douglas. "Home Gyms Muscle Up." *Business Week Online.* 4 December 2006.

Malloy, Courtney L., Ph.D. and Harold N. Urman, Ph.D., "Fitness and Recreation Services." *Vital Research* September 2005.

"Manufacturers Sales by Category Report—2007 Edition." Sporting Goods Manufacturing Association. Available from <http://www.sgma.com>.

"SGMA Recreation Market Report—2001." Sporting Goods Manufacturers Association. May 2002, June 2004, June 2005. Available from <http://www.sgma.com>.

Tuhy, Carrie. "Five for the Money." *Money.* August 1985.

Vermillion, Len. "Custom Made Rollers Shape Treadmill Designs." *Product Design & Development.* May 2005.

Football Equipment

—■—

INDUSTRIAL CODES

NAICS: 33–9920 Sporting Goods Manufacturing

SIC: 3949 Sporting and Athletic Goods Manufacturing

NAICS-Based Product Codes: 33–9920912F, 33–99209146, and 33–99209151

PRODUCT OVERVIEW

American football equipment includes helmets, shoulder pads, pants, leg pads, decals, chinstraps, helmet cleaners, helmet polish, cleats, jerseys, and many other products. Although the equipment undergoes constant improvements, the basic items of sporting gear particular to football were established in the late nineteenth century and the first half of the twentieth century.

The predecessors of American football found their way onto college campuses in the 1800s. American colleges each played their own version of football; they were typically violent hybrids of the games of soccer and rugby. Princeton played a violent kicking game known as Ballown, in which players used their fists and feet to drive the ball forward. Harvard played a football game each Monday that was reportedly so violent that the day became known as Bloody Monday. Darmouth played Old Division Football, a soccer game played with hundreds of freshmen and sophomores. The Oneida Football Club played Boston Game, another violent running and kicking game. As these games developed on American soil, England was in the midst of formalizing the rules to its version of football. In 1863 the Football Association of England codified the rules to English football.

The United States began to formalize the rules to its own version of football in the years after the Civil War. Princeton developed the Princeton Rules in 1867. On November 6, 1869, Princeton played a game with nearby Rutgers that is seen as the first intercollegiate football game. That game bore little resemblance to modern football. Each team had 25 players on the field; there were 11 fielders lined up in their own territory as defenders, while 12 so-called bulldogs moved the ball against the opposing team. Each goal was a score. After each score the teams traded sides on the field. A player could advance the ball only by kicking or batting it with the feet, hands, head, or side. Rutgers won the game with a score of six to four.

Yale, Columbia, Princeton, and Rutgers finalized a set of rules for intercollegiate football in October 1873. The first rules were very close to the rules for soccer, although they did dictate at least one aspect that is part of the modern football game: the kickoff. Subsequent rule revisions helped shape the modern game: the touchdown was introduced in 1876 and the forward pass in 1905.

Walter Camp was a major force in moving the game of American football away from its soccer and rugby roots. Walter Camp was general athletic director and head advisory football coach at Yale University from 1888–1914, and chairman of the Yale football committee from 1888–1912. Camp is credited with a number of significant changes to the game: creating the line of scrimmage; shrinking the number of players from 15 to 11; standardizing the scoring; and creating the quarterback, center, and safety positions. Football's popularity began to extend from the Northeastern United States to the rest of the country at the turn of the twentieth century. The game was still notoriously dangerous; many college players suffered serious injuries and 18 players died in 1904 alone. President Theodore Roosevelt called upon college athletic leaders to institute reforms. The Intercollegiate Athletic Association of the United States was formed in December, 1905. The group instituted new rules and banned the sport's most objectionable practices. In 1910 this group changed its name to the National Collegiate Athletic Association (NCAA). By 1912 rules for the modern game of football were in place.

Many historians point to November 12, 1892 as a crucial date in the history of professional football. On this date William Heffelfinger became the first player to be paid to play in a football game. Leagues began to develop in the United States around the turn of the twentieth century as the sport's popularity increased. But the sport had to resolve serious issues: salaries were rising too quickly, players followed offers from team to team, and college players were allowed to play on professional teams. To resolve this, teams formed the American Professional Football League in 1920. The organization would change its name to the National Football League (NFL) in 1922. The number of franchises increased to 18 in 1924 from the original 10 in 1920. Football's popularity grew in the post-World War II period with athletes such as Johnny Unitas and Bobby Layne helping the sport gain even greater prominence.

The increasing popularity of football on television accompanied by increasing ticket sales and ticket prices in the post-war period created steady growth for both professional and collegiate football. As more money filtered into football programs, more money became available for improved equipment for players, training equipment, and playing fields. In 1960 Lamar Hunt challenged the dominance of the NFL when he created the American Football League (AFL). This development helped create the Super Bowl, the championship game played between the winners of each league (now each division). The first Super Bowl game was played in 1967. The two leagues merged in 1970, shortly after Joe Namath led the New York Jets to the AFL's first Super Bowl victory over the NFL. The merged NFL continued to grow in both revenues and number of teams, reaching 32 teams in 2007.

Football thrived in some places outside the United States. Canada, for example, has a rich and longstanding football tradition of its own. The Grey Cup, the championship trophy of the modern professional Canadian Football League (CFL) has been awarded since 1909. The modern Canadian game is played with slightly different rules than the U.S. game. Differences include twelve players, as opposed to the U.S. game's eleven;

longer end-zones; three downs instead of four; and the rouge, a single point awarded for kicking a ball out of the opponents end zone. Despite the slightly different rules, Canadian football is played with the same equipment as the American game, and indeed many of the players and coaches in the CFL have come from the United States.

Other professional leagues formed in North America since the NFL-AFL merger include the World Football League (WFL) of the 1970s, two international leagues sponsored by the NFL, the short-lived XFL, and at least two professional Arena Football Leagues in which football is played indoors on a hockey-rink sized field.

The popularity of college football also grew rapidly with the help of television in the post-war period. The collegiate game is played by more than six hundred schools. Lower division programs participate in their own playoff system, while schools in the NCAA's top division participate in post-season bowl games and the Bowl Championship Series that attempts to match the two top-ranked teams from the regular season. Thirty-two bowl games were scheduled for Division I schools in 2007. These televised bowl games generate millions of dollars for the participating schools and football programs and provide a prime marketing opportunity for companies wishing to display the newest and best football equipment for the millions of high-school and youth football players and coaches who watch them. In 1994 the University of Michigan's football team entered new territory for sports equipment marketing when Nike signed the Michigan athletic department to an equipment deal that paid the University of Michigan more than $1 million per year. In 2007 Michigan switched from Nike to adidas for a new equipment contract worth $60 million over eight years, making them, again, the most highly paid college program.

MARKET

In the early days of football, players wore almost no protective gear. Some players grew their hair long or tied bandanas around their heads to provide extra cushioning. Some early leather helmets and protective headgear appeared at the turn of the twentieth century. College players were required to wear helmets in 1939; the National Football League made the same demand of its players in 1943. The John T. Riddell Company patented the first plastic helmet in 1939. These first helmets tended to break. Riddell would have difficulty obtaining plastic during the war years.

By the late 1940s Riddell was able to refine its manufacturing techniques. The company changed its helmet design, reshaping the flat top to a teardrop shape. This new design helped reduce concussions and other injuries. A player's head could move side to side; in the old design the top of the player's head had to absorb most of the impact. Color had been used on the leather headgear of the 1920s and 1930s and was an important modification—color helped quarterbacks distinguish receivers far down the field. Color was increasingly becoming important to help distinguish franchises from each other. The Los Angeles Rams were the first team to design a graphic for their helmets. In 1949 colors could be baked directly on to the helmet in the manufacturing process.

Manufacturers were producing protective nose gear as early as 1927. These devices were made of leather and looked somewhat like a faceplate. Vern McMillan invented the first facemask. It was a rubber-covered wire mask on a leather helmet and was used in the 1930s. Riddell has noted how the development of the plastic helmet helped in the improvement of the facemask. The hole drilled for the bolt holding the mask onto the helmet would not expand in rigid plastic as it would if it was drilled through leather. The sides of the plastic helmet would not collapse and press the nut into the wearer's face. The facemask was modified several times in the 1930s and 1940s, using various combinations of leather, wire, and plastic.

Football Market. Football is the most watched sport in the United States. The Super Bowl is the most watched event on television annually; an estimated 90.7 million viewers tuned in to watch in 2006. For its 2005 fiscal year the NFL reported revenues of $5.8 billion, approximately half of which were from broadcast rights. Ticket sales generated $1.4 billion, the balance ($1.1 billion) were proceeds from local broadcasting deals, team sponsorships, and ancillary sources such as the licensing of NFL footage for use in television and film productions. There has always been a good natured tug of war between the American version of football and the version of football played outside the United States, which is what Americans would consider to be either rugby or soccer. The United States tried to export its American version of football with the creation of the WFL in 1990. The goal of the league was to hold football games in Europe and Asia. These were American style football games with slightly altered rules to make the game more enjoyable to soccer and rugby fans. A typical WFL game was attended by 20,000 people, less than one-third of the 68,000 people attending an average NFL game. The WFL was generally ignored in the United States and it took a brief hiatus before returning as the NFL Europe in 1997. The league changed names again to NFL Europa in 2006 before ceasing operations in June 2007.

Sales and Market Share Data. Globally, adidas has approximately 35 percent of the global football market, including boots, balls, shirts, and related equipment. Nike

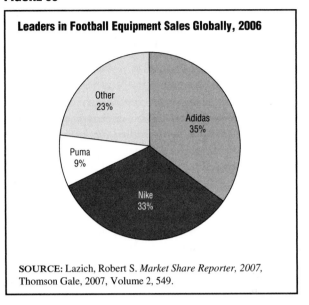

FIGURE 96

Leaders in Football Equipment Sales Globally, 2006

Other 23%
Adidas 35%
Puma 9%
Nike 33%

SOURCE: Lazich, Robert S. *Market Share Reporter, 2007,* Thomson Gale, 2007, Volume 2, 549.

is a close second in the market with a 33 percent market share. Puma is third with a 9 percent market share. These global figures represent football equipment used in American style football, soccer, and rugby.

Tracking a relatively small market such as football equipment in the United States is difficult. Data on football equipment are grouped in with larger data sets by most of the entities collecting such data, including the U.S. Census Bureau. Independent market tracking firms publish in-depth studies on such markets only periodically, which makes tracking of trends difficult. Nonetheless, a sense for the market can be gained by a look at the reports of several organizations, each providing a snapshot of the industry at a particular moment in time.

The Sporting Goods Manufacturers Association estimated that U.S. manufacturers shipped $474 million in football equipment in 2006. The NPD Group noted retail sales of footballs at chain sporting goods stores and mass merchandisers were $5.8 million for the first quarter of 1999. A total of $410,000 in lineman receiver gloves, $171,000 in protective equipment, $26,000 in belt/flag sets, $7,000 in helmets, and $361,000 in other related equipment were sold during this same year. These retail figures refer to consumer retail sales and omit sales made through catalog orders or online. It also omits the large school, institutional, and professional channels.

Some companies have carved out large market shares in various niche categories. Wilson Team Sports reported having 80 percent of the market for footballs as of February 2006. Riddell's contract with the NFL gave it 85 percent of the NFL helmet market according to data published in *Market Share Reporter 2007*. Riddell also holds a dominant share of the market for college and high school helmets. In 2002, according to one estimate, the market for new helmets and reconditioning services was approximately $89 million annually and was expected to reach $111 million by 2010.

KEY PRODUCERS/MANUFACTURERS

Adidas. Adolph Dassler began his shoemaking business in the family home in 1920. The company opened its first shop in 1927. Dassler manufactured special shoes for soccer and track and field, introducing studs and spikes for the first time. Adidas' shoes appeared on athletes at the 1928 Olympic Games in Amsterdam. By 1937 the company was manufacturing 30 different types of shoes for 11 different sports. The Nazis seized the company during World War II but in the post-war period Dassler rebuilt the company and named it adidas. The adidas brand was worn by numerous athletes at Soccer World Cup matches and during the Olympic Games of the 1950s and 1960s.

Having built a successful line of soccer and track shoes the company expanded into other sporting gear. Its popular tracksuits were marketed as early as 1964. By the 1970s the company lost ground in the North American apparel market to Nike. Adidas retained its popularity outside the United States, however. It continued to offer innovations such as the Copia Mundial, the world's best selling soccer shoe in 1979 and the Torsion sole system in 1988. It launched an equipment line in 1991. Adidas has become a major sponsor of soccer and football tournaments, such as the 2003 Women's Football World Cup in the United States and the European Football Championships of 2004 in Portugal. Adidas had revenues of $13.6 billion in 2006.

Nike. Nike, Inc. is based near Beaverton, Oregon, and is the world's leading designer, marketer, and distributor of athletic footwear, apparel, equipment, and accessories. The company began through the efforts of two men: Bill Bowerman, the track coach for the University of Oregon, and Phil Knight, an accounting major at the university and a runner on Bowerman's team. In 1962 Knight approached the Japanese company Onitsuka Tiger about importing high-end athletic shoes into the United States. Knight and Bowerman formed the Blue Ribbon Sports company in 1964. Blue Ribbon Sports began to distribute the shoes from Onitsuka Tiger. By the mid-1960s the Japanese shoes began to challenge Germany's dominance of the U.S. sports shoe market.

The company opened its first retail outlet and became the first distributor of Onitsuka running shoes in the United States. The shoe had a special cushioned wedge heel designed by Bowerman. According to Nike company historians, the name Nike, used first on one line of the company's shoes, was chosen because of its earlier usage, as the name of a Greek Goddess of victory. Graphic artist Carolyn Davidson designed the company's now famous Swoosh logo for a small, one-time fee. In 1972 tennis star, Ilie Nastase, became the first athlete to professionally endorse the new line of Nike shoes. In 1978 Blue Ribbon Sports changed its name to Nike, Inc.

Nike subsidiaries include sporting goods maker Converse Inc.; Nike Bauer Hockey Inc.; Cole Haan, which designs, markets, and distributes fine dress and casual shoes and accessories; Hurley International LLC, which designs, markets, and distributes action sports and youth lifestyle footwear, apparel, and accessories; and Exeter Brands Group LLC, which designs and markets athletic footwear and apparel for the value retail channel. Nike has 28,000 employees and earned revenues of $14.9 billion for the fiscal year ended March 2006.

Riddell Sports Inc. John T. Riddell began a career as a teacher, head football coach, and athletic director at Evanston Township High School in Evanston, Illinois. Riddell soon recognized a problem with the cleats on the football shoes. In rain or otherwise inclement weather, the team's football shoes needed to be refitted with longer cleats. The shoes were not always ready in time. Riddell had the idea for removable cleats and got help from the J.P. Smith Shoe Company to produce them. In 1927 Riddell quit his coaching job and opened the Riddell Corporation two years later.

Riddell manufactured the first molded basketball shoe and the first soft-spike basketball shoe. In 1939, Riddell developed the first plastic football helmet. The first helmets tended to break during games. However, in the post war period the company perfected its manufacturing processes, making the helmet more comfortable and durable. The company continued to develop new products for football in the 1930s and 1940s. During this period Riddell produced the first chin straps, the first plastic facemasks, and the first low cut football shoes. Riddell died in June 1945.

The company continued producing innovative gear, manufacturing the first clear and tubular bar guards, the first nose protector, and the first microfit and air cushioned helmets in the 1950s and 1960s. In the 1980s the company began to produce high school and youth shoulder pads. In 1988 it acquired the Power Athletic Company, a manufacturer of professional shoulder pads. In 1989 Riddell became the official supplier of helmets and shoulder pads to the NFL and in 1990 to the WFL. In 2001 Riddell introduced a new football helmet called the RevolutionT, a new technology designed to reduce the risk of concussion. Riddell also launched the Sideline Response System, a new technology that combines a real-time, on-field head impact telemetry system (HIT

System), team management software, and cognitive testing to provide a new standard of care for the athlete.

In 2006 parent company Riddell Bell merged with Easton Sports, creating the leading company in head protection equipment and a market leader in baseball, football, softball, hockey, auto racing, and motorcycle sports. In addition to its supply agreements with the NFL and the WFL the company supplies approximately half the helmets used in high school and college football leagues. Reconditioning of football helmets is another of Riddell's businesses. Reconditioning involves cleaning, buffing or repainting helmets, and replacing interior straps or pads. Reconditioned helmets are certified as conforming to the National Operating Committee on Standards for Athletic Equipment (NOCSAE) standards.

Under Armour. In 1996 former Maryland football player Kevin Plank designed T-shirts with special microfibers that would keep an athlete cool and dry during a workout. He drew his inspiration from the tight fitting compression shirts that players wore during practices. He tested his new T-shirts on fellow players and then began to market them. The company's T-shirts, hats, and related apparel became known as performance apparel. The new clothing was soon very popular among athletes. In 1997 twelve college teams and ten National Football League teams were wearing Under Armour clothing. The company began advertising and promoting its products through their use in football-themed films such as *Any Given Sunday* and *The Replacements.* In 2000 it began to advertise nationally. By 2007 the company was expected to see sales of $600 million. Under Armour has continued to expand into new markets and at the end of the first decade of the twenty-first century was involved in selling football cleats, headgear, and sporting apparel for women. In 2007 the company had 979 employees and was based in Baltimore, Maryland.

Schutt Sports. Bill Schutt began manufacturing basketball goals and dry line markers out of his hardware store in Litchfield, Illinois. In 1935 he manufactured the company's first faceguard. In 1973 Schutt introduced football faceguards in eight colors and powder coated basketball rims. In 1987 Schutt began marketing the Air football helmet. It also marketed helmets to the basketball market the same year. In 2001 the company marketed the Air Advantage helmet. Schutt introduced the DNA football helmet, featuring SKYDEX padding in 2001. Three inches of this new type of padding replaces 12 to 18 inches of traditional padding. In 2004 Schutt partnered with Concussion Sentinel to offer $20 million of free cognitive testing equipment to high schools and colleges throughout the United States.

Niche Market Players. Within the small niche segments of this market are companies whose names are less familiar than Nike or adidas. Arthron Inc., for example, produces special hip and shoulder padding for athletes. Rogers Athletic manufactures tackling dummies and other training equipment. SpeedCity is a producer of football and sports training equipment and videos to help train athletes to improve 40-yard dash times, lateral quickness, and flexibility. Porta Phone produces wireless headsets used by coaches and their staffs. Bike Athletic manufactures cups, shoulder pads, rib protectors, and football pant inserts.

MATERIALS & SUPPLY CHAIN LOGISTICS

The earliest footballs and helmets were made from simple materials. In the Middle Ages, pig bladders were used as makeshift balls. The bladders were easily acquired from local livestock and were suitable as balls because they were basically round and could be easily inflated and sealed. They were later covered in animal skins or leather and the term pig skins is used to this day when referring to footballs. Producing balls from pig bladders meant that balls could vary in size and durability based on the animal. Rubber balls would not be available until the nineteenth century. Charles Goodyear patented the process for vulcanizing rubber in 1836. In 1855 Goodyear began to produce the first rubber balls that were durable and of a consistent size for use in soccer and football.

The first helmets were also made of leather. Riddell introduced the first plastic helmets that offered players greater head protection but were somewhat unreliable and prone to breaking. As Riddell refined its production techniques the helmets became more durable. Manufacturers improved their designs as well as their materials. Advances in the materials sciences have had an impact on sporting equipment of all types.

How a Football is Made. A regulation American football is 11 inches long and 22 inches in circumference at the center. The ball is inflated to an air pressure of 12.5 to 13.5 pounds per square inch (psi) and weighs 14 to 15 ounces. The football is made from leather panels that are tanned to a natural brown color. The panels are cut into shapes and then sent through a skiving machine to shape the panels into a predetermined thickness and weight. The panels are manufactured in an inside-out manner. One of these lacing panels receives an additional perforation and reinforcements in its center, to hold the inflation valve. A synthetic lining is sewn to each panel. The lining prevents the panel from stretching or growing out of shape during use. Holes are then punched into the panels for use in sewing them together with a hot-wax lock stitch machine. The footballs are then turned right side out. A rubber or plastic bladder is inserted into the football. Finally, quality checks are done and the balls are ready for a manufacturer's label and number.

How a Football Helmet is Made. The manufacturing of a helmet begins with the construction of the shell, made of polycarbonate. Pellets of polycarbonate are sent through an injection-molding machine that heats and molds the substance into the form of a helmet. A drill creates 14 to 15 holes in the mold. Liners are cut and inserted to provide cushioning and support. The lining is attached with airtight seals. The jaw pads, face guards, and chin straps are then attached and the helmet is painted. The helmet then goes through a rigorous quality control testing phase to ensure that it fully conforms to standards established by the National Operating Committee on Standards for Athletic Equipment.

The facemask, which is usually made of plastic or metal bars, attaches to the front of the helmet. There are two types of facemasks: the open cage and the closed cage. Quarterbacks, running backs, wide receivers, and defensive backfield men usually prefer the open cage because the lack of a vertical bar above the nose enables better visibility. The closed cage is usually the choice of linesmen because the two, three, or four horizontal bars help keep other players fingers and hands out of their eyes. In the 1970s vinyl coating was layered onto the bars to protect against chipping and abrasions. Soon, colors were added to the facemasks as another way to distinguish players and teams.

DISTRIBUTION CHANNEL

The market for football equipment can be divided into two categories: the school and institutional channel, by far the larger; and the retail channel. Schools at all levels—grade schools, high schools, and schools of higher education—are potential customers for football equipment.

Most team equipment is purchased through the school and institutional channel. While the institution is the initial purchaser of the equipment it may charge participants in its football program for the equipment used or insist they purchase the equipment outright. While many schools have faced tightening budgets during the first decade of the twenty-first century, they are unlikely to cut deeply into their football programs as a means of savings. These programs tend to be popular and are often large generators of revenue. Equipment manufacturers have sales teams that sell directly to these markets. Riddell, for example, has a sales force of 186 individuals who sell helmets and other football gear to high schools, colleges, and the many youth and professional leagues across the United States. Their staff includes former athletes and coaches who are familiar with Riddell products.

Dixie Sporting Goods has 50 sales professionals in the Southeastern United States to market team uniforms and sports equipment.

The retail market for football equipment is best understood by looking at the sporting goods market. Sporting goods are now sold through numerous retail channels. This has made the market fragmented and very competitive. Some sporting goods stores such as Dick's Sporting Goods and Sports Authority began as regional retailers and then began expanding into other markets in the 1970s. Small independent retailers with sports specific merchandise such as Bass Pro Shops also developed during this period. In the 1990s the discount store industry quickly developed. Wal-Mart and Target were the leading retailers of sporting goods equipment in the early 2000s. According to *Sporting Goods Business*, Wal-Mart sold $8 billion of sporting goods equipment in 2006. Target was second with sales of $4.5 billion. Dick's Sporting Goods, the largest U.S. sporting goods chain, sold $3.1 billion in equipment in 2006. Several trade journals have noted that large discount stores sometimes hesitate to enter into agreements with small manufacturers for various reasons. The company is seen as too new and unproven, or is seen as being unable to manufacture enough goods to fulfill orders.

The Internet has transformed the sporting goods market. The major discount and sporting goods stores have all established an online presence. Some companies such as Moosejaw.com have flourished on the Internet without the benefit of a traditional store. The catalog business for sporting goods also continues to flourish. The Sport Supply Group mails 3 million catalogs and other advertising products to potential clients. It is the largest manufacturer, marketer, and distributor of sporting goods products directly to the institutional and team sports market. Manufacturers have also established retail and online sites to compete in the sporting goods market. Nike has approximately 220 stores and had $1 billion in retail sales. Adidas earned $300 million through its 100 stores. Reebok operated 160 stores and earned $295 million.

KEY USERS

Football remains a popular sport for young men. Approximately 1 million high school boys played football during the 2004/2005 school year, according to the National Federation of State High School Associations. The next most popular sport was basketball, although it had half the participants—545,497 players compared to 1,071,167 for football.

According to the National Collegiate Athletic Association, there were approximately 60,000 football players at the college level in the 2004/2005 school year. These figures remain fairly constant from season to season.

Approximately one million high school boys played football in the 1999/2000 school year and 56,528 men played college football in the 1998/1999 school year. The slight variation can be attributed to changes in the number of high school or college age men each year as well as the state of the football programs at any particular school.

Approximately 2.1 million boys play football frequently according to the National Sporting Goods Association. Its estimate includes school players and those on a park or district team. Football enjoys consistent levels of participation at the high school and college level. This figure is up 4 percent from 2000. However, the number of men in the United States who have ever played the sport has declined. Indeed, participation in many team sports has been declining since the 1990s. According to the Sporting Goods Manufacturers Association, there were 18.3 million people at least six years of age who played football at least once in 2000. By 2004, 16.4 million people had played football. Participation fell more than 10 percent from 1987 to 2004. Touch football participation fell 36 percent over this same period. Many team sports were seeing similar declines in participation in the United States. The number of baseball players fell from 15.1 million in 1987 to 9.7 million in 2004, a decline of nearly 36 percent. The number of volleyball players fell from 36 million to 22.2 million over the same period, a drop of 38 percent. The number of basketball players fell from 35.7 million to 34.2 million players, a decline of 4 percent.

ADJACENT MARKETS

Football equipment is often tracked in the larger context of team sports equipment manufacturing. The market for team sporting equipment other than football is a primary adjacent market to the market for football equipment. Despite football's popularity, baseball and softball equipment represents a much larger share of the team sports equipment market. According to industry tracker SportscanInfo, baseball and softball equipment combined had 46 percent of team sports sales in 2006. Basketball represented 21 percent of team equipment sales. Football had 13 percent of sales. Baseball and softball's commanding share of the team sports market is understandable when one considers the equipment necessary for the game: bats, gloves, helmets, leg guards, chest protectors, batter's guards, catcher's guards, and related equipment. Also, there are more basketball and baseball teams. There were 14,760 high school football teams in the United States in the 2004/2005 season, according to the National Federation of High School Associations. But there were 17,482 basketball teams and 15,161 baseball teams. Similarly, there were 617 football teams at the collegiate level, but 864 baseball and 994 basketball teams. Soccer had 10

FIGURE 97

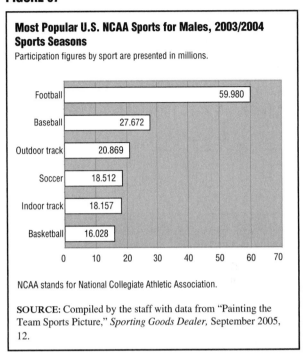

Most Popular U.S. NCAA Sports for Males, 2003/2004 Sports Seasons

Participation figures by sport are presented in millions.

Sport	Participation
Football	59.980
Baseball	27.672
Outdoor track	20.869
Soccer	18.512
Indoor track	18.157
Basketball	16.028

NCAA stands for National Collegiate Athletic Association.

SOURCE: Compiled by the staff with data from "Painting the Team Sports Picture," *Sporting Goods Dealer*, September 2005, 12.

percent of team sales, hockey 4 percent, and lacrosse and volleyball had 3 percent each.

In terms of a spectator sport, football is considered by many to have replaced baseball as the national pastime of the United States. There have been countless theories to explain how this might have happened, however the sport's accessibility to television audiences is most often credited for helping it achieve its widespread success. Football broadcasts feature brief interludes of violent game play separated by brief shots of short-skirted cheerleaders and convenient breaks for advertisers—an ideal combination for television popularity. A baseball season has 162 games, which can drag on even the most devoted fans. The seventeen-week, sixteen-game season of NFL football builds suspense into the outcome of each game. It is unclear when football became America's preferred game, but it happened far earlier than might be imagined. A 1966 *Time* magazine article titled "The National Pastime," noted that 31,000 fans paid to watch a Baltimore Colts scrimmage while only 23,000 fans showed up to watch a baseball game between the Baltimore Orioles and the Detroit Tigers. Four decades later fans regularly lined up to watch their favorite teams practice in the weeks before the season began.

Average attendance at an NFL game was 67,738 in 2006. Total paid attendance in 2005 was 21.79 million, the fourth year of record setting attendance. Major League Baseball (MLB) was second in average attendance, with

31,404 spectators per game in 2006. The popularity of football may seem surprising based on the cost of the tickets. The average cost for a family of four to attend an NFL game was $321.62 in 2004, an 80 percent increase over the cost a decade earlier. That same family had to spend only $164.43 for an MLB game and $263.44 for an National Basketball Association (NBA) game.

Basketball is often the second most popular sport for young men to play in school. Like football, it is fast-paced and exciting. It also has star athletes that attract young men to the game. The game can also be played more easily that football, can be played indoors and outdoors, and requires fewer people and less equipment.

Football Licensing. Football is big business. Plunkett Research estimated that the National Football League generated $5.8 billion in revenues in 2006. Major League Baseball earned $5.2 billion, the National Basketball Association earned $3.1 billion, and the National Hockey League earned $2.2 billion. Much of the NFL's income comes from its licensing and sponsorship deals. Sponsors pay millions for the right to use NFL and Super Bowl logos in their promotions. The NFL has more than 100 licensees, including Electronic Arts, Northwest Airlines, Pottery Barn, Reebok, VF Corp., and Wilson. Billions in fees are earned for the right of the major networks, DirecTV, and ESPN to broadcast the NFL's games.

The National Football League had approximately 39 percent of licensed sports apparel sales in 2006, ahead of Major League Baseball, which took 26 percent of sales and college teams, which collectively took 25 percent of the market. Video games are a new and lucrative market for football teams. *Madden NFL 07* was the best-selling video game in 2006, with sales of 5 million copies. Approximately 2.5 million copies of the *NCAA Football 07* were sold in 2006. By one estimate, a school gets 10 to 25 percent of its licensing revenue from video games. The University of Oregon saw its video game royalties increase from $15,000 in 2000 to $95,000 in 2006. Ohio State and Florida earned revenues of $103,500 from *NCAA Football 2005–06*.

RESEARCH & DEVELOPMENT
The development of the compression apparel market has been one of the major developments in the football equipment and apparel industry. The demand for football gear that keeps players cool and dry has moved into helmets, shoulder pads, and other protective equipment. There is a continual effort to make these products more durable and lightweight. In 2005 Schutt introduced the Typhoon shoulder pad, featuring climate control technology which helps the player better manage body temperature in all types of weather conditions. Russell Athletic has expanded

its Dri Power technology from its sports apparel into its casual apparel lines. Russell also introduced its Xtreme Compression football jersey in the autumn of 2006. The jersey keeps the body cool and dry and because of its compression technology makes a player difficult to grab.

Manufacturers are producing padding that better absorbs and distributes blows to an athlete's head and body. An estimated 1.6 million to 3.8 million amateur and professional athletes suffer a concussion each year, according to the Center for Disease Control and Prevention. The design of the Schutt DNA helmet helps reduce the force of helmet to helmet contact. It was released in 2003. Riddell's Revolution, released in 2002, has a teardrop shape and padding that can be custom fit to the athlete's head. Riddell marketed the $170 helmet as capable of reducing concussions. A study of high school football players in January 2006 showed 5.4 percent of Revolution wearers suffered concussions, as opposed to 7.6 percent among wearers of a more traditionally designed helmet.

CURRENT TRENDS
Arena football is growing in popularity. Viewership of the Arena Football League (AFL) is increasing in part because of a league contract with the NBC that broadcasts the games on Sunday afternoon. From 2002 to 2006, the AFL's total viewership increased from 12 million viewers to 65 million, its team values went up from $12 million to $20 million and its average per game attendance increased from 9,957 to 12,378. When combined with arena football2, the AFL's minor league, Arena Football played in 49 markets in 2007. The game is also gaining fans because of its fast pace and high scoring. It seems to attract young male fans, a highly prized demographic for marketers who sponsor games. Sponsors include Nike, Spalding, Schutt, Upper Deck, and EA Sports (which released an AFL video game to join its NFL video games).

Fantasy football leagues have become increasingly popular. In a Fantasy Football game, a participant becomes the owner of a fantasy team arranged into a league. Owners may acquire a player through a draft or a trade. The owner then scores points based on those players' statistical performances on the field. Fantasy football leagues are not a new phenomenon borne out of the Internet. According to sports lore, the first such league was active in 1963.

Wilfred Bill Winkenbach, part owner of the Oakland Raiders and sports writer Scotty Stirling created the first fantasy football league, known as the Greater Oakland Professional Pigskin Prognosticators League. There were thought to be 15.6 million fantasy football players in 2006, compared to 11.9 million fantasy baseball players and 7.1 million fantasy basketball players. Entering into a league might cost from several hundred dollars to more than one thousand dollars. One estimate suggested that $1 billion was spent on fantasy football leagues in 2006. Such leagues are believed to have boosted the number of football viewers. More than a few NFL players have gone on record to state their dislike of such leagues, since they lead viewers to root for a particular player's statistics instead of the overall team. However, the NFL has embraced the phenomenon and uses many of its own top stars in marketing campaigns to promote its own online fantasy football game at NFL.com.

TARGET MARKETS & SEGMENTATION

The football equipment market is segmented by age range. Grade school aged players make up one segment, high school aged players another. Collegiate level athletes are another large segment, and the large, professional player segment influences all of them. The media exposure received by professional football provides a stage upon which the brand names worn by the players are on view by a large audience. That audience has influence on the equipment used by all the other market segments.

The football equipment market was on solid ground in 2006, according to industry analysts. Average selling prices increased more quickly than in any other team sport. Some of the increase has come from premium lines of equipment being released—sales of footballs costing more than $25 increased three times faster than the sales of footballs costing less than $25. Manufacturers continue to bring technological advances to mouth guards, padding, helmets, and apparel. These improvements not only make a player safer out on the field but more comfortable as well.

The overall football industry remains strong, as well. The National Football League generates billions in revenues and schools spend billions in support of their teams as well. According to an Associated Press poll, Ohio spent $29.6 million on its football team during the 2004/2005 season, the most of any team. It is understandable why so many colleges spend millions on their football teams—football is popular and a major revenue generator. But some may question the fairness of the spending. Rutgers doubled spending on its football program during the 2004/2005 season as it eliminated the men's tennis, swimming, diving, crew, and fencing teams.

RELATED ASSOCIATIONS & ORGANIZATIONS

International Federation of American Football, http://www.ifaf.org

National Federation of State High School Associations, http://www.nfhs.org

National Football League, http://www.nfl.org

National Operating Committee on Standards for Athletic Equipment, http://www.nocsae.org

BIBLIOGRAPHY

Bachman, Rachel. "College Football Video Game Pads Schools Coffers." *Newhouse News Service.* 22 July 2007.

Finlay, Ryan. "Headway Being Made in Concussion Diagnosis." *Arizona Daily Star.* 9 April 2007.

Frick, Bob. "Under Armour: Profits Are No Sweat." *Kiplinger's Stock Watch.* 20 June 2007.

Kerrigan, Andy. "Scoring at Retail." *Sporting Goods Business.* May 2007.

"National Pastime." *Time.* 16 September 1966.

"Professional Football Continues to be the Nation's Favor." Harris Interactive. Available from <http://www.harrisinteractive.com>.

"Record Number of High School Students Participate in Sports." *Research Alert.* 17 November 2006.

"The Retail Matrix." *Sporting Goods Business.* May 2007.

Rovell, Darren. "Michigan Goes Addidas as Nike Runs Out on Big Blue." CNBC. Available from <http://www.cnbc.com/id/19710056>.

"Rutgers Invests in Football but Cuts Other Sports." ESPN. 18 November 2006. Available from <http://sports.espn.go.com>.

"Rutgers: The Birthplace of Intercollegiate Football." Scarlet Knights. 27 July 2007. Available from <http://www.scarletknights.com>.

Ryan, Thomas J. "Tackling the Market: Football Equipment Vendors Make a Play With Technology and Innovation," *Sporting Goods Business.* August 2005.

Schiller, Gail. "And People Say American Idol is a Hit." *Hollywood Reporter.* 2 February 2007.

"Table 1233: Participation in Selected Sports Activities: 2004." *Statistical Abstract of the United States: 2007,* 127th ed. U.S. Department of Commerce, Bureau of the Census. December 2006.

Terry, Robert J. "As Under Armour Gains Market Share Would Be Investors Root for IPO." *Baltimore Business Journal.* 3 December 2004.

"U.S. Sports Industry Overview." Plunkett Research. Available from <http://www.plunkettresearch.com>.

Waltzer, Emily. "Charged Up." *Sporting Goods Business.* May 2006.

SEE ALSO *Athletic Shoes, Men's Apparel*

Frozen Foods

———————■———————

INDUSTRIAL CODES

NAICS: 31–1411 Frozen Fruit, Juice, and Vegetables, 31–1412 Frozen Specialties, 31–1712 Fresh and Frozen Seafood Processing, 31–1813 Frozen Cakes, Pies, and Other Pastries, 31–1822 Flour Mixes and Dough Manufacturing from Purchased Flour

SIC: 2037 Frozen Fruits and Vegetables, 2038 Frozen Specialties, not elsewhere classified, 2053 Frozen Bakery Products, Except Bread, 2092 Fresh or Frozen Prepared Fish, 2097 Manufactured Ice

NAICS-Based Product Codes: 31–141111, 31–14114, 31–1411W, 31–13121, 31–14124, 31–1412W, 31–17122, 31–17123, 31–181301, 31–181302, 31–181303, 31–181303, 31–18130Y, 31–18220261, and 31–18220271

PRODUCT OVERVIEW

Food preservation is as old as humanity, the practice motivated by a surplus of food in the clement seasons and shortages in winter, and also by successful hunting which will produce meat in excess of the amount that can be consumed immediately after the hunt. Preservation by drying, salting, and smoking are ancient practices. Underground storage of vegetables, fruits, and meats is thousands of years old as well. Nine or ten feet underground the temperature is a constant 59 degrees Fahrenheit, not quite refrigerator temperature (35–38 degrees), but still cool. Ancient man knew to harvest ice in winter and to store it under ground in blocks to bring down the temperature

further. Chinese use of iced cellars has been documented back to 1000 BC. If sufficient amounts of ice were stored in winter, some of it will have survived into the warm season to keep stored food cool enough to preserve it. Humanity living in arctic regions froze fish and preserved it in that manner. This situation changed for the first time in the nineteenth century when, stimulated by Napoleon's wars, the French invented canning in glass and the British canning in metal. Preservation of food by artificial freezing did not take hold until the 1930s. The art had to be and was discovered as humanity came to an ever better understanding of the behavior of gases. The knowledge discovered could be put to use only because means of using fossil (and later nuclear) fuels were developed into our modern forms of energy.

This essay deals with freezing as a form of food preservation and with the products of that preservation process—frozen fruits, vegetables, seafood, meats, and various prepared food in finished or semi-finished forms (complete dinners being an example of the finished category and frozen doughs an example of the semi-finished category). Ice cream and its relatives (ices and sherbets) are excluded. In this product category the frozen character of the food is its central and, indeed, defining character whereas, in frozen foods, the freezing is a mechanism of preservation only. Preservation, after all, is most successful when it leaves no traces at all. Thus frozen foods are much to be preferred to foods preserved by salting or spices; these preservatives can never be completely removed. Freezing is effective as preservation because very cold temperatures inhibit or completely stop organic processes that under room temperatures would cause food to spoil. Also excluded from this essay are refrigerated products that do not require freezing; most dairy products belong in this category. Traditionally most dairy products, excluding only cheeses, were consumed shortly after milking the cow or goat had been accomplished. Refrigeration extends the life of these products significantly at much lower temperatures than freezing.

Refrigeration Basics. Refrigeration exploits the physical fact that matter responds to changes in temperature by changing phase. We are most familiar with changes of this sort in water. It crystallizes and solidifies in cooler conditions and vaporizes in the presence of heat. When water loses sufficient heat it freezes; when sufficient heat is present, it vaporizes. Our bodies sweat in order to cool off. They do the job by producing water on the surface of the skin; unless humidity in the air is 100 percent, the sweat will evaporate. Evaporation consumes energy and thus cools our skin. When we hold a glass of iced tea in humid weather, we discover that moisture will form on the outside of the glass. Water vapor condenses on the cold surface of the glass. This means that the glass itself has become warmer by taking energy from the vaporized water; and the moisture in the air has liquefied.

The science underlying refrigeration arose in the seventeenth century as part of the development of thermodynamics (heat-power) in which the behavior of gases was intensively studied and relationships between temperature, pressure, and volume were observed. When under pressure gases take up less room and increase in temperature. Most compressed gases absorb heat as they are decompressed by means of a valve at room temperature.

The inventor of modern refrigeration techniques, the German engineer Carl von Linde (1842–1934), is best known for liquefying air in attempts at producing pure oxygen. Linde compressed air and then cooled it by decompression until it liquefied. Linde also produced the first refrigeration system in 1873 by using dimethyl ether as the refrigerant. Later Linde used ammonia gas, still the dominant refrigerant in commercial food freezing processes but not in domestic uses: ammonia is highly toxic. Ammonia is much easier to liquefy by compression than air.

Using two separate spaces, one in which to compress the gas (a heat-producing process), another in which to decompress the gas (a heat-consuming process), enabled Linde to use energy to create cold. The modern refrigerator, using a less toxic refrigerant than ammonia, is an every-day example of this technology. The hot side of the process is on the outside of the refrigerator (at the back) where the refrigerant gas is compressed and is both heated and liquefied in the process. The hot liquid is first cooled in coils exposed to the air of the kitchen and then decompressed through a valve into a system of coils inside the refrigerator, the cold side of the process. As the refrigerant turns back into gas, it cools the coils and passes back to the compressor outside the refrigerator for another pass.

The earliest machines producing artificial cold were used to make ice in breweries which relied on harvested natural ice for cooling. The technology had to develop and spread before freezing as a technique of food preservation became practical. Clarence Birdseye, an American inventor, is credited with launching frozen foods as a category by building and patenting a machine for rapidly freezing food. He operated his own company in the 1923–1928 period before selling his patents to what later became General Foods. General Foods launched the Birds Eye brand in 1930. The company also pioneered the development of the category by development of refrigerated storage equipment for the retail distribution sector. Birds Eye underwent multiple changes in ownership since 1928 but is still present in the market as a leading producer of frozen foods.

Products of a Modern Industry. Frozen food has unique characteristics, be that as a food or as an industry. It is a product of modernity and relies very heavily on an extensive industrial system which, in turn, depends on modern forms of energy. When food freezes slowly, large ice crystals form within it and, in so doing, the crystals rupture cell membranes of fruits, vegetables, and the tissues of meats. When such foods are thawed out for consumption, they lose their accustomed structure. They are softened, mushy, and also taste differently because chemical changes take place as a consequence of crystalline deformations. To avoid this problem, modern frozen food is *quick-frozen*. When the freezing takes place rapidly and the freezing temperature is very low (-22 to -40 degrees Fahrenheit) the crystals formed are very tiny and cells are left intact. The thawed food therefore tastes very fresh a very long time after it is rapidly frozen.

Quick-freezing is a demanding process, the requirements going well beyond the freezing technology itself. Fruits and vegetables begin decomposing right after they are harvested. The freezing must begin immediately after harvest. To preserve the original color of fresh fruits or vegetables yet also to remove enzymes that would affect the taste later, food to be frozen must first blanched in boiling water or steam—but very briefly so that vitamins in the fruit are not dissolved in water. Freezing takes place immediately after blanching. Blanching is a complex process with critical timing and temperature. This sort of process requires industrial organization and cannot, in practice, be done well at home. While the frozen food is in storage, it must be kept in deepfreeze compartments, thus at just below zero Fahrenheit. Food must be moved from deepfreeze to deepfreeze in vehicles (or ships) with the same freezer capabilities. Seafood, which is very perishable, is sometimes frozen on board ship after being cleaned and eviscerated.

When viewed in this comprehensive manner, it becomes obvious that frozen food is the product of an extensive system of technology. From the moment an organic product is harvested, seafood is caught, or livestock or poultry is butchered to the moment when the consumer opens a package to let it thaw, the preservation requires the constant operation of precision machinery. This machinery must compress and decompress refrigerants at very frequent intervals while keeping the refrigerants from leaking out. This work, as physicists call it, is accomplished by using fossil or nuclear energy sources. Finally, the cooled compartments must be suitably insulated. The entire system depends upon reliable electric energy so that sustained blackouts due to failures of the electric grid through overload, lack of maintenance, or acts of God can cause food to spoil in significant quantities in a matter of a day or two.

MARKET

Measuring the total market for frozen food in all but years ending in 2 or 7, the years for which full Economic Census data are collected, is difficult because the five industries that make up the total (see Industrial Codes) are reported on by the Census Bureau in full and separately only in those years. In the intervening periods, these industries are rolled up into composite industries for which product detail cannot be extrapolated. At the time of writing the *2007 Economic Census* was still being planned. Its results would not be available for another three or four years. Thus the most recent year for which reliable data are available is 2002—and for comparison only data from 1997 are available. Earlier data were reported in different formats.

With these limitations noted, the industry in 2002 had a precisely-measured shipment volume of $31.4 billion, up from $27.3 billion in 1997. This represented an annual compounded growth rate of 2.9 percent, a very energetic rate of growth for a mature industry largely dependent on population growth. Frozen food is a very small percentage of total food manufacturing. All U.S. food shipments in 2002 were $458.2 billion. Frozen foods had a 6.9 percent share of total shipments, up from 6.5 percent five years earlier. Food as a whole grew at an annul rate of 1.7 percent. Thus frozen food's growth was twice as rapid as that of food. To bring these numbers down to the household level, the food industry in 2002 shipped $4.35 worth of food for every man, woman, and child in 2002 each day; and of that total frozen foods were 30 cents per day. Actual expenditures, on average, were higher—these

FIGURE 98

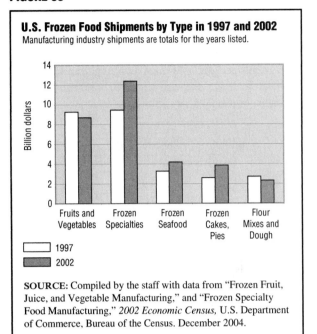

U.S. Frozen Food Shipments by Type in 1997 and 2002
Manufacturing industry shipments are totals for the years listed.

SOURCE: Compiled by the staff with data from "Frozen Fruit, Juice, and Vegetable Manufacturing," and "Frozen Specialty Food Manufacturing," *2002 Economic Census*, U.S. Department of Commerce, Bureau of the Census. December 2004.

foods had to be transported to stores for sale and increased in price through distributors and retailers.

The market is customarily divided into two large and three smaller segments. Fruits and Vegetables, once the largest category, was second largest in 2002, the leading segment being Frozen Specialties. The smaller segments are Frozen Fish; Frozen Cakes, Pies, and Other Pastries; and Flour Mixes and Doughs.

Changes in volume of shipments between 1997 and 2002 reveal interesting shifts in the public's consumption preferences. For instance, Fruits and Vegetables declined in shipments. In 1997 the segment was 33.9 percent of industry shipments; by 2002 its share had shrunk to 27.7 percent. Every category within this segment also showed decline in shipments with the sole exception of orange juice and a miscellaneous catch-all category labeled not specified by kind. Another declining segment was Flour Mixes and Doughs. This segment offers refrigerated or frozen prepared doughs and batters, sweetened or not, for products like biscuits, bread, rolls, pizza, cookies, cakes, and the like. The refrigerated category saw increase between 1997 and 2002 but the frozen category, dominant in 1997, declined enough to cause the entire segment to show negative growth.

The entire frozen foods industry grew—and at a nice rate. Where did this growth take place? It was most pronounced in the largest category, Frozen Specialties. This segment is divided into three categories:

1. Frozen dinners and nationality foods (67%).

2. Other frozen specialties (23%), its three largest product lines being frozen soups, whipped topping, and dairy substitutes.

3. Frozen specialty food manufacturing, not specified by kind, a catch-all category (10%) the components of which are undefined.

This last catch-all category more than tripled in volume of shipments in the 1997–2002 period; second in growth was the other category, increasing by 32 percent, and frozen dinners were last in growth, increasing 17 percent. Among categories within these, meat products and sweet products showed growth, dough-based products (including pizza) exhibited decline. The Frozen Specialty Segment as a whole increased its share of the frozen food industry from 34.6 to 39.3 percent, the largest increase within the industry.

In second rank as measured by growth was Frozen Cakes, Pies, and Pastries, a segment that increased from 9.5 to 12.3 percent of total industry shipments between 1997 and 2002. The largest product category within that segment was frozen pastries (41%) and also had the largest increase: it grew by 60 percent in the period. Frozen soft cakes (24.3% of shipments) increased 44 percent; and frozen pies (23.7%) increased by 9 percent in this period. The largest increase of all was displayed by a category we cannot say much about because the Census Bureau simply calls it frozen bakery products, not specified by kind. This category increased more than three-fold in the period and represented 11 percent of shipments of this segment.

Rounding out the growth pattern was Frozen Seafood. This segment increased its share of the frozen food category from 12 to 13.3 percent in this five-year period. The segment is divided into frozen fish (60%) and frozen shellfish (40%). The frozen fish component increased more rapidly, at 37 percent, shellfish at 15 percent in this period.

Foundational Factors. Since the end of World War II, a major driving force supporting the growth of the frozen food category has been the high quality of its product, its abolition of seasons in food by providing preservation, and its convenience for a population in which the homemaker has altogether changed her style of life. In the early twenty-first century, and beginning somewhat earlier, the major driving force, however, has shifted to innovation aimed at increasing convenience.

In the entire post-war period women gradually became participants in the labor force without, at the same time, relinquishing (or being able to relinquish) their role as primary managers of households. Women thus created a demand for any and all assistance to cope with chores

FIGURE 99

Labor Force Participation Rates for Men and Women in the United States, 1962–2005

Male participation rate
Female participation rate

SOURCE: *Employment and Earnings,* monthly, and *Monthly Labor Review,* U.S. Department of Labor, Bureau of Labor Statistics. January 2006 and November 2005, respectively.

at home—and their earnings provided the wherewithal to purchase the conveniences. In 1959, for instance, only 37.1 percent of adult women participated in the labor force, but this rate grew every year thereafter, based on Bureau of Labor Statistics data, for the next forty years, dipping only slightly in two years (1962 and 1991), reaching a peak participation rate of 60 percent in 1999. Thereafter, however, women's participation rates have declined, reaching 59.6 percent in 2002 and 59.3 percent in 2005. In the period for which we have good data (1997–2002), a steady state in participation had been reached. The growth in frozen foods, therefore, was supported by innovation answering to changes in life-style that had become well-established earlier. This innovation also serves the food industry's desire to grow more rapidly with more profitable products.

Innovation in the food industry has focused on creating added value to the consumer, achieved most efficiently by industrializing food preparation. Put in other words, this has meant delivering more and more food in ready-to-eat or ready-to-heat forms, thus taking labor out of the kitchen. Prepared foods in general can be differentiated better, command higher prices, provide higher margins, and thus higher return on investment. Frozen and refrigerated modes of delivery are very important in achieving this end. It is thus not surprising that frozen food has grown faster than food in general and that, within this category, prepared foods have exhibited growth and frozen commodities (fruits, vegetables) have seen decline. Busy households in which both parents work or the woman is the sole head of household create a receptive market for such innovations.

KEY PRODUCERS/MANUFACTURERS

In the frozen foods industry, consumers are much more aware of brand names than of the corporate entities that control them. People recognize such names as Healthy Choice, Swanson, Lean Cuisine, and Weight Watchers but would be hard put to identify these as being the brands respectively of ConAgra, Inc., Pinnacle Foods Corporation, Nestlé USA Inc., or H.J. Heinz Company. In terms of dominant market share in the United States, however, these companies are the leading producers, in the order shown, in frozen specialties, thus in the frozen dinner and entrée markets. Frozen specialties being the leading segment, leaders in it are also key producers in the industry as a whole. All of these companies feature a much wider range of frozen products in addition and are also major producers of non-frozen foods.

A key producer of frozen vegetables and fruits is Birds Eye Foods—the founder of which, Clarence Birdseye, also launched the entire category of frozen foods in the late 1920s. The company became nationally known in 1930 after it was acquired by a company later called General Foods. Philip Morris acquired General Foods in 1983. Philip Morris also acquired Kraft Foods and placed Birds Eye under its management. In 1993 the Birds Eye operation was sold to Dean Foods Vegetable Company which was then, in turn, acquired by Agrilink in 1998. Agrilink then chose to change its name to Birds Eye Food in recognition of its largest brand. Many other brands in this industry can point back at similar histories as the food industry itself consolidated.

Another important participant in the fruit and vegetable segment is General Mills by way of its Green Giant brand. General Mills is also a leading participant in frozen dough products through its Pillsbury brand. Nestlé's Stouffer's line is another major participant in the segment. An additional twenty companies are recognized suppliers of this category as well.

Kraft Food has the dominant market share in frozen pizza distribution through its DiGiorno and Tombstone brands followed by Schwan Food Company (Red Baron, Freschetta, and Tony's). Kellogg Company, the cereal producer, is the leading manufacturer of frozen waffles. The leading brand of pies, be these pot pies or sweet, is Marie Callender's, owned by ConAgra. The product line was purchased by ConAgra from Marie Callender's, a operator of restaurant chains. The company is now called Perkins and Marie Callender's Inc.

Tyson Foods, Inc. is the leading producer of frozen chicken and specialties. In order of size the three largest producers of frozen red meat products are United Food Group, LLC (Moran brand), Bubba Foods LLC (Bubba Burger), and Philly Gourmet Meat Co. (Philly Gourmet).

Leaders in frozen fish include Chicken of the Sea International, Del Monte, and Groton's.

MATERIALS & SUPPLY CHAIN LOGISTICS

In the logistics of frozen food manufacturing, the product itself, that which is to be preserved, largely defines the location where the freezing must take place. This is as true of fresh fruit and vegetables as of prepared meals or hand-held snacks or of meat products. Freezing itself might be viewed as functionally equivalent to canning, and freezing operations, like canning activities, are integrated into the food production process. The freezing technology in use is installed permanently with the refrigerant sealed into the system. For operations the site needs energy, a heat source for steam generation (in blanching, for instance), water (for washing the product), freezing equipment, and temporary deepfreeze capacity to hold finished product until it is picked up for shipment. Inputs are the products to be frozen, energy, and water. Where cryogenic freezing is used (see below), the refrigerant itself is a continuously consumed input that must be transported in.

Distinctly different freezing systems in use are designed for different applications. These fall into three categories based on the manner in which the cold temperature is transferred to the product. In *air systems* the refrigeration equipment cools a stream of air and uses it to freeze the product. In static or air-blast systems the operator keeps a large space at extremely low temperatures and simply exposes the product to this environment, placed on racks, for the time required for hard freezing. No fans move the air. In tunnel freezers labor or machinery moves the product down a tunnel slowly, the tunnel kept at very cold temperature. Fans move the air ensuring that the trolleys carrying product are touched on all sides. Belt freezers carry the product on a perforated belt with cold air blowing at it from top and bottom. The most effective technique for individually freezing particles are fluid bed systems in which very cold air is blown upward at the product causing it to be suspended in a turbulent air mass continuously swirling and tumbling in the blast.

In contact freezing heat is removed from the product by contact directly with the refrigerant itself or with metal plates in contact with the product. Immersion systems are best for loose product, plate freezers or contact belt freezers work well with packaged products. Pressure applied by the plates to the package speeds up the freezing.

Producers use cryogenic systems when very rapid freezing to very low temperatures is desirable or when freezing must be accomplished at unusual and temporary locations. The refrigerant is brought in and thus no equipment is needed at the site. Capital costs are low, operating costs are high—owing to the expense of the refrigerants and their transportation to the site. Liquid nitrogen and liquid carbon dioxide are used in these applications. Fish and berries are sometimes frozen using cryogenic systems.

Packaging may take place ahead of or after freezing. Thereafter the product remains in deepfreeze compartments throughout its travels to the consumer's shopping cart or use in an institutional setting.

DISTRIBUTION CHANNEL

Distribution of frozen foods is identical to food distribution generally. Thus a three-tier distribution is common in which the producer sells to a wholesaler, the latter to the retailer, and the retailer to the customer. The wholesale level, to be sure, may be wholly-owned by the retailer, common in the case of major food chains. The retailer may be a supermarket or a restaurant.

KEY USERS

Since eating is an unavoidable activity in life—no matter how much those on a diet wish this were not the case—everyone is a user of frozen foods. In the industry itself distinctions are made between consumer and institutional markets, thus bulk buyers and individual- and family-portion purchasers. The dominant consumer of prepared meals is the ordinary grocery shopper. Institutional markets, however, are purchasers of bulk shipments of fruits, vegetables, and meat products requiring additional preparation.

ADJACENT MARKETS

Important adjacent markets are represented by dairy products that require a refrigerated environment but not the use of a deep freeze. Ice cream, of course, is the exception. Ice cream is unique in being, as already noted, the only frozen product intended to be consumed in frozen form. A close relative of ice cream is ice itself, simply frozen water, produced and sold exactly like frozen food but used by the buyer for either temporary cooling of beverages in coolers and/or consumption in the beverage itself.

Adjacent to frozen foods is the full range of dried foods. By removing all water from a product, producers of various dried beverage or soup mixes achieve preservation by removing the indispensable ingredient of life—and also of decay—from the product and sealing out all moisture by packaging. These products are then reconstituted by adding water. Freeze-dried products are an interesting hybrid between the two in that freezing, combined with desiccation, produce a very stable product. It was once thought that freeze-drying would replace most other forms of preservation, but the process has seen major application only in the distribution of coffee.

The largest and most dominant market adjacent to frozen food is canned food, dwarfing the former in sheer volume sold, particularly in fruits and vegetables, but also in seafood and in certain categories of prepared foods like soups and stews.

RESEARCH & DEVELOPMENT

Research & development efforts in this industry are concentrated chiefly on new product development in efforts to penetrate the prepared food segment deeper and deeper with frozen specialties aimed both at meal-time products and quick snacks. Increasingly, as the twenty-first century unfolds, this has taken the form of cooperative research involving the producers of the chief tool consumers use to restore frozen food to appropriate temperatures—the microwave oven—itself a category of product in active development.

Complex frozen meals in which different components require different degrees of heating present problems that technology can overcome. Restoring vegetables on a combined dish to a crisp taste while heating a relatively thick cut of breaded veal through and through cannot be accomplished effectively in the same time and at the same temperature. Enter smart packaging and highly educated microwave ovens. Developments are aimed at producing packaging that a microwave oven is capable of reading in order to program itself to cook the contents of the package just right. This involves targeting heat at different areas of the package for longer or shorter times. Similar issues are involved in creating crisp french fries from frozen fries.

CURRENT TRENDS

The most notable trends touching the frozen foods industry are changes in lifestyle in which the traditional family meal is seemingly disappearing, replaced by a combination of continuous snacking, fast food purchasing, grabbing a meal, and members of a family eating separately as their schedules permit. A counter trend is represented by gourmet food preparation as a hobby or avocation practiced to amaze and to entertain close friends.

Parallel with these trends, a portion of the population is at least professing an interest in healthy eating marked by avoidance of fats, carbohydrates, sugar, and products saturated with bad cholesterol. The frozen food industry is responding to each of these fashions. The industry attempts to provide the experience of a meal but without any of the necessary preparations, planning, and labor. Its leading products are all manner of handheld snacks. It supplies difficult-to-get ingredients for the gourmet chef as well as healthy breakfast foods for children that require minimum time (or skill) to prepare.

TARGET MARKETS & SEGMENTATION

The industry has well-defined segments. These have emerged over time and have developed into the very divisions of the industry as illustrated by segment names: fruits and vegetables; frozen specialties; fish and seafood; cakes, pies, and pastries; and flour and dough products.

The first of these segments represents utility, the fruits and vegetables having a fresher character than their canned counterparts but requiring additional work. Frozen fish and seafood represent safety over fresh-bought goods. Frozen specialties on the one hand and frozen sweets on the other hand cater to the consumers' demand for luxury at great convenience; flour and dough products represent a segment offering the consumer both the convenience of avoiding the more arduous aspects of baking, namely dough preparation, while preserving the pleasure of producing something fresh by leaving the baking and the optional decorating function to the consumer.

RELATED ASSOCIATIONS & ORGANIZATIONS

American Frozen Food Institute, http://www.affi.com/

National Frozen & Refrigerated Foods Association, http://www.nfraweb.org/home.html

National Frozen Pizza Institute, http://www.affi.com/nfpi/default.html

BIBLIOGRAPHY

Canovas, Barbosa, Bilge Altunakar, and Mejia Lorio. "Freezing Fruits and Vegetables." *FAO Agricultural Services Bulletin 158.* Food and Agricultural Organization of the United Nations. 2005.

"Carl von Linde." Chemical Heritage Foundation. Available from <http://www.chemheritage.org/classroom/chemach/gases/linde.html>.

Darnay, Arsen J. and Joyce P. Simkin. *Manufacturing & Distribution USA,* 4th ed. Thomson Gale, 2006, Volume 1, 61–66, 109–131.

"Frozen Fruit, Juice, and Vegetable Manufacturing: 2002." *2002 Economic Census.* U.S. Department of Commerce, Bureau of the Census. December 2004.

"Frozen Specialty Food Manufacturing: 2002." *2002 Economic Census.* U.S. Department of Commerce, Bureau of the Census. December 2004.

"History of Frozen Food." American Frozen Food Institute. Available from <http://www.affi.com/factstat-history.asp>. "History of Frozen Foods: Long and Varied." National Frozen & Refrigerated Food Association. Available from <http://www.nfraweb.org/media/edit1.html>.

Lazich, Robert S. *Market Share Reporter 2007.* Thomson Gale, 2007, Volume 1, 106–112.

SEE ALSO *Canned Foods*

Gasoline

———————————•———————————

INDUSTRIAL CODES

NAICS: 32–4110 Petroleum Refineries

SIC: 2911 Petroleum Refining

NAICS-Based Product Codes: 32–41101111, 32–41101121, and 32–41101Y

PRODUCT OVERVIEW

General Product Characteristics. Gasoline is a liquid formed of selected hydrocarbon compounds, thus of carbon and hydrogen atoms combined in a great variety of ways. The carbon atom has six electrons, which circle the atomic core in two orbits. The innermost orbit of an atom can never hold more than two electrons, while the second orbit maximally holds eight. In the case of carbon, the outermost or second orbit holds only four electrons. It would like to fill in its outer orbit. Having four empty spots, it is called tetravalent, having a valence of four. On the other hand, hydrogen has a single electron and feels an equal need for another electron to fill in its incomplete and only orbit. When carbon and hydrogen meet, carbon lends one of its electrons to fill in hydrogen's void and hydrogen supplies its sole electron to fill one of the voids in carbon. The simplest saturated hydrocarbon is methane, natural gas. It is a single carbon joined with four hydrogen atoms (CH_4). Each hydrogen is fully satisfied by having one of the carbon's actual electrons. The carbon completes its outer orbit with four electrons borrowed from four hydrogens.

Saturated hydrocarbons all have this form. Every empty spot on every participating atom is filled; furthermore, each bond comes from a separate atom. This hydrocarbons class is called the *alkanes*. The number of carbon atoms present is always matched by twice that number of hydrogen atoms plus two. For example, C_2H_6 is ethane, C_3H_8 is butane, and so on. Alkanes are quite stable chemicals because they are saturated. They have no inherent desire to combine any further. Note that in the case of all alkanes beyond methane, when two or more carbons are present, the carbons are joined by single bonds with each atom contributing one electron to and receiving one electron from the other. In a chain of two carbons (C–C) the total valence of the carbons is eight, each having four. The bond between the two carbons satisfies two, leaving a deficit of six. In ethane, this deficit is satisfied by six hydrogen atoms.

When the supply of hydrogen is insufficient to saturate the compound, unsaturated structures form. In these, some of the carbons adhere by double bonds (they share two electrons each) or triple bonds (three each). Compounds with double bonds are called alkenes, while those with triple bonds are called alkynes. Both kinds of bonds are less stable owing to the somewhat awkward orbital geometries involved. For this reason unsaturated hydrocarbons are much more reactive. In the presence of free hydrogen, for instance, they will form alkanes. Yet another category of unsaturated hydrocarbons are the aromatics, which are different from alkenes and alkynes in that their carbons are arranged in ring formations rather than in chains or in branched structures. The most basic aromatic is benzene (C_6H_6), formed of six carbon atoms in a circle and surrounded in a kind of halo by six hydrogen atoms. This class received its name because these substances had quite distinctive smells. Mothballs are an example; they are naphthalene ($C_{10}H_8$), forming two joined rings.

To sum up the basics, there are saturated and unsaturated hydrocarbons. Within each category there are one or more carbons forming the skeletal structure and hydrogen atoms being the decoration. These structures may be chains, branched structures of various complexity, and ring structures. Ring structures may be joined and further elaborated by hanging adornments of more carbon and hydrogen elements.

A barrel of crude oil comes from the ground containing all of the categories of hydrocarbons previously mentioned. Molecules with every conceivable length of carbon structure are present, including traces of methane on up to heavy hydrocarbons with hundreds of chained or branched carbons in the individual compounds. Gasoline is that portion of crude oil made up of hydrocarbons with five to twelve carbons in different arrangements. This small subset, however, is sufficiently great to form more than 500 different compounds. Gasoline is thus a quite diverse community of compounds rather than a single substance, yet each component of it is made up of carbon, hydrogen, and nothing else.

When crude oil is heated, the hydrocarbons within it separate into layers: those with the fewest carbon atoms rising to the top, and those with the most falling to the bottom of a heated column. In simplest terms, the refining of crude consists of thus causing the many thousands of different hydrocarbons to sort themselves by boiling point in towers. Petroleum gases (butane, propane) are taken out at the top, gasoline is drained off at the second level, kerosene and jet fuel drop out at level three, heating oil at level four, and lubricating oils at level five. The heavy residuals are collected at the bottom and contain tars, asphalt, and other heavy substances.

While this description is accurate in broad outline, gasoline refining is more complex in practice. Only 25 percent to 35 percent of the typical barrel of crude is naturally in the gasoline range. Demand for gasoline is greater than for the other naturally occurring fractions in crude oil. For this reason the industry uses additional processes to get more gasoline out of crude. Among these are catalytic cracking, reforming, and polymerization. Cracking, a heat process used in the presence of catalysts, can break down longer hydrocarbon chains. Reforming can change the structure of hydrocarbons so that they have a slower burning rate. This is in part accomplished by adding free hydrogen. Polymerization, also known as alkylation, can cause short chains to form longer chains aided by the presence of acids. These broad methods come in many implementations.

Octane Ratings. When gasoline is used in engines, the gas will ideally burn slowly rather than explosively. Before the sparkplug ignites the gas-air mixture produced by carburetion or fuel injection, the mixture is first compressed in the cylinder by the piston. The composition of the gasoline must be such that very few of its hydrocarbons will ignite and explode while compression is still in process. Those that do are known to be auto-igniting. Premature explosions reduce the power that can be extracted from the gas; explosions also eventually damage the engine. This phenomenon, known as engine knock, is due to the presence of too many straight-chain and not enough branched and ring-shaped hydrocarbons. Straight-line chains ignite from the heat of compression alone. When they dominate, the octane level of the gas is too low.

The industry's use of an octane number is based on experimental observation. The eighth compound in the alkane sequence is C_8H_{18}. Eight carbons give this compound its name, octane. One form of octane, also known as isooctane or as 2,2,4-trimethylpentane, is a highly branched and ideally burning hydrocarbon. It has been given the octane value of 100. The immediately preceding alkane is heptane (7 carbons, 16 hydrogens). It is a straight chain hydrocarbon, is auto-igniting, burns explosively, and has an octane number of 0. All other major groups of hydrocarbons have been assigned octane numbers either higher, lower, or the same as isooctane. Raw gasoline coming from crude distillation towers usually has an octane rating of 70—too low for good performance in modern engines. For this reason refineries employ additional processes to lift the octane. The most powerful engines have the highest compression levels and therefore need the highest octane ratings to perform well.

Octane values are obtained by two laboratory techniques known as Research Method and the Motor Method. In both cases the same fuel is burned using specific methods defined by the American Society for Testing and Materials (ASTM). The two methods produce slightly different results because engine revolutions and other parameters used are differently specified. Octane values as seen by the public are the average of values obtained by the two tests. Octane values of gasoline varieties intended for use in high altitude environments are typically two octane numbers lower than those used in the lowlands. With less oxygen available at high altitudes, the lower octane formulations are equivalent in performance.

Octane number in no way influences the inherent energy contained in gasoline as measured in British thermal units (Btus). When ethanol is mixed with gas, ethanol lowers gasoline's Btu content but raises its octane level. Ethanol octane ratings are at or above 110; for this reason ethanol is a very good anti-knock agent.

Types of Products. Grades of gasoline on the market are defined by octane rating. Regular gasoline has an octane rating of minimally 85 and less than 88. Midgrade gas is at least 88 octane and less than or equal to 90. Premium gas has more than 90 octane but rarely more than 92. High performance engines are designed to achieve a high compression ratio in their cylinders—a means of providing high levels of horsepower. The higher this ratio is—the relation of compressed volume to total cylinder volume—the greater the likelihood that compression will ignite chained compounds in the gas. For this reason high octane gas is recommended for use in such vehicles.

Since the passage of Clean Air Act amendments by Congress in 1995, the Environmental Protection Agency has mandated the use of reformulated gasoline, usually abbreviated as RFG, in areas the EPA has designated as carbon monoxide non-attainment areas of the country, or places where the levels of CO are higher than mandated by law. RFG is also known as oxygenated gasoline, having oxygen content of 2.7 percent or higher by weight. More oxygen in the gas ensures more complete burning of all carbons present—and hence less carbon monoxide in the exhaust. Initially RFG was produced by blending gasoline with methyl tertiary-butyl-ether (MTBE), a product based on methanol and manufactured by the petroleum industry. Subsequently the EPA found that residual levels of MTBE began appearing in bodies of water at unacceptably high levels. Since then ethanol, based on corn in the United States and sugar cane in Brazil, has come to be substituted for MTBE in the production of reformulated gasoline. Gasoline with around 10 percent by volume of ethanol provides the required oxygen content. Mandates for MTBE elimination have been promulgated by state governments, not the EPA, but the process of transition was nevertheless almost completed in the first decade of the 2000s. Approximately 39 percent of all gasoline produced is reformulated gasoline (RFG), sold primarily in major metro areas with air pollution problems.

MARKET

The history of the gasoline market in the United States for the 1980 to 2006 period is summarized in Figure 100,

FIGURE 100

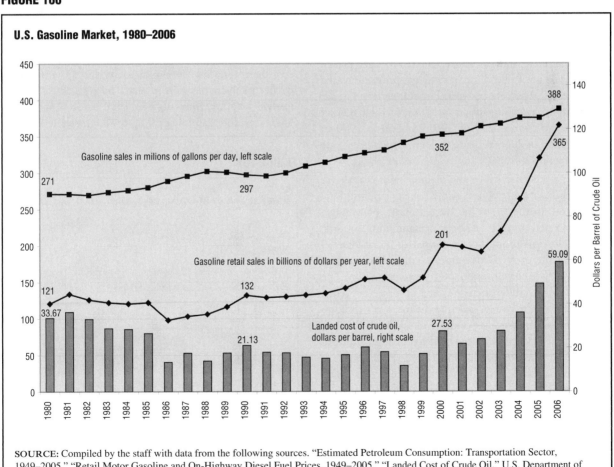

SOURCE: Compiled by the staff with data from the following sources. "Estimated Petroleum Consumption: Transportation Sector, 1949–2005," "Retail Motor Gasoline and On-Highway Diesel Fuel Prices, 1949–2005," "Landed Cost of Crude Oil," U.S. Department of Energy, Energy Information Administration. Various publication dates.

showing consumption per day in millions of gallons of gasoline, total gasoline consumed times average prices per gallon in each year to show total retail sales, and the average cost of a barrel of imported crude landed on U.S. shores.

This data, assembled from statistics maintained by the Energy Information Administration (EIA), an agency of the U.S. Department of Energy, clearly illustrates some of the fundamentals of the gasoline market. A major factor is the unflaggingly steady consumption of gasoline by the U.S. public growing at modest but steady rates in all periods except those in which recessionary forces are at work. Official recessions held sway January to July 1980, July 1981 to November 1982, July 1990 to March 1991, and from March to November of 2001. The last of these recessions did not affect gasoline consumption, at least not as measured one year to the next. In the earlier periods, consumption flattened out. In all other periods, gasoline consumption grew. In the 1980 to 2006 period, consumption increased at an annual, compounded rate of 1.4 percent per year. During that same period, population increased at the rate of 1 percent yearly.

During this quarter century, the price of gasoline averaged $1.35 per gallon, but it was as low as 93 cents in 1986 and as high as $2.58 in 2006—all these being average prices across the nation; they may have been higher or lower in different states. The overall trend in gas prices, however, was up. In the entire period, gas prices increased at the rate of 2.9 percent per year. Physical increases in gas consumption and rising prices produced a total market that advanced from a level of $120.9 billion in 1980 to $365.1 billion in 2006 at the retail level, growing at a rate of 4.3 percent per year.

The graphic also shows the phenomenon that drives the price behavior of gasoline, namely the price of imported crude oil. In the 1980s the United States, on average, satisfied 33 percent of its refinery demand with imports; in the 1990s this proportion had risen to more than 50 percent. In the early 2000s, imports satisfied 63 percent of refinery crude consumption, 66.5 percent in 2006. The price of foreign crude, therefore, had become the defining element of the cost of gasoline. In the 1980 to 2006 period, the lowest crude price came in 1998 ($11.84 per barrel) after the Organization of Oil Exporting Countries (OPEC) increased its production ceiling by 2.5 million barrels per day to 27.5 million barrels. The highest price was recorded in 2006 ($59.09 per barrel), the culmination of shortages caused by natural disaster (the hurricane season of 2005), uncertainties produced by wars in Iraq and Afghanistan, deteriorating United States-Iran relations, and very strong global demand for oil.

Over the period shown in Figure 101 many factors influenced crude prices, including economic cycles, weather, wars and rumors of wars, the collective actions of the producers in tightening or easing supplies, social unrest (e.g., unrest in Venezuela, an OPEC country, in 2002 and 2003), pricing schemes employed, strikes (in Nigeria and Venezuela), major oil spills, and other natural and political events. Neither governments nor participants in the industry, not even major collectives like OPEC, have effective control over the whole range of events affecting crude prices. Many of these entities are able to influence the pricing of gasoline but none of them can control the price due to the number of factors that combine to establish the prices of crude oil and gasoline. According to EIA estimates, the final price of gasoline at the pump is determined 50 percent by crude oil prices (prices that are themselves set by many different factors), 27 percent by refining costs, 15 percent by state, local, and federal taxes, and 8 percent by expenditures on distribution and marketing. Although no one entity controls the price of gas, in public discussion much blame is assigned for rising gas prices and much credit is claimed for falling prices.

In 2005 most gasoline sold in the United States was regular, accounting for 80.6 percent of all gasoline pumped. Midgrade had a 10.2 percent share, and premium a 9.1 percent share. Reformulated gasoline (RFG) was 38.8 percent of all gas sold. All three grades were sold as RFG in non-attainment areas. Figure 101 presents shares by different grades going back to 1990.

The interesting development in this 15-year period has been substantial gain in share by regular gas, up by nearly 14 points, from 66.8 in 1990 to 80.6 percent in 2005. Premium gasoline gave up the largest share between

FIGURE 101

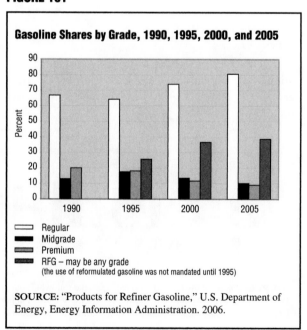

Gasoline Shares by Grade, 1990, 1995, 2000, and 2005

Legend:
- Regular
- Midgrade
- Premium
- RFG – may be any grade
  (the use of reformulated gasoline was not mandated until 1995)

SOURCE: "Products for Refiner Gasoline," U.S. Department of Energy, Energy Information Administration. 2006.

1990 and 2005, 11 points, and midgrade had a middling loss of share, around 3 percent. RFG was mandated in 1995 and, as its use was implemented, reached nearly 39 percent of the market by 2005.

Factors Influencing the Market. As indicated in Figure 100, the chief factor influencing the use of gasoline is the level of economic activity. Economic expansion lifts demand; recessions cause demand to weaken. In its annual forecast of the outlook for energy, domestically and worldwide, the Energy Information Administration looks principally to economic trends across the globe. In periods of global economic downturn, prices of crude oil tend to drop but low prices do not translate automatically into increased consumption. Conversely, very substantial increases in price have but a moderate influence on consumption. Another way to put this is to say that gasoline consumption is relatively inelastic in relation to price. When consumers encounter sharply escalating gas prices, they cope by cutting back on other purchases; when gas prices drop, they may increase leisure-time motoring a little; the more natural response is to spend the saved money on other things. Economic activity, however, directly influences the physical quantity of gasoline used and is therefore the dominant force affecting demand.

The Bureau of Transportation Statistics, an element of the U.S. Department of Transportation, conducts a periodic National Household Travel Survey. Data from surveys conducted in 1983, 1985, 1988, 1991, 1994, and 2001 are available from the EIA (see "Household Vehicles Energy Use," cited below). These data which cover most gasoline use, permit a look at trends below the somewhat abstract level of the economy as a whole. These data indicate (1) more vehicles in use per household and per capita, (2) more miles traveled by each vehicle, but also (3) less gasoline used per 1,000 miles traveled.

Vehicles per household averaged 1.8 throughout the entire 1983 to 1994 period, but then rose to 1.9 vehicles per household in 2001. The number of vehicles per 1,000 population increased from 555 in 1983 to 670 by 2001. Miles traveled annually by each vehicle increased from 9,400 in 1983 to 12,000 in 2001. Gallons of gas used per 1,000 miles of travel, however, decreased from 66.2 in 1983 to 49.5 in 2001. The EIA calls this energy intensity, which has been steadily declining.

Dropping energy intensity suggests that increasing costs of gasoline—costs have been rising steadily since 1986—have as a consequence increasing efficiency in fuel use. This growing efficiency is documented statistically by data on gas mileage achieved and also by growing use of regular gas at the expense of midgrade and especially premium gasoline.

The history of U.S. gas mileage from 1949 through 2004 is shown graphically in Figure 102. The data is drawn from EIA statistics. From the late 1940s to the late 1960s, gas mileage was in steady decline. An awakening took place in the early 1970s when OPEC's oil embargos, aimed at raising crude prices, shook the nation and caused people to stand in line for gasoline. The embargo began in October 1973 and ended in March 1974. The year 1973 coincided with the lowest average vehicle gas mileage in this 55-year period. Passenger cars averaged 13.4 miles to the gallon that year. From that date onward, passenger car mileage increased steadily to reach 22.4 MPG by 2004.

In 1965 EIA separated pickup trucks from the broader category of trucks, vans, and pickups. The agency created a category in which vans, pickups, and sports utility vehicles are grouped. Vans became popular in the 1970s, morphed into SUVs, and competed since then with pickup trucks often used as passenger vehicles. This category largely displaced the gas hogs of the 1950s and 1960s. The category has also increased its gas mileage over time from a bottom of 9.7 MPG in 1967 to a peak of 17.6 MPG in 2001; gas mileage began to drop thereafter and was 16.2 MPG in 2004. The truck category is not strictly comparable with the other vehicular types because its fuel is predominantly diesel. Trucks experienced their historical low in 1981, with 5.3 MPG. They too have improved, if minimally, coming in at 6.7 MPG in 2003 and 2004.

The graphic excludes total gas mileage averages which, in addition to the categories shown, also include buses and motorcycles, the latter from 1989 forward. Until then motorcycles were included in the calculation for passenger vehicles. Gas mileage for all vehicles was 13.1 MPG in 1949, reaching its all-time low in 1973 (11.9 MPG), and its highest level of 17.1 MPG in 2001 and 2004.

The overall improvement in gas mileage closely parallels the results achieved by the van, pickup, and SUV category because this grouping of vehicles is growing whereas the population of passenger cars is essentially flat. Data reported by the Bureau of Transportation Statistics indicate that in the 1980 to 2000 period, passenger car registrations increased at the very modest rate of 0.5 percent per year whereas registrations of vans, pickups, and SUVs increased 5.4 percent per year. In the more recent 1990 to 2000 period, passenger car registrations actually declined, even if at less than 1 percent per year; the other category grew at 5 percent annually. These trends, if extended into the future, signal decreasing gas mileage in the post-2000 period.

Data at this level of resolution were not available at the time of writing for the post-2000 period, but ample information was available in the first decade of the 2000s to show that gasoline prices were up sharply and, based on EIA energy outlook estimates, were expected to continue

FIGURE 102

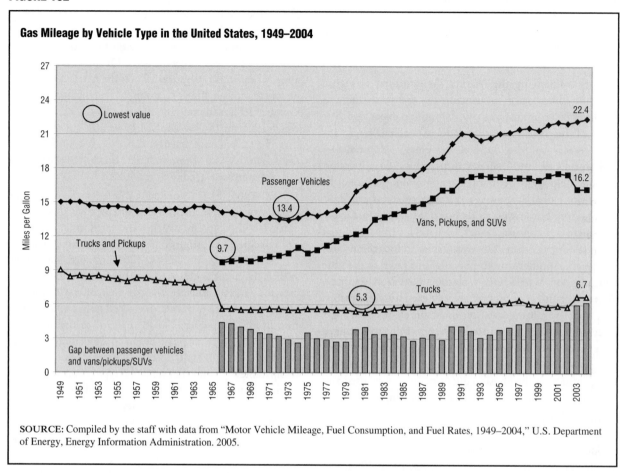

Gas Mileage by Vehicle Type in the United States, 1949–2004

Miles per Gallon

○ Lowest value

22.4

Passenger Vehicles

13.4

16.2

Vans, Pickups, and SUVs

9.7

Trucks and Pickups

5.3

Trucks

6.7

Gap between passenger vehicles and vans/pickups/SUVs

SOURCE: Compiled by the staff with data from "Motor Vehicle Mileage, Fuel Consumption, and Fuel Rates, 1949–2004," U.S. Department of Energy, Energy Information Administration. 2005.

high well into the 2010s. In 2007 an observer of the industry could well wonder just what the consequences may be of gas prices in the range of $3.50 and above. Would $50-plus expenditures at every fill-up cause major shifts in vehicle uses, reduce use of gas-powered vehicles altogether, lead to mandated improvements in fuel efficiency, the adoption of alternative fuels, and massive investments in mass transit, or merely a shift of money from one category to another of the household budget?

KEY PRODUCERS/MANUFACTURERS

Among the world's top twenty oil companies, only seven are privately held. They are shown here in order of size, their location, and overall rank: Exxon Mobil (U.S., No. 2), Royal Dutch/Shell (U.K. and Netherlands, No. 5), BP Plc (U.K., No. 6), Chevron (U.S., No 7), Total (France, tied for No. 8), ConocoPhillips (U.S., No. 12), and Repsol YPF (Spain, tied for No 17). Two other companies are somewhat private. Enti Nazionale is 30 percent privately owned, the balance held by the Italian government; the company is tied for the No. 17 spot with Repsol. Luikol is the Russian oil company with 14.1 percent of its shares privately held; Luikol is ranked No. 20.

The remaining companies include the world's largest, Saudi Arabia's Aramco (No. 1), Petroleus de Venezuela (No. 3), National Iranian Oil (No. 4), Pemex (Mexico, tied for No. 8 with Total), PetroChina (China, No. 10), Kuwait Petroleum Corporation (No. 11), Petramina (Indonesia, tied at No. 13 with Sonatrach), Sonatrach (Algeria, No. 13), Petrobras (Brazil, No 15), Abu Dhabi National (United Arab Emirates, No. 16), and Petroliam (Malaysia, No. 19).

In the U.S. market the top ranking participants, based on refinery capacity as reported by the Energy Information Administration, were ConocoPhillips (12.9% share of total U.S. capacity in January 2006), Exxon Mobile (12.2%), Valero Energy Corporation (10.8%), British Petroleum (6.4%), Chevron USA Inc. (5.8%), Marathon Petroleum Company LLC (5.6%), Sunoco Inc. (5.2%), Flint Hill Resources LP (4.5%), Motiva Enterprises LLC (4.3%), and Citgo Refining (3.7%). Another 59 companies were also active in U.S. gasoline refining in 2006.

Exxon Mobile, the world's largest publicly traded oil giant ($365.4 billion in sales in 2006), is the continuous living representative of the Standard Oil Trust John D.

Rockefeller had assembled in the late nineteenth century, a vast lash-up of companies then principally selling the kerosene fraction of crude oil for lighting. Two important elements of Standard Oil were Standard Oil Company of New Jersey, selling its product under the Esso name and known as Jersey Standard, and Standard Oil of New York, known as Socony. These two companies became Exxon and Mobil a few name changes later. In 1911 trust-busting Teddy Roosevelt broke up Standard Oil into some thirty-four separate entities, Exxon and Mobil being two of them. The Mobil name was introduced by Socony in 1920 as Mobiloil, indicating that kerosene had given way to gasoline, the fuel of transportation. Jersey Standard changed its name to Exxon (a reconfiguration of the Esso name) in 1972. The two companies merged again in 1999.

The world's second-largest private oil company, Royal Dutch Shell, with 2006 sales of $318.8 billion, was formed in 1907 by the merger of Royal Dutch Petroleum and Shell Transport and Trading Company Ltd. of Britain. Both parts of the new entity went back to the nineteenth century, Royal Dutch having developed Malaysian oil deposits. Shell, perhaps surprisingly for many, actually began as a company buying and selling seashells. The quotes around the company's name were part of the name. The business began transporting oil when Marcus Samuel Jr., the founder's son, discovered profits in oil exports when visiting Baku in what is now called Azerbaijan. Shell became a major oil transport company before entering the oil business itself.

In an ironic twist of fate, the original owner of British Petroleum was actually a company called Deutsche Petroleum AG, a German company based in Berlin. In 1907, Deutsche Petroleum established a sales office in London and called it British Petroleum Company. In 1917, the company became part of the Anglo Persian Oil Company (APOC), the operation having been seized during World War I as enemy property. The German original was getting its oil from the Caspian region. APOC had its origins in 1908 with discoveries of oil in Persia (now Iran). The modern BP includes a number of U.S. oil companies it acquired along the way, including Amoco (the former Standard Oil of Indiana), Arco (formerly Atlantic Refining and Richfield Petroleum), and Standard Oil of Ohio (Sohio). The German firm, Aral, is also part of BP. Perhaps looking forward to the challenging future, BP's twenty-first century slogan is "beyond petroleum."

As the name implies, ConocoPhillips is the result of a merger, that of Conoco and Phillips Petroleum. The company had revenues of $188.5 billion in 2006. Conoco and Phillips joined forces in 2002. Conoco began in Ogden, Utah as a coal, oil, kerosene, grease, and candles distributor in 1875, then calling itself Continental Oil and Transportation Company. The original Phillips was Frank Phillips who, with his brother L.E. Phillips, drilled for oil in Bartlesville, Oklahoma and hit a gusher in 1905. The two brothers proceeded to drill 81 productive wells in a row before they hit their first dry hole. Continental developed as a distribution company, and was also owned for a while by Standard Oil. Phillips developed into an oil production and refining company. The Phillips company itself was founded in 1917 with the opening of Phillips' first refinery.

Among the domestic corporations, together controlling more than 70 percent of refining capacity, Valero is a recent entity, having been founded in 1980 and expanding thereafter by acquisition. Flint Hill Resources is a subsidiary of Koch Industries, Inc., a privately held investment company. The company operated as Koch Petroleum Group until 2002 when it acquired its current name. It began in 1940 as Wood River Oil & Refining Company. Motiva Enterprises is even younger than Valero, having been formed in 1998, but with deep roots in the business. It is a joint venture formed by Saudi Refining Inc. and Shell Oil Company USA, the latter itself an affiliate of Royal Dutch Shell. Saudi Refining is a subsidiary of Armaco.

Chevron Corporation, like ConocoPhillips, is the product of two major companies that merged. The first of these was Chevron, originally the Pacific Coast Oil Company, later Standard Oil Company of California, and eventually Chevron. It began in 1879. The other half of the company began in 1901 as The Texas Fuel Company, later The Texas Company, and eventually Texaco. The two companies merged in 2001 as Chevron Texaco. In 2005 the name was simplified to Chevron. Chevron had revenues in 2006 of $204.9 billion.

Marathon Petroleum began in 1887 as The Ohio Oil Company. Between 1889 and 1911 it, too, was part of Standard Oil. In 1911 The Ohio, as Marathon was once known, was spun off again as a consequence of antitrust legislation. The company renamed itself Marathon in 1962. In 1982 it became part of US Steel Corporation but gained its independence again in 2001. In 2006 Marathon had revenues of $64.9 billion. Sunoco, with 2006 sales of $38.7 billion, began in 1886 as the Sun Oil Company in Ohio. The company originated as a diversification effort by the Peoples Natural Gas Company of Pittsburg at a time when oil exploration was still largely centered on Pennsylvania and Ohio. The company expanded by acquisitions and mergers and further diversified into chemicals and other petroleum products, including the production of coke for the steel industry.

MATERIALS & SUPPLY CHAIN LOGISTICS

Gasoline and other fractions of crude oil are always carefully formulated for local conditions and the mix of fuels and lubricants demanded by domestic markets. For this reason oil moves over the greatest of distances in the form of crude and is refined locally. It is much more practical to dispatch one tanker from an oil-rich region than to refine near the well-head and to ship many products in many vessels to many markets. Refining capacity in domestic markets, however, is concentrated near the points where oil occurs or is landed from tankers, with finished products reaching more distant points by pipeline or tanker truck. Refineries have seasons, switching from the production of gasoline to fuel oil and back again. Lubricants are produced throughout the year.

As the twenty-first century advances, a much bigger and general issue is beginning to focus the attention of all those in the business. It is the relatively near-term future. How long will the oil last? Proven world reserves of oil are put at between 1.1 and 1.3 trillion barrels as of 2006. The low number is the estimate published by *World Oil* magazine and the high estimate by the *Oil & Gas Journal*. An intermediate estimate is provided by British Petroleum's *BP Statistical Review*. BP puts world reserves at 1.2 trillion barrels. All three estimates are tracked by the Energy Information Administration. When examined in detail, each of these estimates is heavily hedged with cautions and disclaimers because very substantial portions of each estimate are based on data that are difficult or impossible to independently verify.

Using BP intermediate values for proven world reserves, thus 1.2 trillion barrels, together with the much more accurate statistics measuring total world consumption of oil, 30.857 billion barrels in 2006, reserves will last 39 years, thus until 2045, if we assume no growth whatsoever in either consumption or reserves.

World oil consumption, however, has been growing at the rate of 1.73 percent per year in the 1995 to 2005 period. Assuming such a rate of growth, and no growth in reserves, oil will run out in 29 years instead, thus in 2035.

At the same time, according to BP's compilation for the same period (1995 to 2005), reserves have grown at a rate of 1.6 percent per year. Such data can only be interpreted to mean that the global industry is replacing draw-down for consumption 100 percent by new discoveries or better estimates of actual reserves in the ground and adding additional new reserves at the 1.6 percent rate. In this period, thus, average draw-down was around 30 billion barrels a year and average additions to reserves 32 billion barrels. That rate of addition suggests an endless supply of oil well into the dimmest future unless the recent period is a fluke or reserves are exaggerated.

In this context it is worth noting a paragraph that BP's own report includes as a cautionary note regarding oil reserves, taken from the *BP Statistical Review of World Energy*, issued in June 2006:

> The data series for proved oil and gas reserves [...] does not necessarily meet the definitions, guidelines and practices used for determining proved reserves at company level, for instance, under UK accounting rules contained in the Statement of Recommended Practice, 'Accounting for Oil and Gas Exploration, Development, Production and Decommissioning Activities' (UKSORP) or as published by the US Securities and Exchange Commission, nor does it necessarily represent BP's view of proved reserves by country. Rather the data series has been compiled using a combination of primary official sources and third-party data. Canadian oil sands 'under active development' have been included in proved oil reserves.

If reserve data compiled by BP were as rosy as they appear, the company would probably choose a slogan other than "beyond petroleum" to identify itself. Oil exploration has become tremendously advanced in every aspect since the days of Frank Phillips. The likelihood that major fields, likely to support another century of heavy oil use, have remained undiscovered is very low. BP's reference to Canadian oil sands points at Canada's increase of its oil reserves by 174 billion barrels based on the oil content of the country's tar sands. The oil in these sands is not, strictly speaking, equivalent to pumpable oil as usually understood; the economics of using such oil are certainly less favorable.

DISTRIBUTION CHANNEL

Oil is moved from wells to refineries by oil tankers, pipelines, or tanker trucks. Refineries store the gasoline made in refinery tankage for direct distribution or in pipeline storage tankage for long distance movement. Gasoline is first delivered by tanker trucks or pipelines to bulk terminal storage facilities. Blending of ethanol into gasoline takes place at these facilities, thus close to the customer, because ethanol cannot be pipelined. From bulk terminals gasoline moves by truck to filling stations. These stations, numbering 127,446 in 2002 (according to that year's U.S. Economic Census), are the ultimate retailers of gasoline to the public. Filling stations may be wholly-owned by oil companies, operate as franchisees, or may be independent.

KEY USERS

Using data from the Bureau of Transportation Statistics' 2001 National Household Travel Survey, 88 percent of people aged 15 years or older drive vehicles and thus are key users of gasoline. Approximately 8 percent of households do not own a vehicle. These households tend to have low income, be single-person households, reside in apartments and condominiums (thus often with access to mass transit), and rent rather than own their place of residence.

ADJACENT MARKETS

Markets adjacent to gasoline are all those associated with use of automobiles generally, which tends to include such very diverse areas as the auto industry itself, highway construction, lubricants, tires, construction of outlying suburbs dependent on cars, and convenience stores associated with many filling stations—to name a few. Adjacent in another sense, thus as alternative modes of transportation, are markets, largely in the public sector or controlled by it, providing mass transit. The bicycle market is another alternative industry providing mobility without the need for gasoline. Alternative fuels—in use or just proposed—include petroleum gases like butane and propane, diesel, ethanol derived from agricultural products, and hydrogen. Hydrogen can be produced technically, but not very economically, by hydrolysis of water or (done routinely in fertilizer production) from methane. Electric power, whether produced from fossil fuels, geothermal sources, or uranium represent a feasible alternative if electric vehicles gain traction, which may be possible if gasoline prices reach high-enough levels.

RESEARCH & DEVELOPMENT

Gasoline refining, its use in automobiles, and the impact of its combustion products on the environment have produced wave upon wave of R&D ever since its use as a transportation fuel was adopted in the late nineteenth century. The most recent wave of research has been associated with the production of RFG in which initially MTBE and later ethanol have been used to produce EPA-mandated oxygen levels in this hydrocarbon fuel.

Important R&D efforts were underway within the automobile industry in the first decade of the 2000s aimed, in a sense, at the minimizing or eliminating of gasoline as a fuel, the effort based on concerns over future supplies and fears of increasing dependency on foreign sources of crude. Research in this direction includes tinkering with engines capable of burning pure alcohol, an effort that harks back to 1896 when Henry Ford introduced his first engine, powering the quadricycle, which ran on pure ethanol. Hybrid vehicles that combine gasoline and electric power sources, as well as fuel cell research in which hydrogen would be the chief fuel are other examples. Work is proceeding at perfecting (essentially improving the economics of) the production of liquid fuels from coal.

CURRENT TRENDS

The three most dramatic trends surrounding this product at the end of the first decade of the new century were rising gasoline prices, growing reliance on imported gasoline, and the very public endorsement of ethanol by the George W. Bush administration. The first two trends were viewed as problems, the last as an opportunity to grow our way out of dependence on oil by turning to America's corn farmers. Steeply rising corn prices were producing a negative reaction on the part of livestock raisers, meat producers, and industries reliant on corn for sweeteners, suggesting that converting large segments of corn production to fuel manufacturing could also feature as a problem in the near term, not just in some distant future. In that oil price increases were partially due to international conflicts, in part due to lagging investments in refinery capacity, continuation of these trends was fairly certain.

The environment was not encouraging rapid and massive investments in refinery capacity at a time when ethanol was on the front burner. The end of political conflicts separating the United States from the Muslim world was not readily visible, and even a U.S. pullout from Iraq did not promise rapid return to peaceful oil production in that region as a direct consequence. Current trends, therefore, as seen in 2008, suggests interesting times ahead.

TARGET MARKETS & SEGMENTATION

Gasoline has been traditionally targeted to three groups by performance, thus three types of vehicles: regular gas aimed at the mass market, midgrade gasoline at higher performance vehicles, and premium gas at automobiles with high compression ratios and high horsepower. These segments have been in existence a very long time. Marketing aimed at the segments is thus minimal and largely left to automobile manufacturers who inform users which grade of gasoline will perform best in the new car, pickup, van, or SUV purchased.

RELATED ASSOCIATIONS & ORGANIZATIONS

American Automobile Association (AAA), http://www.aaa.com

American Association of Peak Oil & Gas (ASPO), http://www.peakoil.net

American Association of Petroleum Geologists, http://www.aapg.org

American Petroleum Institute (API), http://www.api.org

Independent Petroleum Association of America, http://www.ipaa.org

The Petroleum Marketers Association of America, http://www.pmaa.org

BIBLIOGRAPHY

Darnay, Arsen J. and Joyce P. Simkin. *Manufacturing & Distribution USA* 4th ed. Thomson Gale, 2006, Volume 1, 442–446.

"Ethanol Timeline." U.S. Department of Energy, Energy Information Administration. Available from <http://www.eia.doe.gov/kids/history/timelines/ethanol.html>.

"Full Text of 2007 State of the Union Speech." MSNBC. 23 January 2007. Available from: <http://www.msnbc.msn.com/id/16672456>.

"Household Vehicles Energy Use: Latest Data & Trends." DOE/EIA-04-64. U.S. Department of Energy, Energy Information Administration. November 2005. Available from <http://www.eia.doe.gov/emeu/rtecs/nhts_survey/2001>.

"International Energy Outlook 2007." DOE/EIA-0484(2007). U.S. Department of Energy, Energy Information Administration. May 2007.

Lazich, Robert S. *Market Share Reporter 2007*. Thomson Gale, 2007, Volume 2, 658–660.

"Number of U.S. Aircraft, Vehicles, Vessels, and Other Conveyances." U.S. Department of Transportation, Bureau of Transportation Statistics. Available from <http://www.bts.gov/publications/national_transportation_statistics/2002/html/table_01_11.html>.

"Quantifying Energy. BP Statistical Review of World Energy." British Petroleum. June 2006. Available from <http://www.bp.com/multipleimagesection.do?categoryId=9011001&contentId=7021619>.

"What is Crude Oil." Chevron Products Company. Available from <http://www.chevron.com/products/learning_center/crude>.

"Where Does My Gasoline Come From." U.S. Department of Energy, Energy Information Administration. Available from <http://www.eia.doe.gov/neic/brochure/gas06/gasoline06.pdf>.

SEE ALSO *Ethanol, Turbines: Wind & Hydropower*

Hand Tools

———————■———————

INDUSTRIAL CODES

NAICS: 33–2212 Hand and Edge Tools, 33–2213 Saw Blade and Hand Saw Manufacturing, 33–3991 Power-driven Hand Tool Manufacturing

SIC: 3423 Hand and Edge Tools, 3425 Hand Saw and Saw Blades, and 3546 Power Driven Hand Tools

NAICS-Based Product Codes: 33–399111, 33–39913, 33–39919, 33–339919, 33–22122, 33–222123, 33–22125, 33–22128, 33–22129, 33–22131, 33–2221302, 33–221303, and 33–222130Y

PRODUCT OVERVIEW

As the term itself implies, hand tools are operated by individuals. In the modern context, the activity itself may be power-assisted, as in the case of an electric saw or drill. In using this term, however, neither in common parlance nor the industrial sphere does anyone mean to imply that a person operating a large machine tool (like a drill press, forging hammer, or a large stationary saw) is using a hand tool. The definition implicitly includes the relative size of the tool and thus the operator's ability to control it by hand. Thus a jackhammer used to break up the pavement is not really a hand tool but a device—a tool used in construction. The jackhammer requires the participation and application of the operator's entire musculature. By contrast a carpenter affixing plywood to a roof using an air-driven hammer, the hammer fed its power from a compressor on a truck located nearby, is considered to be using a power-driven hand tool.

The first major differentiation in this product category is between hand powered and power-assisted tools. Many other distinctions characterize the category as well in various combinations. Tools are classified by the situation in which they are used, thus in domestic or craft activity. There are tools for the casual user and the professional—the differences usually defined by the tool's durability and power; the homeowner presumed to need tools for smaller jobs requiring less skill. Tools are also classified by the many different human goals that they support: lifting, holding, cutting, drilling, affixing, detaching, opening, closing, turning, roughing, smoothing, loosening, heating, measuring, and so on. Tools are associated with functions and are then differentiated within them. Thus, for instance, saws are part of the broader category of cutting tools and are further divided into what they cut: wood or metal. Saws are also designed for different types of cuts ranging from rough to high precision.

The category of work in which the tools are intended to be used also lends its name to the tools. There are agricultural, gardening, automotive, home repair, construction, woodworking, plumbing, electrical, and many other categories. The number of classifications is limited only by what human beings do. There are special tools assisting the philatelist (pincers) and the surgeon (scalpels). Covered in this essay, however, will only be those categories included by the U.S. Census Bureau under its definition of hand tools and power-driven tools—while remaining alert to the fact that the concept of tooling goes well beyond that definition. Neither pincers nor scalpels are included under hand tools by the government but these products are included under other industries. The Census definition used here is tooling used in industrial production activities, but keeping in mind that such tools are also sold to the public at large. At the low end, hand tools sold to consumers are typically much less rugged and have less power than their counterparts used in machine shops and factories. However, in recent decades the differentiation at the upper end of consumer tools is less and less obvious. In the twenty-first century the do-it-yourself person can and does use tools his or her counterpart could only dream about fifty years earlier.

The Census Bureau divides its reporting into two major categories: human-powered and power-assisted hand tools. Distinct classifications are provided for hand and edge tools and saw blades and hands saws, both non-powered. For historical reasons, saws receive special treatment. Power-driven hand tools also receive a classification of their own; within that category there are electrically, pneumatically, and internal combustion engine-driven subcategories. The power-driven category also includes power saws.

Since the Economic Census is the only genuinely reliable microscope that gives us a view into the details of the U.S. economy, these classifications permit us to gauge the overall magnitude of hand tools as manufactured products—the supply side of the equation. The classifications, however, provide less of a view of final use of these products—the demand side. The Census provides much less product-level detail when reporting on wholesale and retail activities. At those levels, down-stream from production, a category like hand tools disappears in the more inclusive category of hardware and farm and garden equipment. It is therefore almost impossible, using Census

data alone, to distinguish between that portion of supply which ends up in the hands of factories or commercial companies and that which finds its way into the hands of the consumer. Hand tools are also too small a category to receive prominent tracking by the government's *Current Industrial Reports* which provide data on import-export trends.

For all of these reasons, data on the domestic production of hand tools in the major categories will be drawn from Census sources. These will include the industrial categories previously listed under the Industrial Codes section. Please note that these categories explicitly leave out significant clusters of tooling such as cutlery and cooking utensils and many categories specific to service and professional activities, including medical and laboratory tools and devices and instruments employed as hand tools in electronics and other industries. The Census data used, however, will be augmented from other sources that illuminate demand and thus trace Census data forward to the point of sale and also account for imports.

MARKET

Using *2002 Economic Census* data as a benchmark, domestic hand tools represented a $10.39 billion industry in which non-powered hand tools and saw blades represented 66.8 percent of shipments and power-driven hand tools 33.2 percent, thus roughly two-thirds of production went into human-powered and one-third into power-assisted products. These data were built by the Census Bureau from a full survey of all producers. Some three years later, as measured by the *Annual Survey of Manufactures*, the

FIGURE 103

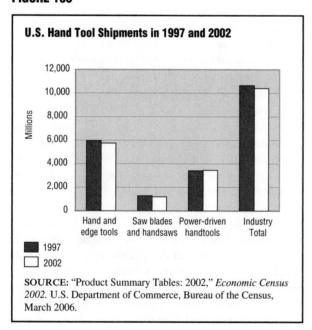

U.S. Hand Tool Shipments in 1997 and 2002

SOURCE: "Product Summary Tables: 2002," *Economic Census 2002*. U.S. Department of Commerce, Bureau of the Census, March 2006.

FIGURE 104

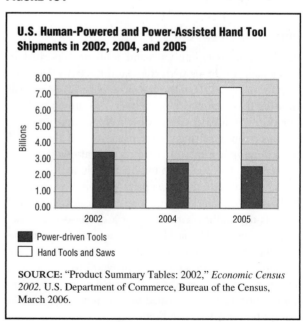

U.S. Human-Powered and Power-Assisted Hand Tool Shipments in 2002, 2004, and 2005

SOURCE: "Product Summary Tables: 2002," *Economic Census 2002*. U.S. Department of Commerce, Bureau of the Census, March 2006.

production had dropped somewhat since 2002 and stood at $10.07 billion, but the respective shares of human-powered and power-assisted tools had changed significantly, representing 74.4 and 25.6 percent of total shipments in 2005—three quarters of the production was hand tools and saws and one-quarter power-driven handtools.

In sharp contrast to these data, which represent domestic supply, the Freedonia Group, an independent research organization, reported quite different numbers about the actual demand for hand tools in a 2005 study widely referred to in this industry. For the 2002 benchmark year, Freedonia estimated total demand to have been $12.1 billion in the United States, nearly $2 billion more than production. Freedonia reported for non-powered tooling demand as $4.86 billion versus the Census Bureau's shipments (supply) of $6.9 billion. The research firm's estimate for the power-driven tools segment was more than double the Census result, $7.24 billion demand versus $3.45 billion supply, the shipments reported by the Census. The research firm estimated that power-driven handtools were 59.8 percent of total demand, non-powered hand tools 40.2 percent. It is well to note that the Census conducted a survey and Freedonia, focusing on demand, carried out an investigation by interviewing associations and companies. Freedonia's definitions of the sectors to be included may have been different from those of the Census Bureau. What is clearly evident from these two sets of data, however, and substantiated by Freedonia's own research, is that substantial portions of domestic demand for power-driven tools were satisfied with imports. In the human-powered hand tool segment, the Census surveys may have included categories not counted by Freedonia. A tentative conclusion is that substantial portions of power-driven tool demand is satisfied with imports. In contrast, substantial portions of domestic production of human-powered hand tools were either exported by their producers or were excluded by Freedonia or its sources.

Freedonia's study, which covered power-driven hand tools globally and in some detail provided estimates of the breakdown between commercial/industrial and consumer demand for power tools. Freedonia's author of the study, Michael A. Deneen, reporting in *Business Economics*, estimated that 70 percent of demand is represented by the commercial/industrial sector and 30 percent by the private consumer or household. Deneen expected this ratio, applicable worldwide, to remain constant into the future. Other than by analogy with power-drive hand tools, no fine breakdown of the ultimate demand for human-powered hand tools is available. A higher portion of that market may well be represented by commercial/industrial demand, at least as hand and edge tools are defined by the Census Bureau. The Census Bureau includes under that category dies and cutting tools that may be used both in hand tools and in machine tools; the Census also includes precision measuring tools. These two categories are likely to be consumed much more in industrial settings; in 2002 they represented nearly $1.5 billion in production. As a rough guess or as a rule of thumb, however, it seems reasonable to assume that all hand tools are distributed between commercial/industrial and consumer markets as are power-driven tools, thus 70/30.

The major distinct end use-markets are industrial production viewed as a whole, the construction industry and consumer do-it-yourself work, motor vehicle repair, and agriculture-forestry-gardening activities. Use-markets, however, are far more extensive than the major activities may indicate. Within the industrial production category are many industrial subdivisions that have their amateur practitioners and hobbyists. Such people also use the industrial or equivalent hand tools. Examples are furniture and cabinet making by hobbyists, boat building and antique automobile restoring, aircraft restoration, home repair and remodeling. Many of these activities take place outside the industrial setting. They are carried out by amateurs who are amateurs in name only as they use the same tools as their professional counterparts. The markets for drills are wherever anyone needs to sink a hole, for saws wherever someone needs to cut metal, plastic, or wood. The individuals engaging in these activities may do so as employees or as independents working on projects of their own.

As a consequence of the pervasiveness of tool use by people, Freedonia found a correlation worldwide between tool use and gross fixed investment. Growing investment correlates with growth in hand tool demand, flat rates with flat or no growth. Other indicators are trends in personal income, level of urbanization, availability of electric power, and the population of automobiles. Using these factors, Freedonia saw the major industrial regions of the world as having mature demand whereas growth was indicated in Asia, particularly China, in Latin America, and in the Africa/Mid-East region.

The U.S. market is mature and, based on long-term trends, is also a declining market for hand tools. The United States has been losing manufacturing employment. Between 1982 and 2004, manufacturing employment has dropped from 17.8 million to 13.4 million according to *Manufacturing and Distribution USA*. Production worker numbers declined in the same period from 12.4 to 9.4 million, suggesting that in 2004, three million fewer production workers needed hand tools than in 1982. Most of these losses (93 percent) were experienced in the 1999 to 2004 period. Other trends have been mixed. The services sector has grown; low interest rates have stimulated remodeling of homes; but personal income in real (inflation-adjusted) dollars has been flat in a consumer market for tools that are saturated in most categories: most people own a drill and the usual kit of hammers, screwdrivers, pliers, and tape measures. The problem is not in ownership, but in finding that tool when needed.

KEY PRODUCERS/MANUFACTURERS

Hand tools manufacturing in the modern context is a relatively easy-to-enter sector. Virtually every country has its own hand-tool making industry. An estimated 50 countries also participate in the power-driven tool industry. For this reason industrial concentration is low and market leaders have relatively small portions of total market. Companies tend to be concentrated in either the non-powered or the power-driven portions of the hand tool industry; however, the leaders typically offer both kinds of tools but dominate in one of the categories.

A selection of key producers is presented under two headings below. U.S. market share leaders in 2005 are presented first based on data collected by Credit Suisse and reported in *Market Share Reporter*. Next, leaders in power-driven hand tools are shown by 2004 market share as presented by Freedonia Group.

Non-Powered Hand Tool Market Leaders. The Stanley Works began in Connecticut in 1843 as a door-bolt company and later developed into the United States' leading producer of non-powered handtools. Stanley had a market share of 21 percent in 2005. Stanley had revenues of $4 billion in 2006, its operations divided into three sectors: Industrial, Consumer, and Security Products. The Industrial sector is a provider of non-powered and power-driven tools in every category. The Consumer sector of the company produces carpentry, electronic, masonry, and mechanics tools as well as measuring instruments. The Security Products sector at Stanley was the direct outgrowth of the company's original involvement with doors. Today Stanley's security systems extend from locks and components up to very sophisticated systems solutions, including sensors and automatic notification protocols.

Danaher held the second largest share of the non-powered tool market in 2004 with 19 percent. The company's sales in 2005 were $7.9 billion, but only a portion of these sales were in the hand tool product lines. Danaher began in the early 1980s as a grouping of companies—named after the Danaher River in Montana where its founders were fishing together and talking of joining their efforts. The company's principal focus is on the support of manufacturing activities by others with products and services in the categories of measurement, motion, water management and related environmental services, product identification, and tooling. Danaher also has activities related to medical technologies.

Snap-On Incorporated was the third place market shareholder in non-powered tools with 11 percent market share. Snap-On had sales of $2.4 billion in 2006. It is a Kenosha, Wisconsin, company operating four divisions: Snap-On Dealer Group sells tools directly to mechanics and repair shops; Commercial and Industrial Group sells tooling and systems to corporations; Diagnostics and Information Group sells diagnostic equipment and services, including management and information services; the Financial Services operates in support of its dealer networks and is jointly owned with CIT Financial Corporation. Founded as the Snap-On Wrench Company in 1920, the company started to sell interchangeable socket and wrench handles invented by its two founders, Joseph Johnson and William Seideman. Direct demonstration at mechanics' shops of the benefits of such tooling turned out to be the best way to sell this innovative product and has remained at the core of Snap-On's strategy.

Cooper Industries had sales of $4.7 billion in 2005. Approximately 85 percent of Cooper's sales are from electrical products, but the company is a leader in non-powered hand tools, with a 7 percent market share. Cooper began in 1833 as a small iron foundry operated by the brothers Charles and Elias Cooper in Mount Vernon, Ohio. It made power and compression equipment for natural gas transmission until the 1960s. After that time, in efforts to diversify, the company entered petroleum and industrial equipment, electrical devices, and also began making electrical and other tools.

Power-Driven Hand Tool Market Leaders. Black & Decker is the global market share leader in power-driven equipment, with 12 percent of the world market in 2004. The company had its start in 1917 when its two founders, S. Duncan Black and Alonzo G. Decker, expanded a small machine shop into a manufacturing plant in Towson, then a small suburb of Baltimore, Maryland. The driving innovation was the invention of a pistol grip and trigger switch on a drill the company was making. Since that time the company has expanded by a combination of innovation and geographical expansion, first to Canada then to England, Australia, and beyond. The company had sales in 2005 of $6.3 billion, of which power tools and accessories accounted for $4.7 billion.

Robert Bosch GMBH, with a 9 percent global market share in power tools in 2004, was founded by Robert Bosch in 1886 as the Workshop for Precision Mechanics and Electrical Engineering, based in Stuttgart, Germany. The company's entry into power tools began in the mid-1930s by deploying its smallest electric motors as power sources for hand-held tools. Among the company's well-known brands are Skil (saws), Dremel (precision tooling), Vermont-American (a producer of bits for tools), and RotoZip (a drill bit, saw, and sawing tool producer). Bosch offers more than 1,000 specific products, including components. In the United States Bosch operates as Robert Bosch Tool Corporation through its division, Bosch Power Tools, located in Mount Prospect, Illinois.

Techtronic Industries was founded in 1985 by Horst J. Pudwill and Roy C.P. Chung to manufacture rechargeable battery packs and appliances. The first production facility began operations in Hong-Kong. The company grew by acquisition and features such brands as Milwaukee (industrial power tools) and the German company AEG (diversified power tools), both acquired in 2005. The Ryobi brand of saws and clean-up devices was acquired from its Australian owners in 2002. Homelite, a producer of chain saws and outdoor clipping equipment was acquired in 2001. Techtronic had a 7 percent global market share in 2004 which apparently increased a year later with the integration of Milwaukee and AEG brands.

Makita Corporation, a Japanese company, had a global market share of 5 percent in the power-driven tool market. It manufactures 320 different types of power-driven hand tools for industrial, consumer, and garden and outdoor applications. Makita began in 1915 by selling and repairing electric motors. Its first power tool was an electric planer.

Hitachi began operations in 1910 as an electrical repair shop and then progressed by making transformers, fans, electric locomotives, elevators, refrigerators, power shovels, transistor radios, and computers. The company thereafter developed into one of the world's electronics giants. The company classifies its power tool activities under the label of DIY (do-it-yourself), which it subdivides into power tools and garden supplies. Hitachi has 54 U.S. subsidiaries. The company's U.S.-based power tool operations are managed by Hitachi Koki U.S.A. headquartered in Atlanta. Power tools are a very small element within this giant's portfolio but represented 4 percent of the global market in 2004.

MATERIALS & SUPPLY CHAIN LOGISTICS

Hand tools cover a spectrum from very simple tools, such as hand-held files made of single pieces of steel, on up to complex machines that require assembly, such as power drills. Human-powered hand tools typically require the joining of a steel tool with a handle made of wood, hard plastic, or a composite of nonmetallic materials. A simple example is a screwdriver made of metal tightly held by a plastic handle. Both components may be fabricated entirely on site from raw materials or, more typically, components may arrive semi-fabricated and are further processed at the final manufacturing stage. Non-powered hand tool producers typically manufacture multiple lines of such

tools and will therefore have complex fabrication and assembly arrangements supported by extensive inventories of componentry and work in progress. Materials will be drawn from multiple vendors and arrive in various stages of completion. Since the most valuable and proprietary component of hand tools is the work piece itself, operations tend to be predominantly metal working factories or associated with such.

In the case of power-driven tools, the power source is the most important component of the tool in terms of complexity and cost. The core may be an electric motor, a pneumatic device, or an internal combustion engine. If the device is electric, it may be cordless and feature chargeable battery packs which the company either specializes in manufacturing, has a license to produce, or purchases from others. These power devices may be purchased or may be the very heart of the production operation. Small makers of power tools may purchase the motor or engine and engage in assembly only or in fabrication and assembly, joining parts to a purchased power-pack. Large producers may make the power-pack and purchase drill bits, blades, saws, chain saws, and other components from others. All manner of combinations are prevalent depending on the history, size, and technological alignments of the producers.

A distinct third category of producer in the hand tools industry is the manufacturer of bits only, including cutting dies, blades, edges, drills, saws, shaped tools for insertion into power equipment, and the like. These companies make the working end of the tool, not the entire tool, and will therefore often be metalworking specialists purchasing steel or exotic metal stocks as raw materials and shaping them in their operations.

The output of this industry ends up in large metropolitan areas, not least outdoors and gardening tooling. Relatively small portions of the industry output are shipped to rural areas or distribution centers in small towns. The hand tools industry's operations tend to be located in or on the edges of large urban areas with access to raw materials, semi-finished goods, and transportation hubs.

DISTRIBUTION CHANNEL

Consumer Products. Virtually all hand tools reaching the consumer markets, whether non-powered or power-driven, move through a three-tier distribution system. The ultimate customer buys from a retailer who makes purchases from a wholesaler who obtains goods from the producer. The retailer may be a Big Box store such as Home Depot or Lowe's, a department store such as Sears/Kmart or Wal-Mart, a hardware store, a lumber yard, a gardening outlet, or a specialized equipment dealer.

Outdoor tools such a chain saws and other gasoline-powered devices such as trimmers, tend to be sold by so-called servicing dealers—for the simple reason that such tools often require servicing; chain saws, for instance, require sharpening. Dealers are typically supplied by independent distributors specializing in power equipment in certain categories, such as green goods, intended for the outdoor trade. Since the appearance of large discounters, a tension has existed between producers who sell both to discounters and to dealers and the dealer chains that distribute their products. Dealers require the higher margins realizable from selling original equipment to subsidize their servicing operations. They resent producers who sell their equipment at a discount. The same equipment will show up at the dealership for servicing. Some companies sell exclusively through dealers and thus cultivate a loyal sales channel at the retail level. Dealer-oriented producers rely on high quality products sold at a healthy margin but effectively supported by excellent services. Other producers use a strategy of offering products at various quality and price points and aiming to achieve their aims through volume.

A three-tier distribution is virtually dictated in this industry by the sheer variety of tools available. The matching of demand at the retail level, particularly with many different types of retailers involved, calls for the additional administrative and sales-outreach provided by the wholesaler in the middle. The exception, however, proves the rule. Large retailers offer store brands made for them directly by the manufacturer based on a product mix specified by the seller. In these cases a two-tier distribution is in effect.

Catalog sales represent another two-tier distribution system, but such sales are limited to categories such as specialized gardening tools and categories of hobbyists tools that are included in the catalog along with many other items not produced by this industry.

Industrial/Commercial Products. Industrial/Commercial hand tool distribution may be three-tier or two-tier depending on the size of the commercial buyer. If the buyer is a self-employed craftsperson, his or her purchases are by definition for commercial use but the buyer will typically purchase the tooling from a retailer. Many retailers provide a commercial discount, however, and the buyer realizes a small benefit over the ordinary homeowner. In large commercial operations and in industrial purchasing, tooling will be bought from a specialized distributor directly. These distributors are organized to meet the requirements of contractors or automotive repair shops or specific industries, carrying very diverse lines of tooling but highly targeted to the mix of businesses that they serve. Online access by the buyer to the distributor's inventory is a common and still growing aspect of this type of distribution as the twenty-first century gets under way. Such access permits an industrial buyer to order items directly as needed with minimal human interaction between buyer and seller.

KEY USERS

The use of hand tools in industrial and commercial contexts is an inherent aspect of all such activity. Tool users and categories of tools emerge into view by simply naming the activity itself. Construction has its tooling as does railroad maintenance, each category often using the same standard tools and some that are unique to the industry. In the consumer category, distinct classes of key users emerge: the Do-It-Yourself (DIY) practitioner, the home gardener and lawn worker, and the person pursuing a hobby or a craft activity that calls for industrial-style tooling. In all of these categories, tooling represents a relatively small proportion of total expenditures on the activity; supplies represent the majority of dollars spent.

This is well illustrated by looking at the DIY market. According to the Home Improvement Research Institute (HIRI), as cited by Credit Suisse, tools are a part of the consumer home improvement market, estimated at approximately $311.1 billion in 2006. Tools were 6 percent of total expenditures; the largest category was lumber and building materials (27%). Another earlier estimate, provided by Darrin M. Brogan, put the 2005 DIY market, a subset of HIRI's home improvement market, at $21.3 billion and estimated total expenditures on hardware and tools at $5.1 billion, just under 24 percent of total.

Among key users at the consumer level are gardening enthusiasts who buy special gardening tools and homeowners who take care of their lawns. This was a market of between $24 to $35 billion dollars depending on the source, the high figure provided by the National Gardening Association. The category is largely dominated by purchasers of fertilizers, seeds, plants, and lawn care services. Equipment may be as high as 25 percent of total, but that includes professional lawn care equipment and riding mowers which do not fit the hand tools category.

The last key user group is represented by the person engaged in a craft as a hobby, purchasing industrial grade equipment, including hand tools, to carry on such activities at home. This user group divides into those engaged in automotive work and those in various kinds of precision carpentry and similar woodwork.

ADJACENT MARKETS

Markets adjacent to hand tools are other tooling markets typically more closely associated with specific activities. As viewed from the perspective of the consumer market, adjacent markets would include cooking, textile, and office tools—thus, kitchen utensils of all kinds, sewing kits and scissors, handheld seamers, irons, textile cleaning tools, computers, rulers, pens, and pencils. Two other adjacent categories are tools used in bodily care, such as razors and devices for shaping or trimming hair, and household cleaning tools including brooms, mops, hand-held vacuums, stain removers, wax appliers, and brushes of all kinds. In part these markets are adjacent because they provide products of similar function, as well as compete with DIY work, outdoors pursuits, or hobbies that use industrial tooling.

Viewed from the industrial perspective, the tools used in laboratories, medicine (including dentistry), precision tools used in making jewelry, and many other categories excluded from the Census definitions being used but included with specific industries, are adjacent in providing functional equivalents.

Within the narrower category under consideration here, adjacent markets are represented by specialty chemicals that in one way or another do the work of hand tools. An example is adhesives that will cause surfaces to adhere in cases where the alternative is nailing. Throughout industry, as well as in the home, advances in materials sciences are impacting tooling, either by eliminating specific activities (e.g., sanding or smoothing) or by providing finished products that do not need dressing or cutting.

RESEARCH & DEVELOPMENT

In this very mature industry, R&D efforts are aimed at product improvements to induce industrial and commercial buyers to upgrade to products that are more efficient and safe and to attract the more conservative and generally inattentive do-it-yourselfer by innovation. Attention is directed at improving the performance of tools at lower power levels; upgrading the performance of 12-volt tools to those of 18-volt tools, for example. Cordless products worked a significant revolution in power tools by making them easier to use without the hassle of managing cords. Efforts to extend the time over which battery packs stay charged and perform at peak is a goal of research.

Ergonomic design is a focus of research as well. Handles and grips developed on the basis of ergonomic studies can not only result in novel forms that draw the customer's interest by looking different, but can also take effort out of using the tool and can introduce higher levels of safety in tool use, a perennial concern to the industry. Multi-purpose tooling has always been popular. Research aimed at speeding up the process whereby one working bit is exchanged for another is an ongoing effort.

CURRENT TRENDS

At the production level, a major but continuing trend is globalization of production whereby particularly power-tool producers are moving their manufacturing operations to low-labor cost markets. In this industry particularly, where barriers to entry are relatively low, a complementary movement is represented by low-labor cost regions also entering the business as exporters and—an example is China—producing for their growing domestic demand as well. This trend is beginning to manifest in the non-powered hand tool industry as well. According to data from Freedonia, the North American market in 2006 accounted for 40 percent of global demand but only produced 30 percent of the product consumed worldwide.

At the retail level in the consumer segment, independent retailers are feeling the competitive pressures of the big box discount houses. The evidence for this is anecdotal—indirectly shown by the high growth rates in these categories achieved by retailers like Home Depot and Lowe's in comparison with total growth rates in the industry. Another trend in the retail sector is the increasing participation of women not only in initiating home improvement projects but also as buyers of hand tools, in part a consequence of growth in numbers of households headed by women.

In part arising from the growth in the DIY market, but a perennial concern of the industry, is emphasis on the safe use of power tools, strongly promoted by the industry associations.

TARGET MARKETS & SEGMENTATION

In the hand tools industry, virtually all markets are not only well defined but are served by specialized tools developed specifically for them. One area where targeting is discernible is the introduction of more and more professional-level tools to the consumer market. This is resulting in a blurring of the distinction between tools in the home and tools in the factory—at least at the higher end of tools sold to the consumer and the do-it-yourselfer. This targeting is almost always accomplished by marketing efforts rather than by product redesign.

A strongly emergent trend is the emergence of women as DIY tool users. This has caused targeting approaches aimed at taking gender bias out of tool promotion and, in some instances, introducing color changes specifically aimed at attracting female buyers.

RELATED ASSOCIATIONS & ORGANIZATIONS

American Hardware Manufacturers Association, http://www.ahma.org

Compressed Air and Gas Institute, http://www.cagi.org

The Hand Tools Institute, http://www.hti.org

Home Improvement Research Institute, http://www.hiri.org/

International Staple, Nail and Tool Association, http://www.ISANTA.org

National Hardware Show, http://www.nationalhardwareshow.com/images/100464/index.htm

Outdoor Power Equipment Institute, http://www.opei.mow.org

Power Tool Institute, Inc., http://www.powertoolinstitute.com

Unified Abrasives Manufacturers Association, http://www.UAMA.org

BIBLIOGRAPHY

Borgan, Darrin M. and Stanton G. Cort. "Industry Corner: DIY Products: A Global Perspective—Do-It-Yourself." *Business Economics*. January 1997.

Carlson, Scott. "Independent Hardware Retailers Lose Market Share to Big-Box Stores." *Saint Paul Pioneer Press.* 9 June 2005.

Darnay, Arsen J. and Joyce P. Simkin. *Manufacturing & Distribution USA,* 4th ed. Thomson Gale, 2006, Volume 2, 891–897, 1208–1212.

Deneen, Michael A. and Andrew C. Gross. "The Global Market for Power Tools." *Business Economics.* July 2006.

Lazich, Robert S. *Market Share Reporter 2007.* Thomson Gale, 2007.

"New and Improved Tools Rule the Market." *Industrial Distribution.* September 2002.

"Product Summary: 2002." *2002 Economic Census.* U.S. Department of Commerce, Bureau of the Census. March 2006.

Sachdev, Ameet. "Hardware Manufacturers, Retailers Now Cater to Women." *Chicago Tribune.* 14 August 2000.

"U.S. Power & Hand Tool Demand." *Do-It-Yourself Retailing.* January 2005.

Zelman, Ivy L., et. al. "Homebuilding: An Investor's Guide." Credit Suisse Equity Research. 13 October 2006.

SEE ALSO *Construction Machinery, Lawn & Garden Tools*

Handbags

■

INDUSTRIAL CODES

NAICS: 31–6992 Women's Handbag and Purse Manufacturing

SIC: 3171 Women's Handbags and Purses

NAICS-Based Product Codes: 31–69920 through 31–69920131

PRODUCT OVERVIEW

A handbag is any type of bag or case that can be carried either by hand or over the shoulder. Such bags are traditionally marketed to women, who use these bags to carry their cosmetics, money, and personal belongings. Handbags and purses may be manufactured from a wide range of materials, including leather, denim, vinyl, and straw. Handbags also may be closed in some way, either by gathering up the straps, or with a snap or clasp. Some industry trackers have begun to include small totes and camera bags when analyzing the handbag category. Such items have traditionally been seen as luggage. However, as more people find cell phones, iPods, and other handheld devices to be necessities, the traditional definition of handbag is being expanded.

There are a number of different handbag styles. The bag's form or its material often dictates its name. According to DesignerHandbags101.com, a box bag is a hard-sided purse, square or rectangle in shape, with a metal, bone, shell, or wooden handle. An envelope purse is a flat, square or rectangular bag with a triangle-shaped top flap that folds over like an envelope. A half moon bag is any bag shaped like a half moon, with or without a handle. A canteen bag is a round, stiff bag that resembles a water flask. A bucket bag is shaped like a bucket, usually has an open top and shoulder strap. A baguette is a small, long, narrow bag that resembles a loaf of French bread. A quilted bag is characterized by its texture—quilted—and is usually carried with a strap or metal chain. An accordion bag consists of several small bags stitched together.

Some types of handbags are better known. A change purse is a small purse just large enough to hold loose change that can be sealed with a zipper, clasp, or snap. It is usually kept in a larger handbag. A clutch is a bag with no handles that must be carried clasped in one hand or under the arm. A satchel is a bag with a wide, flat bottom, zippered or clasped top, and two handles or straps. A tote bag is a medium sized bag with two straps.

The word handbag did not start being used until the twentieth century. The word purse has its origins in the Latin word *byrsa* and the Greek word *bursa*, according to the *Oxford English Dictionary (OED)*. The *OED* defines a purse as "a small bag with straps or drawstrings that allow the bag to be closed and carried."

For centuries both sexes carried their valuables and personal belongings with them rather than on them. Both sexes carried pomanders in the sixteenth century, small mixtures of aromatic substances carried in a little bag. Pomanders were believed to prevent possible infections. Women also carried little bags of lavender to scent handkerchiefs. Women later carried chatelaines, which were a set of chains around the waist that might hold a pencil, keys, or a watch. Men might carry tobacco pouches or large wallets to hold money and documents. Valuables might also be placed into a leather pouch and secured to their belts for extra protection.

People carried their belongings because pockets were a later addition to clothing. The notion that pockets were intended to take the place of bags and satchels can be found in the word's formal definition. The *OED* defines a pocket as a "small bag or pouch worn on the person." Pockets first appeared on men's clothing in the late sixteenth century but were more problematic for women's apparel. Clothing tended to be form fitting in the seventeenth century; pockets would disturb the line of a skirt or dress. Women got around this problem by tying their pockets around waists (some did so until the mid-nineteenth century). The hoop skirt, which came into fashion in the early sixteenth century, was the first dress to incorporate pockets. Slits could be cut into the hoop skirt and pockets could be attached in the ample space beneath. When styles changed and women's apparel became tighter in the eighteenth century, women again began carrying their belongings in little bags called reticules.

Reticules were small, very feminine bags that could be used to hold rouge, face powder, a fan, a scent bottle, or smelling salts. Reticules could be purchased at local markets or estate sales, while many women made their own. Knitting was a popular pastime for women in the eighteenth century, and women could design their reticule to their liking. They could be made of velvet, lace, or other readily available fabrics.

The handbag industry would benefit from both labor and social changes in the twentieth century as women entered the workforce in greater numbers. Women's reticules would need to be larger and sturdier as they took vital accessories into the workplace. While needlepoint and embroidery certainly still existed, the Industrial Revolution meant that goods (including apparel accessories) could be produced more cheaply and quickly. Fabrics and materials became easier to produce and obtain, as retail and distribution lines became more robust. The apparel, leather, and textile industries expanded. Leather and trunk makers began producing small suitcases and leather bags for the first time. The word handbag appears about this time, to describe the leather bag a man might use to carry important papers and belongings.

Women were carrying what looks like a modern purse by the 1920s. This decade is sometimes called the Roaring 20s because of the changes in the entertainment and fashion industries brought on by economic prosperity. Some young women became known as flappers; they wore short skirts, bobbed their hair, drank alcohol, and expressed their disdain for social conventions. Women felt free to behave this way in part because of the recent success of the women's suffrage movement in the United States. The Nineteenth Amendment, which granted women the right to vote, had been passed in June 1919. It was also about this time that some designers began to open their influential fashion houses: Coco Chanel opened her first shop in 1912 in Paris, France; Prada was established in Milan, Italy in 1913 and Gucci opened in Florence, Italy in 1920.

Handbag and purse manufacturing would continue to be driven by fashion and cultural changes. During the 1940s, when many materials were needed for the war effort, handbag manufacturers started using wood rather than metal in the construction of the purse's frame. The apparel industry was under similar restrictions, as nylon and other fabrics were needed for the war. Women's fashions moved from the glamorous, flowing gowns of the 1930s to tailored suits. The war had another effect on the landscape. Millions of women entered the workforce, taking up jobs in factories and offices. Women would need handbags that were simple in design but also durable.

Designers pushed a return to a more feminine look in the years after World War II. This period was another one of economic prosperity for the United States marking the beginning of America's consumer culture as well as a number of important fashion trends such as saddle shoes, poodle skirts, and the stiletto heel. The decade also marked the appearance of two bags that were seen as trend setters and are still coveted by collectors: the Hermes Kelly bag and the Coco Chanel 2.55. Chanel's 2.55 bag, named for its February 1955 release date was a quilted bag with a shoulder strap and was a revolution in the fashion world. The bag was admired for its first-class construction but also for its design. A woman typically had to carry her purse by hand, even if it had a strap. With a strap long enough to fit around the shoulders, a woman now had her hands free. Chanel had to turn down numerous requests to make the bag because of the time spent manufacturing each one. In 1956 Grace Kelly appeared on the cover of *Life* using a Hermès bag to hide her pregnancy after her recent marriage to Prince Rainier of Monaco. The bag was later named the Kelly Bag. There was a six month waiting list to get a bag in 1956. In 2006 the crocodile bag cost between $10,000 and $60,000 and had a 3-year minimum waiting list. The bag reportedly takes 25 hours to manufacture.

The elegant, crafted purse was still very much in vogue in the early 1960s. However, by the mid-point of the decade fashions were changing quickly. The bouffant hairdo popularized by Jackie Kennedy began to fall out of favor with young women by 1967. Mary Quant popularized the miniskirt in 1965. Artists such as Andy Warhol and Roy Lichtenstein were part of the Pop Art movement, in which everyday objects such as Coca-Cola bottles and Campbell's soup cans served as the inspiration for art. Fashion and handbag design seemed to mirror this overall trend. Handbags were often constructed in unusual shapes and made of a wide range of materials: raffia, straw, wicker, vinyl, velvet, fur, crocodile, lizardskin, snakeskin, and patent leather.

The social and political unrest in the United States in the late 1960s continued into the next decade. People seemed to want to forget their problems. Sea horses, Rubiks cubes, and pet rocks were just a few of the novelty items that sparked people's interests in the 1970s. The wish to look beyond the nation's unemployment, high gas prices, and impeachment proceedings are best symbolized in the decade's most enduring symbol: the smiley face. The yellow graphical image was a highly stylized, synthetic version of a human face: just a circle, two dots and curved smile. The commercial use of the image is credited to Harvey Ball, who first used it on a button in 1963. But it wasn't until the early 1970s that brothers Murray and Bernard Spain put the image on mugs, T-shirts, bumper stickers, and purses.

The 1980s brought a return to more traditional, natural styles. The conservative preppy look was very popular: blue blazers, skirts, and button-down oxfords. This decade also is known for its emphasis on luxury. Shows such as *Dynasty* and *Dallas* were popular. The actresses' wardrobes helped make certain styles popular: shoulder pads, more jewelry (both precious and costume), heavily moussed or sprayed hair, and bright colors.

The 1980s also brought another *It* bag, a bag that women just had to have for its elegance and construction. One day in 1984, actress Jane Birkin was sitting next to Jean-Louis Duman, the president of Hermès, on an airplane. Birkin sketched her ideal bag for Duman; that sketch served as an inspiration for the Birkin bag. The bag costs $6,000, although the price can easily reach $80,000 due to its exclusive design. The *Birkin* as it is often called has become a status symbol among wealthy women. Both Martha Stewart and rapper Lil' Kim carried the bags with them when they went to federal court in 2004. The Birkin bag has one of the longest waiting lists of any luxury accessory—reportedly as long as six years.

A few trends caught on in the 1990s: parachute pants, the rise of sports apparel as everyday apparel, and bright colors. The Retro look was in, with fashions from the 1960s and 1970s once again becoming popular. The decade's major fashion trend, however, was grunge fashion. Grunge rock began about 1992, when Nirvana, Pearl Jam, and other bands from the Washington state area captured the interest of the country's young people. The artists and fans kicked off a fashion trend: torn blue jeans, tie-dye T-shirts, flannel shirts, Doc Marten boots, and body piercings. By the late 1990s, grunge music lost its influence. The clean-cut preppy look from the 1980s returned, particularly for teenage boys. Each of these fashions were

reflected in their time by handbags that fit the look of the outfits being worn at the time.

Handbags have been the best-selling fashion accessory for some time. As an accessory, they follow the various trends of the apparel industry. However, as seen above, fashion trends change quickly. The tassels and kiss-lock closures that were popular on handbags in 2006 may not be so in the future. Certain rules seem to remain true, however. Women will always like neutral colored purses that can be worn with a number of outfits. They are also willing to spend heavily on the right purse, a purse with the right combination of style and craftsmanship that appeals to them.

MARKET

U.S. Industry shipments of women's handbags and purses fell from $250.9 million 1997 to $193.1 million in 2002, according to the U.S. Census Bureau. In 2002, approximately 63 percent of the handbags shipped were of leather or mostly leather construction. Only 3.3 percent were made of mostly plastic or vinyl in construction. Other types of bags (excluding those constructed of precious metals) made up the remaining shipments. From 1997 to 2002 the number of handbag manufacturers in the United States fell from 136 to 98, a drop of 27.9 percent.

During this period U.S. manufacturers were moving their operations overseas in an effort to control manufacturing costs. As the number of handbag makers fell in the United States, handbag imports began to creep up, from $1.05 billion in 1997 to $1.3 billion in 2002, according to figures from the International Trade Administration. Imports have increased steadily since then to $2.3 billion in 2006. China is the top provider of luggage to the United States, representing 67 percent of luggage imports in 2006. Italy and France, two countries that are home to many well-known fashion houses, are far behind. Italy was second with a 16 percent market share in 2006, followed by France with a 7.6 percent share. The United States exported only $17.6 million in 2006, mostly to Canada and Japan. However, shipments have improved steadily since 1996, when they were a mere $8.8 million.

Handbag manufacturing in the United States has been on the decline since the early 1980s. Figure 105 presents two decades worth of data from the U.S. Census Bureau on shipments by the industry it calls "Women's Handbag and Purse Manufacturing." The decline has been dramatic. In 2002 industry shipments were less than a fifth what they had been 20 years earlier. The industry is so diminished that for years after 2002, the Census Bureau has combined its reports on this industry with data for another shrinking U.S. manufacturing industries: luggage, personal leather goods, and leather goods. In the meantime, retail sales of handbags and purses have been

FIGURE 105

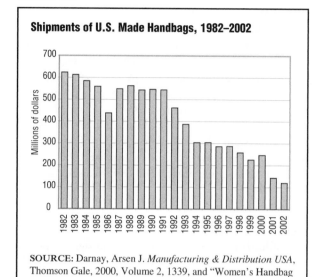

Shipments of U.S. Made Handbags, 1982–2002

SOURCE: Darnay, Arsen J. *Manufacturing & Distribution USA*, Thomson Gale, 2000, Volume 2, 1339, and "Women's Handbag and Purse Manufacturing: 2002," *2002 Economic Census,* U.S. Department of Commerce, Bureau of the Census, January 2005.

healthy. Although handbags are being made in ever fewer numbers in the Untied States, they are being purchased in ever greater numbers. *Accessories Magazine* estimates that retail sales of handbags have climbed from $5.2 billion in 2002 to an estimated $6.9 billion in 2006.

There are several reasons for the increase in retail sales. Purses were once neutral colors such as blue, black or tan. One handbag might coordinate with a number of outfits. Women still follow this rule when making handbag purchases. However, styling has become more dynamic in recent years; women choose handbags that are more unique and express their own individual tastes. The industry has also become so competitive that there are many different styles for a woman to choose from. Due to these trends, women are buying more handbags. According to Coach Inc., in 2005 the typical U.S. woman purchased four handbags, nearly double the number she purchased only

FIGURE 106

Categories of Handbags by Retail Price Range, 2007

Retail Handbag Categories	Retail Price Range
Luxury bags	$301 or more
Upscale bags	$101 to $300
Mid-market bags	$51 to $100
Standard bags	$21 to $50
Low-end bags	$5 to $20

SOURCE: Compiled by the staff with data gathered from a survey of industry trade journals and retail catalogs in the summer of 2007.

five years earlier. In addition to buying a greater number of handbags, the price of the bags purchased appeared to be no obstacle. The price of a luxury bag increased from $500 to $1,000 during the period 2001 to 2005, according to the President and CEO of high-end retailer Bergdorf Goodman.

A bag is considered a luxury or designer bag based on its price and the price will generally vary depending on the materials used to produce the bag. Luxury bags have approximately 43.5 percent of industry sales, according to industry estimates. Figure 106 provides prices ranges for handbags by category widely available in the United States in 2007.

Coach reported that its share of the luxury bag market improved from 17 percent to 23 percent in 2006. Analysts believe it has found success by combining fashion (assorted colors and styles) with function (incorporating laptop and Blackberry storage, for example). Another key to Coach's success is its retail prices. An average bag costs approximately $250, making it a luxury item while remaining affordable to many. It was the second most-reported brand of handbag owned by women in 2006, according to a study by Cowen & Company.

Designers periodically produce exclusive products in the hopes that they will become the next must-have status symbol. Louis Vuitton introduced the $42,000 Tribute Patchwork handbag in late 2006. The Lana Marks Cleopatra Clutch in alligator comes adorned with 1,500 black and white round diamonds. Only one clutch is made for retail each year; it retails for $100,000. The Hermès $148,000 Birkin in croc porosus lisse was the most expensive bag available in 2006 because of the rare leather used in its making and also because of the nine carats of diamonds set in white gold and placed on its clasp and lock. This bag is custom made.

KEY PRODUCERS/MANUFACTURERS

Coach. The industry leader designs and markets a number of apparel accessories, including handbags, outerwear, electronic accessories, luggage and small cases. The company sells its products in nearly all retail channels, including catalogs, online stores and company-operated retail and factory stores in North America and Japan. Its products are also available in department stores and specialty retailers. As of July 1, 2006, it operated 218 retail stores and 86 factory stores in North America; and 118 department store shop-in-shops, retail stores, and factory stores in Japan. Coach, Inc. was founded in 1941 and is based in New York, New York. It has 2,300 employees and earned $2.4 billion in its fiscal year ended July 2006.

Dooney & Bourke. This maker of high-end women's handbags is sold in online stores, in catalogs and in de-

FIGURE 107

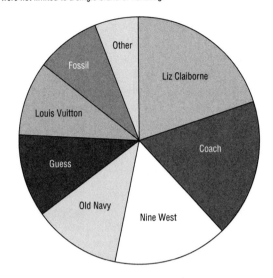

Most Widely Owned Brands of Handbags by Women in the United States, 2006

These data are the self reported results of a survey in which respondents were not limited to a single brand of handbag.

SOURCE: Compiled by the staff with data from Montgomery, Elizabeth and Eric Nanes, *2007 Handbag: U.S. and Japan Markets Look to Different Paths*, Cowen & Company, January 9, 2007, 15.

partment stores. It operates seven of its own stores nationwide, including a flagship location in Manhattan. It also has stores in Tokyo, Japan and Old San Juan, Puerto Rico. In addition to handbags, the company manufactures cellular phone cases, iPod cases, hats, jewelry, luggage, men's and women's apparel, shoes, tote bags, and wallets. It generated $6.2 million in revenues in 2005 and had 104 employees.

Liz Claiborne. The company designs and markets a number of products, including handbags, men's and women's apparel, jewelry, and fragrances. Its products are available in department stores, specialty retailers, catalogs and on-line vendors. It had 17,000 employees and revenues of $5 billion its fiscal year ended December 2006. The company was founded in 1976 and is based in New York.

Louis Vuitton Moet Hennessy (LMVH). LMVH is the world's largest luxury goods company. It makes wines and spirits, perfumes, cosmetics, fashion and leather goods, watches, and jewelry. Some of their well-known brands include Dom Perignon, Moet & Chandon, Christian Dior, Givenchy, Donna Karan and TAG Heuer. The company reportedly had 61,088 employees and revenues of $16.4 billion in 2005.

Nine West Group. The company is known largely for its footwear, but it also has a successful line of handbags and apparel. It operates approximately 600 of its own retail stores in addition to being available at department stores, specialty retailers, and independent shoe stores across the United States. The Jones Apparel Group acquired the company in 1999. Jones Apparel Group generated revenues of $4.6 billion in 2006.

MATERIALS & SUPPLY CHAIN LOGISTICS

Many manufacturers began to move their operations out of the United States in the 1990s. The handbag industry shared many of the same problems experienced by the apparel and leather industries to which it is connected. The industry suffered from high production costs because of the labor-intensive nature of handbag manufacturing. Overseas markets offered a large pool of non-unionized workers and also offered access to more plentiful materials and at lower costs. Also, with the increasing globalization of the international community, it was advantageous for companies to expand into developing markets that were hungry for new products. This was particularly the case for fashion, where certain luxury names appeal to women of all cultures.

Leather production grew in China and the other countries to which U.S. manufacturers were shifting their operations. China produced an average of 4.8 million bovine hides and skins annually from 1984 to 1985, according to World Trade Organization statistics. Bovine hides and skins are processed into leather. By 1990 China produced 9.3 million hides. In 2000, the figure was up to 40.1 million. India's production increased, although not as dramatically as China's. The country manufactured 36.1 million pieces in 1990; a decade later, production was up to 40 million. In terms of the overall leather goods trade (including footwear) the Far East had 15.7 percent of the market from 1984 to 1986, according to World Trade Organization statistics. By the period 1999 to 2001, that figure had more than doubled and stood at 33.7 percent.

Overall costs may be lower for U.S. businesses in these countries, but there are still other considerations. The animal hides must be obtained and then processed into leather in tanneries. From there they must be manufactured into leather goods and shipped back to the United States. This supply process is affected by the quality of hides, the general infrastructure available in the country (from roads to ports and machinery), and potential environmental problems that might be created in the leather treating and manufacturing process. The discovery in the middle of 2007 of large numbers of toys imported from China by Mattel that had been painted with lead-based paints served as a wake-up call to manufacturers in many industries who use contractors in China to produce their products. Internal testing and oversight of contractors and suppliers is a not insignificant cost that must be born by the primary manufacturer.

DISTRIBUTION CHANNEL

Companies generally manufacture handbags along with some other product: luggage, clothing, scarves, or watches, for example. Because of this, handbags benefit from the well-established distribution networks created by the large apparel and leather industries. They are available in nearly every retail channel.

Some specialty retailers went through tough times in the 1980s and early 1990s. By the middle of the decade the country was enjoying strong economic times and many people became affluent. This increased affluence—wealthy consumers willing to spend money—helped to spark change in the retail industry. Specialty retail blossomed and the discount store industry took shape.

Some of these new specialty goods retailers were stores selling luxury goods. Many luxury goods makers decided to expand their product lines in the 1990s to capitalize on consumer interest in luxury items. They moved into retail channels which were previously not available or opened stores selling their own branded goods. Coach Inc., the leader in the luxury handbag category, is an example of a company that found success by expanding into fine apparel, accessories, handbags, and luggage. Handbag makers Kate Spade and Kenneth Cole also opened their first shops in this decade.

The handbag industry became fragmented and increasingly competitive. A woman might purchase a quality purse through a catalog, at a high-end department store, specialty retailer, discount store, or on the Internet. Many of the specialty retailers opened an online site to go along with their regular brick-and-mortar stores. There are numerous third-party vendors that sell designer handbags online as well, often at discount prices. The Internet has been a source of innovation to the handbag trade. Amazon.com launched the site Endless.com in January 2007 in response to shoppers' requests. Women can search for a handbag by designer, color, style, and numerous other criteria. Some Web sites allow women to rent designer bags rather than go to the expense of purchasing them. The Web site Bag, Borrow or Steal allows women to rent bags for $19.95 to $249.95 per month, depending on the subscription plan. One customer on the site described the service as addictive.

In short, handbag makers typically get their product to the consumer in one of four ways: through its own branded stores, any of a number of retail stores, the Internet, or catalogs. According to estimates by Credit Suisse in its *Brandscape 2007* report, department stores represent 30 percent of handbag sales. Specialty stores and mass merchants each account for 20 percent of the retail market for handbags. Specialty chains follow with 15 percent of the market. Other types of stores took the remaining 15 percent of sales.

The competitive market hasn't dissuaded some entrepreneurs have from starting their own handbag businesses. One professor at the Fashion Institute of Technology estimates it takes some good designs and approximately $50,000 for an industrial sewing machine and various materials to get started. It can be formidable to gain a foothold in such a competitive market. However, the rarity of a bag can actually add to its desirability. Kate Spade and Coach are two companies that began as very small operations and then found success through expansion.

KEY USERS

According to Cowen & Company, Japan is the largest market for luxury handbags. Japanese women typically own more handbags than American women (6.10 vs. 5.48 in 2006) and use their handbags more often (2.36 versus 1.92 times per month in 2006). However, Japanese women expressed plans to curtail their handbag spending in 2007. American women 26 to 55 years of age planned to double their purchases in 2007. This age group encompasses those most likely to be working and interested in fashion, and are most likely to be employed and to need a handbag for carrying personal items to and from the workplace. Ownership of handbags varied by education level. In the United States, a woman with a college education owned an average of 6.3 bags in 2006, two to three handbags more on average than her female counterpart without a college degree. Women in the Southwest, Mid-Atlantic, Southeast, and Northeast regions owned more purses than women in the Northeast, West Coast, Plain States, or Mountain regions.

Although women are by far the predominant users of handbags, there may be a growing number of men interested in carrying handbags in Europe, according to a 2006 *Guardian* article. Bags designed for men are usually leather in construction. Some are roomy enough to be mistaken for laptop cases or briefcases, while others definitely are not. American men will probably dig in their heels at the thought of slinging a handbag over their shoulders. More then a few people rolled their eyes when *Metrosexuals*—heterosexual men who devote a great deal of time and money to their appearance and overall lifestyle—began to populate the country in 2002. Many men felt inspired by celebrities Brad Pitt, David Beckham, or the hosts of *Queer Eye for the Straight Guy* to embrace their feminine side. The term also spawned the term *retrosexual*, a term used to describe a man who has no interest in fashion, moisturizers, or culture. With such derisive terms, will American men ever feel comfortable purchasing a handbag for themselves? Will they have to carry around their phone, Blackberry, iPod, and countless other items while women, Louis Vuittons in their laps, smirk from the sidelines? There is anecdotal evidence from some high-end retailers to suggest that male handbag sales are up. The first step in marketing, of course, is a catchy name. The Manbag is perhaps the most common and the most likely to induce cringes; it was first used on an episode of *Friends* in 1999, not long after Prada debuted one of the first handbags for men in 1998.

ADJACENT MARKETS

Handbags are just one product made from finished leather. Other items include outerwear, luggage, footwear, wallets, and upholstered furniture. Leather tanning can be traced back thousands of years. The industry has been on the decline in the United States for a number of reasons, including the cost associated with production. As a result, the high retail cost of quality leather goods is out of reach for many consumers. Also, the changing culture in the United States means certain leather products (saddles, for example) are no longer in high demand. The Census Bureau estimates that the number of establishments devoted to leather production and tanning fell from 357 in 1997 to 262 in 2002. The value of shipments has declined from $3.3 billion in 1997 to $1.9 billion in 2005. Handbags represent only a fraction of the end market for leather. Nearly three quarters of leather production is used to manufacture footwear, garments, and gloves.

Handbags are the best-selling fashion accessory for women, according to *Accessories Magazine*. Industry tracking firm NPD Group reported that the women's apparel market generated sales of $101 billion in 2005. Women's accessories (including hats, scarves, belts, sunglasses and other goods) represented approximately $30.2 billion of that total. As stated, apparel manufacturing is increasingly done overseas. The Census Bureau notes that that there were 6,249 establishments in the United States devoted to making women's, girl's, and infant's apparel in 2002, down from 7,126 in 1997.

RESEARCH & DEVELOPMENT

Designers have introduced a number of new features to handbags. Some purses now include storage specially designed to carry iPods and personal digital assistants. Interior lighting is a popular new feature. Manufacturers are looking into ways to incorporate biometric technology; with such technology a person carrying a handbag would use a thumb or eye scan to access her purse. The technology would be used to deter pickpockets.

Research and Development into new materials is also an area of interest for handbag designers and manufacturers. An increasing awareness of issues surrounding the responsible management of the environment are partially responsible for this effort to identify new materials for use in making handbags. The use of recycled materials is one area of particular focus in this effort.

CURRENT TRENDS

The apparel and handbag industries have struggled with the issue of piracy for some time. Handbags, wallets, and backpacks represented approximately 9 percent ($4.3 million) of all counterfeit goods seized by the U.S. Customs and Border Protection Service for the first six months of 2006. This is a fraction of the roughly $200 billion in total goods thought to be counterfeited each year. However, these are confiscated handbags. The counterfeit handbag trade is estimated to be worth billions. Handbag designers Louis Vuitton, Chanel, and Burberry regularly appear at the top of annual lists of most counterfeited brands. Vietnam, Thailand and China are just a few countries thought to be major producers and consumers of counterfeit handbags and other goods.

Fake goods were once sold on street corners. Now such products are sold on online retail sites, auction site eBay, flea markets, and well-known accessory shops. Some knockoffs are easy to spot. They offer inferior stitching, substandard materials, and, in the worst cases, misspellings on the designer label. Often the price is a giveaway as well. Counterfeit goods might cost one-sixth of the regular retail price; an authentic Chanel handbag retails for approximately $1,500, while a fake may cost approximately $60. Some counterfeit goods, however, are well-made and retail at high prices.

Counterfeit bags sell well for a number of reasons: women get the satisfaction of having a status symbol bag that may pass for the real thing. There's also some psychological satisfaction that the woman has gotten away with something by not having to pay the high cost of the designer label.

One problem in combating the trade in counterfeit goods is that apparel and handbags are not protected under the U.S. Copyright Act. Some have argued that the industry doesn't need such protections, and that it may ultimately stifle creativity in the industry if designers are not allowed to copy designs and find inspirations in each other's work. There are laws in place to protect apparel designers in Europe, although it appears to have had little effect on design copying.

Representative Robert W. Goodlatte (R-Virginia), with six co-sponsors, introduced the Design Piracy Prohibition Act in July 2006. This bill would extend copyright protection to fashion designs for a period of three years. The bill would amend the Copyright Act of 1976. The Act would extend protection to "the appearance as a whole of an article of apparel, including its ornamentation," with "apparel" defined to include "men's, women's, or children's clothing, including undergarments, outerwear, gloves, footwear, and headgear; handbags, purses, and tote bags." In order to receive the three-year term of protection, the designer would be required to register with the U.S. Copyright Office within three months of going public with the design.

TARGET MARKETS & SEGMENTATION

According to the Census Bureau there were 120.9 million women aged 15 years or older in the United States in 2005. This represents a large market for handbag makers. Many women have more than one handbag and acquire more than one per year. A consumer goes through a complex process when she considers a purchase, usually finding some middle ground between practicality and emotional investment.

A survey conducted on drugstore.com offers some insight into women's motivations for purse shopping. Forty-six percent of respondents chose the handbag they currently carry because it is functional and well organized. Only 18 percent claimed to have selected a handbag because it was fashionable or trendy; four percent cared most about color; and seven percent looked for a specific brand name.

RELATED ASSOCIATIONS & ORGANIZATIONS

Accessories Magazine, http://www.accessories magazine.com

Coach Inc., http://www.coach.com

Leather Industries of America, http://www.leatherusa.com

Louis Vuitton, http://www.louisvuitton.com

National Fashion Accessories Association, http://www.accessoryweb.com

BIBLIOGRAPHY

Barkham, Patrick. "Papa's Got a Brand New Bag." *The Guardian.* 29 June 2006.

Chabbott, Sophia, Katya Foreman, and Alessandra Ilari. "Into the Stratosphere: Luxury Handbags Take Prices Ever Higher." *WWD.* 13 February 2007.

Critchell, Samantha. "A Storied History of Handbags, from the Inside Out." *San Diego Union-Tribune.* 8 May 2005.

Hamashige, Hope. "Successful Accessories Businesses Start with Unique Designs," CNN Money. 10 September 2001. Available from <http://www.cnnmoney.com>.

"Handbags: Key Retail Trends 2006." *Accessories Magazine.* 23 March 2007. Available from <http://www.accessoriesmagazine.com>.

"Hides, Skins, Leather, Leather Footwear: Value of Exports." Food and Agriculture Organization of the United Nations. Available from <http://www.fao.org>.

"Historical Statistics for the Industry: 2002 and Earlier Years." *Leather and Hide Tanning and Finishing: 2002.* U.S. Department of Commerce, Bureau of the Census, Economics and Statistics Administration. December 2004.

"Historical Statistics for the Industry: 2002 and Earlier Years." *Women's, Girl's, and Infant's Cut and Sew Apparel Contractors: 2002.* U.S. Department of Commerce, Bureau of the Census, Economics and Statistics Administration. December 2004.

Montgomery, Elizabeth and Eric Nanes. *2007 Handbag: U.S. and Japan Markets Look to Different Paths.* Cowen & Company. 9 January 2007.

Nolan, Kelly. "When it Comes to Accessorizing, It's in the Bag." *DSN Retailing Today.* 27 February 2006.

Prabhakar, Hitha. "World's Most Extravagant Handbags." *Forbes.* 26 March 2007.

Saad, Omar. *Brandscape 2007 Sector Review.* 28 February 2007, 58.

"Top 25 Export Destinations for Handbags." U.S. Department of Commerce, International Trade Administration. 15 February 2007. Available from <http://www.ita.doc.gov>.

"Top 25 Import Destinations for Handbags." U.S. Department of Commerce, International Trade Administration. 15 February 2007. Available from <http://www.ita.doc.gov>.

"Value of Shipments for Product Classes: 2005 and Earlier Years." *Annual Survey of Manufactures: 2005.* U.S. Department of Commerce, Bureau of the Census. November 2006.

"When It Comes to 'Purse-onality' Women Have it in the Bag." *Business Wire.* 14 September 2000.

"Women's Handbag and Purse Manufacturing: 2002" *2002 Economic Census.* U.S. Department of Commerce, Bureau of the Census, Economics and Statistics Administration. December 2004.

SEE ALSO *Luggage; Shoes, Non-Athletic; Women's Apparel*

Helicopters

——■——

INDUSTRIAL CODES

NAICS: 33–6411 Aircraft Manufacturing

SIC: 3721 Aircraft Manufacturing

NAICS-Based Product Codes: 33–64113014 and 33–64113017

PRODUCT OVERVIEW

No one knows how long humans have dreamed of flying, but this concept informed some of Western culture's earliest stories. Around the time BC became AD, the Roman poet Ovid told the tale of Daedalus and his son Icarus. The first storytellers as well as their audiences could see snow persisting on the highest mountain peaks during warm weather; clearly, air did not grow warmer with altitude, so the sun would not have melted the wax holding Icarus's feathers to his wings. This tale appeals to the emotions, not to reason.

A modern audience could note how the genius of Daedalus, using observation to mimic birds' wing shapes, was rendered irrelevant by his son's reckless behavior. Passion defeated by reality was a lesson that early would-be aeronauts repeatedly rediscover, frequently at a high price. This essay summarizes the history of helicopters.

Helicopters are aircraft characterized by large-diameter, powered, rotating blades. Such a craft can lift itself vertically by accelerating air downward at an angle. The helicopter is the most successful vertical takeoff and landing (VTOL) aircraft yet developed due to its relatively high efficiency in performing hovering and low-speed flight missions.

From the ancient literature cited above, we move forward more than fourteen centuries to the first designs of what human-created flight might look like, and waiting for us is Leonardo da Vinci. His earliest written speculation on human-powered flight dates from 1473, when he was twenty-one, and his notebooks demonstrate that this was a lifelong fascination. Many artists of this period designed theater sets; Andrea del Verrocchio, to whom the young Leonardo was an apprentice, made such sets for the Medici family, and illusionist flying machines were often involved.

Leonardo's first sketch of such a device dates to 1478; today, it would be called a hang glider. The first appearance of a helicopter in Leonardo's notebooks dates to 1490, where a depiction of a large central screw-like device was designed to measure about thirteen feet in diameter. Leonardo intended it to be made of reed and covered with taffeta to make a light, resilient wing. This helicopter is shown as powered by four men who ran on the craft's platform around the central shaft, pushing a bar that would cause the spiral to turn. The helicopter would, at least theoretically, bore its way through the air like a giant corkscrew.

Even as a might-have-been, Leonardo's continuous spiral airfoil offers fascinating possibilities. To find those possibilities realized, we must travel forward four centuries more, to a chilly, windy beach called Kitty Hawk in North Carolina. On December 17, 1903, Wilbur and Orville Wright's 750-pound plane launched from a railroad track at less than seven miles per hour, attained an altitude of perhaps ten feet, and landed about twelve seconds later, having traveled 120 feet. This modest beginning led to a new world in which gravity, while impossible to ignore, could be successfully challenged and, less than seventy years later, overcome as humans traveled to and landed on the moon.

A few decades more and we have arrived at the moment where sufficient power combines with essential materials and human daring. The year is 1939; the place, a Connecticut field; and the person in the well-worn fedora, Igor Ivanovitch Sikorsky, born in Russia and emigrated to the United States in his thirties.

Everything about Sikorsky's VS-300 was bare-boned: the cockpit was only a seat in front of the exposed 75-horsepower engine; belts and pulleys drove the blades; the vertical rotor spun at the end of a spar.

This ungainly machine was an attempt to perfect the helicopter, which Sikorsky and others believed would be the aircraft that brought flight to the masses. The vision was one of backyard and rooftop helipads with commuters taking to the air rather than the road. But engineers had yet to perfect the helicopter. Prototypes lifting off the ground proved too cumbersome for regular service.

The idea of the helicopter was inspired not by nature but the screw discovered by Archimedes approximately 2200 years ago. A screw pump can push water up an incline and a propeller screw can push against water to move a ship forward. Why couldn't a large enough screw pull a machine into the air?

Like hundreds of others in aviation, Sikorsky built on the work of others. In 1919 the Spanish aircraft engineer Juan de la Cierva y Cordonia was studying how aircraft stall. As a propeller pulled a plane down the runway or through the air, the rotor turned, producing lift. Even if the engine failed in flight, the rotor would continue to turn, providing enough lift to enable a slow, controlled descent. Cordonia used this phenomenon to create a freewheeling rotor he used as a crucial part of what he called his *autogiro*.

The first sustained helicopter flight was not achieved until 1935, with a coaxial model built by Louis Breguet and René Dorand in France. Building partly on that attempt, Sikorsky took out the tandem rotors that canceled the counter-rotation (known as *torque*) in the French design and used a single main rotor for lift. This greatly simplified the mechanism and made controlling the craft much easier.

The first flight of Sikorsky's VS-300 was something less than astonishing. On Sept. 14, 1939, the craft cleared the ground by just a few inches, probably due to the rotor blowing air downward, and the whole event lasted 10 seconds. This was partly intentional, since the craft had been tethered to the ground in case anything went wrong. For successive flights, the engineers fixed a glitch that shook the machine violently when the rotors whirred at speed and, by November, short one-minute hops were possible.

Test flights through the spring and summer of 1940 helped Sikorsky and his team improve the manner in which aerodynamics applied to the helicopter. They also began practicing some of the three-dimensional feats that make helicopters so useful: landing on a dime, hovering over a single point, even throwing down a rope ladder for a rescue.

Once the control problems were better understood, Sikorsky and his team were able to eliminate first one and then both of the horizontal auxiliary rotors, opting instead for changing the pitch of the main rotor to control longitudinal and lateral motion. The modern helicopter was born.

> The helicopter is probably the most versatile instrument ever invented by man. It approaches closer than any other to fulfillment of mankind's ancient dreams of the flying horse and the magic carpet.—Igor Ivanovitch Sikorsky, comment on twentieth anniversary of the helicopter's first flight Sept. 13, 1959

MARKET

The animation company Hannah-Barbera created *The Jetsons* in the early 1960s. Set in an unspecified future, everyone who wants one has access to a flying car, very much in the spirit of Sikorsky's dream. In the real world, military and other interests reserved helicopters for highly specialized purposes.

Although U.S. forces gained some experience with helicopters late in World War II, the first substantial use of the vertical-takeoff craft came in the Korean War. Between 1950 and 1953 helicopters performed casualty evacuation, search and rescue, troop insertion, cargo transport, and reconnaissance. In 1950 General Douglas MacArthur requested more helicopters for use as organic aircraft within division, corps, and army headquarters units. U.S. Marine Corps units also used helicopters as airlift and combat support. Perhaps the greatest contribution helicopters made to the war effort in Korea came in transporting wounded soldiers to Mobile Army Surgical Hospitals for emergency medical care. By the end of the Korean War, the U.S. military was committed to developing the helicopter's potential for nearly every possible mission.

After the war, helicopter designers concentrated on developing powerful craft that could carry greater payloads over longer distances. Sectors such as oil exploration came to depend on the economical transportation ability provided by helicopter technology. The military concentrated on making helicopters essential to warfare.

The French used helicopters to patrol and dominate large territories in the Algerian War foreshadowing the U.S. Army's airmobile concepts typifying the Vietnam War between 1964 and 1973, when the army created air cavalry divisions with helicopters outfitted to specialize in assault, attack, heavy and medium transport, command and control, search and rescue, and medical evacuation. Even the last images of U.S. involvement in Vietnam included helicopters evacuating embassy personnel and refugees from the roof of the U.S. embassy in Saigon (later, Ho Chi Minh City) as the South Vietnamese government collapsed in March, 1975.

Civilian use of helicopters spread widely after the Vietnam War. The speed, mobility, and vertical takeoff and landing that made helicopters attractive to military forces also appealed to police, emergency services, and firefighters, especially in remote areas. Law enforcement helicopters from federal to local levels assisted ground units in surveillance and pursuit operations. Emergency service helicopters made dramatic rescues of hapless hikers and climbers. Helicopters enhanced firefighting efforts whether in large-scale wildfires or in combating hazardous industrial fires.

Though military and commercial aircraft manufacturers dominate the industry in the early twenty-first century, American companies also produced many aircraft for the general aviation and the helicopter market segments, which included fixed wing aircraft and rotorcraft for business transportation, regional airline service, recreation, specialized uses such as ambulance service and agricultural spraying, and training. American manufacturers historically produced approximately 60 percent of the world's general aviation aircraft and 30 percent of the helicopters.

Sales and exports of U.S. civil helicopters surged in 2005 to record levels, according to the U.S. Aerospace Industries Association (USAIA). In its aerospace industry annual review, the trade group reported civil helicopter sales jumped from $515 million to a record $750 million. The U.S. industry shipped 120 more civil helicopters in 2006 than it did in 2004.

Civil helicopter exports also reached record levels, rising 57 percent to $490 million. "Used civil aircraft exports rose 31 percent from already high levels to $2.8 billion," the USAIA said, "helping exports and the trade balance, but not resulting in new production."

According to Flight's HeliCAS database, a healthy 531 turbine helicopters were civil-registered in 2006. The leading helicopter makers also reported strong order backlogs and were planning higher production rates in 2007. While the helicopter industry was riding the same post-9/11 economic recovery that was boosting other sectors of commercial aerospace, it was also seeing strong growth in the offshore support sector, which was re-equipping

after years of operating aging but depreciated helicopters. Growth in the law enforcement and emergency medical service sectors was also playing a part.

U.S. manufacturers shipped 4,088 units of complete civilian aircraft (fixed wing, powered craft; helicopters; and non-powered types of civil aircraft) in 2002, valued at approximately $34.7 billion. In terms of unit shipments, this figure represented a decrease from 2001, when the industry shipped 4,541 units valued at 41.8 billion, and from 2000 when shipments numbered 5,162 civil aircraft valued at $38.6 billion.

In use as aerial cranes, firefighting, air ambulances, crop-dusting, search and rescue, law enforcement, a host of military purposes, and the transport of the rich and famous, it may be more appropriate to ask where helicopters are not useful than to list where they are.

KEY PRODUCERS/MANUFACTURERS

Five large-scale firms dominated the helicopter production field in the first decade of the twenty-first century, as can be seen in Figure 108. Each is profiled briefly below.

Eurocopter S.A. Europe's largest helicopter maker, Eurocopter, makes a full range of civilian and military helicopters and offers helicopter repair, maintenance, and overhaul services.

This premier manufacturer has four plants, two in France and two in Germany, and many offices worldwide. Eurocopter employed a workforce of 10,822 people pro-

FIGURE 108

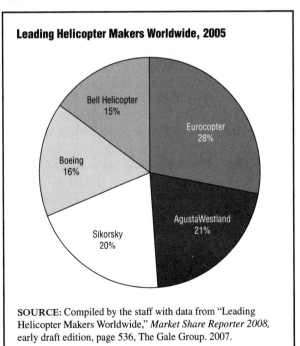

Leading Helicopter Makers Worldwide, 2005

Bell Helicopter 15%
Eurocopter 28%
Boeing 16%
AgustaWestland 21%
Sikorsky 20%

SOURCE: Compiled by the staff with data from "Leading Helicopter Makers Worldwide," *Market Share Reporter 2008*, early draft edition, page 536, The Gale Group. 2007.

ducing 57 percent of the civil market and 25 percent of the military market in the middle of the first decade of the twenty-first century. Eurocopter reported revenues of $2.6 billion in 2002. The European firm, with its vast product range, continues to significantly outsell its rivals, claiming to have captured 52 percent of the market in units in 2005 (twice as much as its nearest rival), and 46 percent by value. It has been first in the U.S. market during the period 2000–2005, with a more than 50 percent market share in the EMS, para-public, utility, and tourism sectors.

Like most of its competitors, Eurocopter increased its investment in the Asia Pacific market to the extent that by 2006, it claimed to have captured half of the civil and para-public market in Japan. Eurocopter announced the creation of a Japanese subsidiary in 2006 to coordinate the commercial network in that region. The European firm also views India as a strategic market, and plans to set up a local organization based in Bangalore.

Bell Helicopter Textron Inc. A subsidiary of Textron, the company makes commercial and military helicopters and tilt-rotor aircraft. Bell's commercial helicopters seat up to 15 passengers and include models designed for transport, emergency medical services, and search and rescue operations. Military models include the venerable UH-1Y *Huey,* a utility helicopter used for personnel and medical transportation; the AH-1Z Super Cobra reconnaissance/attack helicopter; the Eagle Eye Unmanned Aerial Vehicle (UAV); and the V-22 Osprey tilt-rotor (with Boeing). Bell also makes helicopters through joint venture Bell/Agusta Aerospace and provides repair, maintenance, and overhaul services.

Bell representatives claimed that the company achieved a 23 percent increase in civil helicopter shipments in 2005, compared with the previous year. It continues to build its new offerings on the foundations laid by its older models the Bell 210, a derivative of the Huey, which achieved FAA certification in 2005; and the 407X light single, a 407 refitted with Honeywell's new HTS900 engine and which is the basis for Bell's selection for the Armed Reconnaissance Helicopter (ARH) program—a 368-unit order that has given Bell a new lease of life.

Boeing Company. In 1916 William E. Boeing founded the Boeing Company, then called Pacific Aero Products, in Seattle, Washington. Until 1960 Boeing was only a designer and manufacturer of airplanes. After acquiring Vertol Aircraft Corporation, Boeing became a designer and manufacturer of helicopters as well. On September 21, 1961, the CH-47A Chinook helicopter took its first flight.

At the beginning of the twenty-first century, Boeing Company was the world's largest manufacturer of commercial jetliners and military aircraft, and was NASA's leading contractor. In 2006 its total revenues were $61.5 billion. Boeing Integrated Defense Systems, a division of Boeing Company, manufactures the Apache Longbow, the Chinook, and the Osprey helicopters for the military. This division employed 72,000 people and had revenues of $32.4 billion in 2006. Boeing's headquarters are in Chicago, Illinois.

Sikorsky Aircraft Corporation. Though its legendary founder is long gone, helicopters bearing his name still fly the skies. A subsidiary of United Technologies, Sikorsky Aircraft's military helicopters include the Black Hawk, used for troop assault, combat support, special operations, and medevac operations; and the Seahawk, used for submarine hunting, missile targeting, anti-surface ship warfare, and search and rescue.

The company claimed in 2006 to have logged double-digit growth for the prior three years, with dollar sales leaping from $100 million in 2001 to $600 million for civil aircraft deliveries in 2005. Chief Executive Officer Jeff Pino predicted that billings would be close to those of Eurocopter in 2007. Sikorsky continues to pursue new derivatives, such as the S-76D, scheduled for certification by end-2008. With a maximum takeoff weight of almost 6.6 tons, other features include new composite main and tail rotors offering a 2 decibel noise reduction and a new PW210S engine featuring a dual-channel FADEC and 10 to 20 percent extra power in comparison with the current engine.

The company was also excited about the new Thales Topdeck cockpit, which was being installed on civil helicopter for the first time in 2007. Derived from the cockpit developed for the A380 super jumbo, it features four 6 x 8-inch multifunction screens, an integrated interactive flight management system, synthetic vision interface, and a trackball to interact with the moving map display. Also for the first time on a helicopter, the vehicle includes integrated backup instrumentation giving altitude, attitude, and speed.

AgustaWestland N. V. One of the world's largest helicopter manufacturers, this company produces a wide range of high-performance rotorcraft for civil and military markets. Of 92 helicopters it delivered in 2004, 66 went to commercial customers. Formed by combining two leading European helicopter manufacturers, AgustaWestland has operations in Italy (near Milan), the United Kingdom (near Somerset, England), and the United States (Fort Worth, Texas, and Philadelphia, Pennsylvania).

AgustaWestland bought out Bell Helicopter's 25 percent stake in the medium twin-turbine AB139 in November, 2005, to improve support and increase sales. Instead of Bell assembling the helicopter in Amarillo, Texas as originally planned, AgustaWestland was expanding its Agusta Aerospace (AAC) subsidiary in Philadelphia, Pennsylvania, to establish production of the AB139 for the North American market. AAC already manufactures the single-turbine A119 and U.S. production of the AB139 was to begin by the end of 2006. AgustaWestland, based at Cascina Costa, Italy, is the second-largest helicopter manufacturer in revenue terms with sales of €2.54 billion in 2004, but around 90 percent of its business is defense. The company responded to the strong interest from the U.S. market by announcing a second production line in Philadelphia, boosting production capacity to 50 units per year. This line, like its twin in Vergiate near Milan, will be directly supplied with structural elements produced by PZL in Poland and TAI in Turkey.

MATERIALS & SUPPLY CHAIN LOGISTICS

Although not quantified until long after the Wright brothers skidded and soared over Kitty Hawk's sand dunes, the ability to fly was eventually rendered as lift plus thrust having to exceed mass (known on Earth as weight) plus drag. All aircraft design must struggle with this reality. The first powered aircraft were largely paper and thin cables, but engineers continue to explore how machines, especially helicopters, can fly more with less.

Some aircraft of composite materials began to appear in the late 1930s and 1940s; these were usually plastic-impregnated wood materials. The largest and most famous example of this design is the Duramold construction of the eight-engine Hughes flying boat, popularly known as the *Spruce Goose*. A few production aircraft also used Duramold materials and methods.

Fiberglass, fabrics made up of glass fibers, were first used in aircraft in the 1940s and became common by the 1960s. Composite is the term used for different materials that provide strengths, light weight, or other benefits not possible when these materials are used separately. They usually consist of a fiber-reinforced resin matrix. The resin can be a vinyl ester, epoxy, or polyester, while the reinforcement might be any one of a variety of fibers, ranging from glass through carbon, boron, and several other proprietary types.

To these basic elements, strength is often increased by adding a core material, making a structural sandwich. Core materials such as plastic foams (polystyrene, polyurethane, or others), wood, honeycombs of paper, plastic, fabric or metal, and other materials, are surrounded by layers of other substances. This method has been used to create,

for example, Kevlar, used in aircraft panels, and Lucite, superior to glass for aircraft windows and canopies.

By the twenty-first century, almost all helicopter parts include composites. The airframe, or fundamental structure, of a helicopter can be made of either metal or organic composite materials, or some combination of the two. Higher performance requirements encourage the designer to favor composites with higher strength-to-weight ratio, often epoxy (a synthetic resin) reinforced with glass, aramid (a strong, flexible nylon fiber), or carbon fiber. Typically, a composite component consists of many layers of fiber-impregnated resins, bonded to form a smooth panel. Tubular and sheet metal substructures are usually made of aluminum, though stainless steel or titanium is sometimes used in areas subject to higher stress or heat. To facilitate bending during the manufacturing process, the structural tubing is often filled with molten sodium silicate. A helicopter's rotary wing blades are usually made of fiber-reinforced resin, which may be adhesively bonded with an external sheet metal layer to protect edges. The helicopter's windshield and windows are formed of polycarbonate sheeting.

Modern helicopter engines use turbines rather than pistons and are purchased from an engine supplier. The helicopter manufacturer may purchase or produce the transmission assembly, which transfers power to the rotor assembly. Transmission cases are made of aluminum or magnesium alloy.

DISTRIBUTION CHANNEL

Regardless of the kinds of machines in which they are used, most aircraft parts have common origins and distribution channels.

The American aircraft industry can be divided into four segments. In one segment, manufacturers such as Boeing and Lockheed Martin Corp. build the wings and fuselages that make up the airframe. Meanwhile, companies such as General Electric and Pratt & Whitney manufacture the engines that propel aircraft. The third segment covers flight instrumentation, an area where the most profound advances in aviation have taken place. But the fourth segment, broadly defined by the industrial classification "aircraft parts not otherwise classified," includes manufacturers of surface control and cabin pressurization systems, landing gear, lighting, galley equipment, and general use products such as nuts and bolts.

Aircraft manufacturers rely on a broad base of suppliers to provide the thousands of subsystems and parts that make up their products. There are more than 4,000 suppliers contributing parts to the aerospace industry, including rubber companies, refrigerator makers, appliance manufacturers, and general electronics enterprises. This diversity is necessary because in most cases it is simply uneconomical for an aircraft manufacturer to establish, for example, its own landing light operation. The internal demand for such a specialized product is insufficient to justify the creation of an independent manufacturing division.

There is a second aspect to this distribution tier, since aircraft manufacturers have found it cheaper and more efficient to purchase secondary products from other manufacturers, who may sell similar products to other aircraft companies, as well as automotive manufacturers, railroad signal makers, locomotive and ship builders, and a variety of other customers. For example, an airplane builder such as Boeing, Grumman, or Beech might purchase landing lights from a light bulb maker such as General Electric. Such subcontractors supply a surprisingly large portion of the entire aircraft. On the typical commercial aircraft, a lead manufacturer such as McDonnell Douglas may actually manufacture less than half of the aircraft, though it is responsible for designing and assembling the final product.

When a major manufacturer discontinues an aircraft design, as Lockheed did with its L-1011 Tristar, a ripple effect is caused that affects every manufacturer that supplied parts for that aircraft. Therefore, parts suppliers that make up the third tier of distribution strive to diversify their customer base to ensure the decline of one manufacturer will be tempered by continued sales to others. Given the unstable nature of the industry, parts manufacturers also attempt to find customers outside the aircraft business.

In terms of the distribution of helicopters to the end user, most units are produced only after an order has been placed for the vehicle. This is common for large assets that are intended for very specific purposes and therefore often require some level of customization.

KEY USERS

For many practical applications, helicopters are indispensable. They are used to perform important services for cities, industry, and government. Rescue missions and operations depend on the versatility of the helicopter for disaster relief efforts at sea and on land. The Coast Guard uses them regularly, and the ability of the helicopter to hover allows for harnesses to be extended to victims on the ground or at sea, who can then be transported to safety. Helicopters are also useful when rescuing lost or injured hikers or skiers. Hospitals now have helipads so accident victims can be transported as quickly as possible for emergency treatment. Police use them for aerial observation, tracking fleeing criminals, searching for escaped prisoners, or patrolling borders. Police and news agencies use the helicopter to watch for traffic problems in major cities.

Wildlife and forestry employees need helicopters for aerial surveys of animal populations and to track animal movements. Forestry personnel use the helicopter to observe the condition of tree stands and to fight fires. Helicopters transport personnel and equipment to base camps, and spray fires. The agricultural industry engages helicopters to spray fields and to check on and round up cattle.

Helicopters are especially useful to industry, performing jobs that require strength and maneuverability, such as hoisting heavy building materials to the upper levels of a high-rise and hauling awkward or large objects. They have also been used to erect hydro towers and other tall structures. Petroleum industries rely on the helicopter to observe pipelines for damage and to transport personnel to offshore drilling operations.

Helicopters are the prestige vehicle of choice when businesses want to impress clients and employees. Though expensive, helicopter flight is a convenient way to beat the traffic, and downtown businesses in large cities will often have heliports on top of their buildings. Helicopters transport passengers from the airports and are enjoyed recreationally by sightseers and hunters willing to pay for quick transportation to exotic locales.

While helicopters have improved greatly since the first piloted rotary machines of 1907, they are significantly slower than airplanes and cannot reach the same altitudes. Expensive and difficult to fly, helicopters are also highly versatile and can move in ways impossible for fixed-wing craft. This maneuverability makes the helicopter an essential tool for industrial, civil, and military service.

Helicopters in widest use in the U.S. armed services are the Sikorsky UH-60 Black Hawk and the Boeing AH-64 Apache. The Black Hawk, in service since 1978, is designed as a troop carrier and logistical support aircraft, but it can be used for medical evacuation, command and control, search and rescue, armed escort, and electronic warfare missions. The Black Hawk can carry 16 laser-guided Hellfire antitank missiles and a total weapons payload of up to 10,000 pounds of missiles, rockets, cannons, and electronic countermeasures. The helicopter can also transport up to 11 fully equipped soldiers.

Like its predecessor the Black Hawk, the Apache attack helicopter can carry up to 16 missiles as well as 76 aerial rockets for use against lightly armored vehicles, and other soft-skinned targets. The Apache also boasts state-of-the art sensors that can identify targets in all types of weather during the day or night. Both the Black Hawk and the Apache played critical roles in ground attack, troop support, and supply during the Gulf War of 1991 and the Iraq War of 2003. The flexibility and firepower provided by modern military helicopters make them an indispensable part of the U.S. military arsenal.

ADJACENT MARKETS

Even the simplest modern helicopter contains thousands of parts whose peak functioning is essential to a safe landing. Indeed, a significant part of a helicopter's control panel contains instruments indicating whether the other instruments are working correctly; it is not as though, when something goes wrong, the pilot can pull over to the nearest cloud. The categories considered in this section include instrumentation systems and engine instruments. Products produced by these industry sectors are necessary to getting a helicopter off the ground, keeping it in the air, and touching down gently.

Guidance and Control Instrumentation. The products of this industry relevant to this essay include radar systems, navigation systems; flight and navigation sensors, transmitters, and displays; gyroscopes; and airframe equipment instruments.

The main suppliers of search and navigation equipment are the same contractors who supply the larger U.S. aerospace and defense industries, to which search and navigation equipment contribute significantly. Although not necessarily the most prolific producers of search and navigation instruments, many of the largest and most recognizable corporations in the United States have been involved in the business, including AT&T, Boeing, General Electric, General Motors, and IBM.

A substantial majority of the industry's product types fall into the avionics (aviation electronics) classification, which includes aeronautic radar systems, and air traffic control systems.

Historically, the primary customer for industry products has been the U.S. government—in particular, the Department of Defense and the Federal Aviation Administration.

Search and detection systems and navigation and guidance systems and equipment ($29.1 billion worth of shipments in 2001) constitute 91 percent of the total search and navigation market and include the following product groups: light reconnaissance and surveillance systems; identification-friend-or-foe equipment; radar systems and equipment; sonar search, detection, tracking, and communications equipment; specialized command and control data processing and display equipment; electronic warfare systems and equipment; and navigation systems and equipment, including navigational aids.

During the 1970s development of the Global Positioning System (GPS) satellite network began. Inertial navigators using digital computers became common devices in civil and military aircraft. Industry shipment values for the above products totaled $31.9 billion in 2001, an increase over the $29.9 billion shipped in 2000. Employment in the aircraft components sector in the United States also saw growth in the early 2000s. In 2001 the industry's employment base of 153,710 workers was nearly 12,000 people greater than the previous year. Capital investment, which totaled approximately $1 billion in 2000, had remained relatively constant since 1997.

Aircraft Engine Instruments. The main customers of the aircraft engine instruments segment are General Electric, United Technologies, Rolls Royce, and other aircraft manufacturers. This sector produces temperature, pressure, vacuum, fuel and oil flow-rate sensors, and other measuring devices. Growth in this market is linked to aircraft production.

Through the first decade of the twenty-first century, the miscellaneous measuring and controlling devices industry was projected to grow at an annual rate of 3 percent. Aircraft engine instruments were predicted to be one of the industry's faster growing segments. Furthermore, the addition of software and services will contribute to overall industry growth, as will further expansion into overseas markets. The top five export markets in the late 1990s were Canada, Mexico, Japan, United Kingdom, and Germany; these five countries also were the top import countries. Looking into the 2000s, estimates indicated 33 percent of measuring and controlling instrument product shipments would be exported, while 25 percent of U.S. demand would be met by imports.

RESEARCH & DEVELOPMENT

In a sector as competitive as the helicopter industry, it is easy to focus on matters of day-to-day survival. That has never been truer than in the early twenty-first century, when operators have been desperate for aircraft and manufacturers have been under pressure to meet rising demand and demanding production schedules.

What will airframes, power systems, and avionics look like over the next 40 years? That is a more complex question than it would first appear, since even with computer-assisted drawing equipment and high demand, it can take 10 years for a new helicopter to move from the drawing board to the field. Much potential remains for revolutionary advances in the science of vertical flight. Some industry experts feel helicopter technology hasn't advanced significantly since the 1970s; the basic air vehicle performance has remained largely unchanged since the end of U.S. military involvement in Vietnam.

Regardless of how individuals may feel about the need for war, many rotorcraft developments evolve to meet the requirements of the U.S. military. Combat operations in Iraq and Afghanistan, combined with the prospect that they will persist for some time and be coupled with operations elsewhere in the world, confront military and industry leaders with a simple fact: the Pentagon needs an aircraft that can get off the ground and land without requiring a runway but can perform like a fixed-wing airplane in between.

The Bell Helicopter/Boeing V-22 can do that, and was scheduled to go into use in Iraq in 2006, but that process began in the late 1980s. To achieve greater speed, range, and payload, the experts cited the promise of a compound-helicopter design, using an auxiliary propulsion system to supplement the thrust of the rotors for greater forward speed. Fixed wings can provide extra lift. The Piasecki Aircraft, which has long worked on the concept of compound-helicopter design, is preparing its X-49A SpeedHawk for flight tests in 2007.

The Heliplane being developed with funding from the U.S. Defense Advanced Research Projects Agency (DARPA) uses a similar approach. The Salt Lake City-based autogiro maker Groen Brothers Aviation is designing a proof-of-concept, long-range, vertical takeoff and landing aircraft. DARPA's objective is to achieve performance with a rotary-wing aircraft comparable to that of a fixed-wing one.

The Smart Hybrid Active Rotor Control System (SHARCS) integrates actively controlled rotor blades to reduce helicopter noise and vibration. Performance tests of the 6.5-ft rotor were scheduled for early 2007, followed by wind tunnel tests in Milan, Italy. SHARCS is led by the Rotorcraft Research Group at Carleton University in Ottawa, Canada, with funding from the Canadian Natural Sciences and Research Council, AgustaWestland, Manufacturing and Materials Ontario, and Sensor Technology, Ltd.

Developing a low-drag hub alone would be a significant efficiency gain for tomorrow's rotorcraft. Experts said the vertical drag of a rotorcraft's hub is roughly equivalent to the entire drag of a fixed-wing aircraft of a similar gross weight, which greatly limits the performance of helicopters.

Tail rotors are another necessary evil on single main-rotor helicopters. They are critical to controlling torque and directional control, but they also add drag, increase the cost and complexity of maintenance, and generate a lot of noise. Most importantly, tail rotors are a safety weakness. They are a critical flight control that is susceptible to a single-point failure. In the late 1990s, developers and operators of unmanned air vehicles (UAVs) realized they were losing aircraft to single-point flight-control failures. They revised their designs to make them doubly, and lately, triply redundant. As a result, it is difficult to lose a UAV to a flight control failure.

Tomorrow's aircraft also are likely to be less expensive to operate. Since military operators are shifting life-cycle costs to aircraft suppliers through long-term support contracts, suppliers are motivated to maximize aircraft reliability. The United Kingdom has led in this contracting area, but the United States is expanding its use of this practice.

CURRENT TRENDS

Manufacturing processes and techniques will continue to change in response to the need to reduce costs and the introduction of new materials. Automation may further improve quality (and lower labor costs). Computers will become more important in improving designs, implementing design changes, and reducing the amount of paperwork created, used, and stored for each helicopter built. Also, industrial robots that can wind filament, wrap tape, and place fiber will permit fuselage structures to be made of fewer, more integrated pieces. Advanced, high-strength thermoplastic resins promise greater impact resistance and repairability than current materials such as epoxy and polyimide. Metallic composites such as aluminum reinforced with boron fiber or magnesium reinforced with silicon carbide particles also promise higher strength-to-weight ratios for critical components such as transmission cases while retaining the heat resistant advantage of metal over organic materials.

FIGURE 109

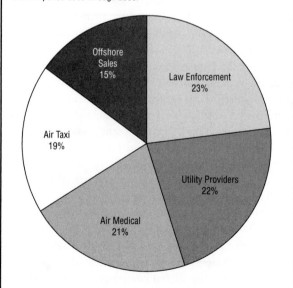

U.S. Turbine Helicopter Sales by Category of End User, 2000–2005

Market shares are shown based on average sales by end users category over the period 2000 through 2005.

- Offshore Sales 15%
- Law Enforcement 23%
- Air Taxi 19%
- Utility Providers 22%
- Air Medical 21%

SOURCE: Compiled by the staff with data from *Helicopter Market Newsletter Turbine,* Helicopter Association International. 2006.

TARGET MARKETS & SEGMENTATION

Helicopters have proven themselves so crucial to so many markets, they would seem to almost sell themselves. In truth there is a great deal of competition among helicopter manufacturers and each must maintain a close relationship with a large variety of institutional buyers.

The major target markets for helicopters are the agencies and organizations that use helicopters as a part of their very function. They include the military, law enforcement, tourism, firefighting, agriculture, construction, the wealthy, and hospitals. Because of the unique functionality offered by the helicopter, it has a built in audience and thus, a capitive market. Even the hazards inherent in all aircraft, particularly acute with helicopters since they often fly at relatively low altitude, with little time to react to a sudden event, do not deter from the popularity of this most useful machine. Air shows remain the most widely attended venues for customers old and new to marvel at the latest advances in aircraft design. Other ways in which helicopters are promoted include through the listings of helicopter charter companies in telephone directories and the plethora of Internet Web sites devoted to the sale, maintenance, supply and repair of helicopters.

RELATED ASSOCIATIONS & ORGANIZATIONS

American Helicopter Society, http://www.vtol.org

Competition Helicopter Association, http://torchs.org/clubs/clubs.htm

Helicopter Club of Great Britain, http://www.hcgb.co.uk

Helicopter Foundation International, http://www.hfi.rotor.com

Naval Helicopter Association, http://www.navalhelicopterassn.org

Popular Rotorcraft Association, http://www.pra.org

Rotor Rats, http://torchs.org/clubs/clubs.htm

Twirly Birds, http://www.twirlybirds.org

Whirly-Girls International, http://www.whirlygirls.org

BIBLIOGRAPHY

"Advanced Technology Prompts Reevaluation of Helicopter Design." *Aviation Week & Space Technology.* 9 March 1987, 252.

"Airframes: Sleeker, Safer, Friendlier." *Rotor & Wing.* 15 May 2007.

Allen, Matthew. *Military Helicopter Doctrines of the Major Powers, 1945–1992.* Greenwood Press, 1993.

Basic Helicopter Handbook. IAP Inc. 1988.

"Boeing in Brief." April 2007. Available from <http://www.boeing.com/companyoffices/aboutus/brief.html>.

Boyne, Walter J., and Donald S. Lopez, eds. *Vertical Flight: The Age of the Helicopter.* Smithsonian Institution Press, 1984.

Brown, Stuart F. "Tilt-rotor Aircraft." *Popular Science.* July 1987, 46.

Davis, Kenneth C. *Don't Know Much About History,* Rev. ed. Harper Collins Publishers Inc., 2003.

Fay, John. *The Helicopter: History, Piloting, and How It Flies,* 4th ed. Hippocrene, 1987.

Francis, Devon F. *The Story of the Helicopter.* Coward-McCann, 1946.

Futrell, Robert Frank. *The United States Air Force in Korea, 1950–1953,* Rev. ed. U.S. Department of the Airforce, Office of Air Force History. 1983.

"Graphite Tools Produce Volume 'Copter Parts." *Design News.* 17 February 1986, 30.

"History." Available from <http://www.boeing.com/history/chronology/>.

"Integrated Defense Systems." Available from <http://www.boeing.com/companyoffices/aboutus/brief/ids.html>.

Momyer, William W. *Airpower in Three Wars: World War II, Korea, Vietnam.* U.S. Department of the Air Force. 1978.

Musquere, Anne. "High Times for Helo Makers: This Year's Heli-Expo Saw Record Attendance and Sales." *Interavia Business & Technology.* Spring 2006, 14.

Nicholl, Charles. *Leonardo da Vinci: Flights of the Mind, A Biography.* Viking Penguin, 2004.

"Researchers Work on Noise Reduction in Helicopters." *Research & Development.* January 1986, 55.

"Rotary-Wing Technology Pursues Fixed-Wing Performance Capabilities." *Aviation Week & Space Technology.* 19 January 1987, 46.

Seddon, J. *Basic Helicopter Aerodynamics.* American Institute of Aeronautics & Astronautics. 1990.

Smith, Bruce A. "Helicopter Manufacturers Divided on Development of New Aircraft." *Aviation Week & Space Technology.* 29 February 1988, 58.

"U.S. Civil Helo Sales Hit Record Levels in 2005." *Rotor & Wing.* 15 February 2006.

SEE ALSO *Airplanes, Jet Aircraft*

Ice Cream

———■———

INDUSTRIAL CODES

NAICS: 31–1520 Ice Cream and Frozen Dessert Manufacturing

SIC: 2024 Ice Cream and Frozen Desserts

NAICS-Based Product Codes: 31–152001, 31–152002, 31–152003, 31–152004, 12–152005, and 31–15200Y

PRODUCT OVERVIEW

Ice Cream and Related Products. Strictly speaking ice cream is a frozen dairy product with at least 20 percent of its content consisting of milk solids and 10 percent butterfat. The product is required to weigh at least 4.5 pounds to the gallon. Ice cream is sweetened with sugar. Ice cream may contain egg whites and will contain a stabilizer. This is the product people call regular ice cream. Most ice cream made in the United States is regular, but a wide variety of similar products will be found in grocers' and supermarkets' freezers. The Census Bureau treats these variants as part of the same industry and calls them frozen desserts. The full range consists of ice cream; frozen custard; low fat or non-fat ice cream, sometimes called ice milk or dietary frozen dessert; sherbet; water ice; frozen yogurt; frozen pudding; and mellorine. These products differ from one another by fat content or the source of the fat. Legal definitions vary from state to state but are very similar. Ice cream is further subdivided into packaged forms and novelty products: predominantly ice cream bars, fruit bars, and ice cream sandwiches.

Frozen custard differs from ice cream in that it contains egg yolks. Custards are also cooked products whereas ice cream batter is a pasteurized and homogenized mixture. Frozen puddings contain flour in addition to egg yolks. Custards and puddings have a higher egg content and a more dense structure than ice cream. Low fat ice creams have reduced levels of butterfat; their chief ingredient is skim milk. No-fat ice creams are produced using whey or non-fat dry milk products. So-called lite varieties are aimed at the dieting public; they may be low in fat, have no fat, have reduced amounts of sugar or may be sweetened by artificial sweeteners.

According to the National Yogurt Association, citing the official definition of the Food and Drug Administration (FDA), yogurt is milk (or other dairy products) fermented with the benign bacteria Lactobacillus bulgaricus and Streptococcus thermophilus. The bacteria feed on milk sugars and release lactic acid. The acid causes yogurt's curdled character and creates an environment in which pathogens cannot thrive. Since yogurt can be produced by fermenting whole as well as skim milk, frozen yogurt products come in regular and low-fat varieties. The benefits of yogurt come chiefly from the bacterial cultures that they carry. These aid digestion, promote the immune system, increase calcium content, and have other systemic benefits. Frozen yogurt, however, does not offer dietary benefits as such.

Sherbets are slightly sour frozen fruit ices and come in milk or water-based forms. The word itself, derived from Turkish and Persian roots means drink and originally signified a fruit-drink. Sherbets contain milk solids whereas water ices don't. The term ice sherbet means a water ice and is not a sherbet, strictly speaking.

Mellorine is something of a hybrid product which combines milk-solids and only a minimum of milk fats. The fats in the dessert must be largely derived from animal or vegetable sources such as cottonseed oil or soy. The origins of this variety are obscure, but the product appears to have been introduced in the 1950s. In those days the emergence of mellorine encouraged cotton growers to hope for large new markets for their products. But mellorine never took off and is fading away with less than half of one percent of market share, and that minute share is also declining. The original motive for the formulation of this product was to take cost out of ice cream by substituting cheap sources of fat for more expensive ones.

Origins. Ice cream plays such a starring role in the enjoyment of food across the world that it is almost obligatory to romanticize its origins, to see it shrouded in mystery, and to detect its appearance thousands of years ago. Alexander the Great was not the first but a later aficionado of the product. Cliff Lowe, however, writing for inmammaskitchen.com, notes that ancient references to ice cream are really references to mixes of flavored ice. Ancient humans had indeed discovered the pleasure of eating chipped ice juiced up with fruits and honey, but what they ate was a slushy mix rather than a cream. Properly speaking ice cream appears to have originated around the year 1700 in France and Italy. Recipes for ice cream batter first appeared, and then multiplied, in French recipe books around that time. The invention of the current form of ice cream appears to have required the widespread use of metal cookware made of silver and pewter. Such utensils transmit temperature quickly. People made the first batches of real ice cream using two bowls. The larger bowl held ice mixed with salt, the smaller bowl inside the larger bowl held the batter. To make ice cream the cook whipped the batter while simultaneously agitating the ice-salt mixture by moving both bowls up and down rapidly. Salt causes ice to melt. That process consumes heat. A bowl suspended in an ice-salt mixture will be cooled as the salt melts ice and draws the energy for this melting from the batter, causing the batter to freeze. The whipping of the cream introduces air into the mixture. Ice cream was no sooner invented on the Continent than it appeared in the Colonies too. There is documentation to the effect that Jefferson and Washington both served ice cream on special occasions and took interest in ice cream recipes.

Making ice cream was an arduous, lengthy process; it took approximately 40 minutes of awkward whipping and agitation. Its preparation required physical effort. Nancy Johnson, of New Jersey, is credited with inventing the first hand-cranked ice cream machine, likely modeled on a butter churn. The invention dates to 1843. On September 9th of that year Nancy Johnson received Patent No. 3254 for her invention. The first commercial producer

FIGURE 110

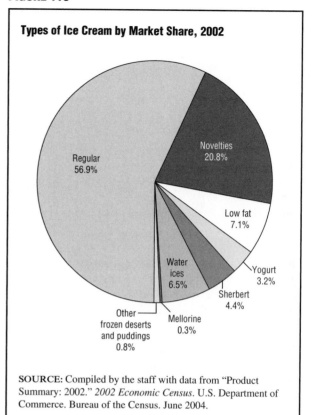

Types of Ice Cream by Market Share, 2002

Regular 56.9%
Novelties 20.8%
Low fat 7.1%
Yogurt 3.2%
Sherbert 4.4%
Mellorine 0.3%
Water ices 6.5%
Other frozen deserts and puddings 0.8%

SOURCE: Compiled by the staff with data from "Product Summary: 2002." *2002 Economic Census.* U.S. Department of Commerce. Bureau of the Census. June 2004.

of ice cream emerged soon after. He was Jacob Fussell, of Baltimore, Maryland. Fussell was a dairy farmer who discovered that demand for his milk was declining and parlayed this challenge into founding the first ice cream company. It became a national enterprise; eventually he sold it to the Borden Milk Company.

The Manufacturing Process. Industrial production of ice cream may be carried out in batch or continuous modes. The same basic steps are followed either way. These steps are blending of the batter, pasteurizing by heat, homogenizing the blend by pressure, cooling and flavoring the mix, freezing, packaging, and hardening. Milk either arrives already refrigerated or is held at such a temperature in cans awaiting the blending process. Milk is blended with egg whites, sugar, and stabilizers for a brief period of time. Pasteurization takes place at a temperature of 182 degrees Fahrenheit. It is accomplished by contacting the batter with plates washed by very hot water on one side. At the pasteurization temperature all bacteria in the mixture perish.

Raw milk, as it comes from cows, will naturally separate into butterfat and skim milk if left to itself. Homogenization causes fat particles to break apart into tiny particles. The process is accomplished by forcing the milk, still hot from pasteurization, through tiny orifices. Under high pressure the milk and the particles are effectively mixed in the homogenization chamber. The pasteurized and homogenized mixture is then allowed to cool and age. On the way to continuous or batch freezers, feeders add flavoring which mixes and blends in as the batter flows.

The actual creation of ice cream itself begins at this quite advanced stage. The mixture is pumped through freezers maintained at -40 degrees Fahrenheit and rapidly cools. But the cream, at this point, is still a cold semi-liquid paste. Air is introduced into the cooling mixture during freezing. The industry calls air the overrun. The amount of air injected has a direct relationship to the quality of the ice cream. Low overrun produces denser, richer ice cream, but some air entrainment is needed to give ice cream its creaminess. If the recipe calls for the inclusion of bits of fruit, chocolate, nuts, or candied mixtures, the cold paste passes through feeders that insert and mix these ingredients into the flow. The finished product is then dispensed into packaging carried by a moving belt, be these cartons or tubs. As the packages are filled they are automatically closed, and the belt carries the products into a freezing chamber.

The freezing chamber, also called the hardening room, is maintained at a temperature of -30 degrees Fahrenheit. The packaged ice cream must reside in such a space until it has reached at least -10 degrees of temperature. In large applications the freezing chamber is a tunnel through which the packaging moves for two or three hours before being extracted for transport to refrigerated warehouses ahead of retail distribution. On exit from the hardening room or tunnel, the ice cream is rock hard.

Variations on this process, following the initial freezing step, produce ice cream bars, sandwiches, and other novelty products by the deployment of specialized equipment that doles out the semi-liquid ice cream, inserts wooden or plastic handles for a bar, or sandwiches the ice cream between baked sheets before sealing them in individual sacks.

MARKET

Industrial Production. The ice cream and frozen dessert manufacturing industry shipped product valued at $9.01 billion in 2005, up from $8.2 billion in 2002. The market was $5.9 billion in 1997 and $5.2 billion in 1992. The growth rates in this period varied a great deal having been 2.1 percent per year on a compounded basis between 1992 and 1997, 7 percent per year in the period from 1997 to 2002, and slowing again to an annual rate of 3.2 percent per annum between 2002 and 2005.

In the 14-year period from 1992 to 2005, industry shipments showed negative growth in four years (1994, 1997, 1999, and 2003) and wildly different growth in ten others. Between 1997 and 1998 the industry grew 1.6 percent, for instance. Between 2002 and 2003 it grew at a heady 21.6 percent. Unless these data are the consequence of survey anomalies, one is tempted to suppose that ice cream reflects changing public moods. For instance, the period from 1992 to 1999 exhibited rather energetic growth in the economy (it was the so-called dot-com boom). During that time ice cream turned in a rather sluggish performance, growing at a compounded rate of only 0.4 percent per year. The 1999 to 2005 period, by contrast, was more troubled and somber. The economy began to show a softening in 2000 and gross domestic product turned negative in early 2001. That same year brought the 9/11 shock, military activities overseas, and in 2002 the beginning of a jobless economic recovery. Yet during this period, ice cream sales advanced at a very brisk rate of 8.6 percent per year.

Four factors influence demand for ice cream: weather, pursuit of health, flavor, and impulse-buying. Hot weather favors ice cream consumption. The desire to eat in a healthy manner either dampens ice cream buying or favors the low-fat or no-fat brands. Novel flavors can lift demand and, in effect, make small companies into dominant forces. Impulse buying is observed as a common behavior—so that a careful ice cream buyer will occasionally throw all inhibitions to the wind. Over the long run the industry is steadily growing.

FIGURE 111

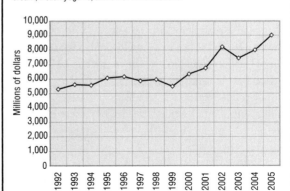

U.S. Ice Cream Shipments 1992–2005

Shipments are presented in millions of dollars and include all types of ice cream, frozen yogurts, and other frozen deserts.

SOURCE: Compiled by the staff with data from "General Statistics: 2002," *2002 Economic Census*, and *Annual Survey of Manufactures*, U.S. Department of Commerce, Bureau of the Census. June 2004 and March 2006.

One of the motives behind food selection decisions—to eat healthier foods—does find measurable support in the data available between 1997 and 2002. In that period regular ice cream lost a share of the market declining from 60.9 to 56.9 percent of total shipments. Yogurt and mellorine lost share as well. Sherbets, water ices, and low fat ice cream gained in share, 1.8, 1.3, and 0.1 percent respectively. But frozen novelties exhibited the largest growth, a 3.2 percent gain in share, suggesting that impulse buying on the one hand balanced healthy-eating tendencies on the other.

Retail Perspectives. Based on data from the U.S. Department of Agriculture (USDA), which independently monitors this industry from the dairy perspective, ice cream production in 2004 was 1.5 billion gallons, representing approximately 8 percent of total U.S. milk production. The value of this production, excluding water ices (on which USDA does not report) was approximately $4.50 per gallon at the point of shipment. For the same year, the International Dairy Foods Association (IDFA) estimates that total sales of ice cream and frozen desserts at retail were $21.4 billion, suggesting a nearly three-fold increase in value from the point of production to the point of sale. The products of the industry sold, on average (including water ices), for $13.30 per gallon.

The retail market is of necessity an estimate because retail sales of ice cream and related products are combined by the Census Bureau with the sales of other commodities in reporting on such categories of distribution as wholesalers, grocers, and restaurants; the last category includes ice cream shops along with fast-food enterprises and luxury restaurants. Bulk sales of ice cream reach ice cream shops and restaurants; such shipments are a relatively small proportion of total production (approximately 19%) but command high prices, anywhere from $35 to $150 per gallon. Expressed on a per-gallon basis, ice cream novelties will typically cost between $13 and $40 per gallon. Ice cream in 16 ounce containers will retail for around $30 on a per-gallon basis; in 56 ounce containers ice cream sells for $10 to $13 per gallon. The most popular size, the half-gallon container, sells for $6 to $9 per gallon. Popular, big-name brands have the highest prices, while store-brands come in at the lowest end of the price scale.

Flavors. The IDFA, citing The NDP Group's National Eating Trends Services, reports that vanilla leads the flavor parade; 20 percent of consumers opt for the flavor. Next in line is chocolate with a 12.9 percent share, neapolitan with 4.8, strawberry with 4.3, and cookies-and-cream with 4 percent of the market. Baskin-Robbins' famed 31 flavors, or at least 26 of them, have to compete for the

remaining 68 percent—as well as with the flavors offered by more than 360 producers in the United States.

KEY PRODUCERS/MANUFACTURERS

Based on the *2002 Economic Census*, 365 companies participated in the industry in 2002, operating 407 establishments. The total number of corporate participants was down from 410 in 1997; establishment counts were down from 451. Just under one-third of establishments in 2002 (31%) were of some size, employing twenty or more people. These data indicate an industry of many small independent operations serving local or regional markets. Many are family owned, many are local dairies. Just a handful of companies have a commanding national or international market. Most of these latter have been assembled under the fold of two major food companies: Unilever N.V. and Nestlé S.A. Yet others with national name recognition are also owned by food companies. An example is Baskin-Robbins owned by Dunkin' Brands (of Dunkin' Donuts fame) which is itself owned by several investment companies (Bain Capital, Thomas Lee Partners, and the Carlysle Group). Another is Healthy Choice, owned by ConAgra.

Based on data assembled by *Market Share Reporter*, in 2005 Nestlé was the leading company in the United States with 23 percent of the market, followed by Unilever with 22 percent, and in 2002 Unilever was the top producer worldwide (16.3%) followed by Nestlé (11%). In 2007 both of these companies maintained their dominance, controlling half the market in the United States and more than one-third of the market globally. Nestlé owns Dreyer's and Häagen-Dazs, the second and third-ranked brands in the United States in 2006 according to *Dairy Field*. Dreyer's is known as Edy's in markets west of the Rocky Mountains and the brand is usually referred to as Dreyer's/Edy's. Nestlé is a leading brand in ice cream novelties in its own right, and novelties are the most rapidly growing category in ice cream.

Unilever owns Breyers, the top brand in the United States in 2006, acquired in 1993. That same year Unilever also acquired the Sealtest line and merged these into its wholly-owned Good Humor-Breyer's Ice Cream Company. Good Humor had also been an independent company and was the originator of the first-ever ice cream bar. Unilever acquired Good Humor in 1961. By the time of Breyer's acquisition, Unilever had also purchased Gold Bond Ice Cream in 1989. The company acquired Ben & Jerry's in 2000, but has been operating that acquisition as a separate company rather than as part of the Good Humor-Breyer's combination. Ben & Jerry's was the fifth-ranking U.S. brand in 2006. Like Nestlé, Unilever represents a family of ice cream brands but, unlike Nestlé, has no ice cream called by its own name.

Ice cream making is a relatively low-cost-of-entry type of business and continues to attract entrepreneurs who hope to make their mark by some unique differentiation. Two of the leading brands of the early twenty-first century began in the second half of the twentieth century. Häagen-Dazs being a new brand arising from a family already in the ice cream business since the 1920s and Ben & Jerry's, a start-up company.

The inventor of Häagen-Dazs, Reuben Mattus, a New Yorker, introduced a new high-quality brand of ice cream with just three flavors (vanilla, chocolate, and coffee) in 1961. The brand took off. It had two important features to promote it. It aimed at a market largely neglected then, a luxury market for richly textured, dense ice cream; it also had an eye-catching name that Mattus invented, giving it an umlaut, to suggest old-world quality and craftsmanship. The market was ready for such a product. The brand's reputation spread across the East Coast. In 1976 Doris Mattus, Mattus' daughter, formed the first Häagen-Dazs ice cream shop. The franchise soon became national and, after Mattus sold the business to Pillsbury, became a world-wide brand. General Mills, later the owner of Pillsbury, sold the Häagen-Dazs line to Nestlé.

Bennett Cohen and Jerry Greenfield, childhood friends, began Ben & Jerry's in 1977 in Burlington, Vermont. They were in their twenties at the time, wanted to enter the food business, and chose the ice cream business because it cost less to enter than bagel-making. They learned to make ice cream in a correspondence course—but their achievement was clearly not learned from books but flowered from talent and flair. They built their franchise by combining a great instinct for novel flavors, showmanship, and community involvement. They were members of the baby boom and well in tune with it. Within ten years they had built a $32 million business—which was just a beginning.

Equally representative of key producers and leaders in this industry are Blue Bell Creameries, Wells' Dairy, Inc., and Turkey Hill Dairy. In rankings of U.S. ice cream brands these three companies held fourth, sixth, and seventh rank in 2006. All three are family-owned. Blue Bell, based in Brenham, Texas, celebrated its 100th birthday in 2007. Blue Bell has been operated by the Kruse family since 1919. The Wells' Dairy's brand is the Blue Bunny. Introduced in 1935, it comes in nearly 500 flavors. The company is based in Le Mars, Iowa, a town that styles itself the Ice Cream Capital of the World, claiming to be the community with the largest ice cream production within its borders. The company has been continuously owned by the Wells family since 1913. Turkey Hill was originally a farm owned by Armour Frey in Lancaster County, Pennsylvania. It was transformed into a dairy in 1947 when Frey's sons bought out the business. Turkey Hill began manufacturing ice cream in 1980.

MATERIALS & SUPPLY CHAIN LOGISTICS

The ice cream industry's chief raw material is fresh milk, a commodity consumed by the population and reaching its market in refrigerated tanker trucks everywhere. At the same time, the finished product, just like its main input, requires refrigerated transportation to the point of sale. The product, therefore, is significantly dependent on technically advanced transportation systems in any case. For this reason ice cream is produced both at significant distances from population centers but near dairies, and in the center of urban areas relying on milk reaching it from the countryside.

DISTRIBUTION CHANNEL

Ice cream reaches the final consumer by two very distinct channels, one being the grocery store or supermarket serving those intending to eat the product at home, and the ice cream shop or restaurant where product is purchased for immediate consumption. According to the International Dairy Foods Association, as measured in dollar expenditures, expenditures away from home represent two-thirds of all ice cream purchases, one-third reaching the consumer in packages.

On a volume or gallonage basis, the divide appears to be the other way around, most ice cream being packaged and purchased for at-home consumption. This indicates that portion-bought ice cream is the most expensive. If the ice cream is made directly at the site where it is sold in cones and cups, a one-tier distribution is practiced (producer to consumer). Most ice cream reaches the consumer by means of a three-tier distribution in which producers sell to a frozen foods wholesale merchant, the merchant supplies the retailer, and the retailer sells to the ultimate user.

KEY USERS

Ice cream is eaten by many, often as a dessert or as a refreshment on a hot day. The product line includes varieties even for those who, being lactose-intolerant, can only enjoy water ices, but that product is firmly part of the ice cream universe as a whole.

ADJACENT MARKETS

Most adjacent to ice cream—indeed almost indistinguishable from the category when consumed in an ice cream shop—is the ice cream cone itself. Cones and waffles are bakery products. In the history of ice cream, the origin of the cone is typically said to be at the World Fair of 1904.

According to Cliff Lowe's well-researched history, an ice cream vendor named Charles Menches found himself booth-to-booth to a waffle vendor named Ernest Hawmi. When Menches ran out clean ceramic cups he found a solution by getting waffles from his neighbor. Hawmi formed cone-shaped waffles on the spot to satisfy this sudden demand. Lowe, however, points out that a New York-based Italian named Marchiony sold ice cream in waffle cups in the late 1800s and early 1900s and, indeed, obtained a patent for such cups in 1903.

Adjacent markets to ice cream are also other forms of dessert, most closely adjacent being refrigerated creams, puddings, and whipped toppings. A market for home ice cream making machinery is another adjacent market. Such equipment may be hand-cranked or electrically operated, may be designed to be used with ice and salt or to be placed in the refrigerator's freezing compartment. High-end ice cream makers cooled by liquid gases are also available.

RESEARCH & DEVELOPMENT

Most efforts of this nature in the field of ice cream are centered on product development, strongly focused on impulse products in the novelty category. Efforts are also expended on producing ice creams and frozen desserts that have low fat and low sugar content yet produce the taste sensations associated with high-end luxury, regular ice creams.

CURRENT TRENDS

The dominant trends in this industry are toward richer and better-tasting products on the one hand and healthier frozen desserts on the other. These trends are somewhat in conflict. If the recent decades of ice cream history are representative of the future, major success will reward those who bet on richer and more delicious products, as exemplified by the triumphs in market-penetration achieved by Häagen-Dazs and Ben & Jerry's.

In the food industry as a whole, snack foods are under intense development because evolving life styles are marginalizing food consumption at regularly scheduled family meals. The search for new snack foods is reflected in the ice cream industry in the rapid growth of ice cream novelties. Currently available market data also suggest that healthy ice cream will, at the very least, establish a major segment for itself and, aided by innovation, will hold its own.

TARGET MARKETS & SEGMENTATION

The three major markets are regular ice cream, novelties, and healthy formulations—targeted at the general buyer who is interested in taste alone and feels little guilt for indulgence, at the impulse buyer looking for five minutes of pleasure, and at those minding their dietary budgets. Within the health-minded segment, non-fat and low- or no-sugar products represent one extreme. Within regular ice cream, high-end luxury brands represent the other. The same divide is present in novelties which come in regular, sherbet, and low-fat varieties. There is, thus, a high, a middle, and a low end, each with its clientele.

RELATED ASSOCIATIONS & ORGANIZATIONS

International Association of Ice Cream Vendors, http://www.iaicv.org/indlinks.htm

International Dairy Foods Association (IDFA), http://www.idfa.org/about/index.cfm

National Ice Cream Retailers Association, http://www.nicyra.org

National Yogurt Association, http://www.aboutyogurt.com/lacYogurt/facts.asp

BIBLIOGRAPHY

"Ben Cohen—Co-founder of Ben & Jerry's Ice Cream." Ben & Jerry's. Available from <http://www.benjerry.com/our_company/about_us/our_history/benbio.cfm>.

"A Brief History of Ice Cream." Zinger's Homemade Ice Cream. April 2006. Available from <http://www.zingersicecream.com/history.htm>.

Dairy Products 2002 Summary. U.S. Department of Agriculture. April 2003.

Darnay, Arsen J. and Joyce P. Simkin. *Manufacturing & Distribution USA,* 4th ed. Thomson Gale, 2006, Volume 1, 927–929.

"Ice Cream Manufacture." University of Ghuelph. Available from <http://www.energiemanagertraining.com/dairy/pdf/Ice_Cream03.pdf>.

"Just the Facts: Ice Cream Sales and Trends." International Dairy Foods Association. Available from <http://www.idfa.org/facts/icmonth/page2.cfm>.

Lazich, Robert S. *Market Share Reporter 2006.* Thomson Gale, 2006, Volume 1, 80–81.

Lowe, Cliff. "The History of Ice Cream, How to Make Ice Cream at Home: Ice Cream Makers Past & Present." inmammaskitchen.com. Available from <http://www.inmamaskitchen.com/FOOD_IS_ART_II/food_history_and_facts/ice_cream.html>.

Petrak, Lynn. "The Cold Truth: Although Sales Dip Slightly, Diverse Products Keep Rolling Out to a Splintered Marketplace." *Dairy Field.* January 2007.

———. "Freezing Points: Better-for-you and Decadent Ice Cream Products Continue on Parallel Growth Tracks as the Category Churns Ahead." *Dairy Field.* January 2006.

Scott, Mark and Cassidy Flanagan. "Ice Cream Wars: Nestlé vs. Unilever." *BusinessWeek.* 24 August 2007. Available from

<http://www.businessweek.com/globalbiz/content/aug2007/gb20070824_230078.htm>.

Smith, Pamela Accetta. "What's in a Name?" *Dairy Field.* February 2007. Available from <http://www.dairyfield.com/content.php?s=DF/2007/02&p=15>.

Wilson, Steve. "Doc Wilson's Ice Cream Page." Doc Wilson's Ice Cream Site. Available from <http://www.users.nwark.com/~piperw/icpage.htm.>.

SEE ALSO *Milk & Butter, Snack Foods*

Industrial Finishes & Coatings

———◆———

INDUSTRIAL CODES

NAICS: 32–5510 Paint and Coating Manufacturing

SIC: 2851 Paints and Allied Products

NAICS-Based Product Codes: 32–55104 through 32–55104265, 32–55107 through 32–55107061, and 32–5510A through 32–5510A041

PRODUCT OVERVIEW

The three major subdivisions of the Paint and Allied Products industry, as defined by the U.S. Bureau of the Census, are: (1) architectural coatings, (2) product finishes for original equipment manufacturers, and (3) special purpose coatings, including all marine paints. Architechtural coatings, commonly known as house paints, are not covered in this essay. Here, the focus is on the finishes and coatings used in making other products.

Product finishes for original equipment manufacturers (OEM). These products consist of all types of internal or surface coverings typically applied by manufacturers on the assembly line while making a product. An everyday example is the exterior finish on automobiles. A less obvious example is the internal coating applied to food cans to protect the metal from the food and the food from the metal, as well as the material applied to the outer coating of the can to protect it from rusting and to receive the inks of a label or brand image. The application of OEM finishes is an intrinsic part of manufacturing. These finishes are sprayed on as either liquids or powders. This group of products will be referred to as product finishes or as OEM finishes.

Special purpose coatings. This group of products, also called specialty coatings, is applied after the manufacturing process. They comprise a distinct product subdivision because they are applied away from the assembly line or the component manufacturing system. Marine paints are the only exception; whether applied during manufacturing of new vessels or during the process of maintaining older ships, all marine paints are labeled special purpose coatings by the U.S. Census Bureau. That is not the case in autos, however. Automotive finishes applied on the assembly line are known as product finishes, but automotive coatings applied after the manufacturing process in repair or paint shops are special coatings. Special coatings are applied to provide special protection to existing products, in maintenance or repair operations, or as markings. For instance, paints used in traffic signs and markings are considered special coatings.

The product finishes industry arose only after industrialization began and brought with it the mass production of products on a large scale. Product finishes were first used in the transportation industry. Two of today's top product finishes and special coatings producers were present at the marriage of coatings and the transportation industry. PPG Industries, founded in 1883 as Pittsburgh Plate Glass Company, has roots in North America in the automobile center itself, Detroit, Michigan, while Akzo Nobel began as Sikkens in Western Europe, more specifically, in Arnhem, the Netherlands. In Detroit in 1902, the Ditzler brothers manufactured color varnishes and sold them to carriage makers and fledgling automakers. Fast drying super durable undercoats were developed to meet Detroit automakers' needs. In Arnhem, during the same time period, father and son Sikkens developed fast drying car lacquers that later protected aircraft, railroad equipment, and farm machinery and equipment. In 1928 PPG purchased Ditzler and introduced solventborne synthetic enamels, which became the standard in product finishes because they created strong, smooth, and glossy finishes. The 1950s brought innovations like powder coating finishes. The 1960s brought more innovation as product finish producers realized that electronics were a growth area. The 1990s brought high solids solventborne and waterborne products. The new century brought high performance finishes and growth in the special coatings market due to innovations known as smart coatings.

The transportation industry remains the single most important user of finishes and coatings. Trends often start in the transportation industry and then migrate to other industries. For instance, trends in metallics coatings began in the transportation industry and traveled to appliances and then electronics. In terms of relative size, product finishes represent the larger of the two markets, with special coatings being the smaller market.

MARKET

The market for paints and coatings of all types in the United States was $20.8 billion in 2006 as reported by the Census Bureau in its *Current Industrial Reports* series. OEM finishes represented 29 percent, and special coatings 21 percent of this total. The combined value of these two categories had shipments in 2006 of $10.4 billion. Shipments in this combined grouping were $8.6 billion in 1997. Industrial finishes and coatings thus grew at a rate of 2.1 percent per year, underperforming the durable goods sector of the U.S. economy, which was growing at an annual rate of 2.6 percent. Durables are selected for comparison because these products are used to coat them. Figure 112 shows the big picture for the period 1997 to 2006.

Detailed data on components categories within the two major groups, OEM Product Finishes and Special Finishes, were available only from 2001 and beyond. An analysis of product categories is presented in Figure 113 for two years, 2001 and 2006. Twelve subcategories divide product finishes and eight divide the special finishes product cluster. Despite such detail, the tabulation is still highly aggregated because the subcategories, each explode into multiple products.

OEM Product Finishes. In 2006 product finishes accounted for 58 percent of the industrial finishes and coatings grouping. In the 2001 to 2006 period, the

FIGURE 112

U.S. Paint and Allied Products Industry Shipments, 1997–2005

◆ Product Coatings (original equipment manufacturers)
■ Special-purpose Coatings

SOURCE: Compiled by the staff wtih data from "Paint and Allied Products: 2005," *Current Industrial Reports*, U.S. Department of Commerce, Bureau of the Census. June 2006.

growth of this category was unremarkable at 1.5 percent annually, produced by some rapidly growing and some declining categories. The largest category in shipments was automotive finishes, including autos, light trucks, vans, sports utility vehicles, and parts for all these products. Next in size were finishes for wood furniture and wood products. Metal building products, which include sidings, were third. The smallest category with, furthermore, the greatest drop in shipments, was coatings for electrical insulation products.

All but one of the subcategories under product finishes, are liquid coatings. The exception is powder coatings. This powder subcategory had the most rapid growth in the OEM group, its shipments advancing at 3.9 percent yearly. The next two subcategories ranked by growth were paper, board, and foil finishes (3.8% per year) and container and closure finishes (2.8%). Growth was paced by a product category offering technological/environmental advantages and two others that largely represent commercial and consumer packaging products—corrugated containers and canned goods.

The powder coatings category (14.5% of OEM and 8.4% of industrial coatings), represents a group within a group because powdered coatings are used precisely in those OEM categories that show up, in Figure 113, with negative growth rates on the liquid side. Negative growth in liquid finishes is in part balanced by positive growth in powdered finishes. This shift is particularly notable in coatings for electrical insulation products which appear to be shifting strongly from liquid to powder.

FIGURE 113

U.S. Industrial Finishes & Coatings 2001 and 2006 by Subcategory

Product Category	Shipments (million dollars)		Annual Growth (percent)	Percent in 2006 of	
	2001	2006		Category	Total
OEM Product Finishes, Total	**5,601**	**6,035**	**1.5**	**100.0**	**58.3**
Auto, light trucks, vans, SUVs	1,204	1,350	2.3	22.4	13.0
Other transportation (truck, air, rail)	455	374	-3.8	6.2	3.6
Appliance, heaters, air-conditioning	177	97	-3.6	1.6	0.9
Wood furniture and other wood products	581	676	3.1	11.2	6.5
Non-wood furniture	479	396	-3.7	6.6	3.8
Metal building products	582	650	2.2	10.8	6.3
Containers and closure finishes	438	482	1.9	8.0	4.6
Machinery including road building	468	536	2.8	8.9	5.2
Paper, board, foil finishes	115	139	3.8	2.3	1.3
Electrical insulating coatings	29	19	-7.4	0.3	0.2
Powder coatings	721	872	3.9	14.4	8.4
All other or not specified	412	444	1.5	7.4	4.3
Special Finishes, Total	**3,243**	**4,324**	**5.9**	**100.0**	**41.7**
Industrial (interior)	212	475	17.4	11.00	4.6
Industrial (exterior)	542	707	5.5	16.3	6.8
Traffic marking paints	280	330	3.3	7.6	3.2
Auto and other machinery repainting	1,672	2,312	6.7	53.5	22.3
Marine (ship, offshore)	277	243	-2.6	5.6	2.3
Marine (yacht, pleasure)	(D)	9	-	0.2	0.1
Aerosol paints for products	(D)	228	-	5.3	2.2
Not Specified	58	21	-18.1	0.5	0.2
Total Industrial Finishes and Coatings	**8,844**	**10,359**	**3.2**	**-**	**100.0**

(D) indicates data suppression by the U.S. Census Bureau.

SOURCE: Compiled by the staff with data from "Paints and Allied Products 2002 and 2005," *Current Industrial Reports,* U.S. Department of Commerce, Bureau of the Census, March 2004 and June 2006.

Powder coatings were introduced in the 1950s. The dry pigment is applied to surfaces by blowing the powder. The powder is formulated so that resins it incorporates cause it to adhere uniformly to the receiving surface. The powder is then immediately baked in high-temperature ovens to a hard finish. Because heat is needed to set the coating, powder coatings are also known as thermosets.

Special Finishes. The lines between categories are not always clear. Thus, certain products that should appear under the OEM grouping show up as special finishes. Most notably coatings, finishes, and paints used in marine products, ships, yachts, oil rigs, and shore facilities are functionally misplaced. Whether the paint is applied to a brand new ship or to an old one, the paint used is part of special finishes. The category also includes paints that are put in aerosol cans and sold to consumers directly for use in homes but is included under the industrial category.

The largest special finish subcategory is coatings of all kinds used in repainting and maintenance of autos, trucks, and machinery in general. This category represented 53.5

percent of the special finishes group and 22.3 percent of industrial finishes and coatings overall. Second and third in shipments were industrial finishes for exteriors (16.3%) and interiors (11%). The U.S. Census Bureau describes these finishes as "Industrial new construction and maintenance paints...especially formulated coatings for special conditions of industrial plants and/or facilities requiring protection against extreme temperatures, chemicals, fumes, etc."

Special finishes, the smaller part of industrial coatings, 41.7 percent of the broad category, had much better growth, with shipments increasing 5.9 percent per year, faster than the Gross Domestic Product (GDP), which grew at 5.4 percent and more than twice as fast as shipments of durable goods at 2.7 percent. The auto/machinery finishes category alone had annual growth of 6.7 percent, industrial interior finishes a dramatic growth of 17.4 percent, and industrial exterior coatings 5.5 percent per year. In general, during this period, finishes for maintenance were beating out finishes for new construction.

KEY PRODUCERS/MANUFACTURERS

The companies present in the early years of the development of the industrial paint and coatings market—Akzo and PPG—are still at the top, along with Sherwin-Williams.

Akzo. Although Akzo is number one in the world, with headquarters in The Netherlands and coatings sales worldwide in the $7.2 billion range, North America is important in the company's strategy for growth. On the last day of 2006, Akzo announced that it would make Nashville its North American powder coatings headquarters to focus on appliances and furniture. Akzo divides its portfolio into products for industrial applications such as powder, wood, coil, and specialty coatings; automotive refinishes; and marine, protective, and aerospace coatings. In Western Europe, to focus on wood finishes, Akzo acquired ICI Group's wood business in 2005. In North America, it acquired The Flood Company's wood coatings with its patented Penetrol technology in September 2006. Akzo reiterated in December 2006 that the company is continuing to target coatings market acquisitions.

PPG Industries. PPG Industries has been a leader in making product finishes and special coatings since its acquisition of Ditzler in 1928. Headquartered in Pennsylvania, PPG is a diversified manufacturer that recently married its historic interest in glass to coatings, capturing the emerging architectural glass market. PPG is ranked number two in the world with estimated 2005 global coatings sales of $5.61 billion. PPG divides its portfolio into applications for auto; auto refinish; industrial; architectural; aerospace and packaging. PPG announced in January 2007 the purchase of the industrial coatings unit of a South American paint maker. PPG recorded an after-tax charge of $106 million in the third quarter of 2006 in part related to environmental remediation at a former chromium manufacturing site in New Jersey. PPG has also acquired a Singaporean leader in powder coatings for the flooring industry, opened a Taiwanese coatings facility for small parts, and expanded an existing Chinese facility for automotive and industrial coatings.

The Sherwin-Williams Company. One of the founding firms of the United States finishes and coatings industry, Sherwin-Williams is considered number three in the world. Sherwin-Williams reported 2005 sales of $7 billion with global coatings sales of $5.55 billion, approximately 79 percent of the total, an increase of roughly $1 billion over the previous year. Sherwin-Williams divvies up its portfolio more simply than others, into automotive finishes; industrial and marine coatings; industries; and special-purpose coatings for the automotive-aftermarket, industrial-maintenance and traffic-paint markets. Sher-win-Williams is not diversified. It has always emphasized paint and coatings and is generally accepted to be the largest producer of paints and coatings in the United States. *Fortune* magazine named Sherwin-Williams to their 100 Best Companies to Work For list.

ICI Paints. A once prominent company was ICI Group. At the end of 2005 ICI Paints, headquartered in the United Kingdom, had estimated 2005 global sales of $2.972 billion and was number five in the world. Since 2003, in the wake of a devastating profit warning, ICI hammered its balance sheet back into shape with a string of well-executed sell-offs. This allowed ICI to pay off £900 million in debt and put £230 million into the company's pension black hole. As a result, ICI was in termoil in December 2006, when speculation was rife that Akzo Nobel of The Netherlands would mount a raid on ICI. It appears ICI is ripe for takeover.

Other prominent companies include DuPont Coatings & Color Technologies Group, Wilmington, Delaware, with estimated 2005 global coatings sales of $4.2 billion. BASF Coatings AG, Munster, Germany, had estimated global 2005 coatings sales of $2.8 billion, while Valspar Corp., Minneapolis, Minnesota estimated 2005 global coatings sales at $2.4 billion. Up and comers in the industry include Nippon Paint Co. Ltd., Osaka, Japan, with estimated 2005 global coatings sales of $1.7 billion. Nippon announced that it acquired Rohm and Haas Company's automotive coatings business in December 2006 and intends to create a leading position in the North American plastic automotive parts coatings market, a market that grew 25 percent in the five-year period from 2001 to 2005.

MATERIALS & SUPPLY CHAIN LOGISTICS

The cost to purchase the materials needed to manufacture all paints and allied products increased 10 percent on average between 1997 and 2002. The largest categories of materials purchased are:

- Resins (including alkyd, acrylic plastics, vinyl, epoxy and polyester)

- Organic and inorganic color pigments (including chrome colors, zinc oxide, iron oxide, metallics and predispersed colorants) and titaniuim dioxide pigments

- Solvents (hydrocarbon, alcohol, ester, ketone and glycol)

Resins can be either natural or synthetic, oil- or water-based, and are the material used as the binder to suspend pigments evenly. Binders make finishes and coatings adhere to the surface. Binders determine performance properties such as hardness, color retention and durability. Alkyd resin is a synthetic resin used in solvent-based finishes and coatings; acrylic resins are an expensive synthetic polymer used to make high-performance finishes and coatings; vinyl is a clear synthetic resin used in water-based finishes and coatings. In 2002 the entire $19.9 billion paint and allied coatings industry spent $2.2 billion on this important class of materials.

Pigments are a basic component of product finishes and special coatings. They are powdery substances that provide whiteness or color and hiding power. They can be organic or inorganic, and their quality and cost vary widely. High performance, high-end pigments are needed in finishes and coatings. High performance pigments like silica and silicates provide excellent durability; zinc oxide pigments resist mildew and corrosion and rust. Spending on this class of materials increased 6 percent in the period between 1997 and 2002, from $1.5 million to $1.6 million. Titanium dioxide is an expensive bright white pigment that is the most important color-producer needed to make product finishes and special coatings. In 2002 the paint and coatings industry purchased $868 million worth of titanium dioxide.

Solvents are a liquid in which pigments are dissolved or dispersed. Solvents include hydrocarbons like toluene and xylene; alcohols like butyl, ethyl, and isopropyl; esters include ethyl acetate and butyl acetate; ketones are methyl ethyl ketone and methyl isobutyl ketone; and glycols. Spending by the entire paint and allied products industry on this class of materials increased 10 percent in the period between 1997 and 2002, from $753 million to $837 million.

DISTRIBUTION CHANNEL

Single tier and two-tier distribution systems coexist in these two subdivisions of the paint and allied products industry. Single tier distribution is characteristic of the OEM product finishes group. Finishes and coatings are sold by the manufacturer directly to the industrial user who will incorporate them into their own production system. Because 38 percent of product finishes are purchased by transportation-related industries, the makers of OEM finishes are intimate with the assembly line processes of this industry. This intimacy frequently leads to custom-designed product finishes and other cooperative innovations.

Two-tier distribution systems are characteristic of the special coatings subdivision of the paint and allied products industry. These coatings are sold through specialized coatings, sealants, and adhesives distributors who sell to the smaller commercial automobile paint and repair shops—the automotive aftermarket. The complexity of this industry at the technical level provides a natural niche for distributors in the middle between producers and industrial consumers.

KEY USERS

Key users of product finishes—38 percent—are the transportation-related industries who make equipment on a production line: light duty, parts, heavy duty, aircraft and railroad, machinery and equipment. Other key users are the building and construction industries, especially companies that make aluminum extrusions.

Key users of special coatings—53.5 percent—are in the transportation industry aftermarket. These are the many commercial operations engaged in repair and maintenance. Other key users of special coatings are industrial facilities.

This industry's own trade association, the National Paint and Coatings Association, which boasts a 70 percent participation rate within its industry, classes key users of product finishes and special coatings as the makers of 13 products: homes, buildings, factories, bridges, ships, automobiles, buses, furniture, appliances, machinery, metal food cans, highway safety markings, and aircraft.

ADJACENT MARKETS

Adjacent markets to finishes and coatings are those that provide the functional services of these chemicals in a different way. The most important category in the future could be engineered plastic surfaces that do not require coatings because the functionalities of coatings—such as color, reflectivity, resistance to abrasion, and protection of the underlying substance—are inherent in the plastic. Engineered plastics with such characteristics are still very expensive and are used only in limited applications such as in the form of small internal parts intended to reduce friction. Two other products the markets for which are adjacent to the industrial finishes market are lubricants and exotic metal alloys. Lubricants eliminate the need for coatings intended to protect internal moving parts. Exotic metal alloys make finishes unnecessary in many cases, by producing surfaces that will not oxidize. Rolls-Royce, the manufacturer of aircraft based in the United Kingdom, said coatings now account for 30 percent of production costs against 5 percent forty-five years ago because internal coatings are now crucial to the effective operation of all aircraft components.

Demand for product finishes and special coatings can be tied to the thirteen sectors listed under key users, who are also adjacent markets. Transportation related industries account for 38 percent of product finishes and 53.5 percent of special coatings; thus, what happens in the auto industry has an important and direct influence on demand for special coatings. Building and construction industries use nearly 11 percent of product finishes for aluminum extrusions, so what happens in that industry impacts the paints and coatings industry, making it an adjacent market. Special coatings related to facilities construction and maintenance are growing at a rate of 17 percent annually (for interior coatings), which affects special coatings growth.

RESEARCH & DEVELOPMENT

U.S. product finishes and special coatings grew to $10.3 billion per year in 2006 primarily due to chance discoveries by creative and often unconventional ideas that were frequently labeled impossible when first suggested. This interest in doing things that have never been done before remains a hallmark of finishes and coatings manufacturers who employ 25 percent of their workforce in research and development. Without a doubt, both product finishes and special coatings makers—and resins manufacturers—are devoted to research and development in resins because resins are at the heart of coatings. Resins comprise approximately 30 percent of a product's composition. They are the substance that holds everything together, and the most expensive class of materials needed—$2.2 billion were spent on resins in 2002. Resins determine durability, gloss, hardness, and longevity. Newer resins that inhibit rust, for instance, are a paramount research and development goal at Krylon Industrial Coatings.

Research and development is typically focused on growing market segments. Since growth in the product finishes subdivision has essentially been flat, research and development is now focused on special coatings with their double-digit growth in both big segments; the automotive refinishing aftermarket and the facilities coatings. New markets for architectural glass and environmental regulations also contribute to R&D.

New markets contribute to research and development. There is growing demand in the building and construction industry for energy efficient, high performance glass products. Architectural glass has thin, transparent coatings applied to enhance performance. PPG, for instance, has melded its historic interest in glass to contribute to this growing market segment. It has added glass coating capabilities to its Wichita Falls, Texas plant. The new plant began operations in 2006 and is PPG's sixth glass coating line. PPG anticipates this fast-growing segment of the architectural glass industry will continue to increase demand for special purpose coatings.

Regulations contribute to research and development. The European Union mandated reductions in the use of hexavalent chromium and issued directives to the automotive industry, the household appliance industry, and the construction industry. Regulatory pressure to find substitutes for hexavalent chromium affects the steel industry that supplies the automotive industry and the household appliance industry with galvanized thin sheet steel. Thin sheet steel is coated with a thin layer of zinc to make it resistant to corrosion. Hexavalent chromium is the alloy used to galvanize and harden thin sheet steel and to make it corrosion resistant using a process that involves chromium (VI)-passivation. Henkel develop a completely chromium-free passivation system for galvanized steel called Passerite 5004 that offers the same level of corrosion protection. Voestalpine, an Austrian-based steelmaker, was using Passerite 5004 on all four of its galvanized thin sheet production lines as of 2006. It has completely replaced hexavalent chromium.

CURRENT TRENDS

Three developments in the market for product finishes and special coatings stand out as trends in the market today. These include the continued commitment to consumers in North America, the ongoing debate over powder versus liquid applications, and the trends toward changes based on styling and design, especially the increased use of metallics.

Product finish manufacturers are still committed to steady growth in North America. The commitment to North America is demonstrated by capital expenditures for new plant construction in Indiana and Michigan. Vitracoat America Inc., built a 14,000 square foot powder coatings plant in October 2006 in Elkhart, Indiana. Akzo Nobel announced in August 2006 that it will add an 11,000 square foot addition to its Pontiac, Michigan, automotive plastics coatings plant—although much of the automotive industry is cutting costs, plastics remain important. On the last day of 2006, Akzo announced that it will make Nashville, Tennessee, its North American powder coatings headquarters to focus on appliances and furniture.

The growth of powder coatings relative to liquid coatings is another trend in the market today. Overall, powder coatings represents 14.4 percent of the product finishes subdivision in 2006, up from 13 percent the year before, and lead growth in the group. The debate surrounding the vivid growth patterns on each end of the spectrum in powder coatings will continue. For instance, automotive powders decreased by 63 percent while appliance powders increased by 17 percent during the same time period. The entire powder product class had been predicated to grow after the U.S. Environmental Protection Agency implemented phased reductions in volatile organic compounds (VOCs), culminating in 2003 with total reductions of 45 percent. While benefits of powder coatings include that they are solvent-free, producing no VOC emissions, and that they are available in a variety of colors and textures, powder coatings growth has been lower than some in the industry had anticipated. Mushrooming growth may still be ahead. In 2005 Akzo constructed a new powder coatings plant in Russia; in 2006 the company created Akzo Nobel Powder Coatings in the Middle East near Cairo, Egypt, and built a powder coatings plant in China. Akzo also has headquarters in Tennessee to focus on anticipated growth in powder coatings for use on appliances and furniture.

Another emerging trend is the increased use by industrial consumers to use product finishes—especially metallics—to differentiate their products in a competitive marketplace. Known as the style and design trend, this trend is expected to continue to drive growth in sales. The style and design trend shows an increased use of vivid colors and textures to establish a unique style for the items on which they are applied. An early example being the metallic effects in the all important transportation sector. Metallic effects were first used in car paints with metallic silver comprising as much as 37 percent of new car registration in the 2000 to 2006 period. Other metallic effects growing in the transportation sector include metallic grays, blues, and blacks.

After metallics got wheels in the transportation industry, they traveled to appliances and then electronics. Metal effects pigments cost more, but it appears manufacturers are willing to pay more for the stylish results they produce. For example, classic organic pigments range from $7.50 to $20/kilogram (kg) while metallic pigments cost $22 to $58+/kg. High performance pigments have been formulated for design effects that include interference/color shifting and luminescent/phosphorescent. Metallic effects include sparkle, pearlescence, liquid metal effects, and combinations of these. For instance, LuminOre has a cold alloying process for composite metals. While the process involves only a single coating, a matrix of metals can be combined to achieve an endless array of metal colors including brass, aluminum, and even white gold. The style and design trend is expected to continue to stimulate growth for product finishes and special coatings.

TARGET MARKETS & SEGMENTATION

With the product finishes subdivision having shown a nearly flat growth rate in recent years, manufacturers are targeting the two biggest special coatings segments—the automotive aftermarket and facilities coatings (industrial interior and exterior finishes).

The automotive aftermarket represented 53.5 percent of the value of special coatings shipments in 2006. The annual growth rate since 2001 of nearly 7 percent is expected to continue to hold. Facilities coatings represented 23 percent of the value of special coatings shipments in

2006. Shipments in combination (interior and exterior) increased at a 9.4 percent annual rate from 2001 to 2006. Growth is projected to continue. Two examples of how these segments with high growth rates are targeted in marketing are through the expansion of existing lines and the creation of smart coatings.

Expanding an existing product line. Sherwin-Williams' automotive finishes segment launched Planet Color, a collection of optically enhanced automotive coatings for the custom finishing aftermarket. Planet Color has color names to reflect moods like *Bikini Brites* consisting of wild, electric colors, *Rugged* consisting of coatings that create texture and dimension, and *Alloy* coatings that blend natural elements to reflect gold, silver, bronze, and copper. Sherwin-Williams is targeting the dominant automotive aftermarket segment, with its large market of custom repair and painting outlets, its high growth rate, and its $2.3 billion-a-year market.

Creating smart coatings. To be smart, a coating has to be able to be active by responding to environmental changes. Most smart coatings are derived from titanium dioxide pigments, which is known for its photocatalytic characteristics. As a result coatings with suitably prepared titanium dioxide particles can be made smart. With the help of nanotechnology, which increases the surface area of titanium dioxide particles, energy from ambient light can be used as an energy source for photocatalysis. This chemical reaction can cause the breakdown of organic toxins and odors. It gives coatings capabilities such as deodorization, water purification and removal of environmental pollutants like nitrogen and sulfur dioxides. Titanium dioxide-sourced photocatalysis is now being used to make surfaces superhydrophilic so that they have self-cleaning properties and antimicrobial capabilities. Sherwin Williams is using such smart coatings to target the pharmaceutical and food industries, sectors where microbial growth on floors is problematic. The company added a silver-based antimicrobial protection product to its existing FasTop product range of flooring coatings. This smart coating protects against a broad spectrum of bacteria and mold and is targeted at the facilities coatings segment.

RELATED ASSOCIATIONS & ORGANIZATIONS

The Association of Home Appliance Manufacturers, http://www.aham.org

National Paint and Coatings Association, http://www.paint.org

National Coil Coating Association, http://www.coilcoating.org

BIBLIOGRAPHY

"Akzo Breaks Ground on Nashville Powder Coatings Center." *Chemical Week.* 11 October 2006, 30.

"Akzo Nobel Acquires The Flood Company." *Coatings World.* September 2006, 10.

"Akzo Nobel Expands in North America." *Coatings World.* August 2006, 10.

"Akzo Nobel Moves." *Chemical Business Newsbase.* 31 December 2006.

"Just as ICI Starts to Paint a Pretty Financial Picture it May Disappear." *Sunday Business (London).* 11 January 2007.

"Nippon Paint Acquires Rohm and Haas Auto." *Coatings World.* December 2006, 10.

"Paint and Allied Products: 2005." *Current Industrial Reports.* U.S. Department of Commerce, Bureau of the Census. June 2006.

"Pigments: Bright or Bleak?" *ICIS Chemical Business Americas.* 22 January 2007.

"The Powder Coatings Market." *Coatings World.* Available from <http://www.coatingsworld.com>.

"PPG Industries has Acquired an Architectural and Industrial Coatings Business in Brazil for an Undisclosed Sum." *ICIS Chemical Business Americas.* 22, January 2007.

"PPG Purchases Land in Tianjin and Expands Production." *Chemical Business Newsbase.* 6 November 2006.

"PPG to Record $106 Million in Environmental, Legal Charges." *Chemical Week.* 4 October 2006, 4.

"The Shapemakers Solution: Aluminum Extrusion." The Aluminum Extruders Council. Available from <http://www.jobshop.com/techinfo/papers/alumextpaper.shtml>.

"Smart Coatings Are in the Spotlight: Smart Coatings Are Receiving a Lot of Attention for Their Ability to Increase the Functionality of Paint and Boost Sales." *Coatings World.* June 2006, 18.

"Vitracoat Begins Construction of New Plant." *Finishing Today.* October 2006, 9.

SEE ALSO *Paints*

Inks & Toners

―――――■―――――

INDUSTRIAL CODES

NAICS: 32–5910 Printing Ink Manufacturing

SIC: 2893 Printing Ink

NAICS-Based Product Codes: 32–5910E, 32–5910E111, and 32–5910E121

PRODUCT OVERVIEW

Ink is a solution used in writing and printing. Much of the $4 billion worth of ink manufactured each year in the United States goes unnoticed. Newspapers and books are printed with ink using the conventional lithographic offset method. At the local grocer, packaged foods have a U.S. Department of Agriculture-required nutritional facts label printed with ink using the conventional flexographic press method. Magazines are printed with ink using the conventional gravure press method.

Inks are categorized by the presses upon which they are used. The three conventional classes of ink mentioned above are lithographic offset inks, flexographic inks, and gravure inks. The fourth conventional product class is inks related to letterpress printing. These four conventional methods of printing all require ink and all require pressing. Printing presses are pieces of equipment that use a plate to press ink against a substrate, traditionally paper. The characteristic that is shared by the four traditional classes of ink is that a plate comes in contact with the substrate to transfer words and images.

The fifth class of printing ink is distinct. It does not require contact with the substrate to produce words or images. This newest of ink classes is nonimpact digital ink. Printing with digital ink does not require a plate to press the ink against the paper. Digital ink is typically used in smaller printing equipment introduced in the late 1980s. Smaller, more affordable digital ink printers created a consumer market for ink commonly referred to as the SOHO (small office and home office) market. Larger institutions also use digital inks in their growing networks of small printers so the SOHO market should not be confused as being restricted to small settings.

Nonimpact digital ink is experiencing double digit growth while the four conventional ink classes are either stagnant or shrinking. Moreover, digital ink is distributed to the SOHO consumer market in a non-conventional manner.

MARKET

The market for printing ink of all types in the United States in 2002 measured $4 billion in value of products shipped by manufacturers. The value of all shipments is provided by the U.S. Census Bureau in its *2002 Economic Census.* In 2002 the four traditional ink product classes represented 94 percent of the ink industry based on dollar value of shipments. Digital inks comprised the remaining 6 percent, or $242 million of the total $4 billion per year ink market. Because digital inks are the fastest growing class of inks, they are the primary focus of this essay. A review of the four traditional ink product classes is first provided, along with an overview of the nonimpact digital ink industry.

Printing Inks for Lithographic and Offset Printing. These inks represent almost half, or 42 percent, of the value of all inks in the marketplace. Manufacturers' shipments of lithographic and offset inks stayed even over the five-year period from 1997 to 2002, at $1.68 billion each year. Books and newspapers are printed with offset lithography ink. Even given the increased use of electronic media and the affiliated forecasts that the book and newspaper industry will not survive the new technology era, offset lithography remains the single largest class of inks because offset presses are designed for efficiency in high quality, high volume printing.

Offset lithography uses a large printing plate on which words and images are ink-receptive while the remainder of the plate is water-receptive. Ink is transferred from the plate to a rubber blanket; the rubber blanket presses the words and images to paper. A great advantage is that the words and images transferred (offset) from plate to blanket to paper are not reversed. Offset printing is the most common form of high volume commercial printing.

Printing Inks for Flexographic Presses. Flexographic printing inks represent 20 percent of the value of inks in the marketplace. Manufacturers' shipments for these inks increased slightly over the five-year period from 1997 to 2002, from $718 million to $777 million, an increase of 7 percent.

Most commonly used in the packaging industry, flexographic presses use a roller system with a flexible rubber-like plate. Quick-drying ink is applied to a raised pattern of words and images on the plate, which is rolled over the substrate. The fast-drying characteristic of flexographic inks makes these inks ideal for printing on substrates such as plastics and foils. Flexography is primarily used to print packaging materials such as plastic bags, milk and beverage cartons, nutrition facts labels, and candy and food wrappers. Its fast drying time and somewhat lower quality is ideal for these disposable packaging products.

Printing Inks for Gravure Printing. This class of inks represents 13 percent of the value of inks in the marketplace. Manufacturers' shipments declined during the five-year period from 1997 to 2002, from $573 million in 1997 to $496 million in 2002, a drop of 13 percent. Because it takes a long time to set up gravure presses, gravure inks are used only for high-quality projects such as glossy magazines, of which *National Geographic* is an example. Gravure inks are also used for printing postage stamps and paper money for the U.S. Treasury.

In gravure printing, words and images consist of small holes etched onto a cylinder, typically by a diamond tipped or laser etching machine. The small holes, called cells, are filled with ink, then a rubber-covered roller presses paper, or any other substrate, onto the surface of the cylinder plate. Because the cells are filled with ink, gravure transfers more ink to the substrate than other printing processes and is noted for its remarkable density range.

Gravure inks are the first choice for fine art and photography reproduction. Because of its high set up costs, gravure printing is used primarily for press runs in excess of one million copies. *Vanity Fair* is an example; it is printed by RR Donnelley, a long-time purchaser of gravure printing inks. RR Donnelly purchases gravure inks to print six out of the top ten glossy magazines in the United States.

Printing Inks for Letterpress Printing. This class of inks represents 4 percent of the value of inks in the marketplace. Manufacturers' shipments of letterpress inks shrunk in the five-year period from 1997 to 2002, from $184 million to $135 million, a drop of almost 25 percent. Letterpress printing consists of raised words and images (known as type) on a plate. The plate is locked onto a flat surface, inked, and the substrate is pressed against the inked type to produce the impression.

Letterpress was once the only kind of printing in the world and was the primary means of mass communication for over 500 years. It was invented in 1450 by a German named Gutenberg as an alternative to calligraphy and is commonly thought to be one of the inventions that revolutionized communications. Although letterpress inks represent a small sector of the ink market, purists adore the slightly embossed look that results from the direct impression of raised inked type upon paper. Letterpress printing inks are used to produce a high-end look.

Printing Inks for Nonimpact Digital Applications. This class of inks is the fastest growing of the ink types and is a distinct category. Its use does not involve a large plate to press ink against the substrate, or paper. In 1997 manufacturers' shipments of nonimpact digital printing inks were $96 million per year, or 3 percent of all ink shipments—by far the smallest of the ink classes. By 2002 shipments of digital printing inks surpassed the demand for letterpress inks reaching a value of $242 million while shipments for letterpress inks stood at $135 million that year.

In the five-year period from 1997 to 2002, digital inks doubled their share of the printing ink market, increasing from 3 percent in 1997 to 6 percent in 2002. While the rest of the printing ink industry was relatively flat, digital inks grew by over 40 percent, from $96 million to $242 million.

Digital ink is distributed to the consumer market via a distinctly different distribution channel than the distribution channel used for the four types of conventional inks. Digital inks are used by the owners of small and affordable printers while conventional inks are generally used by the owners of gigantic printing presses. Small and affordable printers created the consumer printing aftermarket that involves repeated purchases of digital ink in the form of replaceable and disposable cartridges.

Nonimpact digital inks are subdivided into two product classes: Inkjet inks and electrophotographic printing inks. The latter is known generally by the term toner. Both types are quick drying and stored in disposable cartridges.

Inkjet printers use a printing process where liquid ink is propelled or jetted at a substrate to form images or words. Quality depends on the relationship between the ink, the print head, and the paper. Two different print head designs are used to jet ink to paper: thermal and piezoelectric. The thermal method forces droplets of ink out of the print head by heating a resistor to cause an air bubble to expand; when the bubble collapses, the ink droplet is jetted off the print head onto the paper. The piezoelectric method charges crystals that expand to jet droplets of ink onto the paper. The thermal method is a popular technology used by Hewlett Packard and Canon; the piezoelectric method is used by Epson.

Electrophotographic printing inks are used in laser printers. Laser printers use static electricity to print words and images on paper. First, a laser beam light creates electrostatically charged words and images on a roller drum. The drum is then rolled through a reservoir of electrophotographic printing ink (toner), which is picked up by the charged portions on the drum. Finally, the toner ink is transferred to the paper through a combination of heat and pressure. In addition to printers, photocopy machines use electrophotographic printing inks in this same process, which was developed by Xerox in 1971.

KEY PRODUCERS/MANUFACTURERS

In 2005 the printing ink industry experienced an unprecedented wave of consolidation, unprecedented raw material price increases, and unprecedented overseas competition. Overall, the U.S. printing ink market is flat and some product segments are shrinking.

During the first nine months of 2006, according to the National Association of Printing Ink Manufacturers (NAPIM), U.S. ink sales rose 5 percent compared to the previous year. NAPIM's *2005 State of the Industry Report* showed that U.S. ink manufacturers had average returns on net assets of 3 percent. Paltry margins within the industry explain the intense industry-wide interest in the digital ink segment with its 40-plus percent growth rate. Manufacturers not traditionally involved in digital inks are looking for ways to break into this fastest growing of the ink categories, primarily through acquisitions.

FIGURE 114

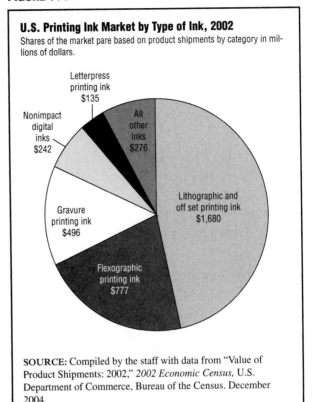

U.S. Printing Ink Market by Type of Ink, 2002

Shares of the market pare based on product shipments by category in millions of dollars.

- Letterpress printing ink $135
- Nonimpact digital inks $242
- All other inks $276
- Gravure printing ink $496
- Lithographic and off set printing ink $1,680
- Flexographic printing ink $777

SOURCE: Compiled by the staff with data from "Value of Product Shipments: 2002," *2002 Economic Census*, U.S. Department of Commerce, Bureau of the Census. December 2004.

The bright spot in the printing ink industry is this very particular growth in nonimpact digital inks: inkjet inks and toner inks. *Print Week* reported that in 2005, the proportion of digital color inks was just 4 percent by volume, but 12 percent by value of shipments. Not only are digital inks the fastest growing portion of the ink market, they also represent a segment of the market that offers manufacturers the opportunity for greater profit margins. This is yet another reason for conventional ink makers to see digital ink—whether black or color—as a potentially attractive opportunity.

Leading manufacturers of digital inks, relatively speaking, have a lot of experience with digital ink products and technically advanced product lines. They also have something the conventional ink manufacturers do not: an understanding of the consumer segment aftermarket created by the introduction of small affordable printers in the late 1980s. The top three manufacturers of digital ink are Nazdar, Triangle Digital INX, and Von Son.

Nazdar. In February 2006 North America's largest ink manufacturer, Nazdar, announced the acquisition of Lyson, Ltd., in the United Kingdom, and its subsidiary, Lyson, Inc., in Chicago, Illinois. Lyson is known for its upscale line of inkjet inks. Lyson became part of the Nazdar Inks and Coatings Division, based in Shawnee,

Kansas. In addition to its 120,000 square foot plant in Shawnee, Nazdar has manufacturing facilities in Chicago, Illinois, Atlanta, Gerogia, and Toronto, Canada. The Lyson product line includes piezo and thermal digital inks for all major inkjet print technologies: photographic, graphic, grand format, office recycling, textile and industrial. Lyson also caters to the consumer aftermarket for replaceable and dispoable ink cartridges.

Triangle Digital INX. In early 2006 INX International Ink Co. formed a joint venture with inkjet ink specialist Triangle Digital, creating Triangle Digital INX. Triangle Digital LLC was founded in 2002 and gained a reputation with outdoor durable digital inks, which they had been manufacturing since 1993. Triangle targeted the digital market well before conventional inkmakers, developing digital inks for all printhead types that use a patented ink dispersion technology that packs up to twice as much pigment in every drop of ink, giving stronger cyans, richer magentas, a jetter black, and a truer yellow. INX International is the third largest producer of conventional ink in North America and part of Sakata Inx worldwide operations. Sakata traces its heritage as an ink provider to 1896 and focuses on commercial and package printing.

Van Son. Van Son, the Dutch-based ink maker, is one of the few European conventional ink companies involved in making and selling digital inks for both the industrial and the consumer aftermarket segments. The company focuses on inks for home printers, particularly those used for photographs from digital cameras. Van Son predicts this market will grow to an enormous size and have developed an own-brand photo-printing kit for digital camera owners. Van Son acknowledges the difficulties of trying to market ink to consumers without a recognized brand name; much of the marketing of its products to end-users is done by distributors and retail chains. Van Son also promotes their products through computer and specialist photographic magazines. They are a respected conventional ink maker that got into the consumer aftermarket relatively early, with ArtColour technology inks that can improve the visual dots per inch beyond the actual rating of the printer.

Other small companies in the news with innovative digital ink applications include Collins Ink in Lebanon, Ohio, which focuses on injet inks for industrial applications, and Squid Ink in Spring Lake Park, Minnesota, which specializes in the making ink jet inks for the packaging industry.

MATERIALS & SUPPLY CHAIN LOGISTICS

Raw materials make up the bulk of the cost of ink. The $4 billion per year U.S. ink industry spent half of that—$2.1 billion—on materials used in manufacturing in 2002. The printing ink industry purchases many of the same materials that paint makers purchase. However, because the printing ink manufacturing industry is only one-quarter the size of the paint industry, ink makers do not have many opportunities to purchase large quantities of raw materials and negotiate bulk price savings on the primary components of ink.

The three primary components needed to make ink are: the vehicle (or carrier), the pigment, and additives. The vehicle is approximately 75 percent of the product by weight. The pigment—primarily carbon black, since black is the most used ink color—is 20 percent by weight. The additive package is 5 percent by weight. The vehicle defines the ink as either water-based or solvent-based.

The vehicle is a carrier fluid that keeps ink in a liquid state and carries the pigment to the substrate, or paper. It also determines characteristics such as dry rate, gloss level, adhesion rate, and scuff resistance. Vehicle fluid is designed to evaporate as ink dries, leaving the ink either on or in the substrate. To assist with drying, vehicle co-solvents are used, usually glycol or glycerin. The pigment in black ink is always carbon black. In color inks, titanium dioxide is the white pigment used to adjust color ink characteristics. Additives control things such as drop formation, print head corrosion, pH level, fade resistance, and color brilliance.

Overall, the cost to purchase the materials needed to manufacture inks decreased 9 percent between 1997 and 2002. This decrease was due primarily to decreases in two of the smaller categories of materials consumed by the print industry: hydrocarbon oils and solvents expenses decreased from $137 million in 1997 to $113 million in 2002, a drop of 17 percent, and wood chemicals (such as wood rosin and turpentine) expenses decreased from $19 million in 1997 to $10 million in 2002, a drop of 47 percent. The largest categories of materials needed to manufacture inks are:

- Pigments (organic and inorganic) including carbon black pigments

- Paints, varnishes, lacquers, shellacs, japans, enamels (and all ink vehicles and varnishes)

- Plastics resins (granules, pellets, powder, liquid)

Pigments are a basic component of ink. In 2002 the entire $4 billion ink manufacturing industry spent $658 million on pigments. Spending on just color pigments increased 10 percent in the period between 1997 and 2002, from $515 million to $536 million.

In April 2006 *Purchasing* magazine reported global consumption of pigments in the $8 to $10 billion range, with the ink industry comprising the largest market for color pigments. Color pigments represent 25 percent of the cost of materials consumed each year by the ink industry. This can be understood in terms of consumer preference: organic pigments made from salts of nitrogen-containing compounds such as yellow lake and peacock blue offer a wide spectrum of vibrant colors. Some examples of inorganic pigments are chrome green, Prussian blue, and cadmium yellow.

Taken together, color pigments and carbon black pigments represent 31 percent of the cost of materials consumed each year by the ink industry. Carbon black is essential for making black ink. It is an oil by-product made by shooting a hot mist of oil particles into a very hot flame in the absence of oxygen, an expensive energy-intensive process with a limited number of makers. The resulting powder is commonly called carbon black, a generic name for grades such as acetylene black, flame black, and rubber black. Ink makers must purchase the high grade special blacks which comprise only 10 percent of carbon production worldwide. The rest of the world's carbon black production is used in the automotive industry to strengthen tires.

The few remaining carbon black makers raise prices at regular intervals. In June 2006 *Ink World* reported that Degussa, a German company, increased furnace black prices by 8 percent in November of 2005 and by 9 percent in May of 2006, for a total 17 percent price increase.

Paints, varnishes, and lacquers are the vehicles that keep ink fluid so it can be carried to the substrate. This category represents 9 percent of the cost of materials consumed each year by the ink industry. In 2002 the ink manufacturing industry spent $196 million on these chemical vehicles.

Plastics resins promote adhesion. They are often used to create quick drying inks which give good gloss. Also, they improve print quality by affecting adhesion and resistance properties. Among other ink ingredients, the cost of plastics resins has gone down because the cost of styrene has dropped; styrene is their principal component. This category represents 5 percent of the cost of materials consumed each year by the ink industry. In 2002 ink manufacturers spent $113 million on plastics resins.

DISTRIBUTION CHANNEL

Manufacturers of large printing presses that use conventional inks set a standard in the ink industry that determined how inks were sold and distributed. Printing press business owners operated presses that were often made by the same company that made inks. It was not unusual for purchase or lease agreements on press equipment to be linked to a contractual obligation to purchase ink. This practice continues to this day. In-plant printing equipment often belongs to the equipment maker who is also the ink maker who services the press to keep it running smoothly.

Printer manufacturers, whether large industrial or small digital SOHO, are vertically integrated; they make the equipment and they make the ink. For most of the history of ink industry sales and distribution, printing press manufacturers had a privileged position: "use my equipment, buy my ink." To guarantee access to the lucrative ink sales aftermarket, where most of the profit was made, printing equipment makers often sold or leased equipment at or below market rates. This became known as blending equipment and consumables.

Because digital printers were the lineal descendent of the printing presses, SOHO printer makers initially assumed ink purchasing decisions would be made in deference to them. When manufacturers began selling smaller and affordable digital ink printers in the late 1980s for SOHO use, they counted on the ink cartridge aftermarket for much of their profit margin.

Digital ink printers have gone down in price and are often sold at a breakeven point, or even a loss to the manufacturer. This is done because on some models, the printer is a loss leader, the ink and toner supplied to the machine over its lifetime being the profit point on such machines. However, this pattern may be changing.

For a time, printer manufacturers succeeded in blending equipment and consumables and had a relative monopoly on the sale of replaceable and disposable ink and toner cartridges. Hewlett Packard did this by making equipment in which only its ink and toner cartridges worked. Epson still makes photo quality printers that use only Epson inks because only Epson uses piezoelectric technology. Cartridges were not interchangeable between printer brands and models, and this was just the way manufacturers of printers and ink wanted it.

The digital ink market changed the distribution channel because it changed the nature of the lucrative ink sales aftermarket and created a new distribution channel. Owners of small affordable printers must repeatedly purchase ink and toner cartridges in order to continue to operate their equipment. Because ink cartridges are designed by printer makers to be disposable, new cartridges must be purchased to replenish the ink supply. Often a separate disposable cartridge is used for each of the major ink colors, which are referred to as CMYK for cyan, magenta, yellow, and key (printing terminology for black). These new cartridges are expensive, and the need to repeatedly replenish the ink supply increases the consumer cost of using the printing equipment.

The distribution channel for consumer aftermarket ink and toner cartridges exploded. The broadened distribution channel offers consumers four routes to purchase cartridges required to operate low cost printers. Competition within the broadened distribution channel resulted in competitively-priced ink cartridges of at least four origins: original equipment manufacturers (OEM) cartridges, OEM compatible cartridges, remanufactured cartridges, and refilled cartridges.

Original equipment manufacturers cartridges are produced by the printer manufacturers. OEM compatible cartridges are manufactured with new drums and working parts to the OEM's specification by another company and sold under a separate brand name.

Remanufactured cartridges involve an industrial process whereby spent inkjet or toner cartridges are disassembled and cleaned, refilled, engineered as new, and then marketed as own-brand cartridges. Remanufactured cartridges compete on price and quality, and are billed as the environmentally friendly choice because without remanufacturing, empty cartridges end up in landfills. Refilled cartridges can be either original equipment manufacturers cartridges or OEM compatible cartridges. These are filled with ink at a retail location. Common cartridge refillers are franchises such as Australia-based Cartridge World Inc., Canada's Island Ink-Jet Sytems Inc., and Caboodle Cartridge Inc., of California. In 2004 the Wall Street Journal reported the size of the U.S. market for ink cartridges was $4.15 billion in 2003, up from $3.7 billion in 2002.

As a result of the broadened distribution channel that offers consumers a range of competitively-priced ink and toner cartridges, makers of digital printers have seen their share of worldwide cartridge shipments and revenue decline. While original equipment manufacturers cartridge sales still dominate, their monopoly is weakening.

KEY USERS

Key users of conventional ink are the four kinds of printing presses and the industries that typically use those presses: offset lithographic inks are used by newspaper and book publishers, flexographic inks are used by all industries in product packaging, gravure inks are used by glossy magazines and by advertisers who can afford glossy catalogues. Letterpress inks are commonly use for chic stationary and invitations.

Key users of nonimpact digital inks are primarily consumers who own small office and home office printing equipment. Larger institutions also use digital inks in their growing networks of small printers.

ADJACENT MARKETS

A primary adjacent market related to the use of conventional ink is the substrate such as paper, plastic, foil, board, and corrugated cardboard, upon which conventional presses print. The primary adjacent market related to nonimpact digital ink is the manufacture of printers, led by Hewlett Packard, Canon, Epson, and Lexmark. Adjacent to the printer market is the manufacture of computers, which are used to operate printers that use digital ink.

An important adjacent market to digital inks is paper used by the owners of small office and home office printers, typically reams of 8½ inch by 11 inch paper. Since computers and printers contributed to the creation of the SOHO market, another adjacent market is paper. Even though in the 1990s, visionaries promised a paperless society due to the availability of computers, printers, and digital inks, small printers churn out more than 1.2 trillion sheets in North America every year. Notable is the wide range of paper designed for printing photographs with digital ink. Photo paper is available in gloss levels ranging from glossy, semi-gloss, semi-matte, and matte.

RESEARCH & DEVELOPMENT

The raw materials needed to manufacture ink are expensive. Consequently, ink makers commit research and development dollars to finding substitutes for the products they need: pigments, carrier vehicles, and plastics resins. It is hard to find a lower-cost, yet high-quality raw material to substitute for these basic ingredients. Ink manufacturers used research to develop various waterborne vehicles. One problem of waterborne inks is that they tend to slow down drying time and thus the printing process as a whole.

Because the aftermarket has historically been the source of lucrative sales and profits, ink makers commit research and development to finding ways to more completely blend consumables with equipment. For instance, ink giant Sun Chemical launched a new business unit, SunJet, based in the United Kingdom, specifically to focus on inkjet use in packaging. Then, in September of 2006, in a classic example of a blending of consumables and equipment, Sun sold its share in Colour Valid Group to a manufacturer of press control systems based in Germany, and agreed to jointly sell on-press colored ink management solutions through its new SunJet unit.

Many research and development topics are off limits due to confidentiality agreements, the president of Collins Inks told *Ink World* in July 2006. Metallics, however, play an important role in digital ink sales and are the focus of research and development for new formulations that successfully utilize them.

Ink World reported in June 2006 that metallics pigment prices are up: copper pigment was up 2.5

times in April 2006 over January 2005; zinc pigment was up 3.0 times; and aluminum pigment was up 0.66 times in the same period. Increased prices often reflect increased demand. The president of Collins Inks explained that as little as three years ago it was impossible to run silver pigment inks through the printhead of a digital printer, but formulators have found ways to do it.

CURRENT TRENDS

The rapidly growing digital ink segment created two trends. The first is an expanded distribution channel that offers consumers a range of competitively-priced ink cartridges. The second trend is the increased quality of photographic inkjet products.

The existence of a distribution channel offering consumers a range of competitively-priced ink cartridges created a need for aftermarket trade associations to protect this new market segment. One example is the European Toner & Inkjet Remanufacturers Association (ETIRA), established in 2003. ETIRA represents the interests of approximately 1,400 inkjet and toner cartridge remanufacturers. ETIRA deals with the alleged quality disparities between remanufactured cartridges and name brand cartridges by reporting independent consumer surveys.

A 2003 study by the reputable German test organization Stiftung Warentest confirmed that non-original equipment manufacturers cartridges are not only cheaper, but are also of equal quality. A February 2005 performance study by the national Dutch independent consumer organization Consumentenbond concluded that "contrary to what printer manufacturers tend to say, printers using non-original cartridges do not give perceptibly more complaints than printers using original cartridges." Remanufactured cartridges are often 20 to 30 percent cheaper than original equipment manufacturers' cartridges.

The increased quality of photographic inkjet products is another current trend. Chicago, Illinois, based Lyson is known for its upscale line of inkjet inks, especially photo graphic inkjet products. Its technically advanced photographic inkjet product line includes products such as: Lysonic Archival Inks, Fotonic Photo Inks, Quad Black Inks, Small Gamut Inks, Cave Paint/Photochrome Inks and Daylight Darkroom.

Daylight Darkroom is a Lyson product line dedicated to digital black and white photograph printers. It is so technically advanced that it "gives silver a run for its money," according to the May/June 2005 issue of *Photo Techniques.* Daylight Darkroom involves four to seven colors of black ink that replace the color ink cartridges in an existing printer with Lyson cartridges containing blacks and grays at various densities. Colors such as black shaded cyan, magenta, and brown are included. The result is lab

quality black and white prints. A basic starter kit with the software, a set of cleaning cartridges, a set of ink, and a 50 sheet box of Lyson photopaper sells for $510. Kodak has called photography inkjet the new frontier.

TARGET MARKETS & SEGMENTATION

The digital ink market is segmented into a consumer side and an industrial side. Some firms are going to continue to serve the consumer market because they have an understanding of the market. Conventional ink makers will continue to serve the industrial market because they prefer the business-to-business relationship.

Manufacturers such as Lyson, Nazdar, Triangle, and Van Son, who have experience with digital ink products and an understanding of the consumer segment will continue to target it. For instance, Lyson sells original equipment manufacturer replacement ink cartridges through its office recycling inks product line. Lyson also has an expanded range of recyclable refill products for the office formulated to provide exact color matching of original equipment manufacturers originals, an area where many third-party remanufacturers fail. Lyson also sells bulk digital inks for use in filling both compatible and remanufactured inkjet cartridges. Triangle developed a revolutionary bulk ink delivery system called the EasyFill Pro that allows the consumer to avoid repeatedly replenishing their supply of original equipment manufacturer's cartridges, and saves money with bulk purchasing of digital inks.

RELATED ASSOCIATIONS & ORGANIZATIONS

European Toner and Inkjet Remanufacturers' Association, http://www.etira.org

Flexographic Technical Association, http://www.flexography.org

Gravure Association of America, http://www.gaa.com

National Association of Printing Ink Manufacturers, http://www.napim.org

Packaging Label and Gravure Association Global, http://www.plga.com

BIBLIOGRAPHY

"2006 Year in Review." *Ink World.* December 2006.

"Carbon Black." *Encyclopedia of American Industries,* 4th ed. Thomson Gale, 2005, Volume 1, 562–564.

"Cheaper Oil will Deflate Elevated Carbon Black Prices." *Purchasing Magazine.* 14 December 2006. Available from <http://www.purchasing.com>.

"Debunking the Myths of Digital Inks." Marrutt Digital Solutions. Available from <http://www.marrutt.com/digital-ink-myths-2.php>.

"The Future for Ink and Laser Cartridges." *Recharger Magazine.* 3 November 2005. Available from <http://www.rechargermag.com/articles/37174>.

"Ink 101, Part II: Types of Printing Inks." *Paper, Film & Foil Converter.* 1 October 2006.

"The Inkjet Report: As Inkjet Printing Continues Its Rapid Growth, Ink Manufacturers Are Continually Creating New Technologies that Would Have Been Unthinkable Just a Few Years Ago." *Ink World.* July 2006, 20.

"Opportunities Are Springing Up in the Ink Jet Ink Market." *Ink World.* July 2002, 20.

"Printing Ink Manufacturing 2002." *2002 Economic Census.* U.S. Department of Commerce, Bureau of the Census. November 2004.

"Review: Dedicated Digital B&W System Gives Silver a Run for its Money." *Photo Techniques.* May/June 2005.

"What's That Stuff?" *Chemical and Engineering News.* 16 November 1998.

SEE ALSO *Copiers, Copy & Printer Paper, Premium Paper, Printers*

Jet Aircraft

————————◆————————

INDUSTRIAL CODES

NAICS: 33–6411 Aircraft Manufacturing

SIC: 3721 Aircraft Manufacturing

NAICS-Based Product Codes: 33–64111 through 33–6411100, and 33–64113 through 33–64113021

PRODUCT OVERVIEW

The earliest written story about human-powered flight was from the Roman poet Ovid, whose tale of the inventor Daedalus and his headstrong son Icarus dates to around the time BC was becoming AD.

Fourteen centuries later, Leonardo da Vinci made hundreds of sketches of birds both in flight and on the dissection table. Two related sketches stand out. One, made when da Vinci was twenty-six, shows what today would be called a hang glider. Another, dated a dozen years later, shows four men seated on a rotating platform while they work levers to spin a screw-shaped airfoil, suggesting modern helicopters.

Whether these devices were built, and if so, whether they worked, are beside the point. Da Vinci provided a way to approach the problem of projecting the human body into the atmosphere under its own power. When that imaginative leap was taken, all that remained was for engineering to solve the problem.

Wilbur and Orville Wright were the first to do so at a beach called Kitty Hawk in North Carolina on December 17, 1903. It took internal combustion, daring, and wings largely made of paper, but from then on, humans could fly. That feat became addictive for many. The only questions left were how far and how fast; issues that remain relevant as humans consider traveling across and outside the solar system.

Changes in language are one way to trace the swift progression from a modified glider to the first jet engine. Da Vinci and the Wright brothers both used the term flying machine, which was probably printed in English in the twentieth century's first decade; by 1910, the song "Come, Josephine in My Flying Machine (Up She Goes!)" was wildly popular. Aeroplane first appeared around 1896 and became first air-plane, then airplane by 1927 and was reduced to plane shortly thereafter.

Jet engines work according to Sir Isaac Newton's Third Law of Motion, which states that every force acting on a body produces an equal and opposite force. The jet engine draws in some of the air through which the aircraft is moving, compresses and mixes it with fuel, ignites it, and the expanding gases power a turbine that produces sufficient force to thrust the plane forward. The force produced by such engines is expressed as pounds of thrust, a term that refers to the number of pounds the engine can move. This force in the earliest jet engines achieved speeds beyond the wildest dreams of pilots whose aircraft speed was limited by whirling wooden blades.

Dr. Hans von Ohain and (then) Royal Air Force Lieutenant Frank Whittle are both recognized as being the co-inventors of the jet engine. Each appears to have worked separately, knowing nothing of the other's efforts. Von Ohain is generally recognized as the designer of the first operational turbojet engine. Whittle was the first to register a patent for the turbojet engine in 1930, while von Ohain was granted a patent for his turbojet engine in 1936. However, von Ohain's jet was the first to fly in 1939, followed by Whittle's in 1941. By 1951 a jet aeroplane appeared in science fiction; three years later, the use of jet in casual conversation was widely accepted.

After achieving success with the Whittle engine, the British promptly shipped a prototype to their allies in the United States, where General Electric (GE) began producing copies. The first American jet engine, produced by GE, took flight in a plane constructed by Bell Aircraft late in 1942. Although use of jets was limited during World War II, by the end of the war all three countries (Germany, United Kingdom, and the United States) had begun to utilize elite squadrons of jet-powered fighter planes.

One of those whose research paralleled technological achievements was the Austrian physicist Ernst Mach (1838–1916). He discovered how airflow becomes significantly disturbed at the speed of sound, which in the lower atmosphere is 1,126 feet per second, approximately 770 miles per hour. As an object exceeds this speed it generates a shock wave that would be later called a sonic boom. Early researchers, one team of whom was German, thought this then-amazing velocity would be a good goal; accordingly, the term mach became a jet's first hurdle. Human exuberance being what it is, twice the speed of sound was dubbed mach 2, three times, mach 3, and so on.

The concept of a sound barrier is semantic rather than physical; nevertheless, it was surpassed on October 14, 1947, by the American X-1 rocket plane piloted by Charles E. Yaeger. This led to advances in aircraft that created a certain irony: air battles in the Korean War were often conducted by jet planes moving at such high velocities that relatively few were intentionally shot down. Half a century later, the U.S. space shuttles reenter the atmosphere at over mach 20, a phenomenon that might have surprised even the person whose name supplied the concept. Contemporary commercial jet engines, up to eleven feet in diameter and twelve feet long, can weigh over 10,000 pounds and produce more than 100,000 pounds of thrust.

MARKET

American aircraft companies build and sell airplanes for three markets: the military; commercial aviation; and general aviation, which includes business aviation. From the end of World War II until the collapse of the Soviet Union in 1989, American military services had an unending appetite for sophisticated aircraft, which American firms attempted to satisfy. The end of the Cold War, which reduced military spending around the world, provided the greatest challenge for American aircraft manufacturers, who had grown accustomed to lucrative contracts from the U.S. Department of Defense (DOD).

Developing commercial aircraft posed significantly greater risks than those of military aircraft. The development process for a passenger airliner capable of carrying several hundred people was lengthy and costly, requiring manufacturers to anticipate the needs of airlines far in advance, and to gamble large amounts of money on the product's success. For this reason, Boeing canceled its development of a super jumbo aircraft. Manufacturers found a more stable market by designing new or modifying existing aircraft in response to the demands of carriers, who typically asked for improved fuel efficiency and more seating.

Due to the risks involved, commercial aircraft manufacturers tended to modify existing airframes rather than reinventing; most existing commercial airliners changed little in the last half of the twentieth century. However, some exciting new aircraft developments occurred in the areas of speed, range, capacity, and fuel efficiency. Many manufacturers by the early 2000s worked cooperatively, jointly developing a design and dividing work among partners if the design was successful. The merger of Boeing Company of Seattle, Washington, and the McDonnell Douglas Company of St. Louis, Missouri in 1997, resulted in economies of operation in many areas. Boeing concentrated on long-range, fuel-efficient planes with a slightly higher passenger capacity, justifying this with the ratio of development costs. Airbus continued to pursue the super jumbo concept, announcing the new A380 in 2007.

American manufacturers historically produce approximately 60 percent of the world's general aviation aircraft and 30 percent of the helicopters. The major U.S. manufacturers of general aviation aircraft are the Beech Aircraft Corp., Fairchild Aircraft Inc., the Cessna Aircraft Co., Gulfstream Aerospace, and Learjet Inc.

Most aircraft manufacturers derive much of their profits from producing replacement and upgrade parts for their airplanes. Since large commercial jets represent such a large investment—a new twin-engine passenger jet may cost several hundred million dollars—airlines try to keep them in the air for many years. Moreover, the Federal Aviation Authority (FAA) sets stringent guidelines on repair and replacement procedures for passenger aircraft.

By 2003 the aircraft industry was struggling in the wake of downturns in the air transportation market. The leading U.S. airlines lost more than $7 billion in 2001 and more than $3 billion through the first half of 2002. A slack economy, a decline in travel following the attacks of September 11, 2001, and heightened competition from discount airlines, contributed to the air transportation sector's woes. In December of 2002, United Airlines, which accounted at the time for some 20 percent of U.S.

flights, filed for bankruptcy after losing $4 billion over two years and laying off 20,000 employees.

U.S. manufacturers shipped 4,088 units of complete civilian aircraft (fixed wing, powered craft; helicopters; and non-powered types of civil aircraft) in 2002, valued at approximately $34.7 billion. In terms of unit shipments, this figure represented a decrease from 2001, when the industry shipped 4,541 units valued at 41.8 billion, and from 2000 when shipments numbered 5,162 civil aircraft valued at $38.6 billion.

The Aerospace Industries Association forecasted that shipments of complete civil aircraft would total 2,751 in 2003, with an estimated value of approximately $25 billion. Some 275 airliners were expected to account for the majority of this total ($18 billion). According to *Standard & Poor's Industry Surveys*, Avitas, Inc. expected aircraft orders to fall from an estimated 816 in 2001 to 561 in 2002, after which levels would steadily improve, reaching an expected 973 by 2005. During the same timeframe, Avitas expected aircraft deliveries to fall from an estimated 1,148 in 2001 to 941 in 2002 and 707 in 2003. After 2003 deliveries were expected to improve slowly through 2005, when levels were forecast to reach 829.

FIGURE 115

Commercial Jet Transportation Market Worldwide, 2005–2014

From 2005–2014 an estimated 7,259 commercial jet transports will be sold valued at $472.6 billion.

Airbus
45.5%

Boeing
54.5%

SOURCE: Compiled by the staff from Lazich, Robert S., "Commercial Jet Transporation Market Worldwide, 2005–2014," *Market Share Reporter 2007*, Thomson Gale, Volume 2, 498. 2006.

FIGURE 116

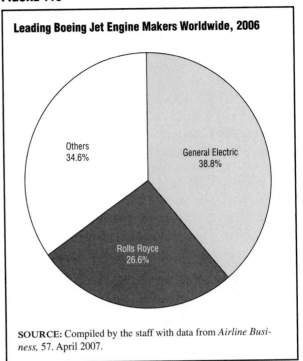

Leading Boeing Jet Engine Makers Worldwide, 2006

Others
34.6%

General Electric
38.8%

Rolls Royce
26.6%

SOURCE: Compiled by the staff with data from *Airline Business*, 57. April 2007.

KEY PRODUCERS/MANUFACTURERS

Boeing Company. In 1997 Boeing Company was the world's largest manufacturer of commercial jetliners and military aircraft, as well the leading contractor for the National Aeronautics and Space Administration (NASA). The company employed more than 200,000 people in 1999 in more than 60 countries worldwide. However, by 2002 it had reduced its workforce to 166,000. Company revenues were $54.1 billion in 2002, representing a nearly 83 percent drop since 2001. During the late 1990s, production problems resulted in lost aircraft orders from companies such as British Airways, United Parcel Service, and Airbus Industries. These problems cost Boeing millions of dollars and threatened its standing as the top manufacturer.

Airbus S.A.S. Airbus was able to surpass Boeing's orders in 2001 and 2002. According to *Air Transport World*, Airbus reported net orders of 233 planes in 2002, compared to Boeing's 176. However, Boeing was still the market leader in deliveries, with 56 percent of the total market share. Airbus began as a five-nation European consortium named Airbus Industrie. It was conceived as a European answer to America's domination of the large commercial transport market. By the early 2000s, Airbus was restructured into a corporation named Airbus S.A.S. The company's majority shareholder (80 percent) is the European Aeronautic Defense & Space Co. BAE SYS-

TEMS of the United Kingdom holds the remaining 20 percent interest.

Boeing and Airbus disagreed on the future needs of the industry. Boeing canceled plans to develop a larger-capacity 747 jumbo jet, while Airbus continued to perfect its design of a new super-jumbo jet. In the early months of 2007, the A30 ordered by the Australian carrier Qantas began touring the world. Eight stories tall, with a wingspan of one hundred yards, this double-decker craft can be configured to carry between 500 and 1,000 people for up to 9,300 miles without refueling. This behemoth is also the most fuel-efficient, least polluting, and quietest civilian air transport ever to fly. Regular routes for the A30 are planned to open toward the end of 2008.

Northrop Grumman Corp. Based in Los Angeles, California, this company employed 96,800 people in 2001. The company is responsible for the design, development, and manufacture of aircraft (including the less-than-perfectly-concealed Stealth Bomber), aircraft sub-assemblies, and electronic systems for the military. In addition to building ships, the company also designs, develops, operates, and supports computer systems. Sales totaled $17.2 billion in 2002. That year, Northrop Grumman saw its net income fall 85 percent, reaching $64 million.

Learjet Inc. In 1969 aviation engineer Bill Lear introduced the first private jet, which Cessna and Beech later imitated. Learjet Inc. is currently a subsidiary of Bombardier and builds high-performance business jets, the limousines of the skies. A pioneer in the business jet industry, the company has built nearly 2,000 aircraft at its Wichita, Kansas, plant since its first jet rolled off the assembly line in 1964. Current Learjet models include the Learjet 31A (light jet), Learjet 45 (super-light jet), and Learjet 60 (midsize jet). The Bombardier Completion Center in Tucson, Arizona, delivers and refurbishes interiors for Learjet 31A, Learjet 60, and Challenger 604 aircraft. The center also conducts Learjet 45 factory completions. A plant in Canada assembles wings for the Learjet 45. The United States accounts for most of Learjet's sales.

Cessna Aircraft. This company is one of the most famous names in small planes. A subsidiary of Textron, Cessna manufactures business jets, utility turboprops, and small single-engine planes. Best known for its small prop planes, Cessna is also a leading maker of business jets, making nine variations of its popular Citation jet. Its utility turboprop plane, the Caravan, has freight, bush, amphibious, and commercial (small connecting flights) applications. Cessna's single-engine planes are typically used for per-

sonal and small-business purposes. As it prepared to enter the twenty-first century, Cessna remained the largest private aircraft manufacturer in the United States. With its line of cargo craft and advanced private jets, including the new Citation X, Cessna still offered the broadest product range in the industry. With the company's relationship to owner Textron on solid ground, Cessna looked certain in the first decade of the twenty-first century to remain America's leading small aircraft manufacturer.

Beech Aircraft Corporation. Best known for its line of Beechcraft propeller and jet airplanes, Beech Aircraft Corporation is one of several American manufacturers of small aircraft. Beech competes with Cessna, Piper, and Lear for shares of such markets as private pilots, small air taxi services, corporate customers, and military forces. Beech also manufactures a variety of aircraft parts and special systems for larger companies, principally McDonnell-Douglas.

In 1990 Beech recorded its best year, turning out 433 aircraft and collecting $1.1 billion in sales. Also in 1990, the new Starship model won certification. In 1992, Beech's 60th anniversary year, the company's 50,000th aircraft rolled out of the factory. That same year, however, a sales slump, attributed to a 10 percent federal luxury tax, caused the company to cut back production and lay off 180 administrative staff.

Due to its 1900 and Jayhawk projects, Beech remains the largest of the small aircraft manufacturers, though Cessna builds more private aircraft. It offers a complete line of advanced aircraft, from the single-engine Bonanza, to the twin-engine Baron and Super King Air series, to the futuristic Starship. The bulk of Beech's more recent success, however, lies with its Beechjet and 1900 airliner. Barring any severe depression in small aircraft markets, Beech is likely to retain its leading position in this sector of the aviation industry.

MATERIALS & SUPPLY CHAIN LOGISTICS

The number of units produced per year by the U.S. civilian aircraft industry is approximately one-thousandth that of the automotive industry. The U.S. aerospace industry shipped a total of 4,068 civil and 450 military aircraft in 2005 according to the U.S. Department of Commerce's International Trade Administration. On average, commercial transport aircraft cost $300 per pound. The use of relatively advanced materials, combined with the production of intricate component forms without net-shape processes, contribute to this higher cost per pound. The airframes of commercial aircraft are currently largely aluminum (approximately 70–80%) with smaller weight fractions of steel, titanium, and advanced composites. The gas turbine engines that power these aircraft use alloys of nickel (approximately 40%), titanium (approximately 30%), and steel (approximately 20%), with the balance being advanced composites and aluminum.

A combination of composite materials and computer design has grown from the occasional application for a nonstructural part (such as baggage compartment doors) to the construction of complete airframes. For military applications, these materials have the advantage of being less detectable by radar.

Some aircraft of composite materials began to appear in the late 1930s and 1940s; these were usually plastic-impregnated wood materials. The largest and most famous example of this design is the Duramold construction of the eight-engine Hughes flying boat, popularly known as the *Spruce Goose*. A few production aircraft also used Duramold materials and methods.

Fiberglass, fabrics made up of glass fibers, were first used in aircraft in the 1940s and became common by the 1960s. Composite is the term used for different materials that provide strengths, light weight, or other benefits not possible when these materials are used separately. They usually consist of a fiber-reinforced resin matrix. The resin can be a vinyl ester, epoxy, or polyester, while the reinforcement might be any one of a variety of fibers, ranging from glass through carbon, boron, and several other proprietary types.

To these basic elements, strength is often increased by adding a core material, essentially making a structural sandwich. Core materials such as plastic foams (polystyrene, polyurethane, or others), wood, honeycombs of paper, plastic, fabric or metal, and other materials, are surrounded by layers of other substances to create a structural sandwich. This method has been used to create, for example, Kevlar, used in aircraft panels, and Lucite, superior to glass for aircraft windows and canopies.

One advantage of composites is being able to take a wide variety of shapes, accomplished by various methods. The simplest is laying fiberglass sheets inside a form, infusing the sheets with a resin, letting the resin cure, and polishing the result; this is how synthetic canoes are constructed. More sophisticated techniques involve fashioning the material into specific shapes with complex machinery. Some techniques use molds; others employ vacuum bags that allow atmospheric pressure to force parts into the desired shape.

Composite materials allow jet aircraft engineers to design lighter, stronger, and cheaper streamlined parts simply not possible when using metal or wood. Composites use has spread rapidly throughout the industry and will probably continue to be developed in the future.

DISTRIBUTION CHANNEL

Aircraft parts manufacturing can be seen as predating the invention of powered aircraft. The Wright Brothers' first airplane, little more than a propeller-driven kite, was equipped with cables, chains, and an engine built by others. In one sense, Orville and Wilbur Wright invented nothing; they merely designed and assembled their aircraft from existing parts. However, this view is too simplistic, since the Wrights put these parts together in a way no one had done before.

The American aircraft industry can be divided into four segments. In one segment, manufacturers such as Boeing and Lockheed Martin Corp. build the wings and fuselages that make up the airframe. Meanwhile, companies such as General Electric and Pratt & Whitney manufacture the engines that propel aircraft. The third segment covers flight instrumentation, an area where the most profound advances in aviation have taken place. Lastly the fourth segment, broadly defined by the industrial classification aircraft parts not otherwise classified, includes manufacturers of surface control and cabin pressurization systems, landing gear, lighting, galley equipment, and general use products such as nuts and bolts.

Aircraft manufacturers rely on a broad base of suppliers to provide the thousands of subsystems and parts that make up their products. There are more than 4,000 suppliers contributing parts to the aerospace industry, including rubber companies, refrigerator makers, appliance manufacturers, and general electronics enterprises. This diversity is necessary because in most cases it is simply uneconomical for an aircraft manufacturer to establish, for example, its own landing light operation. The internal demand for such a specialized product is insufficient to justify the creation of an independent manufacturing division.

There is a second aspect to this distribution tier, since aircraft manufacturers have found it cheaper and more efficient to purchase secondary products from other manufacturers, who may sell similar products to other aircraft companies, as well as automotive manufacturers, railroad signal makers, locomotive and ship builders, and a variety of other customers. For example, an airplane builder such as Boeing, Grumman, or Beech might purchase landing lights from a light bulb maker such as General Electric. Such subcontractors supply a surprisingly large portion of the entire aircraft. On the typical commercial aircraft, a lead manufacturer such as McDonnell Douglas may actually manufacture less than half of the aircraft, though it is responsible for designing and assembling the final product.

When a major manufacturer discontinues an aircraft design, as Lockheed did with its L-1011 Tristar, a ripple effect is caused that affects every manufacturer that supplied parts for that aircraft. Therefore, parts suppliers that make up the third tier of distribution, strive to diversify their customer base to ensure the decline of one manufacturer will be tempered by continued sales to others. Given the unstable nature of the industry, parts manufacturers also attempt to find customers outside the aircraft business.

KEY USERS

Twenty-first century Americans, long used to constant technological innovations, accept flying as a necessary fact of life despite a small percentage that still refuse to board a plane. A century ago there was no airline industry, even though the military made use of aircraft in its operations by 1910. Beginning with the 1920s, with the ranks of trained pilots enlarged with aviator veterans of World War I, flying was increasingly used to deliver the U.S. mail, to manage crops, and to provide thrill seekers with excitement at state fairs. When a particularly plucky Midwestern postal pilot named Charles Lindbergh flew his *Spirit of St. Louis* across the Atlantic Ocean in 1927, a veritable craze began for organized passenger flights.

The first air passenger services did not begin until 1937, when the emergence of the DC-3 and Electra enabled airlines to make money from passenger services alone and end their reliance on airmail. Many of the early passengers became airsick, so for decades, flight attendants could only be hired after completing nursing training. Even then, they could not be taller than five feet, two inches due to the headroom in the vehicle's passenger cabins.

Air travel continued to grow rapidly in the 1960s in business and leisure travel, and the transition to jets was virtually complete by the middle of the decade. Yet the potential market remained largely untapped; as late as 1962, two-thirds of the American population had never flown. In 1961, Eastern inaugurated an hourly, unreserved shuttle service connecting New York, Washington, D.C., and Boston. There were major safety advances, although traffic growth strained the air traffic control system, and there was a growing conflict between increasing volume and safety. Busy airports experienced conflicts between scheduled and business aircraft operations.

Despite financial pressures, air transportation was a critical part of the economy at the end of the twentieth century. Domestic traffic, 321 million passengers in 1984, rose to 561 million by 1998. United had a fleet of more than seven hundred airliners, and American and Delta had more than six hundred each by 2000. The mass market for leisure and recreation depends greatly on the ability of consumers to access particular locations by air. Gamblers cycle in and out of McCarren International Airport in Las Vegas, for example, at a level of more than 20 million ar-

rivals a year, and spend an average of three to four days in town. Ski resorts in Utah, New Mexico, and Colorado cater to short-term visitors who jet in and out when both snow and flying conditions are attractive. Even avid golfers and fishermen or hunters take advantage of low fares to squeeze in a few days of recreation in places such as the Southwest or Alaska, locations known for their allure.

Added to these specialized recreational activities are the many packaged junkets put together by the airline and tourist industries to attract short-term vacationers with a week or less of leisure time for fully organized getaways at resorts. Highly efficient jet travel on planes that carry large numbers of passengers to various specialized destinations has helped to create a mass consumer industry of recreation and leisure that combines vacations with sport activities for all seasons.

Perhaps the strongest indicator of the importance of the industry came after the hijacking of four airliners on September 11, 2001, in terrorist attacks against the United States. The federal government canceled all air traffic for several days. Even after resumption, traffic dropped sharply, since much of the public was reluctant to fly again. Congress immediately appropriated approximately $15 billion in direct grants and loan guarantees to scheduled carriers, since many tottered on the brink of financial collapse. Manufacturers also suffered, because many airlines postponed or canceled orders in response to the drop in passenger traffic and reduced schedules. Flight delays, overcrowding, overbooking, and cancellations are only some of the incidents that traumatize passengers. Nevertheless, the basic pattern of frequent air travel trips to pursue business, tourism, recreation, and leisure activities remains in place. Long-term growth in the early twenty-first century looks favorable, and although various carriers continued to dodge in and out of bankruptcy, by 2007 air traffic neared pre-9/11 levels.

ADJACENT MARKETS

Even a single-engine, single-wing airplane contains thousands of parts whose peak functioning is essential to a safe landing, and jet aircraft require much more sophisticated instrumentation. Indeed, a significant part of all aircraft control panels contain instruments indicating whether the other instruments are working correctly. Three main adjacent categories include instrumentation systems, engine instruments, and fuel. Products produced by these industry sectors are necessary to getting aircraft off the ground, keeping it in the air, and landing at controllable speeds.

Guidance and Control Instrumentation. The products of this industry include radar systems and navigation systems; flight and navigation sensors, transmitters, and displays; gyroscopes; airframe equipment instruments; and speed, pitch, and roll navigational instruments.

The main suppliers of search and navigation equipment are the same contractors who supply the larger U.S. aerospace and defense industry, to which search and navigation equipment contribute significantly. Although not necessarily the most prolific producers of search and navigation instruments, many of the largest and most recognizable corporations in the United States have been involved in the business, including AT&T, Boeing, General Electric, General Motors, and IBM.

A substantial majority of the industry's product types fall into the avionics (aviation electronics) classification, which includes aeronautic radar systems, air traffic control systems, and autopilots. Historically, the primary customer for industry products has been the U.S. government—in particular the Department of Defense (DoD) and the Federal Aviation Administration (FAA).

Search and detection systems, as well as navigation and guidance systems and equipment ($29.1 billion worth of shipments in 2001) constitute 91 percent of the total search and navigation market. They include the following product groups: light reconnaissance and surveillance systems; identification-friend-or-foe equipment; proximity fuses; radar systems and equipment; sonar search, detection, tracking, and communications equipment; specialized command and control data processing and display equipment; electronic warfare systems and equipment; and navigation systems and equipment, including navigational aids for aircraft.

During the 1970s the Global Positioning System satellite network first came under development. Inertial navigators using digital computers also became common on civil and military aircraft.

Industry shipment values for the above products totaled $31.9 billion in 2001, an increase over 2000 levels of $29.9 billion. In 2001 the industry's employment base of 153,710 workers was an increase from the previous year's count of 145,990 workers. Capital investment, which totaled approximately $1 billion in 2000, has remained relatively constant since 1997.

Aircraft Engine Instruments. The main customers of the aircraft engine instruments segment are General Electric, United Technologies, Rolls Royce, and other aircraft manufacturers. The sector shipped temperature, pressure, vacuum, fuel and oil flow-rate sensors, and other measuring devices. Growth in this market is linked to aircraft production.

Through the middle of the first decade of the twenty-first century, the miscellaneous measuring and controlling devices industry was projected to grow at a rate of 3 percent annually. Aircraft engine instruments were

predicted to be one of the industry's faster growing segments. Furthermore, the addition of software and services will contribute to overall industry growth, as will further expansion into overseas markets. The top five export markets in the late 1990s were Canada, Mexico, Japan, United Kingdom, and Germany; these five countries also were the leading importing countries. Looking into the 2000s, estimates indicated that 33 percent of measuring and controlling instruments product shipments would be exported, while 25 percent of U.S. production would be imported.

Aviation Fuel. The most common fuel is an unleaded/paraffin oil-based fuel classified as JET A-1, which is produced to an internationally standardized set of specifications. In the United States only, a version of JET A-1 known as JET A is also used. The only other jet fuel commonly used in civilian aviation is called JET B, a naphtha-kerosene mixture especially effective in cold-weather conditions. However, JET B's lighter composition makes it more dangerous to handle, and it is therefore restricted to areas where its cold-weather characteristics are absolutely necessary.

Jet fuels are sometimes classified as kerosene or naphtha-type. Kerosene-type fuels include Jet A, Jet A1, JP-5, and JP-8. Naphtha-type jet fuels include Jet B and JP-4. Both JET A and JET B can contain additives including antioxidants, antistatic agents, corrosion inhibitors, and fuel system icing inhibitors. The annual U.S. usage of jet fuel in 2006 was 21 billion gallons.

RESEARCH & DEVELOPMENT

In the wake of the September 11, 2001 terrorist attacks, the U.S. federal government ordered airlines to ensure that existing cockpit doors on commercial aircraft would be locked at all times and secured with extra bars and barriers. It also developed a standard redesigned, reinforced cockpit door that airlines were required to install on all aircraft by 2003.

With an annual research budget exceeding $1 billion for its aeronautical division, NASA contributes substantially to advances in aircraft technology. NASA has assisted the general aviation industry in the United States in such areas as developing new wing and blade designs—including the civil tilt-rotor project—and cockpit technology for business and commuter aircraft. NASA plans to develop aircraft that meet the world's new environmental and safety standards.

The Kyoto Protocol to the United Nations Framework Convention on Climate Change developed in 1997 was the first international treaty to set standards for greenhouse gas emissions—primarily carbon dioxide—by countries ratifying it. Although as of December 2006, the United States had yet to accept this treaty's limitations, NASA plans to develop aircraft that meet those environmental and safety standards.

Boeing is another leading aircraft technology researcher. Each year the company devotes between $1.5 billion and $1.8 billion for research and development (R&D). In the mid-1990s, the majority of the company's research funds went to developing its 777. In the early 2000s, with the delivery of its 777s, Boeing turned to refining its existing aircraft and designing new planes. In cooperation with NASA and several universities, Boeing began to develop a blended-wing-body (BWB) plane. The BWB's advantages include superior fuel economy, lower production costs, greater capacity, and greater range than the conventional aircraft of the 1990s. The BWB's capacity comes from the design of the wings, which hold seats for passengers. Researchers estimate that the plane could be ready by 2015. Meanwhile, Boeing plans to meet the fast-approaching requirements for environmentally friendly aircraft with its 717–200, which features reduced emissions and lower noise levels than its rivals. Test flights of the 717–200 began in early 1998.

In the mid-1990s, United States and Russian researchers jointly studied how to develop new supersonic civil aircraft. Although supersonic projects had largely ended in 1978, both countries renewed their interest. The U.S. component of the research team consisted of NASA, Boeing, Rockwell-Collins, Pratt & Whitney, and General Electric. The Russian component of the team included Tupolev, the developer of the Tu-144 supersonic jet. The collaborators went to work rebuilding the plane's engine to use the plane in studying the ozone layer and sonic-boom problems.

In contrast to the huge government-sponsored research programs of the aerospace conglomerates, the R&D efforts of the makers of ultralights and kit planes, designed to be assembled by the user, were lean but smart. The popular kit designs offered by Burt Rutan and others in the 1970s offered advanced materials such as exotic composites, plastic foams, and fiberglass and epoxy laminates. Also featured in these designs were canards, small wings placed at the nose of the aircraft, and winglets, fins at the end of the main wing, both of which increased efficiency and stability. Computer modeling enabled designers to incorporate advanced wing shapes into designs the ordinary enthusiast could build at home. At least one company has adopted these technologies to produce an inexpensive, six-passenger business turboprop (less than $1 million, compared to $3 million and up for competitors).

A significant experimental aircraft was the Gossamer Condor. In 1977 it enabled the first human-powered flight. In 1986 came perhaps Rutan's greatest achievement—the Voyager, the first aircraft to circle the world without refueling. By the early 2000s Rutan's conceptions of lightweight craft with intercontinental range had found a military application in the U.S. armed forces—highly capable drones, used effectively during the hostilities in Afghanistan. High-altitude drones with extended range were also expected to acquire satellite-like global or regional communications roles in the new century. Rutan's designs and principles have found their civil application in the Beech Starship, a small business turboprop, and in a small jet fighter/trainer.

Instrumentation is another area of continuing research. A computerized display of flight information, the Electronic Flight Information System (EFIS), has promised to improve the decision-making abilities of pilots by providing an integrated, improved display of navigational, meteorological, and aircraft performance information in the cockpit. State-of-the-art airliners and business craft, such as the Boeing 757 and 767, the Airbus A-310, and the Beech Starship, are equipped with this system.

The Global Positioning Satellite (GPS) system, first developed for use by the U.S. military in the early 1970s, relies on groupings of satellites to provide extremely precise location information (including altitude) to receiving units within airplanes. Pilots using GPS navigational systems could fly in a straight line from airport to airport and could even fly an instrument approach using GPS navigational systems (precise within 50 feet). This change could reduce both travel time and fuel costs, especially for regional airlines. The growing number of regional jets had the potential for allowing for the return of more direct flights. Newer, more fuel-efficient planes with longer ranges, such as the Boeing 777, were also expected to impact the airlines.

CURRENT TRENDS

Civil aircraft production is controlled by the commercial market, supplying the jets and turboprops used by the world's passenger and cargo airlines. As of 2005, just two manufacturers—Boeing in the United States and Airbus S.A.S. in France—controlled nearly the entire market for commercial aircraft for more than a decade. This market domination was secured by manufacturing medium and large jets for 100 or more passengers, the industry's most lucrative and capital-intensive segments. Aircraft manufacturers noted the increasing demand for large-capacity, wide-body planes and expected that the average number of seats per plane would increase to 240 by 2015. They expected Asian countries would help drive this trend with a 356-seat average capacity per plane by 2015. Airbus has constructed its A380, whose interior can be configured to carry between 500 and 1,000 passengers, with an initial nonstop route from Los Angeles, California to Melbourne, Australia, scheduled to begin in 2008.

The huge costs and risks of aircraft manufacturing encouraged business consolidation and a proliferation of international joint ventures in what has been termed a borderless industry. Few countries could be considered self-sufficient in production, and even for those that could, most competitors in the industry pursue multiple cross-border ventures in order to keep costs down and draw on the special competencies and efficiencies of firms around the globe.

Globally, the industry experienced continued growth in the early 2000s. Boeing's World Air Cargo Forecast predicted an annual expansion rate of 6.2 percent through 2023, tripling the levels of overall air traffic. Strong growth was reported in international trade, with the most reported in the Asia-Pacific region. Traffic in North America and within Europe was expected to see below average increases. The U.S. firms Cessna Aircraft Co. and Raytheon led the continuing strong surge in sales in the general aviation segment. The U.S. industry reached $147 billion in 2003 sales. That year, Boeing, with 280 units, and Airbus, with 300 units, produced a combined $33 billion in aircraft. These had a per-unit value of $50 million or more, according to *Fortune.* For the first half of 2004, the companies delivered a combined 312 aircraft. According to researchers from the Teal Group, $421 billion in aircraft will be built between 2004 and 2012.

Environmental groups in the United States, Europe, and Australia have focused on noise pollution. The U.S. Airport Noise and Capacity Act of 1990 required U.S. airlines to make their fleets meet quieter noise specifications. Smaller business jets were exempt from this rule. The International Civil Aviation Organization (ICAO) imposed similar standards.

Heavily congested airports have suggested the need for 600–800 seat, ultra-high-capacity aircraft (UHCA or VLCT, very large commercial transport). Airbus began research on such a project, estimated to cost between $6 billion and $8 billion. Boeing also began research for its proposed UHCA, the 747-X. The potential market for these aircraft was projected at between 400 and 500 aircraft by 2010. In the early 2000s Boeing studied development of smaller capacity, but higher speed, transports than the proposed UHCAs.

A concept for a 300-seat supersonic airliner, dubbed the *Orient Express,* has been the subject of a study group comprised of engineers and others from Boeing, Aerospatiale, British Aerospace, Japan Aircraft Development Corp., Tupolev, and Alenia. Traveling at Mach 3, or three times the speed of sound, the aircraft would cut travel time between Tokyo and Los Angeles to 4 hours, from the cur-

rent 10. Fares were projected to eventually fall to a level just 20 percent higher than those for conventional flight.

Two types of vertical takeoff and landing (VTOL) aircraft also were being developed to serve inner-city airports. Ishida Corp. of Japan (in collaboration with United Kingdom and U.S. firms) is developing the 14-passenger TW-68. With wings that rotate 90 degrees, the craft would allow vertical takeoff and landing. Due to traffic congestion, Boeing projected a need for thousands of civil tilt-rotor aircraft (such as the Bell/Boeing V-22) in the first few decades of the new century.

TARGET MARKETS & SEGMENTATION

The opportunity to fly for business and the perceived right to fly to increased exotic vacation possibilities have led the aircraft industry to a simple, if perhaps immodest goal: to get everyone in the air at some point in life, and the more often the better. Beginning in the late 1990s online travel agencies such as Orbitz, Priceline.com, Travelocity, and Expedia.com used highly complex algorithms to scan myriads of combinations for discount hotel rooms, car rentals, and air flight times and days that would, these agencies claimed, provide the consumer with the supreme package for cheapest available services, accommodations, and travel.

However, these algorithms have been analyzed, with the mildly disturbing conclusion that thorough examination of all available variables would take longer than the probable age of the known universe. The agencies ignore this massive fact by choosing what appears to be several of the top possibilities and presenting them in a few seconds on the computer screen, thereby motivating the user to reach for a credit card. Airlines contribute this process partly because competition constantly increases and because, thanks to various technologies and new ways of ordering data, flights from almost anywhere to almost anywhere are increasingly likely. Distance is no longer the barrier it was as little as fifty years ago.

Parents of the Baby Boom generation (those born between 1945 and the early 1960s) who flew in or before the mid-twentieth century recall a different travel experience than post-9/11 airports offer: friends and relatives could walk you to the gate and welcome you there when you returned; you could smoke on planes and even bring your own liquor; people dressed up as if for a business meeting; meals were served and they often tasted good. Ironically, it was the success and safety of air travel, plus lowered costs and quicker journey times, that led to lowering the level of gentility in the flight experience.

Despite delays, congestion, and sometimes embarrassingly intimate searches for the sake of security, most people who fly usually get where they intended to go in pretty much the time promised. Flying is cheaper than a train, quicker and safer than driving. Airlines work ceaselessly to ensure that, sooner or later, everyone will fly.

RELATED ASSOCIATIONS & ORGANIZATIONS

Aerospace Industries Association of America, http://www.aia-aerospace.org

Air Line Pilots Association, International, http://www.alpa.org

Buffalo Rocket Society, http://www.buffalorocketsociety.org

F-4 Phantom II Society, http://www.f4phantom.com

Fellowship of Christian Airline Personnel, http://www.fcap.org

Giant Scale Warbirds Association, http://www.giantwarbirds.org

International Airline Passengers Association, http://www.iapa.com

Negro Airmen International, http://www.blackwings.com

United Flying Octogenarians, http://www.unitedflyingoctogenarians.org

World Airline Historical Society, http://www.wahsonline.com

BIBLIOGRAPHY

Airbus S.A.S. "2001 Commercial Results Consolidate Airbus' Position as World's Leading Aircraft Manufacturer." Airbus. Available from <http://www.airbus.com>.

"Airbus Sees Strong Cargo Market." *The Journal of Commerce Online.* 30 March 2004.

Asimov, Isaac. *Asimov's Guide to Science.* Penguin Books Ltd., 1975.

Barnes-Svarney, Patricia, ed. *The New York Public Library Science Desk Reference.* Macmillan, 1995.

"Boeing Commercial Airplanes." *Airfinance Journal.* April, 2005.

"Bright Outlook in 2004." *Airline Business.* 1 February 2004.

"China to Buy 400 Aircraft in Five Years. *Alestron.* 3 July 2001.

The Compact Oxford English Dictionary, 2nd ed. Oxford University Press, 1989.

"Component Tracking." *Flight International.* 13 April 2004.

Draper, Deborah J., ed. *Business Rankings Annual.* Thomson Gale, 2004.

Flores, Jackson. "Forecasts." *Flight International.* 27 April 2004.

Global Market Forecast 2004–2023. Airbus S.A.S. 2004–2005. Available from <http://www.airbus.com.>.

"India Will Buy 570 Aircraft—Forecast." *Airline Industry Information.* 22 June 2005.

Kingsley-Jones, Max. "Production." *Flight International.* 13 December 2004.

Lazich, Robert S. *Market Share Reporter, 2007.* Thomson Gale, 2007, Volume 2, 493–499.

Materials Research Agenda for the Automobile and Aircraft Industries. National Materials Advisory Board (NMAB). 1993.

Moxon, Julian. *How Jet Engines Are Made.* Threshold Books, 1985.

Napier, David H. "2001 Year-End Review and 2002 Forecast: An Analysis." Aerospace Industries Association, 2002. Available from <http://www.aia-aerospace.org>.

Nicholl, Charles. *Leonardo da Vinci: Flights of the Mind, A Biography.* Viking Penguin, 2004.

Ott, James. *Jets: Airliners of the Golden Age.* Pyramid Media Group, 1990.

Pattillo, Donald M. "Air Transportation and Travel." *Dictionary of American History,* 3rd ed. Charles Scribner's Sons. 2003, Volume 1, 82–86.

Peace, P. *Jet Engine Manual.* State Mutual Book & Periodical Service, 1989.

Taylor, Alex. "Lord of the Air." *Fortune.* 10 November 2003.

"Teal Group Analysts Predict that 6,743 Commercial Aircraft, Valued at $421 Billion, Are to Be Built Between 2003 and 2012." *Airfinance Journal.* September 2003.

"U.S. Firms Start Year Well." *Flight International.* 3 May 2005.

SEE ALSO *Airplanes, Helicopters*

Jewelry

---◼---

INDUSTRIAL CODES

NAICS: 33–9911 Jewelry (except Costume) Manufacturing and 33–9914 Costume Jewelry and Novelty Manufacturing

SIC: 3911 Jewelry, Precious Metal, and 3961 Costume Jewelry

NAICS-Based Product Codes: 33–9911, 33–99111 through 33–991117, and 33–99140 through 33–99140YWY

PRODUCT OVERVIEW

Jewelry is one of the oldest forms of personal embellishment and is a part of nearly all cultures. Jewelry use includes as a currency, as a functional object, as symbolism, and as an adornment.

Jewelry has been used as a form of currency. Early cultures in Africa, the Middle East, and the Americas maintained wealth in the form of jewelry stored on their person. Jewelry in the form of currency included slave beads and wedding dowries. Wedding dowries, for instance, may contain jewelry given to a female throughout her life from the time that she is a child through the time when she is to be married. The accumulated dowry is given to the man or his family after the wedding. In other cultures, women wear all the gold and jewelry they possess at all times to symbolize their worth.

Jewelry is also highly functional. Items like broaches, pins, hair combs, seal rings, and pill and snuff boxes on chains or rings have practical uses. Viking women wore functional brooches as part of their normal dress. Viking female dress did not change much for roughly 200 years. Women wore a conservative wool pinafore over a long pleated linen shift. The pinafore had shoulder straps fastened by a pair of brooches. The design of the brooches, just as of the clothes, was standardized.

Much jewelry is worn for symbolism. Symbolic jewelry signals the membership of a particular group, religious or secular. Religious examples include the crucifix or the Star of David worn around the neck. Classic symbolic jewelry includes birthstone, engagement, and wedding rings. Jewelry is especially adept at symbolizing those that belong to the upper and ruling classes. European royalty wore their richest jewels during the Medieval and Renaissance eras to demonstrate wealth, power, and lineage. One way to demonstrate lineage is with cameos and other portrait jewelry. While European royalty wore jewelry in everyday life, jewels were prominently worn in portraits painted by royal painters so that the beauty, power, and wealth of royalty could be seen by a greater audience and praised and remembered by generations to come.

In the 1920s costume jewelry became popular. It was sold along with an outfit, or costume, with which it was meant to be worn. This introduced inexpensive jewelry to the masses. Costume jewelry tends to imitate precious stones and metals at a fraction of the cost. Base metals, such as tin and lead, were used and made to look like precious metals such as gold. To imitate expensive stones, manufacturers used cut glass, enamel, and, later, acrylic and other hard plastics. This made it possible for consumers to buy jewelry on a whim and base the choice on aesthetics and design rather than the value of the materials alone.

In the first decade of the twenty-first century, most jewelry is used for symbolism or adornment. For many consumers, time and money will be committed to purchasing rings symbolizing marital status. A common guideline is that a man should spend three months salary on a woman's diamond engagement ring.

Diamonds are graded according to four classifications, commonly called the Four Cs, for clarity, cut, color, and carat weight. Clarity is judged to be anywhere from flawless to very slight inclusions to imperfect. Cut influences diamond brilliance. Cuts with many facets reflect more light more brilliantly. Color varies anywhere from colorless to yellow, although diamonds have been known to have many colors. One of the rarest colors is pink. Carat weight denotes stone size. A larger carat weight results in a higher price.

Often consumers, mostly female, will purchase rings, earrings, and necklaces for adornment. With the advent of costume jewelry, the market offers a wide spectrum of choices. Rings, earrings, and necklaces are available in the marketplace for under $10 and for over $10,000.

MARKET

In the United States in 2002, 1,962 establishments were primarily involved with the manufacture of jewelry and 655 establishments manufactured costume jewelry. Most jewelry makers were located in New York, California, and Rhode Island, while most costume jewelry makers were located in California and Rhode Island. In 2002 jewelry manufacturers shipped $5.5 billion worth of product. Costume jewelry manufacturers shipped $797 million worth of product in 2002, down 35 percent from $1.2 billion in 1997. Jewelry manufacturer shipments increased 19 percent during the same timeframe, from $4.6 billion in 1997 to $5.5 billion in 2002.

The costume jewelry industry is divided into three categories by the Census Bureau: non-precious metal jewelry (65%), other jewelry/costume novelties (16%), and costume jewelry not-specified-by kind (19%). Precious jewelry is categorized by the primary metal with which it is made. Jewelry made of gold and platinum accounted for nearly two thirds (63%) of the precious jewelry industry based on the value of manufacturers shipments in 2005. Jewelry made of silver accounted for 15 percent, and jewelry made of other metals and types accounted for 12 percent of 2005 shipments. The remaining 10 percent was made up of shipments of stamped metal coins and jewelry not-specified-by-kind.

Shipment data detailed enough to track changes in the trends for different types of jewelry are collected by the Census Bureau every five years in its economic census cycle. Using data for the two years 1997 and 2002, growth of 19 percent is seen for shipments of fine jewelry, from total industry shipments of $4.6 billion in 1997 to $5.5 billion in 2002. Jewelry made of gold and platinum grew 12 percent, from $3.2 to $3.5 billion. Most of the gold and platinum jewelry growth was for wedding rings. Most Americans will marry and many will re-marry. Manufacturers' shipments of gold and platinum wedding ring sets rose sharply between 1997 and 2002, from $419 million to $729 million, or 74 percent.

Jewelry made of silver grew 93 percent between 1997 and 2002, from $399 million to $767 million. Silver is valued for jewelry. It is comparatively scarce, brilliantly silver-white colored, malleable, and resists oxidation. Silver jewelry grew in popularity in part because it is more affordable than gold and platinum, for which it often substitutes. Most silver jewelry sports a numeric hallmark, which indicates metal content. For instance, sterling silver is defined as containing 92.5 percent silver and 7.5 percent of an alloy metal, usually copper to make it stronger. Sterling silver products therefore tend to sport a tiny hallmark number of 925 on the back or inside of the piece to denote its extremely high quality. Jewelry silver is defined as containing 80.0 percent silver and 20 percent copper. Silver jewelry products tend to therefore sport a small hallmark number of 800, or higher, to attest to its high percentage of the more valuable metal.

Jewelry made of other metals and types grew 43 percent between 1997 and 2002, from $452 million to $648 million. While the three major categories of jewelry each grew 12, 93, and 43 percent, respectively, notable was the doubling in value of shipments of silver jewelry and of wedding ring sets.

KEY PRODUCERS/MANUFACTURERS

Tiffany & Co. In 1837 Tiffany and Young started selling stationery and fancy goods on Broadway Avenue in New York, New York. Every item was sold at a non-negotiable selling price, a radical business model that made headlines in local newspapers. By the 1850s Tiffany silver designs captured the attention of people around the world and Charles Tiffany assumed control, renaming the company Tiffany & Co. Tiffany designs, such as the Tiffany diamond setting, are American standards. Tiffany contributed to the establishment of the 925 sterling silver standard. It has long attracted some of the world's foremost designers.

In 1867 Tiffany became the first American firm to win a medal for the excellence of its silverware at the Paris Exposition Universelle. Tiffany products have been purchased by museums across the world since the Boston Museum of Fine Arts first did so in 1873. In 1956 Parisian master jeweler Jean Schlumberger started designing for Tiffany, and in 1995 a retrospective of his work was on display at the Louvre museum in Paris, France. Other famous designers include Elsa Peretti, Paloma Picasso, and Frank Gehry. Frank Gehry, arguably the world's greatest living architect, designed six collections priced from $100 to $7,800.

Jewelry represents the majority of Tiffany sales at 82 percent, with silverware, watches, and other categories rounding out the balance. The United States represents the bulk of Tiffany sales, with 60 percent of sales in 2005. Japan is the next largest market for Tiffany at 20 percent. The Asia Pacific and European regions amount to 8 percent and 6 percent respectively. The flagship store in New York represented 10 percent of total company sales in 2005. In the 1987 initial public offering, Tiffany stock was offered at $23 per share. In September 2007 shares of Tiffany stock hovered around $50 per share.

De Beers Institute of Diamonds. Prolonged battles during the 1880s over the diamond fields on the southern portion of the African continent contributed to the formation of De Beers Consolidated Mines Limited. The company was established in 1888 and listed in 1893 on the Johannesburg Stock Exchange. De Beers grew both organically with energetic mine production levels and through aggressive acquisition tactics of other firms in the region. In its early years De Beers controlled 90 percent of worldwide production. In later years De Beers created the well-known "A Diamond is Forever" advertising campaign.

With a workforce of almost 23,000 employees and 10,000 contractors in 25 countries, De Beers in 2007 accounted for roughly 40 percent of worldwide diamond production and 45 percent of worldwide diamond distribution. More than 19,000 employees work in Africa. In 2005 De Beers posted a pre-tax profit of $1.07 billion.

Swarovski Group. Headquartered in Wattens, Austria, Swarovski had worldwide sales in 2005 of $2.14 billion with 17,000 employees. It had production facilities in 16 countries, including the United States. In 2005 it operated 565 Swarovski Own Shops.

In 1892 Daniel Swarovski perfected a design for a machine that cut crystal faster and with greater precision than the manual process of the time. Over time Swarovski business units grew to include Tyrolit (abrasive tools and cutting tools), Swarovski Optik (optical instruments), Swareflex (reflective and luminous road markings), and Signity (genuine and synthetic gemstones). Swarovski crystals for consumers are produced under two distinct labels. The Daniel Swarovski couture collection includes jewelry, accessories, and home accessories. Swarovski, the international brand, produces jewelry and accessories of cut crystal, gifts, and interior décor pieces made of elaborately faceted crystal.

MATERIALS & SUPPLY CHAIN LOGISTICS

Jewelry is generally categorized based on the material from which it is made: gold, platinum, or silver. Gemstones are typically embedded into or mounted upon the gold, platinum, or silver jewelry. By general consensus, precious gemstones are diamonds, rubies, emeralds, and sapphires. Everything else is considered semi-precious. This includes gemstones like amethyst, citrine, and garnet. Turquoise is considered semi-precious, as is onyx, aquamarine, and pearls. Synthetic gemstones and pearls are used in costume jewelry.

For fine jewelry, precious metals represent 32 percent of the cost of materials purchasing, while precious, semi-precious, and synthetic gemstones and pearls represent 22 percent. Two preferred materials for fine jewelry making are gold and diamonds. Gold and diamonds are the standard bearer for wedding ring sets. Both gold and diamonds are mined.

Gold Mining. Gold is found in its metallic form in various geological formations. Since it does not tarnish, it is easily identified. Gold is mined primarily in South Africa and Russia. Gold mining companies mine gold in one of two ways. Open-pit mining, also called cyanide heap leaching, requires blasting rock from the earth. The broken rock, or ore, is hauled to a processing facility where it is crushed. Crushed ore is sprayed with cyanide, which seeps through the pile to dissolve gold particles. The gold liquid is collected from the bottom of the heap and smelted, which removes remaining impurities by applying intense heat. Open-pit or cyanide heap leaching mining is cost effective. It is estimated that two-thirds of mined gold is obtained by open-pit mining. The remaining third is shaft-mined.

Shaft-mining is the traditional, and more costly, mining technique for gold. It is less invasive, more time consuming, and not widely used by modern gold miners. Shaft-mining requires life support (pumping in cool air, pumping out gasses and water), complex logistics, and high risk levels. Shafts are carved from rock to locate rich gold veins. Gold is extracted from the vein by digging, hammering, or small-scale blasting.

Gold is also mined in the United States; total 1989 production was 363 million ounces. Approximately 15 percent of U.S. gold comes from copper, lead, and zinc mines. Where these base metals are deposited, either in veins or as scattered mineral grains, minor amounts of gold are commonly deposited with them. When the predominant metal is mined, the gold is recovered as a byproduct. The largest single source of byproduct gold in the United States is the porphyry deposit at Bingham Canyon, Utah.

Diamond Mining. In the eighteenth century diamonds had been found only in India and Brazil and were very rare. In the mid-nineteenth century, diamond mining began in Africa extensively throughout the southern part of the continent. Although the rarity of diamonds diminished due to a greater amount of stones mined, diamonds retained value due to the tightly controlled selling and trading practices of DeBeers.

Diamonds are commonly mined by a method called pipe mining and are extracted from volcanic pipes. Approximately 250 tons of ore, or broken pieces of rock, must be mined in order to produce a one-carat gem-quality polished diamond. Once the rock that bears diamonds is located, it is transported to a processing facility where it is separated from the rock. Alluvial mining of diamonds is also possible but it is generally not as profitable for large mining companies. Alluvial mining involves extracting diamonds from riverbeds or ocean beaches. Once the diamond is separated from the surrounding mineral formation, it is cut and polished.

Cutting can reduce the size of a diamond by half. Diamonds are typically cut across the grain by a thin metal disc coated with diamond dust. Facets are ground individually. Facets are applied by a revolving mechanism made of iron and coated with oil and diamond dust. It is held against the revolving machine, or turntable, as it revolves at a high speed. A faceted diamond has an average of fifty-eight facets, depending on the cut. Facets are valued when they are clean, symmetrical, and sharp.

Less than half of mined diamonds are gem quality. Those that are not gem quality are often used for cutting tools and abrasives. World diamond production figures are presented in Figure 117, which depicts the proportion of mined diamonds used as gemstones and the proportion used in industrial applications. One example of an industrial application is Swarovski's Tyrolit business unit. It makes abrasive and cutting tools that involve diamond dust.

DISTRIBUTION CHANNEL

After gold and diamonds are mined, they are shipped to jewelry manufacturers throughout the world, including many of the 1,962 U.S. establishments that make fine jewelry. Manufacturers cast gold, platinum, and silver, and set diamonds and other stones into jewelry like rings, earrings, bracelets, necklaces, and watches.

The retail distribution channel for finished jewelry is bifurcated into jewelry retailers and specialty jewelry retailers. Some companies, including Tiffany & Co., overlap into both categories.

As of 2000, jewelry stores represented the chief distribution channel for jewelry, accounting for 60 percent of all jewelry sales in that year. Large U.S. jewelry retailers

FIGURE 117

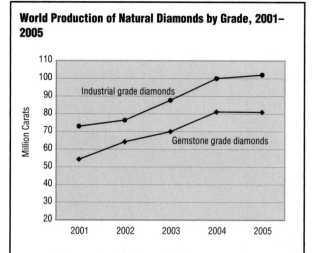

World Production of Natural Diamonds by Grade, 2001–2005

SOURCE: "Industrial Diamonds Statistics and Information," *2005 Minerals Yearbook, Volume 1 – Metals and Minerals,* U.S. Department of the Interior, United States Geological Survey, May 2006.

with national chains are Kay Jewelers (with origins as far back as 1916) and Zales (1924). Kay Jewelers operates 868 stores and is known by its advertising slogan, "every kiss begins with Kay." Jared Galleria of Jewelry and JB Robinson are both under the Kay umbrella. Zales emerged from bankruptcy in 1993 to post $2.8 billion in 2000. Zale Corporation operates approximately 2,250 retail locations throughout the United States, Canada, and Puerto Rico. Its brands include Zales Jewelers, Zales Outlet, Gordon's Jewelers, Bailey Banks & Biddle Fine Jewelers, Peoples Jewelers, Mappins Jewelers, and Piercing Pagoda.

The U.S. retail jewelry distribution channel is broad. *Market Share Reporter 2007* reported the jewelry and watch sales figures of 15 companies. The channel included, in order of sales: Wal-Mart, Sterling, Zale, QVC, Tiffany & Co., JCPenney, Sears Roebuck & Co., Finlay Fine Jewelry, Helzberg Diamond Shops, Fred Meyer Jewelers, Costco, Home Shopping Network, Target Stores, and two television retailers. In 2006 Wal-Mart had $2.7 billion in jewelry sales.

Online jewelry sales are increasing. According to *Market Share Reporter 2007,* online sales increased from $2.8 billion in 2004 to a projected $6.0 billion in 2008. Zales, for instance, operates online at www.zales.com, www.gordonsjewelers.com, and www.baileybanksandbiddle.com.

KEY USERS

In 1999, 43 percent of the U.S. adult population bought jewelry. Compared to men, women have a slightly higher purchase rate, 48 percent for women and 36 percent for men. Men have a lower purchase rate incidence, but they spend considerably more money on their jewelry purchases than women do. Most of the jewelry purchases by men are gifts. The highest jewelry purchase rate is among younger adults 18 to 24 years old.

Key users of jewelry are people planning to wed who want to symbolize their love with wedding ring sets, typically gold and diamond. Other key users of jewelry include groups who are marking life transitions. For instance, class rings symbolize the completion of high school, college, or university. Identity and medical bracelets are used to help people in the medical profession care for people in a prompt and appropriate manner.

ADJACENT MARKETS

Gold is a valuable metal. It is a stable chemical that does not easily combine with other substances, nor does it tarnish or corrode. This makes it ideal for a number of industrial uses. More than 80 percent of the gold mined worldwide is used for jewelry. The remaining 20 percent is used in industrial applications. World gold prices increased 45 percent between 2005 and 2006. Rising prices often indicate increased demand in an adjacent market. The adjacent market for gold used in jewelry is gold used in industrial applications. Industrial uses for gold include fine wire for electronics, as a protective coating for spacecraft and satellites, and in industrial and medical laser systems.

India is the largest worldwide consumer of gold. Demand in India for gold is anticipated to increase from 800 tons per year in 2000 to 981 tons by 2010 and 1,152 tons by 2015. The United States is the second largest worldwide gold consumer.

The fashion industry and the jewelry industry are closely adjacent markets. Trends in apparel lead to trends in jewelry. New jewelry collections are rolled out as often as new apparel collections, sometimes four times per year.

RESEARCH & DEVELOPMENT

Research and development in jewelry relates primarily to gemstones. Gemstones are artificially manufactured and artificially farmed.

Artificially manufactured gemstones are either synthetics or simulants. Synthetic manmade gemstones are laboratory grown. They are of the same or extremely similar chemical composition as their natural equivalents. Most synthetic stones are produced using a high temperature process. Synthetically grown gemstones have, in essence, the same appearance as well as the same optical, physical, and chemical properties as the naturally occurring gemstones they represent.

Simulated stones are also manmade, but of a lesser quality than synthetic gemstones. Simulants have an appearance that is similar to a natural gemstone, but different optical, physical, and chemical properties. Simulant gemstones produced in the United States include coral, cubic zirconia, turquoise, malachite, and birthstones such as garnet, amethyst, aquamarine, emerald, ruby, peridot, sapphire, and topaz.

Consumer acceptance of synthetic and simulant gemstones has increased in recent years. The recognition of synthetics and simulants for their own merits and not as inexpensive substitutes is perhaps the main reason. United States law requires that synthetic and simulant gemstones be clearly marked so that they are not confused for natural gemstones. The value of synthetic gemstone production in 2004 was $30.4 million and $100 million for simulants.

Human intervention is also used to produce cultured pearls. Most pearls used for jewelry today are cultured pearls. Research and development related to cultured pearls is attributed to Japanese researchers. Cultured pearls are developed through pearl farming. Cultured pearls are generally larger and of a more consistent size and color. They are made through the insertion of an artificial nucleus or shell bead directly into the tissue of an oyster. This process is referred to as grafting.

Pearl farming is a relatively simple form of aquaculture because oysters do not require artificial nourishment. Pearl farming, does, however, require a long-term investment of time and money. Only 5 to 10 percent of each pearl crop will be the high quality gems that generate 90 percent of the profit. Pearls are farmed throughout the world. In the United States, pearls are farmed in Tennessee, Arkansas, Louisiana, and Alabama.

CURRENT TRENDS

The current trend is greater concern about social and environmental aspects of diamond and gold mining. The plot of the 2006 movie *Blood Diamond,* starring Leonardo DiCaprio, brought issues surrounding conflict diamonds to the attention of the general public. Conflicts in Sierra Leone during the 1990s resulted in the coining of the phrase blood diamond.

Conflict diamonds come from regions that are controlled by forces, splinter groups, or factions that are opposed to the legitimate and internationally recognized governments in the areas. Conflict diamonds are sold to generate funds to support illegal activities, such as rebel military forces, and their profits contribute to prolonged brutal wars in parts of Africa. Legitimate diamonds play a role in the prosperity and economic development of many parts of Africa.

The United Nations, the diamond industry, and other organizations created a system to prevent conflict diamonds from entering the legitimate diamond supply chain. The system, known as the Kimberley Process Certification Scheme (KPCS) was implemented in 2003. KPCS participating governments guarantee that shipments of rough diamonds are exported in secure containers accompanied by a uniquely numbered, government validated certificate stating that they are conflict free. KPCS participating governments also agree not to import rough diamonds without approved KPCS certificates.

Two diamond industry programs supplement KPCS with self-regulation. These are the International Diamond Manufacturers Association and the World Federation of Diamond Bourses. These two bodies effectively represent all diamond processors and traders. Self-regulation involves warranties that accompany invoices covering the sale of rough diamonds, polished diamonds, and finished diamond jewelry. Warranties apply to rough diamonds mined after 2002 and the products fabricated from them.

In 2005 the Council for Responsible Jewelry Practices (CRJP), a non-profit organization, was founded in part by the Jewelers of America, a trade organization. CRJP promotes and develops responsible social, ethical, and environmental business practices throughout the gold jewelry industry. It held its first meeting in London, United Kingdom, in 2006. The Council believes gold should be extracted and processed in a manner that respects the needs of current and future generations. It supports transparency in gold mining, procuring, and selling.

TARGET MARKETS & SEGMENTATION

The market is segmented into a low-end and a high-end. On the low-end, manufacturers produce costume jewelry. Stores such as Wal-Mart cater to consumers who may be interested in costume jewelry. On the high-end, manufacturers produce luxury jewelry items and target consumers who have enough disposable income to purchase them.

RELATED ASSOCIATIONS & ORGANIZATIONS

British Jewelers Association, http://www.bja.org.uk

Council for Responsible Jewelry Practices, http://www.responsiblejewellery.com

Diamond Facts, http://www.diamondfacts.org

Gemological Association and Gem Testing Laboratory, http://www.gem-a.info

International Diamond Manufacturers Association, http://www.worlddiamondcouncil.com

Jewelers of American, http://www.jewelers.org

Jewelers Vigilance Committee, http://www.jvclegal.org

The National Craft Association, http://www.craftassoc.com

No Dirty Gold, http://www.nodirtygold.org

Society of American Silversmiths, http://www.silversmithing.com

Society of North American Goldsmiths, http://www.snagmetalsmith.org

BIBLIOGRAPHY

Alastair, Duncan. *Masterworks of Louis Comfort Tiffany.* Harry N. Abrams and the Smithsonian Institution Press. September 1998.

Brown, Bina. "Jewelry Demand Helping Gold Price." *CNN.* 14 August 2006.

Campbell, Greg. *Blood Diamonds: Tracing the Deadly Path of the World's Most Precious Stones.* Westview Press, 2004.

"Conflict Diamonds." United Nations, Security Council Affairs Division. 21 March 2001. Available from <http://www.un.org/peace/africa/Diamond.html>.

"Costume Jewelry and Novelty Manufacturing: 2002." *2002 Economic Census.* U.S. Department of Commerce, Bureau of the Census. 2004.

"Eliminating Conflict Diamonds." World Diamond Council. Available from <http://www.diamondfacts.org>.

"The Fashion Industry and New York City." Garment Industry Development Corporation. Available from <http://www.gidc.org/industry.html>.

Hamilton, Adam. "Gold Stock Investing." Zeal Speculation and Investing. 7 June 2002. Available from <http://www.zealllc.com/2002/goldstk101.html>.

Haws, Maria. "The Basic Methods of Pearl Farming: A Layman's Manual." U.S. Department of Agriculture, Center for Tropical and Subtropical Aquaculture. March 2002.

"Jewelry (except costume) Manufacturing: 2002." *2002 Economic Census.* U.S. Department of Commerce, Bureau of the Census. 2004.

"Jewelry Sales Reach $39.8 Billion." United Marketing Press Release. 17 April 2001. Available from <http://www.unitymarketingonline.com>.

Kanfer, Stefan. *The Last Empire: De Beers, Diamonds, and the World.* Farrar, Straus, Giroux, 1993.

Kirkemo, Harold, Newman, William L. Ashley, Roger P. "Gold." U.S. Department of the Interior, United States Geological Survey. Available from <http://pubs.usgs.gov/gip/prospect1/goldgip.html>.

"Mining Diamonds." Costello's Australian Jewellers. Available from <http://www.costellos.com.au/diamonds/mining.html>.

"A Monopoly Isn't Forever." *CNN.* 30 September 2000.

Perlez, Jane and Kirk Johnson. "Behind Gold's Glitter: Torn Lands and Pointed Questions." New York Times. 24 October 2005. Available from <http://www.nytimes.com/2005/10/24/international/24GOLD.htm>.

"Quarterly Retail E-commerce Sales, 4th Quarter 2004." U.S. Department of Commerce, Bureau of the Census. 24 February 2005. Available from <http://www.census.gov/mrts/www/data/html/04Q4.html>.

"Synthetic and Simulant." U.S. Department of the Interior, United States Geological Survey. Avaliable from <http://minerals.usgs.gov/minerals/pubs/commodity/gemstones/sp14-95/synthetic.html>.

Kitchen Cabinets

INDUSTRIAL CODES

NAICS: 33–7110 Wood Kitchen Cabinets and Countertop Manufacturing

SIC: 2434 Wood Kitchen Cabinet Manufacturing, 2541 Wood Partitions and Fixtures Manufacturing

NAICS-Based Product Codes: 33–71101 through 33–71101121, 33–71104 through 33–71104121, 33–71107 through 33–71107121, 33–7110A through 33–7110A121, 33–7110E through 33–7110E121, and 33–7110H through 33–7110H100

PRODUCT OVERVIEW

Kitchen cabinets are storage units used to house, organize, and protect anything typically found in the kitchen. This would include dishes, pots and pans, non-perishable food, and other similar items. While usually a permanent fixture in a kitchen, stand-alone hutches and armoires used to be quite common in the kitchens of old. These stand-alone pieces of furniture served the same storage function as wall mounted cabinets do today. Kitchen cabinets are the most prominent feature of the kitchens in most homes and the most costly component of a typical residential kitchen remodel, surpassing even appliances as an investment. Accounting for nearly half of an average remodeling budget, they can be the homeowner's most significant expenditure in the kitchen.

Kitchen cabinets are available in a wide array of styles, sizes, and colors. The least expensive way to purchase cabinets is to buy stock cabinets, or those varieties that are on-hand. They are factory-built in a small selection of pre-finished styles and in standard dimensions. Semi-custom cabinetry is factory-built in standard sizes, but leaves the consumer many choices of style, finish and functionality. Custom cabinetry is built, sometimes on-site, to a particular kitchen's dimensions and the style may be unique.

Wood cabinets are the most popular, compared with other materials from which cabinets are made, primarily laminates and metals. Wood cabinets that are not solid wood or hewn from a single piece of lumber are either plywood, many layers of wood bonded together perpendicularly, or veneers, a thin sheet of real wood bonded to the face of another, less expensive material such as particleboard or multi-density fiberboard (MDF). Thermofoil, a resin that permanently hardens after heat-curing, and melamine resin, formed when melamine reacts to formaldehyde, take a distant second and third of the kitchen cabinet market, with 6.4 and 1.2 percent respectively.

MARKET

The Bigger Picture. The kitchen cabinet industry is generally considered a subset of the furniture industry, so it is somewhat subject to trends in the larger industry. For instance, the availability and cost of raw materials like lumber will influence many areas of the larger furniture market. Kitchen cabinetry is an independent enough segment that it, along with office furniture, are the only two components of the U.S. furniture industry that are growing overall and are predicted to continue to do so. Overall, the cabinet industry in the United States is extremely strong, weathering the downturns in both the housing and furniture markets. While new home construction was expected to drop in 2007 to 1.408 million, down from its high in 2005 of 2.155 million, and major residential furniture makers like Broyhill, Pulaski, and others are battling foreign competition and closing down production lines, cabinetry manufacturing saw an unprecedented 9-year growth streak over the last decade that only began to abate in late 2006.

According to an April 2006 Forest Products Survey, the kitchen cabinet industry is expected to claim $16 billion of the U.S. forest products market by 2008. The Ohio-based Freedonia Group research firm concurred with this assessment with its projection that the repair and remodeling segment of the market will constitute $16.7 billion of the cabinet industry during this same period. By contrast, *Kitchen and Bath Business* predicted a more modest $14 billion. Industry data on kitchen cabinetry, however, often include information about countertop manufacturing and sometimes bathroom vanity production as well, since the industries are so closely related.

The cabinet industry in late 2006 ended a tremendous growth period of 127 consecutive months of increasing sales. The end of the streak as industry insiders called it, is a necessary economic correction, according to experts, brought on by rising interest rates, record high home costs, and a rising inventory of unsold new homes. It follows that the industry should plateau and stabilize, after this period of dizzying growth rates. January 2007 figures report a decrease in cabinet sales of 12.7 percent and some cabinet manufacturers have resorted to minor, short-lived layoffs. These are comparatively minor disturbances compared to the furniture market as a whole. In fact, some cabinet makers are building new plants, namely American Woodmark in Virginia, Merillat in New Mexico, and Kraftmaid in Utah. On the whole, the industry is very strong and has a reputation for reacting nimbly to changing market forces by adjusting production strategies, investing in technology, responding quickly to customer desires, and reducing lead times.

The Big Picture. Residential applications represent the largest end use of kitchen cabinets, claiming 70 percent of industry sales. Demand for kitchen cabinets was predicted to rise as residential consumers continue to repair and refurbish private homes rather than sell. Once considered an enclave to retreat from the world, one's home is now considered more a gathering place for loved ones, especially in the wake of the post 9/11 cocooning trend—the tendency to stay at home more and go out less. As such, according to *Builder* magazine, buyers are looking for showcase kitchen and eating areas. One major component, both stylistically and functionally, of the showcase is cabinetry.

FIGURE 118

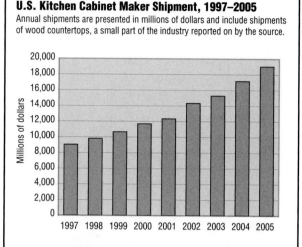

U.S. Kitchen Cabinet Maker Shipment, 1997–2005

Annual shipments are presented in millions of dollars and include shipments of wood countertops, a small part of the industry reported on by the source.

SOURCE: Compiled by the staff with data from the "Wood Kitchen and Countertop Manufacturing: 2002," *2002 Economic Census,* and "Statistics for Industry Groups and Industries: 2005, *Annual Survey of Manufactures,* U.S. Department of Commerce, Bureau of the Census. December 2004 and November 2006 respectively.

Furthermore, even with the addition of nearly 11.4 million new homes over the past decade, the average age of the U.S. home was 31 years in 2005, up from an average 23 years in 1985, and well past the 25 year old milestone generally recognized as a homeowner's initial yen to remodel. Industry forecasters from Harvard University's Joint Center for Housing Studies in *Foundations for Future Growth in the Remodeling Industry,* predicted a very encouraging 45 percent rise in spending on repair and remodeling between 2005 and 2015, even after adjusting for inflation over the same period.

Kitchen cabinet demand continues to be strong in the new home construction sector, as well, even with the industry waning, because it benefits from the trend toward larger kitchens that rely on greater cabinet space to enhance utility. While the new home sector slowed in the middle of the first decade of the twenty-first century, builders accounted for a slightly smaller than potential portion of the cabinet purchases, but expect their cabinet buying to increase as the new housing market recovers. In the meantime, remodeling is on the rise with homeowners confident that they will recover their investments when the time comes to sell their houses.

KEY PRODUCERS/MANUFACTURERS

Masco Corporation. The leading producer of kitchen cabinets in the United States is Taylor, Michigan-based Masco Corporation, which captured 21 percent of the market in 2006. Although the cabinet industry was experiencing a temporary dip in demand, Masco still led the pack with $3.1 billion in sales for the second quarter of 2007 alone, down from $3.4 billion for the same period the year before. This figure was in keeping with an overall industry decline of approximately 10 percent and referred only to cabinet sales, not the total sales of all its holdings.

Masco's acquisitions over the last two decades encompass every brand that leaps to mind: Kraftmaid (1990), Merrilat (1985), and Mill's Pride (1999), among others. Industry experts attribute Masco's success to its ability to choose acquisitions wisely and then allow the new holding to run autonomously, rather than subsuming its assets and liquidating the undesirable remainder. In the instance of Kraftmaid, for example, Masco allowed Kraftmaid to invest some $10 million it would not have had otherwise, in new technologies and leaner manufacturing processes that wildly improved its productivity and ability to capture the market.

Masco's cadre of cabinet manufacturers only adds to its command of the home improvement market. Before it became a cabinet giant, it was already a strong force in the plumbing supply industry with its Delta and Peerless brands. Its cabinet and plumbing brands together make it the largest manufactured goods supplier to Home Depot, a leading national retail outlet in the do-it-yourself market. Masco also holds interests in the installation industry, which serves the new home building market by installing cabinetry, fireplaces, gutters, bath accessories, garage doors, shelving, windows, and paint. Rounding out Masco's product lineup are its Decorative Architectural division, which owns well-known paint brand Behr Process Corporation; hardware interests, including Brainerd and Liberty in the United States, and Avocet in Europe; and its Specialty Products division, which manufacturers windows, doors, power tools, fasteners and many other home supply products. All told, Masco owns nearly fifty companies based all over the globe that serve almost every aspect of homebuilding and remodeling.

Fortune Brands. The second largest producer of cabinets for the kitchen is Deerfield, Illinois-based Fortune Brands, capturing 14 percent of the cabinet market in 2006. Fortune, with $8 billion in sales per year for all of its brands, is an entirely different sort of company than Masco. Where Masco's strength is its breadth of offerings within the homebuilding and home-improvement industries, Fortune is widely diversified across multiple industries including cabinets and furniture, liquor and wine, and golf supplies and apparel. As unlikely a family as this may seem, it provides Fortune with a sort of cushion against downturns in any one of the industries. While the housing market is in a slump, Fortune's other products and brands keep the company's earnings steady.

Fortune's cabinet companies include Aristokraft, Omega, and Diamond, but it also holds interest in some other areas of home supply with Moen plumbing and fixtures, Waterloo tool storage, Therma Tru doors, and Simonton windows. All of Fortune's holdings combined reported net sales of $2.35 billion in the second quarter of 2007, up 4 percent from same quarter the previous year; however, this is due to Fortune's non-cabinet producing entities. Fortune's home improvement holdings brought $232 million in net sales that quarter, down from $248 million the previous year.

American Woodmark Corporation. The next leading producer of cabinets for the U.S. market is American Woodmark Corporation with a 7 percent share of the market. American Woodmark reported fourth-quarter sales of $166.1 million for its 2007 fiscal year ending April 30, a 23 percent decrease compared to the same period in 2006. The company said its core product sales dropped by 15 percent "as remodeling sales were roughly flat and new construction sales accounted for the decline."

American Woodmark's range of products is exclusively kitchen and bath cabinetry. The American Woodmark brand is sold only through Home Depot, while its Shenandoah brand is available only through Lowe's. One other brand, Timberlake Cabinetry, is available through many builders, building supply companies, dealers, and

FIGURE 119

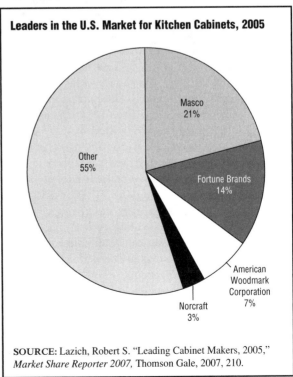

Leaders in the U.S. Market for Kitchen Cabinets, 2005

Other 55%
Masco 21%
Fortune Brands 14%
American Woodmark Corporation 7%
Norcraft 3%

SOURCE: Lazich, Robert S. "Leading Cabinet Makers, 2005," *Market Share Reporter 2007*, Thomson Gale, 2007, 210.

distributors nationally. It is pitched as a high-quality, budget-minded offering.

The three top manufacturers share 42 percent of the market, with the remaining 58 percent held by companies that command less than three percent of the market each—most less than one percent.

MATERIALS & SUPPLY CHAIN LOGISTICS

Wood Is Everything. Raw material for kitchen cabinetry is almost entirely wood. Some 95 percent of cabinets rely exclusively on wood, irrespective of hardware. Non-wood materials, such as decorative laminates, glass, and metal account for the other 5 percent of the market. Consequently, kitchen cabinetry depends largely on lumber suppliers in the manufacturing process. The cabinet market is subject to the trends associated with reliance on a natural resource like wood.

The largest suppliers of wood are Weyerhaeuser, Canfor, and West Fraser Timber, with 7.2, 5.2, and 4 billion board feet of lumber produced per annum respectively. Canfor and West Fraser are Canadian companies. Lumber production within the United States is concentrated in four states: California, Idaho, Oregon, and Washington from where 83 percent of U.S. lumber originates. In terms of sales Weyerhaeuser was still the largest company with $17.27 billion in the first quarter of 2005 alone. The second largest U.S. lumber company was Louisiana-Pacific with $2.02 billion in sales. Nexfor Inc. rounds out the top three with $1.36 billion in sales for the same period.

Process. According to *Wood and Wood Products* magazine, completing a cabinet from raw material to ready-to-ship can require some 250 processes. Raw materials are cut, assembled, and finished sometimes all at one plant, sometimes at different locations. Since the advent of lean manufacturing practices, time-to-market has plunged dramatically. Large quantities of inventoried stock demand expensive storage space. New manufacturing practices have reduced these inventories and the costs associated with them. Cabinets that once crossed miles of factory flooring as they were taken through numerous assembly, finishing, and packing tasks, now move smoothly from station to station and onto a waiting truck. By implementing lean manufacturing practices, KraftMaid has reduced its lead time from three weeks to one week and increased production by 250 percent, from 4,000 cabinets to 10,000 cabinets in a single day.

After orders are received and entered into a company's database, they are assigned a place on the production line. Components, once cut from raw lumber, are gathered and assembled. Once assembled, the unfinished product is moved to the finishing phase, after which they are loaded directly for delivery. The same is generally true of thermofoil and melamine cabinets.

Finishing of wood cabinets consists of applying stains, mechanical or manual color consistency procedures, and mechanical spraying of sealer and topcoats. This is followed by air drying, then oven curing, which prevents cracking and lengthens the life of the cabinet. Sometimes, at the end of production, factory-applied wormholes, compression marks, and oversanding are added to meet the demand for a distressed, or slightly aged and worn, look to the cabinet.

Thermofoiling is an alternate technique wherein MDF blanks (doors or drawer fronts) are routed with a given design, then moved to a "clean room" where adhesive is applied and dried. Then the piece is draped with the foil (basically vinyl), heated to reactivate the adhesive, compressed to bond the material to its face, trimmed, inspected, and assembled.

DISTRIBUTION CHANNEL

Imports and Exports. The period 1995 through 2004, on the whole, was a successful time for the foreign and domestic cabinet manufacturing industry. While the U.S. cabinet industry was beginning the largest growth streak in its history, foreign manufacturers, particularly those in Canada, were also expanding their exporting activities in the United States. Over the same 9-year growth streak for U.S. manufacturers, cabinetry imports more than quadrupled, rising 350 percent. U.S. manufacturers were also exporting their goods at higher rates during the same period, tripling the value of cabinets shipped from $19.2 million in 1995 to $58.5 million in 2004. Despite this growth in exports, imports far out number exports and this trade imbalance may foreshadow troubles for the domestic industry in the future.

Finished Cabinets to Consumers. The Kitchen Cabinet Manufacturers Association lists nine points of purchase for consumers: builders, remodelers, retail showrooms, architects/designers, wholesale distributors, home centers such as Lowe's or Home Depot, direct-to-consumer, lumber yards, and via Multi-Family building projects. Among companies that claim more than $25 million in sales annually, retail dealers represented the largest outlet of kitchen cabinets in the market, claiming 44 percent of sales in 2006. Home centers continued to lose market share, falling off 4.4 percent to 16.6 percent in 2006, a drop from 20 percent on 2005 figures. The market share held by distributors was up from 2005 to 2006, from 15 percent to 22 percent. Builders' position in the market slipped somewhat during this period, from 17 percent in 2005 to 15.3 percent in 2006, owing in large degree to the housing slump.

Sales by outlet for smaller companies, those with sales of less than $25 million annually, were dominated by retail dealers, which accounted for 75 percent of sales. Sales made directly to consumers were the next largest category in 2006, representing 7.3 percent of sales as did builders who represented another 7.3 percent of sales. Finally, distributors and home centers accounted for 4 to 5 percent of sales for these smaller cabinet makers.

Customer Control. One growing trend in the distribution of kitchen cabinets is what is known as customer-controlled ordering. The use of sophisticated Web sites by cabinet manufacturers allows customers to log onto the site, make selections, and place orders themselves, directly with the company.

Once only used by designers, kitchen planning software is becoming more available than ever as either downloadable software from a Web site, as is the software available from the retailer IKEA, or purchased directly at a store for home use. Such software allows customers to plan their own kitchens without the necessity or cost of a designer. Some kitchen planning software enables a direct interface with a cabinet company's inventory and ordering systems through which purchases can be made without having to even deal with a salesperson. Planning software is also available increasingly for in-store use in conjunction with a representative at dealers, distributors, and home centers.

KEY USERS

The largest outlet for kitchen cabinetry, consuming more that half of the total production volume, is the residential market. More than 55 percent of all kitchen cabinetry goes to owner-occupied dwellings and single family homes. Commercial and institutional applications, residential additions and alterations, maintenance of existing residential and non-residential structures, and other real estate outlets constitute a combined additional 18 percent of the total output. Homes and other residential dwellings account for nearly three-quarters of total U.S. kitchen cabinet purchases.

Other end users include manufactured and mobile homes (1.8%), the retail trade (1.4%) and multifamily dwellings (1.2%). The remaining consumers of kitchen cabinets are spread across a wide range of applications from recreational vehicles, food service and drinking venues, hotels, hospitals and other health practitioners' offices, the recreation industry, the government, nursing homes, and several other outlets, each capturing less than one percent of total production.

ADJACENT MARKETS

Bathroom Vanities and Related Cabinetwork. Although the same companies that produce kitchen cabinetry often also produce bathroom cabinetry, it represents a minor but notable subset. According to the July 2007 edition of *Kitchen and Bath Business*, of the more than $10 billion in sales of cabinets by manufacturers, 86 percent was cabinets and just over 10 percent was generated by the sale of bathroom vanities. The remaining market share was spread among countertops, millwork, and other products.

Hardware. Cabinets without hardware, that is drawer pulls and doorknobs, are available, but to complete the look and functionality of most cabinets and cater to individuals' tastes, most users finish their cabinetry choices with a selection of hardware. The choice of hardware gives the end user an opportunity for customization. In some cases, the installation of new hardware on old cabinets is a method used to update the look of a kitchen with little expense. Installing hardware on door and drawer-fronts also saves wear on cabinet and drawer edges by concentrating contact on the pulls themselves rather than on the cabinetry finish.

Popular styles in hardware follow styles in the cabinets themselves. While cabinets are nearly always wood, almost anything goes with drawer pulls and door knobs. Materials used for the hardware include stainless steel, pewter, brass, nickel, and other metals in any size, shape, color or finish; glass and ceramic in every hue; and others. Once cabinet producers offered fewer, simpler selections, but this has been changing as customers are using pulls and knobs as creative outlets and expressions of originality. In fact, novelty and organic shapes were popular in cabinetry hardware in the later part of the first decade of the twenty-first century.

Kitchen Countertops. Often milled by the same lumber companies and assembled by some cabinet manufacturers, kitchen countertops are a tagalong to the kitchen cabinet industry. Many people will replace countertops without replacing cabinets at the same time, however, the opposite is less often true: if a homeowner goes to the trouble of replacing all the cabinetry, the countertop generally gets upgraded at the same time. In fact, since it is a work surface, it will generally need replacing at more frequent intervals than cabinets. While kitchen countertops represent a completely separate industry, especially owing to the fact that while 95 percent of cabinets are wood, countertops are made of many different materials, they are often included along with cabinetry, flooring, vanities, and other items in data sets that track remodeling, repairing, and other home improvement trends.

RESEARCH & DEVELOPMENT

Environmental Impact. The most important area of research and development for the kitchen cabinet industry happens further up the supply chain at the production of raw materials. With the ever-growing concern over deforestation and other environmental issues related to industrial production, the pressure for alternatives and more sustainable hardwoods continues. One major initiative, still in its infancy is the Germany-based Forest Stewardship Council (FSC) program, a global voluntary certification program that is roughly equivalent to organic certification in the food industry. FSC lumber comes from foresters that practice environmentally sound and sustainable agricultural and harvesting methods, treat and pay workers fairly, and charge an ethical amount for their timber. These and other checks and balances are then independently verified by third parties. FSC lumber is enormously popular in Europe and is beginning to make gains in the United States. This is not only due to the public's increasing desire to make environmentally responsible purchasing decisions, but also because FSC compliance is actually profitable, especially since some large American cabinet outlets are onboard: Home Depot, Lowe's, and Office Depot all carry FSC products. One factor undermining the effectiveness of the FSC program is the fact that some U.S. manufacturers, in an attempt to buy the most inexpensive wood possible, purchase foreign hardwoods imported from companies that do not practice sound forest management.

The largest U.S. forest products company, Weyerhaeuser, participates with CERFLOR, Brazil's forest certification program and the European Programme for the Endorsement of Forest Certification (PEFC). It is developing more sustainable hardwoods, such as their Lyptus trademarked brand of lumber grown in Brazil, where sustainability is less of an issue than in colder regions. Lyptus grows quickly and can be ready for reharvest in 14 to 16 years. It is also grown among more traditional hardwoods in forests that replicate more realistic and ecologically diverse environments. Diversity reduces stress on the environment by preserving natural habitats and supporting the most natural complexity of a given ecology. Lyptus and other alternatives are not grown as a one-species orchard, but as single components of a regular forest system. However, sustainability is an issue that affects all timber uses—flooring, countertops, furniture, building materials, paper, and many other outlets, not just cabinetry.

Volatile Organic Compounds (VOCs). Since the early 1990s cabinet makers have striven to remove harmful VOCs from their production process. VOCs are found in spray finishes and adhesives and affect air quality and worker health and safety. Cabinet manufacturers, much like wood floor manufacturers, are working to use less harmful UV finishing and other techniques that produce fewer toxic off-gassing.

Alternate Applications. Another catalyst expanding the kitchen cabinet industry is the willingness of consumers to choose cabinets for other areas of their homes, such as media and laundry rooms. The concurrent trend of designers repurposing kitchen cabinetry for the specific needs of other rooms in the home, in turn piques consumers' interest in improving the aesthetics and functionality of rooms other than the kitchen. This self-propelling trend of user willingness and designer retooling helps fuel the momentum.

Number three on *Builder* magazine's top ten list of elements of home design in 2004 was the laundry room. This begins with a bit more space to fold the laundry, then expands to include a hobby or craft (storage intensive activities), then perhaps evolves into another place to log onto the computer and shop, pay bills, or check stocks. *Builder* handily captions it as "wash, spin, sew, surf." All of this necessitates a greater use of cabinetry to house the accoutrements in a room designed for true multi-tasking. Moreover, work areas in home offices; mud rooms or breezeway storage areas; home crafting centers; transitional workspaces near the kitchen, such as computer areas used for and in the kitchen; and even fitted bedroom components, are applications helping kitchen cabinets cross over to other areas of the home.

CURRENT TRENDS

Design. Traditional cabinetry still comprises the largest share of cabinetry manufactured for the U.S. market claiming 63 percent in 2006. The next most popular style is contemporary, which is gaining on traditional. It once represented 10 percent of market but by 2006 claimed 16 percent and rising. Contemporary in the broadest sense, represents a continuum of style choices, but can be defined loosely by clean, open lines and brighter colors on one end of the spectrum to tailored, but warm, and even deep-toned, and inviting on the other. Kitchen cabinets favor the warm end of the spectrum where it meets the more traditional old world style. Where contemporary was once thought of as bright and harsh, it now enjoys a much wider span of possibilities as palettes deepen, and less cluttered designs are emerging that favor warmer woods. Designers are even recycling older door styles and pairing them with contemporary finishes to bridge the gap and soften the look. This is helping the kitchen become contemporary, without losing its sense of tradition. Glass and aluminum accents on doors are seen in abundance, with clear, opaque, and patterned glass inserts making for limitless possibilities. Overall, contemporary will be expected to make inroads into the traditional market as well as stealing more market share from its next biggest style competitor, country, which claimed 12 percent of the market in 2006.

Finishes. Thermofoil manufacturers see white as a continuing trend, no matter what is happening in the larger wood cabinet market, mostly because they are still successfully selling it. However, where white thermofoil was once very popular in the residential market, it is losing ground there while making up volume in the general and medical office markets. Some tone-on-tone white finishes remain popular in residential installation because they are neutral canvases easily used as a backdrop for any number of decorating styles. However, the residential market is, by and large, seeing darker finishes and glazes on woods that give kitchen cabinetry a furniture-like finish. Consumers are also mixing different finishes and colors for a more eclectic look that mimics the rest of the home.

The darker wood species have been the trend in the early 2000s include black walnut, wenge, rosewood, and ipê, a dense, tropical, hardwood from Brazil. Nonetheless, the more tried and true oak and maple woods are definitely maintaining a hold on the market. Notably, trends in the larger furniture industry tend to precede those same trends in the cabinet market eventually, so the forces behind painted, lacquered, and dark stained finish choices in home furnishings, will afterward rise to prominence in the cabinet arena as well.

Funtionality/Features. Kitchens are bigger than they ever have been and consequently cabinet buyers are asking for more than they ever have. Copious use of open space has replaced the long, narrow galley kitchen of yore. Islands and double islands are a popular trend. Islands provide additional work surfaces while filling some of large spaces buyers like. Another trend is the more strategic use of cabinets: while kitchens are bigger, filling them with large banks of wall cabinets is not more popular. Instead designers are using cabinets as architectural elements such as in columns or as arch supports, and stacking them vertically to the ceiling, without the benefit of a soffit, such as a 48 inch cabinet topped with a 12 inch cabinet and finished with decorative molding.

Features that are in demand are those that maximize functionality. These include pull out shelves, tilt-outs that utilize shallow spaces with hinged drawer fronts that attach to the frame, two-tiered drawers with smaller pull-outs behind a single-drawer front, and multi-tiered vertical pull-outs that hold small items like spices or canned goods in small, otherwise unusable space. In answer to the call for no space used unwisely, Top Drawer Components produces a bank of drawers that fits into a corner, complete with 90 degree mitered drawer fronts rather than the usual flat front.

Business Relationships. Partnering of builders and cabinet makers is another trend seen in the industry. To attract more buyers, some builders are beginning to make the entire range of a given cabinet supplier's offerings available, instead of the smaller selection that is customary. Homeowners then deal directly with suppliers for the final products to be installed in the home as the finishing work is done. By taking themselves out of the process, builders give homeowners greater choice and control over customization, increasing the buyers' satisfaction with their new homes. At the same time they save themselves the aggravation of selling, ordering, and delivering the cabinets to the buyers of their homes. This is a trend seen in other areas of home buying as well. It is not uncommon for builders to ask their new home clients to deal directly with suppliers of lighting and flooring materials.

TARGET MARKETS & SEGMENTATION

Kermit Baker, director of the Remodeling Futures Program at the Harvard Joint Center for Housing Studies, explained at the National Association of Homebuilders semi-annual Construction Forecast Conference in late 2004, that the nation's residential remodeling industry should be expected to grow by 5 percent annually over the next several years. Of the estimated $138 billion spent on home remodeling in 2004, $60 billion went for interior space remodels and additions, including kitchen and bath upgrades and alterations. Remodeling and repair on cabinets specifically will claim some 70 percent of the homeowner's budget with the average homeowner buying 21 cabinets.

As to exactly which 21 cabinets arrive in the purchaser's home, the data is sharply divided. Of companies whose sales are more than $25 million per year, cabinet buyers favored stock cabinets giving them 49 percent of the market in 2005. In the same category, 32.5 percent of the market share went for semi-custom cabinetry, with the remaining 18.5 percent captured by fully custom cabinetry. For companies whose sales were under $25 million per year, the percentages were very different. Among these smaller companies, custom cabinets lead the pack snagging nearly 65 percent of the market, where stock and semi-custom cabinets split the remaining market, with 14 and 21 percent respectively.

A home's age is a key factor in remodeling decisions and in the decision to purchase new cabinetry. Homes built in the 1970s are now hitting the typical remodeling age of over 25 years: not only is the kitchen out of date, but it no longer functions well. In fact, the age of the average U.S. home is actually going up, making remodeling increasingly necessary.

Also, the topmost spenders are supporting the growth of the whole category. The number of households spending $25,000 or more per year on remodeling doubled from 16 percent in 1995 to 31.2 percent in 2003. A homeowner's reluctance to buy a new home during a period in which the housing market is in flux is only likely to fuel their desire to remodel their current one.

Immigrant homeownership is another prominent factor in the remodeling surge, with Hispanics on the leading edge of home improvement spending. "This trend will only intensify," stated Baker, "as minorities increasingly become homeowners. They are expected to account for nearly half of the increase in the home-owning population by 2015."

Many factors, including aging homes, the high cost of new homes, the wide availability of high quality cabinets, rising mortgage rates, and a general desire to upgrade and improve will likely support the kitchen cabinet industry over the next decade.

RELATED ASSOCIATIONS & ORGANIZATIONS

Hardwood Plywood and Veneer Association (HPVA), http://www.hpva.org

Kitchen Cabinet Manufacturers Association (KCMA), http://www.kcma.org

National Kitchen and Bath Association (NKBA), http://www.nkba.org

Wood Component Manufacturers Association (WCMA), http://www.woodcomponents.org

BIBLIOGRAPHY

"2005 Looking Solid as '04 Closes Strong." *Kitchen & Bath Design News*. March 2005, 1.

Adams, Larry. "How KraftMaid Doubled Production." *Wood & Wood Products*. November 1999, 45.

"American Woodmark 4Q Sales Fall 23%." *Wood & Wood Products*. July 2007, 1.

"Ask Natural Life: Answers to Your Questions About Healthy, Sustainable Living." *Natural Life*. March-April 2005, 2.

Baxter, Steve. "Thermofoils Expand Designers Choice's Door Selection." *Wood & Wood Products*. May 2001, 55.

"Builder Wants Suppliers to Join His Team." *Chilton's Hardware Age*. April 1995, 33.

"Cabinet Sales Up 9.2% in April." *Wood & Wood Products*. July 2003, 18.

Christianson, Rich. "Chins Up! Cabinet Industry Is Down, but Hardly Out." *Wood & Wood Products*. March 2007.

———. "Masco Amasses an Impressive Collection of Cabinet Companies." *Wood & Wood Products*. November 1999, 11.

Curry, Pat. "Top 10 Elements of Style: Trends in Home Design Reflect a Consumer Desire for Less Maintenance and More Free Time." *Builder*. January 2004, 8.

Darnay, Arsen J., and Joyce P. Simkin. *Manufacturing & Distribution USA*, 4th ed. Thomson Gale, 2006, Volume 2, 1612–1616.

"Design Trends in Kitchen Cabinets: A Brief Look at What's Hot in Kitchen Design." *Wood & Wood Products*. February 2007, 3.

"Forest Products Society." *Forest Products Journal*. April 2006.

"Growth Forecast for Nation's Residential Remodeling Industry Over Next Few Years." *Kitchen & Bath Design News*. December 2004.

Koenig, Karen M. "The Streak Goes On: But After 107 Months of Cabinet Growth, How Much Longer Will It Last?" *Wood & Wood Products*. April 2005, 8.

Lantz, Gary. "Certified Wood: Eco-fad or Everlasting? Want to Back Sustainable Forests and Natural Capital When You Reach for Lumber? A Look at What Those Symbols Mean and Where the Movement's Headed." *American Forests*. Spring 2005, 4.

McQueen, Gregg. "Living Lean." *Industrial Maintenance & Plant Operation*. August 1999, 30.

Partsch, Bill. "Slow and Steady." *Kitchen & Bath Business*. 1 July 2007.

Reep, Sarah. "It's Time to Update with Contemporary Influence." *Kitchen & Bath Design News*. May 2006, 2.

"Service with Style: As Today's Homebuyers Turn to Custom-Looking Cabinets and Increased Design Options, Big Builders Are Looking for Cabinet Suppliers to Streamline the Process." *Builder*. August 2003, 3.

"The U.S. Furniture Industry: Yesterday and Today... Will There Be a Tomorrow?" *Wood Digest*. June 2007, 20.

SEE ALSO *Countertops, Wood Flooring*

Lawn & Garden Tools

INDUSTRIAL CODES

NAICS: 33–3112 Lawn and Garden Tractor and Home Lawn and Garden Equipment Manufacturing

SIC: 3524 Lawn and Garden Equipment Manufacturing

NAICS-Based Product Codes: 33–31121, 33–31123, 33–31127, 33–3112W, and 33–3111J

PRODUCT OVERVIEW

The lawn and garden tools industry talks about its product, in the aggregate, as *green goods*—this despite the fact that a rather large proportion of the actual objects sold are various shades of red, sometimes yellow, and occasionally black. But the phrase, of course, refers to the great outdoors. The term is also intended to convey that lawn and garden products are a category of appliance. The trade talks about their indoor counterparts as *white goods*—despite modern preferences for indoor appliances that come in a variety of bright colors to match the décor of kitchens and laundry rooms of the United States.

From the consumer's perspective, everything in the garage or outdoor shed falls into the category of lawn and garden tools, be the specific object a lawnmower, a power trimmer, a shovel, a rake, or a set of garden shears. In the corporate sector, the category means power-driven tools. Yet another way of classifying things is that adopted by the U.S. Census Bureau, the nation's collector of economic data. The Census Bureau divides this category into power-driven lawn and garden equipment, in which category it includes mowers, tractors, and snow blowers. It classi-

fies chainsaws as power-driven hand tools produced by a different industrial category. Furthermore, at least some of the implements (like shovels and rakes, for instance), are classified as hand and edge tools. Finally, the Census Bureau classifies commercial-grade mowers and tractors under farm machinery and equipment. The discussion presented in this essay will be focused on power-driven outdoor tools used by the consumer, including chainsaws and also commercial turf maintenance equipment. The main reason for this delimitation is the availability of good statistical information and also because data from the industry itself aligns well with this approach.

The largest single segment of this category is lawn care, and the dominant product is the lawnmower. Mowers come in an ascending hierarchy of overall expense, beginning with the simple gas-engine-driven mower. Just a few such mowers are electrically operated, usually by battery power, preferred by those who dislike the noise of the engines, the smell of hydrocarbon fumes, or wish to avoid the environmental impact of relatively uncontrolled gasoline combustion. Mowers are usually hand-propelled, but the cutting action is supplied by the engine. Some mowers are also machine-propelled so that the user need only guide but need not push the device. Riding mowers, also called lawn tractors, are the next step up. Rather more powerful and comfortable garden tractors occupy the summit; some of these have special features such as the ability to pivot around their center-point for maximum access to hemmed-in locations; they also feature gearing in reverse. Supplementing these basic lawn tools are edge and weed trimmers, lawn vacuums and sweepers, leaf blowers, and power spray devices to clean up the sidewalks and the driveway. Most of these devices can be purchased

529

powered by batteries driving electric motors or by gasoline engines. People who have large amounts of waste matter to reduce to manageable heaps, or for composting, can also buy shredders or choose mowers specifically designed to mulch the grass.

Snow blowers, sometimes called snow throwers, provide an outdoor device of equivalent heft and power to the homeowner in winter. These come in small, often electrically-driven packages on up to massive, self-propelled machines powered by gasoline engines.

A smaller but still sizeable category of products serve in assisting the homeowner in the management, not least the removal, of trees and large bushes. The principal product is the engine-driven chainsaw. The homeowner who must have everything will likely also acquire a log splitter, a stump cutter, and have a chipper or wood-waste shredder. For trimming and dressing bushes, he or she will also have a trimmer, typically powered by a rechargeable battery pack.

For soil preparation the homeowner has a choice of tillers and cultivators powered so that the effort to turn, loosen, and prepare the soil will require virtually no muscle-power. These devices may be quite small and intended to serve a person squatting down by a flower bed or may be quite large devices with wheels intended to be walked behind. Lawn aerators belong in this category as well. These devices are more typically used by commercial services; they are usually employed just once a year, in spring. Homeowners, however, may rent them from supply houses.

Commercial turf care equipment includes all of the products mentioned above, with the difference that the tools are designed for constant use, are often very large, and have special features. Commercial service providers also use golf cars and other self-propelled utility vehicles in getting their work done; such vehicles are, therefore, also included as products of this industry.

Finally, all of these products have attachments and accessories purchased as options and are supported by servicing dealers who, for instance, will sharpen or replace lawnmower blades or chainsaw chains. The core product underlying all of these products is the gasoline engine. Only a few companies make engines; the majority of producers buy engines from others.

MARKET

The last economic census completed at the time of this writing, conducted in the year 2002, showed total industry shipments for powered lawn and garden tools of $8 billion. Products aimed for the household accounted for $6.5 billion and commercial turf equipment for $1.55 billion of that total, or 80.7 and 19.3 percent respectively. Subsequent annual surveys conducted by the government

in the 2003 to 2005 period, the *Annual Survey of Manufactures*, and trend projections thereafter indicate a 2008 market of approximately $10.14 billion. Between 1997 and 2002, the industry declined at an annual rate of 1.8 percent, but losses were not uniform. Within the consumer category, the largest segment in 1997, non-powered lawnmower shipments declined at a rate of 2.8 percent per year, but all other segments registered growth at an annual rate: riding mowers increased at 1.4 percent, parts shipments at a rate of 1.6 percent, and other equipment (hand held tools and snow equipment) at a rate of 3.7 percent per year. Growth in the commercial turf equipment market was also positive; it grew at a rate of 3 percent per year. Projections suggest that the entire industry's annual growth rate will have been just under 4 percent per year between 2002 and 2008.

As a consequence of the decline in non-riding lawnmower shipments in the period from 1997 to 2002, riding mowers became the largest single segment in the consumer-oriented part of this industry. In 2002 riding equipment represented 43.7 percent of total shipments, just a nose ahead of non-riding equipment, representing 43.4 percent of shipments. Parts represented 11.6 and the rapidly growing "other equipment" category 1.4 percent of shipments.

The lawn and garden equipment sector is a well-established and mature industry responding only slowly to broader forces in the economy. The market tends to soften in economic down-turns and to flourish or languish in response to trends in housing. If the devil is in the details,

FIGURE 120

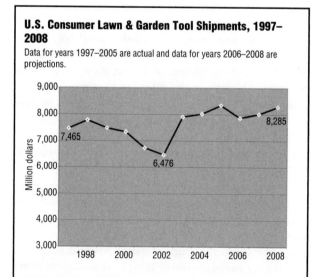

U.S. Consumer Lawn & Garden Tool Shipments, 1997–2008

Data for years 1997–2005 are actual and data for years 2006–2008 are projections.

SOURCE: Compiled by the staff with data from the following sources, *Annual Survey of Manufactures 2003 and 2005*, U.S. Department of Commerce, Bureau of the Census, November 2004 and November 2006.

FIGURE 121

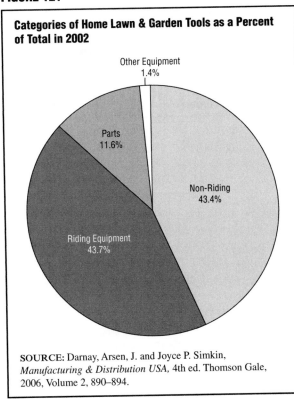

Categories of Home Lawn & Garden Tools as a Percent of Total in 2002

Other Equipment
1.4%

Parts
11.6%

Non-Riding
43.4%

Riding Equipment
43.7%

SOURCE: Darnay, Arsen, J. and Joyce P. Simkin, *Manufacturing & Distribution USA*, 4th ed. Thomson Gale, 2006, Volume 2, 890–894.

the details illuminating the closing years of the twentieth and the opening years of the twenty-first century suggest both a slight reaction to the economic downturn that came in 2001 and also a process of upgrading: expenditures shifted to the more expensive product segment within this industry's consumer/household segment, illustrated by the growth in riding mowers and, especially, in the sale of other equipment—the tillers, shredders, leaf blowers, and similar hand-held categories. The functions performed by these tools, of course, could be performed with much less expensive tools like shovels and rakes. Meanwhile the commercial sector, which flourishes with the growth of services provided to homeowners and institutions, illustrates that a segment of the public used more commercial services and, for that reason perhaps, did not replace that old mower. This mixed picture results in modest growth projections from 2002 forward.

Census Bureau data are excellent for gauging the production picture but less helpful in tracing the flow of goods from producers downward to the point of sale. At the wholesale level of the market, Census Bureau data blend the details of this industry with that of farm equipment and report the results only for Farm and Garden Machinery and Equipment Merchant Wholesalers. Details for product categories are not reported so that the wholesale level is entirely swamped by the farm machinery category. At the same time, the Census Bureau data for

the retail level, reported for the category Outdoor Power Equipment Stores shows only a portion of the industry's shipments as moving through this specialized channel, 68 percent of shipments in 2002. A presumed 32 percent of the total volume shipped is thus delivered to market by big box retailers such as Home Depot and Lowe's as well as the department stores. But even these data, scant as they are, support the presumption of a general upgrading process by the consumer. In 1997 the specialized outdoor power equipment stores accounted for a smaller portion of shipments, 54.5 percent, with 45.5 percent left to larger retailers more likely to offer a discount. This suggests that, by 2002, more consumers bought from specialists who offer the more pricey products (like garden tractors) that also require higher levels of service.

KEY PRODUCERS/MANUFACTURERS

In 2002, according to the Census Bureau counts, the U.S. lawn and garden tools industry had 128 participating companies. The top 35 participants, listed in Figure 122, are arranged alphabetically by name and by seven major product categories. This is not an exhaustive list of producers; readers who do not see their favorite brand may discover that the brand is owned by one of the companies listed or is the offering of a smaller participant.

By the very nature of the products sold in this industry, the most expensive and complex component of a lawn and garden tool is its gasoline engine or its electrical driver, including the motor, battery pack, and charging mechanism. Historically the industry had been structured around a few engine builders who did not make the tools themselves and a large number of companies that built the tools around a purchased engine. This has changed, but it was once quite common for people to say that their lawnmower was a Briggs & Stratton when, in fact, the maker of the mower had bought a Briggs & Stratton engine. Even well into the first decade of the twenty-first century, this pattern still largely holds. Briggs & Stratton remains the world's leading producer of small engines; the majority of lawnmowers sold, under whatever brand name, have such engines. Tecumseh Power Company, another U.S. engine producer, is second largest in the U.S. market. Two other U.S. companies participate in a leading role. Both sell more powerful engines: Kohler Company and Caterpillar, Inc. The latter is an important supplier of heavy-duty engines for heavier equipment used in some of the commercial turf products.

Competition in the engine market intensified in the 1980s with the entrance of Japanese companies into the market, Honda and Kawasaki. Both offered engines for lawn and garden tools adapted to this market from the small engines they used to power motorcycles. Honda departed from the long-established pattern in the industry

FIGURE 122

Key Participants in the Lawn & Garden Manufacturing Business by Type of Equipment Produced

Company	Engines	Walk-behind lawn-mowers	Lawn tractors	Snow-blowers	Edgers/ trimmers	Chain-saws	Commercial turf equip.
American Honda Motor Co., Inc.	X	X		X	X		
Ariens Company		X		X			
Black & Decker		X			X	X	
Briggs & Stratton	X						
Bush Hog, LLC							X
Caterpillar, Inc.	X						
Dolmar GmbH					X	X	
Echo, Inc.					X	X	
Exmark Manufacturing Co., Inc.							X
Global Garden Products		X	X		X	X	
Hoffco, Inc.		X					
Husqvarna Outdoor Products		X	X		X	X	
Husqvarna Professional Outdoor Products Inc.		X	X	X	X	X	X
Hustler Turf Equipment		X		X			
Jacobsen, A Textron Company							X
John Deere Company		X	X	X	X	X	X
Kawasaki Motor Company	X				X		
Kohler Company	X						
Kubota Tractor Corporation			X	X			X
Lastec, Inc.							X
MTD Products Inc.		X	X	X	X	X	
Positec USA, Inc.					X		
Redmax/Komatsu Zenoah America, Inc.	X				X	X	
Robin America, Inc.	X						
Shindaiwa, Inc.	X				X	X	
Simplicity Manufacturing, Inc.		X	X	X			
Snapper, Inc.		X	X	X			
Solo Incorporated					X	X	X
Stihl Incorporated					X	X	
Tanaka America Inc.					X	X	
Techtronic Industries, NA, Inc.					X	X	
Tecumseh Power Company	X						
The Toro Company.		X	X	X	X		X
Walker Manufacturing Company				X			X
Woods Equipment Company				X			X

SOURCE: Compiled by the staff from a study of the individual company literature as well as information available from the Lawn and Garden Dealers Association.

by offering both engines and lawn tools under its own brand. American Honda Motor Co., based in Swepsonville, North Carolina, now produces 75 percent of all of Honda's small engines and powered lawn and garden tools on U.S. soil. Three other Japanese companies are key participants in the engine market. Redmax is the American entity organized by Komatsu Zenoah and is a participant, also, in a variety of lawn and garden tools like trimmers and chainsaws. Robin America, Inc. was organized by Fuji Heavy Industries to distribute the Subaru line of Robin engines. Shindaiwa, Inc., like Redmax, is an engine maker that also sells trimmers and similar tools but not lawnmowers. The leading supplier of electrically powered walk-behind lawnmowers is Black & Decker, the world's leading producer of electrically-powered tools.

Not surprisingly, Briggs & Stratton began to diversify, perhaps in response to Honda's and other engine makers' entry into the market. In 2004 the company acquired Simplicity Manufacturing, Inc., a company that had already grown by acquisition and had a comprehensive line of products including commercial turf maintenance products.

The world's largest producer of lawnmowers in 2005 was MTD Products, a U.S. company, with a 25 percent share of the market, followed by Husqvarna, a Swedish company, with 20 percent. Huqvarna was then still owned by Electrolux Home Products. In that year Murray, Inc., held a 15 percent share. Murray is not listed in Figure 122 because it filed for bankruptcy and, at time of writing, was attempting to reorganize. The Toro Corporation held the fourth largest share at 11 percent. Snapper, Inc. (owned by Simplicity Manufacturing Inc.), perhaps best known for its riding mowers, still held a 2 percent share of the lawnmower market in the United States in 2005. Leaders in the lawn tractor/garden tractor category in 2005 were John Deere, Electrolux Home Products, owners of American Yard Products, MTD, Simplicity/Snapper, and Toro, in that order.

The largest participants in the lawn and garden equipment market tend to offer a wide range of products, thus mowers as well as tractors, snow blowers, as well as hand-operated equipment like trimmers, leaf blowers, and chainsaws. Among the major participants, Husqvarna, John Deere, and Toro also offer commercial turf maintenance equipment, but, by and large, that market is dominated by specialists who have their roots in farm or construction equipment manufacturing or have specialized, since their inception, in heavy-duty professional equipment. Examples of such companies are Bush Hog, LLC, Exmark Manufacturing, Walker Manufacturing, and Woods Equipment Company.

Trends in the consumer portion of this market, the dominant portion, representing 81 percent of all equipment sales, has been in the direction of consolidation and, occasionally, divestiture. An example of the latter has been the purchase of Husqvarna by Electrolux Home Products in 1997 and its divestiture of the same property in 2006. More typical has been aggregation of assets—and most notably brand names—by acquisition. The brand names continue to be maintained although ownership and marketing arrangements change. Indeed, it is often difficult, when looking at a subsidiary, to discover that it is actually owned by someone else. Consumer-oriented companies have acquired capabilities in the commercial sector to round out their offerings; but the same trends in reverse, with commercial companies entering the consumer market, are not discernible.

MATERIALS & SUPPLY CHAIN LOGISTICS

The Outdoor Power Equipment Institute (OPEI) published a study in 2002 in which it analyzed inputs to the industry, dividing these into materials on the one hand and components on the other. The Institute measured these as percentages of the total value of shipments of which they were made a part. Data for the year 2000 indicated that raw materials represented 24.1 percent of total shipments in that year, up from 13.7 percent in 1996 and 9.6 percent in 1992. Purchased components in 2000 were 38.7 percent of shipments, down from 45.8 percent in 1996 and 47.6 percent in 1992. These data, of course, reflect the trend noted above—consolidation. Producers were buying fewer components and buying more materials—making more of the finished product in-house from scratch. What, then, are these materials and components?

The largest material categories were steel (41.5%), plastics (24.3%), shipping cartons (15.5%), other materials (7.9%), paint (4.5%), and aluminum (1.7%). Not surprisingly, engines lead acquired component costs, representing 47.5 percent of total. The next four categories are transmissions (13.8%), cables and controls (6.7%), engine parts (6.2%), and tires (2.7%). This listing, also provided by OPEI, indicates that even grass catching bags for lawnmowers and seats for riding mowers are purchased components, also cutting blades and fuel tanks. These patterns very much support what closer familiarity with the industry reveals: most companies actually fabricate only a portion of the tool itself. In the case of lawnmowers, typically, the deck itself and the handles are made in-house from incoming materials, the deck painted in the factory. The rest of the componentry arrives ready-made and is assembled and then packaged for delivery.

Production facilities in this industry are usually located on the outer perimeters of metropolitan areas. Large component producers like Briggs & Stratton have very extensive and well-developed supply chains so that a producer has ready access to engines or parts located anywhere near a reasonably-sized city. In that steel products—of which steel sheet represents the largest percentage followed by stampings and forgings, as shown by Census Bureau data—are the largest inputs next to engines and engine components, from a logistical vantage point location in the vicinity of a steel distribution center is ideal for lawn and garden tool producers.

DISTRIBUTION CHANNEL

Based on surveys of producers conducted by the Outdoor Power Equipment Institute, the United States was a net exporter of lawn and garden tools in 2000 by a substantial margin. In that year domestic producers exported $867 million in such goods while imports amounted to $359 million—for a rare foreign trade surplus of just over half a billion dollars. The highest export rates were achieved by producers of commercial turf equipment (who exported 32.9% of their production) and hand-held equipment makers (33.6%), approximately one-third of their shipments. In the consumer products category 10.4 percent of goods made went into foreign markets. Canada purchased the largest share of exports.

Producers in this industry use both two- and three-tier distribution. In the first case goods move from the producer directly to the retailer and then to the customer, be that consumer a member of the public or an institutional buyer. In doing its distribution analysis, OPEI established five categories of identified buyers. These were: (1) small retailers, (2) general merchandise, hardware, and auto chains, (3) discount houses, (4) home improvement, home center, and building supply stores, (5) wholesalers/distributors, and (6) a very small Other category. The first four are different kinds of retail operation. If the results for these categories are combined, one sees that commercial lawn and garden tool producers, (but excluding hand-held goods) sold 84.8 percent of their goods directly into the retail channel, suggesting the dominance of the two-tier distribution system. This was also true of producers of hand-held tools who sold 84 percent of their product directly to the retail channel and commercial turf maintenance equipment builders who sold 81 percent of products directly to retail stores. Wholesaler/distributor participation was thus relatively minor in this industry, amounting to 13.6 percent for commercial, 15.2 percent for hand-held, and 16.4 percent for professional tools. The rest went into the Other channel. A surprisingly large proportion of product went to small retailers with fewer than fifteen stores each. In the consumer sector, 33.8 percent went to such stores; in the hand-held sector, 29.2 percent, and in the commercial turf maintenance sector a very high 79.8 percent. These percentages are net of exports, and thus reflect only domestic sales.

The predominance of small retailers reflects the fact that many of the leading firms have endeavored to develop their own dealer networks; some directly own such retail outlets. Direct sales are also driven by decades-long consolidation in the retail sector and the development of very large chain stores. These operations buy product under contract directly from producers and thus bypass the intervening wholesale channel. Distributors typically handle the more expensive products like lawn and garden tractors and also sell these to their networks of small independent dealers. In the consumer sector just over half of the product made (51%) and in the hand-held sector a slightly higher proportion (54.8%) is sold to large chains. In commercial turf maintenance sector an insignificant portion (1.2%) reaches chains.

KEY USERS

Key users of lawn and garden tools are homeowners, accounting for 81 percent of the market. The remaining 19 percent of production is purchased by institutional buyers. These buyers may be divided into: (1) commercial service organizations that use such equipment to deliver services to others, not least homeowners; (2) rental organizations who offer such equipment on a short-term-use basis; (3) institutional facility operators who employ staffs to maintain extensive terrains, including owners of golf courses, recreational areas, resorts, ball parks, industrial parks, hotels with extensive grounds, cemeteries, and owners or managers of similar facilities; and (4), most likely the largest market for commercial equipment, the public sector comprising such units as towns, cities, counties, highway departments, port authorities, park and forest services, elements of State and Federal governments, and, indeed, any agency responsible for the management of public lands.

Key users may also be defined by the surface area of the lawn requiring servicing. Users of walk-behind lawnmowers typically have modestly sized lawns less than half of an acre in extent. Homeowners with half an acre and up to two acres may use riding mowers or lawn tractors if the terrain permits their easy use. Garden tractors are usually deployed on lawns of two acres or greater. These, of course, are rules of thumb applicable to the household where one person will typically do the yard work. In institutional settings, equipment of all sizes may be employed. Thus a large crew using industrial-grade push mowers may be utilized or, depending on terrain, one or two large devices may be used. Service organizations working on residential estates typically arrive with one or two large self-propelled machines, several push mowers, and assorted hand-held equipment. Three or more laborers will fan out to do work, each person doing a different job.

ADJACENT MARKETS

Adjacent markets may be viewed as products adjacent to the core tools described here, thus lesser tools and greater tools. Using that point of view, non-powered tools represent one extreme. In the 1950s and earlier periods, the majority of homeowners used rotary push-mowers entirely powered by human muscle. Such equipment is not only still available but growing in popularity in that segment of the market interested in environmental issues, good exercise, relative silence, and possibly nostalgia and memories of having done grandma's or the neighbor lady's yard as a teenager. At the other extreme, beyond the commercial turf maintenance equipment, farm machinery represents equipment functionally equivalent to lawn and garden tools but writ large.

Another view of adjacency is to ask, What competes with what? In the early twenty-first century, with the most populous generation of the United States getting ever older—the Baby Boom generation—the aging population itself is influencing markets adjacent to particular slices of lawn and garden tooling. For instance, power-driven tillers are favored by people still eager to prepare their own garden soil but with less physical effort. In adopting powered-equipment, such people displace non-powered hand tools. The relative growth in riding equipment is in part due to people seeking greater comfort, in part due to trends in residential construction: new construction has featured progressively increasing square footage; such housing is often sited on greater acreage as well. The shift toward riding mowers has reduced demand for walk-behind mowers. Landscaping may be viewed as an adjacent market in those cases where people reorganize their lawns, replacing grass with shrubbery, ground cover, and flower beds—thus eliminating the need to mow.

Paralleling the aging of the baby boom has been increasing participation by women in the work force and thus a reduction in the time families can devote to housekeeping chores. The two trends have converged and have lead to the increased use of professional lawn care services, itself an adjacent market—although, of course, a market that itself uses the higher end of the lawn and garden equipment. This is reflected in greater growth experienced by commercial turf equipment over walk-behind lawnmowers. Elderly people, even with small yards, engage professionals to mow and fertilize their lawns and to clear away snow in the winter. Working couples busy conveying children to multiple outside activities have less time; some have opted to hand over routine lawn chores to companies that do the work when no one is home.

RESEARCH & DEVELOPMENT

Significant research and development efforts have gone into making gasoline engines used in lawn and garden equipment compliant with Environmental Protection Agency regulations. This effort has been on-going since the 1990s, but has not ended as the twenty-first century marches on. The State of California, traditionally a leader in prodding and leading the nation to achieve ever higher air purity (motivated by West Coast smog), has issued more stringent regulations under the auspices of its California Air Resources Board (CARB) applicable to all equipment sold in the state. Leading engine makers are conformant to both EPA and CARB rules. Efforts in this direction have focused on improved carburetion to minimize the emission of carbon monoxide and unburned hydrocarbons, reduced surface areas within the combustion chamber itself, helpful in reducing hydrocarbon emissions, and improved control of oil so that less of it is burned in the combustion chamber, reducing smoke—a particular problem in the more powerful 2-cycle engines that burn a gas-oil mixture. Greater fuel efficiency is another R&D goal in part motivated by environmental

concerns: the less fuel combusted, the less pollution; attempts in this direction also provide a consumer benefit. Attempts at complying with environmental regulations have raised engine prices because it is easier to comply with carbureted engines, thus by using more expensive 4-cycle engines.

A significant irritation for the user of powered-equipment is the effort required to get the smaller engines started. A good deal of R&D effort has gone into giving consumers easy-to-start engines. Briggs & Stratton, for example, features three different engine types that accomplish this end, called ReadyStart, FreshStart, and Z-Start. Better engine shielding to reduce noise and lighter engines to make hand-held devices easier to use are also under continuous development.

Powered lawn and garden tools can injure users if they are improperly operated. Producers and industry groups in the lawn and garden industry are expending continuous R&D effort to achieve higher safety in two ways: through the tightening and refinement of safety standards and their implementation in actual equipment designs.

Rather exotic R&D is also underway in this industry and may, in the longer term, become widely applied. One example is an effort by Udi Peless and Shai Abramson, two Israeli engineers, to market their invention, RoboMower, a completely automated, electrically driven lawn mower so independent that, once programmed, it can undock itself from its charging station, mow the lawn every week, and then return to its dock to replenish its batteries. The device is on sale in the United States through Sears.com. Lest it be thought that RoboMower is an absolutely unique product, there is also the automatic LawnBott offered by Kyodo America Industries Co. Ltd. LawnBott is guided by a perimeter cable defining the area it is supposed to mow. It detects its own boundaries by communicating with the cable. If its owner wishes to dispense with the cable, a well-defined fence will do as well. LawnBott sells for $1,849 and is made in Italy.

CURRENT TRENDS

The predominant trend in the mature lawn and garden tools industry is slow growth, stability, incremental technological change, corporate consolidation, and possibly a tendency by product users to replace lower-value with higher-value equipment, the last trend in part dictated by environmental pressures. This is an industry still dominated by U.S. manufacturers who export more of their product than they import, with a handful of major producers enjoying a global market. In general, one might say, the industry reflects the nature of the plant it is primarily intended to manage—the grass: it comes in its season and, at regular intervals, grows dormant, to reemerge again as green as before.

The industry underwent some competitive turbulence in the 1970s and 1980s with the entry of Japanese companies into the market, most notably Honda. However, with Honda's centralization of its small engine and lawn and garden tools manufacturing in the United States, U.S. manufacturing dominance has not been affected. Entry into the tools segment by the leading engine producer, Briggs & Stratton has transformed the once hierarchical character of this industry—its division into engine-producing and equipment-producing halves. Consolidation has reduced the number of companies, but the proliferation of many distinct brands has remained.

TARGET MARKETS & SEGMENTATION

In the lawn and garden tools sector, the product itself actually defines its own market and segment. Thus, for example, low-cost lawnmowers are targeted for the low end of the consumer market and are treated as commodities. The addition of features—such as, for lawn mowers, bigger engines, mulching features, glass-catching equipment, self-propelled walk-behinds—are intended for the higher end of the market and expected to appeal to those seeking ease, comfort, or even prestige. Lawn work frequently performed by males and thus the old adage "the bigger the boy the bigger the toy" applies in this industry. Riding mowers, lawn tractors, and garden tractors are in part targeted to appeal to the male ego—and, indeed, are often purchased for uses on quite small properties if the egos are big enough. Commercial equipment, in turn, is targeted to buyers who view the equipment in terms of equipment life, efficiency, and economy of use.

RELATED ASSOCIATIONS & ORGANIZATIONS

American Society of Agricultural Engineers, http://www.asae.org

Equipment Engine Training Council, http://www.eetc.org

Lawn and Garden Dealers Association, http://www.lgda.com

North American Equipment Dealers Association, http://www.naeda.com

Outdoor Power Equipment Aftermarket Association, http://www.opeaa.org

Outdoor Power Equipment & Engine Service Association, http://www.opeesa.com

Outdoor Power Equipment Institute, http://www.opei.mow.org

The Professional Landcare Network –PLANET, http://www.landcarenetwork.org

Professional Lawn Care Association of America, http://www.plcaa.org

BIBLIOGRAPHY

Darnay, Arsen J. and Joyce P. Simkin. *Manufacturing & Distribution USA,* 4th ed. Thomson Gale, 2006, Volume 2, 1029–1033.

DeRosa, Angie. "Hardware Show Abounds with Innovation." *Plastics News.* 29 May 2006.

Lazich, Robert S. *Market Share Reporter 2007.* Thomson Gale, 2007.

Murray, Charles J. "Mowing on Autopilot." *Design News.* 26 June 2006.

"Product Summary: 2002." *2002 Economic Census.* U.S. Department of Commerce, Bureau of the Census. March 2006.

Profile of the Outdoor Power Equipment Industry 2002. Outdoor Power Equipment Institute. Old Town Alexandria, Virginia. Undated.

"The Share-of-Market Picture for 2005." *Appliance Magazine.* September 2006.

SEE ALSO *Hand Tools, Construction Machinery*

Lighting

INDUSTRIAL CODES

NAICS: 33–5110 Electric Lamp Bulb and Parts Manufacturing, 33–5121 Residential Electric Lighting Fixture Manufacturing, 33–5122 Nonresidential Electric Lighting Fixture Manufacturing, 33–5129 Lighting Equipment Manufacturing, not elsewhere classified

SIC: 3641 Electric Lamps, 3645 Residential Lighting Fixtures, 3646 Commercial Lighting Fixtures, 3648 Lighting Equipment, not elsewhere classified

NAICS-Based Product Codes: 33–51101, 33–51103, 33–5110W, 33–51211, 33–51214, 33–5121W, 33–51221, 33–51222, 33–5122W, 33–51291, 33–512941, 33–51294Y

PRODUCT OVERVIEW

Modern lighting systems are classified by the device that converts electric power into light, thus by the technology behind the bulb or the tube. The three major product groupings are incandescent lighting, represented by the ordinary household light bulb, plasma lighting, represented by fluorescent lights, and LEDs (light-emitting diodes), semiconductor light sources still in development. The category is also known as solid state lighting.

Incandescents. Incandescent lighting is produced by a filament made of tungsten, a metal. The filament is mounted as a connector between the two poles of an electric circuit. Current enters the filament at one end and leaves at the other through a wire. Tungsten does not conduct electricity well. It resists the current's flow. This resistance produces a very high temperature—approximately 4,600° Fahrenheit and higher. A small portion of the electrical force, maximally 10 percent, turns into light; the rest is released as heat. Tungsten has the highest melting point of all elemental metals (6,192° Fahrenheit) and the second highest melting point of all elements; only carbon has a higher melting point. Tungsten is thus an ideal filament as it can be formed into a very thin wire and heated to a very high temperature while keeping its shape. The longer the filament the more power the lamp can produce; the higher the melting point, the more light the filament will emit. In a 60 watt bulb the filament, if unwound, would be more than six feet in length. To put that much wire into a small bulb, it has to be coiled tightly and the coils coiled in turn.

In the production process, the manufacturer draws all the air out of the bulb or replaces it with an inert gas. An argon-nitrogen mixture is typical. Vacuum bulbs are used for lower wattage lamps (40 watts and under); higher wattage bulbs are filled with gas. Tiny portions of the super-heated filament boil away at high temperature while the bulb is on. In vacuum bulbs these particles deposit on the glass, at its thick end if it is screwed in upside down, at its neck if it is installed in an upright position. Over time the glass darkens and the filament grows thinner—until at last it breaks. When it does, the electric power produces a temporary arc, sometimes an audible pop, and the bulb dies. The clouded appearance of burned-out bulbs is produced by the dying arc. Producers introduced inert gas fillings to prolong filament life. The heated gas inside the enclosed bulb creates a convection current. Heated air rises, cooler air falls, hence a current. The current carries the tiny particles of filament and will deposit some of them back on the filament again so that fewer particles end up on the glass.

The most efficient incandescent lamp is the halogen lamp developed by General Electric in 1959. It did not become widely available until later. This lamp features a hard, quartz glass tube filled with inert gases with a small amount of halogen gas (bromine or iodine). Quartz glass is used because it can withstand the high heat to which the glowing filament may bring it, maximally 1,652° Fahrenheit. In a halogen lamp tungsten boiling off will form a compound, tungsten bromide; the compound will circulate in the gas. This volatilized metal is most likely to redeposit again onto the filament at those points where it is hottest—precisely those points where it is also thinnest. At the point of redeposit, the tungsten separates from the bromine, the bromine returning to the current. This technology provides long filament life but at the cost of very hot operations, limiting the deployment of halogen lights. They are widely used in automotive lighting.

Plasma or Arc Lights. The role that the tungsten filament plays in incandescent bulbs is taken over either by mercury or sodium in these lights. Sodium suggests salt, but the element is actually an alkali metal. These metals cannot be formed into a wire and are present in the glass tube as vapors suspended in the carrier, an inert gas. The two electric poles, positive and negative, are far apart—unlike in incandescent bulbs where they are physically linked by the filament. When the light is turned on, electrons come from the cathode and seek to reach the far-away anode of the circuit, traveling through the gas. In effect an arc is set up between the poles. Many electrons moving in the tube energize the gas, turning it into a plasma, thus into a mass of ionized gas, meaning a gas with many free electrons—hence the name plasma light. As the electrons speed through the tube, they collide with molecules of mercury or sodium and cause these to be excited. Excitement means that an incoming electron from the cathode causes one of the mercury molecules' (or sodium molecules') electrons to jump from its normal orbit into a higher orbit. This unstable situation soon causes the metal's electron to drop back down to its ordinary orbit, but as it does so, it releases the energy that kicked it upward—as a photon. This transaction—the leap into a higher orbit, the drop back, the release of a photon—is the source of light in every kind of lighting. In plasma lights mercury or sodium are the important actors.

Plasma lights using mercury, of which the most common is the ordinary fluorescent light, all produce ultraviolet (UV) light, invisible to humans. To serve a useful purpose, UV light must be converted to visible light. This is accomplished by coating the glass tubes of the lamp with a substance called phosphor. The material is made of rare earth elements, sixteen metals in the periodic table. Phosphor is not phosphorus although the latter's name has been appropriated because it is the element that naturally glows. Phosphor is capable of reacting with UV light and transforming most of it into visible light, the rest into heat (infrared radiation). Thus invisible light coming off mercury is rendered as visible light by means of the phosphor coating. Mercury lights are very efficient because most of the incoming energy is converted to light, very little to heat, but the frequency of this light is toward the shorter waves (green-blue) thus producing less color. This phenomenon is covered in more detail later.

Sodium-based lamps produce visible light directly but also in a very narrow frequency range (yellow-orange) so that objects of other colors appears as shades of grey. Such light is referred to as monochromatic. Sodium lamps are coated with indium tin oxide which permits visible light to exit but infrared radiation (heat) to be reflected back. Sodium lamps are the most efficient sources of light, but the monochromatic light is unsuitable for normal lighting in a home.

Creating Arcs. In fluorescent lights electric current moves between separated poles in an arc. Neural gas facilitates the movement of electrons from atom to atom across the tube's length. Electrical current flowing from the outlet comes in at 120 volts and is not suitable to start the light up or to maintain it as it runs. To start the lamp, thus to seed the interior of the tube with electrons, low power, known as cathode voltage, is needed at the coils, the latter usually made of tungsten. Cathode voltage is 20 volts. To get such power requires reducing line voltage. To produce the arc itself once the lamp has been started requires 200 volts or higher, calling for increasing the line voltage. The *ballast* used in fluorescent lights is the transformer that does both jobs. The older form, known as magnetic ballast, does the job with coils of wire and magnets. The advanced electronic ballast uses solid state (silicon) circuits to achieve the same end. Electronic ballasts can also modulate the frequency of the current and can thus reduce the flickering associated with conventional fluorescent light. Advanced ballasts are used in the type of compact fluorescent lights (CFLs) that can be screwed into lamps like incandescent bulbs.

Special Switches. Modern three-way lamps can produce light at three intensities. They have two filaments which combine to produce three levels of brightness. One, for example, will have a 30 and a 70 watt filament. When both are on they produce 100 watts. Alternatively a lamp with a 50 and a 100 watt filament will produce three levels of light at 50, 100, and 150 watts. A switch inside the lamp determines which filament should be turned on for the two lower settings and when to turn on both to get the maximum amount of light. Touch lamps feature switching activated by sensing the temperature of the finger touching the lamp. Body temperature is almost always higher than the ambient temperature in which the lamp rests. Alternatively, such lamps can detect the electrical capacitance of the skin (the electric force it stores) and contrast it to the capacitance of the lamp's surface. Dimmer switches dole out current from the electric line to the device to be dimmed, thus reducing the voltage that reaches the bulb. Dimmer switches for fluorescent lights do not work on incandescent lights—or vice versa. Fluorescent dimmers depend on the type of ballast used and must be chosen for compatibility. Some fluorescent lights will not work with any dimmer switch. Dimming halogen lights reduces the power that reaches them, thus also the heat they generate. Dimming such lights defeats their purpose, which is to increase filament life while keeping the glass clear of tungsten deposits—both dependent on heat.

LED Lights. The newcomer to lighting is the light-emitting diode or LED. It emerged in the 1990s and is still in early stages. A diode is a kind of valve that permits current to flow in one direction only through what is known as a p-n junction, where p stands for positive and n for negative. The junction is where p and n meet; in that gap energized electrons move around, meaning that they increase or decrease in energy. Spots where an electron was but is no longer are referred to as electron holes. When another electron encounters such a hole, it drops into it, releasing a photon. LEDs thus work very similarly to fluorescent lights. Photons are emitted through a phosphor coating converting UV radiation into visible light. LED lights have reached a luminous efficiency better than any incandescent lamp but below that of fluorescent lamps, but the technology is still young.

Color. The color output of lamps is measured by the Color Rendering Index (CRI) maintained by the International Commission on Illumination. CRI values range from 0 to 100, in which a CRI of 0 means pure monochromatic, black and white lighting and a CRI of 100 is equivalent to color as seen in sunlight. Incandescent bulbs produce a CRI of 100, ordinary fluorescents a CRI of 63 (faces look paler), and low pressure sodium lamps used in street lights produce a CRI of nearly 0 (in its yellow-orange light people and objects are mostly in shades of gray). Our eyes are able to see only a very narrow band of the electromagnetic radiation, those with waves between 0.4 to 0.7 microns (millionth of a meter). In increasing wave lengths the colors are violet, indigo, blue, green, yellow, orange, and red. Thus the greater the wavelength the warmer the color.

The color of a photon depends on its level of energy, and that level is dependent on the distance by which the electron, displaced from its natural orbit, has jumped. In a superheated tungsten filament, all kinds of photons within the visible range of light are produced—as well as photons in the invisible infrared range. In effect the tungsten emits maximally 10 percent of its input energy as light (all colors), the rest as heat. The desirable color-rendering of incandescents also makes such lamps the least efficient. Looked at objectively, the common bulb might be described as a heater that incidentally produces light in that its function is to give light, it is inefficient.

In a fluorescent lamp, by contrast, molecules of mercury vapor are not, in effect, tortured by massive flows of current, as the tungsten filament is. Instead a single electron in the arc collides with a single electron of the mercury floating in the pressurized argon gas. When the mercury's electron drops back to is normal orbit, it also gives up a photon, but it is almost always in a narrow slice of the UV frequency. The light output therefore is unlike sunlight, which is our standard for real light. The same circumstances limit the light produced by sodium lamps to a narrow range of visible light.

Applications. Residential lighting accounts for approximately one-third and all other kinds—commercial, institutional, and industrial lighting—for two-thirds of all energy consumed in lighting. The residential segment is the dominant consumer of incandescent lighting and although, to be sure, fluorescent lights are present in many home shops, compact fluorescents are penetrating the market slowly, and LED lamps are appearing in the home, principally as flashlights.

The commercial market—using that designation, as does the U.S. Department of Energy (DOE), to refer to all other sectors—relies principally on fluorescent lighting. About three quarters of all lighting provided in nonresidential structures is fluorescent, thus making fluorescents the largest category of lighting. Fluorescent lights are the most cost-effective in terms of electrical consumption over the life of the bulb and they produce reasonably decent white light. Since this sector also includes hotels, motels, and communal living facilities, the sector is also a user of incandescent lighting, which is more pleasing to people.

Outdoor lighting uses mercury vapor, metal halide, and sodium lamps, all part of the plasma lighting category. The color rendering of these lights is inferior to fluorescent lamps but they are very efficient, have much longer life, and lowest overall costs. All of them require noticeably long periods to reach full luminosity, ranging from 2 minutes after turn-on for metal halide lamps to as much as 15 minutes for low pressure sodium lamps. It takes time to create the necessary plasma ionization before the arc that produces the light can take hold. Mercury vapor and high pressure sodium lamps are used in street lighting, metal halide lamps are favored for lighting stadia and warehouses. Low pressure sodium lamps are used in parking lots, warehouses, and tunnels. The difference between high and low pressure sodium lights is visible in their color; high pressure sodium lamps provide a whiter light because they contain mercury alongside sodium; low pressure sodium lamps produce a yellow light.

Most automobiles are equipped with halogen incandescent bulbs. Some feature so-called high intensity discharge (HID) lamps, usually with metal halide implementations working at high pressure and high heat to vaporize the metals used to produce light in a wider range of frequencies than mercury or sodium can deliver. The HIDs used in autos have rapid turn-on features, unlike larger metal halide lamps that need a warm-up.

Lamp Efficiency and Economics. The light production from a lamp is measured in lumens, one lumen being the light thrown by a candle on a square foot of surface located one foot from the candle. The efficiency of a lamp can be measured by the watts of electricity it takes to produce the same illumination. Let us assume we wish to have 1,800 lumens, the high end of an incandescent 100 watt bulb. The same illumination can be produced by a 70 watt halogen lamp and a 50 watt compact fluorescent.

According to the DOE's Energy Information Administration, electrical power cost 9.86 cents per kilowatt hour (a thousand watts) in 2006. To operate for 750 hours, the incandescent bulb will consume 75, the halogen 52.5 and the CFL 37.5 kilowatt hours—the numbers derived by multiplying 750 by the watt rating and dividing by 1,000 to obtain kilowatts. These numbers, times the average cost of electricity, produce operating costs of $7.40, $5.18, and $5.04 respectively for the three types of bulbs. To this operating cost we must add the acquisition cost of the bulb itself.

An operating time of 750 hours was chosen for this example because an incandescent lamp has a minimum life of that length. That bulb will cost $0.50, the halogen lamp $7.00, the CFL $7.50. It would thus seem that the incandescent provides the best cost-efficiency if the bulb cost must be amortized over 750 hours. In actuality, however, the halogen bulb will have a minimum life of 3,000 hours and the CFL 8,000 hours. For 750 hours, the halogen lamp itself will cost only $1.75 ($7 x (750/3,000)). Using the same approach, the CFL lamp will only cost $0.70 in the first 750 hours. Total costs therefore will be $7.90 for the incandescent, $6.93 for the halogen, and $5.74 for the CFL bulb. The CFL comes out as the winner.

Lamp efficiency is usually stated as lumen production per watt of electricity—a measure we inverted above to derive an economic comparison. Figure 123 shows the efficiency ranges of different types of lights, along with rated life hours, showing the minimum and maximum values in both categories and a ranking in each. The lamps are arranged by minimum efficiency.

Low pressure sodium lights are the most energy efficient and also have the third-highest rated bulb life. LEDs have the longest life at 35,000 hours. Incandescent bulbs come in last in both energy efficiency and in rated life. In this product category however, as in many others, sheer technical effectiveness and cost-benefit do not adequately capture the reality of the product. If we were to arrange this table by color-rendering index, incandescent bulbs

FIGURE 123

Efficiency of Lamp Types by Power Used and Rated Hours

Type of Lamp	Lumens per Watt			Rated Life Hours		
	Low	High	Rank Efficiency	Low	High	Rank Life Hours
Sodium, low pressure	183	200	1	18,000	n/s	3
Sodium, high pressure	150	150	2	20,000	24,000	2
Metal halide	50	90	3	7.500	20,000	6
Mercury vapor	50	55	4	16,000	24,000	4
Fluorescent	45	100	5	20,000	30,000	2
Compact Fluorescent	35	60	6	8,000	10,000	5
LED	30	50	7	35,000	50,000	1
Halogen	24	35	8	3,000	4,000	7
Incandescent	10	18	9	750	2,000	8

LED stands for light emitting diode.
n/s stands for not shown.
The ranking by Life Hours has two types of lamps ranked as second as they have the same life expectancy.

SOURCE: Compiled by the staff with data from the U.S. Department of Energy. January 2007.

would be the best (at 100 CRI) and low-pressure sodium lamps the worst (at around 0 CRI); the other categories would also stay in the same places but in reverse order. In residential settings color matters and thus trumps efficiency and cost.

The resource consequences of lumen efficiency can also be calculated for each type of light from the data in Figure 123. The theoretical maximum lumen-yield of a watt of electricity is 683 lumens per watt—although that light will be monochromatic. The energy efficiency of different lamps can be calculated by dividing lumens per watt by 683 and multiplying that number by 100 to obtain energy efficiency as a percentage. The low pressure sodium lamp, for example, with 183 lumens per watt produces an efficiency of 26.8 percent, meaning that nearly 27 percent of the incoming electrical energy is translated into light, 63 percent lost as heat. At the lowest end of efficiency, an incandescent bulb producing 10 lumens per watt, the efficiency is 1.5 percent, with nearly all incoming power leaving as heat.

MARKET

The lighting sector in the U.S. economy, as it is presented by the U.S. Census Bureau, consists of four industries. Of these Electric Lamp Bulb and Parts Manufacturing provides the core product and three other industries manufacture the fixtures intended to hold the light-producing element and often also the ballast of fluorescent lights. The Census Bureau divides the field into residential and nonresidential light fixtures and into another category it labels Lighting Equipment Manufacturing not elsewhere classified. This last industry classifies outdoor lighting as one major component and flashlights and other non-residential portable lights as the other—thus the largest and the smallest fixtures are combined in one industry.

To get some perspective on the role of each user sector, data from the DOE are illuminating. In order of importance, and measured in total electricity consumed in lighting, the top sector is commercial (51%) followed by residential (27%), industrial (14%), and stationary outdoor lighting (8%). The nonresidential sector, therefore, excluding outdoor lighting, is 65 percent of all electrical consumption for lighting.

The sector taken as a whole represented a market, at the production level, of $12.1 billion in 2005, down slightly from a level of $12.3 billion in 1997. The lighting sector as a whole is a mature industry. Looking at component industries, there is more variability seen in performance. Lamp sales have been declining at the steepest rate (3.8% per year). Sales of residential fixtures increased at a nominal 0.9 percent per year, essentially flat. Commercial, industrial, and institutional fixture sales declined at 1.4 percent per year. Outdoor and portable lighting showed

the only growth, roughly matching that of durable goods in the economy, increasing at 3.8 percent per year. The performance of the sector's component industries is shown graphically in Figure 124.

Despite continuing innovation in lighting, the technological thrust has been to produce more efficient and particularly longer-lasting light bulbs. These more efficient and longer-lasting devices have much higher prices but provide substantial savings in use. The feedback to residential consumers, however, works rather poorly. Bulbs represent a tiny portion of total household spending. In order to realize how high the savings are, people would have to keep track of bulb life and make extensive calculations to determine the level of energy savings they have achieved. Most consumers do not bother. Acceptance of high-priced bulbs is therefore slow. Competition, however, is intense. Market research firms who follow these products, Mintel International Group Ltd. being an example, thus conclude that the light bulb market has traditionally low growth. Competition causes prices to fall. Acceptance of new products leads to lower replacement rates. As a consequence unit sales to satisfy a given rate of demand decline with bulb efficiency. Demand is largely a function of household formations, which have been advancing at the rate of 1.4 percent per year in the 1997 to 2005 period.

Within the residential market, fixed light fixtures, thus ceiling- or counter-mounted devices, representing 59 percent of the market in 2005, exhibited positive growth at 3.5 percent per year—yet slower than new housing completions in the same period of 4.1 percent yearly. The other categories of residential lighting, however, namely portable devices, have been declining at a rate of 4.7 percent per year. The reason for this decline may be loss of earning power and a busier lifestyle, which impacts home decorating projects.

Within the commercial/institutional/industrial sector, only industrial lighting fixtures (representing just under 19% of the market) exhibited growth at 5.2 percent per year. Other users have purchased fixtures at declining rates of nearly 4 percent per year. The sharp dip in fixture consumption in this sector beginning in 2001 and lasting until 2003 was most likely the consequence of the recession that began in 2001, putting the brakes on corporate expenditures.

Outdoor and portable lighting (excluding lamps in the home but including flashlights) have shown a contrary pattern, quite possibly mirroring a sense of insecurity in the public in the aftermath of terror attacks and other dramatic incidents of public violence, such as massacres at schools. This category exhibited largely flat performance in the 1998 to 2001 period, as shown in Figure 124. Sales rose sharply between 2001 and 2002 and have been climb-

FIGURE 124

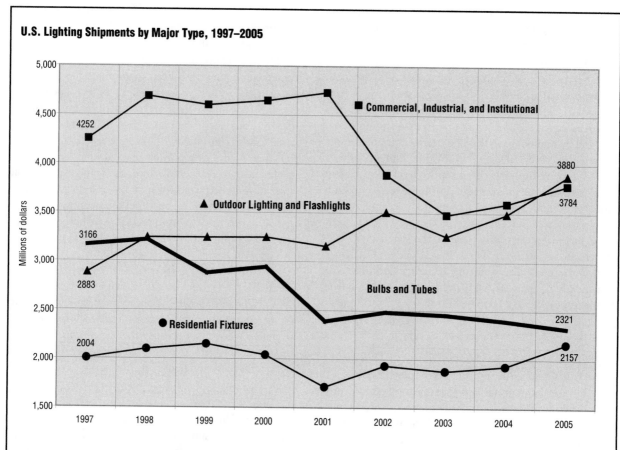

U.S. Lighting Shipments by Major Type, 1997–2005

Commercial, Industrial, and Institutional

Outdoor Lighting and Flashlights

Bulbs and Tubes

Residential Fixtures

4252
3166
2883
2004
3880
3784
2321
2157

SOURCE: Compiled by the staff with data from "Value of Product Shipments: 2005," *Annual Survey of Manufactures*, U.S. Department of Commerce, Bureau of the Census. November 2006.

ing since. Outdoor lighting grew at 1 percent in the 1998 to 2001 period but at the rate of 2 percent in the 2001 to 2005 period. Flashlight sales declined in the 1998 to 2001 period at nearly 3 percent per year and grew at a heady 11 percent yearly between 2001 and 2005.

KEY PRODUCERS/MANUFACTURERS

The top producers of lamps in the United States are General Electric, Osram Sylvania Inc., Philips Lighting Company, and Feit Electric Company, Inc., thus a publicly held U.S. corporation, a German multinational, a Dutch multinational, and a privately held U.S. company.

General Electric Company. Thomas Edison formed General Electric Company (GE) in 1890 as the umbrella organization for various enterprises. Edison was one of the inventors of the incandescent lamp. He filed his patents for the device in 1879, a year after the British inventor Joseph Wilson Swan obtained his patent for a similar device Swan had first introduced in 1860. Edison did not succeed

in defending his patents in the United States, where the U.S. Patent Office declined to recognize his design, saying that it relied on the earlier invention of another pioneer, William Sawyer. Edison also could not prevail in Britain, but he produced his own invention, which was superior to others, under contractual agreements and thus launched an industry. GE has gone far beyond light bulbs in its lengthy history. In its official filings with the Securities and Exchange Commission for 2006 (in the company's 10-K report) GE reported sales of $163.4 billion. Of that total its industrial segment accounted for 20.5 percent of revenues. Lighting products barely received a mention at the tail end of a discussion of a sub-segment, Commercial & Industrial Products. Nevertheless, General Electric had the dominant market share in light bulbs and tubes in the United States in 2006, holding in excess of 70 percent of the market.

Osram Sylvania Inc. This company is fully owned by Osram GMBH, headquartered in Munich, Germany.

Osram, with 2006 revenues of €4.6 billion, is part of Siemens. Approximately 43 percent of Osram's sales were realized in the Americas, much of the total by Osram Sylvania. In the period 1909 to 1993, Sylvania, which began as Hygrade Incandescent Lamp Company in Salem, Massachusetts, operated as a manufacturer of light bulbs and other electrical products. Sylvania's founder began his career by buying burned out light bulbs, removing their filaments, and producing new bulbs from old. Between 1959 and 1993 Sylvania was owned by General Telephone & Electronics (GTE). Osram purchased GTE's lighting operation in 1993. Osram Sylvania has a market share of approximately 8 percent in the United States.

Philips Lighting Company. Part of Royal Philips Electronics, Philips Lighting Company is based in The Netherlands. That global company reported €27 billion in revenues in 2006 of which lighting products represented €5.5 billion. Approximately 29 percent of the company's revenues were earned in North America. The company's U.S. share of the lamp market was approximately 4 percent.

Feit Electric Company, Inc. This company is a privately held bulb producer headquartered in Pico Rivera, California. Feit began operations in 1978 as a manufacturer of fluorescent lamps. Its own brand of products commands a 1 percent share of the lamp market. Feit is also involved in substantial private label lamp production for others.

Light fixture manufacturing is an extremely fragmented market with many small participants. In the Census Bureau's *2002 Economic Census*, 1,092 companies were reported to be participants in the industry. These companies operated 1,155 establishments of which 699 had fewer than 20 employees. The market is thus populated by many small producers domestically. In addition, valuable lamps—chandeliers, for instance—are imported predominantly from Europe.

Cooper Lighting Industries Ltd. This company, based in Peachtree City, Georgia, is representative of the handful of diversified and large producers of lighting. Cooper grew by acquisition of lamp companies and had sales in 2005 of $4 billion. Cooper combines the capabilities of nine separate companies, including Halo Lighting and McGraw Edison, the latter having acquired Halo.

Acuity Brands, Inc. This company is a publicly traded diversified fixtures producer with 2006 sales of $2.4 billion, of which 72 percent were realized in a wide line of lighting fixtures serving industrial, commercial, institutional, municipal, and residential markets. The balance of Acuity's products were chemicals.

Two other large participants in the market in 2006 were Hubble Lighting, Inc., with electric fixtures representing $1.6 billion of its total sales of $2.4 billion in that year, and Genlyte Group Inc., which had 2006 sales of $1.48 billion. Genlyte concentrates on light fixtures for a wide market.

MATERIALS & SUPPLY CHAIN LOGISTICS

If there is a critical material consumed in lighting, it is tungsten. The metal is used, although in different ways, in all kinds of lamps except LEDs, although in somewhat different ways. Tungsten wire is the second highest material input to the lamp and bulb manufacturing industry, glass holding the top rank as an input. The United States no longer has active tungsten mining operations although reviving closed mines was under discussion in the latter half of the first decade of the twenty-first century. Producers obtain the metal from scrap and from imports, principally from Canada. The world's largest producer, with the largest reserves of the metal, is China. China accounted for 84.6 percent of world mine production and 62 percent of world reserves in 2006 according to the U.S. Geological Survey. China's own consumption of tungsten, however, has increased significantly so that the country was actually importing tungsten scrap to alleviate local shortages there. Canada, with 9 percent of world reserves, is the closest source of tungsten for U.S. producers. More than half of all tungsten is used as an alloying agent to harden other metals in critical wear applications; tungsten production in machinery applications, therefore, drives the demand, not lighting.

Glass bulbs and tubes are manufactured in so-called ribbon machines, originally invented by Corning Glassworks. Molten glass runs on conveyor belts equipped with tiny openings through which glass is blown by machine into molds to produce blanks in various sizes and configurations. These high-precision, high temperature operations are centrally located and feed lamp producers pre-coated blanks ready for assembly. This portion of lamp production is part of the Pressed and Blown Glass and Glassware Manufacturing industry (NAICS 32–7212) centered in Ohio, New York, Pennsylvania, West Virginia, and South Carolina.

In the lighting fixture segments of the industry a great diversity of semi-finished materials are fabricated and purchased components are assembled into fixtures of all kinds ranging from small flashlights on up to street lights resting high on poles made of aluminum and steel tubing. Apart from the residential sector in which the single most important input, as measured in dollars, is paperboard packaging material for shipments of relatively small items, the largest input to lighting fixtures are specialty transformers and fluorescent ballast manufactured by electronics components producers in industries upstream, as it were, from fixture manufacturers. Specialty transformers and ballasts were also the second highest inputs for residential fixture manufacturers.

Lamp and bulb production in the United States is concentrated in Connecticut, Illinois, California, and New Jersey, the states shown in rank order based on shipments in 2002. Lighting fixture production is much more widespread. The top producing states, in order of 2002 shipments, were California, Illinois, Pennsylvania, New Jersey, Ohio, and New York, thus concentration matches population densities across the country.

DISTRIBUTION CHANNEL

Lighting plays so universal a role in human affairs that distribution of lighting products takes place through multiple channels. Landscape and garden lighting fixtures, for example, may be found in garden centers, endoscopes for lighting the interior of the body are delivered by the medical supply chain, auto lights (and their replacements) are sold through automotive dealerships, municipalities and highway departments purchase outdoor lighting systems by public procurement, and the list could be extended to many more niche markets.

Residential bulbs are sold through hardware stores, drug stores, food stores, department stores, and mass merchandisers. Low-end fixtures are typically also available in such outlets. Higher end fixed lights and portable lamps are sold in furniture stores and specialized lamp stores. These channels are supplied by distributors who, in turn, rely on electrical or specialized lighting wholesalers.

Specialized lighting distributors play an important role in servicing builders and contractors who work on behalf of commercial, industrial, and institutional clients. These wholesalers typically offer very extensive lines of lighting fixtures, systems, and components ranging from interior to outdoor products. Some wholesalers are further specialized to serve important lighting markets such as theatrical and movie lighting applications.

KEY USERS

In that all people need lighting—even the sightless do so on behalf of those close to them—we are all key users. The context of lighting, however, the places to be illuminated and for what purpose—creates quite different user profiles. Residential and similar lighting (hotels, resorts) is dominated by incandescent lamps because a full range of colors is important to simulate daylight. Lighting fixtures, similarly, are used as means of expressing aesthetic values. In more practical situations where lighting becomes a functional means to an end, economic efficiency becomes the crucial element. Fluorescent lighting is a good compromise in that it produces good lighting at low cost at the sacrifice of CRI. Outdoor lighting of streets, buildings, and parking lots to prevent accidents or to discourage crime require minimum color rendering but consume high amounts of energy. It is costly to light the night when confining walls are absent and do not help by reflecting the light back. These applications use the most cost-effective sources of light with the least pleasing appearance.

ADJACENT MARKETS

Mood lighting provided by candles represents the closest adjacent market to utilitarian electric light. Rare indeed is the household where candles affixed to candlesticks are entirely missing and where, on festive occasions, the lights are not dimmed and shadowy but smiling faces are lit by flickering but very warm lights, and candle flames reflect from raised glasses.

Gas lights used in outdoor lighting represent an alternative. Such lighting is used both in residential and municipal settings to provide a traditional feel to certain areas. Oil lamps are used in camping situations and when the lights fail.

In that lighting is a central aspect of decoration, all interior furnishings, from carpets to drapes to wall coverings, from furniture to paintings to sculptures to knick-knacks, are adjacent markets when lighting is chosen to illuminate such surroundings and, sometimes, new lighting installations lead to changes in furnishings.

RESEARCH & DEVELOPMENT

The central focus of R&D in lighting is on solid state lighting (SSL) based on light-emitting diode technology. In describing its programs of R&D support for this technology, the Department of Energy states: "No other lighting technology offers the Department and our nation so much potential to save energy and enhance the quality of our building environments." LEDs have emerged as a technology capable of providing very high efficiencies in converting energy into light (rather than heat) while providing a superior CRI approaching, and eventually matching, that of incandescent bulbs.

DOE expends approximately $1.3 billion annually supporting research on building efficiencies. Of that a portion is dedicated to supporting lighting innovations by sponsoring research by universities, corporations, associations, and national laboratories. An example of such research was the development of an organic light-emitting diode lamp by Universal Display Corporation producing light at 45 lumens per watt, matching fluorescent lamps, but with a CRI of 78 (versus 63 for fluorescents). The efficiency and color production of such lights is expected to improve; they already have the longest lamp life on record, and their performance is not affected, like all other lights are, either by vibration or by on-and-off switching. In just the LED category alone, in 2007 DOE was supporting 43 projects, and sponsoring others with GE, Philips, and Osram Sylvania. These companies were financially participating in the sponsored research with R&D expenditures of their own.

DOE is also sponsoring research intended to improve conventional lighting processes, focusing attention on higher-efficiency filaments for incandescent lamps, next-generation fluorescent products, and multi-photon phosphor research. The last category represents basic research on phosphors used to coat fluorescent lights with the aim of doubling their output of visible light and producing an ideal multi-colored spectrum of light.

R&D in this field is aimed at major improvements in lighting efficiency on the one hand, promised by the exploitation of semiconductor technology, the aim being energy conservation, hence DOE's participation. On the other hand, the aim is to improve the color-rendering index of efficient light sources to aid their acceptance by the public.

CURRENT TRENDS

Perhaps the two most important issues impacting lighting are global warming, viewed as, at least in part, human-caused by emission of carbon dioxide, and looming future shortages of petroleum and, in due time, all other hydrocarbon fuels as global resources are drawn down. Efficient lighting is one way to reduce carbon emissions and to conserve fuels.

Public efforts to forcibly cause the population to conserve energy used in lighting have surfaced in the latter years of the first decade of the twenty-first century. As reported by Diane Katz in *Michigan Science* in August 2007, "the incandescent light bulb no longer will be sold in Australia, Canada, Cuba or Venezuela within five years. Similar phase-outs are pending in California, New Jersey and several other states as well as the European Union."

Such legislative efforts are matched, as already noted by extensive and in part publicly-supported activity to provide the consumer with products that deliver the same warm light in a new package, eventually at the same low acquisition price.

TARGET MARKETS & SEGMENTATION

The subdivision of the market into major user segments with different requirements has been noted at various points above. Under this heading we need only to note in addition that lighting products must be sold not only to their ultimate end-use consumers but also to at least two professional categories that importantly influence what kind of products are purchased: architects and engineers. Producers in the industry therefore expend marketing and sales efforts to reach these professions in efforts to keep them abreast of new developments.

RELATED ASSOCIATIONS & ORGANIZATIONS

The Illuminating Engineering Society of North America, http://www.iesna.org

International Association of Lighting Designers, http://www.iald.org

National Electrical Manufacturers Association, http://www.nema.org/about

BIBLIOGRAPHY

"The Arc Lamp." Industrial Electronic Engineers, Inc. An IEE Archives Department Exhibition. Available from <http://archives.iee.org/about/Arclamps/arclamps.htm.>.

Darnay, Arsen J. and Joyce P. Simkin. *Manufacturing & Distribution USA,* 4th ed. Thomson Gale, 2006, Volume 2, 1387–1397.

"Did Thomas Edison Really Invent the Light Bulb?" Demand Entertainment, Inc. Available from <http://www.coolquiz.com/trivia/explain/docs/edison.asp>.

"Fluorescent Lighting." HyperPhysics. Available from <http://hyperphysics.phy-astr.gsu.edu/hbase/electric/lighting.html>.

Goodman, Marty. "History of Electric Lighting Technology." Available from <http://www.sheldonbrown.com/marty_light_hist.html>.

Katz, Diane S. "Fluorescent Revolution." *Michigan Science.* August 2007.

Lazich, Robert S. *Market Share Reporter 2007.* Thomson Gale, 2007, Volume 1, 1863.

"Light Bulbs—US." Mintel International Group Ltd. Available from <http://www.mindbranch.com/listing/product/R560-700.html>.

"Lighting Research and Development." Building Technologies Program. U.S. Department of Energy. 30 August 2006. Available from <http://www.eere.energy.gov/buildings/tech/lighting>.

"Measuring Light Source Life." Building Technologies Program. U.S. Department of Energy. Available from <http://www.netl.doe.gov/ssl/usingLeds/general_illumination_life_measuring.htm>.

"Product Summary: 2002." *2002 Economic Census.* U.S. Department of Commerce, Bureau of the Census. March 2006.

"Tungsten." *Mineral Commodity Summaries.* U.S. Department of the Interior, Geological Survey. January 2007.

"Value of Product Shipments: 2005." *Annual Survey of Manufactures: 2005.* U.S. Department of Commerce, Bureau of the Census. November 2006.

Linens

INDUSTRIAL CODES

NAICS: 31–4129 Household Textile Product Mills, not elsewhere classified

SIC: 2392 Housefurnishings, not elsewhere classified

NAICS-Based Product Codes: 31–41291, 31–41293, 31–41295, 31–41296, 31–4129W

PRODUCT OVERVIEW

The oldest meaning of linen is cloth woven from the fibrous portions of the stem of the flax plant, the oldest source of textile fibers. Industrial-style production of flax yarn goes back at least 5,000 years. Neolithic grave remains hold residues suggesting that prehistoric humanity already harvested and processed flax stems to make clothing as far back as 8000 years before the current era. People started weaving cloth from wool a little later in history—in Nordic climates. Flax is native to a wide global region extending from India to the Mediterranean. It produces the strongest and longest plant fibers and is more than twice the strength of cotton. The fabric is soft, light, and absorbs moisture better than any other on the market. Its natural color is ivory to grey with some kinds tinged a light tan. The characteristic white clothing of the ancients in the Mediterranean regions, from Egypt to Europe, were made of linen yarn, not least the famous Roman togas. Flax has yielded to cotton over time. Cotton is easier to produce and weave. Flax, however continues to be the source material for a variety of useful products. In addition to textiles it is used as the strong fiber in currency, in cigarette paper, and its seed yields linseed oil.

The dominant role linen played in the history of textiles—its excellent performance in garments, its ability to absorb moisture easily, its softness, its durability, its light color suggesting cleanliness and purity—made it the fabric of choice for direct contact with the body. For this reason linen acquired a more generic meaning long ago. The word is used to designate textiles used in the bedroom, bathroom, kitchen, and in the dining room regardless of the fibers actually employed to weave modern sheets, covers, slipcovers, mattress protectors, pillowcases, comforters, towels, washcloths, shower curtains, napkins, and table covers. The word linen is also still used for undergarments, a meaning present in the word lingerie, despite the fact that most such goods are now made of cotton, synthetics, and of silk.

In this essay we shall focus attention on linens as a generic category, thus on bed, bath, and beyond, but excluding underwear. The delimitation of the subject in this manner is in part motivated by industrial reporting conventions that define a major industry as Household Textiles nec. The nec (not elsewhere classified) addendum suggests that the industry produces miscellaneous goods—which these products certainly are not. Considering the amount of time we spend each night in bed and washing and drying after waking, the household textiles category is a rather major part of our daily life. Linen as a fabric is a relatively small proportion of this industry—as it is also a small percentage of the apparel industries. As appropriate, linens will be discussed, if only peripherally, also as they occur in the apparel category.

Textile Fibers. Textiles form a kingdom divided into natural and man-made categories. Natural fibers come from animals and plants whereas man-made fibers are classified as inorganic and organic.

The major animal fibers are sheep's wools, cashmere from goats, and silk from the silkworm. Wool is the largest category within this subdivision; silk is the most expensive. Plant-based textiles derive from cotton, flax, sisal, jute, hemp, and bamboo. Cotton is King, of course—a phrase introduced as the title of a book by David Christy in 1855. Cotton has had to yield the global throne since those day but remains the sovereign among natural fibers. Hemp and flax are sometimes confused because both are tall plants with fibrous stems from which textiles are made. Hemp and marijuana both belong to the Genus *Cannabis*; hemp used for its fiber, however, *Cannabis sativa*, has a very low content of tetrahydrocannabinol, the psychoactive drug, compared with *Cannabis indica*, marijuana itself. Flax belongs to the Genus *Linum*.

Inorganic fibers, made of carbon, ceramics, and glass are used in industrial products principally, including insulation materials. Organic fibers fall into two further categories. Of these cellulosic fibers are made from wood. Rayon, once thought of as artificial silk, is the best-known category. Non-cellulosic organic fibers come from petroleum and include polyester, nylon, olefins, and acrylic. The first two account for the greatest volume. In the modern world organic (also called synthetic) fibers are the new king of the Age of Oil. They represented approximately 58 percent of all fibers made in the first decade of the 2000s; cotton accounted for 38 percent; all of the other fibers combined, including wool, silk, and linen claimed for a mere 4 percent of fiber consumption.

Within the household textiles industry, the proportions are more skewed toward natural fibers—in large part because the products are in close contact with the skin and have superior moisture-absorption characteristics. Based on 2006 import data—and imports dominate this industry—cotton accounted for 64.0 percent, man-made fibers for 34.8 percent, linen for 0.8, silk for 0.3, and wool for 0.1 percent of fiber consumption.

Ironically, perhaps, linen, the flax-based fiber, played a larger role in all other categories of textiles than in the linens category. Overall its share of textile imports was 4.2 percent, in apparel 2.5 percent, and highest in floor coverings with a 21.5 percent share, second only to wool at 21.9 percent—owing to its great strength combined with its inherent softness. Unlike other categories of textiles, however, total consumption (in contrast with imports) of floor coverings is dominated by man-made fibers and imports do not play as large a role as in the other categories.

Warp, Woof, and More. The industrialization of textiles in the modern sense, dating back to the mid-eighteenth century, represented, in effect, the birth of industrial civilization and also foreshadowed the age of the computer—because weaving is a fundamentally digital technology in which the warp represents the *0* and the woof the *1*.

On a weaving loom the warp is formed by lines of thread held rigidly by the machinery. Although warp threads are all aligned in parallel, each is affixed to a movable part of the loom (known technically as the *Bolus hook*) and can thus be lowered or raised individually. The woof—which is as often referred to as the weft—is a line of thread interwoven with the warp at right angles. If every other warp-holder is lowered during a single pass of the weft, the weft will run under the first, over the second, under the third, over the fourth, and so on. After one pass the upper-lying warp is lowered, the lower level raised, and the weft passes back. As this process continues, an even network of warp and weft produces a web of fabric. This type of weave is known as plain weave. The thread count is calculated by counting the number of threads running in both directions in one square inch of fabric. A good sheet will have a thread count of 180 to 200. The extreme ranges are 80 to 700. The higher the thread count the softer the fabric.

Much as in computers where the artful manipulation of 0s and 1s can create intricate images on a display screen, so also in weaving the ability to manipulate the warp in different ways produces the many types of fabric on the market. The most common four types are plain weave, basket, twill, and satin. As already discussed above, plain weave is a simple web. In basket weave multiple threads of weft, always the same number, pass over and under the same number of warp threads on each pass and, on returning, pass under those they ran over and over those they ran under on the previous pass. This forms a checkerboard pattern. The twill weave is achieved by passing a weft under a single warp and then over the next two, repeating this pattern with a step or shift at each pass so that a diagonal marking of the fabric results. Variations on this theme have produced well-known varieties like denim, gabardine, tweed, and serge. In satin weaves a smooth finish is achieved by moving four weft strands beneath a single warp or the other way around. The multiple threads are said to float over the single thread. Varying the number of floating threads produces different effects. Satin weave is employed on silk. When cotton or linen are used instead, the weave is known as sateen. Warp and weft, of course, need not be the same thickness or color. The vast range of fabrics on the market is built up from variations on a few parameters.

As the nineteenth century was blinking itself awake, in 1801, Joseph Marie Jacquard in France invented a loom in which punch cards activated the lowering and raising of individual warp threads automatically in complex patterns remembered by the cards. The punch cards were large, the width of the loom, and made of cardboard, but they anticipated the much smaller punch cards to be used first in calculators and then in computers. A particular weave pattern, once "programmed" onto punch cards, could be stored for later use. The invention introduced the *Jacquard weave*, the most complex kind. In the modern version Jacquard weaving is, needless to say, computer-controlled. The highest end of tablecloth, damask, is produced by Jacquard weaving. A lustrous silk satin or linen fabric is produced with a raised design usually executed in the same thread. Very complex and colorful fabrics can be produced using this general technology.

Mechanical knitting and crocheting methods produce non-woven fabrics. Of this type, uncut pile fabrics are used in toweling; a network of threads is further filled by knitting loops to the web, loops appearing on both sides. Cut pile fabrics are made the same way but closed loops are cut open, usually on one side only. Such fabrics are used in carpeting. When producers need padding and want to save money they use felts made of wool. Wool is ideal for non-woven applications because moistened fibers, when agitated, will randomly lock together because the fibers have tiny scales. Felts, however, pull apart easily as well. Shower curtains, a product category included under household textiles, are typically fiber-reinforced or fiber-filled vinyl sheets in which the filler is held by but also reinforces the plastic.

Major Product Categories. In industrial reporting it is customary to divide household furnishings into three large groupings and then to combine the remainder into an All Other category—principally because the major components of this last grouping are sold in many different contexts.

The three major categories are sheets and pillowcases (15.5% of the market in 2005), bedspreads and bedsets (3.3%), and towels and washcloths (6.3%). These categories correspond to traditional departments or sections of retail stores. The largest product category in this industry, however, is represented by pillows of all kinds (22.6% of shipments); but because pillows are sold in all manner of contexts and in several different departments, the product is placed in the All Other bin for statistical purposes. Comforters are the second largest single functional category (17.8% of shipments)—then sheets and pillowcases, towels and washcloths, followed by table covers and napkins (4.9%), mattress covers and protectors (4.5%), shower curtains (4.0%), bedspreads and bedsets, and blankets and quilts (3%). The remaining shipments of the

industry are not specified by kind and may be anything else the industry reports under this category of goods.

Bedsets—in commerce usually referred to as bedding or comforter sets—are matched products for formal bedrooms that come in four to nine pieces including comforters, quilts, cushions, shams, tube-shaped neck rolls, sheets, pillow cases, and bed skirting or dust ruffles that hang all around the bed nearest to the floor. In crib sets bumpers are included to prevent babies rolling against the bars of the crib.

Another view of this industry is that its products are principally used in the bedroom or in resting situations where pillows are used for comfort. Nearly two-thirds of shipments are associated with sleep-related products, approximately 10 percent with the bathroom, and 5 percent with the dining room. If we add up the shipments of the three major industry groups which produce textiles—textile mills, textile product mills, and apparel—we see an industry of $109.5 billion in 2005. Of that total household furnishings represented a mere 5.3 percent that year. In looking at that percentile, it is worth a moment's reflection to note that we spend approximately 33 percent of our time on earth in bed—but since that activity is solitary and largely unconscious, we spend enough for comfort but not a lot—in contrast to apparel—on show.

MARKET

In the textile industry, as in many others in which foreign producers have come to dominate the market, highly reliable industrial production statistics provided by the U.S. Bureau of the Census provide but a partial view of total activity. To get at the market itself, *apparent consumption* must be calculated. This is achieved by obtaining data on domestic shipments, subtracting from that total the portion shipped out of the country as exports, and adding product coming into the country as imports. In 2005, the last year for which all of the necessary data were available at time of writing, apparent consumption in the United States at the production level (not retail) was $12.2 billion, up from $10.3 billion in 2000, representing an annual growth rate of 3.4 percent per year. The rate of growth fell below that of Gross Domestic Product in this period (4.8% per year) as well as below that of non-durable goods (3.8% per year).

In this same period, domestic shipments declined from a level of $7.1 billion in 2000 to $5.8 billion in 2005, an erosion of 4.1 percent per year. Exports declined from an estimated $471 million to $368 million, dropping at the rate of 4.8 percent annually. In this same period, however, imports grew at a heady rate of 13 percent per year from a level of $3.7 billion in 2000 to $6.8 billion in 2005. Net domestic production (shipments less exports) represented 64 percent of apparent consumption in 2000 but had fallen to 44 percent by 2005. These relationships are presented graphically in Figure 125.

Trade Imbalance and Consequences. The rapid replacement of domestic manufacturing by imports has been stimulated by the shift in production from the relatively high-wage area of the United States to low-wage industries elsewhere. The top suppliers of U.S. textile imports in 2007, shown in order, were China, Mexico, India, Indonesia, and Vietnam—top-ranked China shipping nearly five times more product than second-ranking Mexico. Consumers in the United States, in consequence, have had plentiful products competitively priced, but significant erosion of domestic buying power has been one price we have had to pay.

In this industry alone, some 160 production plants disappeared as a consequence of this shift in the 2000–2005 period accompanied by the loss of 21,000 jobs in the household textiles industry alone. Employment in this industry was declining at a rate of 8.7 percent per year. The domestic industry has suffered instability as a consequence, with well-known firms being merged, acquired, and experiencing bankruptcies. These losses have had their greatest impact on southern states. According to data published by the National Council of Textile Organizations (NCTO), 80 percent of all textile-related plant closings in the 1997 to 2007 period took place in North and South Carolina, Georgia, Virginia, and Alabama. North Carolina lost the most plants, 39% of the total compiled by NCTO.

FIGURE 125

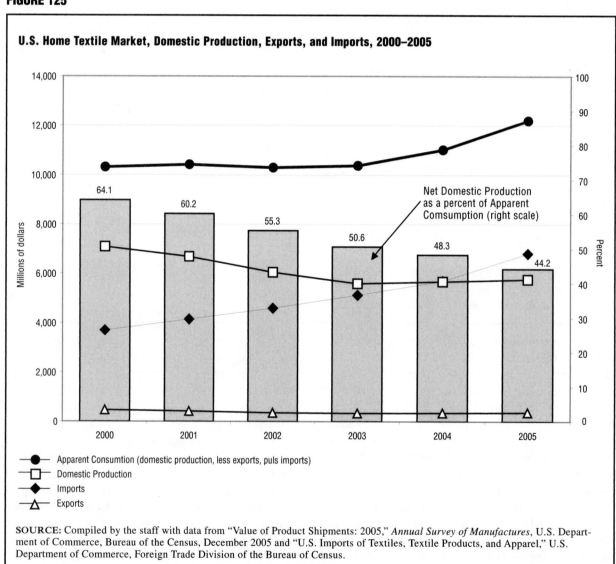

U.S. Home Textile Market, Domestic Production, Exports, and Imports, 2000–2005

Net Domestic Production as a percent of Apparent Comsumption (right scale)

- Apparent Consumtion (domestic production, less exports, puls imports)
- Domestic Production
- Imports
- Exports

SOURCE: Compiled by the staff with data from "Value of Product Shipments: 2005," *Annual Survey of Manufactures*, U.S. Department of Commerce, Bureau of the Census, December 2005 and "U.S. Imports of Textiles, Textile Products, and Apparel," U.S. Department of Commerce, Foreign Trade Division of the Bureau of Census.

Retail Estimates. Based on projections provided in *Manufacturing and Distribution USA*, which provides data for NAICS 44-2299, All Other Home Furnishings Stores, retail sales in the 2000 to 2005 period were advancing at the rate of 6 percent per year. Product details for this or any other retail industry are not reported comprehensively enough to trace products down to the retail level, but in rough terms the industry here discussed as linens represented, at retail, approximately $20 billion in consumer expenditures.

KEY PRODUCERS/MANUFACTURERS

What could be fairly described as a major foreign trade assault on U.S. textile producers throughout the 1990s and the first decade of the 2000s has transformed the domestic industry supplying household textiles. Among the top six producers, three have been in bankruptcy in the first decade of the twenty-first century. Two have responded to severe pressures by selling themselves to foreign companies. One has licensed out its product lines to another in the process of being acquired by a bankruptcy turn-around specialist, retaining fabric production activities only. One has reorganized itself into a private corporation with outside investors. Only one company—it had always been privately held—has remained above the fray by specializing in niche markets and focusing on research-based products.

Milliken & Company. The acknowledged leader in the fine table covers and napkins category—along with other textile lines—is Milliken & Company, founded in 1865 as a wool fabrics firm in Portland, Maine by Seth Milliken and William Deering. With Deering soon after departing to start another enterprise, Milliken developed into a diversified textiles manufacturer involved in carpets, apparel, specialty and industrial textiles, dust-control products, automotive textiles, chemicals, and interior and household textiles. This last category includes wall coverings as well as table linens. Hoover's, the Dun & Bradstreet publisher, estimates Milliken's 2005 sales at $3.3 billion; of that only a portion was earned in selling linens.

Springs Global US, Inc. The largest U.S. company in this industry appears to be Springs Global US, Inc. based on its leading role, as indicated by *Market Share Reporter*, in multiple product segments. Hoover's estimated Springs Global's sales at $2.7 billion in 2005. The company was the top producer of sheets and pillowcases, comforters, and bedding, second in bath towel manufacturing, and third in mattress fabrics. The company acquired Burlington Industries' bedding brands after the latter filed for bankruptcy. In 2006, having closed a number of production plants, Springs Global (earlier it was Springs

Mills, later Springs Industries) merged with Cotenimas of Brazil, South America's top-ranked textile company. Springs Global began in 1887 in Fort Mill, South Carolina.

WestPoint Home, Inc. This company formed in 2005 and is the inheritor of an enterprise once known as WestPoint-Stevens and earlier as WestPoint-Pepperel. J.P. Stevens and Pepperel were once independent textile producers for the household market. The company is now a privately held operation formed jointly by its predecessor and American Real Estate Partners LP. WestPoint holds top share in blankets and bathing towels, is second in sheets, pillowcases, and bedding, and second in comforters. Its estimated sales in 2006 were approximately $440 million.

Dan River Inc. With sales as estimated by Hoover's of $227 million, Dan River holds third rank in bedding and fourth place in comforters. One of the oldest of the leading companies, Dan River began in 1882 in Danville, Virginia as a cotton mill. Over a century later, in 2004, the company was forced to file for bankruptcy, another casualty of an avalanche of imports. Dan River worked its way out of Chapter 11 by being acquired by Gujarat Heavy Chemicals Ltd., an Indian firm.

International Textile Group (ITG). Burlington Industries, Inc., once a major factor in household textile products, also filed for bankruptcy and was eventually acquired by W.L. Ross & Company. W. L. Ross, a billionaire, specializes in rescuing bankrupt companies. Ross formed International Textile Group in 2004. Burlington became a part of that group along with Cone Deming and Carlisle Finishing. By that time Burlington had sold off its bedding brands to Springs Global but remains active in the household business as a fabrics supplier. ITG reported sales of $720.9 million in 2006. Of that total $36.4 million were generated by the company's interior furnishings fabrics segment.

MATERIALS & SUPPLY CHAIN LOGISTICS

Two hundred years before the current era, merchants guided pack animals westward on a route from China to the Mediterranean. The route soon came to be known as the Silk Road. The animals laboring along it carried precious cloth for sale to ancient elites in Mesopotamia, Egypt, and later in the Roman Empire. Although the high value of silk has always made it the most profitable textile to transport, textiles have always moved long distances from their production to points of use, whether as fibers, thread, cloth, or finished goods. Eli Whitney's invention of the cotton gin in 1793 enabled low cost spinning of cotton fiber. The use of cotton in clothing predates the discovery of silk and goes back 7,000 years. Cotton thread became a major export product from America to Europe and was the most important raw material in the world's first high tech industry. Precisely because no effective machine has ever been invented for spinning flax fiber into yarn—and manual methods are still required for fine linen—flax is a minor fiber whereas cotton is second only to synthetics. Nevertheless, linens have always traveled long distances as well. Manmade fibers, similarly, begin in oil fields concentrated in just a few areas of the world. The oil moves across oceans and crosses continents in pipelines before fractions of it are extracted in refineries and turned into fibers.

The textile industry has a three-tier structure. Textile mills spin thread; thread is sold as a final product to product mills which weave it into fabric; fabric is the final product purchased by those who produce consumer, commercial, and industrial products in apparel industries and other textile mills using purchased materials. The very fact that all three industrial segments are viable and operate independently indicates that textiles generally are not limited by the logistics of materials.

In 2005 thread and yarn mills operating in the United States demanded 8.4 hours to produce $1,000 worth of product, fabric mills 8.9 hours, other textile product mills (in which our products fall) demanded 9.9 hours, and apparel production required 10.3 hours of labor. Thus in moving from the lowest to the highest tiers of the industry, labor-intensity increases—and this for obvious reasons: much more automation can be applied to the mass production of uniform outputs. Where labor inputs are low and technology can be applied effectively, the United States is competitive. Thus the U.S. textile industry is the fourth largest exporter of textiles in the world—but these products are threads, yarns, and fabrics. Conversely, the United States is the largest importer of finished textiles at the upper tier. These relationships illustrate that the logistical structure of the industry is dominated by the distribution of low-cost labor rather than materials.

DISTRIBUTION CHANNEL

Distribution of household textile products parallels that for textiles generally. The products are sold by department stores, mass merchandisers, warehouse clubs, and specialized retailers. The largest of the specialists is Bed, Bath & Beyond, Inc., a company with sales in Fiscal Year ending March 2007 of $6.6 billion, operating 888 stores across the country. Factory outlet stores with limited product offerings also distributed the goods of the industry.

Products of the industry typical reach retailers through independent home furnishings wholesalers and,

to a lesser extent, furniture wholesalers. Large chain operations tend to operate their own distribution centers and, in procuring goods, deal directly with manufacturers and importers. Large institutional buyers of such textiles, like the military, use public procurement methods.

KEY USERS

The products of this industry are used by everyone, but certain user groups within the industry have distinguishable profiles. Fine table linens and bedsets used in bedroom decoration are purchased more or less routinely only by the more affluent segments of the population and, within that segment, those portions of it that often entertain formally and live in homes intended in part for display. Most established families, however, with some modicum of means, will own at least one set of formal table linens. The industry also serves large commercial and institutional markets: hospitality, medical facilities, the military, and other institutions that house people, including jails and prisons. Hotels, motels, and resorts are large users of bedding linens, bedsets, and bathroom toweling. Hospitals and related communal institutions (from nursing homes to hospices) are a market similar to hospitality. Medical buyers have a utilitarian orientation and do not buy decorative textiles.

ADJACENT MARKETS

The largest adjacent markets to household linens are furniture markets, especially bedroom and dining room furniture, as well as cabinetry and shelving to hold the products in readiness before use on the bed or in the bathroom. Rugs and carpeting are related textile products; not surprisingly, many producers are also makers of these product lines. Laundry machinery and chemicals in the home and laundry services purchased by institutional users of the product category are adjacent markets. Textile-based, often fire-resistant wall-coverings are an important category producers in this industry manufacture for the hospitality market.

RESEARCH & DEVELOPMENT

Most research and development in this field is focused on synthetic fabrics, particularly, in this industry category, on developing synthetics with higher rates of moisture absorption and thus good performance in the laundry as well. Most synthetics have half or less the moisture absorption capabilities of plant-based textiles like cotton and linen—characteristics important in bedding and toweling applications. Work on polyester promises the best results for matching cotton's performance, and work along these lines was advancing in the first decade of the twenty-first century with positive results.

CURRENT TRENDS

The most important trend in this industry in the twenty-first century is the transformation of the domestic industry from a production into a distribution sector. The general shape of this transformation is a shift of manufacturing to other countries, with domestic firms maintaining some production but benefiting principally by stamping their well-known brands on goods made elsewhere and using their well-established distribution networks to sell the product to U.S. consumers.

On the materials side, the major trend is the competition between cotton, the largest natural fiber, and polyester, the largest synthetic fiber. The long-term trend in cotton has been eroding market share, with cotton yielding share to polyester. The latter material, however, is strongly influenced by oil prices which have been rising sharply in the later years of the first decade of the twenty-first century. According to analyses conducted by the U.S. Department of Agriculture, upward pressure on polyester prices has slowed the erosion of cotton in the market. The longer-term outlook appears to depend on the future performance of the petroleum industry worldwide. In the very long run, however—barring exceptionally large new oil discoveries somewhere—cotton may once more emerge as the king of textiles.

TARGET MARKETS & SEGMENTATION

The principal buyers of household textiles—as well as textiles generally—are women. Pricier products are targeted at fashion-conscious home decorators, with the message suggesting that these products will enable the buyer to create pleasing home environments and attractive settings for communal entertaining. Appeals to prestige play a role. Top-of-the-line products are promoted as gifts to newlyweds. In commodity-style products (e.g., sheets and pillow cases) the marketing is based on quality, including that of the fabric itself and its thread count. Performance is important to the buyer in these categories, not least the feel of the fabric—its smoothness or softness or ability to absorb moisture—and its durability. Products in many shades of color are provided to let the consumer integrate them into the home's color scheme. The performance of the product in the laundry is part of the performance appeal. Low-end products are promoted on the basis of cost-benefit: good quality but affordable.

RELATED ASSOCIATIONS & ORGANIZATIONS

American Fiber Manufacturers Association, Inc., http://www.afma.org/afma/afma.htm

American Textile Machinery Association, http://www.atmanet.org/home.aspx

National Cotton Council of America, http://www.cotton.org

National Council of Textile Organizations, http://www.ncto.org

BIBLIOGRAPHY

Braudel, Fernand. *The Wheels of Commerce.* Harper & Row, 1982.

"CEO Blames Imports, Not Merger, for Springs Closings." *The State.* 4 August 2007.

"Cotton and Wool Yearbook." U.S. Department of Agriculture, Economic Research Service. Available from <http://usda.mannlib.cornell.edu/MannUsda/viewDocumentInfo.do?documentID=1282>.

Darnay, Arsen J. and Joyce P. Simkin. *Manufacturing & Distribution USA,* 4th ed. Thomson Gale, 2006, Volume 1, 233–237.

Johnson, James, et. al. "The United States and World Cotton Outlook." *Agricultural Outlook Forum 2006.* 17 February 2007. Available from <http://www.usda.gov/oce/forum/2006%20Speeches/Commodity%20PDF/cotton%20outlook%202006.pdf>.

"King Cotton." Civil War Potpourri. 16 February 2002. Available from <http://www.civilwarhome.com/kingcotton.htm>.

Lazich, Robert S. *Market Share Reporter 2007.* Thomson Gale. 2007, Volume 1, 204–206.

"Plant Closings and Job Losses." National Council of Textile Organizations. Available from <http://www.ncto.org/ustextiles/closings.asp>.

"Textile Materials and Technologies." Cornell University, Art, Design, and Visual Thinking. Available from <http://char.txa.cornell.edu/media/textile/textile.htm>.

"Types of Weaves." Textile Exchange. Available from <http://www.teonline.com/types-of-weaves.html>.

"Value of Product Shipments: 2005." *Annual Survey of Manufactures.* U.S. Department of Commerce, Bureau of the Census. November 2006.

Luggage

INDUSTRIAL CODES

NAICS: 31–6991 Luggage Manufacturing

SIC: 3161 Luggage Manufacturing

NAICS-Based Product Codes: 31–69910 through 31–699101F4

PRODUCT OVERVIEW

Luggage refers to the various types of bags and containers that travelers use to transport their belongings. The term includes suitcases, backpacks, computer cases, briefcases, tote bags, carry-on bags, and garment bags. Suitcases are usually rectangular bags 24 to 36 inches in height that open on hinges like a clamshell. They may be manufactured from leather, metal, plastic, or textiles; the material will dictate if the suitcase is hard-sided, soft-sided, or semi-soft sided. Suitcases have a handle so they may be carried; many have wheels, which means a particularly heavy bag may be rolled. Some suitcases possess locks, which may be opened with keys or a combination. Suitcases are sometimes referred to as *Pullmans* or *Pullman cases*. This term was used in the early days of mass travel, when suitcases were designed to fit under seats in Pullman sleeping cars—George Pullman invented the railroad sleeping car in 1857; the cars held his name until 1980.

A garment bag typically holds one or two items on hangers, while a garment carrier holds three or four. Tote bags are small bags worn around the shoulder; they are often sold as accessory pieces within larger luggage collections. Carry-ons are less than 22 inches tall and can be easily stored under an airline, train, or bus seat. Sports bags may vary in size; designs vary based on their function. A briefcase is a small, flat case originally designed to carry important papers. Backpacks are small- to medium-sized bags with two shoulder straps with which the bag can be carried on the back; backpacks are popular with students and campers. Computer cases are used to hold laptops and other portable electronic devices. All of these products may be manufactured from leather, but other textiles and metals are proving to be popular because they offer a weight advantage, are attractive, and are durable.

Early History. The word luggage first appears in English in 1596, according to *The Oxford English Dictionary*. It comes from the word *lug,* which means to haul or to drag. Such a word is appropriate for the earliest forms of luggage. Trunks and cases are some of the oldest methods of transporting belongings; their origins go back thousands of years in China and Egypt. Early travelers would have transported their belongings in heavy trunks for their ocean trips or in carts and caravans across long distances. Trunks were typically made of wood with metal trim. The interior was lined with paper; by the early nineteenth century, fabric was used. The trunk proved durable on long journeys. There are a number of trunk styles, including Jenny-Lid trunks, steamer trunks, Barrel-Staves, Bevel-tops, and Dome-Top trunks. The flat-top lid style was the most popular; one explanation for this is that ship, stagecoach, and train porters could not properly stack the trunks with curved lids. Flat-top trunks could be slid in and out of baggage holds and not be damaged.

Trunk manufacturers in the early twentieth century were small, regional operations with just a handful of employees. A few of the prominent makers were M.M. Secor, Seward Trunk & Bag Co., Louis Vuitton, Hartmann, and Shwayder Trunk Manufacturing Company (predecessor to luggage industry leader, Samsonite). Some of these companies expanded into side industries, such as making small bags, cases, and related leather goods.

Post World War II. The luggage industry is, for obvious reasons, deeply tied to the travel industry. Changes in the United States after World War II benefited both industries. By the 1950s couples were getting married and moving to the suburbs; people were starting to live further away from family members. Wages increased, which granted consumers more disposable income. The overall leisure and vacation industry in the United States became much more robust. This decade also marked the beginning of America's love affair with the automobile. Travel in the late nineteenth and early twentieth centuries meant an expensive, difficult trip by rail or ship. In the post World War II period, Americans could pack their bags and take to the open road.

Increasingly, Americans were taking to the open skies as well. Commercial air travel existed before World War II but was out of reach for the vast majority. Commercial airlines were helped along in the 1950s and 1960s by technological improvements in jet-engine design, improvements that grew out of lessons learned by the airforce during the war. Airlines offered air coach class seating to compete with coach seating in the rail industry.

The newness of air travel lent it an air of sophistication, which further made it desirable to many Americans. By the 1970s larger aircraft were being used, helping bring down ticket prices and making air travel more affordable. The largest aircraft of the time, the Boeing 747, was first flown commercially in 1970. It was the first aircraft designed with overhead compartments for luggage storage.

Business travel was on the rise as well during this period. The economic expansion that began in the United States after World War II continued through the 1960s. As the economy and population expanded, so too did companies. The decade of the 1960s is often seen as the birth of modern Corporate America. Novels such as *The Man in the Gray Flannel Suit* dramatize the situation many men found themselves in at this time: going off to work every day in their gray flannel suit, briefcase in hand, and conforming to a professional standard. Briefcases began to sell well during this period, as did luggage in general. Quantus Airlines of Australia was the first airline, in 1971, to introduce a business class section on its planes.

In response to this growing market, luggage makers began offering new suitcase designs and styles. Suitcases were typically constructed using heavy metals and woods. Manufacturers now experimented with lighter woods and

metal-blends to help make suitcases lighter. They also explored new colors, although black and gray remained the most popular. Samsonite released the first suitcase with wheels in 1975.

Luggage makers typically extolled the durability of their products. "Tough Enough to Stand On" was the slogan used by Shwayder in some early advertisements for his suitcases; the slogan ran with a picture of Shwayder, his father, and his three brothers standing on one of the company's suitcases. But in the 1950s and 1960s luggage makers started copying larger advertising trends and marketed their products as being part of a consumer lifestyle. A Hartmann luggage ad from 1968 shows a sharply dressed woman with matching luggage and the suggestion: "Maybe we should call ourselves Hartwomann." Hartwomann is a play on the company name, of course, but the *we* asks the female viewer to make the same identification: I, too, am a Hartwomann. Another advertisement followed that was even edgier, featuring an attractive woman with her luggage and the line "Hartmann Will Never Be an Old Bag."

In another example of how luggage advertising was trying to attract new consumers is a Samsonite advertisement of the time in which an array of suitcases appears beneath the suggestion "Choose Your Luggage like You Would Your China." The message behind such advertising is clear: luggage is increasingly a necessity not a luxury for the American home.

The 1980s and Beyond. The late 1970s were a hard period for the luggage business. Sudden increases in oil prices caused an economic slowdown and airline hijacking put an additional burden on all travel related businesses. It was not until the mid-1980s that the economy began to grow again and it did at a rapid pace, ushering in a period of economic growth that lasted in the United States until the end of the century, with only minor slowdowns in the early 1990s.

These economic cycles were important in driving for the luggage industry. The baby boom generation reached adulthood during the 1970s and 1980s. This led to an increase in the number of people likely to travel, with disposable income enough to travel for leisure as well as for work. In addition, members of this age group were more likely than their parents to purchase luggage for their children, for example, as a college graduation gift.

Producers of luxury goods, including high-end luggage makers, saw what was happening: there was a growing market of consumers who could afford luxury products, many of whom wished to buy products as a status symbol. The luxury goods industry boomed. A product becomes a luxury good because it has a design, a level of durability, and an overall quality that is superior to the equivalent mass market product. With this status also comes a higher degree of desirability among many shoppers.

Some luxury goods makers ventured into the luggage market for the first time during this period. Montblanc, known for its luxury pens, did so in 1995. Established luggage firms enjoyed improved sales. Trunk maker Louis Vuitton advertised its handmade suitcases and monogrammed specially numbered trunks. Luxury firms began selling suitcases in eye-catching styles, colors, and expensive fabrics. They also offered clever features such as better storage and recessed locks, features that other luggage makers would copy. Even established luggage makers recognized consumers' desire for products that combined function and fashion. Hartmann Inc. recruited fashion designers Gloria Vanderbilt and Roy Halston to design luggage for them.

Luggage makers were increasingly aiming products at women. They had several reasons for doing so. Women were believed to make 80 percent of all luggage purchases. Women generally spend more money on luxury items than men do. But it was also recognition of women's growing role in the business world. More women were executives and were traveling for business. They were willing to buy luggage that was marketed to them and addressed their specific needs. The Man in the Gray Flannel Suit with his briefcase of the 1960s had, to some extent, changed into the Woman with the Louis Vuitton luggage by the late 1980s.

The luggage industry, along with other manufacturing industries in the United States, went through a great deal of change in the 1990s. The industry saw a number of sales and consolidations. Hartmann Inc. was bought by Browns-Forman, known for its alcoholic beverage products. Samsonite struggled after its parent company declared bankruptcy. The American Tourister name remains an important brand name with the American public, but the company was bought out by Samsonite. Atlantic Luggage was taken over by Travelpro International. Luxury luggage makers such as Victorinox and Tumi have moved into department stores (a market once dominated by Atlantic Luggage) and taken market share.

The manufacturing of luggage began to move out of the United States during the 1990s as did manufacturing activities generally. The industry was struck hard by the downturn in traveling which occurred after the terrorist attacks against the United States in September 2001. Industry sales in the United States declined in both 2002 and 2003, rebounding slightly in the following years.

The luggage industry remains intrinsically tied to the overall travel industry. Globalization is a driving force behind growth in the travel industry. However, the industry faces many challenges as well; challenges which make the future difficult to predict. The travel industry continues to struggle with how to ensure passenger security, how to respond to terrorist attacks, how to handle possible pandemics, and how to manage volatile fuel costs.

MARKET

A report from Global Industry Analysts estimated the global retail bag and luggage market to have been worth $20.8 billion in 2004. The source expects the industry to grow 4 to 5 percent annually through 2010 and thus reach $27 billion in 2010. Growth will be dependent on the stability of the travel industry, the affordability of air travel, the development of new air routes, and the health of the overall economy.

Suitcases, Pullman cases, casual bags, and garment bags represented approximately half of global luggage sales in 2004. Sports bags, backpacks, and daypacks accounted for another 27 percent of sales. Fifteen percent of sales were sales of business bags, which include briefcases and laptop cases. Other types of luggage accounted for the remaining 8 percent of sales. The United States was the largest market worldwide for luggage with sales of $7.8 billion (38% of the global market). Figure 126 presents global luggage sales by region in 2004.

FIGURE 126

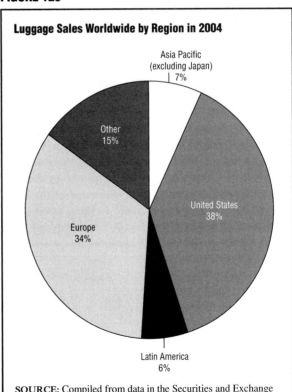

Luggage Sales Worldwide by Region in 2004

Asia Pacific (excluding Japan) 7%

Other 15%

United States 38%

Europe 34%

Latin America 6%

SOURCE: Compiled from data in the Securities and Exchange Commission Filing by Samsonite Corporation filed with the U.S. Security and Exchange Commission on May 18, 2006.

The luggage market is very fragmented. Samsonite has been a leader in the industry for many years in each of the major markets, followed by regional players that have single digit market shares. These smaller companies are often successful by appealing to some niche category in the field, such as the luxury sector, or the travel-retail sector.

Samsonite claimed a 27 percent share of the luggage market in the United States in 2004, based on sales of hardside and softside luggage and garment bags. Other players were Travelpro, Atlantic Luggage, Jansport, Targus, Tumi, Swiss Army, and East Pak. The United States is the largest consumer market for luggage on the global stage and it imports an increasing proportion of its luggage requirements from abroad. In the late 1990s, 53 percent of the leather and leather products sold in the United States were imported, according to the Census Bureau. By the early twenty-first century, imports were estimated to make up over 90 percent of luggage sold in the United States. In 2006, 81 percent of those imports were coming from China.

The luggage industry can be difficult to track using government figures. One must look at the general luggage category (bags made of leather) but also consider cases made of metals and textiles to get a truly accurate picture. The leather luggage industry was in decline for much of the late 1990s. The Census Bureau reported 230 establishments devoted to leather luggage manufacturing in 2002, down from 278 in 1997. The value of shipments fell as well. Luggage manufacturers shipped $1 billion worth of goods in 1997; by 2002, that value was nearly cut in half. Shipments of textile-based suitcases fell 78 percent between 1997 and 2002, and textile-based garment bags fell 84 percent. Leather-based business cases fell 62 percent. Luggage shipments fell to a recent low of $492.1 million in 2003. There are several reasons for this decline. The industry was simply following larger manufacturing trends and consolidating operations and moving operations out of the United States. As well, the industry was suffering the effects of the terrorist attacks of September 11, 2001. The overall travel industry would not begin to return to pre-2001 levels until 2004. In 2005 U.S. luggage shipments were valued at $639.1 million.

KEY PRODUCERS/MANUFACTURERS

The largest luggage manufacturer worldwide is Samsonite which had 20 percent of the global market in the early 2000s, based on industry sources. VIP was second largest with a 6 percent share of the market. By luggage category, Samsonite claimed to control 22 percent of hard and soft luggage and garment bag sales in Europe. Samsonite is the leading luggage player in France, Germany, and the United Kingdom. Other players include Antler, Delsey, Roncato, Stratic, and Rimova. In the Asian Pacific market, Samonsite again had a leading position, with a 27.5 percent share of soft and hard luggage and casual bag sales. Major players in this region include VIP, Eminent, and Crown. Profiles on Samsonite and three other prominent players in the luggage industry follow.

Samsonite Corporation. This industry leader was the largest producer of luggage, computer cases, outdoor and casual bags worldwide in 2004. The company licenses its brand names to third parties for use on products that include travel accessories, leather goods, handbags, clothing, and furniture. The company also manufactures products under various brand names, including Samsonite, American Tourister, Lacoste, and Samsonite Black Label. The company was incorporated in 1987 and is headquartered in Denver, Colorado.

Samsonite was started by Jesse Shwayder. He founded the Shwayder Trunk Manufacturing Company in Denver, Colorado in 1910. Shwayder and his ten employees marketed trunks and small hand luggage in the Western United States. Jesse Shwayder introduced a new style of luggage, Samsonite Streamlite, in 1941. The product was named after the Biblical giant and was intended to convey the product's strength and resilience. Streamlite luggage was tapered in shape and was made by covering a wooden frame with vulcanized fiber. The company had made suitcases for many years. It was not until this period of time that the company changed its manufacturing process to help make suitcases that were identical in design. In short, Shwayder Trunk finally started making sets of luggage.

Shwayder thought it time to update the look of the suitcase to capitalize on the growing air travel market of the late 1950s. Consumers were interested in light cases, as were airlines. He abandoned the typical wood construction in favor of sheet metal and magnesium combined with injection moulded ethyl cellulose. In 1958 he created the Silohuette. It was far lighter than a wooden suitcase and more sleekly designed. It also possessed some sensible features, such as recessed hardware to protect the locks and hinges from the wear and tear that all bags receive during travel. The company marketed the Classic Attaché in the early 1960s, which was a hit with businessmen. In 1965 Shwayder officially changed the company name to Samsonite. Firmly established as the leading maker of hard luggage and briefcases, the company expanded into the soft luggage market in the 1970s. The company made several noteworthy acquisitions in the 1980s. In 1993, Samsonite acquired American Tourister, another prominent name in the luggage field.

American Tourister. Sol Koffler started the company in 1933, and at the time it was called the American Luggage Works. The Rhode Island-based company sold 5,000 suitcases in its first year. A few years later, Koffler used new machinery to streamline the manufacturing process to produce a suitcase that was more uniform in construction. The process allowed the suitcase to be more easily shaped and increased its durability. Koffler's new suitcase called the American Tourister proved to be very successful. The company made other innovative manufacturing changes in the coming decades; Koffler was the first luggage maker to produce an all-vinyl case. In 1954 it improved the chemical manufacturing process so that American Tourister cases were virtually indestructible. Such claims were used in the company's advertising, and were boosted by reports of American Tourister luggage surviving accidents intact. American Tourister produced a series of highly successful television commercials which featured owner testimonials interspersed with shots of a gorilla kicking and stomping on the suitcases to prove their durability.

Koffler sold the company in 1978 to Hillenbrand Industries. It was sold again in 1993 to holding company Astrium International, which also owned Samsonite. Efforts were made to keep these two companies separate. This was feasible because Samsonite's target audience was interested in hard-luggage and briefcases while American Tourister products performed well in the soft-luggage sector. In 1995 Astrum split into Samsonite and Culligan Water Technologies Inc. Samsonite took American Tourister with it. Samsonite consolidated much of American Tourister's operations. As of the late 1990s, American Tourister has existed as a brand only.

Travelpro International. Founded in 1987 and headquartered in Florida, this company is best known for its Rollaboard product. Rollaboards are the small carry-on bags with telescoping (collapsible) handles that can be wheeled easily through airports and onto planes. The rollaboards were invented by Bob Plath, company founder and former Northwest Airlines pilot. Travelpro luggage was reportedly used by over 425,000 airport personnel worldwide. Travelpro gained more control of the market with its merger with Atlantic Luggage.

Hartmann, Inc. Bavarian Trunk maker Joseph Hartmann founded this company in Wisconsin in 1877 and moved it from Milwaukee to Racine a few years later. In 1905 Hartmann announced plans to build luggage so fine it would stand as a symbol of excellence. By the 1930s the company offered over 800 different styles of trunks and suitcases. The company started releasing lighter luggage in the 1940s and 1950s after switching from aluminum and steel to basswood framing. Lenox Incorporated purchased Hartmann, Inc. in 1983; Lenox was then acquired by Browns-Forman the same year. Browns-Forman had a number of subsidiaries, and Hartmann makes up around

2 percent of the company's revenue. In early 2007 there were persistent rumors that the company was once again going to be sold.

Louis Vuitton. A young Loius Vuitton apprenticed himself to Monsieur Marechal in 1835, who was making trunks for the court of Empress Eugenie, wife of Napoleon III. Louis Vuitton started his trunk manufacturing business in 1854; he would expand into London in 1885. The company advertised the craftsmanship of the trunks they offered and such trunks became popular with the public. The company would expand into luggage and handbags in the 1930s. Louis Vuitton developed the distinctive LV monogram after his products began to be copied by counterfeiters; fake Vuitton luggage and leather goods are still regularly seized by U.S. Customs officials. In the twenty-first century the company still manufactures its products much as it did in the nineteenth century; a suitcase reportedly takes 15 hours to produce and a trunk 60 hours. The Louis Vuitton brand is one of the most recognizable brands in the luxury goods industry more than 150 years after the founding of the company.

MATERIALS & SUPPLY CHAIN LOGISTICS

The luggage industry uses a variety of leathers, textiles, metals, and plastics in the manufacturing process. Piece fabrics are used to manufacture the case. Metals are needed for locks, zippers, and hardware. Plastics might also be needed for hardware or linings. Various chemicals are also needed to treat the fabrics to increase their durability.

In the 1990s the luggage industry, like most manufacturing industries, was looking for ways to cut costs. Luggage manufacturing is costly in part because it is labor-intensive, much like the larger apparel and leather industries to which it belongs. But the cost of raw materials used in the manufacturing process is also an important contributor to overall cost. Basic commodity prices have been on the rise during the period from the late 1990s through 2005. The expansion of industrialization around the world through the growth of globalization has caused demand for raw materials to rise sharply. This, in turn, has caused prices for those materials to become volatile causing sourcing problems for many manufacturing industries, luggage being one of them.

Leather luggage makers spent $572 million on materials used in manufacturing in 1997 to produce shipments valued at $1.43 billion. Materials consumed represented 40 percent of shipment values that year. In 2002 materials consumed had fallen as a percentage of the value of shipments to 27 percent or $154.9 million for shipments valued at $571.9 million. Although the value of shipments of leather luggage was down considerably (-60%) between

1997 and 2002, the cost of the materials consumed in the production of those shipments was down even further (-73%).

DISTRIBUTION CHANNEL

Luggage makers started moving operations out of the United States in the 1990s. Overseas operations were attractive to companies because overall manufacturing costs were cheaper than in the United States: reasons for this include lower wages to workers, cheaper and more accessible materials, and lower rents on facilities. U.S. luggage and leather makers also moved into overseas markets because the developing countries into which they were moving offered new markets into which to sell their products. China, India, and Indonesia are a few of the countries that have growing numbers of wealthy citizens. These affluent individuals are hungry for high-end electronics, automobiles, and leather goods.

Far less manufacturing of luggage was done in the United States in the early twenty-first century than had been done during the twentieth century. U.S. luggage makers import much of their product from third-party manufacturers in Asia and Europe. Luggage companies then market these goods under various labels to appeal to a certain demographic: affluent travelers or young, price-conscious shoppers, for example. The Travel Goods Association, in analyzing Census figures, estimates that 96 percent of luggage, briefcases and computer cases, 97 percent of travel and sports bags, and 98 percent of luggage locks sold in the United States are imported.

Samsonite, as the largest luggage maker and the only one with a real global presence, has the most mature distribution networks. In fiscal year 2006 Samsonite reported purchasing 89 percent of its soft-sided goods from third-party retailers in Eastern Europe and Asia. The remaining 11 percent came from company-operated facilities. The company sells in stores across the globe, and is available in all retail channels. It also sells products through approximately 284 Samsonite-operated retail stores in North America, Europe, Asia, and Latin America; 150 shop-in-shop corners principally in Asia; and 173 franchised retail stores principally in India.

The luggage industry is labor-intensive, but not capital-intensive; in other words, it is not difficult to enter the market if one has access to enough workers and the textiles and fabrics needed for the manufacturing process. This is a major reason for the number of small, niche-market players in many countries. Small companies have the ability to stake out a place in the market but lack the distribution networks that Samsonite has to become a global player. This situation creates a competitive market, but it creates problems as well. For example, India has

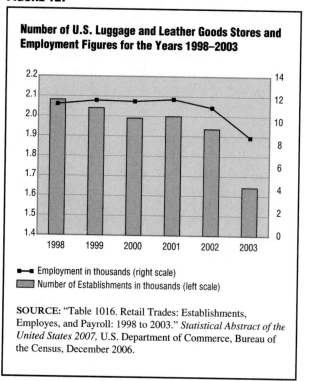

FIGURE 127

Number of U.S. Luggage and Leather Goods Stores and Employment Figures for the Years 1998–2003

- Employment in thousands (right scale)
- Number of Establishments in thousands (left scale)

SOURCE: "Table 1016. Retail Trades: Establishments, Employees, and Payroll: 1998 to 2003." *Statistical Abstract of the United States 2007*, U.S. Department of Commerce, Bureau of the Census, December 2006.

a growing class of young people who are well educated and wealthy (much of this wealth, it should be noted, has been generated through outsourcing contracts with the United States). Its luggage market was forecasted to grow approximately 50 percent between 2007 and 2010. Nearly half of its luggage market is, however, unorganized (generic suitcases), while prominent firms such as VIP and Samsonite control the organized sector. Luggage is sold through small, mom and pop stores; high-end leather and luggage retailers hesitate to open stores in India because of uncertainty in the country's distribution networks.

Luxury goods makers started to diversify in the 1990s. They began to sell through a wider range of retail channels. The luggage retail picture became even more complex when some leather and luggage goods makers opened stores selling their own branded goods. Specialty retailing became very popular during the 1990s (specialty stores carry limited merchandise devoted a particular industry or a series of related ones). Leather accessories maker Coach opened stores to sell its handbags and leather goods. Wilson Leather was formed in 1988, and opened outlets in many malls. Samsonite opened its own stores as well. Consumers can now purchase luggage online through the online version of popular retailers or through independent vendors such as eBags or LuggageFactory.com.

There were 1,977 luggage and leather goods stores in the United States in 2002, according to the Census

Bureau. These stores employed more than 12,000 people. According to *Market Share Reporter 2007*, Coach stores held the largest market share (30%) for luggage and leather goods sold in specialty stores during 2003. Figure 127 presents the leaders in specialty store sales of luggage and leather goods for that year. Market research firm, IBISWorld estimated that revenues for these retail outlets would climb from $1.6 billion in 2003 to $1.9 billion in 2010.

KEY USERS

Travelers are, of course, the key users of luggage. After the terrorist attacks against the United States in 2001 and subsequent attacks in Europe the following year, the travel industry experienced a severe downturn. Recovery began in 2003. The Travel Industry Association estimated that 168 million business trips were made in the United States in 2004 and 490 million leisure trips. In both categories travel was up over levels seen in 2002. But the way Americans spend their leisure time has changed in recent years. More people take vacations closer to home rather than flying to distant locations, shunning airport delays and expenses. Part of this shift has to do with a decline in the number of Americans taking the traditional two-week vacation every year. According to the Travel Industry Association, Americans are much more likely to take a number of long weekend trips during the year instead of a single, longer trip. There were 225 million weekend leisure trips made in 2005, up 10 percent over the previous five years. This trend has helped shape the luggage industry. Consumers want small, easy-to-carry, overnight bags, not large, heavy suitcases.

Another user group of products produced by the luggage industry is children. The growing number of school-age children in the United States means increased sales for backpacks and bookbags. Business travelers are another user group upon which the industry relies. The improvement in the business travel category may translate into higher garment bag sales while business travelers also use computer cases for laptops that they take on trips of all kinds, whether by air or simply on foot down to the local coffee shop.

ADJACENT MARKETS

The travel industry is the driver behind luggage manufacturing. There were 806 million international tourist arrivals around the world in 2005, the third year of increase in a row, according to the World Tourism Organization. Africa increased the most over 2004, up 9 percent, followed by Asia and the Middle East with 8 percent each. Tourist arrivals were up only 6 percent in the Americas. Balancing national security with international tourism has been problematic for the United States. In January 2007 travel industry officials met with the Bush administration

to ask that the government revise its policies regarding international travel to the United States. They noted that the United States will need to stop "treating tourists like terrorists" if it wishes to see tourism increase. Overseas travel to the United States fell 17 percent from 2002 to 2007; the country's share of the international travel business has fallen from 9 percent to 6 percent. In another report on the travel business, Euromonitor's *World Travel Market 2006* found that total business arrivals to the United States fell by 10 percent to 7 million over the 2004 to 2005 period. Over the same period the number of business visitors to Europe grew by 8 percent to 84 million.

The luggage industry is also part of the larger leather goods market. According to the Census Bureau, manufacturers of all leather goods shipped $2 billion worth of goods annually from 2002 to 2005. Leather goods include apparel accessories, handbags, and footwear. High-end items are part of the luxury market. The industry still performs well, although it is down from the levels in enjoyed in the 1980s. According to studies done by Bain & Company, the luxury goods market globally was worth around $200 billion in 2006. Global leather goods sales jumped 18 percent in 2005.

RESEARCH & DEVELOPMENT

Consumers of luggage in the early twenty-first century wanted bigger, leaner, and lighter. Manufacturers are experimenting with various materials and designs in the hope of producing a product that can offer great storage options while maintaining a sleek profile and being as lightweight as possible.

New models of luggage released in early 2007 offer clever tweaks to recent luggage designs. One model allows a suitcase to be wheeled at a traveler's side rather than behind him or her, producing less stress on the shoulder. Another suitcase had a built-in seat that could be pulled from its side. A traveler with this piece of luggage can turn his suitcase into a stool, perfect for more comfortably weathering the long lines at airport check-in counters. Suitcases that can be wheeled are very common. However, in new models the wheels are covered to protect them from damage and work is ongoing to find ever more flexible wheel designs.

Adding features is another way in which manufacturers work to improve their products. One example of such features is a model that includes built-in speakers and an iPod port. Several luggage makers also offer combination garment bag/suitcases, in which the garment bag is rolled around a hard case. Manufacturers are experimenting with microchips and special ID tags to track lost luggage. With airline security a continuing issue, some makers are exploring security alarms for suitcases. One European suitcase model, used to transport rare materials, features a strong deterrent to potential thieves: an 80,000 volt shock to anyone who tampers with it.

CURRENT TRENDS

In December 1972 the U.S. Federal Aviation Administration (FAA) issued security guidelines to screen passengers and luggage. Passengers would soon be walking through metal detectors and have their carry-on baggage x-rayed. These measures were instituted to combat the rising number of hijackings worldwide during the 1970s. Early airport security equipment could screen for a gun or knife, but they did not have the ability to screen for explosives or possible toxins that a terrorist might try to bring aboard.

The issue of airline security drew national attention after the terrorist attacks of September 11, 2001. Among the efforts to tighten national security in the aftermath of these attacks was a mandate by the American Transportation and Safety Board that all baggage be screened for explosives by December 31, 2002. The Transportation Security Administration (TSA) had not, as of early 2007, met this goal for a number of reasons: expense, lack of manpower, and poor training of TSA personnel. Explosive detection equipment has been installed in some airports, but the equipment is expensive, bulky, and time-consuming to operate.

The TSA issued stricter guidelines regarding carry-on luggage in August 2006. Airlines are trying to enforce baggage restrictions; the maximum size carry-on bag for most airlines is 45 linear inches (the total of the height, width, and depth of the bag). This creates occasional conflicts because some passengers are forced to check oversized bags at the gate. With tighter rules in place, passengers are now checking 20 percent more luggage than they did prior to 2001. With this increase in checked luggage has come an increase in lost and delayed luggage. This irksome situation, however, may have a silver lining for some entrepreneurial individuals. David Sempler, president of the Air Travelers Association, noted that the delivery of mishandled luggage has the potential to fuel growing businesses. Universal Express and Luggage Express are two firms that contract with airlines to deliver recovered bags to passengers' homes and hotels.

TARGET MARKETS & SEGMENTATION

Luggage manufacturers market a number of labels to appeal to different categories of customers. Samsonite, for example, markets its signature Samsonite label to middle and upper income shoppers. Samsonite's Lark brand is aimed at the executive traveler. Lacoste is aimed at affluent customers. Samsonite's American Tourister label is aimed at lower and middle-income shoppers. Its Trunk & Co. brand is aimed at young people.

Other companies also offer different luggage lines to different customer groups. Atlantic Luggage claims to be the largest supplier to the department store channel; its luggage is aimed at those who are looking for quality items that are reasonably priced. Travelpro has a loyal following among airline personnel because of its Rollaboard product.

Tumi is a leading manufacturer of luxury bags and travel accessories. Most of its revenue comes from its airport locations. The company targets professionals between the age of 30 and 55, men 20 to 35 years of age and women who use luxury items. The company claimed that 65 to 70 percent of its luggage was sold to men. It, too, offers a number of products to compete in various markets. They include the T-Tech (urban-casual-business, day and travel bags for young professionals), Couriers (day, business, and travel bags for professional women), Accent (mixed-material briefcases for young men) and Superlights (nylon briefcases for young men).

RELATED ASSOCIATIONS & ORGANIZATIONS

American Dealers Luggage Association, http://www.luggagedealers.com/rcairlines.htm

National Luggage Dealers Association, http://www.nlda.com

Travel Goods Association, http://www.travel-goods.org

Travel Industry Association, http://www.tia.org

BIBLIOGRAPHY

Abdend, Jules. "Luggage, Accessories Set to Tackle Tough Times." *Bobbin.* December 1993.

Cordle, Ina Paiva. "Boca Company Expands Control Over Luggage Delivery." *Miami Herald.* 24 August 2006.

"India, Russia, China Seen Booting Luxury Market." *China Economic Net.* 25 March 2007.

Keefe, Cathy. "Weekend Getaways Grow in Popularity." U.S. Department of Commerce, International Trade Administration. 27 November 2006. Available from <http://www.tia.org>.

Lazich, Robert S. *Market Share Reporter 2007.* Thomson Gale, 2007, Volume 2, 689.

Lipke, David. "Tumi Taps Chu for Creative Role." *Daily News Record.* 15 January 2007, 1.

"Luggage Makers Are Striking Gold." *India Business Insight.* 23 January 2007.

"Luggage Manufacturing: 2002." *2002 Economic Census.* U.S. Department of Commerce, Bureau of the Census. December 2004.

Rast, Charlie. "Top 25 Import Sources for Luggage." U.S. Department of Commerce, International Trade Administration. 15 February 2007. Available from <http://ita.doc.gov/td/ocg/imp316991.htm>.

Rozario, Kevin. "Leaving the Baggage Behind." *Duty-Free News International.* 15 July 2003, 30.

"Samsonite Corp." Information filed with the Security and Exchange Commission. 18 May 2006. Available from <http://www.secinfo.com/dVut2.v789.htm>.

Sen, Arushi. "VIP to Expand Hard Luggage Portfolio." *Hindu Business Online.* 18 July 2006.

"State of the U.S. Travel Goods Market 1992–2005." Travel Goods Association. Available from <http://www.travel-goods.org>.

"Value of Product Shipments: 2005" *Annual Survey of Manufactures.* U.S. Department of Commerce, Bureau of the Census. November 2006.

Walters, Helen. "Tumi's New Itinerary." *Business Week Online.* 24 January 2007.

Wardell, Jane. "U.S. Business Tourism Suffers on Travel Restrictions." *USA Today.* 6 November 2006: 1.